Bacterial Adhesion
to Animal Cells and Tissues

Bacterial
Adhesion
to Animal Cells and Tissues

Itzhak Ofek
Department of Human Microbiology
Sackler Faculty of Medicine
Tel-Aviv University
Tel-Aviv, Israel

David L. Hasty
University of Tennessee College of Medicine
and Department of Veterans Affairs Medical Center
Memphis, Tennessee

Ron J. Doyle
Department of Microbiology and Immunology
University of Louisville School of Medicine
Louisville, Kentucky

ASM PRESS

Washington, D.C.

Front cover figure: Adhesion of bacteria to substrata is a remarkably specific phenomenon. The scanning electron micrograph on the cover illustrates type 1 fimbria-mediated adhesion of uropathogenic *Escherichia coli* to murine bladder epithelial cells. An epithelial cell near the center of the field is covered by hundreds of *E. coli* cells, while the five surrounding cells are virtually devoid of adherent bacteria. To date, there is no molecular explanation for this mosaic pattern of *E. coli* adhesion. Both commensal and pathoadaptive alleles of the adhesin subunit of type 1 fimbriae, FimH, can be found within the overall *E. coli* population. The strain illustrated in this micrograph expresses a pathoadaptive allele of FimH that is typical of uropathogenic *E. coli*. Strains expressing commensal alleles of FimH adhere in extremely small numbers to uroepithelial cells. See chapters 4 and 6 for more information. Micrograph courtesy of D. L. Hasty (University of Tennessee and Veterans Administration Medical Center, Memphis, Tenn.), E. V. Sokurenko (University of Washington, Seattle, Wash.), and Lou Boykins (University of Memphis Integrated Microscopy Center, Memphis, Tenn.). Tim Higgins (Illustrator, University of Tennessee Health Science Center, Memphis) contributed to the cover design.

Back cover figure: Surface view of the FimH molecule (purple), crystallized in the presence of the FimC chaperone (yellow) and a molecule of C-HEGA (green) occupying the mannose-binding site of the lectin. The figure was created with the PyMOL molecular graphics system program of W. L. Delano (http://www.pymol.org) and is provided courtesy of Stefan Knight, Department of Molecular Biology, Swedish University of Agricultural Sciences, Uppsala Biomedical Centre, Uppsala, Sweden.

Copyright © 2003 ASM Press
American Society for Microbiology
1752 N St., N.W.
Washington, DC 20036-2904

Library of Congress Cataloging-in-Publication Data

Ofek, Itzhak.
  Bacterial adhesion to animal cells and tissues / Itzhak Ofek, David L.
    Hasty, and Ron J. Doyle.
      p. ; cm.
  Updates: Bacterial adhesion to cells and tissues / Itzhak Ofek and Ronald J. Doyle. 1994.
  Includes bibliographical references and index.
  ISBN 1-55581-263-5 (hardcover)
    1. Bacteria—Adhesion. 2. Cell adhesion molecules.
[DNLM: 1. Bacterial Adhesion. 2. Adhesins, Bacterial. QW 52 O31b 2003] I. Hasty,
David L. II. Doyle, Ronald J. III. Ofek, Itzhak. Bacterial adhesion to cells and tissues.
IV. Title.
QR96.8.O45 2003
616'.014—dc21

                                                                    2003005915

All Rights Reserved
Printed in the United States of America

Address editorial correspondence to ASM Press, 1752 N St., N.W., Washington, DC
20036-2904, U.S.A.

Send orders to: ASM Press, P.O. Box 605, Herndon, VA 20172, U.S.A.
Phone: 800-546-2416; 703-661-1593
Fax: 703-661-1501
E-mail: books@asmusa.org
Online: www.asmpress.org

DEDICATION

We dedicate this volume to our coauthor and great friend, Ron Doyle. We first became aware that Ron was suffering from amyotrophic lateral sclerosis (Lou Gehrig's disease) in the fall of 2000, when he told us of some weakness in his right arm. The rapidity with which he succumbed to this terrible disease makes us wish we could change our field of research to search for a treatment or cure. The grace with which he accepted his fate makes us wish to have a fraction of his strength of character. With the untimely death of Ron Doyle on 18 January 2002, microbiology lost one of its leaders and we, and many, many others, lost a valued friend and colleague. We miss him tremendously.

Ron was born in Calvert City, Ky., and attended Northeast Louisiana University, Monroe, La., where he studied chemistry and played basketball. He obtained his Ph.D. in microbiology at the University of Louisville Medical School in 1967. He then moved to Roswell Park Institute in Buffalo, N.Y., for postdoctoral studies in protein chemistry. In 1969 he moved back to Kentucky and started his 33-year career in the Department of Microbiology at the University of Louisville, where he became full professor in 1979. During Ron's career in Louisville, he was heavily involved in teaching assignments and directed a very active research laboratory. Ron also served in a variety of administrative posts, including associate dean for research in the University of Louisville School of Dentistry, a Sigma Xi National Lecturer, an American Society for Microbiology division chair, and a Canadian Society of Microbiology section chair. He was also elected to membership in ASM's prestigious American Academy of Microbiology. To say that Ron liked to travel is a major understatement. He traveled all over the world, becoming involved in international research and teaching projects. He lectured widely throughout the world and participated in joint research projects with scientists in various countries. He was an honorary member of the Israeli Society of Microbiology and the Romanian Academy of Medicine.

Although Ron traveled to many different countries around the world, everyone who knew him could attest to the fact that Israel was his favorite. In fact, his determination to write this book on bacterial adhesion, and the 1994 Ofek–Doyle book, was very possibly because it gave him a good excuse to return to Israel year after year. His first contact with Israel was during a 1984 sabbatical at the Weizmann Institute of Science. This initial visit led to continuing collaborations with scientists at the Weizmann Institute, Tel Aviv University, the University of Jerusalem, and Bar-Ilan University. Several of the collaborative projects were supported by joint U.S.-Israel grants. Ron also initiated research projects with scientists from the Arab community of Israel, focusing on rapid diagnosis of mycobacteria, funded by a grant from the Israeli Ministry of Health. Ron went to Israel to write and work and experience its cultures and its history even during times when almost no one else would come.

Ron's early studies focused on cell wall constituents, particularly those of gram-positive bacteria, such as peptidoglycan, teichoic acid, and teichuronic acid. His work greatly advanced our knowledge about the function of these constituents in cell growth and division. His interest in bacterial adhesion originated in the late 1960s, when it became clear that bacterial lectins were involved in the mechanisms used by oral bacteria to cause disease. His interest in this subject was undoubtedly influenced by his earlier studies of lectin-sugar interactions. Ron was an early proponent of using mathematical approaches developed for ligand-receptor interactions for studying bacterial attachment to substrata. Thanks to his contributions, adhesion data can now be analyzed using Scatchard, Hill, and related plots to estimate the affinity, cooperativity, and other parameters of the adhesion. Ron was also one of the first to point out the importance of hydrophobicity in the adhesion process. This was a concept initially dismissed as irrelevant by many researchers in the field, but Ron stood his ground and has long since been vindicated, as the importance of the hydrophobic effect became widely accepted. In more recent years, Ron collaborated with scientists in Mexico, Israel, Romania, and Bulgaria to develop rapid diagnostic tools involving lectin recognition of microbial surfaces. Ron also worked on antiadhesion therapies for bacterial infections. As a variation on this theme, he found that fluoride, at concentrations present in fluoridated tap water, suppresses the ability of streptococci to express adhesins and other virulence factors.

These are a few examples of Ron's contributions to the field of microbiology, many of which have had a marked impact on the field. Ron's legacy includes some 200 primary publications and many reviews and books. His books dealt with a great diversity of subjects, including a number of different areas of current microbiological research and also with the far-reaching consequences of microbiology on culture and history. Family, friends, and colleagues all miss Ron, as do the many dozens of scientists and students from around the world who were the beneficiaries of his generosity. Microbiology has lost a pioneering spirit, and we have lost a very dear friend.

ITZHAK OFEK AND DAVID HASTY
September 2002

CONTENTS

PREFACE

It will be clear from our Dedication that the preface to this book is being written one year after the death of Ron Doyle. Ron played an instrumental role in much of the writing of the text. Even throughout 2001, the year in which he was most afflicted by the debilitating effects of amyotrophic lateral sclerosis, he worked with us on various sections of this book. On several occasions, we visited Ron in Louisville or he visited us during Ofek's sabbatical in Memphis. He had even planned one final trip to Israel for the fall of 2001 to continue working on the book, as well as to visit the country he loved. He continued to play a very important role in writing this book and remained much more active than we had any right to expect, almost until the end of his life.

This text represents an effort to update a 1994 text, *Bacterial Adhesion to Cells and Tissues,* by Ofek and Doyle. The literature surveyed for the previous publication ended on 1 January 1992. The literature surveyed in this volume ended, with some exceptions, on 1 January 2002. Thus, the period covered was a full decade of very active investigation in the field of bacterial adhesion. Many achievements were made over this 10-year period, as evidenced from the many publications cited. We have endeavored to cover as many important issues as possible, but a complete overview of the field became impossible. Although a wholehearted effort was made to review all of the pertinent publications, undoubtedly there were some that escaped our attention, and we apologize in advance to the investigators whose work we did not include. The chapters cover general principles, methodology, characteristics of target cells and tissues, characteristics of bacterial surfaces and the regulation of surface protein expression and biogenesis, and several common themes that emerged during the last decade or two and are under very active investigation. The final chapter compiles, primarily in table form, a list all of the pathogenic bacterial species that were tested for their ability to adhere, the test substrata used, and the adhesins involved in the cases where they were known. It will only take a cursory glance to see which species have received the most attention and to indicate, perhaps, which species offer an opportunity for more active investigation.

We express our gratitude to a number of people who have contributed to our project in various ways. Large parts of this book were written in countries other than the United States and Israel due to the hospitality and inspiration of many of our international colleagues, and our very special thanks go to them: Giuseppi Teti and colleagues at the University of Messina, Messina, Italy; Karen Krogfelt at the Statens Seruminstitut, Copenhagen, Denmark; Hany Sahly at the University of Kiel, Kiel, Germany; and Thomas Hartung at the University of Konstanz, Konstanz, Germany. The hospitality of Erika Crouch, Washington University Medical School, St. Louis, Mo., is also gratefully acknowledged. We are grateful to Kelly Cowan (Department of Microbiology, Miami University, Oxford, Ohio) for writing the Mathematical Analyses section of chapter 2. Naomi Balaban (Department of Human Microbiology, Tel Aviv University, Tel Aviv, Israel), Kevin McIver (Department of Microbiology and Immunology, University of Texas, Southwestern, Dallas, Tex.), and Mark Schembri (Department of Microbiology, Technical University of Denmark, Lyngby, Denmark) were kind enough to read and offer advice on various sections of chapter 4. Evgeni V. Sokurenko provided help with the discussion of probiotics as antiadhesion therapeutic agents in chapter 11. A special debt of gratitude goes to Memphis coworkers, especially Loretta Hatmaker, Tim Higgins, Susan Price, and Linda Snyder, for the variety of important ways in which they helped.

Finally, we acknowledge granting agencies (NIH, Department of Veterans Affairs, U.S.-Israel Bi-National Science Foundation, Ocean Spray Cranberry Co.) for funding work on mechanisms of bacterial adhesion and antiadhesion-based therapy in our laboratories.

ITZHAK OFEK AND DAVID HASTY
January 2003

BASIC CONCEPTS IN
BACTERIAL ADHESION

I

The study of bacterial adhesion to animal cells can be traced to the turn of the 20th century, with the first published results being those of Guyot (32) on hemagglutination (Table 1.1). Although relatively quiescent for several decades, the science of bacterial adhesion began to show signs of emerging from its lag phase in the late 1940s and early 1950s. Electron microscopic studies by Houwink and van Iterson (38) revealed thin, straight filaments which were thought to be adhesive structures. These structures were termed pili by Brinton (4, 5), who studied their biophysical properties and who suggested that they were important adhesive organelles. Elegant studies defining the specificities of bacteria for hemagglutination by Duguid and colleagues (20–22) showed unequivocally that these structures, now termed fimbriae, mediated adhesion to animal cells. By the early 1970s, work by Gibbons and coworkers on streptococcal adhesion (23, 24, 30) established the concept of tissue tropism and initiated an exponential phase of growth of this emerging field. By the end of that decade, a number of milestone studies could be identified, showing that (i) there are specific structures, called adhesins, that are responsible for bacterial attachment; (ii) cloning of genetic elements encoding fimbrial adhesins; (iii) the relationship of adhesion to bacterial tissue tropism and to the pathogenesis of experimental infections; (iv) the identity of certain specific receptor residues of host cells and tissues; and (v) a new way of treating infections based on antiadhesion therapy. These milestone studies then set the stage for a burst of publications during the 1980s that refined and extended the early concepts on adhesion. Thus, during the first several decades of intensified investigation of bacterial adhesion mechanisms, several fundamental principles were established. In the following section, we summarize some of the most important of these principles.

THE PROCESS OF ADHESION

Pathogenic bacteria are negatively charged at pHs encountered in physiologic media. Similarly, adhesive substrata on animal cells and tissues possess negative surface characteristics. If the bacteria are to adhere, an energy barrier must be overcome. Adhesion is thus an energy-dependent event. The energy barrier must be overcome regardless of the mechanism of adhesion. The DVLO theory (17, 99) accounts for the repulsive forces encountered between similarly charged surfaces and is useful in describing bacterium-substratum complexes. Figure 1.1 shows an idealized representation of a bacterium approaching a site on a cell or tissue.

1

TABLE 1.1 Some of the milestones in bacterial adhesion from 1900 to 2001

Year	Observation	Reference(s)
1908	Discovery that *E. coli* causes hemagglutination	32
1949	Electron microscopy of filaments; speculation of adhesive role	38
1955	Discovery that mannose inhibits hemagglutination	9
1955	Relationship of fimbriation and hemagglutination to adhesion to epithelial cells	22
1959	Phase variation of fimbriation	4
1961	Chromosomal location of fimbriation factor	6
1965	Structure of type 1 fimbriae	5
1971	Tropism of oral bacteria	30
1973	Antiadhesin-based vaccine	76
1975	Identification of LTA as streptococcal adhesin	64
1977	Identification of surface sugars as receptors for adhesion	61
1977	Discovery that purified fimbriae hemagglutinate	77
1978	Hydrophobic effect in interaction with phagocytic cells	98
1978	Soluble fibronectin-bacterium interaction	43
1979	Discovery that subinhibitory concentrations of antibiotics modulate adhesin expression	63, 90
1979	Use of type 1 fimbrial adhesin as an adhesin-based vaccine	82
1979	Cloning of adhesin-encoding gene cluster	50
1979	Discovery that infection in humans depends on adhesion	78, 79, 81, 100
1979	Phagocytosis mediated by fimbrial adhesin	75, 83
1979	Receptor analogs for treatment of infections	2
1980	Glycolipids as fimbrial receptors	44
1982	Requirement of multiple adhesins for streptococcal adhesion	53
1982	Hydrophobic effect in adhesion of oral bacteria	54
1983	Fibronectin as cellular receptor for adhesion	84
1985	Identification of adhesin as a minor component of fimbrial structure	48
1989	Biogenesis of fimbriae	45
1999	First determination of adhesin three-dimensional structure	8

The adhesion process is thought to occur in distinct steps. The bacteria must first overcome repulsive forces that separate negatively charged particles, and this is thought to be accomplished by hydrophobic determinants that effectively lead to weak, reversible adhesion. Although the hydrophobic effect is sometimes referred to as nonspecific, this is probably an oversimplification. A theorem in organic chemistry is that "like dissolves like." In this regard, any hydrophobic molecule should "dissolve" (or interact with) any other hydrophobic molecule. The hydrophobic interaction (or hydrophobic effect) is therefore conventionally considered to be nonspecific, by definition. This convention, however, does not take into account the fact that fatty acids often bind to certain host proteins in a saturable and inhibitable manner, suggest-

ing that there is a degree of specificity even in hydrophobic interactions. The term "hydrophobin" is used to refer to a molecule or site on bacterial surfaces that are engaged in the adhesion process by interacting via hydrophobic moieties (72–74). In the relatively few studies examining the initial stage of adhesion, specific molecules have been identified on the bacterial surface that mediate the reversible interaction. For example, lipoteichoic acid (LTA) on the surface of group A streptococci and the type IV bundle-forming family of fimbriae produced by several different human pathogens (see chapter 4) are mediators of reversible adhesion. Firm adhesion of bacteria is the result of one or more succeeding steps in which a second tier of stereospecific adhesins bind to complementary receptors on host surfaces. The physical forces that drive both types

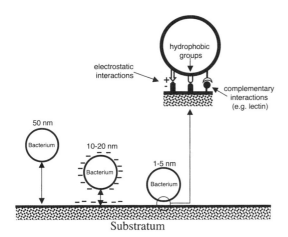

FIGURE 1.1 Description of the electrical double layer and bacterial adhesion. In this diagram, three distinct interaction regions are depicted. One region (>50 nm) reflects van der Waals' attractions only. A closer region (10 to 20 nm) involves both van der Waals' attractions and Coulombic forces, and it is in this region that maximum repulsion due to the net negative charges of the opposing surfaces occurs. Because the repulsive forces increase in proportion to the diameter of the particles approaching each other, fimbriae or other polymers having a smaller diameter can be a very effective means of overcoming the barrier. A third, even closer region (>2 nm) requires complementary binding sites, which may involve hydrophobin-hydrophobin, lectin-carbohydrate (complementary), and charge-charge (electrostatic) interactions. The presence of hydrophobic sites may stabilize other interacting sites. Hydrophobic sites and/or numerous charge-charge sites contribute to form, over time, a virtually irreversible adhesion. (Adapted from reference 7.)

of adhesive interactions include ionic and Coulombic interactions, hydrogen bonding and the hydrophobic effect, and coordination complexes involving multivalent metal ions.

The multistep kinetic model we proposed 10 years ago (37) was based largely on studies with oral streptococci and group A streptococci. Adhesion of *Streptococcus sanguis* to saliva-coated hydroxylapatite (SHA) was found to depend on multiple interacting sites (13–15, 53, 54). It was suggested that adhesion to SHA was cooperative in that the binding of a few bacteria tended to promote the binding of additional bacteria. This cooperative adhesion was dependent on both the

hydrophobic effect and nonhydrophobic interactions. Other work has supported the view that *S. sanguis* adheres to SHA via multiple sites (27, 31, 51, 52). The *S. sanguis*-SHA complex seems to proceed by two distinct kinetic steps, the first of which is readily reversible and is mediated by the hydrophobic effect. The second step is mediated primarily by nonhydrophobic interactions and can only follow successful completion of the first steps. Once the second reaction has taken place, the cell becomes firmly bound and this binding is only very slowly reversible. Detachment would be expected only when all adhesin-receptor interactions are coordinately reversible, the probability of which is low. The observation that agents which interfere with ionic or hydrogen bonds (salts or high temperatures) or with the hydrophobic effect (chaotropes or low temperature) inhibit adhesion supports the validity of the multiple-site model (13–15, 53, 54, 106). Furthermore, it has been shown that adhesion of *S. sanguis* occurs in two distinct kinetic steps whereas desorption appears to occur in a single step (13, 14). This behavior is also consistent with the multiple-site model.

The adhesion of group A streptococci may also obey the two-step kinetic model. One step would involve hydrophobic interactions between LTA and the fatty acid-binding domains on epithelial cells (Fig. 1.2). The correlation between hydrophobicity of the streptococci and the ability of the bacteria to adhere to epithelia supports this premise. A time-dependent, virtually irreversible adhesion of group A streptococci to fibronectin was observed by Speziale et al. (87). They found that proteolysis of the bacteria reduced the adhesion to fibronectin. Preparations of a tryptic digest of the bacteria which contained very small amounts of LTA were capable of inhibiting adhesion. It was postulated that the adhesion of group A streptococci to epithelial cells follows a two-step reaction sequence (37). LTA mediates the first step of streptococcal adhesion to fibronectin on animal cells (11), and M protein (12) or fibronectin bind-

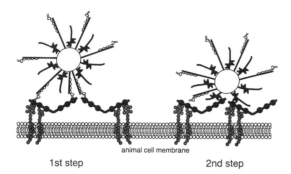

1st step   2nd step

FIGURE 1.2 Hypothetical two-step model for the interaction of bacteria with host substrata. The model suggests that there are at least two sequential steps leading to firm adhesion. The different steps probably involve two different adhesins and receptors but could involve different binding sites on a single molecule. The first step may be relatively weak and reversible and is probably accomplished by adhesins that extend some distance from the surface of the bacterium in order to bridge the charge repulsion of the opposing surfaces. In this model, successful completion of the first step is predicted to be requisite and facilitatory for the second step, which perhaps would involve more stereospecific interactions. Successful completion of this second step would probably bring the organism past the charge repulsion barrier, within a very few nanometers of the host cell surface, and increase the strength of adhesion, making the interaction essentially irreversible. This model should be considered generic, but the figure is based on one hypothetical mechanism for adhesion of group A streptococci to host cells. The first step occurs when LTA (complexed with a surface protein) binds to fibronectin, and the second step occurs as protein F subsequently binds to fibronectin. Reactions of other bacteria with other substrata, such as *S. sanguis* binding to the dental pellicle, should follow similar steps. Although the model is largely derived from results with *S. sanguis* and *S. pyogenes* (37), its principal features should be valid for any bacterium-host cell interaction.

ing protein (e.g., protein F [33, 34, 97]) mediates the second step of adhesion.

In most cases the reversible step of adhesion has been completely overlooked because nonadherent bacteria are removed by numerous washing steps, procedures which also remove loosely adherent bacteria. Indeed, the reversible step of adhesion is not easily studied, and experiments must be carefully designed for this purpose (see chapter 2). It must be pointed out, however, that it was recently shown that the application of shear force, per se, does not necessarily lead to removal of bacteria that are in the early stages of adhesion. In fact, under certain circumstances, shear stress was found to enhance the binding activity of a fimbrial adhesin, presumably by the induction of conformational changes in the ligand binding site of the adhesin (94). Because adhesion of bacteria to surfaces almost always occurs in the presence of shear forces, their effects must be taken into consideration when designing or interpreting adhesion experiments. Understanding the forces and specific molecules that are involved in the different steps of adhesion is important, since the information may have significant consequences in the design of effective antiadhesive therapeutic approaches (see chapter 10).

NATURE OF ADHESINS AND THEIR COGNATE RECEPTORS

When considering adhesin-receptor interactions, the definitions of the words "adhesin" and "receptor" must be carefully understood. The literature frequently interchanges the words. The adhesin would bind the receptor regardless of the origin of the receptor. In this book, the word "adhesin" is referred to as the binding entity originating from the bacterium, regardless of its chemical nature (e.g., polysaccharide or protein). To date, the adhesive interactions of almost 300 human- and animal-pathogenic bacterial species have been investigated (41, 56) (see chapter 12). Based on these studies, three principal types of adhesin-receptor interactions have been described (Table 1.2; Fig. 1.3). The first type of adhesive interaction is due to the binding of lectins with carbohydrate structures. Lectin-carbohydrate types of recognition can occur when the lectin is on the bacterial surface or on the host surface. The type in which the lectin is on the bacterial surface predominates and undoubtedly contains the largest group of adhesins so far described among bacterial pathogens. Although comprising a smaller number of cases, there are nevertheless examples of other lectin-carbohydrate adhesive

TABLE 1.2 Types of adhesin-receptor interactions in bacterial adhesion to mucosal surfaces

Type of interaction	Bacterial adhesin (example)	Receptor on epithelial cell (example)	Reference(s)
Lectin-carbohydrate	Lectin (type 1 fimbriae)	Glycoprotein (uroplakin on bladder cells)	104
	Capsule (hyaluronic acid of *S. pyogenes*)	Lectin (CD41 on epithelial cells)	16
Protein-protein	Fibronectin binding proteins (F protein of *S. pyogenes*)	Fibronectin (fibronectin on respiratory cells)	33, 34
Hydrophobin-protein	Glycolipid (LTA of *S. pyogenes*)	Lipid receptors? (fatty acid binding site in fibronectin on host cells)	11, 37
	Lipid binding proteins (surface protein of *Campylobacter* spp.)	Membrane lipids (phospholipids and sphingolipids of cells)	91, 92

interactions in which the lectin is on the mucosal surface. The second type, of which a significant number of cases are known, involves recognition between a protein on the bacterium and a complementary protein on the mucosal surface. The third type, and the least well characterized, comprises the binding interactions that occur between hydrophobins, frequently involving hydrophobic moieties of proteins interacting with lipids. The hydrophobic protein and the lipid can be found on either the bacterial or the host cell surface. The contribution of hydrophobicity to bacterial adhesion to mucosal surfaces is probably underestimated because it is often responsible for the initial,

weak and reversible interaction that is so difficult to measure. A key feature of bacterial adhesins is that they are associated with surface structures (57) (see chapter 4). Typical examples are the fimbrial adhesins of *Escherichia coli,* consisting of macromolecular assemblies of several hundreds of subunits of several different proteins (28).

LECTINS AS ADHESINS
Because the lectin family of adhesins is shared by many bacterial pathogens, it has been studied the most extensively. In gram-negative bacteria, the lectins may be presented as parts of fimbriae, as amorphous or seemingly amor-

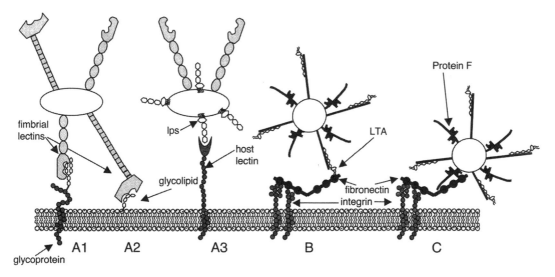

FIGURE 1.3 Diagrammatic representation of bacterial adhesion involving lectin-carbohydrate interactions (A1, bacterial lectin-host glycoprotein; A2, bacterial lectin-host glycolipid; A3, host lectin-bacterial LPS) (A), hydrophobin-protein interactions (B), and protein-protein interactions (C). For more details, see Table 1.2 and the text.

phous structures, or as other types of outer membrane components (58). In gram-positive bacteria, lectins may be found in the peptidoglycan matrix or may protrude through the cell wall from the cytoplasmic membrane.

The lectin adhesins are usually classified by sugar specificity. Specificity can be determined by inhibiting the adhesion with either simple or complex carbohydrates that compete with the binding of the adhesins to host cells. In general, the affinity of simple sugars (e.g., mono- or disaccharides) to the lectin adhesins is low (i.e., the concentration required for 50% inhibition is in the millimolar range). An increase of several orders of magnitude in the inhibitory potency can be achieved by suitable chemical derivatization. For example, a marked increase in affinity of the inhibitors for the bacterial lectins can be obtained by attaching them to polymeric carriers to form multivalent ligands (46). Methylumbelliferyl or *para*-nitro-phenyl derivatives of alpha-mannosides are 500 to 1,000 times more inhibitory than the methyl-alpha-mannoside (26) (also see chapter 11). However, the sugar specificity is usually defined by the simplest saccharide that is able to inhibit the adhesion in the millimolar range.

Some bacterial lectin-adhesins have been identified that recognize terminal saccharide structures, but many recognize internal sequences as well. For example, the tip adhesin PapG of P fimbriae recognizes internal Gal(α1→4)Gal sequences in glycolipids on host cell surfaces (88, 89). The glucan binding lectin of some oral streptococci recognizes 6 to 10 internally linked (α1→6) glucose residues (19). Another interesting feature of the carbohydrate-lectin interaction is that, in at least some cases (e.g., *Helicobacter pylori*), the binding of the bacterial lectin to glycolipids appears to be dependent on the linkage of fatty acids to the sugar moieties (Table 1.3).

TABLE 1.3 Examples of carbohydrates as attachment sites for bacteria colonizing mucosal surfaces[a]

Organism	Target tissue	Carbohydrate structure	Form[b]
E. coli			
Type 1	Urinary tract	Man(α1→3)Man(α1→3)Man(α1→6)Gp	
P	Urinary tract	Gal(α1→4)Gal	GL
S	Neural	NeuAc(α2→3)Gal(β1→3)GalNAc	GSL
CFA/1	Intestinal	NeuAc(α2→8)	Gp
CS3	Intestinal	GalNAc(β1→4)Gal	Gp
K1	Endothelial	GlcNAc(β1→4)GlcNAc	Gp
K99	Intestinal	NeuGc(α2→3)Gal(β1→4)Glc	GL
H. pylori	Stomach	NeuAc(α2→3)Gal	GL
		Lewis-b blood group	Gp
		Glucose-fatty acid	GL
		Lactosyl ceramide	GL
N. gonorrhoeae	Genital	Gal(β1→4)Glcβ	GL
		NeuAc(α2→3)Gal(β1→4)GlcNAc	Gp
P. aeruginosa	Intestinal	Gal(β1→3)GlcNAc	Gp
		Fucose	Gp
		Mannose	Gp
	Respiratory	GalNAc(β1→4)Gal	GL
H. influenzae	Respiratory	GalNAc(β1→4)Gal	GL
	Respiratory	GalNAc(β1→4)Gal	GL
S. pneumoniae	Respiratory	GlcNAc(β1→3)Gal	Gp
M. pneumoniae	Respiratory	NeuAc(α2→3)Gal(β1→4)GlcNAc	Gp
S. suis	Respiratory	Gal(α1→4)Gal	Gp
K. pneumoniae	Respiratory and intestinal	Gal(α1→4)Gal	Gp

[a]Based on references 41, 58, and 80.
[b]Predominant forms: Gp, glycoproteins; GL, glycolipids; GSL, glycosphingolipids. In all cases, the simplest carbohydrate structure inhibiting adhesion is shown.

The study of bacterial lectin-adhesins, especially those that are associated with multisubunit fimbriae, has been hampered because of the difficulties in obtaining the lectins in a pure and soluble form. Whole fimbriae are purified with relative ease, but fimbrial lectins have not been purified in a soluble and functional form from purified fimbriae. Recently, however, major breakthroughs have been achieved in the preparation of soluble fimbrial lectins. By preparing fusion proteins of the amino-terminal regions of the P fimbrial lectins, PapGI, PapGII, or PapGIII, with a fragment of staphylococcal protein A, the Gal(α1→4)Gal-binding domain was clearly identified (35). The three fusion proteins exhibited distinct fine sugar specificities identical to those of the parent fimbriae (see below). Furthermore, a soluble and functional form of FimH was purified by expressing the lectin either as a chimeric protein with maltose binding protein (93) or as a complex with the FimC chaperone. The FimH-FimC complex has been crystallized with a mannose inhibitor, and the three-dimensional structure has been solved (Fig. 1.4) (8, 39). It is anticipated that many more of the fimbrial lectins will now be purified by using similar techniques and that their combining sites will be identified and characterized.

A great deal of effort has been given to the definition of the saccharide specificities of bacterial lectin-adhesins. In the past, monosaccharide and disaccharide inhibitors were used to group lectin-bearing bacteria into distinct adhesive classes. For example, in the earliest studies, enterobacteria were divided into only two groups based on the sensitivity of hemagglutination reactions to mannose: mannose sensitive (MS) and mannose resistant (MR). As more saccharides were used as either inhibitors or adhesive substrata and as increasing numbers of isolates were tested, the MR group was subdivided into several very distinct categories. Some of these broad categories, for instance the Gal(α1→4)Gal binding P fimbriae of uropathogenic *E. coli,* were subsequently subdivided into several different classes based on the specific location of the

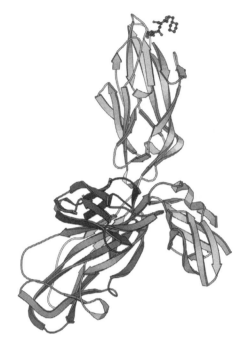

FIGURE 1.4 Three-dimensional structure of a fimbrial adhesin. The FimH lectin adhesin is presented at the tips of type 1 fimbriae of *E. coli.* Purification of soluble FimH was made possible by overexpressing FimH in a complex with its periplasmic chaperone, FimC. Although the fine sugar specificity of this FimH complex has not been defined, a ligand binding pocket present on the surface of the lectin domain of FimH is capable of accommodating a monomannose unit. The saccharide indicated in the figure, cyclohexylbutanoyl-*N*-hydroxyethyl-D-glucamide, was cocrystallized with the FimH-FimC complex (8). This has now been confirmed by cocrystallization with mannoside (39). (Courtesy of Stefan Knight.)

digalactoside within the oligosaccharide moieties of glycolipid (88, 89) (Fig. 1.5).

From the foregoing, it is clear that the sugar specificity of a bacterial lectin must include both the primary sugar specificity and the fine sugar specificity. The primary sugar specificity defines the simplest carbohydrate structure that inhibits lectin-mediated adhesion, whereas the fine sugar specificity defines the structure of the most potent carbohydrate inhibitor. Many bacterial variants or clones or even species may have the same primary sugar specificity but may otherwise differ significantly in the fine sugar specificity of their lectin adhesins. For example, MS adhesion is the property of type 1 fimbriae, which initially were thought to be an adhesive class that was functionally homogeneous among many enterobacterial species. However, distinctions have recently been identified within the strains that exhibit MS adhesion via type 1 fimbriae (see Table 6.1). Whereas all type 1-fimbriated strains are sensitive to D-mannose, intergenus differences (e.g., *E. coli* versus *Salmonella*) were noted when aromatic mannosides were used as inhibitors (62). It has recently been found that other mannose derivatives can also be used to detect even intraspecies differences, such as differences among *E. coli* isolates (see Table 6.2) (86). Careful analysis of the fine sugar specificities of lectin-bearing bacteria is therefore an extremely useful tool in the characterization of bacterial clones or variants associated with particular types of infections.

BACTERIAL POLYSACCHARIDES AND LIPOPOLYSACCHARIDES AS ADHESINS

Until recently, only lectins expressed on the surface of macrophages had been found to interact with complementary carbohydrates on bacterial surfaces (65). In all of these cases, the carbohydrate structures recognized by the macrophage lectins were contained in either the capsular polysaccharides or the lipopolysaccharides (LPS) of the outer membrane of gram-negative bacteria. During the last 5 years, a number of studies have shown that surface carbohydrates, such as LPS, or the lipooligosaccharides, which simply have fewer repeating units, can also serve as adhesins

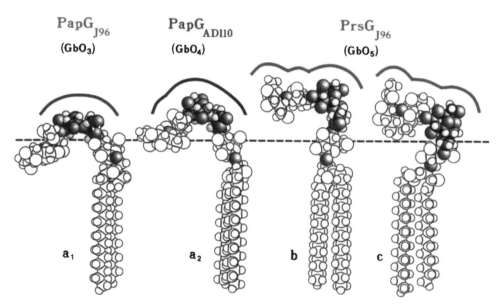

FIGURE 1.5 Schematic illustration of the three classes of globoseries glycolipid receptors for PapG adhesins of *E. coli* P fimbriae. The simplest carbohydrate structure for the PapG adhesins is the Gal(α1→4)Gal moiety, but flanking saccharides are important in the binding of different classes of the adhesins. Class I, represented by the PapG adhesin cloned from *E. coli* J96, binds most effectively to the globotriaosyl ceramide (GbO$_3$). The class II adhesin is represented by the PapG adhesins cloned from *E. coli* IA2 and AD110 and binds globotetraosyl ceramide (GbO$_4$). Class III adhesins, represented by another PapG cloned from *E. coli* J96, binds best to the Forssman glycolipid (GbO$_5$). The classes of adhesins affect both tissue and host tropism, with class II predominating in human urinary tract infections due to the predominance of the globotetraosyl ceramide at the mucosal surface of the human urinary tract (see chapter 6). (Reprinted from N. Strömberg, P. G. Nyholm, I. Pascher, and S. Normark, Saccharide orientation at the cell surface affects glycolipid receptor function, *Proc. Natl. Acad. Sci. USA* **88:**9340–9344, 1991, with permission from the publisher.)

which mediate binding of the microorganisms to nonprofessional phagocytes, such as mucosal epithelial cells, as well as to mucus constituents, presumably by recognizing lectins on the host cell (40).

Rough mutants lacking the O side chain of LPS bind to the epithelial cells to a lesser degree than do the parent strains, and isolated LPS can inhibit the binding. Further evidence for the adhesin activity of LPS was obtained in a study showing that isolated *Vibrio mimicus* LPS caused hemagglutination of rabbit erythrocytes (1). It has also been proposed that the heptose-3-deoxy-D-manno-2-octulosonic acid (KDO) present in the inner core region of LPS is recognized by a lectin-like receptor on the plasma membrane of rat hepatocytes (67). The inner-core LPS of *Pseudomonas aeruginosa,* having a terminal glucose residue, was found to be necessary for the bacteria to bind to and be internalized by corneal epithelial cells (105). When the delipidated core region of LPS was used as an inhibitor of the association of the bacteria with corneal cells, the magnitude of inhibition was only 50%, based on CFU counts. In general, bacteria which lose repeating carbohydrate units from LPS tend to be more hydrophobic (72), suggesting that the increased adhesion of LPS mutant may depend on the hydrophobic effect rather than on protein-carbohydrate interactions. Furthermore, unlike phagocytic cells or hepatocytes, an animal lectin as an integral component of the membrane of mucosal epithelial cells has not yet been characterized (29).

Because most of the studies cited above showed that the core region of LPS was implicated in mediating adhesion, it is possible that a host-derived component that specifically binds this region of LPS is involved. Indeed, it has been known for some time that lung surfactant protein D (SP-D) binds the core region of LPS (42). Although SP-D was found in large amounts in lung secretions, various amounts of SP-D mRNA were also expressed in a number of other tissues (47). Therefore, the possibility that SP-D may become immobilized on certain host mucosal cells and mediate the adhesion of bacteria by recognition of the LPS core region cannot be excluded.

PROTEIN-PROTEIN INTERACTIONS

The protein-protein type of interaction in bacterial adhesion is probably best exemplified by the interaction of fibronectin binding proteins and fibronectin on the animal cell surface. Fibronectin or fibronectin-like domains are present on a great variety of animal cells and in tissues. The number of bacteria known to express fibronectin binding proteins is growing (36). However, the mere fact that a fibronectin-binding protein is present does not indicate that this protein functions as an adhesin. In fact, in only few studies has the interaction of a fibronectin binding protein with fibronectin been documented clearly to directly mediate adhesion. Whether this type of protein-protein interaction is a common mechanism through which bacteria adhere to animal cells remains to be determined. In one bacterial species that has been well studied, *Streptococcus pyogenes,* more than six different fibronectin binding proteins have been identified (25). One of the fibronectin binding proteins, protein F1, binds fibronectin via domains that are 37 amino acids in length and repeated two to six times. Interestingly, the fibronectin binding sites span the C terminus of one repeat and the N terminus of the following repeat (66). These domains, termed RD2 domains, bind to the 29-kDa N-terminal fragment of fibronectin. Another region in protein F1, termed UR, is located immediately N-terminal to RD2 and binds not only to the N-terminal 29-kDa fragment of fibronectin but also to the adjacent 40-kDa collagen binding domain.

Protein-protein interaction may involve bacterial proteins that bind to components of the extracellular matrix other than fibronectin. These matrix components include the collagens, elastin, fibrinogen, laminin, thrombospondin, and vitronectin. Although these proteins are normally found beneath the epithelial cells of mucus membranes (see chapter 3), they can become exposed by wounding the mucosal barrier by various means. There are many different bacterial proteins that bind one or more components of the extracellular matrix, so many that the term MSCRAMMs (for "microbial surface components recognizing adhesive matrix molecules") (68) has become a commonly used descriptor.

HYDROPHOBINS AS BACTERIAL ADHESINS

The term "hydrophobin" was given to any surface component that promotes surface hydrophobicity and adhesion to surfaces (73). Although only a very few hydrophobin adhesins have been characterized, the role of the hydrophobic effect in adhesion has been well documented (72). As discussed above, the use of hydrophobins as adhesins may be a common mechanism to overcome repulsive forces. There are a number of methods to determine hydrophobicity (see chapter 2), but none of these methods is specific for a particular hydrophobin (18). It may also be possible that the contribution of an adhesin to hydrophobicity will be evident when one method is used but not when a different method is used. Nevertheless, adhesion to hydrocarbon droplets (see chapter 2) has become the test of choice to demonstrate the contribution of an adhesin to hydrophobicity (74).

COMPLEXITIES OF POTENTIAL RELATIONSHIPS BETWEEN ADHESINS AND RECEPTORS

The identity of the bacterial adhesins and their cognate receptors on host cells has been determined in a considerable number of pathogenic organisms (reviewed in references 56, 57, and 103) (see chapter 12). The complexities of adhesin-receptor relationships exhibit some general features (Table 1.4). A particular receptor may contain more than one attachment site specific for two or more adhesins. This is illustrated by the Dr blood group, a glycoprotein, which acts as a receptor on host cell membranes for three different clones of E. coli, each of

which produces a distinct adhesin that binds to a different region of the Dr blood group molecule. Another feature is that two different pathogens, each expressing structurally distinct adhesins, can exhibit the same receptor specificity. This is the case with Staphylococcus aureus and Streptococcus pyogenes, both of which bind the amino-terminal region of fibronectin on mucosal cells. The adhesin on S. aureus is a fibronectin binding protein, whereas one of the fibronectin-specific adhesins of S. pyogenes is LTA (Table 1.4).

The finding that several different respiratory tract pathogens recognize the disaccharide GalNAc($\beta1\rightarrow4$)Gal is yet another example of adhesin-receptor complexity (Table 1.4). It has been suggested that the GalNAc($\beta1\rightarrow4$)Gal sequence is preferentially accessible in the glycolipids of the respiratory epithelium, allowing firm binding of a diverse group of respiratory pathogens bearing the complementary adhesins. In some cases, however, distinct adhesins have the same specificity but are carried by different bacteria that colonize different tissues and animal hosts. For example, the Gal($\alpha1\rightarrow4$)Gal-specific lectins occur in uropathogenic P-fimbriated E. coli, the pig pathogen Streptococcus suis (95), and respiratory and enteropathogenic fimbriated Klebsiella pneumoniae (69).

Conversely, the same bacterial adhesin can bind to several distinct receptors on different cell types. Such receptors are called isoreceptors. Several glycoproteins ranging in size from 45 to 110 kDa have been described as receptors for type 1 fimbriae on different cell types (Table 1.4). All these isoreceptor glycoproteins have a common oligomannose-containing attachment site for FimH, the adhesin moiety of type 1 fimbriae.

Another complexity is illustrated by the filamentous hemagglutinin adhesin of Bordetella pertussis, which contains multiple domains, each with a distinct receptor specificity. This hemagglutinin, which has been cloned and sequenced, contains at least three domains. One of these domains contains an arginine-glycine-aspartic (RGD) sequence which binds the bacteria to a

TABLE 1.4 Examples of various complexities of receptor-adhesin relationships[a]

Type[b]	Bacterial clone[c]	Host target cell[c]	Adhesin[c]	Receptor[c]	Attachment site of receptor[c]
A	E. coli, UTI	Erythrocytes	Dra fimbriae	Dr-BG	Dra specific
	E. coli, ETEC	Erythrocytes	AFA II	Dr-BG	AFA II specific
	E. coli, pigs	Erythrocytes	F 1845 fimb	Dr-BG	F 1845 specific
B	S. pyogenes	Mucosal	LTA	FN	Amino terminal
	S. aureus	Mucosal	FBP	FN	Amino terminal
C	E. coli, Gal(α1→4)Gal	Uroepithelial	P fimbriae	Fso G	P-BG
	Pyelonephritis	P fimbriae	Fso F/H, FN	ND
D	E. coli	Erythrocytes	Type 1 fimbriae	66-kDa Gp	Oligoman
	PMN	Type 1 fimbriae	CD11/18 Gp	Oligoman
	Macrophages	Type 1 fimbriae	CD 48 Gp[d]	Oligoman
	Uroepithelial cells	Type 1 fimbriae	Uroplakin	Oligoman[e]

[a]Adapted from references 55 to 60.

[b]A, Target host cell expresses one receptor molecule that contains three attachment sites for three different adhesins produced by three clones of bacteria. B, Two bacterial species express two distinct adhesins that bind the bacteria to the same receptor molecule on the target host cell. C, the same bacterial clone produces a fimbrial structure composed of two subunits, each of which binds the bacteria to distinct receptors on the host target cell. D, the same adhesin binds the bacteria to similar attachment sites contained in different receptor molecules (isoreceptors) expressed by various host target cells.

[c]BG, blood group; FN, fibronectin; FBP, fibronectin-binding protein; Gp, glycoprotein; LTA, lipoteichoic acid; Oligoman, oligomannoside; PMN, polymorphonuclear leukocytes; UTI, urinary tract infection; ND, not defined;

[d]Data from reference 3.

[e]Data from reference 104.

CR3 integrin present on pulmonary macrophages (70). The hemagglutinin also contains two carbohydrate binding domains, one of which is specific for galactose (96) and the other is specific for sulfated sugars (49).

MULTIPLE ADHESINS

Human bacterial isolates are confronted with numerous environmental challenges, one of which is the necessity that they adhere to one or more of the various mucosal surfaces in the body between their portal of entry into the host and the site where physiological conditions favor colonization. The various habitats can be considered either conducive to commensalism, where the colonizing bacteria rarely reach sufficient density to cause disease, or conducive to pathologic colonization, where such a critical mass of bacteria is more readily achieved. To maintain themselves in different environments such as the upper respiratory system, the gastrointestinal system, or the urogenital system, bacteria must be able to interact with different receptors on the distinct surfaces. To achieve this goal, evolutionary pressures have selected clones of bacterial species that are able to express

more than one type of adhesin. Indeed, as shown in chapter 12, many bacterial species have been found to express more than one type of adhesin, and in some cases more than 10 have been described. The expression of these adhesins is under the control of complex regulatory systems that is discussed later (see chapter 4). In the relatively few cases in which it has been tested, it was found that most bacterial cells preferentially produce only one type of adhesin at any one time (59). In one study, for instance, fewer than 9% of fimbriated cells coexpressed both type 1 and P fimbriae simultaneously. Other studies have shown that some E. coli strains are capable of coexpressing type 1 and S fimbrial lectins. In group A streptococci, M protein and protein F1 adhesins are differentially expressed and environmental conditions that upregulate the expression of one of these adhesins downregulates expression of the other (97). There are other cases, however, in which equal amounts of several different adhesins are expressed, such as with Fusobacterium nucleatum (101). It is obviously of great interest to gain a detailed understanding of the factors of the environment and/or the bacterial genome that contribute to the regulation of multiple adhesin expression.

The chemical, functional, and ultrastructural diversity of adhesins expressed by a single bacterial clone undoubtedly endow the organisms with the ability to colonize a range of substrata. Traditionally, investigators have emphasized studies of a particular phenotype expressing one type of adhesin. The detection of any particular phenotype is dependent on the presence of specific receptors on the target substratum chosen and on other factors such as the genotypic background of the strain and the specific growth conditions. As growth conditions and test substrata are varied, the chances of detecting more than one type of adhesin increase. For *E. coli,* it is not unreasonable to suggest that any single clone is capable of producing at least two different adhesins. This is probably true for most, if not all, bacteria. Multiple adhesins have been reported in at least 15 different pathogenic species (see chapter 12).

ROLE OF ADHESION IN INFECTIONS

The likelihood that differences in the adhesive capacity of bacteria would be reflected in their tissue tropism and, especially, their pathogenic potential was an early speculation. Indeed, the relationship between adhesion and tissue tropism or pathogenesis was undoubtedly the primary driving force behind the impressive development of this field of science during the mid-1970s.

The earliest studies of the role of adhesion in infections were only able to correlate the relative ability of bacteria to adhere to cells in vitro with the relative infectivity of the strains in vivo (55, 60). In most cases, there was little or no information about the specific adhesins or receptors involved in these interactions. Whereas the use of adhesin-expressing and adhesin-negative variants was a minor improvement in the assays performed, these strains were, nevertheless, nonisogenic pairs.

It was only when it became possible to clone and express adhesin genes and to engineer adhesin-negative recombinant strains that it became relatively easy to obtain a clear correlation between adhesion and infectivity (Table 1.5). This relationship became even clearer when the infectivity of the pathogen in hosts that express different levels of the adhesin receptor, as exemplified for the infectivity of K99 *E. coli* in piglets, which express the K99-specific receptor, was compared to the infectivity of K99 in adult pigs, which do not express this receptor (Table 1.5). Perhaps

TABLE 1.5 Examples showing relationship of adhesion to infectivity

Bacterial species	Adhesin or receptor	Manipulation	Relative infectivity compared to parent	In vitro adhesion	Reference
Adhesin manipulation					
E. coli	FimH	Inactivation[a]	Low	Low	10
E. coli	FimH	Inactivation[a]	Low	Low	85
E. coli	FimH	3-Man+[b]	Low	Low	85
E. coli	FimH	1-Man+[b]	High	High	85
S. pyogenes	M protein	Inactivation[a]	Low	Low	12
S. pyogenes	Capsule	Inactivation[a]	Low	High	102
Receptor manipulation					
E. coli	NeuGC(α2→3)Gal(β1→4)Glc	Variant/piglets[c]	High	High	
	NeuAc(α2→3)Gal(β1→4)Glc	Variant/adult pigs[c]	Low	Low	

[a]Inactivation of the gene encoding the bacterial adhesin.
[b]Bacterial variants that express trimannose or monomannose alleles of FimH.
[c]Hosts that express receptor variants.

the most impressive examples highlighting the role of adhesion in the infectious process come from the application of adhesion inhibitors, which prevented infection (see chapter 11). It should be noted, however, that not all studies have found a correlation between the ability of bacteria to adhere to cells in vitro and their ability to cause an infection in vivo (71).

REFERENCES

1. **Alam, M., S.-I. Miyoshi, K.-I. Tomochika, and S. Shinoda.** 1996. Purification and characterization of novel hemagglutinins from *Vibrio mimicus*: a 39-kilodalton major outer membrane protein and lipopolysaccharide. *Infect. Immun.* **64:**4035–4041.
2. **Aronson, M., O. Medalia, L. Schori, D. Mirelman, N. Sharon, and I Ofek.** 1979. Prevention of colonization of the urinary tract of mice with *Escherichia coli* by blocking of bacterial adherence with methyl alpha-D-mannopyranoside. *J. Infect Dis.* **139:**329–332.
3. **Baorto, D. M., Z. Gao, R. Malaviya, M. L. Dustin, A. van der Merwe, D. M. Lublin, and S. N. Abraham.** 1997. Survival of FimH-expressing enterobacteria in macrophages relies on glycolipid traffic. *Nature* **389:**636–639.
4. **Brinton, C. C., Jr.** 1959. Non-flagellar appendages of bacteria. *Nature* **183:**782–786.
5. **Brinton, C. C., Jr.** 1965. The structure, function, synthesis and genetic control of bacterial pili and a molecular model for DNA and RNA transport in Gram-negative bacteria. *Trans. N. Y. Acad. Sci.* **27:**1003–1054.
6. **Brinton, C. C., P. C. Gemski, Jr., S. Falkow, and L. S. Baron.** 1961. Location of the piliation factor on the chromosome of *Escherichia coli. Biophys. Biochem. Res. Commun.* **5:**293–299.
7. **Busscher, H. J., and A. H. Weerkamp.** 1987. Specific and non-specific interactions in bacterial adhesion to solid substrata. *FEMS Microbiol. Rev.* **46:**165–173.
8. **Choudhury, D., A. Thompson, V. Stojanoff, S. Langermann, J. Pinkner, S. J. Hultgren, and S. D. Knight.** 1999. X-ray structure of the FimC-FimH chaperone-adhesin complex from uropathogenic *Escherichia coli. Science* **285:**1061–1065.
9. **Collier, W. A., and J. C. deMiranda.** 1955. Bacterien—Haemagglutination. III. Die Hemmung der Coli—Haemagglutination durch Mannose. *Antonie Leeuwenhoek* **21:**133–140.
10. **Connell, H., W. Agace, P. Klemm, M. Schembri, S. Mårild, and C. Svanborg.** 1996. Type 1 fimbrial expression enhances *Escherichia coli* virulence for the urinary tract. *Proc. Natl. Acad. Sci. USA* **93:**9827–9832.
11. **Courtney, H. S., D. L. Hasty, and I. Ofek.** 1990. Hydrophobicity of group A streptococci and its relationship to adhesion of streptococci to host cells, p. 361–386. *In* R. J. Doyle and M. Rosenberg (ed.), *Microbial Cell Surface Hydrophobicity.* American Society for Microbiology, Washington, D.C.
12. **Courtney, H. S., M. S. Bronze, J. B. Dale, and D. L. Hasty.** 1994. Analysis of the role of M24 protein in streptococcal adhesion and colonization by use of Ω-interposon mutagenesis. *Infect. Immun.* **62:**4868–4873.
13. **Cowan, M. M., K. G. Taylor, and R. J. Doyle.** 1986. Kinetic analysis of *Streptococcus sanguis* adhesion to artificial pellicle. *J. Dent. Res.* **65:**1278–1283.
14. **Cowan, M. M., K. G. Taylor, and R. J. Doyle.** 1987. Role of sialic acid in the kinetics of *Streptococcus sanguis* adhesion to artificial pellicle. *Infect. Immun.* **55:**1552–1557.
15. **Cowan, M. M., K. G. Taylor, and R. J. Doyle.** 1987. Energetics of the initial phase of adhesion of *Streptococcus sanguis* to hydroxylapatite. *J. Bacteriol.* **169:**2995–3000.
16. **Cywes, C., I. Stamenkovic, and M. R. Wessels.** 2000. CD44 as a receptor for colonization of the pharynx by group A streptococcus. *J. Clin. Investig.* **106:**995–1002.
17. **Derjaguin, B. V., and L. Landau.** 1941. Theory of the stability of strongly charged lyophobic soils and of the adhesion of strongly charged particles in solutions of electrolytes. *Acta Physicochim. (USSR)* **14:**633–662.
18. **Doyle, R. J.** 2000. Contribution of the hydrophobic effect to microbial infection. *Microbes Infect.* **2:**391–400.
19. **Drake, D., K. G. Taylor, A. S. Bleiweis, and R. J. Doyle.** 1988. Specificity of the glucan-binding lectin of *Streptococcus cricetus. Infect. Immun.* **56:**1864–1872.
20. **Duguid, J. P., and R. R. Gillies.** 1957. Fimbriae and adhesive properties of dysentery bacilli. *J. Pathol. Bacteriol.* **74:**397–411.
21. **Duguid, J. P., E. S. Anderson, and I. Campbell.** 1966. Fimbriae and adhesive properties in salmonellae. *J. Pathol. Bacteriol.* **92:**107–138.
22. **Duguid, J. P., I. W. Smith, G. Dempster, and P. N. Edmunds.** 1955. Non-flagellar filamentous appendages ("fimbriae") and hemagglutinating activity in *Bacterium coli. J. Pathol. Bacteriol.* **70:**335–358.

23. **Ellen, R. P., and R. J. Gibbons.** 1972. M protein-associated adherence of *Streptococcus pyogenes* to epithelial surfaces: prerequisite for virulence. *Infect. Immun.* **5:**826–830.

24. **Ellen, R. P., and R. J. Gibbons.** 1973. Parameters affecting the adherence and tissue tropisms of *Streptococcus pyogenes. Infect. Immun.* **9:**85–91.

25. **Finlay, B. B., and M. G. Caparon.** 2000. Bacterial adherence to cell surfaces and extracellular matrix, p. 67–80. *In* P. Cossart, P. Boquet, S. Normark, and R. Rappuoli (ed.), *Cellular Microbiology.* ASM Press, Washington, D.C.

26. **Firon N., S. Ashkenazi, D. Mirelman, I. Ofek, and N. Sharon.** 1987. Aromatic alphaglycosides of mannose are powerful inhibitors of the adherence of type 1 fimbriated *Escherichia coli* to yeast and intestinal epithelial cells. *Infect. Immun.* **55:**472–476.

27. **Fives-Taylor, P. M., and D. W. Thompson.** 1985. Surface properties of *Streptococcus sanguis* FW 213 mutants non-adherent to saliva coated hydroxyapatite. *Infect. Immun.* **47:**752–759.

28. **Gaastra, W., and A. M. Svennerholm.** 1996. Colonization factors of human enterotoxigenic *Escherichia coli* (ETEC). *Trends Microbiol.* **4:**444–452.

29. **Gabius, H. G.** 1997. Animal lectins. *Eur. J. Biochem.* **243:**543–576.

30. **Gibbons, R. J., and J. van Houte.** 1971. On the formation of dental plaques. *J. Periodontol.* **44:**347–360.

31. **Gibbons, R. J., I. Etherden, and E. C. Moreno.** 1983. Association of neuraminidase-sensitive receptors and putative hydrophobic interactions with high-affinity binding sites for *Streptococcus sanguis* C5 in salivary pellicles. *Infect. Immun.* **42:**1006–1012.

32. **Guyot, G.** 1908. Uber die bakterielle haemagglutination. *Zentbl. Bakteriol. Abt. I Orig.* **47:**640–653.

33. **Hanski, E., J. Jaffe, and V. Ozeri.** 1996. Proteins F1 and F2 of *Streptococcus pyogenes.* Properties of fibronectin binding, p. 141–150. *In* I. Kahane and I. Ofek (ed.), *Toward Anti-Adhesion Therapy for Microbial Diseases.* Plenum Press, New York, N.Y.

34. **Hanski, E., P. A. Horwitz, and M. G. Caparon.** 1992. Expression of protein F, the fibronectin-binding protein of *Streptococcus pyogenes* JRS4, in heterologous streptococcal and enterococcal strains promotes their adherence to respiratory epithelial cells. *Infect. Immun.* **60:**5119–5125.

35. **Hansson, L., P. Wallbrandt, J.-O. Andersson, M. Byström, A. Bäckman, A. Carlstein, K. Enquist, H. Lönn, C. Otter, and M. Strömqvist.** 1995. Carbohydrate specificity of the *Escherichia coli* P-pilus PapG protein is mediated by its N-terminal part. *Biochim. Biophys. Acta* **1244:**377–383.

36. **Hasty, D. L., H. S. Courtney, E. V. Sokurenko, and I. Ofek.** 1994. Bacteria-extracellular matrix interactions, p. 197–211. *In* P. Klemm (ed.), *Fimbriae. Adhesion, Genetics, Biogenesis and Vaccines.* CRC Press, Inc., Boca Raton, Fla.

37. **Hasty, D. L., I. Ofek, H. S. Courtney, and R. J. Doyle.** 1992. Multiple adhesins of streptococci. *Infect. Immun.* **60:**2147–2152.

38. **Houwink, A. L., and W. van Iterson.** 1949. Electron microscopical observations on bacterial cytology. II. A study on flagellaton. *Biochim. Biophys. Acta* **5:**10–44.

39. **Hung, C. S., J. Bouckaert, D. Hung, J. Pinkner, C. Widberg, A. DeFusco, C. G. Auguste, R. Strouse, S. Langermann, G. Waksman, and S. J. Hultgren.** 2002. Structural basis of tropism of *Escherichia coli* to the bladder during urinary tract infection. *Mol. Microbiol.* **44:**903–915.

40. **Jacques, M.** 1996. Role of lipo-oligosaccharides and lipopolysaccharides in bacterial adherence. *Trends Microbiol.* **4:**408–409.

41. **Karlsson, K.-A.** 1995. Microbial recognition of target-cell glycoconjugates. *Curr. Opin. Struct. Biol.* **5:**622–635.

42. **Kuan, S. F., K. Rust, and E. Crouch.** 1992. Interaction of surfactant protein D with bacterial lipopolysaccharides: Surfactant protein D is an *Escherichia coli*-binding protein in bronchoalveolar lavage. *J. Clin. Investig.* **90:**97–106.

43. **Kuusela, P.** 1978. Fibronectin binds to *Staphylococcus aureus. Nature* **276:**718–720.

44. **Leffler, H., and C. Svanborg-Eden.** 1980. Chemical-identification of a glycosphingolipid receptor for *Escherichia coli* attaching to human urinary-tract epithelial cells and agglutinating human-erythrocytes. *FEMS Microbiol. Lett.* **8:**127–134.

45. **Lindberg, F., J. M. Tennent, S. J. Hultgren, B. Lund, and S. Normark.** 1989. Pap D, a periplasmic transport protein in P-pilus biogenesis. *J. Bacteriol.* **171:**6052–6058.

46. **Lindhorst, T. K., C. Kieburg, and U. Krallmann-Wenzel.** 1998. Inhibition of the type 1 fimbriae-mediated adhesion of *Escherichia coli* to erythrocytes by multiantennary alpha-mannosyl clusters: the effect of multivalency. *Glycoconjugate J.* **15:**605–613.

47. **Madsen, J., A. Kliem, I. Tornoe, K. Skjodt, C. Koch, and U. Holmskov.** 2000. Localization of lung surfactant protein D on

mucosal surfaces in human tissues. *J. Immunol. Methods* **164:**5866–5870.

48. **Maurer, L., and P. E. Orndorff.** 1985. A new locus, *pilE,* required for the binding of type 1 piliated *Escherichia coli* to erythrocytes. *FEMS Microbiol. Lett.* **30:**59–66.

49. **Menozzi, F. D., R. Mutombo, G. Renauld, C. Gantiez, J. H. Hannah, E. Leininger, M. J. Brennan, and C. Locht.** 1994. Heparin-inhibitable lectin activity of the filamentous hemagglutinin adhesin of *Bordetella pertussis. Infect. Immun.* **62:**769–778.

50. **Mooi, F. R., F. K. De Graaf, and J. D. van Embden.** 1979. Cloning, mapping, and expression of the genetic determinant that encodes for the K88ab antigen. *Nucleic Acids Res.* **6:**849–865.

51. **Morris, E. J., and B. C. McBride.** 1984. Adherence of *Streptococcus sanguis* to saliva-coated hydroxyapatite: evidence for two binding sites. *Infect. Immun.* **43:**656–663.

52. **Morris, E. J., N. Ganeshkumar, and B. C. McBride.** 1985. Cell surface components of *Streptococcus sanguis:* relationship to aggregation, adherence, and hydrophobicity. *J. Bacteriol.* **164:**255–262.

53. **Nesbitt, W. E., R. J. Doyle, K. G. Taylor, R. H. Staat, and R. R. Arnold.** 1982. Positive cooperativity in the binding of *Streptococcus sanguis* to hydroxylapatite. *Infect. Immun.* **35:**157–165.

54. **Nesbitt, W. E., R. J. Doyle, and K. G. Taylor.** 1982. Hydrophobic interactions and the adherence of *Streptococcus sanguis* to hydroxylapatite. *Infect. Immun.* **38:**637–644.

55. **Ofek, I., and E. H. Beachey.** 1980. General concepts and principles of bacterial adherence in animals and man, p. 1–27. *In* E. H. Beachey (ed.), *Bacterial Adherence. Receptors and Recognition,* series B, vol. 6. Chapman & Hall, Ltd. London, United Kingdom.

56. **Ofek, I., and R. J. Doyle.** 1994. *Bacterial Adhesion to Cells and Tissues.* p. 321–512. Chapman & Hall, New York, N.Y.

57. **Ofek, I., and R. J. Doyle.** 1994. *Bacterial Adhesion to Cells and Tissues.* p. 54–93. Chapman & Hall, New York, N.Y.

58. **Ofek, I., and R. J. Doyle.** 1994. *Bacterial Adhesion to Cells and Tissues.* p. 94–135. Chapman & Hall, New York, N.Y.

59. **Ofek, I., and R. J. Doyle.** 1994. *Bacterial Adhesion to Cells and Tissues,* p. 239–320. Chapman & Hall, New York, N.Y.

60. **Ofek, I., and R. J. Doyle.** 1994. *Bacterial Adhesion to Cells and Tissues.* p. 1–15. Chapman & Hall, New York, N.Y.

61. **Ofek, I., D. Mirelman, and N. Sharon.** 1977. Adherence of *Escherichia coli* to human

mucosal cells mediated by mannose receptors. *Nature* **265:**623–625.

62. **Ofek, I., D. L. Hasty, S. N. Abraham, and N. Sharon.** 2000. Role of bacterial lectins in urinary tract infections: molecular mechanisms for diversification of bacterial surface lectins. *Adv. Exp. Med. Biol.* **485:**182–192.

63. **Ofek, I., E. H. Beachey, B. I. Eisenstein, M. L. Alkan, and N. Sharon.** 1979. Suppression of bacterial adherence by subminimal inhibitory concentrations of beta-lactam and aminoglycoside antibiotics. *Rev. Infect. Dis.* **1:**832–837.

64. **Ofek, I., E. H. Beachey, W. Jefferson, and G. L. Campbell.** 1975. Cell membrane-binding properties of group A streptococcal lipoteichoic acid. *J. Exp. Med.* **141:**990–1003.

65. **Ofek, I., J. Goldhar, Y. Keisari, and N. Sharon.** 1995. Nonopsonic phagocytosis of microorganisms. *Annu. Rev. Microbiol.* **49:**239–276.

66. **Ozeri, V., A. Tovi, I. Burstein, S. Natanson-Yaron, M. G. Caparon, K. M. Yamada, S. K. Akiyama, I. Vlodavsky, and E. Hanski.** 1996. A two-domain mechanism for group A streptococcal adherence through protein F to the extracellular matrix. *EMBO J.* **15:**989–998.

67. **Parent, J. B.** 1990. Membrane receptors on rat hepatocytes for the inner core region of bacterial lipopolysaccharides. *J. Biol. Chem.* **265:**3455–3461.

68. **Patti, J. M., B. L. Allen, M. J. McGavin, and M. Höök.** 1994. MSCRAMM-mediated adherence of microorganisms to host tissues. *Annu. Rev. Microbiol.* **48:**585–617.

69. **Przondo-Mordarska, A., D. Smutnicka, H. L. Ko, J. Beuth, and G. Pulverer.** 1996. Adhesive properties of P-like fimbriae in *Klebsiella*-species. *Zentbl. Bakteriol.* **284:**372–377.

70. **Relman, D. A., M. Domenighini, E. Tuomanen, R. Rappuoli, and S. Falkow.** 1989. Filamentous hemagglutinin of *Bordetella pertussis:* nucleotide sequence and crucial role in adherence. *Proc. Natl. Acad. Sci. USA* **86:**2637–2641.

71. **Robins-Browne, R. M., A. M. Tokhi, L. M. Adams, and V. Bennett-Wood.** 1994. Host specificity of enteropathogenic *Escherichia coli* from rabbits: lack of correlation between adherence in vitro and pathogenicity for laboratory animals. *Infect. Immun.* **62:**3329–3336.

72. **Rosenberg, M., and R. J. Doyle.** 1990. Microbial cell surface hydrophobicity. History, measurement and significance, p. 1–37. *In* R. J. Doyle and M. Rosenberg (ed.), *Microbial Cell*

Surface Hydrophobicity. ASM Press, Washington, D.C.

73. **Rosenberg, M., and S. Kjelleberg.** 1986. Hydrophobic interactions: role in microbial adhesion. *Adv. Microb. Ecol.* **9:**353–393.

74. **Rosenberg, M., R. B.-N. Greenstein, M. Barki, and S. Goldberg.** 1996. Hydrophobic interactions as a basis for interfering with microbial adhesion, p. 241–248. *In* I. Kahane and I. Ofek (ed.), *Toward Anti-Adhesion Therapy for Microbial Diseases*. Plenum Press, New York, N.Y.

75. **Rottini, G., F. Cian, M. R. Soranzo, R. Albrigo, and P. Patriarc.** 1979. Evidence for the involvement of human polymorphonuclear leukocyte mannose-like receptors in the phagocytosis of *Escherichia coli*. *FEBS Lett.* **105:**307–312.

76. **Rutter, J. M., and G. W. Jones.** 1973. Protection against enteric disease caused by *Escherichia coli*—a model for vaccination with a virulence determinant? *Nature* **242:**531–532.

77. **Salit, I. E., and E. C. Gotschlich.** 1977. Hemagglutination by purified type 1 *Escherichia coli* pili. *J. Exp. Med.* **146:**1169–1181.

78. **Satterwhite, T. K., D. G. Evans, H. L. DuPont, and D. J. Evans, Jr.** 1978. Role of *Escherichia coli* colonisation factor antigen in acute diarrhoea. *Lancet* **ii:**181–184.

79. **Scheld, W. M., J. A. Valone, and M. A. Sande.** 1978. Bacterial adherence in the pathogenesis of endocarditis. Interaction of bacterial dextran, platelets, and fibrin. *J. Clin. Investig.* **61:**1394–1404.

80. **Sharon, N., and H. Lis.** 1996. Micribial lectins and their receptors, p. 475–506. *In* J. F. Montreuil, G. Vliegenthart, and H. Schachter (ed.), *Glycoproteins*. Elsevier Publishing Co., Amsterdam, The Netherlands.

81. **Silverblatt, F. J.** 1974. Host-parasite interaction in the rat renal pelvis: a possible role for pili in the pathogenesis of pyelonephritis. *J. Exp. Med.* **140:**1696–1711.

82. **Silverblatt, F. J., and L. S. Cohen.** 1979. Antipili antibody affords protection against experimental ascending pyelonephritis. *J. Clin. Investig.* **64:**333–336.

83. **Silverblatt, F. J., J. S. Dreyer, and S. Schauer.** 1979. Effect of pili on susceptibility of *Escherichia coli* type 1 pili and capsular polysaccharides on the interaction between bacteria and human granulocytes. *Scand. J. Immunol.* **20:**299–305.

84. **Simpson, W. A., and E. H. Beachey.** 1983. Adherence of group A streptococci to fibronectin on oral epithelial cells. *Infect. Immun.* **39:**275–279.

85. **Sokurenko, E. V., V. Chesnokova, D. E. Dykhuizen, X.-R. Wu, K. A. Krogfelt, M. A. Schembri, I. Ofek, and D. L. Hasty.** 1998. Pathogenic adaptation of *Escherichia coli* by natural variation of the FimH adhesin. *Proc. Natl. Acad. Sci. USA* **95:**8922–8926.

86. **Sokurenko, E. V., V. Chesnokova, R. J. Doyle, and D. L. Hasty.** 1997. Diversity of the *Escherichia coli* type 1 fimbrial lectin: differential binding to mannosides and uroepithelial cells. *J. Biol. Chem.* **272:**17880–17886.

87. **Speziale, P., M. Höök, L. Switalski, and T. Wadström.** 1984. Fibronectin binding to a *Streptococcus pyogenes* strain. *J. Bacteriol.* **157:**420–427.

88. **Striker, R., U. Nilsson, A. Stonecipher, G. Magnusson, and S. J. Hultgren.** 1995. Structural requirements for the glycolipid receptor of human uropathogenic *Escherichia coli*. *Mol. Microbiol.* **16:**1021–1029.

89. **Strömberg, N., B.-I. Marklund, B. Lund, D. Ilver, A. Hamers, W. Gaastra, K.-A Karlsson, and S. Normark.** 1990. Host-specificity of uropathogenic *Escherichia coli* depends on differences in binding specificity to Galα1-4Gal-containing isoreceptors. *EMBO J.* **9:**2001–2010.

90. **Svanborg-Eden, C., T. Sandberg, K. Stenqvist, and S. Ahlstedt.** 1979. Effects of subinhibitory amounts of ampicillin, amoxycillin and mecillinam on the adhesion of *Escherichia coli* bacteria to human urinary tract epithelial cells: a preliminary study. *Infection* **7:**S452–S455.

91. **Sylvester, F. A., D. Philpott, B. Gold, A. Lastovica, and J. F. Forstner.** 1996. Adherence to lipids and intestinal mucin by a recently recognized human pathogen, *Campylobacter upsaliensis*. *Infect. Immun.* **64:**4060–4066.

92. **Szymanski, C. M., and G. D. Armstrong.** 1996. Interactions between *Campylobacter jejuni* and lipids. *Infect. Immun.* **64:**3467–3474.

93. **Thankavel, K., B. Madison, T. Ideda, R. Malaviya, A. H. Shah, P. M. Arumugam, and S. N. Abraham.** 1997. Localization of a domain in the FimH adhesin of *Escherichia coli* type 1 fimbriae capable of receptor recognition and use of a domain-specific antibody to confer protection against experimental urinary tract infection. *J. Clin. Investig.* **100:**1123–1136.

94. **Thomas, W. E., E. Trintchina, M. Forero, V. Vogel, and E. V. Sokurenko.** 2002. Bacterial adhesion to target cells enhanced by shear force. *Cell* **109:**913–923.

95. **Tikkanen, K., S. Haataja, C. François-Gerard, and J. Finne.** 1995. Purification of a galactosyl-α1-4-galactose-binding adhesin from

the gram-positive meningitis-associated bacterium *Streptococcus suis*. *J. Biol. Chem.* **270:** 28874–28878.

96. **Tuomanen, E., H. Towbin, G. Rosenfelder, D. Braun, G. Larson, G. C. Hansson, and R. Hill.** 1988. Receptor analogues and monoclonal antibodies that inhibit adherence of *Bordetella pertussis* to human ciliated respiratory epithelial cells. *J. Exp. Med.* **168:** 267–277.

97. **van Heyningen, T., G. Fogg, D. Yates, E. Hanski, and M. Caparon.** 1993. Adherence and fibronectin binding are environmentally regulated in the group A streptococci. *Mol. Microbiol.* **9:**1213–1222.

98. **Van Oss, C. J.** 1978. Phagocytosis as a surface phenomenon. *Annu. Rev. Microbiol.* **32:**19–39.

99. **Verwey, E. J. W., and J. T. G. Overbeek.** 1948. *Theory of Stability of Lyophobic Colloids.* Elsevier Publishing Co., Amsterdam, The Netherlands.

100. **Watt, P. J., and M. E. Ward.** 1980. Adherence of *Neisseria gonorrhoeae* and other *Neisseria* species to mammalian cells, p. 251–288. *In* E. H. Beachey (ed.), *Bacterial Adherence.* Chapman & Hall, New York, N.Y.

101. **Weiss, E. I., B. Shenitzki, and R. Leibusor.** 1996. Microbial coaggregation in the oral cavity, p. 233–240. *In* I. Kahane and I. Ofek (ed.), *Toward Anti-Adhesion Therapy of Infectious Diseases.* Plenum Press, New York, N.Y.

102. **Wessels, M. R., and M. S. Bronze.** 1994. Critical role of the group A streptococcal capsule in pharyngeal colonization and infection in mice. *Proc. Natl. Acad. Sci. USA* **91:**12238–12242.

103. **Wilson, M.** 2002. *Bacterial Adhesion to Host Tissues. Mechanisms and Consequences.* Cambridge University Press, Cambridge, United Kingdom.

104. **Wu, X.-R., T. T. Sun, and J. J. Medina.** 1996. *In vitro* binding of type 1-fimbriated *Escherichia coli* to uroplakins Ia and Ib: relation to urinary tract infections. *Proc. Natl. Acad. Sci. USA* **93:**9630–9635.

105. **Zaidi, T. S., S. M. Fleiszig, M. J. Preston, J. B. Goldbeg, and G. B. Pier.** 1996. Lipopolysaccharide outer core is a ligand for corneal cell binding and ingestion of *Pseudomonas aeruginosa. Investig. Ophthalmol. Visual Sci.* **37:** 976–986.

106. **Zhang, X.-H., M. Rosenberg, and R. J. Doyle.** 1990. Inhibition of the cooperative adhesion of *Streptococcus sanguis* to hydroxylapatite. *FEMS Microbiol. Lett.* **71:**315–318.

METHODOLOGICAL APPROACHES TO
ANALYSIS OF ADHESINS AND ADHESION

2

Ideally, in vivo experiments should complement in vitro approaches to the characterization of adhesin-receptor interactions and their role in the infectious process. In this respect, adhesion experiments are not exceptions. The majority of the available methods are used for in vitro experiments, but in vivo experiments with appropriate animal models are becoming more and more common. If antiadhesive therapies are to become commonplace, model systems are essential. The study of mechanisms of adhesion vary from simple methods providing limited information to more complex methods that are difficult to perform and analyze but provide more meaningful information. Step-by-step methodological details for many of these techniques are presented in a variety of other texts (4, 17, 32, 38, 129). In this chapter, we discuss some of the most important methods, emphasizing the types of knowledge gained from the use of a specific method.

There is often some confusion in the use of terms such as specificity and nonspecificity in describing the mechanisms of bacterial adhesion, due in part to the fact that the same basic physicochemical forces are responsible for both specific and nonspecific binding: the so-called Van der Waals' and electrostatic forces, and hydrogen bonding. It has been argued that the primary determinant of specific binding is the occurrence of highly localized chemical groups which together form a stereochemical combination, whereas in nonspecific binding stereospecificity is not involved (13).

PARTITIONING OF BACTERIA

The net surface charge of the bacteria and the presence of lipophilic residues endow members of bacterial populations with the ability to partition between two immiscible liquids or between nonpolar substrata and the aqueous phase. There are a number of partition systems, including microbial adhesion to hydrocarbon, hydrophobic interaction chromatography, salt aggregation test, contact angle measurement, two-phase partitioning, and adhesion to plastics.

The choice of a particular system is guided mainly by the ease of performing the test and by the reproducibility of the quantitative data obtained. The microbial adhesion to hydrocarbons test (MATH test), using partitioning of bacteria between hexadecane and aqueous phases, has been the most widely employed (36, 105).

Cell surface hydrophobicity can be defined most simply as the relative tendency of a microorganism to adhere to a nonpolar material compared to water (32, 33, 35). The partition techniques are used to obtain information about the accessibility of lipophilic residues (i.e., the hydrophobin) on the bacterial surface and their role in adhesive interactions, as discussed in chapter 1. The MATH test is performed by overlaying a suspension of bacteria with a hydrocarbon, such as hexadecane (Fig. 2.1). Following vigorous vortexing, the mixture is allowed to settle. After a few minutes, the hydrocarbon droplets with adsorbed bacteria rise to the top, reducing the optical density of the aqueous phase. The lower the optical densities of the aqueous phase, the higher the hydrophobicity of the bacterial population.

The following set of experiments helps to clarify the value of the MATH assay (21, 93). When *Streptococcus pyogenes* is harvested during the exponential phase of growth, most of the cells possess a hydrophilic hyaluronic acid capsule which masks the accessibility of the surface hydrophobin, lipoteichoic acid (LTA) (Fig. 2.2). As a result, only the small fraction of the cell population that is unencapsulated will partition at the hexadecane-aqueous interface, and so the optical density recorded will be high. However, when encapsulated exponential organisms are treated with hyaluronidase to remove the capsule or are allowed to grow to the stationary phase, by which time they have become unencapsulated naturally, the lipophilic LTA complexed to surface proteins becomes exposed in such a way as to mediate adhesion of the bacteria to hexadecane. In this case, therefore, the optical density recorded will be low. If the unencapsulated stationary-phase bacteria are treated with trypsin, the LTA-protein complexes are removed. Under these conditions, most of the cells are not able to adhere to hexadecane droplets, and so the optical density readings will be high. These studies show that for the streptococcal cells to partition to the nonpolar phase, the lipophilic residues must not only be present on the streptococcal surface but must also be properly exposed.

The hexadecane method is very useful in selecting for mutants or recombinant strains that are more hydrophobic (or hydrophilic) because the partitioning is very efficient. In interpreting hydrocarbon-aqueous experiments, however, it is important to realize that a value such as "45% adhesion to hexadecane" does not mean the bacteria are half as hydrophobic as those giving a value of "90% adhesion to hexadecane" but that roughly half of the cells of the test population have an exposure of hydrophobins passing the threshold necessary for allowing binding to the oil droplet in a given period. It is also important to note that inactivation of a gene reducing the proportion of the bacterial population capable of binding to hexadecane may be the result of either reduced expression of a hydrophobin (e.g., LTA) or increased expression of a hydrophilic component (e.g. hyaluronate).

Overall, the hexadecane method has been extremely useful to study the exposure of hydrophobins and selection of mutants (35, 36). In addition, the hexadecane method allows kinetic measurements of the rate of binding to hexadecane (75). The slope of the rate curve follows first-order kinetics and represents the affinity of the organisms for the hydrophobic material. Some bacteria are highly hydropohobic and may partition from distilled water. Otherwise, it is necessary to suppress the bacterial surface charge to optimally partition at hydrocarbon-aqueous interfaces. This can be accomplished by increasing the ionic strength of the aqueous medium or by lowering its pH (37). For example, adding amphipathic cationic agents (e.g., cetyl-pyridinium chloride) can potentiate microbial adhesion to hydrocarbons.

Another partitioning system utilizing two immiscible aqueous phases, polyethylene glycol (PEG) and dextran, has been described previously (80, 86). Hydrophilic bacteria partition into the lower (dextran) phase, whereas hydrophobic bacteria are found in the upper (PEG) phase. The information gained is basically similar to that obtained using the adhesion-to-hydrocarbon test. In this system, the

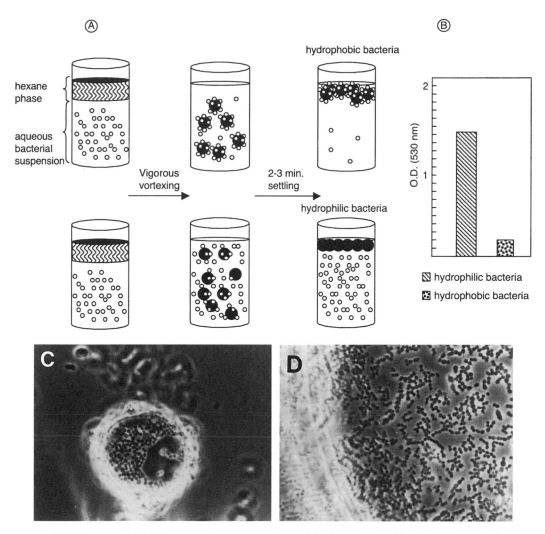

FIGURE 2.1 (A) Diagrammatic representation of the interaction of hydrophobic and of hydrophilic bacteria with hexadecane droplets. The bacterial suspension is overlaid with hexadecane, usually at a 1:10 ratio of hexadecane to aqueous solution (e.g., 0.1 ml of hexadecane plus 1.0 ml of bacterial suspension). After vigorous vortexing for 1 min, the hexadecane breaks down within the bacterial suspension into droplets of various sizes. Bacteria will attach in numbers that are relative to the degree of surface hydrophobicity. After the mixed phases are allowed to separate for a few minutes, the hexadecane droplets will rise to the top, carrying the bound hydrophobic bacteria. Hydrophilic bacteria will remain in relatively greater numbers in the lower (aqueous) phase because they do not bind to the hydrocarbon. (B) Spectrophometric determination of the optical density of the aqueous phase after allowing phase separation to occur will allow the determination of the relative hydrophobicity of bacterial populations. The less turbid aqueous phase contains the bacteria with the most hydrophobic surfaces. This assay can be used to fractionate hydrophobic and hydrophilic bacteria in a mixed population. (C and D) Light micrographs (low [C] and high [D] magnifications) of bacteria bound to hexadecane droplets.

PEG can be palmitylated to measure the hydrophobic partitioning of bacteria which exhibit a very hydrophobic surface. Positively charged PEG (e.g., sulfonated PEG) has also been useful in assessing the surface charge of particles by partitioning.

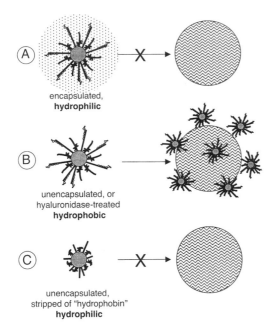

(A) encapsulated, **hydrophilic**

(B) unencapsulated, or hyaluronidase-treated **hydrophobic**

(C) unencapsulated, stripped of "hydrophobin" **hydrophilic**

FIGURE 2.2 Schematic diagram showing the effects of exposure of hydrophobin and protein adhesins (or masking by capsule) on the binding of streptococci to hexadecane droplets. (A) The hyaluronate capsule masks the LTA hydrophobin, and as a result the bacteria cannot adhere to the hexadecane droplets. (B) The hydrophobin is exposed in naturally unencapsulated streptococci (e.g., stationary phase), and so the bacteria are able to bind to hexadecane droplets. The same would be true for bacteria treated with hyaluronidase to remove the capsule. (C) Removal of the hydrophobin by enzymatic digestion of the bacteria makes the bacteria hydrophilic and, therefore, unable to bind to hexadecane. See the text and reference 93 for more detail.

The salt aggregation test is simple, and bacteria are "salted out" by increasing the concentration of ammonium sulfate, essentially analogous to "salting out" of proteins (76, 106). With high densities of bacteria, the salting out can result in visible aggregates and the rate of aggregation can be quantified by turbidimetry. When a low density of bacteria is used, the bacteria are deposited onto a plastic surface and the bound bacteria can then be quantified by enzyme-linked immunosorbent assay (ELISA) (115).

Agarose beads conjugated to hydrophobic probes such as phenyl or octyl groups have been widely used to fractionate proteins. The method has been adapted to fractionate bacteria into subpopulations possessing various degrees of hydrophobicity (119). Although adhesion to agarose-conjugated hydrocarbons can also be used to fractionate hydrophobic and hydrophilic bacteria, it is not an efficient method to separate multiple subpopulations that vary in their level of hydrophobicity. The method is also not useful when the bacteria express lectins specific for sugar residues associated with the agarose beads. Alternatively, the experiment may be performed in the presence of an appropriate saccharide to prevent potential lectin-agarose bead interactions.

In the contact angle technique, a water droplet is applied to the surface of a dried lawn of bacteria. The angle formed where the water contacts the organisms is proportional to the surface hydrophobicity of the bacteria. This is an indirect measurement of hydrophobicity; it requires expensive equipment, and the need to prepare a dried bacterial film makes the assay relatively complicated. However, the contact angle method has been useful to predict the interaction of a wide range of bacterial species with animal cells in vitro. Bacteria that exhibit increased hydrophobicity, as determined by the contact angle technique, are more readily engulfed by phagocytes, consistent with the notion that hydrophobicity is important in adhesion (127). A variation on this technique measuring surface tension has recently been suggested (50).

When a colony containing a certain proportion of hydrophobic bacteria is pressed gently onto the surface of polystyrene (or other plastic), the bacteria exhibiting a threshold of hydrophobin expression will adhere to the plastic, will withstand washing, and can be readily visualized using standard staining techniques (e.g., methylene blue or gentian violet). The method is extremely useful in screening a large number of colonies for mutants that are either much more or much less hydrophobic than the parent organism (104). In some instances, the plastic used has been in the form of microspheres (70). Adhesion to such microspheres can be determined by light microscopy, among many other techniques (e.g., ELISA and use of radiolabeled bacteria). A disadvantage of this technique is that plastics differ from one manufacturer to another and also from batch to batch.

In each of these test systems, the positive role of surface hydrophobicity in bacterium-animal cell interactions has been confirmed (30, 35, 80).

AGGLUTINATION REACTIONS

Historically, agglutination of erythrocytes (e.g., hemagglutination) was the first test showing interactions between bacteria and animal cells (see Table 1.1). Very early, this simple type of assay was expanded to include bacterium-induced agglutination of other types of particles, such as yeast cells or receptor-coated particles (87, 88) (Table 2.1). Nevertheless, erythrocytes remain the primary cells of choice for agglutination tests to monitor adhesins of pathogens. The most important reason for this is probably that they are convenient to assay for lectin adhesins produced by a large number of bacterial species. Also, the numerous species of animals that are readily available yield an equal number of erythrocytes, each presenting species-specific glycoproteins or glycolipids on their surfaces (112, 113). This results in the availability of a remarkable diversity of carbohydrate structures.

Hemagglutination reactions have been responsible for the initial identification of many lectin adhesins and were important in the characterization of adhesin-saccharide specificities, through the use of carbohydrate inhibitors of the reactions. A positive hemagglutination reaction is very likely to indicate the involvement of a bacterial lectin adhesin (88). As far as is known, there are no examples of protein-protein-type adhesive interactions in bacterium-induced hemagglutination. There are, however, a limited number of examples in which hemagglutination is promoted through a hydrophobic effect that may involve protein-protein interactions, as is the case in *Pseudomonas aeruginosa* (45), or through secreted lipids that induce erythrocyte fusion events (61, 125).

Frequently, nonerythrocyte particles can be employed in bacterial agglutination tests (Table 2.1). One of the primary reasons for the popularity of particle agglutination tests is because there is no need to separate adherent and nonadherent bacterial cells from the reaction mixtures. Furthermore, in most cases there is no need for specialized equipment since reactions can be performed by mixing

TABLE 2.1 Types of target cells and particles commonly used in agglutination reactions with bacteria

Target cell for agglutination	Quantitation method	Data presentation	Reference(s)
Erythrocytes	Minimal bacterial density exhibiting visible agglutination	Titer of twofold dilutions of standard suspensions	39, 47, 48
	Intensity of hemagglutination caused by a single density of bacteria	0, ±, +, ++, +++, etc.	
Yeast cells	Minimal bacterial density exhibiting visible agglutination	Titer of twofold dilutions of standard suspension	
	Aggregation rate measurements using single densities of bacteria and yeast cells in an aggregometer	Increase in light transmission per unit of time (rate)	8, 92
Receptor-coated beads	Intensity of agglutination caused by a single density of bacteria	0, ±, +, ++, +++, etc. (empirical scoring)	28
Bacteria	Intensity or rate of coaggregation	0, ±, +, ++, +++, etc., or increase in light transmission per unit of time (rate)	11, 66, 67, 84

small amounts of the reactants on glass slides (48). However, the optimal conditions for the agglutination reactions must be established for each bacterium-particle pair, including the pH and composition of buffer, the temperature, and the degree of agitation.

Agglutination reactions are always a secondary effect of adhesion. There is a critical density of adhesin molecules on the bacterial surface that is required to induce agglutination (39). Assuming that the bacteria possess the critical number of adhesins, it has been calculated that a minimal density of approximately 10^7 bacteria/ml and a 2% suspension of particles are required to easily visualize agglutination. It should be noted that whereas the agglutination reactions are easy to perform, used alone they provide little information about the specificity involved. However, by using a battery of erythrocytes lacking specific blood group antigens, it is possible to gain information about the blood group specificity of the reaction. For example, most pyelonephritogenic isolates of *Escherichia coli* readily agglutinate p^+ erythrocytes, but not p^- erythrocytes (59, 73). This selective hemagglutination facilitated the determination of the receptor specificity of the fimbriae (i.e., P fimbriae) as the disaccharide $Gal(\alpha 1 \rightarrow 4)Gal$, which is shared by all P blood group antigens.

In contrast to erythrocyte surfaces, yeast cell surfaces are thought to be more simply arrayed. The yeast surfaces are composed mostly of mannan and glucan and therefore may be used only for bacteria that present lectins specific for these polysaccharides. Commonly, yeast cells are used to monitor the presence of type 1 fimbriae and to test the fine sugar specificity of the type 1 fimbrial lectin (43, 44). Receptor-coated bead assays are valuable, but they require the availability of purified, well-characterized molecules (28).

Agglutination (or hemagglutination) can be induced by whole bacteria and by certain purified lectin adhesins (48). In some cases, the isolated bacterial lectin acts in a univalent manner and requires cross-linking to achieve agglutination, especially when the active lectin is located only, or primarily, at the fimbrial tip (118, 125).

Dental plaque is a biofilm, composed of multiple genera of bacteria, that forms on teeth at the gingival margin and is an important cause of oral diseases such as periodontal disease. It has become useful to study intergenera, bacterium-bacterium interactions to understand this important type of adhesion phenomenon. The bacterial interaction is called coaggregation, and although plaque on teeth is multigeneric, the phenomenon in vitro is usually tested by mixing one genus with another and measuring the intensity or rate of aggregation (66, 67, 84).

ADHESION TO ANIMATE AND INANIMATE SURFACES

Unlike for the agglutination reactions, appropriate methods must be devised for removal of nonadherent bacteria and quantitation of remaining adherent bacteria when dealing with animate and inanimate surfaces. The various methods commonly used to separate adherent from nonadherent bacteria are listed in Table 2.2. For dense particles, such as hydroxylapatite beads, the separation is simplified because these dense particles settle quite rapidly and shearing forces are minimized. In contrast, shear forces are more significant in the separation of nonadherent from adherent bacteria by other methods. Shear forces may have damaging effects on ligands, receptors, or ligand-receptor complexes and may therefore remove bound bacteria. Recently, the issue of shear forces and their effect on weakly bound bacteria has been dealt with by manipulating washing procedures (23, 98). Flow cytometry has also been used to study bacteria under low shear forces (Fig. 2.3) (111). Bacteria adherent to medical devices are frequently subjected to shearing forces in vivo. A special instrument employing a flow cell method has been developed to study bacterial adhesion to medical devices (14). Although shear forces have long been thought to exclusively remove bacteria from substrata, it has recently been demonstrated that in some cases

TABLE 2.2 Methods commonly used to separate nonadherent bacteria from bacteria adherent to animate surfaces

Target surface	Method	Comments
Immobilized animal cells (e.g., phagocytes, tissue culture cells) and receptor-coated inert surfaces	Washing with buffer	This method is easy to perform. One can distinguish between weak and strong adhesion either by controlling the number of washes (23) (Fig. 2.2) or by inverting the assay plate and applying differential centrifugal forces (98). The strength of adhesion in this latter case can be calculated when the centrifugal force is known.
Cells in suspension	Differential centrifugation or filtration through selective-pore-size membranes	Differential centrifugation may also be performed using a density gradient to allow better separation of adherent and nonadherent bacteria. In some cases, magnetic forces can be used to separate appropriately coated iron particles carrying adherent bacteria.
Dense receptor-coated low particles (e.g., hydroxyapatite beads)	Inversion and decanting following settling of particles	In general, this procedure produces relatively low shearing forces (108).
Phagocytes	Stimulation of cells (in this particular case, there is no need to separate adherent and nonadherent cells)	This method is widely used for phagocytic cells and mast cells, which on interaction with bacteria are readily stimulated to secrete oxidative products and cytokines that can be measured (81, 101). Usually there are minimal or no shearing forces.

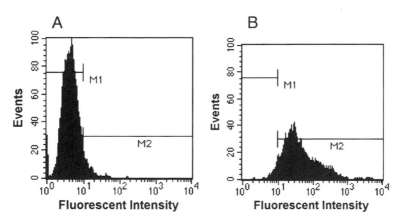

FIGURE 2.3 Flow cytometric analysis of adhesion of fluorescein-labeled *S. pyogenes* to HEp-2 tissue culture cells. The flow chamber is gated to detect epithelial cell-sized particles as a function of the level of fluorescence. The figure shows light scattering of epithelial cells alone (A) and epithelial cells 10 min after being mixed with the bacteria (B). Data were extracted from fluorescence analysis of epithelial cells and depicted as dot plots of forward scatter versus side scatter. A shift of the peak of fluorescence intensity to the right indicates increase in fluorescence intensity of the epithelial cells caused by adhesion of bacteria to the cells. The appearance of a shoulder on the right of the histogram indicates that a subpopulation of cells binds an increased number of bacteria. (Reprinted from reference 111 with permission from the publisher.)

shear forces actually enhance the effectiveness of bacterial adhesins (see chapter 1).

Immobilized Receptors and Their Analogs

The coating of inert surfaces to provide a substratum for bacterial adhesion has been used to mimic the natural environment, to screen various potential receptors, and to develop a detailed characterization of adhesin-receptor interactions (Table 2.3).

Animal Cells

Ultimately, the science of adhesion requires demonstration of bacterial attachment to animal cells within the host. The type of animal cells used in in vitro experiments should be representative of those that are usually colonized during the natural course of infections caused by the test bacteria. Authors of the early studies that employed hemagglutination reactions to characterize different adhesins realized that erythrocytes offered unusually diverse targets for adhesion (see above) but were not the normal target cells for most bacterial pathogens. Some of these investigators also used animal cells derived from tissues that were normally colonized by the pathogen in parallel with the erythrocyte agglutination experiments (40). Animal cells used for studies of bacterial adhesion may be cells in suspension which either have been scraped off mucosal surfaces, in the case of nonphagocytic cells, or have been obtained from the blood or

TABLE 2.3 Methods commonly used to separate nonadherent bacteria from bacteria adherent to inanimate surfaces

Category	Surface	Comments
Mimicking natural environment	Hydroxylapatite coated with buffer or saliva	Related to bacterial adhesion to teeth. Specificity may be established by using inhibitors (i.e., antibodies, saccharides, chaotropes).
	Implants or prosthetic devices (contact lenses, peritoneal dialysis tubing, catheters, etc.)	Surfaces coated with natural fluids may have selectivity for certain microorganisms.
Screening for adhesin expression and receptor specificity	Beads or plastics (96-well plates, etc.)	Much information can be obtained with respect to the binding characteristics of microbial adhesion. For example, when coated with defined glycoconjugates, they can be used to monitor carbohydrate binding properties of microbial surfaces.
Identification of receptors	Nitrocellulose sheets	Macromolecules derived from surfaces commonly colonized by test organisms may interact with microbial surface constitutents after separation of host cell components by PAGE and blotting onto nitrocellulose or a variety of other membranes. This method is useful to identify and define the properties of macromolecues that specifically interact with microbial surfaces (Fig 2.4) (130).
	Silica gels	Glycolipids are readily separated on silica gels (TLC). Microbial adhesins can then bind to the separated glycolipid. This method is especially useful in determining the specificities of certain microbial lectins (Fig. 2.5) (62).
	Receptor depletion by plating onto antireceptor antibodies	This method is used mainly for identifying a receptor on phagocytic cells by using several monoclonal antibodies against defined cell surface glycoproteins (103).

other phagocyte-rich sites, such as the lungs or the peritoneal cavity. Often, tissue culture cells have been used because an immobilized monolayer is simple to use and because they provide a constant supply of relatively uniform cells. In other cases, adhesion to excised tissues has been studied in an effort to test mucosal surfaces that have undergone minimal manipulation. One very commonly used procedure to search for cellular receptors is to separate protein and glycoprotein extracts by sodium dodecyl sulfate-polyacrylamide gel electrophoresis (SDS-PAGE), blot them to

nitrocellulose, and react the blots with bacteria or purified adhesins (Fig. 2.4). It has also proved very useful to separate extracts of cellular glyco-lipids by thin-layer chromatography (TLC) and react bacteria or adhesins with the separated glycolipids (Fig. 2.5).

All mucosal surfaces are bathed with mucus secreted by specialized epithelial cells and glands (see chapter 3), but studies that have focused on adhesion of bacteria to mucus-covered epithelium in vitro are limited. The role of mucus or mucins in adhesion has usually been studied by immobilizing the com-

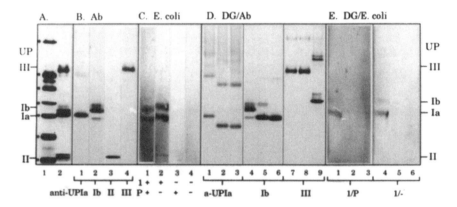

FIGURE 2.4 Identification of *E. coli* receptors by using blots of isolated membrane glycoproteins. Asymmetric unit membrane plaques were purified from bladder uroepithelial cells and separated by SDS-PAGE. These membrane plaques are enriched for uroplakin (UP) proteins Ia, Ib, II, and III. (A) Electrophoretic pattern of bovine urothelial plaque proteins stained with silver nitrate (lane 2). Lane 1 contains molecular weight markers. (B) Identification of uroplakin proteins by immunoblotting with antibodies (Ab) against synthetic peptides mimicking proteins Ia (lane 1), Ib (lane 2), II (lane 3), and III (lane 4). (C) Binding of radiolabeled *E. coli* to uroplakins. After blotting SDS-PAGE-separated uroplakins to nitrocellulose, membranes were treated as follows: those in lane 1 were incubated with *E. coli* J96, which expresses both type 1 and P fimbriae; those in lane 2 were incubated with *E. coli* SH48, which expresses only type 1 fimbriae; those in lane 3 were incubated with *E. coli* HU849, which expresses only P fimbriae; those in lane 4 were incubated with *E. coli* P6678-54, which expresses neither type 1 nor P fimbriae. (D) Uroplakins were deglycosylated by treatment with control buffer (lanes 1, 4, and 7) or with endo H to remove high-mannose-type sugars (lanes 2, 5, and 8) or endo F (lanes 3, 6, and 9). Following deglycosylation, uroplakins were separated by SDS-PAGE and blotted to nitrocellulose, and the membranes were probed by using uroplakin-specific antibodies. Note the removal of ~3-kDa equivalents of the high-mannose-type sugars from uroplakin Ia and Ib by endo H and the removal of ~20-kDa equivalents of sugars, most probably the complex type, from uroplakin III. (E) Binding of radiolabeled *E. coli* to deglycosylated uroplakins. Before SDS-PAGE and blotting, uroplakins had been treated with control buffer (lanes 1 and 4), endo H (lanes 2 and 5), or endo F (lanes 3 and 6). The membranes were incubated with radiolabeled *E. coli* J96 (lanes 1 to 3), or *E. coli* SH48 (lanes 4 to 6). The binding of type 1-fimbriated *E. coli* to uroplakin Ia and Ib was abolished by endo H and endo F treatment, suggesting the involvement of a high-mannose-type saccharide in the binding. (Reprinted from reference 130 with permission from the publisher.)

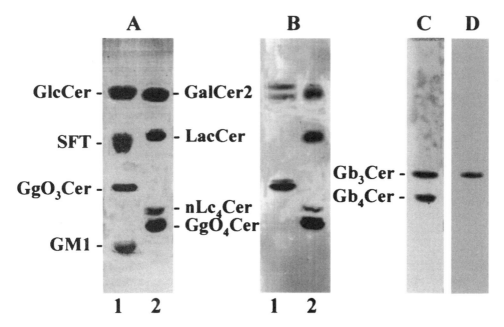

FIGURE 2.5 Identification of fimbrial glycolipid receptors by TLC. Purified glycolipids were separated by TLC and stained with orcinol (A and C), or probed with F1C fimbriae purified from *E. coli* and immunostained with antifimbrial antibodies (B and D). GlcCer, glucosylceramide; SFT, sulfatide; GgO₃Cer, asialo-GM₂ ganglioside; GM1, sialylated GM₁ ganglioside; GalCer2, galactosylceramide; LacCer, lactosylceramide; nLc₄Cer, paragloboside; GgO₄Cer, asialo-GM₁ ganglioside; Gb₃Cer, globotriaosylceramide; Gb₄Cer, globotetraosylceramide. Using this assay and other binding assays, the disaccharide sequence GalNAc(β1→4)GalB of asialo-GM₂ and asialo-GM₁ could be identified as the high-affinity binding epitope for F1C fimbriae of uropathogenic *E. coli*. (Reprinted from reference 62 with permission from the publisher.)

plex mixture on plastic substrata (20). The role of mucus secretions in bacterial adhesion is a particularly overlooked area that requires increased attention (see chapter 3).

There are qualitative and quantitative aspects of the study of adhesion to animal cells or other substrata. Some of the qualitative aspects are whether adhesion is localized or diffuse or is localized to specific structures (e.g., cilia) or to specific cells within a complex epithelium (31). These aspects can only be resolved by microscopical techniques. These techniques vary from simple phase-contrast microscopy of living cells, light microscopic evaluation of stained preparations, immunofluorescence, or one of several electron microscopic techniques, such as scanning electron microscopy. Microscopic techniques can also provide useful quantitative

information, but evaluation of a large number of cells is usually quite labor-intensive.

Quantitative techniques are detailed in Table 2.4. Unlike measurement of adhesion of bacteria to inert surfaces, quantitation of adhesion to animal cells may reflect either numbers of bacteria adherent to individual cells (i.e., if a microscopic method or fluorescence-activated cell sorting [FACS] is used) or an average of the number of bacteria bound to the whole population of cells present. Quantitative techniques that measure average numbers of bacteria bound to the whole population of cells will often include bacteria that are adherent to cell-free areas of the substratum in the count. Thus, the method used for quantitation will obviously affect the interpretations that can be made. Probably the only methods that provide both quantitative and

TABLE 2.4 Methods commonly used to quantify adherent bacteria

Method	Comments
ELISA	Because bound bacteria must be fixed in some way to resist multiple washing steps, one must select an antibody against a bacterial antigen that is not affected by fixation or these washings. An antibody directed against an antigen that is uniformly expressed in the bacterial population is required. The system is good for bacteria bound to inert surfaces as well as bactera bound to host cells (7).
Light microscopy	This method is used exclusively to measure numbers of bacteria bound to cells. It is useful to determine the pattern of bacteria bound to individual cells (e.g., localized versus diffuse adherence) or to extracellular products such as adsorbed serum components or extracellular membrane components secreted by host cells, etc. (89). A major advantage of the method is that it provides a direct measure of bacteria bound to specific cellular sites. A disadvantage is that it is labor-intensive.
Fluorescence (immunofluorescence microscopy [including confocal], fluorescent DNA and RNA probes, flow cytometry or FACS, luminometer, image analysis)	Useful for examination of adhesion to immobilized cells or cells in suspension. Also valuable for the study of invasion and involvement of cytoskeletal components (18, 82, 96, 100). Especially useful when examining adhesion of specific bacteria to surfaces already bearing other bacterial species (e.g., colonic mucosa). Direct counting or flow cytometry can be used to measure numbers of bacteria. FACS does not require that adherent and nonadherent populations be separated (Fig. 2.5) (111). Cells are excluded by size, using volume-gating procedures. Image analysis (9, 49) and luminescence (116) have also been used.
Scanning electron microscopy	Effective to quantitate bacteria bound to structures such as tissue explants and to focus on adhesion only to surfaces normally colonized by the test bacteria. An example is adhesion of uropathogenic *E. coli* to mucosal surface as opposed to cut edges of tissue (Fig. 2.6) (53, 54, 64).
Measurement of metabolites (CO_2, free radical production)	Used successfully to study adhesion of *H. pylori* by monitoring urease activity (12, 114) and of *H. influenzae* by monitoring acid phosphatase activity (15). The target animal cell can be lysed, enabling the selective monitoring of enzymes associated with intact bacteria (69, 97).
Methods dependent on radiolabeled bacteria	Many investigators have adopted this method because it is sensitive enough to detect a small number of bacteria (5, 20, 27, 54, 58, 66, 67, 77, 85, 98, 107, 109, 113). An appeal of the method is that it is relatively easy to obtain radiolabeled bacteria. Another advantage is that it is possible to monitor the simultaneous adhesion of different bacterial species or strains to the same substrata when the bacteria are differentially radiolabeled. Problems associated with this method are in the proper use and disposal of radioactive materials and the associated record keeping.
Viable counts (CFU)	To get an accurate count, it is necessary to ensure that bacteria are appropriately dissociated (e.g., no clumps, chains, or aggregates). The method detects only viable bacteria.
Growth assay	This method is based on the correlation between the number of adherent, living bacteria and the mass obtained over a defined period in a suitable culture medium. Theoretically, growth of bacteria arising from as few as one to five live bacteria can be measured. It does not require specialized probes (e.g., antibodies) but requires up to 3 to 5 h of culture of the adherent bacteria before turbidity can be determined (96, 122). The results can be standardized to obtain average numbers of bacteria.
Biomechanical measurements	A quartz crystal changes its oscillation frequencies when organisms attach to the surface (95). Surface plasmon resonance spectroscopy measures affinity constants for ligand–ligand interactions (3).

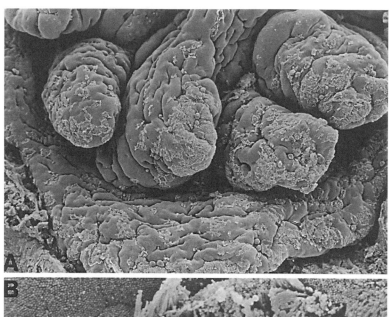
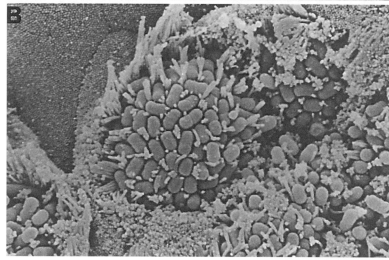

FIGURE 2.6 Scanning electron micrographs of human small intestinal mucosa infected with enteropathogenic *E. coli.* (A) Large areas of mucosal surface can be observed easily at low magnification. (B) Finer details of adhesion can be observed at higher magnification. Here, the bacteria can be seen to adhere intimately, dramatically altering the brush border of the entero-cytes. (Reprinted from reference 64 with permission from the publisher.)

qualitative information are some of the microscopic techniques, such as scanning electron microscopy (Fig. 2.6) (64). Atomic force microscopy is a relatively new microscopic technique that does not require processing of tissues through fixatives and dehydrating agents, thus allowing one to analyze living tissues (68).

As indicated above, one of the most common approaches to the study of bacterium-animal cell interactions has been the addition of suspended bacterial cells to tissue cultured monolayers. It should be pointed out, how-ever, that the reverse may also be studied. Many bacteria can be made to form monolay-ers on tissue culture plate surfaces, thereby providing a substratum to which suspended animal cells may adhere (74, 85, 120). This technique is especially useful when animal cell suspensions that cannot be immobilized are used. This methodological variant avoids the problems of separating unbound bacteria by differential centrifugation when the assay is performed in the presence of an agglutinating material. Numbers of adherent cells can be measured by microscopic counts (120), by measuring alkaline phosphatase released by lysing animal cells (85), or, in the case of erythrocytes, by releasing hemoglobin of the bound erythrocytes (58).

QUANTITATION OF ADHESION

After successfully removing the nonadherent bacteria from a test system, a number of methods may be employed to quantitate the bound bacteria (Table 2.4). When quantitating bacteria bound to receptor-coated surfaces or immobilized target animal cells (e.g., tissue culture cells [see below]), the method of choice is the ELISA. In this assay, the adherent bacteria are fixed to the substratum and an antiserum which reacts specifically with any component on the bacterial surface may be used as the first antibody. One advantage of this method is that neither the target cells nor the bacteria are treated before the adhesion assay is completed. A disadvantage of the method is that it is unable to distinguish viable from nonviable bacteria, since it depends only on an intact surface adhesin. The method is highly reproducible and easy to perform and is not limited to a particular phase of growth (7, 41). The ELISA values obtained may also be converted to number of bound bacteria by relating them to a standard curve generated with a known number of immobilized bacteria.

MATHEMATICAL ANALYSES OF QUANTITATIVE ADHESION DATA

Many of the methods outlined above are quantitative in that they enumerate a certain density of bacteria that adhere to a substratum. Such methods provide useful information, but they essentially ignore the fact that adhesion involves multiple equilibria, similar to most biological ligand-receptor interactions. To fully appreciate the operative mechanisms sufficient to design appropriate methods to modulate adhesion, one must eventually treat adhesion according to the general rules established for equilibria. In this way, one can develop reasonable estimations of, for example, attachment rates, numbers of combining

sites, and association and dissociation constants, providing important additional insights into the mechanisms involved.

In approaching important subjects such as affinity, two basic types of experiments can be performed. In the first and more commonly used approach, various densities of bacteria are added to defined substrata containing constant receptor densities and the binding is measured after the components have had sufficient time to reach an equilibrium. Equilibrium isotherms and their adaptations are commonly used to determine K_A (the equilibrium or affinity constant), numbers of binding sites, and the presence of multiple or interacting classes of sites. These analyses are based on the Langmuir isotherm

$$U/B = 1/KN + U/N$$

where U is the number of unbound cells, B is the number of bound cells, N is the number of binding sites, and $K = K_A$, the (average) affinity constant. Data are plotted in the form of U/B versus U and yield a straight line with a slope of $1/N$ and a y axis intercept equal to $1/KN$ (34). This approach is used to describe fully reversible (25) attachment to independent and noninteracting sites (34). Data obtained from these same assays can be plotted using the Scatchard equation

$$B/U = K(N - B)$$

Nonlinear plots are obtained from this equation, indicating that there is more than one class of receptors on the substratum. These equilibrium data can also be displayed in a Hill plot, which will reveal whether there is negative or positive cooperativity between receptor sites (55). For a more detailed discussion of the equilibrium approach, see reference 34.

The second approach to obtaining the averaged affinity (K_A) and, in addition, valuable information about complex or multiple-step adhesion reactions is to monitor the kinetics of binding. In contrast to the approach used for equilibrium experiments, these experiments employ a single concentration of ligand and a single concentration of receptor, and they document the extent of attachment over time. This approach will yield k_a and k_d, rate constants for adsorption and desorption, which in turn provide K_A via the equation $K_A = k_a/k_d$. In fact, the first (equilibrium) experiment must be performed to obtain the kinetic constants, as described below. Although it seems like more work, the kinetic approach must be used when there is no certainty that the binding being studied occurs via first-order kinetics involving a single step. This is commonly the case with bacterial attachment. Also, obtaining K_A by the kinetic approach does not require that the reaction be fully reversible (25), a condition seldom fulfilled in microbial attachment reactions. The kinetic approach allows the identification of deviations from so-called first-order attachment reactions and can therefore be used to determine whether multiple adhesions are operative (26).

The methods for obtaining kinetic constants, and ultimately K_A, have been described in detail (26). Briefly, an equilibrium experiment (see above) is performed to obtain the denominator for θ, the fractional saturation, also known as "coverage". θ is an indicator of the relative density of bacteria employed in a kinetic experiment with respect to the maximum number of bacteria accommodated on the substratum. The maximum number of accommodated bacteria constitutes the denominator of θ, and the numerator is the largest number of bacteria which are actually bound when using the single density of bacteria chosen for the study at hand. It is generally best to use a density of bacteria which results in a θ of <0.5.

Kinetic experiments yield numbers of bound (B) and unbound (U) bacteria at times t. The kinetic adaptation of the Langmuir isotherm seems very complex but makes fewer assumptions than the original Langmuir equation:

$$-\ln\frac{(U_t - U_f)/U_0 - U_f)}{1 - [1 - (U_t/U_0)]\theta}$$
$$= k_aNt[U_0/(U_0 - U_f) - \theta$$

The equation requires the numbers of unbound bacteria at time t (U_t), at infinity (U_f), and at the very beginning to the experiment (U_0). The left hand side of the above equation is plotted against time and yields a straight line. K_a is then calculated from the slope of the line (m):

$$m = k_aN[U_0 - U_f) - \theta]$$

When the line is not straight, it may be due to changes in the binding as the substratum becomes more covered. In that situation, the experiment should be repeated at a θ of <0.20.

This approach also allows the calculation of a desorption rate constant from the same experimental data, using the equation

$$k_d = k_a(U_0 - B_f)(1 - \theta)/B_f$$

where B_f is the number of cells bound at equilibrium. The average affinity constant, K_A, can be calculated using the two kinetic constants, as indicated above. However, the validity and usefulness of this quantity are uncertain if the kinetic plots have revealed nonlinear or two-phase adsorption. In those cases, the kinetic constants themselves are more reliable. For instance, if there is more than one discernable phase in adsorption or desorption, the average K_A (expressed as k_a/k_d) would obscure important details in the overall reaction, modeled in Fig. 2.7, in which "Bacterium----Substratum★" represents some intermediate state of loose adhesion which is difficult to directly assess experimentally. By obtaining desorption rate constants near the beginning of the adhesion reacton and also after the partners have reached equilibrium (i.e., at the "end" of the adhesion assay), the less accessible rate constants can be inferred. In this way, the existence of distinct phases of attachment, perhaps mediated by different surface structures, can be visualized (24).

The usefulness of the kinetic approach is extended when experimental variables are manipulated. For example, studying the kinetics of adhesion for isogenic mutants of bacter-

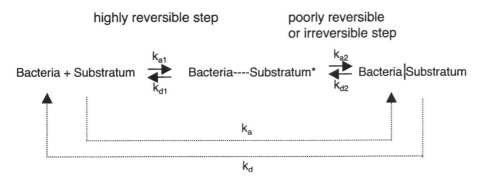

FIGURE 2.7 Kinetic method for evaluating bacterial adhesion to a substratum. The model depicts two steps of adhesion, each with separate association and dissociation constants. One represents weak, reversible adhesion, and the other represents firm adhesion. See the text for additional details.

ial strains, or in the presence of antibody to a receptor or ligand, can reveal the relative contribution of single macromolecules. In addition, some laboratories have developed methods to greatly facilitate the study of kinetic constants through the use of a continuous flow system coupled with computerized image capture and analysis. For a detailed analysis of such systems, see references 83 and 117.

DIFFERENTIATION BETWEEN ADHESION AND INTERNALIZATION

Phagocytic cells are committed to bind and phagocytose bacteria, whereas nonphagocytic cells are not. Nevertheless, some bacteria can be internalized by nonphagocytic cells, including epithelial cells. This internalization process has also been called "invasion" by numerous authors. We think, however, that "internalization" is a more appropriate term and effectively distinguishes the entry of bacteria into nonphagocytic cells from the very different pathogenic processes that are involved in true invasive infections. The internalization process is complex and will be dealt with in more detail in chapter 7. However, one clear requirement for the study of internalization is that one must be able to distinguish between attached and internalized bacteria.

Because internalization by phagocytic cells has been studied for a very long time, meth-

ods that were initially developed to selectively quantitate bound and ingested bacteria in the phagocytic process were adapted to quantitate the internalization of bacteria by nonprofessional phagocytic cells. One of the primary differences, however, is that the process of internalization takes several hours with nonprofessional phagocytes whereas ingestion normally occurs in minutes with phagocytes. Methods have been developed to easily differentiate between extracellularly bound and internalized bacteria (Table 2.5). The primary method of choice is based on selective antibi-

TABLE 2.5 Methods commonly used to distinguish between extracellularly bound and internalized bacteria

Type of cell	Methods
Phagocytic cells	Selective killing of extracellular bacteria by antibiotic
	Metabolic incorporation of radio-labeled metabolites that cannot penetrate phagocytes
	Fluorescence labeling of extracellular bacteria with antibodies
	ELISA specific for extracellular bacteria
	Expression of green fluorescent protein
	Electron microscopy
Nonphagocytic cells	Antibiotic killing of extracellular bacteria
	Electron microscopy
	Expression of green fluorescent protein

otic-mediated killing of bacteria remaining adherent to the exterior of the cell (118). Assays are performed as described for adhesion of bacteria to animal cells, and then, following removal of unbound bacteria, incubation is continued for several more hours, after which an antibiotic that does not penetrate animal cell membranes is added. The internalized bacteria are then released by lysing the animal cells with detergents at concentrations that do not affect the viability of the bacteria. CFU present within the cells are then enumerated by conventional techniques. The percentage of the bound bacteria that become internalized during the extended incubation period can also be determined. For these determinations, it is essential to set up parallel experiments in which the total number of viable bacteria is determined immediately following separation of the unbound bacteria and before adding the antibiotic.

A number of assumptions must be made in order to properly interpret the results of internalization assays. (i) There is neither intracellular growth nor killing. Although neither growth nor killing is likely to occur within 2 h of incubation with nonprofessional phagocytes, a proper control has yet to be devised. (ii) All externally exposed bacteria are killed by the antibiotic. In this regard, it should be realized that there are data indicating that bacteria bound to cells are somewhat more resistant to antibiotics than are freely suspended organisms (90). (iii) The bacteria, internalized or merely adherent, do not produce a pore-forming toxin which may facilitate the entry of antibiotics into the cells.

Studies of adhesion and ingestion by phagocytic cells require a similar approach in that antimicrobial agents are used to kill extracellular bacteria and bacteria attached to the cells (78, 79, 99, 123). It should be noted that some of the antimicrobial agents used successfully to kill only extracellular bacteria on nonprofessional phagocytes may also slowly penetrate phagocytes. Because phagocytic cells internalize and kill bacteria within a relatively short period (e.g., < 2 h), several other methods were devel-

oped to enumerate attached and internalized bacteria (94). One method employs bacteria metabolically radiolabeled with [^{3}H]uracil, which does not penetrate phagocytic cells. Enumeration of the attached versus ingested bacteria is possible by pulse-labeling the bacterium-phagocyte mixture and then recording the decrease in uptake of radioactivity by the phagocytic cells (19). This method, however, requires metabolically active bacteria, a criterion that does not always coincide with the expression of specific components mediating attachment of the organisms to phagocytes (e.g., type 1 fimbriae are expressed at increased levels during late exponential or stationary phase when the bacteria are not growing very actively). Furthermore, bacteria attached to, but not ingested by, phagocytes may have an altered ability to incorporate the radioactive label.

Fluorescein-labeled antibodies, which do not penetrate phagocytes, were used to selectively distinguish attached from ingested bacteria (9, 29). The total number of bacteria in Giemsa-stained monolayers and the number of fluorescent bacteria associated with phagocytes are determined by light and fluorescence microscopy, respectively. The method is somewhat cumbersome and subjective. In another method, an ELISA allowing the separate quantitation of bacteria attached to and bacteria ingested by phagocytic cells was described (7, 91). In this method, bacteria extracellularly attached to a phagocytic monolayer after various periods of incubation are quantitated by adding antimicrobial antibodies, which do not penetrate phagocytes, and then adding horseradish peroxidase-labeled anti-rabbit immunoglobulin G and a suitable substrate. Under these conditions, the decrease in the amount of substrate reduced as a function of time of incubation of the bacterium-phagocyte mixture reflects internalization of bacteria by phagocytes. By combining the ELISA technique with viable-count measurements, the rate of killing to the rate of ingestion could be determined.

Green fluorescent proteins and related molecules are now widely used as fluorescent tags to probe for bacteria or to probe for genes that are expressed only when the bacteria invade animal cells. The *gfp* gene encodes green fluorescent protein, a protein that fluoresces green when exited by blue light. The expression of the protein is placed under the control of a promoter whose activation is dependent on the entry of the bacteria into the host cell, and this reporter element is introduced randomly into the bacterial genome. Cells containing fluorescent bacteria (e.g., those whose genes were turned "on" inside the mammalian cell) can be collected by FACS, the bacteria can be isolated, and the activated genes can be identified (126).

GENETIC AND BIOCHEMICAL TECHNIQUES FOR IDENTIFICATION OF ADHESINS AND THEIR RECEPTORS

The identification of adhesins and their receptors has been the target of intense research, because the information will eventually provide means of interfering with the infectious processes, discussed in detail in chapter 11. There are numerous methods to achieve the goal of identifying the adhesins and their cognate receptors. In this section, an attempt is made to develop a scheme that is logical to follow and useful in situations where neither adhesin nor receptor is known or when only one is known. Even though a great many adhesins and receptors have already been identified and their interactions have been well characterized, current trends demand that more adhesins and their receptors be identified. These current trends include (i) emergence of new pathogens (e.g., *E. coli* O157:H7), (ii) reemergence of quiescent pathogens (e.g., *Streptococcus pyogenes*), (iii) appearance within a single species clone of multiple adhesins with distinct specificities, and (iv) recognition that allelic variants of adhesins may give rise to subtle receptor specificity changes. Indeed, during the 1980s a large number of publications appeared describing potential new adhesins for just one single species, *E. coli* (56, 60, 65, 102, 124).

Just within the roughly 30-year time frame of the development of the science of bacterial adhesion, several new pathogens have been identified, including *Helicobacter pylori, Legionella pneumophila,* and *Vibrio vulnificus.* The schemes outlined in Tables 2.6 and 2.7 have been followed successfully to identify some of their adhesins and receptors (42). Once initial knowledge about the identity of the adhesin and/or its receptor is obtained,

TABLE 2.6 Suggested progression of steps to be taken for the identification of new bacterial adhesins

Step	Approach
If receptor is unknown	
1	Inhibition assays using
	Bacterial surface components after extraction and fractionation
	Antibodies against defined surface components (16)
2	Treatment assays: treat bacteria with enzymes (e.g., proteases or glycosidases) and assay for effects on adhesion
3	Screening assays
	Screen a large number of erythrocytes for hemagglutination reaction
	Screen a large number of glycoconjugates for their ability to bind the test bacteria
4	Positive hemagglutination or binding suggests lectin, prompting study of the inhibitory effects of saccharides
5	Genetic assays: screen transposon insertion mutants for lack of adhesion (this is practical only if there is an assay procedure convenient for screening of large numbers of mutants, e.g., binding of erythrocytes or agglutination reactions)
If receptor is known	
1	Blotting assays: extract surface components, separate on gel, and blot with receptor or analog (98, 130)
2	Perform an assay of receptor binding (e.g., using receptor-coated particles) and screen fractions generated in step 1 for binding or agglutination
3	Genetic assays:
	Screen bank of transposon mutants for binding of receptor
	Prepare gene bank and screen lysates for clones expressing receptor binding activity or activity inhibiting adhesion or ligand binding (51)

TABLE 2.7 Suggested progression of steps to be taken for the identification of new adhesin receptors

Step	Approach
If adhesin is unknown	Follow inhibition and screening assays as described in Table 2.6
	Separate membrane glycoproteins on SDS gels and glycolipids on thin-layer plates, and blot with radiolabeled whole bacteria (62, 98, 130)
	Chemically identify reactive bands, subject to amino-terminal amino acid sequencing, etc.
	Chemically modify receptive surfaces (e.g., use periodate to oxidize carbohydrates) (46)
	Use antibodies against known membrane components (110)
If adhesin is known	Same as above steps, but using adhesin instead of whole bacteria (27)

succeeding steps generally follow the schemes outlined in Tables 2.6 and 2.7. This may allow localization of the adhesin-combining site (72), cloning of the chromosomal or plasmid locus to study regulation and biogenesis of the adhesin (107, 128), and determination of possible allelic variants with subtle differences in receptor specificities (121, 122).

MEASUREMENT OF POSTADHESION SIGNALING EVENTS

A number of bacteria express adhesins which interact with receptors on phagocytic cells and thereby trigger the cells to release proinflammatory agents. It has also been recently found that adhesins of certain bacteria also trigger nonprofessional phagocytes to release proinflammatory cytokines (2). These issues are discussed in more detail in chapter 8. The methods to assay for many of these adhesion-dependent events are well developed, and kits for the assay of most cytokines are available. The primary drawback of these experiments is that the process for releasing cytokines usually requires incubation of the bacteria with the target cells for up to 24 h. For this reason, it is essential to include antibiotics in the reaction mixture to prevent overgrowth of bacteria. Even then, it is not clear if the stimulation of

cytokine expression or secretion is primarily a postadhesion event stimulated by direct interaction of the adhesin with its receptor or a secondary event triggered by a product released by the bacteria during the incubation period. A control employing a mutant deficient in the adhesin may be run in parallel to determine if the signal is generated by an adhesive event (23).

Another approach is to separate the adhesion process from the signaling event. This may be achieved by incubating the animal cells with a relatively high density of bacteria in the presence or absence of an adhesin inhibitor for a short period, separating adherent and nonadherent bacteria as described in Table 2.3, and further incubating them for 24 h to measure cytokine release into the bulk medium. The cytokine release may be dependent not only on adhesion but also on invasion processes. Methods described in Table 2.5 to kill adherent bacteria may be used during or after the adhesion process and may be followed by further incubation of the animal cells to detect cytokine release. In addition, an invasin-deficient mutant may be used.

NATURAL AND EXPERIMENTAL INFECTIONS

It is difficult to study the adhesion process during experimental infections and much more difficult, if not impossible, to study adhesion during the course of a natural infection. Nevertheless, a number of methods are available by which one may be able to evaluate the involvement of adhesion in the infectious process (Table 2.8). These methods invariably involve quantitation of the ability of bacteria to colonize and/or cause disease at one or more sites in the experimental animal. These abilities, in turn, are dependent on many virulence factors, only one of which is the adhesin in question.

Adhesins in a Natural Infection

One of simplest tests to determine whether an adhesin is expressed in vivo during a natural infection is to assay for antibodies against the

TABLE 2.8 Methods used to assess the role of adhesion in infection

Expression of adhesin and its receptor in natural infections
    Assay for antiadhesin antibodies in patient sera
    Assay expression of adhesin by bacteria in clinical specimens
    Perform epidemiological survey for presence of adhesin gene in clinical isolates
    Determine the presence and density of receptor in infected sites and the correlation with natural infection
Experimental infections
    Monitor phenotypic expression of adhesin by immunohistological techniques at site of infection
    Perform experimental infection in animals deficient in expression of adhesin receptor
    Perform experimental infection with specific inhibitors of adhesins
        Antiadhesin-based vaccine
        Antireceptor antibodies
        Receptor analogs
        Other inhibitors
    Perform experimental infection with adhesin-deficient mutant

adhesin in patient sera (22, 71). Another relatively easy assay is to screen isolates from an infection for mRNA for an adhesin by reverse transcription-PCR. Reverse transcription PCR analysis could be difficult to analyze in cases where the gene of interest is present in both commensal and pathogen alike and/or present in different species (57). Nevertheless, if it is determined that the infecting organisms are genotypically positive, the isolates taken directly from the infected area can then be tested for adhesin expression by either direct assay or microscopic examination of specimens. This is useful only for specimens that contain large numbers of the infecting organism and for which the infecting organism comprises a large fraction of the population obtained (63). For instance, intestinal infections would be difficult to analyze in this way, because the infectious organisms may constitute only a small proportion of the total bacterial cells present. Sites that are normally sterile, such as the urinary bladder, would be more likely to have a very high proportion of the infecting organism in the sampled popula-

tion. Should it be necessary to cultivate the organisms in the laboratory between the time they are isolated from the site of infection and the time they are assayed for presence of the adhesin, there will always be a question about whether laboratory media affected gene regulation and subsequent adhesin expression.

Adhesins in an Experimental Infection

The role of adhesion in experimental infection can be assessed by (i) monitoring the time interval after infection for the phenotypic expression of an adhesin by immunohistochemical staining of infected tissues, (ii) performing experimental infections of an animal known to be deficient in adhesin receptor (see Table 1.3), (iii) infecting animals and using specific inhibitors of adhesion (antibodies against adhesin [71], antibodies against receptor [1], or receptor analogs [6]), or (iv) infecting animals with parent and mutant strains that are deficient in adhesin or that express different adhesin alleles (10).

REFERENCES

1. **Abraham, S. N., J. P. Babu, C. Giampapa, D. L. Hasty, W. A. Simpson, and E. H. Beachey.** 1985. Protection against ascending pyelonephritis with bacterial adhesin-specific or host cell receptor-specific monoclonal antibodies. *Infect. Immun.* **48:**625–628.
2. **Agace, W. W., S. R. Hedges, M. Ceska, and C. Svanborg.** 1993. Interleukin-8 and the neutrophil response to mucosal gram-negative infection. *J. Clin. Investig.* **92:**780–785.
3. **Amano, A., T. Nakamura, S. Kimura, I. Morisaki, I. Nakagawa, S. Kawabata, and S. Hamada.** 1999. Molecular interactions of *Porphyromonas gingivalis* fimbriae with host proteins: kinetic analyses based on surface plasmon resonance. *Infect. Immun.* **67:**2399–2405.
4. **An, Y. H., and R. J. Friedman.** 2000. *Handbook of Bacterial Adhesion. Principles, Methods, and Applications.* Humana Press, Totowa, N.J.
5. **Ardehali, R., and S. F. Mohammad.** 1993. ^{111}Indium labeling of microorganisms to facilitate the investigation of bacterial adhesion. *J. Biomed. Mater. Res.* **27:**269–275.
6. **Aronson, M., O. Medalia, L. Schori, D. Mirelman, N. Sharon, and I. Ofek.** 1979. Prevention of colonization of the urinary tract

of mice with *Escherichia coli* by blocking of bacterial adherence with methyl alpha-D-mannopyranoside. *J. Infect. Dis.* **139:**329–332.

7. **Athamma, A., and I. Ofek.** 1988. Enzyme-linked immunosorbent assay for quantitation of attachment and ingestion stages of bacterial phagocytosis. *J. Clin. Microbiol.* **26:**62–66.

8. **Bar-Shavit, F., R. Goldman, I. Ofek, N. Sharon, and D. Mirelman.** 1980. Mannose binding activity of *Escherichia coli,* a determinant of attachment and ingestion of the bacteria by macrophages. *Infect. Immun.* **29:**417–424.

9. **Barthelson, R., C. Hopkins, and A. Mobasseri.** 1999. Quantitation of bacterial adherence by image analysis. *J. Microbiol. Methods* **38:**17–23.

10. **Bieber, D., S. W. Ramer, C.-Y. Wu, W. J. Murray, T. Tobe, R. Fernandez, and G. K. Schoolnik.** 1998. Type IV pili, transient bacterial aggregates, and virulence of entero-pathogenic *Escherichia coli. Science* **280:**2114–2118.

11. **Boughey, M. T., R. M. Duckworth, A. Lips, and A. L. Smith.** 1978. Observation of weak primary minima in the interaction of polystyrene particles with nylon fibres. *J. Chem. Soc. Faraday Trans.* I **74:**2200–2208.

12. **Burger, O., I. Ofek, M. Tabak, E. Weiss, N. Sharon, and I. Neeman.** 2000. A high molecular mass constituent of cranberry juice inhibits *Helicobacter pylori* adhesion to human gastric mucus. *FEMS Immunol. Med. Microbiol.* **29:**295–301.

13. **Busscher, H. J., M. M. Cowan, and H. C. van der Mei.** 1992. On the relative importance of specific and non-specific approaches to oral microbial adhesion. *FEMS Microbiol. Rev.* **88:**199–210.

14. **Busscher, H. J., and H. C. van der Mei.** 1995. Use of flow chamber devices and image analysis methods to study microbial adhesion. *Methods Enzymol.* **253:**455–477.

15. **Chance, D. L., T. J. Reilly, and A. L. Smith.** 1999. Acid phosphatase activity as a measure of *Haemophilus influenzae* adherence to mucin. *J. Microbiol. Methods* **39:**49–58.

16. **Cisar, J. O., E. L. Barsumian, S. H. Curl, A. E. Vatter, A. L. Sandberg, and R. P. Siraganian.** 1981. Detection and localization of a lectin on *Actinomyces viscosus* T14V by monoclonal antibodies. *J. Immunol.* **127:**1318–1322.

17. **Clark, V. L., and P. M. Bavoil.** 1994. Bacterial pathogenesis, part B. Interaction of pathogenic bacteria with host cells. *Methods Enzymol.* **236:**1–642.

18. **Cloak, O. M., G. Duffy, J. J. Sheridan, D. A. McDowell, and I. S. Blair.** 1999. Development of a surface adhesion immunoflu-orescent technique for the rapid detection of *Salmonella* spp. from meat and poultry. *J. Appl. Microbiol.* **86:**583–590.

19. **Cohen, P. S., S. Renate, J. Syter, and K. Vosbeck.** 1985. Quantitation of the rate of the ingestion of *Escherichia coli* by human polymorphonuclear neutrophils using ^{3}H-uracil. *J. Microbiol. Methods* **3:**223–235.

20. **Cohen, P. S., and D. L. Laux.** 1995. Bacterial adhesion to and penetration of intestinal mucus *in vitro. Methods Enzymol.* **253:**309–314.

21. **Courtney, H. S., D. L. Hasty, and I. Ofek.** 1990. Hydrophobicity of group A streptococci and its relationship to adhesion of streptococci to host cells, p. 361–386. *In* R. J. Doyle and M. Rosenberg (ed.), *Microbial Cell Surface Hydrophobicity.* ASM Press, Washington, D.C.

22. **Courtney, H. S., J. B. Dale, and D. L. Hasty.** 1996. Differential effects of the streptococcal fibronectin-binding protein, FBP54, on adhesion of group A streptococci to human buccal cells and HEp-2 tissue culture cells. *Infect. Immun.* **64:**2415–2419.

23. **Courtney, H. S., I. Ofek, and D. L. Hasty.** 1997. M protein mediated adhesion of M type 24 *Streptococcus pyogenes* stimulates release of interleukin-6 by HEp-2 tissue culture cells. *FEMS Microbiol. Lett.* **151:**65–70.

24. **Cowan, M. M.** 1995. Kinetic analysis of microbial adhesion. *Methods Enzymol.* **253:** 179–189.

25. **Cowan, M. M., K. G. Taylor, and R. J. Doyle.** 1986. Kinetic analysis of *Streptococcus sanguis* adhesion to artificial pellicle. *J. Dent. Res.* **65:**1278–1283.

26. **Dahlquist, F. W.** 1978. The meaning of Scatchard and Hill plots. *Methods Enzymol.* **48:**270–279.

27. **Dean-Nystrom, E. A.** 1995. Identification of intestinal receptors for enterotoxigenic *Escherichia coli. Methods Enzymol.* **253:**315–324.

28. **de Man, P., B. Cedergren, S. Enerback, A. C. Larsson, H. Leffler, A. L. Lundell, B. Nilsson, and C. Svanborg-Edén.** 1987. Receptor-specific agglutination tests for detection of bacteria that bind globoseries glycolipids. *J. Clin. Microbiol.* **25:**401–406.

29. **Dilworth, J. A., J. O. Hendly, and G. L. Mandell.** 1975. Attachment and ingestion of gonococci by human neutrophils. *Infect. Immun.* **11:**512–416.

30. **Donlon, B., and E. Colleran.** 1993. A comparison of different methods to determine the hydrophobicity of acetogenic bacteria. *J. Microbiol. Methods* **17:**27–37.

31. **Donnenberg, M. S., and J. P. Nataro.**

1995. Methods for studying adhesion of diarrheagenic *Escherichia coli*. *Methods Enymol.* **253:**324–335.

32. **Doyle, R. J.** 2000. Contribution of the hydrophobic effect to microbial infection. *Microbes Infect.* **2:**391–400.

33. **Doyle, R. J.** 2001. Microbial growth in biofilms, Part B. Special environments and physicochemical aspects. *Methods Enzymol.* **337:**1–439.

34. **Doyle, R. J., J. D. Oakley, K. R. Murphy, D. McAlister, and K. G. Taylor.** 1985. Graphical analyses of adherence data, p. 109–113. *In* S. Mergenhagen and B. Rosan (ed.), *Molecular Basis of Oral Microbial Adhesion.* ASM Press, Washington, D.C.

35. **Doyle, R. J., and M. Rosenberg.** 1990. Microbial cell surface hydrophobicity: History, measurement, and significance, p. 361–386. *In* R. J. Doyle and M. Rosenberg (ed.), *Microbial Cell Surface Hydrophobicity.* ASM Press, Washington, D.C.

36. **Doyle, R. J., M. Rosenberg, and D. Drake.** 1990. Hydrophobicity of oral bacteria, p. 387–419. *In* R. J. Doyle and M. Rosenberg (ed.), *Microbial Cell Surface Hydrophobicity.* American Society for Microbiology, Washington, D.C.

37. **Doyle, R. J., and M. Rosenberg.** 1995. Measurement of microbial adhesion to hydrophobic substrata. *Methods Enzymol.* **253:**542–550.

38. **Doyle, R. J., and I. Ofek.** 1995. Adhesion of microbial pathogens. *Methods Enzymol.* **253:**1–600.

39. **Duguid, J. P., and R. R. Gillies.** 1957. Fimbriae and adhesive properties of dysentery bacilli. *J. Pathol. Bacteriol.* **74:**397–411.

40. **Duguid, J. P., and D. C. Old.** 1980. Adhesive properties of *Enterobacteriaceae,* p. 185–217. *In* E. H. Beachey (ed.), *Bacterial Adherence. Receptors and Recognition,* ser. B., vol. 6. Chapman & Hall, London, United Kingdom.

41. **Dunne, W. M., Jr.** 2000. Evaluating adherent bacteria and biofilm using biochemical and immunochemical methods, p. 273–283. *In* Y. H. An and R. R. Friedman (ed.), *Handbook of Bacterial Adhesion.* Humana Press, Inc., Totowa, N.J.

42. **Evans, D. G., and D. J. Evans.** 1995. Adhesion properties of *Helicobacter pylori. Methods Enzymol.* **253:**336–360.

43. **Firon, N., I. Ofek, and N. Sharon.** 1983. Carbohydrate specificity of the surface lectins of *Escherichia coli, Klebsiella pneumoniae,* and *Salmonella typhimurium. Carbohydr. Res.* **120:**235–249.

44. **Firon, N., S. Ashkenazi, D. Mirelman, I. Ofek, and N. Sharon.** 1987. Aromatic alpha-glycosides of type 1 fimbriated *Escherichia coli* to yeast and intestinal epithelial cells. *Infect. Immun.* **55:**472–476.

45. **Garber, N., N. Sharon, D. Shohet, J. S. Lam, and R. J. Doyle.** 1985. Contribution of hydrophobicity to hemagglutination reactions of *Pseudomonas aeruginosa. Infect. Immun.* **50:**336–337.

46. **Gbarah, A., C. G. Gahmberg, I. Ofek, U. Jacobi, and N. Sharon.** 1991. Identification of the leukocyte adhesion molecules CD11 and CD18 as receptors for type 1-fimbriated (mannose-specific) *Escherichia coli. Infect. Immun.* **59:**4524–4530.

47. **Goldhar, J.** 1994. Bacterial lectinlike adhesins: determination and specificity. *Methods Enzymol.* **236:**211–231.

48. **Goldhar, J.** 1995. Erythrocytes as target cells for testing bacterial adhesions. *Methods Enzymol.* **253:**43–50.

49. **Grivet, M., J. J. Morrier, C. Souchier, and O. Barsotti.** 1999. Automatic enumeration of adherent streptococci or actinomyces on dental alloy by fluorescence image analysis. *J. Microbiol. Methods* **38:**33–42.

50. **Hanlon, G. W., C. J. Olliff, J. A. Brant, and S. P. Denyer.** 1999. A novel image-analysis technique for measurement of bacterial cell surface tension. *J. Pharm. Pharmacol.* **51:**207–214.

51. **Hanski, E., P. A. Horwitz, and M. G. Caparon.** 1992. Expression of protein F, the fibronectin-binding protein of *Streptococcus pyogenes* JRS4, in heterologous streptococcal and enterococcal strains promotes their adherence to respiratory epithelial cells. *Infect. Immun.* **60:**5119–5125.

52. **Hasty, D. L., and H. S. Courtney.** 1996. Group A streptococcal adhesion. All of the theories are correct, p. 81–94. *In* I. Kahane and I. Ofek (ed.), *Toward Anti-Adhesion Therapy for Microbial Diseases.* Plenum Press, New York, N.Y.

53. **Hasty, D. L., and E. V. Sokurenko.** 2000. The FimH lectin of *Escherichia coli* type 1 fimbriae. An adaptive adhesin, p. 481–515. *In* R. J. Doyle (ed.), *Glycomicrobiology.* Klewer Academic/Plenum Publishers, New York, N.Y.

54. **Hazlett, L. D.** 1995. Analysis of ocular microbial adhesion. *Methods Enzymol.* **253:**53–66.

55. **Hill, A. V.** 1913. The combinations of haemoglobin with oxygen and with carbon monoxide. *Int. Biochem. J.* **7:**471–480.

56. **Honda, T., M. Arita, and T. Miwatani.** 1984. Characterization of new hydrophobic pili of human enterotoxigenic *Escherichia coli:* a pos-

sible new colonization factor. *Infect. Immun.* **43**:959–965.

57. **Johnson, J. R.** 1995. Epidemiological considerations in studies of microbial adhesion. *Methods Enzymol.* **253**:167–179.

58. **Kahane, I.** 1995. Adhesion of mycoplasmas. *Methods Enzymol.* **253**:367–373.

59. **Kallenius, G., R. Møllby, S. B. Svensson, J. Winberg, A. Lundblad, S. Svensson, and B. Cedergren.** 1980. The p^k antigen as a receptor for the haemagglutination of pyelonephritogenic *Escherichia coli. FEMS Microbiol. Lett.* **7**:297–302.

60. **Karch, H. K., J. Heesemann, R. Laufs, A. D. O'Brien, C. O. Tacket, and M. M. Levine.** 1987. A plasmid of enterohemorrhagic *Escherichia coli* O157:H7 is required for expression of a new fimbrial antigen and for adhesion to epithelial cells. *Infect. Immun.* **55**:455–461.

61. **Kawai, Y., and I. Yano.** 1983. Ornithine-containing lipid of *Bordetella pertussis,* a new type of hemagglutinin. *Eur. J. Biochem.* **136**:531–538.

62. **Khan, A. S., B. Kniep, T. A. Oelschlaeger, I. Van Die, T. Korhonen, and J. Hacker.** 2000. Receptor structure for F1C fimbriae of uropathogenic *Escherichia coli. Infect. Immun.* **68**:3541–3547.

63. **Kisielius, P. V., W. R. Schwan, S. K. Amundsen, J. L. Duncan, and A. J. Schaeffer.** 1989. In vivo expression and variation of *Escherichia coli* type 1 and P pili in the urine of adults with acute urinary tract infections. *Infect. Immun.* **57**:1656–1662.

64. **Knutton, S.** 1995. Electron microscopical methods in adhesion. *Methods Enzymol.* **253**:145–158.

65. **Knutton, S., D. Lloyd, and A. McNeish.** 1987. Identification of a new fimbrial structure in enterotoxigenic *Escherichia coli* (ETEC) serotype O148:H28 which adheres to human intestinal mucosa: a potentially new human ETEC colonization factor. *Infect. Immun.* **55**:86–92.

66. **Kolenbrander, P. E.** 1995. Coaggregations among oral bacteria. *Methods Enzymol.* **253**:385–387.

67. **Kolenbrander, P. E., K. D. Parrish, R. N. Andersen, and E. P. Greenberg.** 1995. Intergeneric coaggregation of oral *Treponema* spp. with *Fusobacterium* spp. and intrageneric coaggregation among *Fusobacterium* spp. *Infect. Immun.* **63**:4584–4588.

68. **Kotra, L. P., N. A. Amro, G.-Y. Liu, and S. Mobashery.** 2000. Visualizing bacteria at high resolution. *ASM News* **66**:675–680.

69. **Kreft, B., O. Carstensen, E. Straube, S. Bohnet, J. Hacker, and R. Marre.** 1992. Adherence to and cytotoxicity of *Escherichia coli* for eukaryotic cell lines quantified by MTT (3-[4,5-dimethylthiazol-2-yl]-2,5-diphenyltetrazolium bromide). *Zentbl. Bakteriol.* **276:** 231–242.

70. **Lachica, R. V., and D. L. Zink.** 1984. Determination of plasmid-associated hydrobicity of *Yersinia enterocolitica* by a latex particle agglutination test. *J. Clin. Microbiol.* **19:** 660–663.

71. **Langermann, S., S. Palaszynski, M. Barnhart, G. Auguste, J. S. Pinkner, J. Burlein, P. Barren, S. Koenig, S. Leath, C. H. Jones, and S. J. Hultgren.** 1997. Prevention of mucosal *Escherichia coli* infection by FimH-adhesin-based systemic vaccination. *Science* **276**:607–611.

72. **Lee, K. K., W. Y. Wong, H. B. Sheth, R. S. Hodges, W. Paranchych, and R. T. Irvin.** 1995. Use of synthetic peptides in characterization of microbial adhesins. *Methods Enzymol.* **253**:115–132.

73. **Leffler, H., and C. Svanborg-Eden.** 1980. Chemical-identification of a glycosphingolipid receptor for *Escherichia coli* attaching to human urinary-tract epithelial cells and agglutinating human-erythrocytes. *FEMS Microbiol. Lett.* **8**:127–134.

74. **Leong, J. M., L. Moitoso de Vargas, and R. R. Isberg.** 1992. Binding of cultured mammalian cells to immobilized bacteria. *Infect. Immun.* **60**:683–686.

75. **Lichtenberg, D., M. Rosenberg, N. Sharfman, and I. Ofek.** 1985. A kinetic approach to bacterial adherence to hydrocarbon. *J. Micbiol. Methods* **4**:141–146.

76. **Lindahl, M., A. Faris, T. Wadström, and S. Hjerten.** 1981. A new test based on 'salting out' to measure relative surface hydrophobicity of bacterial cells. *Biochim. Biophys. Acta* **677**:471–476.

77. **Ljungh, Å., and T. Wadström.** 1995. Binding of extracellular matrix proteins by microbes. *Methods Enzymol.* **253**:501–514.

78. **Maaloe, O.** 1974. On the dependence of the phagocytosis stimulating action of immune serum on complement. *Acta Pathol. Microbiol. Scand.* **24**:33–46.

79. **MacKaners, J. B.** 1960. The phagocytosis and inactivation of staphylococci by macrophages of normal rabbits. *J. Exp. Med.* **112**:35–53.

80. **Magnusson, K. E.** 1994. Testing for charge and hydrophobicity correlates in cell-cell adhesion. *Methods Enzymol.* **228**:326–334.

81. **Malaviya, R., and S. N. Abraham.** 1995.

Interaction of bacteria with mast cells. *Methods Enzymol.* **253**:27–43.

82. **Manning, P. A.** 1995. Use of confocal microscopy in studying bacterial adhesion and invasion. *Methods Enzymol.* **253**:159–166.

83. **Meinders, J. M., H. C. van der Mei, and H. J. Busscher.** 1992. *In situ* enumeration of bacterial adhesion in a parallel plate flow chamber: elimination of in focus flowing bacteria from the analysis. *J. Microbiol. Methods* **16**: 119–124.

84. **Metzger, Z., L. G. Featherstone, W. W. Ambrose, M. Trope, and R. R. Arnold.** 2001. Kinetics of coaggregation of *Porphyromonas gingivalis* with *Fusobacterium nucleatum* using an automated microtiter plate assay. *Oral Microbiol. Immunol.* **16**:163–169.

85. **Meyer, D. H., and P. M. Fives-Taylor.** 1995. Adhesion of oral bacteria to soft tissue. *Methods Enzymol.* **253**:373–385.

86. **Miorner, H., G. Johansson, and G. Kronvall.** 1983. Lipoteichoic acid is the major cell wall component responsible for surface hydrophobicity of group A streptococci. *Infect. Immun.* **39**:336–343.

87. **Mirelman, D.** 1986. *Microbial Lectins and Agglutinins. Properties and Biological Activity.* John Wiley & Sons, Inc., New York, N.Y.

88. **Mirelman, D., and I. Ofek.** 1986. Microbial lectins and agglutinins, p. 1–19. *In* D. Mirelman (ed.), *Microbial Lectins and Agglutinins. Properties and Biological Activity.* John Wiley & Sons, Inc., New York, N.Y.

89. **Mobley, H. L. T., G. R. Chippendale, and J. W. Warren.** 1995. In vitro adhesion of bacteria to exfoliated uroepithelial cells: criteria for quantitative analysis. *Methods Enzymol.* **253**: 360–367.

90. **Nickel, J. C., I. Ruseska, J. B. Wright, and J. W. Costerton.** 1985. Tobramycin resistance of *Pseudomonas aeruginosa* cells growing as a biofilm on urinary catheter material. *Antimicrob. Agents Chemother.* **27**:619–624.

91. **Ofek, I.** 1995. Enzyme-linked immunosorbent-based adhesion assays. *Methods Enzymol.* **253**:528–536.

92. **Ofek, I., and E. H. Beachey.** 1978. Mannose binding and epithelial cell adherence of *Escherichia coli. Infect. Immun.* **22**:247–254.

93. **Ofek, I., E. Whitnack, and E. H. Beachey.** 1983. Hydrophobic interactions of group A streptococci with hexadecane droplets. *J. Bacteriol.* **154**:139–145.

94. **Ofek, I., J. Goldhar, Y. Keisari, and N. Sharon.** 1995. Nonopsonic phagocytosis of microorganisms. *Annu. Rev. Microbiol.* **49**: 239–276.

95. **Otto, K., H. Elwing, and M. Hermansson.** 1999. Effect of ionic strength on initial interactions of *Escherichia coli* with surfaces, studied online by a novel quartz crystal microbalance technique. *J. Bacteriol.* **181**:5210–5218.

96. **Papaioannou, W., D. van Steenberghe, J. J. Cassiman, J. Van Eldere, and M. Quirynen.** 1999. Comparison of fluorescence microscopy and culture assays to quantitate adhesion of *Porphyromonas gingivalis* to mono- and multi-layered pocket epithelium cultures. *J. Periodontol.* **70**:618–625.

97. **Peck, R.** 1985. A one-plate assay for macrophage bactericidal activity. *J. Immunol. Methods* **82**:131–140.

98. **Prakobphol, A., H. Leffler, and S. J. Fisher.** 1995. Identifying bacterial receptor proteins and quantifying strength of interactions they mediate. *Methods Enzymol.* **253**:132–142.

99. **Ran, J. A., C. Watanakunakorn, and J. P. Phair.** 1971. A modified assay of neutrophil function: use of lysostaphin to differentiate defective phagocytosis from impaired intracellular killing. *J. Lab. Clin. Med.* **78**:316–387.

100. **Reinhard, J., C. Basset, J. Holton, M. Binks, P. Youinou, and D. Vaira.** 2000. Image analysis method to assess adhesion of *Helicobacter pylori* to gastric epithelium using confocal laser scanning microscopy. *J. Microbiol. Methods* **39**:179–187.

101. **Rest, R. F.** 1995. Association of bacteria with human phagocytes. *Methods Enzymol.* **253**: 12–26.

102. **Rhen, M., P. Klemm, and T. K. Korhonen.** 1986. Identification of two hemagglutinins of *Escherichia coli*, N-acetyl-D-glucosamine-specific fimbriae and a blood group M-specific agglutinin, by cloning the corresponding genes in *Escherichia coli* K-12. *J. Bacteriol.* **168**:1234–1242.

103. **Rodzinski, E., and E. Tuomanen.** 1995. Adhesion of microbial pathogens to leukocyte integrins: methods to study ligand mimicry. *Methods Enzymol.* **253**:1–12.

104. **Rosenberg, M.** 1981. Bacterial adherence to polystyrene: a replica method of screening for bacterial hydrophobicity. *Appl. Environ. Microbiol.* **42**:375–377.

105. **Rosenberg, M., D. Gutnick, and E. Rosenberg.** 1980. Adherence of bacteria to hydrocarbons: a simple method for measuring cell surface hydrophobicity. *FEMS Microbiol. Lett.* **9**:29–33.

106. **Rozgonyi, F., K. R. Szitha, S. Hjerten, and T. Wadström.** 1985. Standardization of salt aggregation test for reproducible determination of cell-surface hydrophobicity with special

reference to *Staphylococcus* species. *J. Appl. Bacteriol.* **59**:451–457.

107. **Schifferli, D. M.** 1995. Use of Tn*phoA* and T7 RNA polymerase to study fimbrial proteins. *Methods Enzymol.* **253**:242–258.

108. **Schilling, K. M., R. G. Carson, C. A. Bosko, G. D. Golikeri, A. Bruinooge, K. Hoyberg, A. M. Waller, and N. P. Hughes.** 1994. A microassay for bacterial adherence to hydroxyapatite. *Colloids Surfaces Ser. B* **3**:31–38.

109. **Schilling, K. M., and R. J. Doyle.** 1995. Bacterial adhesion to hydroxylapatite. *Methods Enzymol.* **253**:536–542.

110. **Schrager, H. M., S. Alberti, C. Cywes, G. J. Dougherty, and M. R. Wessels.** 1998. Hyaluronic acid capsule modulates M protein-mediated adherence and acts as a ligand for attachment of group A streptococcus to CD44 on human keratinocytes. *J. Clin. Investig.* **101**:1708–1716.

111. **Sethman, C. R., R. J. Doyle, and M. M. Cowan.** 2002. Flow cytometric evaluation of adhesion of *Streptococcus pyogenes* to epithelial cells. *J. Microbiol. Methods* **51**:35–42

112. **Sharon, N., and H. Lis.** 1989. *Lectins.* Chapman & Hall, London, United Kingdom.

113. **Sharon, N., and I. Ofek.** 1995. Identification of receptors for bacterial lectins by blotting techniques. *Methods Enzymol.* **253**:91–98.

114. **Simon, P. M., P. L. Goode, A. Mobasseri, and D. Zopf.** 1997. Inhibition of *Helicobacter pylori* binding to gastrointestinal epithelial cells by sialic acid-containing oligosacchrides. *Infect. Immun.* **65**:750–757.

115. **Singh, J., and R. J. Doyle.** 1993. Salt enhanced lectinosorbent assay. *J. Microbiol. Methods* **17**:61–65.

116. **Siragusa, G. R., K. Nawotka, S. D. Spilman, P. R. Contag, and C. H. Contag.** 1999. Real-time monitoring of *Escherichia coli* O157:H7 adherence to beef carcass surface tissues with a bioluminescent reporter. *Appl. Environ. Microbiol.* **65**:1738–1745.

117. **Sjollema, J., H. J. Busscher, and A. H. Weerkamp.** 1989. Real-time enumeration of adhering microorganisms in a parallel-plate flow cell using automated image analysis. *J. Microbial. Methods* **9**:73–78.

118. **Small, P. L. C., R. R. Isberg, and S. Falkow.** 1987. Comparison of the ability of enteroinvasive *Escherichia coli*, *Salmonella typhimurium*, *Yersinia pseudotuberculosis*, and *Yersinia enterocolitica* to enter and replicate within HEp-2 cells. *Infect. Immun.* **55**:1674–1679.

119. **Smyth, C. J., P. Jonsson, E. Olsson, O. Soderlind, R. Rosengren, S. Hjerten, and T. Wadström.** 1978. Differences in hydrophobic surface characteristics of porcine enteropathogenic *Escherichia coli* K88 antigen as revealed by hydrophobic interaction chromatography. *Infect. Immun.* **22**:462–472.

120. **Sokurenko, E. V., and D. L. Hasty.** 1995. Assay for adhesion of host cells to immobilized bacteria. *Methods Enzymol.* **253**:220–226.

121. **Sokurenko, E. V., V. Chesnokova, D. E. Dykhuizen, I. Ofek, X.-R. Wu, K. A. Krogfelt, C. Struve, M. A. Schembri, and D. L. Hasty.** 1998. Pathogenic adaptation of *Escherichia coli* by natural variation of the FimH adhesin. *Proc. Natl. Acad. Sci. USA* **95**: 8922–8926.

122. **Sokurenko, E. V., V. A. McMackin, and D. L. Hasty.** 1995. Bacterial adhesion measured by growth of adherent organisms. *Methods Enzymol.* **253**:519–528.

123. **Solberg, C. O.** 1972. Protection of phagocytized bacteria against antibiotics. *Acta. Med. Scand.* **191**:383–387.

124. **Tacket, C. O., D. R. Maneval, and M. M. Levine.** 1987. Purification, morphology, and genetics of a new fimbrial putative colonization factor of enterotoxigenic *Escherichia coli* O159:H4. *Infect. Immun.* **55**:1063–1069.

125. **Tsivion, Y., and N. Sharon.** 1981. Lipid-mediated hemagglutination and its relevance to lectin-mediated agglutination. *Biochim. Biophys. Acta* **642**:336–344.

126. **Valdivia, R. H., and S. Falkow.** 1997. Fluorescence-based isolation of bacterial genes epxressed within host cells. *Science* **277**: 2007–2011.

127. **Van Oss, C. J., and C. F. Gillman.** 1972. Phagocytosis as a surface phenomenon. I. Contact angles and phagocytosis of non-opsonized bacteria. *Reticuloendothel. Soc. J.* **12**:283–292.

128. **Vetter, V., and J. Hacker.** 1995. Strategies for employing molecular genetics to study tip adhesins. *Methods Enzymol.* **253**:229–242.

129. **Wilson, M.** 2002. *Bacterial Adhesion to Host Tissues. Mechanisms and Consequences.* Cambridge University Press, Cambridge, United Kingdom.

130. **Wu, X.-R., T.-T. Sun, and J. J. Medina.** 1996. *In vitro* binding of type 1-fimbriated *Escherichia coli* to uroplakins Ia and Ib: relation to urinary tract infections. *Proc. Natl. Acad. Sci. USA* **93**:9630–9635.

TARGET TISSUES FOR
BACTERIAL ADHESION

3

A thorough understanding of cell and tissue biology is critical to the identification of the mechanisms of bacterial adhesion that operate within the host, especially in order to develop methods to selectively prevent pathogen adhesion. In this chapter, we discuss some basic elements of cell biology that are relevant to bacterial adhesion. There is a voluminous literature on cell biology, and a comprehensive review is clearly beyond the scope of this book. For more detailed information, the reader is encouraged to look to one of the many excellent anatomy, physiology, and cell biology texts (see, e.g., references 1, 55, and 70).

GENERAL STRUCTURE OF ANIMAL TISSUE SURFACES

With the exception of infections that result from penetrating wounds, bacteria first contact epithelial cells (e.g., skin keratinocytes, buccal squamous epithelium, or intestinal enterocytes) or their products (e.g., mucus or fingernails). The epithelial cells of the skin, the upper and lower gastrointestinal tract, the urinary tract, etc., exhibit all of the general components of cell membranes. While they also possess many unique specializations, they exhibit certain basic characteristics that make them an epithelium. There are essentially

three basic characteristics of an epithelium (Fig. 3.1): (i) the cell layer has a free apical surface, and the superficial cells are keratinized (e.g., the skin) or are nonkeratinized and coated with a layer of mucus (e.g., the colon); (ii) the contiguous cells are joined by junctional complexes (Fig. 3.2); and (iii) the cell layer is attached basally to an extracellular matrix specialization termed the basal lamina, which itself sits on other extracellular matrix components. Some epithelia are "stratified," and in these epithelia only the basal layer of cells is attached to the basal lamina (Fig. 3.1A). Examples of this include the skin and the buccal and pharyngeal epithelia. There is an apical layer that borders the free surface, and there are various numbers of intermediate layers that vary in an organ-specific manner. Other epithelia are "simple," in that they are composed of a single layer of cells that rest on a basement membrane (Fig. 3.1B). In most cases, the free surfaces of these epithelia are coated with serous or mucous secretions that can come from major salivary glands (e.g., the parotid gland) or minor unnamed glands or from specialized cells within the epithelial sheet (e.g., goblet cells). Below the basement membranes is a lamina propria, which contains elements of the connective tissues (e.g., collagen, fibronectin, and elastin), various cell

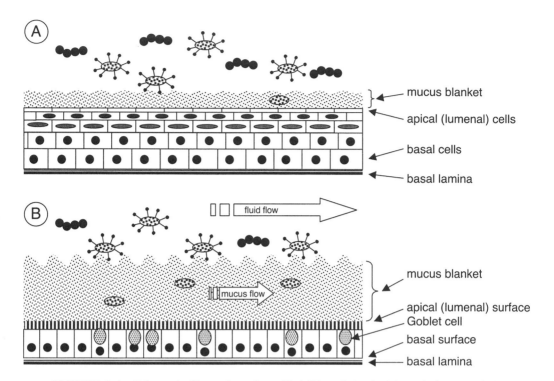

FIGURE 3.1 Schematic illustration of stratified (A) and simple (B) epithelia. Stratified epithelia, such as stratified squamous epithelium of the pharyngeal mucosa (A), have up to 20 or more layers of cells. The basal layer of cuboidal cells lies on a basal lamina composed primarily of type IV collagen, laminin, heparan sulfate proteoglycans, fibronectin, and other extracellular matrix components. The cells are joined by a variety of junctional complexes composed, for instance, of tight junctions, desmosomes, and gap junctions (see also Fig. 3.2). The cells differentiate and change shape as they come closer to the luminal surface. The apical (luminal) surfaces of the squamous cells are covered with a mucus blanket, except for the skin, where the superficial cells are protected by becoming keratinized. In the buccal and pharyngeal epithelium, the mucus coat is not very thick. Nevertheless, to become attached, bacteria must be able to bind to mucus components, and before they are able to attach directly to epithelial cells, they must be able to penetrate or destroy the mucus blanket. Simple epithelia, such as the simple columnar epithelium lining most of the lower gastrointestinal tract (B), have only a single layer of cells. The majority of the cells are enterocytes, but mucus-producing goblet cells are interspersed with them. Specialized areas of the GALT called Peyer's patches are also found along the length of the intestine (see Fig. 3.3). The apical surface of each cell borders the lumen, and the basal layer rests on a basal lamina. The mucus blanket covering most of the intestinal surfaces is very thick, ranging up to 500 μm, several times thicker than the height of the cells. As in panel A, to gain attachment, bacteria must have adhesins for the mucus components. To enter the mucus blanket, bacteria frequently must be able to cycle between adhesive and nonadhesive phases. Once within this blanket, the bacteria can withstand being swept away by the much more rapid flows of fluid within the intestinal lumen (indicated by different-sized arrows).

types (e.g., lymphocytes, macrophages, mast cells, and fibroblasts), and neural and vascular elements (Fig. 3.2). Adhesion of bacteria to these subepithelial components becomes important if the epithelial barrier is breached by wounding or the area is accessed by invasion of the bacteria through the epithelial sheet. The intestinal epithelium, a "simple" epithelium, is used as an example in further discussions of some of the cell and matrix

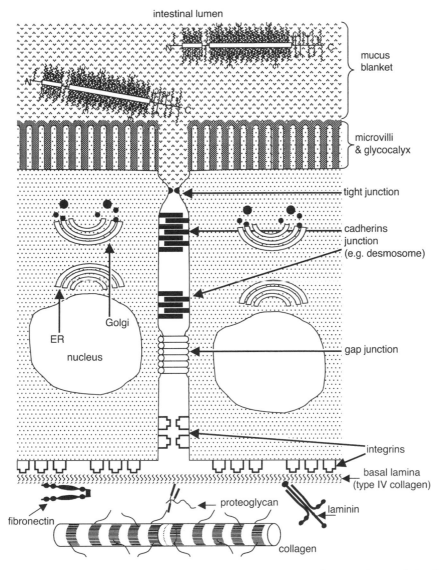

FIGURE 3.2 Schematic illustration of the basic components of intestinal epithelial cells (enterocytes). The apical surface of enterocytes have microvilli (i.e., the brush border). The microvilli are filled with contractile elements of the cytoskeleton and coated with a glycocalyx (see the text). Covering the microvilli and glycocalyx is a mucus blanket that can be up to 500 μm thick (see the text for a description of its composition). The cells contain typical organelles, such as the nucleus, endoplasmic reticulum, and Golgi apparatus. They are joined by junctional complexes such as tight junctions, desmosomes, and gap junctions. The basal surface of the cells rests on a basal lamina composed of type IV collagen, laminin, heparan sulfate proteoglycans, fibronectin, and other extracellular matrix components. Fibrillar collagen, as well as other elements not illustrated (blood vessels and a variety of cell types, such as macrophages and mast cells), are found beneath the basal lamina. The cells are joined to the basal lamina by integrins. Cadherins, integrins, and connexins are localized preferentially but not exclusively to the basolateral membranes.

components encountered by bacteria when colonizing and infecting a host.

SOME BASIC CHARACTERISTICS OF AN EPITHELIAL SURFACE: THE INTESTINAL MUCOSA

The total mucosal surface in the adult human gastrointestinal tract measures approximately 200 to 300 m², constituting the largest surface area of the body that is in contact with the external world. The intestinal epithelium is a single cell layer, consisting primarily of absorptive enterocytes. However, there are a wide variety of additional cell types that differ between different sections of the gastrointestinal tract. These additional cell types include secretory goblet cells, undifferentiated crypt cells, Paneth cells, M cells, enteroendocrine cells, and intraepithelial lymphocytes. Beneath this cell layer and its basement membrane are found the basic elements of the lamina propria mentioned above. Enterocytes are joined into a sheet of contiguous cells by junctional complexes, including tight junctions, adherens junctions, desmosomes, and gap junctions (Fig. 3.2). It is the components of the multi-stranded tight junctions (e.g., occludin and claudin) that form a generally leakproof seal between the cells. The basal plasma membrane is attached to the basal lamina, which is composed primarily of type IV collagen, laminin, fibronectin, and proteoglycans. The basal lamina provides a point of attachment, enabling polarization of the cells. The apical surface exhibits microvilli that are covered by a glycocalyx and mucous blanket (Fig. 3.2). The glycocalyx is a layer rich in carbohydrates that exhibits a dense staining in electron micrographs when cells are treated with saccharide-reactive reagents such as ruthenium red or lectins. Within the glycocalyx are found disaccharidases, peptidases, and other, nonenzymatic molecules. Covering the glycocalyx is a thick layer of mucus produced by goblet cells, which increase in number from the proximal duodenum to the distal colon. Intraepithelial lymphocytes can be found on the luminal side of the basement membrane,

between the enterocytes. In general, the mucus secreted by goblet cells (described in more detail below) forms a thick covering over the apical surfaces of enterocytes. Thus, any bacterial adhesion that occurs in these regions is normally associated, at least initially, with mucus components. In fact, studies of *Helicobacter pylori* have revealed that most of the bacteria colonizing the gastric mucosa reside within the mucus layer over the epithelium (7, 22). Rates of multiplication within the mucus layer are probably higher because of increased nutrients, and mucus flow is significantly slower than that of the fluid phase in the lumen of the stomach as well as that in the lower parts of the gastrointestinal tract.

Areas where the mucus layer of the gastrointestinal tract is essentially absent are the areas of Peyer's patches, prominent components of the gut-associated lymphoid tissue (GALT). GALT is, in aggregate, one of the largest lymphoid organs in the body. Peyer's patches are groups of lymphoid follicles in the small intestinal mucosa. There are three major domains within a Peyer's patch: the follicular area, the parafollicular area, and the follicle-associated epithelium. This epithelium of the Peyer's patch is composed primarily of M cells (Fig. 3.3), but there are a few columnar epithelial cells, intraepithelial lymphocytes, a few goblet cells, and some tuft cells. M cells are scattered over the domes that cover lymphoid follicles. They possess short, irregular "microfolds" (hence the term "M cell") and are surrounded by enterocytes with numerous tall microvilli. Of all the intestinal epithelial surfaces, the apical surfaces of M cells are the least heavily covered by mucus. The glycocalyx of M cells is also much less dense than on adjacent enterocytes. M cells are therefore in a prime position for antigen sampling and for interaction with bacteria (27, 50, 63) (Table 3.1). The basolateral membrane of M cells envelops lymphoid cells, but macrophages, plasma cells, and neutrophils are also present. In fact, exposure to the intestinal microbiota is essential for development of GALT; GALT is undeveloped in germfree animals.

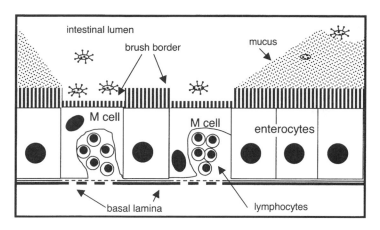

FIGURE 3.3 Schematic illustration of a segment of the follicle-associated epithelium of a Peyer's patch. M cells have much shorter surface projections than do the surrounding enterocytes, and the mucus blanket is essentially absent. Absence of the mucus blanket allows microbes relatively free access to the apical membrane of M cells. Leukocytes are present within a compartment between the M-cell basolateral membrane and the basal lamina. The basal lamina underlying M cells is incomplete.

TABLE 3.1 Examples of bacteria taken up by M cells[a]

Bacillus Calmette-Guérin
Brucella abortus
Campylobacter jejuni
Escherichia coli RDEC-1
Escherichia coli O124:K72
Mycobacterium paratuberculosis
Streptococcus pyogenes
Streptococcus pneumoniae
Shigella flexneri
Salmonella enterica serovar Enteritidis
Salmonella enterica serovar Typhi
Salmonella enterica serovar Typhimurium
Vibrio cholerae
Yersinia enterocolitica
Yersinia pseudotuberculosis

[a]Adapted from reference 50.

ORGANIZATIONAL FEATURES OF CELLULAR MEMBRANE CONSTITUENTS

All animal cell membranes have common compositional and organizational features. (i) The major membrane lipids are arranged in a planar bilayer configuration that is predominantly in a "fluid" state under physiological conditions (11, 18). The membrane lipids are commonly composed of glycerolphospholipids, sphingolipids, and sterols. Although there is fluidity, there is also a polarity that is maintained under most normal conditions. For example, phosphatidylserine is found in the inner but not the outer membrane leaflet, except in a cell undergoing apoptosis. (ii) The membrane lipid bilayer contains integral (or intrinsic) membrane constituents composed of glycolipids, glycoproteins, proteoglycans, lipid-linked proteins, and mucin-like glycoproteins (Fig. 3.4). (iii) Other proteins and glycoproteins are bound to the surface of the plasma membrane (external and cytoplasmic faces) by weak ionic interactions, hydrogen bonding, or the hydrophobic effect. These surface-associated molecules are referred to as peripheral or extrinsic components. (iv) In many animal cells, especially superficial layers of epithelial cells, there is a substantial layer of carbohydrate-containing materials of variable thicknesses on the external surface of the plasma membrane. One layer of this kind, the mucous blanket, arguably reaches its greatest extent when overlying the intestinal epithelium. Another "layer" is composed of extracellular matrix macromolecules (e.g., collagen, fibronectin, and laminin) that underlie epithelia and surround many other types of cells. Although some of them lie at a significant distance from the cell surface in molecular terms, they do come into intimate contact with membrane receptors in many instances and can have dramatic effects on cellular physiology.

The distinction between membrane constituents as integral, peripheral, or belonging to the cell coat or extracellular matrix is based to some extent on the method required to dissociate the constituent in question from the

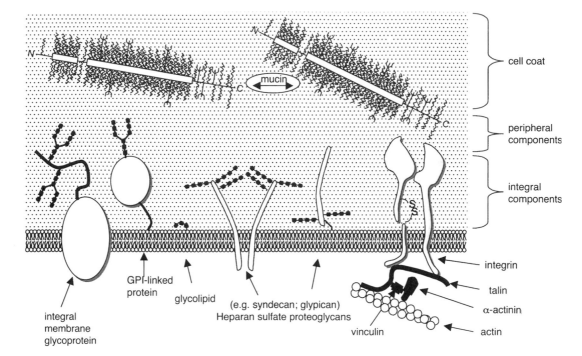

cell coat

peripheral
components

integral
components

integrin

talin

α-actinin

actin

integral
membrane
glycoprotein

GPI-linked
protein

glycolipid

(e.g. syndecan; glypican)
Heparan sulfate proteoglycans

vinculin

mucin

FIGURE 3.4 Schematic composite illustration of the various components of animal cell membranes. The components illustrated vary depending on the cell type and on whether the membrane under discussion is the apical or basolateral membrane. Other than the lipid bilayer, the integral membrane components consist of integral membrane glycoproteins (e.g., integrins) and proteoglycans (e.g., syndecan), GPI-linked glycoproteins (e.g., CD14) and proteoglycans (e.g., glypican), and glycolipids (e.g., globoseries glycolipids). Peripheral components consist of glycoproteins such as fibronectin and laminin. The cell coat consists of the much more heavily glycosylated mucins and other components of the mucus blanket (see the text). Fibronectins and laminins would be present in greater amounts on basolateral surfaces, while mucins would be present in greater amounts on apical surfaces. The integral membrane proteins are linked to the actin cytoskeleton via linking elements such as talin, α-actinin, and vinculin.

cell membrane. The integral constituents may be released only after disruption or perturbation of the phospholipid bilayer, usually by detergents (53). This is because one or more segments of the polypeptide chain contain stretches of hydrophobic amino acids and they are physically embedded in the lipid bilayer. Many of these integral membrane proteins have multiple transmembrane segments that span the bilayer. Other integral membrane proteins do not have a segment of its polypeptide embedded in the bilayer but are covalently attached to a glycosylphosphatidylinositol (GPI) lipid anchor which is embedded in the bilayer. The GPI anchor contains two

fatty acids, inositol, N-acetylglucosamine, and mannose. This type of integral membrane protein can be found on either cytoplasmic or extracellular membrane faces, with the cytoplasmic forms often having different types of lipid groups, such as farnesyl or prenyl, attached to their cytoplasmic domains. Both types are released only by detergents or by other lipid-solubilizing agents, such as release of GPI-anchored proteins by cleaving the phosphate-glycerol bond with phospholipase C.

Peripheral membrane proteins (Fig. 3.4) can usually be dissociated from membranes by less drastic means, such as changing the pH, increasing salt concentrations, or using chelating agents (e.g., EDTA) or chaotropes (e.g., guanidinium chloride). There is no general method, however, to selectively release any of these components. The conditions required to release specific integral and peripheral proteins require optimization.

As mentioned above with regard to lipids, one key feature of the membrane is its asymmetry. For nonglycosylated lipids, the asymmetry is only partial, in that every phospholipid is present on both sides of the bilayer but in different amounts. In human erythrocytes, for example, lipids with positively charged head groups (e.g., phosphatidylethanolamine and phosphatidylserine) are predominant in the internal leaflet facing the cytoplasm (57). The asymmetry with respect to proteins, glycoproteins, and glycolipids appears to be absolute: every molecule of a given membrane constituent has the same orientation across the bilayer, with the carbohydrate moieties of the glycosylated compounds always exposed on the outer surface or within cytoplasmic vesicles (i.e., away from the cytoplasm). For further information about the general organizational characteristics of the animal cell membrane, the reader is referred to textbooks and reviews (see, e.g., references 1 and 55).

Although the above characteristics are shared by many animal cell types, there are variations among the different cell and tissue types. Because the epithelial cells of the mucosal surfaces are the first cell type to come into contact with foreign elements, including bacteria, these are the cells initially colonized during the natural course of most infections. Thus, in the following discussion we focus on the molecular organization of epithelial cells (Fig. 3.2 to 3.4). A major characteristic of this type of cell is the polarization of apical versus basolateral surfaces, such that the apical surface is mainly covered by mucus as a cellular coat and the basal surface is bound to a basal lamina. The integral membrane components are also distinctly polarized. For example, integrins are preferentially expressed on basolateral surfaces. This distinction is important because polarization of the cells means that receptors for bacterial adhesins are polarized as well. For example, the enteropathogenic bacteria *Shigella* and *Listeria* are capable of invading intact enterocytes only from the basal side. To gain access to the basal side, they must first adhere to and invade neighboring M cells, whose distribution of receptors is different from that of enterocytes and which are not covered by a thick layer of mucus (102).

INTEGRAL MEMBRANE CONSTITUENTS

General Features

As mentioned above, glycolipids, glycoproteins, GPI-linked glycoproteins, proteoglycans, and mucins belong to the integral membrane constituents. With respect to their organization in the cell membrane, these components share several properties (Fig. 3.4). Of especial interest is their mobility within the plane of the membrane, made possible by the fluidity of the lipids of the membrane bilayer. Their movement may be restricted by interactions with extrinsic proteins, other intrinsic proteins, and cell-cell junctions (34, 64). As a result, the topography of integral glycoproteins is nonrandom and some of these constituents are confined to specific regions of the membrane.

Much interest has recently been devoted to the fact that many integral membrane constituents are associated with specific membrane lipids organized into domains whose composition is different from other parts of the membrane. These domains have been called lipid "rafts" (8). Lipid rafts are cholesterol- and glycosphingolipid-rich domains characterized by a high degree of acyl chain order and melting temperatures that are higher than those of other membrane regions. GPI-anchored and palmitoylated or myristoylated proteins apparently prefer the environment of lipid rafts, possibly moving into these areas only after aggregation by antibodies or other ligands. Glycosphingolipids are confined primarily to the extracellular leaflet of the bilayer, but the arrangement of molecules in rafts and whether they are confined to the extracellular leaflet are not yet fully understood. Rafts are also associated in the literature with caveolae. The indentations that mark these "tiny caves" form when a protein, caveolin, binds to the cytoplasmic side of rafts. The "caves" can often be identified in electron micrographs, but because of the dynamic nature of caveola formation, there may be functional caveolae that are not indented. Caveolin is usually acylated, but this is apparently not required for caveolin to associate with rafts. Interestingly, CD48, a mannose-containing GPI-anchored protein and a receptor for type 1-fimbriated *Escherichia coli* has been localized within caveolae of bone marrow-derived mast cells (85) (Table 3.2). Caveolae appear to be involved in bacterial entry into these cells, because agents that disrupt caveolae block entry and markers of caveolae were actively recruited to sites of bacterial entry (86). The way in which the GPI-anchored CD48 fimbrial receptor activates entry is not known, since it does not have a cytoplasmic, cytoskeleton binding domain. Because caveolae do not typically fuse with lysosomes, the bacteria contained within caveolar vesicles may be protected and thereby may escape or delay being killed.

Even though the lipids and proteins within rafts are thought to be less fluid than other regions, most membrane constituents are relatively free to move (87, 88). The concept of membrane fluidity has been very important in describing many phenomena associated with the cell surface. An early example proving this mobility is the phenomenon of "patching" and "capping" of integral membrane macromolecules following their interaction with multivalent extracellular agents (49, 68). This has also been shown for the receptors of adhesins. For example, lipoteichoic acid, an adhesin of gram-positive bacteria, binds to receptors on polymorphonuclear leukocytes, which, in the presence of anti-lipoteichoic acid, can be seen to patch and cap (Fig. 3.5).

All glycolipids of the cell membrane are constituents of the outer half of the bilayer, where they comprise 30 to 60% of the total lipid content (Fig. 3.4). The ceramide group of the glycolipids is responsible for anchoring the molecules to the membrane by intercalating within the lipid bilayer, whereas the oligosaccharide chains extend to the outer surface. Chemically, the glycolipids are either neutral or acidic. The acidic glycolipids contain sialic acids (i.e., gangliosides), which contribute significantly to the net negative charge of the animal cell surface. When isolated in pure form from the cell membrane, glycolipids tend to form micelles. When mixed with animal cells, the glycolipids can coat most cells and assume an orientation comparable to the original one, i.e., with the fatty acid chains intercalated into the bilayer and the oligosaccharide portion exposed to the outer

TABLE 3.2 Some important integral membrane components of epithelial cells that act as receptors for bacterial adhesins

Type of membrane component	Cell type	Adhesin	Bacterium
Glyolipids			
Globoseries glycosphingolipids	Uroepithelial cells	P fimbriae	*E. coli*
Glycoproteins			
Integrins	M cells, enterocytes	IpaB, IpaC	*Shigella*
		Invasin	*Yersinia*
Heparan sulfate proteoglycan	Epithelial cell	Opa proteins	*N. gonorrhoeae*
E-cadherin	M cells, enterocytes	Internalin A	*L. monocytogenes*
Uroplakin	Uroepithelial cells	Type 1 fimbriae	*E. coli*
Glycolipoprotein (GPI linked)			
CD48	Mast cells	Type 1 fimbriae	*E. coli*

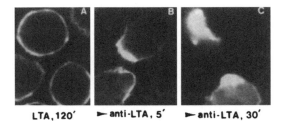

LTA,120′   ►anti-LTA, 5′   ►anti-LTA, 30′

FIGURE 3.5 (A) Uniform binding of lipotei-choic acid (LTA) to polymorphonuclear leukocytes in the absence of a cross-linking agent. (B and C) Capping that occurs 5 to 30 min after exposure to anti-lipoteichoic acid antibodies. (Reprinted from reference 13 with permission from the publisher.)

surface (9, 77). This behavior is similar to that of the bacterial glycolipids, such as lipopolysaccharides of gram-negative bacteria and lipoteichoic acids of gram-positive bacteria (107). Many investigators take advantage of this property for experiments in which the type of oligosaccharide and the number of molecules to be inserted in the animal cell can be controlled (108).

Virtually all proteins of animal membranes are glycosylated (79–81). In many respects, membrane glycoproteins are similar to soluble glycoproteins. For example, they do not differ significantly in their overall amino acid composition and carbohydrate content and they contain the same monosaccharide constituents: the hexoses D-galactose and D-mannose, the methylpentose L-fucose, N-acetyl-hexosamines, and the sialic acids. D-Glucose, found in the collagens, is not known to occur in membrane glycoproteins. The uronic acids, D-glucuronic acid and L-iduronic acid, are confined to proteoglycans such as heparan sulfate. Several types of carbohydrate-peptide linkages are common to both groups of glycoproteins: the N-glycosyl bond between N-acetyl-D-glucosamine and asparagine (N-acetyl-D-glucosaminyl-asparagine) and the O-glycosidic bond between N-acetyl-D-galactosamine and a serine or threonine of the protein polypeptide backbone. A distinctive feature of the primary amino acid sequence of membrane glycoproteins is the presence of a region(s) rich in hydrophobic amino acids. The hydrophobic region(s) is responsible for anchoring the molecules in the lipid bilayer. Most integral glycoproteins extend across the entire thickness of the bilayer at least once, so that they are exposed to both the internal and the external environments. Some of the glycoproteins, such as glycophorin in human erythrocytes (56, 97) and the E2 glycoprotein of Semliki Forest virus (32), span the membrane only once, whereas others traverse it more than once (90). The oligosaccharide moieties are always exposed on the cell surface (or on the inner surface of cytoplasmic vesicles). The internal segment of the membrane-spanning intrinsic glycoprotein may be in close contact with some of the proteins located on the cytoplasmic face of the membrane. The intrinsic glycoproteins may therefore play an essential role in the processes that require communication between the outside of the cell and its interior, such as transport phenomena and the responses of cells to external stimuli, such as hormones, toxins, bacterial adhesins, and other cells (24, 64, 65). Unlike glycolipids, the integral glycoproteins in isolated form do not reassociate with or coat cell membranes in a manner identical to that in their native state, but they can be reassembled into liposomes (21). In such artificial instances, however, the orientation of the reconstituted glycoproteins is symmetrical in that half of the molecules insert with the glycosylated part facing outward and half facing inward. There are a number of integral membrane components of epithelial cells that are of especial importance for bacterial adhesion (Table 3.2). The most salient examples of these components which have been studied with respect to their role in bacterial adhesion are further discussed below.

Integrins

One class of integral membrane proteins is referred to as the integrins (41, 42, 74). The integrins are glycosylated membrane-spanning heterodimers. The largest subunit of integrin heterodimers is an α subunit of roughly 100 kDa, and the smaller is a β subunit of roughly 80 kDa. There are at least 16 different α subunits and 8 β subunits, and heterodimeric integrins are classified as $\alpha_{IIb}\beta_3$, etc. More than 20 different integrins have been identified so far. A few α subunits may combine with any one of the β subunits. Each subunit has a transmembrane segment, and all have a short cytoplasmic tail, except for the β_4 integrin subunit, in which the cytoplasmic C-terminal domain is much larger, equal in size to the extracellular domain. The extracellular domains of integrins are located preferentially, but not exclusively, on basolateral surfaces of epithelial cells, where they bind to basal lamina components and other extracellular matrix macromolecules (e.g., fibronectin, laminin, and collagens [see below]), while the cytoplasmic domains interact with cytoskeletal components and other cytoplasmic signaling partners. Integrins act as receptors for a number of pathogens which bind either to the oligosaccharide side chains or to specific amino acid domains of integrins (Table 3.2). Once enterocytes have become damaged or destabilized, whether by the action of pathogenic bacteria or via other insults, these basolateral integrins may redistribute to apical surfaces (102).

Proteoglycans

Another important group of integral membrane glycoproteins that are prominently involved in binding microbes are some of the members of the heparan sulfate proteoglycan family (5, 43). There are at least five distinct classes of heparan sulfate proteoglycans, with syndecan and glypican being the main integral membrane components in the group. Syndecans are transmembrane molecules, while glypicans are GPI anchored. Other members (perlecan, agrin, and type 18 collagen) are extracellular matrix components. Because the extracellular forms are commonly associated tightly with integrins or other integral membrane adhesion molecules, Iozzo (43) has called them pericellular, rather than extracellular, proteoglycans. Each member of the

family has a core protein to which is covalently bound a variable number of heparan sulfate side chains. The core proteins differ quite significantly in sequence and size (from ~30 to ~500 kDa), and the heparan sulfate side chains can also vary widely (from ~5 to ~70 kDa). There are four different syndecans that have three to five heparan sulfate chains attached to the protein (6, 20). Chondroitin sulfate chains can also be present. A protease cleavage site between the transmembrane domain and the extracellular domain of the core protein allows for shedding of the heparan sulfate-containing proteoglycan from the cell surface. The cytoplasmic domains are capable of interacting with cytoskeletal elements and are involved in signaling events (69). There are six mammalian glypicans, distinguished by a cysteine-rich extracellular domain and the presence of two or three heparan sulfate chains attached to the core protein near its attachment to the GPI anchor (23). Although amino acid sequences of the core proteins vary, the position of the 14 cysteine residues is conserved, suggesting a conserved tertiary structure.

Many bacteria, parasites, and viruses bind to cells via heparan sulfate proteoglycans (71) (Table 3.3). In some cases, this binding appears to trigger a signaling cascade mediating invasion. Because heparan sulfate binds so many different ligands (e.g., growth factors, cytokines, and cytokine receptors), one can imagine that bacterial binding to this cell surface molecule could have myriad effects.

TABLE 3.3 Examples of bacteria that bind to heparan sulfate proteoglycan receptors

Species	Reference(s)
Bordetella pertussis	36
Borrelia burgdorferi	44
Chlamydia trachomatis	10, 91
Escherichia coli	25
Haemophilus influenzae	66
Helicobacter pylori	3
Listeria monocytogenes	2
Mycobacterium spp.	59
Neisseria gonorrhoeae	104
Staphylococcus aureus	51
Streptococcus spp.	109

Cadherins

More than 80 human cadherin genes have been sequenced (98). They play important roles in tissue differentiation. E-cadherin, P-cadherin, and N-cadherin are the most widely expressed. They each are composed of approximately 700 amino acids, with a large extracellular domain, a single transmembrane domain, and a short cytoplasmic domain that interacts with cytoskeletal elements. The extracellular portion of the molecules contains domains (called cadherin domains) which mediate calcium-dependent homophilic interactions between cadherin molecules on separate cells. Tepass et al. (98) have defined seven different cadherin subfamilies on the basis of the numbers and sequence of cadherin domains. Virtually all cells express one or another of the cadherins; E-cadherins (also called uvomorulins) are present on many different types of epithelial cells, N-cadherins are expressed primarily by nerve and muscle cells, and P-cadherins are expressed by placenta and epidermal cells. Intestinal cells are formed into a contiguous epithelium by several types of junctions, including laterally localized E-cadherins holding epithelial sheets together. E-cadherin dimers bind head-to-head to E-cadherin dimers on neighboring cells. They do not bind to cells that express only P-cadherin. E-cadherin is a receptor for internalin of *Listeria monocytogenes*.

Glycocalyx

The cellular glycocalyx was described first by Bennett (4) and later by Ito (45, 46) as an electron-dense mat of filamentous material attached to the surfaces of cat intestinal epithelial cells. The glycocalyx is a general term for a complex, highly glycosylated material composed in part of the integral glycoproteins, proteoglycans, and glycolipids mentioned above. It is also contributed to by a dense coat of short (100 to 500 nm in length) nonsecreted mucin molecules that are anchored in the cell membrane by a transmembrane segment of the core protein. Hundreds of oligosaccharide chains bearing neutral, sialylated, or sulfated sugars can be attached to a single core protein (35). The

oligosaccharide chains are extremely variable but are undoubtedly important for the biological function of these cell surface molecules. The high density and diversity of saccharides in this cellular compartment make it an ideal receptive surface for lectin-bearing bacteria (see chapter 1) (61).

PERIPHERAL MEMBRANE CONSTITUENTS

The peripheral membrane proteins and glycoproteins are anchored to the surface of the membrane by weak ionic interactions or by hydrogen bonding with integral constituents of the cell membranes, e.g., glycoproteins, glycolipids, or polar head groups of phospholipids. Some proteins are associated mainly with the cytoplasmic face of the membrane. Others interact with elements of the cytoskeleton of the cell, thereby providing a link between integral membrane constituents and the interior of the cells. Because of the difficulties mentioned earlier in distinguishing between peripheral glycoproteins of the cell surface and glycoproteins of the extracellular matrix, very few compounds have been unequivocally identified as peripheral glycoproteins. Some key examples of peripheral and extracellular components important for bacterial adhesion are discussed below. In fact, matrix components have become so important as receptors for bacteria that an acronym for microbial surface components recognizing adhesive matrix molecules, MSCRAMM, has come to be commonly used.

Extracellular Matrix Components

COLLAGENS

There are currently 20 or more different molecules that are called collagen, and, in fact, collagen is the most abundant protein in animals (40, 60, 62, 67, 106). It may not be surprising that the vast majority of the collagenous protein in the body is composed of the first three collagens described, types I, II, and III. These collagens are made up of a helical association of α chains that polymerize in a staggered fashion to yield the distinct, 67-nm repeat, collagen banding pattern seen in electron micrographs. These and, indeed, the other collagens contain a repeating amino acid pattern of Gly-Pro-X. Type I and III collagen are present in such tissues as the dermis, cornea, tendons, ligaments, and bone. Type II collagen is present primarily in cartilage. Several other collagen types, such as type VI or IX collagen, do not form fibrils but are associated with the fibrillar collagens, linking them and other macromolecules in the extracellular matrix. Type IV collagen forms a two-dimensional sheet that is a primary structural framework of basement membranes that lie, in addition to a few other locations, beneath all epithelia (19). Other components of basement membranes include laminin, proteoglycans, and several other molecules (e.g., entactin and fibronectin) that differ somewhat from tissue to tissue.

FIBRONECTINS

Fibronectins are virtually ubiquitous adhesive glycoproteins that are found in tissue matrices and in blood and that were first known by such other names as the LETS protein (for "large external transformation sensitive") of normal but not transformed cancer cells and CIG (for "cold-insoluble globulin") of plasma. Fibronectins are disulfide-linked dimeric glycoproteins of approximately 450 kDa. They are com-

posed of three different types of structural units, termed type I, type II, and type III repeats (Fig. 3.6). An important characteristic is the multiadhesive capacity of fibronectin, due in part to the fact that the type I, II, and III polypeptide units form several different domains that bind to various receptors, such as collagen, proteoglycans, cells (integrins), and bacteria, and to the fact that it is a dimeric molecule. Indeed, one excellent method for fibronectin purification from plasma is to take advantage of the collagen and gelatin binding domain by using gelatin-Sepharose affinity chromatography. The fibronectin molecules can be conveniently cleaved into individual functional domains (e.g., cell binding, collagen binding, and bacterium binding) because they are separated in the intact molecule by protease-sensitive sites. The RGD seqeuence contained within one of the type III repeat modules is the binding site for integrins. Alternative splicing of RNA transcripts leads to 20 or more different fibronectin monomers. Plasma fibronectin differs from tissue fibronectin in that it lacks one or two of the type III repeats.

LAMININS

Laminins are another type of multiadhesive glycoprotein found in the extracellular matrix, primarily in basement membranes. They are trimeric molecules of more than 800 kDa (A

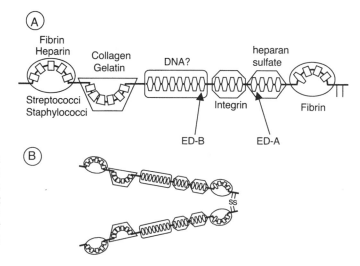

FIGURE 3.6 Schematic illustration of the fibronectin monomer, its type I (□), II (▱) and III (○) domains, and an indication of some of their binding activities. ED-A and ED-B indicate the "extra" domains that are spliced into or out of some fibronectin variants. (B) Disulfide-linked dimer of fibronectin.

chain, ~400 kDa; B1 and B2 chains, ~200 kDa each) and have binding sites for type IV collagen, heparan sulfate, entactin, collagens, and integrins. Purified laminin is cross-shaped in the electron microscope. Laminin interacts with several different pathogenic species, including *E. coli* (89), viridans streptococci (92), and group A streptococci (93).

VITRONECTIN

Vitronectin is one of a variety of adhesive macromolecules that interact with integrins (100). It is a blood glycoprotein that is identical to S protein, which is a component of the complement system. It also binds to collagen (33), thrombin-antithrombin III complexes (99), type 1 plasminogen activator inhibitor (78), and heparin. Vitronectin is thought to be a receptor for group A streptococci (52, 103).

THE CELL COAT

Most cells are coated with a layer of glycoproteins that are bound to the cell surface by noncovalent bonds with an integral constituent. The coat can be of variable thickness and chemical complexity. It may take the form of a matrix, as in mature cartilage, where it is composed of a collagen-proteoglycan-hyaluronic acid complex (Fig. 3.7) associated with the cell membranes of chondrocytes. Other cells, such as mucosal cells of the respiratory and gastrointestinal tracts, are coated with a layer rich in highly sialylated, high-molecular-weight glycoproteins known as mucins (Fig. 3.8). Mucins are undoubtedly important in promoting and/or inhibiting bacterial adhesion to mucous membranes of the host, but the contribution of these complex mixtures of macromolecules to the adhesion of bacteria has often been overlooked. For this reason, mucins are discussed in some detail below. The different types of coats produced by specialized cells at specific sites in the host are well adapted to their specific physiological functions. Thus, the matrix of cartilage is able to withstand stress and to bear considerable loads whereas the mucins form viscous solutions that perform protective, lubricant, and transport roles in the appropriate tissues. Regardless of the thickness or composition of the cell coat, for a bacterium to colonize tissues, it must either bind to or penetrate through the cell coat.

THE MUCUS BLANKET

As already mentioned, many studies of bacterial adhesion essentially ignore the potential contributions of the mucus blanket. This is

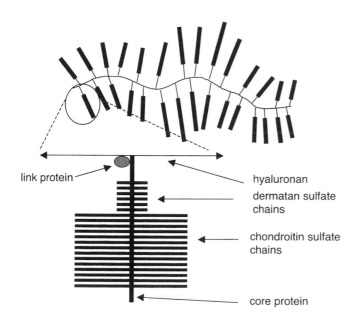

link protein

hyaluronan

dermatan sulfate chains

chondroitin sulfate chains

core protein

FIGURE 3.7 Schematic illustration of cartilage proteoglycans linked to hyaluronan. Chondroitin sulfate and dermatan sulfate glycosaminoglycan chains are covalently linked to serine residues of core proteins. These are, in turn, linked to hyaluronan (hyaluronic acid) by link proteins to form the cartilage proteoglycan, also called aggrecan.

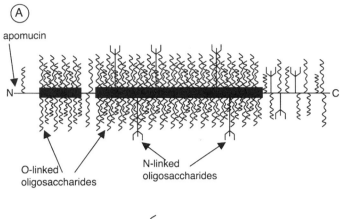

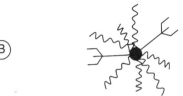

FIGURE 3.8 Schematic illustration of a mucin monomer. (A) Large numbers of O-linked and a few N-linked oligosaccharide chains are bound to the apomucin protein. (B) Cross section showing that the oligosaccharide chains take on a radial arrangement.

undoubtedly due in large part to its extremely complex nature (73). In this section, we focus primarily on intestinal mucus as but one example of a layer of secreted and bound or trapped materials that exists on virtually every mucosal surface. For this reason alone, the mucus blanket should be taken into greater consideration in analyses of bacterial adhesion mechanisms. One normal physiological function of this mucus blanket is to form a protective barrier between environmental insults taken in through the mouth or nose, which clearly includes bacterial invaders, and the more delicate cells of the mucous membranes. In most instances, bacteria that colonize any mucosal surface must first come into contact with and negotiate with the mucus blanket.

Mucus is a viscous material that coats virtually the entire respiratory and gastrointestinal tracts. It is constituted mainly by molecules of mucin glycoprotein, but it also contains fairly high concentrations of lipid vesicles, antibodies, ions, dietary products, entrapped microbes, and water. It can be produced by cells of distinct, named mucus glands (e.g., the sublingual gland) or by the goblet cells that are interspersed among the enterocytes in the gastrointestinal epithelium. The proportion of goblet cells increases along the course of the small and large intestines. Even though it is made up primarily of soluble, secreted products, the mucus coat has viscoelastic properties that normally ensure that, as it slowly moves along the epithelial surface, it still remains strongly associated with the epithelium despite the variety of shearing forces that can come into play.

The basic building block of a mucin is a polypeptide chain encoded by one of several different genes (for instance, humans have at least eight of these genes [35]). All the proteins have large numbers of threonine, serine, and proline residues. Up to 80% of the weight of mucin monomers is made up of oligosaccharides in which proximal GalNAc residues are O linked to threonine and serine residues of the core protein, called the apomucin (Fig. 3.8). There are also N-linked oligosaccharides, but these make up a much smaller fraction of the weight (<5%). In between the bottlebrush O-linked saccharide regions that have little secondary structure, there are "beads" composed of hydrophobic amino acids. There are also cysteine-rich regions near the C terminus. Both types of regions contribute to intermolecular cross-linking.

While a mucin monomer may be 1,000,000 to 2,000,000 Da, cross-links are formed between monomers to increase the effective size to 10,000,000 Da (37, 83, 101). The involvement of disulfide bonds is strongly suggested by the fact that dithiothreitol breaks down most mucus blankets. The pattern of glycosylation varies with the epithelial site of synthesis (48).

There are so-called unstirred layers at the surface of essentially all cells, and the presence of mucus simply increases the thickness of this layer. The mucus blanket can measure up to 200 to 500 μm in width. The width varies between different anatomical regions and is a function of the different characteristics of the mucins secreted, the rates of secretion and shedding, the fraction of cell population that are goblet cells, the pH, and the concentrations of cations. An outer lipid layer forms on the luminal surface of many mucus blankets. Often it is cationic surfactants, such as dipalmitoyl phosphatidylcholine, whose positively charged head groups bind to the highly negatively charged mucins while aliphatic chains form a continuous hydrophobic monolayer facing the outer surface of the gel (54).

The negative charge of the mucin glycoproteins contributes to the creation of a localized acidic microenvironment, presumably by impeding the diffusion of hydrogen ions (16, 75). Because of the density of negative charges, the mucin fibers absorb large amounts of water and swell very rapidly as they are being secreted from the goblet cells, and they can increase their volume several hundred-fold in a few milliseconds (96). Also because of the negative charges, there is a tremendous impact of pH and cation concentration on the degree of either swelling or condensation of the mucus blanket and, consequently, on viscosity. The nature of the mucus blanket also varies dramatically between different anatomical regions, and there are pathologic conditions in which it is compromised.

The high concentrations of antibodies in the mucus blanket can block adhesins or create aggregates that are too large to penetrate the gel and approach cell surfaces. The viscosity of intestinal mucus inhibits the motility of some bacteria. Most bacteria inhabit the outermost layer of mucus because of its viscosity and binding activities (17, 58, 84). Some bacteria are able to degrade mucins by using specific glycosidases and utilize them as nutrients (39). This aids the bacteria, but it apparently also aids the host, since the ceca of germ-free animals contain excessive amounts of mucus, which are dramatically reduced by adding commensal organisms (94). The viscosity of saliva is lower, and the saliva is readily penetrable by motile bacteria (12); however, certain components of salivary mucus can also agglutinate streptococci (14, 95).

EXPRESSION OF RECEPTORS ON CELL SURFACES

On living tissues, there is a dynamic change in the production and expression of any particular cell membrane constituent as a function of age and the physiological state of the cell. Changes are also bound to occur in cells exposed to the actions of drugs, some of which may affect the biosynthesis and expression of cell membrane constituents. Such changes have been best documented in carbohydrate residues of glycoproteins and glycolipids (15, 28–31, 38, 105, 110), largely due to the availability of specific lectin and gycosidase probes (26, 72, 80, 81). Moreover, the proximity of other cell surface constituents may interfere with the ability of receptors to bind the ligand. When comparing the binding of a given ligand to cells under different physiological conditions, it is therefore not always possible to determine whether any difference observed is due to a change in the number of receptor molecules, alterations in the recognition site, or interference by other surface constituents. Receptors for bacterial adhesins are probably no exception. Unfortunately, factors that specifically govern the presentation and distribution of receptors that bind bacteria have not been studied in depth.

An excellent example illustrating the

importance of changes in adhesin receptor specificity in the outcome of infections comes from the binding of *E. coli* bearing K99 fimbriae to glycolipids. These bacteria bind to glycolipids containing *N*-glycolylneuraminic acid, but they do not bind to those that contain *N*-acetylneuraminic acid (47). These two glycolipids differ by only a single hydroxyl group present in the acyl substituent on the 4-NH group of the former compound but absent from the latter. *N*-Glycolylneuraminic acid is found on intestinal cells of newborn piglets but disappears as the animals grow. This glycolipid is not normally found in humans. There is strong evidence, therefore, that this single hydroxyl group explains why *E. coli* K99 can cause frequently lethal diarrhea in piglets but not in adult pigs or in humans. This also illustrates that developmental changes in receptor structures can affect adhesion and susceptibility to infection.

REFERENCES

1. **Alberts, B., A. Johnson, J. Lewis, M. Raff, K. Roberts, and P. Walter.** 2002. *Molecular Biology of the Cell.* Garland Publishing Co., New York, N.Y.
2. **Alvarez-Dominguez, C., J. A. Vazquez-Boland, E. Carrasco-Marin, P. Lopez-Mato, and F. Leyva-Cobian.** 1997. Host cell heparan sulfate proteoglycans mediate attachment and entry of *Listeria monocytogenes,* and the listerial surface protein ActA is involved in heparan sulfate receptor recognition. *Infect. Immun.* **65:**78–88
3. **Ascencio, F., L. Fransson, and T. Wadström.** 1993. Affinity of the gastric pathogen *Helicobacter pylori* for the N-sulphated glycosaminoglycan heparan sulphate. *J. Med. Microbiol.* **38:**240–244.
4. **Bennett, H. S.** 1963. Morphological aspects of extracellular polysaccharides. *J. Histochem. Cytochem.* **11:**14–23.
5. **Bernfield, M., M. Gotte, P. W. Park, O. Feizes, M. L. Fitzgerald, J. Lincecum, and M. Zako.** 1999. Functions of cell surface heparan sulfate proteoglycans. *Annu. Rev. Biochem.* **68:**729–777.
6. **Bernfield, M., R. Kokenyesi, M. Kato, M. T. Hinkes, J. Spring, R. L. Gallo, and E. J. Lose.** 1992. Biology of the syndecans: a family of transmembrane heparan sulfate proteoglycans. *Annu. Rev. Cell Biol.* **8:**365–393.
7. **Blaser, M. J.** 1997. *Helicobacter pylori* eradication and its implications for the future. *Aliment. Pharmacol. Ther.* **11:**103–107.
8. **Brown, D. A., and E. London.** 1998. Functions of lipid rafts in biological membranes. *Annu. Rev. Cell Dev. Biol.* **14:**111–136.
9. **Callies, R., G. Schwarzmann, K. Radsak, R. Siegert, and H. Weigandt.** 1977. Characterization of the cellular binding of exogenous gangliosides. *Eur. J. Biochem.* **80:**425–423.
10. **Chen, J. C. R., J. P. Zhang, and R. S. Stephens.** 1996. Structural requirements of heparin binding to *Chlamydia trachomatis.* *J. Biol. Chem.* **271:**11134–11140.
11. **Cherry, R. J.** 1979. Rotational and lateral diffusion of membrane proteins. *Biochim. Biophys. Acta* **559:**289–327.
12. **Cone, R. A.** 1999. Mucus, p. 43–64. *In* P. L. Ogran, J. Mestecky, M. E. Lamm, W. Strober, J. Bienenstock, and J. R. McGhee (ed.), *Mucosal Immunology.* Academic Press, Inc., New York, N.Y.
13. **Courtney, H., I. Ofek, W. A. Simpson, and E. H. Beachey.** 1981. Characterization of lipoteichoic acid binding to polymorphonuclear leukocytes of human blood. *Infect. Immun.* **32:**625–631.
14. **Courtney, H. S., and D. L. Hasty.** 1991. Aggregation of group A streptococci by human saliva and its effect on streptococcal adherence to host cells. *Infect. Immun.* **59:**1661–1666.
15. **Critchely, D. R.** 1979. Glycolipids as membrane receptors important in growth regulation, p. 63–101. *In* R. O. Hynes (ed.), *Surfaces of Normal and Malignant Cells.* John Wiley & Sons, Inc., New York, N.Y.
16. **Daniel, H., B. Neugebauer, A. Kratz, and G. Rehner.** 1985. Localization of acid microclimate along intestinal villi of rat jejunum. *Am. J. Physiol.* **248:**G293–G298.
17. **Denari, G., T. L. Hale, and O. Washington.** 1986. Effect of guinea pig or monkey colonic mucus on *Shigella* aggregation and invasion of HeLa cells by *Shigella flexneri* 1b and 2a. *Infect. Immun.* **51:**975–978.
18. **Edidin, M.** 1974. Rotational and translational diffusion in membranes. *Annu. Rev. Biophys. Bioeng.* **8:**165–193.
19. **Erickson, A. C., and J. R. Couchman.** 2000. Still more complexity in mammalian basement membranes. *J. Histochem. Cytochem.* **48:**1291–1306.
20. **Esko, J. D., and U. Lindahl.** 2001. Molecular diversity of heparan sulfate. *J. Clin. Investig.* **108:**169–173.
21. **Eytan, G. D.** 1982. Use of liposomes for

reconstruction of biological function. *Biochim. Biophys. Acta* **694**:185–202.

22. **Falk, P. G., A. J. Snyder, J. L. Guruge, D. Kirschner, M. J. Blaser, and J. I. Gordon.** 2000. Theoretical and experimental approaches for studying factors defining the *Helicobacter pylori*-host relationship. *Trends Microbiol.* **8**:321–329.

23. **Filmus, J., and S. B. Selleck.** 2001. Glypicans: proteoglycans with a surprise. *J. Clin. Investig.* **108**:497–501.

24. **Finean, J. B., R. Coleman, and R. H. Mitchell.** 1984. *Membranes and Their Cellular Function,* 3rd ed. Blackwell Scientific Publications, Oxford, United Kingdom.

25. **Fleckenstein, J. M., J. T. Holland, and D. L. Hasty.** 2002. Interaction of an outer membrane protein of enterotoxigenic *Escherichia coli* with cell surface heparan sulfate proteoglycans. *Infect. Immun.* **70**:1530–1537.

26. **Flowers, H. M., and N. Sharon.** 1979. Glycosidases-properties and application to the study of complex carbohydrates and cell surfaces. *Adv. Enzymol.* **48**:29–95.

27. **Frey, A., K. T. Giannasca, R. Weltzin, P. J. Giannasca, H. Reggio, W. I. Lencer, and M. R. Neutra.** 1996. Role of the glycocalyx in regulating access of microparticles to apical plasma membranes of intestinal epithelial cells: implications for microbial attachment and oral vaccine targeting. *J. Exp. Med.* **184**:1045–1059.

28. **Fukuda, M., and M. N. Fukuda.** 1989. Changes in cell surface glycoproteins and carbohydrate structures during the development and differentiation of human erythroid cells. *J. Supramol. Struct.* **8**:313–324.

29. **Gahmberg, C. G.** 1977. Cell surface proteins: changes during cell growth and malignant transformation, p. 371–421. *In* G. Poste and G. L. Nicholson (ed.), *Cell Surface Reviews.* North-Holland, Amsterdam, The Netherlands.

30. **Gahmberg, C. G.** 1981. Membrane glycoproteins and glycolipids: structure, localization and function of carbohydrates, p. 127–160. *In* J. B. Finean and R. H. Mitchell (ed.), *Membrane Structure.* Elsevier/North-Holland, Amsterdam, The Netherlands.

31. **Gahmberg, C. G., and L. C. Andersson.** 1982. Surface glycoproteins of malignant cells. *Biochim. Biophys. Acta* **651**:65–83.

32. **Garoff, H.** 1979. Structure and assembly of the Semliki Forest virus membrane. *Biochem. Soc. Trans.* **7**:301–306.

33. **Gebb, C., E. G. Hayman, E. Engvall, and E. Ruoslahti.** 1986. Interaction of vitronectin with collagen. *J. Biol. Chem.* **261**:16698–16703.

34. **Geiger, B.** 1983. Membrane cytoskeleton interactions. *Biochim. Biophys. Acta* **737**:305–341.

35. **Gendler, S. J., and A. P. Spicer.** 1995. Epithelial mucin genes. *Annu. Rev. Physiol.* **57**:607–634.

36. **Geuijen, C. A. W., R. J. L. Willems, and F. R. Mooi.** 1996. The major fimbrial subunit of *Bordetella pertussis* binds sulfated sugars. *Infect. Immun.* **64**:2657–2665.

37. **Gum, J. R., Jr.** 1995. Human mucin glycoproteins: varied structures predict diverse properties and specific functions. *Biochem. Soc. Trans.* **23**:795–799.

38. **Hakomori, S.** 1981. Glycosphingolipids in cellular interaction, differentiation and oncogenesis. *Annu. Rev. Biochem.* **50**:733–764.

39. **Hoskins, L. C., M. Agustined, W. B. Mckee, E. T. Boulding, M. Kriaris, and G. Niedermeyer.** 1985. Mucin degradation in human colon ecosystems. Isolation and properties of fecal strains that degrade ABH blood group antigens and oligosaccharides from mucin glycoproteins. *J. Clin. Investig.* **75**:944–953.

40. **Hulmes, D. J.** 2002. Building collagen molecules, fibrils, and suprafibrillar structures. *J. Struct. Biol.* **137**:2–10.

41. **Hynes, R. O.** 1992. Integrins: versatility, modulation and signaling in cell adhesion. *Cell* **69**:11–25.

42. **Hynes, R. O.** 2002. Integrins: bidirectional, allosteric signaling machines. *Cell* **110**:673–687.

43. **Iozzo, R. V.** 2001. Heparan sulfate proteoglycans: intricate molecules with intriguing functions. *J. Clin. Investig.* **108**:165–167.

44. **Isaacs, R. D.** 1994. *Borrelia burgdorferi* bind to epithelial cell proteoglycans. *J. Clin Investig.* **93**:809–819.

45. **Ito, S.** 1965. The enteric surface coat on cat intestinal microvilli. *J. Cell Biol.* **27**:475–491.

46. **Ito, S.** 1969. Structure and function of the glycocalyx. *Fed. Proc.* **28**:12–25.

47. **Karlsson, K. A.** 1998. Meaning and therapeutic potential of microbial recognition of host glycoconjugates. *Mol. Microbiol.* **29**:1–11.

48. **Karlsson, N. G., A. Herrmann, H. Karlsson, M. E. Johansson, I. Carlstedt, and G. C. Hansson.** 1997. The glycosylation of rat intestinal Muc2 mucin varies between rat strains and the small and large intestine. A study of O-linked oligosaccharides by a mass spectrometric approach. *J. Biol. Chem.* **272**:27025–27034.

49. **Karnovsky, M. I., and E. P. Unanue.** 1973. Mapping and migration of lymphocyte surface macromolecules. *Fed. Proc.* **32**:55–59.

50. **Kato, T., and R. L. Owen.** 1999. Structure

and function of intestinal mucosal epithelium, p. 115–132. *In* P. L. Ogra, J. Mestecky, M. E. Lamm, W. Strober, J. Bienenstock, and J. R. McGhee (ed.), *Mucosal Immunology*. Academic Press, Inc., New York, N.Y.

51. **Liang, O. D., F. Ascencio, L. Fransson, and T. Wadström.** 1992. Binding of heparan sulfate to *Staphylococcus aureus*. *Infect. Immun.* **60:**899–906.

52. **Liang, O. D., K. T. Preissner, and G. S. Chhatwal.** 1997. The hemopexin-type repeats of human vitronectin are recognized by *Streptococcus pyogenes*. *Biochem. Biophys. Res. Commun.* **234:**445–449.

53. **Lichtenberg, D., R. J. Robson, and E. A. Dennis.** 1982. Solubilization of phospholipids by detergents: structural and kinetic aspects. *Biochim. Biophys. Acta* **737:**285–304.

54. **Lichtenberger, L. M.** 1995. The hydrophobic barrier properties of gastrointestinal mucus. *Annu. Rev. Physiol.* **57:**565–583.

55. **Lodish, H., A. Berk, S. L. Zipursky, P. Matsudaira, D. Baltimore, and J. Darnell.** 1999. *Molecular Cell Biology*. W.H. Freeman & Co., New York, N.Y.

56. **Marchesi, V. T., H. Furthmayr, and M. Tomita.** 1976. The red cell membrane. *Annu. Rev. Biochem.* **45:**667–698.

57. **Marinetti, G. V., and R. C. Crain.** 1978. Topology of amino-phospho-lipids in the red-cell membrane. *J. Supramol. Struct.* **8:**191–213.

58. **McCormick, B. A., P. Klemm, K. A. Krogfelt, R. L. Burghoff, L. Pallesen, D. C. Laux, and P. S. Cohen.** 1993. *Escherichia coli* F-18 phase locked 'on' for expression of type 1 fimbriae is a poor colonizer of the streptomycin-treated mouse large intestine. *Microb. Pathog.* **14:**33–43.

59. **Menozzi, F. D., J. H. Rouse, M. Alavi, M. Laude-Sharp, J. Muller, R. Bischoff, M. J. Brennan, and C. Locht.** 1996. Identification of a heparin-binding hemagglutinin present in mycobacteria. *J. Exp. Med.* **184:**993–1001.

60. **Miller, E. J., and S. Gay.** 1992. Collagen structure and function, p. 130–151. *In* I. K. Cohen, R. F. Diggelman, and W. J. Lindblad (ed.), *Wound Healing: Biochemical and Clinical Aspects*. The W. B. Saunders, Co., Philadelphia, Pa.

61. **Mirelman, D., and I. Ofek.** 1986. Introduction to microbial lectins and agglutinins, p. 1–19. *In* D. Mirelman (ed.), *Microbial Lectins and Agglutinins*. John Wiley & Sons, Inc., New York, N.Y.

62. **Myllyharju, J., and K. I. Kivirikko.** 2001. Collagens and collagen-related diseases. *Ann. Med.* **33:**7–21.

63. **Neutra, M. R., and J.-P. Kraehenbuhl.** 1999. Cellular and molecular basis for antigen transport across epithelial barriers, p. 101–114. *In* P. L. Ogran, J. Mestecky, M. E. Lamm, W. Strober, J. Bienenstock, and J. R. McGhee (ed.), *Mucosal Immunology*. Academic Press, Inc., New York, N.Y.

64. **Nicolson, G. L.** 1976. Trans-membrane control of the receptors on normal and tumor cells. I. Cytoplasmic influence of cell surface components. *Biochim. Biophys. Acta* **457:**57–108.

65. **Nicolson, G. L.** 1979. Topographic display of cell surface components and their role in transmembrane signaling. *Curr. Top. Dev. Biol.* **3:**305–338.

66. **Noel, G. J., D. C. Love, and D. M. Mosser.** 1994. High-molecular-weight proteins of nontypeable *Haemophilus influenzae* mediate bacterial adhesion to cellular proteoglycans. *Infect. Immun.* **62:**4028–4033.

67. **Ottani, V., M. Raspanti, and A. Ruggeri.** 2001. Collagen structure and functional implications. *Micron* **32:**251–260.

68. **Raff, M. C., and S. dePetris.** 1973. Movement of lymphocyte surface antigens and receptors: the fluid nature of the lymphocyte plasma membrane and its immunological significance. *Fed. Proc.* **32:**48–54.

69. **Rapraeger, A. C.** 2000. Syndecan-regulated receptor signaling. *J. Cell Biol.* **149:**995–998.

70. **Ross, M. H., L. J. Romrell, and G. I. Kaye.** 1995. *Histology. A Text and Atlas*. Lippincott Williams & Wilkins, Philadelphia, Pa.

71. **Rostand, K. S., and J. D. Esko.** 1997. Microbial adherence to and invasion through proteoglycans. *Infect. Immun.* **65:**1–8.

72. **Roth, J.** 1980. The use of lectins as probes for carbohydrates-cytochemical techniques and their application in studies on cell surface dynamics. *Acta Histochem. Suppl.* **22:**113–121.

73. **Roussel, P., G. Lamblin, M. Lhermitte, N. Houdret, J. J. Lafitte, J. M. Perini, A. Klein, and A. Scharfman.** 1988. The complexity of mucins. *Biochemie* **70:**1471–1482.

74. **Ruoslahti, E.** 1991. Integrins. *J. Clin. Investig.* **87:**1–5.

75. **Said, H. M., R. Smith, and R. Redha.** 1987. Studies on the intestinal surface acid microclimate: developmental aspects. *Pediatr. Res.* **22:**497–499.

76. **Schulster, D., and A. Levitski.** 1980. *Cellular Receptors for Hormones and Neurotransmitters*. John Wiley & Sons, Inc., New York, N.Y.

77. **Sedlacek, H. H., J. Stark, F. R. Seiler, W. Ziegler, and H. Wiegandt.** 1976. Cholera

toxin induces redistribution of sialoglycolipid receptor at the lymphocyte membrane. *FEBS Lett.* **61**:272–276.

78. **Seiffert, D., and D. J. Loskutoff.** 1996. Type 1 plasminogen activator inhibitor induces multimerization of plasma vitronectin. A suggested mechanism for the generation of the tissue form of vitronectin *in vivo. J. Biol. Chem.* **271**:29644–29651.

79. **Sharon, N.** 1981. Glycoproteins in membranes, p. 117–182. *In* R. Balian, M. Chabre, and P. F. Devaux (ed.), *Membranes and Intercellular Communications.* North-Holland, Amsterdam, The Netherlands.

80. **Sharon, N., and H. Lis.** 1981. Glycoproteins: research booming on long-ignored, ubiquitous compounds. *Chem. Eng. News* **59**:21–24.

81. **Sharon, N., and H. Lis.** 1982. Glycoproteins, p. 1–144. *In* H. Neurath, and R. L. Hill (ed.), *The Proteins,* vol. V, 3rd. ed. Academic Press, Inc., New York, N.Y.

82. **Sharon, N., and H. Lis.** 1989. Lectins as cell recognition molecules. *Science* **246**:227–234.

83. **Sheehan, J. K., D. J. Thornton, M. Somerville, and I. Carlstedt.** 1991. Mucin structure. The structure and heterogenicity of respiratory mucus glycoproteins. *Am. Rev. Respir. Dis.* **144**:S4–S9.

84. **Sherman, P., N. Fleming, J. Forstner, N. Roomi, and G. Forstner.** 1987. Bacteria and the mucus blanket in experimental small bowel bacterial overgrowth. *Am. J. Pathol.* **126**:527–534.

85. **Shin, J. S., and S. N. Abraham.** 2001. Co-option of endocytic functions of cellular caveolae by pathogens. *Immunology* **102**:2–7.

86. **Shin, J. S., Z. Gao, and S. N. Abraham.** 2000. Involvement of cellular caveolae in bacterial entry into mast cells. *Science* **289**:732–733.

87. **Singer, S. J.** 1974. The molecular organization of membranes. *Annu. Rev. Biochem.* **43**:805–833.

88. **Singer, S. J., and G. L. Nicolson.** 1972. The fluid mosaic model of cell membranes. *Science* **175**:710–731.

89. **Speziale, P., M. Höök, T. Wadström, and R. Timpl.** 1982. Binding of basement membrane protein laminin to *Escherichia coli. FEBS Lett.* **146**:55–58.

90. **Steck, T. L.** 1978. Band 3 protein of the human red cell membrane: a review. *J. Supramol. Struct.* **8**:311–324.

91. **Stephens, R. S., K. Koshiyama, E. Lewis, and A. Kubo.** 2001. Heparin-binding outer membrane protein of chlamydiae. *Mol. Microbiol.* **40**:691–699.

92. **Switalski, L. M., H. Murchison, R. Timpl,** R. Curtiss, and M. Höök. 1987. Binding of laminin to oral and endocarditis strains of viridans streptococci. *J. Bacteriol.* **169**:1095–1101.

93. **Switalski, L. M., P. Speziale, M. Höök, T. Wadström, and R. Timpl.** 1984. Binding of *Streptococcus pyogenes* to laminin. *J. Biol. Chem.* **259**:3734–3738.

94. **Szentkuti, L., H. Rieesel, M. L. Enss, K. Gaertner, and W. VonEngelhardt.** 1990. Pre-epithelial mucus layer in the colon of conventional and germ-free rats. *Histochem. J.* **22**:491–497.

95. **Tabak, L. A.** 1995. In defense of the oral cavity: structure, biosynthesis and function of salivary mucins. *Annu. Rev. Physiol.* **57**:547–564.

96. **Tam, P. Y., and P. Verdugo.** 1981. Control of mucus hydration as a Donnan equilibrium process. *Nature* **292**:340–342.

97. **Tanner, M. J. A.** 1978. Erythrocyte glycoproteins. *Curr. Top. Membr. Transp.* **11**:279–325.

98. **Tepass, U., K. Truong, D. Godt, M. Ikura, and M. Peifer.** 2000. Cadherins in embryonic and neural morphogenesis. *Nat. Rev. Mol. Cell Biol.* **1**:91–100.

99. **Tomasini, B. R., and D. F. Mosher.** 1988. Conformational states of vitronectin: preferential expression of an antigenic epitope when vitronectin is covalently and noncovalently complexed with thrombin-antithrombin III or treated with urea. *Blood* **72**:903–912.

100. **Tomasini, B. R., and D. F. Mosher.** 1990. Vitronectin. *Prog. Hemostasis Thromb.* **10**:269–305.

101. **Toribara, N. W., J. R. Gum, Jr., P. J. Culhane, R. E. Lagace, J. W. Hicks, G. M. Petersen, and Y. S. Kim.** 1991. MUC-2 human small intestinal mucin gene. Structure, repeated arrrays and polymorphism. *J. Clin. Investig.* **88**:1005–1013.

102. **Tranh van Nhieu, J., and P. J. Sansonetti.** 2000. Cell adhesion molecules and bacterial pathogens, p. 97–111. *In* P. Cossart, P. Boquet, S. Normark, and R. Rappuoli (ed.), *Cellular Microbiology.* ASM Press, Washington, D.C.

103. **Valentin-Weigand, P., J. Grulich-Henn, G. S. Chhatwal, G. Muller-Berghaus, H. Blobel, and K. T. Preissner.** 1988. Mediation of adherence of streptococci to human endothelial cells by complement S protein (vitronectin). *Infect. Immun.* **56**:2851–2855.

104. **van Putten, J. P. M., and S. M. Paul.** 1995. Binding of syndecan-like cell surface proteoglycan receptors is required for *Neisseria gonorrhoeae* entry into human mucosal cells. *EMBO J.* **14**:2144–2154.

105. **Varki, A.** 1999. Exploring the biological roles of glycans, p. 57–68. *In* A. Varki, R. Cummings, J.

Esko, H. Freeze, G. Hart, and J. Martin (ed.), *Essentials of Gycobiology*. Cold Spring Harbor Laboratory Press, Cold Spring Harbor, N.Y.

106. **Vuorio, E., and B. deCrombrugghe.** 1990. The family of collagen genes. *Annu. Rev. Biochem.* **59:**837–872.

107. **Wicken, A. J., and K. W. Knox.** 1981. Composition and properties of amphiphiles, p. 1–7. *In* G. D. Shockman and A. J. Wicken (ed.), *Chemistry and Biological Activities of Bacterial Surface Amphiphiles.* Academic Press, Inc., New York, N.Y.

108. **Wiegandt, H., S. Kanda, K. Inoue, K. Utsumi, and S. Nojima.** 1981. Studies on the cell association of exogenous glycolipids. *Adv. Exp. Med. Biol.* **152:**3433–3352.

109. **Winters, B. D., N. Ramasubbu, and M. W. Stinson.** 1993. Isolation and characterization of a *Streptococcus pyogenes* protein that binds to basal laminae of human cardiac muscle. *Infect. Immun.* **61:**3259–3264.

110. **Yamakawa, T., and Y. Nagai.** 1978. Glycolipids at the cell surface and their biological functions. *Trends Biochem. Sci.* **3:**128–131.

ADHESINS AS BACTERIAL CELL SURFACE STRUCTURES: GENERAL CONCEPTS OF STRUCTURE, BIOGENESIS, AND REGULATION

4

A central tenet of bacterial adhesion is that in order for molecules to function as adhesins, they must be presented at the bacterial surface in such a manner that the binding domain is capable of docking with the complementary host cell receptor. In most cases, the adhesins are assembled on and attached to the surface. In some instances, it has been found that the adhesins are initially secreted in a soluble form and then associate with the bacterial surface to function as molecular bridges between the bacterium and host cell (5, 135, 136). Whether the adhesin is an integral component of the bacterial surface or a bridging molecule, it must be anchored on the bacterial surface in some fashion before it can participate in adhesive processes. In this chapter, concise reviews of the general architectural plan of gram-positive and gram-negative cell surfaces are presented. Within this context, a few specific examples are used to present basic concepts of the genetics of adhesin expression and adhesive organelle biogenesis. More comprehensive discussions of bacterial cell surfaces can be found elsewhere (15, 16, 44, 91, 95).

GENERAL FEATURES
There are a number of common structures and surface components in both gram-positive and gram-negative bacteria, and each cell type

also has distinguishing characteristics (Fig. 4.1 and 4.2). Although some features of bacterial cell surfaces remain relatively constant, it must be kept in mind that the surface composition varies during different phases of growth and is highly dependent on the composition of the milieu in which the bacteria are grown. Whether it is laboratory agar or broth or host tissues or fluids, the growth medium composition can have dramatic effects directly or indirectly on adhesin expression and/or function.

Both gram-positive and gram-negative bacteria possess peptidoglycan, a stress-bearing and shape-determining structure. Peptidoglycan is sometimes referred to as murein. The entire insoluble network surrounding the cytoplasmic membrane of bacteria is called the sacculus. The peptidoglycan layer is considerably thicker in gram-positive ($\sim$20 to 40 nm) than in gram-negative ($\sim$5 to 10 nm) bacteria. In addition to peptidoglycan, the cell wall of gram-positive bacteria contains covalently bound teichoic acid or teichuronic acid. In gram-positive bacteria, the osmotic pressure inside the cell is equivalent to approximately 20 to 25 atm, roughly equivalent to 1 mol of colligative molecules per liter. The high turgor of gram-positive cells in most environments requires that the cells withstand the internal pressure without bursting. The thick

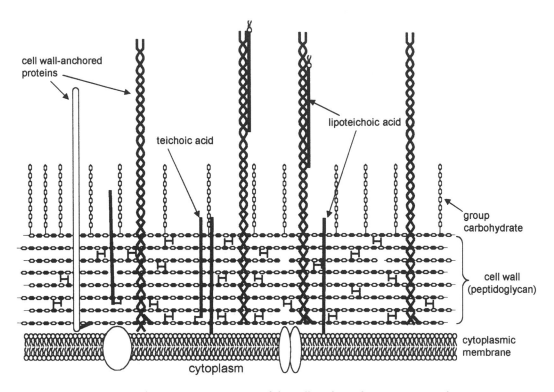

FIGURE 4.1 Schematic representation of the cell surface of a gram-positive bacterium. External to the cytoplasmic membrane is a multilayered peptidoglycan matrix composed of *N*-acetylglucosamine (○) and *N*-acetylmuramic acid (●). Many surface proteins are covalently anchored to muramic acid (see Fig. 4.3) but extend through the peptidoglycan to be presented at the bacterial surface. Adhesion functions may be ascribed to virtually any surface structure or macromolecule. In electron micrographs, the most prominent adhesins appear to be in the form of fibrils and filamentous structures. Lipoteichoic acid can be found within the cytoplasmic membrane but is also secreted through the cell wall and surface protein layer, often becoming associated with proteins and reoriented to contribute to the hydrophobic characteristics of the cell surface.

cell wall is able to accommodate the high internal pressure of the bacterial cell (68). In gram-negative bacteria, the peptidoglycan is surrounded by a second membranous layer, called the outer membrane, whereas the peptidoglycan of the gram-positive cell is generally exposed to the external milieu. The cell wall of the gram-negative bacterium is classically regarded as peptidoglycan plus outer membrane, although some maintain that the outer membrane should not be classified as a wall component.

Peptidoglycan typically contains nonreducing *N*-acetylglucosamine (GlcNAc) residues. Many peptidoglycans also contain unsubsti-

tuted glucosamine and muramic acid (114). These nonreducing residues could conceivably serve as ligands for animal cell surface proteins, although there is presently no evidence for their function in adhesion. Its main function in adhesion is to support or form a matrix for adhesin stabilization and presentation. There is a great deal of heterogeneity in the composition of peptidoglycans from various bacteria. Most of the heterogeneity arises in the peptide portion of the peptidoglycans. The peptides are very short sequences of 3 or 4 amino acids in both the D and L configurations. Cross-linking of the amino acids by a single amino acid or a short peptide also con-

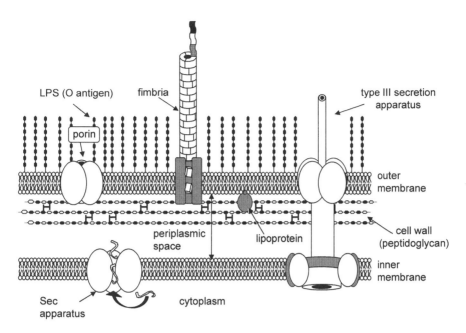

FIGURE 4.2 Schematic representation of the cell surface of a gram-negative bacterium. External to the cytoplasmic membrane is a periplasmic space, a few layers of peptidoglycan matrix, and an outer membrane in which adhesins are embedded. Gram-negative bacterial adhesins are most often found on fimbriae, although fibrils, flagella, OMPs, and capsule can also possess adhesin characteristics. The adhesive subunit of fimbriae may be found at the fimbrial tips (illustrated) or at multiple sites along the length (not illustrated). In most cases, adhesins are integral components of the outer membrane, but examples of secreted components that become associated with integral membrane components do exist.

tributes to peptidoglycan heterogeneity (114). In the gram-negative bacterium, protein molecules covalently attached to peptidoglycan may extend into and partially stabilize the outer membrane. These proteins, frequently referred to as Braun lipoproteins (22), have no apparent role in adhesion.

GRAM-POSITIVE BACTERIA

The adhesins of the gram-positive bacteria are anchored on the surface by four different mechanisms (Table 4.1). One prominent mechanism for anchoring gram-positive surface proteins, the cell wall-anchoring mechanism, involves the anchoring of surface proteins to the cell wall by cleavage from its intramembranous domain and subsequent covalent linkage to the cell wall by a transpeptidation reaction (Fig. 4.3). Another mechanism, the transmembrane mechanism, is the anchoring of a protein in the

cytoplasmic membrane of the bacterium by the presence of a hydrophobic, membrane-spanning domain. A third mechanism is the association of adhesins with surface proteins, while the fourth mechanism is the association of adhesins with surface glycolipids. The best described and probably the most prominent mechanism for anchoring adhesins is the cell wall-anchoring mechanism. A few examples of the surface molecules anchored by covalent linkage to the cell wall are given in Table 4.2 (for more details, see reference 91). Because this is such a common mechanism for anchoring adhesins, it is described in more detail below.

Gram-Positive Cell Wall

The glycan of the gram-positive peptidoglycan consists of a repeating disaccharide: N-acetylmuramic acid-($\beta1\rightarrow4$)-N-acetylglucosamine (-MurNAc-GlcNAc- [GN-MN in

TABLE 4.1 Mechanisms for anchoring adhesins on bacterial cell surfaces

Mechanism	Example	Reference
Gram-positive bacteria		
Cell wall-anchored proteins	Protein F1 (*S. pyogenes*)	54
Transmembrane proteins	Elastin binding protein (*S. aureus*)	39
Associated with surface proteins	Lipoteichoic acid (*S. pyogenes*)	101
Associated with surface glycolipids	Internalin B (*L. monocytogenes*)	63
Gram-negative bacteria		
Fimbriae (pili)	Type 1 fimbriae (*E. coli*)	130
OMP	Opa50 (*N. gonorrhoeae*)	134
Glycolipid	LPS (*H. pylori*)	40
Associated with unknown surface structures	Fucose-sensitive lectin (*P. aeruginosa*)	136

Fig. 4.3]). In this way, gram-positive peptidoglycan is similar to that of most other bacteria. The number of disaccharide units in each mature glycan chain varies considerably, depending on the bacterial species, and can be more than 50. A short, four- or five-residue peptide is linked to the D-lactyl moiety of the MurNAc component of the disaccharide, creating the so-called muramyl peptides. These MurNAc-linked peptides are themselves cross-linked to other MurNAc-linked peptides of other glycan chains via another peptide bridge. In *Staphylococcus aureus,* for example, this connecting peptide is a pentaglycine peptide. The peptide structures are the components of cell walls that differ most among the various species. Cell wall construction begins when a muramyl peptide consisting of

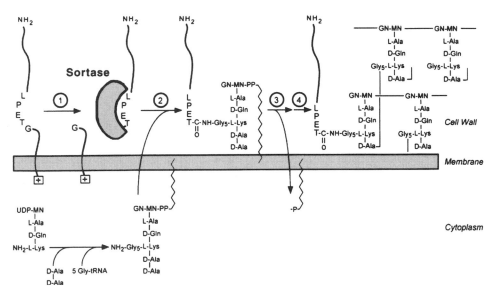

FIGURE 4.3 Hypothetical model for the anchorage of surface proteins to cell wall peptidoglycan. The process is divided into four steps. In step 1, the full-length precursor of a surface protein is exported via an N-terminal leader peptide; in step 2, the protein is prevented from being released extracellularly by a hydrophobic membrane-spanning domain and a charged cytoplasmic tail; in step 3, the protein is cleaved between Thr and Gly of the LPXTG motif by the sortase enzyme; and in step 4, the newly cleaved carboxy terminus of Thr is linked to the amino group at the end of a cell wall cross-bridging peptide. (Reprinted from reference 91 with permission from the publisher.)

TABLE 4.2   Examples of gram-positive proteins covalently anchored to cell walls

Organism	Protein	Function	Reference(s)
Enterococcus faecalis	Aggregating substance, Asa1	Adhesion; promotes invasion	102
Streptococcus pyogenes	Immunoglobulin binding proteins	Binding immunoglobulins	90
	M protein	Adhesion; evasion of phagocytosis	36, 37
	Protein F1 (PrtF1)	Adhesion	54
	Serum opacity factor	Adhesion	35, 71, 112
Staphylococcus aureus	Protein A	Binding immunoglobulins	127
	Fibronectin binding protein	Adhesion	128
	Collagen binding protein	Adhesion	107
Staphylococcus epidermidis	Fibrinogen binding protein	Adhesion	82, 96
Streptococcus pneumoniae	Neuraminidase	Invasion	25
Streptococcus mutans	Glucan binding protein C	Adhesion	120
Streptococcus agalactiae	Immunoglobulin binding protein	Binding immunoglobulins	55, 61
Listeria monocytogenes	Internalin A	Adhesion and internalization	46

the disaccharide linked to the pentapetide and five glycine residues is transferred across the cytoplasmic membrane and cross-linked to the glycine residue of an existing chain of peptidoglycan via its terminal alanine amino acid by a transpetidase reaction (Fig. 4.3).

Gram-positive bacteria typically have several other components cross-linked into the cell wall peptidoglycan. For instance, most gram-positive species express teichoic acid (commonly polyglycerol or polyribitol phosphate chains) that is covalently linked to peptidoglycan (Fig. 4.4A). Lipoteichoic acids (Fig. 4.4B) are intercalated into the cytoplasmic membrane via their fatty acid moieties but are otherwise not covalently linked to the cell wall. In this way, the sacculus, a gigantic, chain mail-like macromolecule that surrounds the bacterial cell, is formed. β-Lactam antibiotics inhibit the transpeptidation reaction carried out by penicillin binding proteins, leading to the release of soluble muramyl peptides of various sizes.

Anchorage of Gram-Positive Surface Proteins to the Cell Wall

As one can see from the number of examples given in Table 4.2, many protein adhesins are anchored to the gram-positive surface by covalent linkage to the cell wall peptidoglycan. Before they can be anchored to the cell wall, the proteins are brought to the surface by secretion via a typical Sec-dependent pathway that requires an N-terminal leader peptide. Several other common characteristics can be identified in these gram-positive surface proteins (Fig. 4.5). Almost all contain N-terminal repeating domains which are commonly critical for the ability of the protein to function as an adhesin. The repeat domains can vary from a few to several hundred amino acids. Strain-to-strain variation in the exact number of tandem repeat domains has been observed for many surface proteins, suggesting that gene segments encoding these domains undergo recombination and/or duplication events. Further, there is usually a proline-rich stretch of amino acids immediately preceding the wall-anchoring region. The increased numbers of proline residues are thought to generate random coils in the protein that facilitate its ability to traverse the peptidoglycan network (43). All of the wall-anchored proteins also have hydrophobic, C-terminal membrane-spanning domains and a short, charged cytoplasmic tail, which are responsible for transiently retaining the protein in the cytoplasmic membrane. In this transiently stabilized position, the wall-anchored proteins present a cell wall-sorting signal, the LPXTG (Leu-Pro-X-Thr-Gly) motif, that lies just external to the cytoplasmic membrane (Fig. 4.3). The LPXTG motif is then recognized by

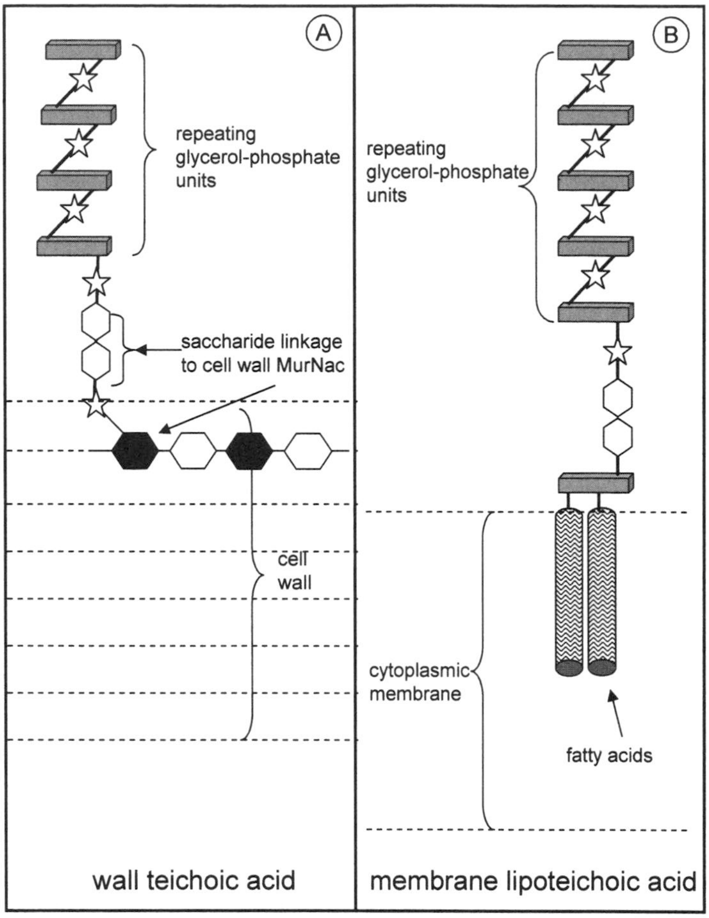

FIGURE 4.4 Schematic diagrams of wall teichoic acid (A) and membrane lipoteichoic acid (B) contained in the cell walls of gram-positive organisms. Teichoic acid is composed primarily of repeating glycerol-phosphate units linked via a saccharide moiety to *N*-acetylmuramic acid (MurNAc). The numbers of glycerol-phosphate units, the substitutions of glycerol (not shown), and the specific structure of the saccharide linkage moiety are species specific. Lipoteichoic acid is similar, in that a poly-glycerol-phosphate chain is linked via a saccharide moiety to its anchor. In this case, however, the anchor is not peptidoglycan but, instead, consists of fatty acid moieties that are embedded in the outer half of the cytoplasmic membrane.

(A)

repeating glycerol-phosphate units

saccharide linkage to cell wall MurNac

cell wall

wall teichoic acid

(B)

repeating glycerol-phosphate units

cytoplasmic membrane

fatty acids

membrane lipoteichoic acid

FIGURE 4.5 Schematic illustration of features common to many wall-anchored surface proteins. All feature an N-terminal leader sequence followed by regions either without (open areas) or with (hatched areas) repeating domains, an LPXTG sorting signal, and a charged C-terminal cytoplasmic tail. Some proteins contain a dozen or more repeated domains, and the number of repeats of the proteins expressed by different clones within one species is often variable. In some cases, proline-rich areas are present distal to the LPXTG motif and may aid the protein in traversing the thick peptidoglycan layer.

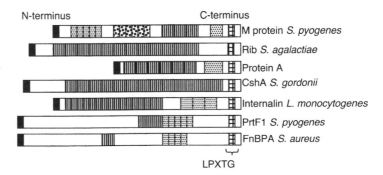

N-terminus

C-terminus

M protein *S. pyogenes*

Rib *S. agalactiae*

Protein A

CshA *S. gordonii*

Internalin *L. monocytogenes*

PrtF1 *S. pyogenes*

FnBPA *S. aureus*

LPXTG

an extracellular enzyme that has been called the "sortase." The sortase cleaves the nascent protein between Thr and Gly of the LPXTG motif. The carboxyl group of the now-C-terminal Thr is then joined to the terminal amino group of the branch peptide of the nascent muramyl petide by a transpeptidase reaction. The branch peptide glycine residue that is used to anchor the surface protein would otherwise be used in formation of the interpeptide bridge of the peptidoglycan.

This seemingly elaborate mechanism is apparently designed to ensure the stable association of secreted surface proteins with the bacterial cell. Most of the proteins anchored in this way have only a single hyrophobic stretch of amino acids spanning the cytoplasmic membrane and would be significantly less stable than others that span the membrane more than once, such as the elastin binding protein of *S. aureus,* which has two or three hydrophobic domains that anchor it in the membrane (Fig. 4.6). Proteins anchored by covalent linkages to the cell wall can be released intact only by enzymatic degradation of the peptidoglycan. For example, the glycine endopeptidase, lysostaphin, releases proteins that are homogeneous in size because it cleaves the bond between the protein and the branch peptide. Muramidase, on the other hand, cleaves within the glycan chains of peptidoglycan and therefore releases proteins that are heterogeneous in size, with the variation being dependent on the amount of peptidogycan remaining attached at the C terminus of the protein.

Many protein adhesins are anchored at the gram-positive bacterial surface by this mechanism. Therefore, it should not be surprising that a sortase-defective mutant would grow normally in vitro but would exhibit reduced virulence in a mouse model (79, 80). This suggests that targeting the sortase by antimicrobial agents would probably follow the same rules as the use of antiadhesion agents in the treatment of bacterial infections with regard to the emergence of resistant strains. Because the bacteria would become ineffective pathogens but would not be killed, strains resistant to this type of antimicrobial agent would be expected to spread much more slowly than would resistance to antibiotics (see chapter 11).

The morphological appearance of gram-positive surface proteins varies from short filaments, commonly called 'fuzz,' to dramatically longer fibrils and filaments. For instance, *Streptococcus pyogenes* is one of the organisms with a surface fuzzy layer (Fig. 4.7A). Some of these surface proteins bind several different

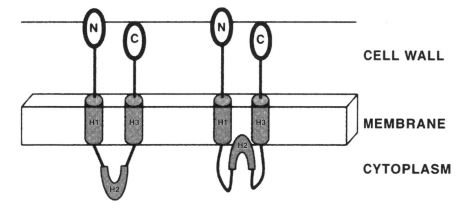

FIGURE 4.6 Schematic illustration of the *S. aureus* elastin binding surface protein, EbpS. This is an example of a surface protein that is not anchored via an LPXTG motif and sortase enzyme but, instead, is anchored via two or three hydrophobic domains thought to be oriented in one of two possible ways with regard to the cytoplasmic membrane. (Reprinted from reference 39 with permission from the publisher.)

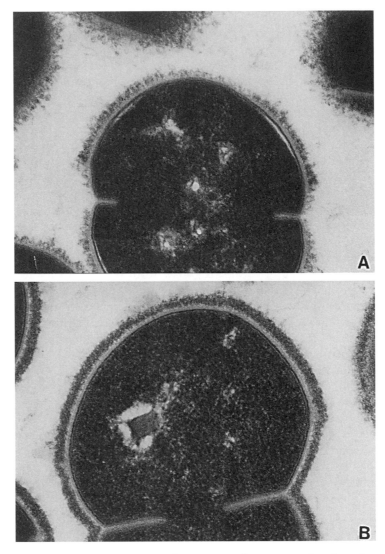

FIGURE 4.7 Electron micrographs of *S. pyogenes*, illustrating surface fuzz. Streptococci were grown overnight in broth and then incubated in the absence (A) or presence (B) of plasma. *S. pyogenes* binds several different serum and plasma proteins (e.g. fibrinogen, fibronectin, and albumin), which dramatically accentuate the presence of surface fibrils. Magnification, ×75,000.

plasma proteins, and when *S. pyogenes* is incubated in blood or grown in broth containing plasma, the appearance of surface 'fuzz' is dramatically enhanced (Fig. 4.7B). A variety of oral streptococci, such as *S. sanguis, S. gordonii,* and *S. oralis,* have much longer filaments, some of which are sufficiently long to be termed fimbriae (138). The 259-kDa fibronectin binding protein of *S. gordonii,* CshA, forms a fibril of approximately 60 nm in length, while the 261-kDa Fap1 protein of *S. parasanguis* FW213 forms 216-nm fimbriae (87). The 60-nm CshA fibrils of *S. gordonii* DL1 are very sparsely distributed on the cell surface (86a), but other oral streptococci carry a much denser array of surface appendages. Fibrils can be either peritrichously arranged (Fig. 4.8A) or arranged in lateral tufts on the cell surface (Fig. 4.8B). The composition and functions of the fibrils on these organisms have not been identified. However at least one fibril class carried on *S. sanguis* GW2 is closely related to the CshA fibrillar antigen, since cells of GW2 react with antiserum raised against the amino acid repeat region of CshA (41a). There are likely to be multiple classes of fibrils on oral streptococci with multiple, perhaps diverse, functions.

Regulation of Expression of Gram-Positive Adhesins

The expression of many sortase-mediated, wall-anchored adhesins is regulated by complex mechanisms involving global regulators. The specific regulatory mechanisms differ from adhesin to adhesin and from species to species. While a comprehensive presentation of the state of knowledge of the regulation of bacterial virulence factors is beyond the scope of this book, two examples are briefly described here, utilizing data on *S. aureus* and *S. pyogenes* to focus on gram-positive adhesins (Table 4.3).

STAPHYLOCOCCUS AUREUS

Regulation of expression of virulence factors by bacterial pathogens is essential for their survival in multiple unique microenvironments. It is frequently coordinated according to a precise temporal pattern of expression. A good example of the mechanisms for regulating virulence factor production are those of *S. aureus* (Fig. 4.9A). These bacteria produce multiple virulence factors that include adhesins and secreted toxins. While several surface protein adhesins, such as clumping factor A (ClfA), are expressed throughout the growth cycle, most adhesins, such as the fibronectin binding adhesins FnbpA (128) and FnbpB (64), the ClfB fibrinogen-binding adhesin (93), and protein A (56), are expressed from the early exponential to the midexponential phase of growth, when cell density is low (76). As the growth cycle progresses and cell density increases in the late exponential phase, most adhesins are downregulated. At

the same time, the expression of many of the secreted virulence factors, such as cytotoxins, superantigens, and proteases, is upregulated. Orchestration of this pattern of expression across the *S. aureus* growth cycle is regulated by reciprocally active global regulatory systems, some of which respond to quorum-sensing molecules (4, 6, 7, 30, 97).

One of the key molecules regulating the switch (phase variation) from surface adhesin expression to toxin production is an RNA molecule, called RNAIII, which is an important regulatory element of the *agr* (for "accessory gene regulator") cascade (Fig. 4.9B). *agr* transcription is driven by two divergent promoters, P-2 and P-3. P-2 is the promoter for transcription of the polycistronic *agrB, agrD, agrC,* and *agrA* mRNA, RNAII (99). AgrC is the receptor-histidine kinase of the two-component signaling pathway. AgrB and AgrD generate and secrete a cyclic thiolactone bond-containing autoinducing peptide (AIP). Binding of AIP to the extracellular domain of the transmembrane protein AgrC leads to autophosphorylation of an AgrC histidine. AIP accumulates extracellularly, and levels sufficient to stimulate AgrC are reached only as the organisms reach a density that is typically found late in the exponential phase of growth. Transphosphorylation of the cytoplasmic response regulator, AgrA, with the help of another of the global regulatory components of *S. aureus,* SarA (for "Staphylococcal accessory regulator A"), activates P-3, resulting in the production of RNAIII, which acts as a regulatory RNA molelcule as well as encoding for the delta-hemolysin (*hld*) (Fig. 4.9B).

RNAIII is the major effector of the dozen or more *agr* target genes. It appears that RNAIII may have subdomains with different regulation-related functions. It is predicted that RNAIII can either take on an untranslatable configuration or bind to other mRNAs, generating different effects depending on the particular mRNA involved. A 5′ domain of RNAIII that is complementary to the leader region of *hla* (alpha-hemolysin gene) can bind

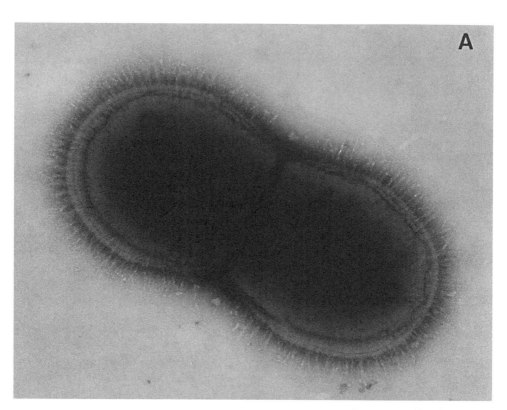

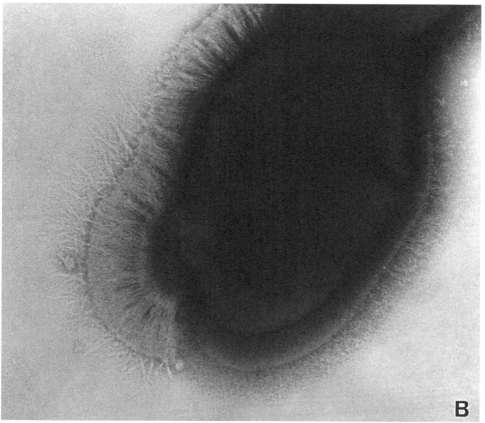

TABLE 4.3 Examples of regulation of expression of adhesins in gram-positive bacteria[a]

Bacterium	Regulatory gene	Mechanism	Effect
S. aureus	RNAIII of the Agr system	Quorum-sensing-mediated transduction	
		On RNAIII transcribed	Surface protein genes (e.g., spa) downregulated
		Off RNAIII not transcribed	Surface protein genes upregulated
S. pyogenes	Mga	DNA binding protein	Upregulation of multiple surface proteins

[a]See the text for more details.

to that mRNA and block the formation of pairing of nucleotides that inhibit translation of alpha-hemolysin (89, 98). A 3′ domain of RNAIII complementary to the spa (protein A) mRNA leader, on the other hand, binds in a translation-inhibitory conformation. Molecular mechanisms such as these for the stimulation of alpha-hemolysin production and concomitant inhibition of protein A production may also apply to the other genes affected by the P-3 transcript. Translation of the delta-hemolysin would require RNAIII taking on a translatable conformation by an as yet undetermined mechanism.

While data are accumulating regarding the switching "off" of surface protein expression by quorum-sensing mechanisms, relatively little is known about the initial stimulus for initiation of surface protein expression that occurs in the early exponential phase. The sarA locus is another well-studied member of the growing list of S. aureus global regulatory systems and a candidate for upregulating the expression of at least some surface proteins (31). SarA is a DNA binding protein that has differential effects on virulence factor expression. For instance, the sarA locus is involved in upregulation of fnbpA but represses the expression of spa. Some of the effects of SarA

are due to its upregulation of transcription of both agr operons (33, 99), but SarA can also affect the expression of virulence factor genes directly. Since both agr and sarA inhibit the expression of at least one surface protein, protein A, it was logical to predict an activator of spa expression. SarS, a homolog of SarA, appears to be at least one of the predicted activators of protein A synthesis (32). Whether it affects the expression of other surface proteins has not been studied. It is logical to predict that other activators of surface protein expression will be identified in the future.

At least one other parallel quorum-sensing system, the RAP-TRAP system, also affects the expression of adhesins and toxins across the exponential phase of growth. A ~33-kDa secreted autoinducer protein called the RNA3-activating protein (RAP) accumulates in the medium as the cell density increases during the exponential phase of growth (7). When RAP reaches a threshold concentration, it induces the phosphorylation of a target protein called the target of RNA-activating protein (TRAP). Phosphorylation of TRAP leads to activation of the agr locus through an as yet unknown mechanism. Interestingly, it has been shown that S. aureus virulence can be inhibited by a heptapeptide called the

FIGURE 4.8 Differing morphologies of structures on the surfaces of oral streptococci. (A) A dividing cell of S. sanguis strain GW2 negatively stained with 1% (wt/vol) methylamine tungstate. A dense fringe of peritrichous, short fibrils (75.3 ± 22 nm long) is visible, but no fibrils can yet be seen on the newly developing cell wall near the septum. (B) S. oralis CN 3410 stained with 1% methylamine tungstate. One side of the cell carries a dense tuft of fibrils of two lengths. Longer, sparser fibrils (289 ± 15 nm long) project through a very dense fringe of shorter tuft fibrils (159 ± 5 nm long). An indistinct fuzz covers the rest of the cell surface. Magnifications: ×73,500 (A) and ×85,700 (B). Micrographs were provided courtesy of Pauline Handley.

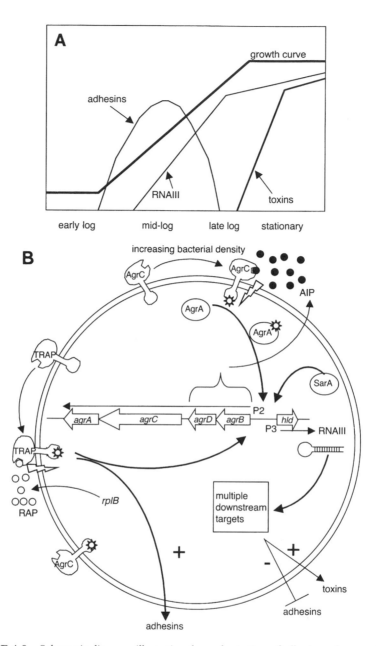

FIGURE 4.9 Schematic diagrams illustrating the orchestration of adhesin and secreted toxin expression across the *S. aureus* growth curve and regulatory mechanisms. (A) Schematic illustration of the early log, mid-log, late log, and stationary phases of the growth curve, in relation to the expression of adhesins, the transcription "effector" RNAIII, and secreted toxins. Adhesins begin to be expressed in early-log phase, peak by mid-log phase, and begin to be turned off by late-log phase by the expression of the *agr* locus and the RNAIII effector. (B) Schematic illustration of the Agr and RAP-TRAP quorum-sensing transcripton regulation systems of *S. aureus*. As the cell density increases, the levels of the autoinducing molecules AIP and RAP reach a threshold and activate the expression of *agr*. AgrB and AgrD combine to express and secrete AIP, which binds to the sensor protein AgrC. AgrC is phosphorylated, and the phosphate group is transferred to the response regulator AgrA, which activates the transcription of RNAII and RNAIII. RNAII encodes an additional autoinducing peptide, sensor, and response regulator, while RNAIII is the primary activator or suppressor of transcription of multiple targets. (Panel A modified from reference 97.)

RNAIII-inhibiting peptide, which appears to act by competitively inhibiting RAP binding and, hence, TRAP phosphorylation (8, 9, 49). Inhibition of TRAP phosphorylation leads, by an as yet unknown mechanism, to inhibition of bacterial adhesion (10, 11).

STREPTOCOCCUS PYOGENES

The first *S. pyogenes* genome that was completed contains roughly 100 genes with homologies to transcriptional regulators. The most thoroughly studied of the regulatory proteins of *S. pyogenes* is the multiple gene regulator, Mga. Mga was called VirR when it was first identified (129) and was later called Mry (108). Mga is a global positive transcriptional regulator that increases the transcription of a number of genes encoding primarily surface proteins, such as M and M-related proteins, serum opacity factor, and a collagen-like protein, SclA (Fig. 4.10). There are two different *mga* genes, which have approximately 70% homology. The two *mga* genes are associated with specific serotypes but effectively complement each other functionally (3). While low-level transcription of *mga* occurs in the absence of Mga binding, full expression requires Mga binding to the *mga* promoter (86). Expression of proteins of the Mga regulon is important in adhesion and colonization, internalization by nonphagocytic cells, and evasion of the immune system and appears to be important in a variety of types of infections caused by group A streptococci. Mga is a DNA binding protein, and its DNA binding capacity is essential for its function (86). Specific Mga binding sequences have been identified within the promoters of the genes it regulates, and comparison of these sites has allowed assignment of a concensus Mga binding sequence (83). Autoregulation of *mga* transcription also involves Mga binding within its own promoter, but the activation mechanism in this case appears to be somewhat different from the mechanism activating other Mga-targeted genes (85). Four potential helix-turn-helix (HTH) motifs are present in Mga on the basis of homologies to other regulatory proteins (29, 108, 109; K. S. McIver, unpublished results). HTH-4 is identical among various serotypes and appears to be essential for binding to all Mga binding sites. Thus, HTH-4 may be the major DNA binding domain. HTH-3 appears to augment Mga binding and may allow modulation of the binding of Mga to its promoter targets. Because transcription of the *mga* operon is stimulated by elevated CO_2 concentrations and other environmental factors (26), Mga has long been thought to represent a response regulator of a two-component signal transduction cascade. However, no sensor protein has yet been identified. Mga is expressed in the exponential phase of growth and is downregulated in stationary phase (84), possibly by Nra/RofA (see below) and/or Rgg/RopB. Although originally identified as a regulator of the SpeB cysteine proteinase (77), Rgg is now known to influence the expression of several *S. pyogenes* genes, such as the M protein adhesin, the collagen-like protein Scl1, and the cytolytic toxins, streptolysin S and streptolysin O (27, 28). The regulatory effects appear to be effected primarily by the interaction of Rgg with other important regulatory systems, such as *mga*, *csrRS* (see below), and *fasBCA* (see below) (27, 28).

RofA (Fig. 4.10) is a member of a family of regulatory proteins that is called the RofA-like protein (RALP) family (51). While there is only one Mga gene in each serotype of group A streptococci, several different RALP genes may be found in each strain. RofA inhibits the transcription of *mga* (13) but is a positive activator of itself and *prtF,* the gene encoding the fibronectin-binding adhesin protein F (45). Nra is another member of the RALP family (~60% identity of *nra* to *rofA*) but appears to act primarily as a negative regulator. Its targets include the gene for a second fibronectin binding protein, *prtF2, cpa* (which encodes a collagen binding protein) (110), *sof49,* and *fbp54* (70). Secreted toxins are also regulated by Nra (Fig. 4.10); they are encoded by *speB, speA,* and *sagA* (88). It is possible that the downregulation of certain virulence fac-

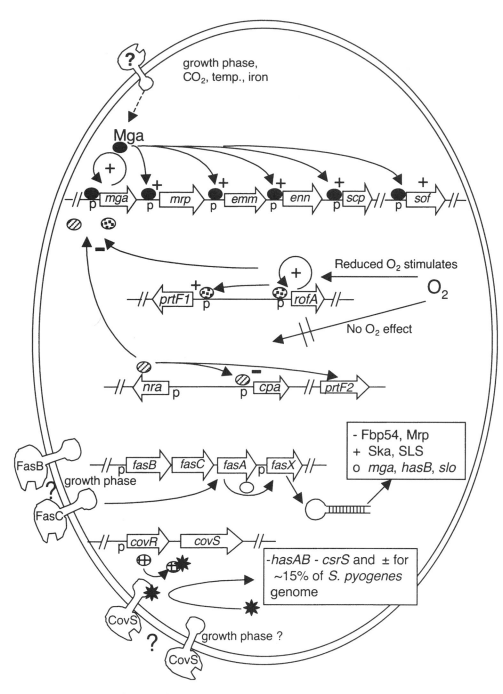

FIGURE 4.10 Schematic illustration of several different *S. pyogenes* regulatory systems. The *mga* locus was the first described virulence factor regulon and remains the best characterized. Mga is a DNA binding protein that activates the transcription of a number of surface proteins, some of which are known to serve as adhesins. Mga is self-regulating and is suppressed by Nra and RofA. The FasBCAX locus upregulates a fibronectin binding adhesin, Fbp54, and Mrp. FasB and FasC appear to serve as sensor molecules, with FasA being the response regulator. The *fasX* transcript does not encode a protein but serves as a regulatory RNA molecule. Thus, the *fasBCAX* locus has some similarities to the *agr* locus. CovR/CovS (also CsrRS) is another two-component regulatory system in which CovS is the membrane-bound sensor molecule and CovR is the response regulator. Recent evidence suggest that this global regulatory system affects up to 15% of the *S. pyogenes* genome.

tors by Nra contributes to the avoidance of host cell damage and thereby increases persistence within nonphagocytic cells.

Of a dozen or more potential two-component regulatory systems identified in the first *S. pyogenes* genome completed, only two sensor kinase-response regulator pairs have been studied in any detail. These are the FasBCAX (69) and CsrRS (74) operons. Fas (for "fibronectin/fibrinogen binding/hemolytic activity/streptokinase regulator") has homolgies to the Agr system of *S. aureus*. The *fas* operon contains two potential histidine kinase sensors, *fasB* and *fasC,* both having roughly 50% homology to the *agrC* sensor of *S. aureus* (Fig. 4.10). The *fasA* gene has a similar level of homology to the *agrA* response regulator. A 3′ region transcribed separately from *fasBCA* contains a 300-nucleotide transcript, which does not appear to encode true peptide sequences but is hypothesized to encode a nontranslated regulatory RNA that is the primary effector component of the *fas* operon, similar to the Agr system. *fasX* transcription is dependent on the FasA response regulator. Inactivating either of these genes leads to a phenotype that is also very similar to that of Agr mutants, in that adhesion to extracellular matrix components is increased and secretion of toxins is decreased. Binding to fibrinogen and fibronectin, but not type I collagen, was increased in a *fas* mutant, along with increased transcription of *fbp54,* encoding a fibronectin binding protein, and *mrp,* encoding a fibrinogen binding protein (69). The secreted virulence factors streptokinase and streptolysin S were markedly decreased. Despite the various levels of homology to the *agr* operon, *fas* is not yet known to be involved in quorum-sensing regulation.

The CsrRS (for "capsule synthesis regulator regulator/capsule synthesis regulator sensor") operon was first described as a locus that inhibited the expression of the hyaluronic acid capsule (74) (Fig. 4.10). CsrR binds in the *has* promoter (14), and inactivation of *csrR* led to increased capsule expression. It was later shown that this locus also affected the expression of several other virulence factors (42); in

that report the locus was called CovR/S (for "control of virulence"). More recently, a genomic approach was taken to begin to understand the spectrum of genes that may be regulated by the CovR/S operon (50). Results suggested that CovR/S is truly a critical regulatory system and that it may influence up to 15% (ca. 300 genes) of all chromosomal genes of the *S. pyogenes* strain analyzed.

It must be noted here that at present the systems described above are incompletely understood, at best. In most instances, our understanding derives from the study of one or a very few clones of each species. It is also clear that bacterial clones for each species vary considerably with regard to gene content, allelic diversity, and gene expression levels. Thus, the extent to which the information provided here have broad implications is unknown and will remain unknown for some time. Nevertheless, a great deal information has accumulated in the last decade with regard to the mechanisms by which gram-positive bacteria maintain themselves within the distinct microenvironments they encounter during their infection of a host or during their spread between hosts.

GRAM-NEGATIVE BACTERIA

General Characteristics of the Gram-Negative Cell Wall and Membranes

Outer membranes of gram-negative bacteria cover the peptidoglycan layer, creating a periplasmic space, and the adhesins of these microorganisms are most commonly anchored to the outer membrane (Fig. 4.2). The outer membrane is a phospholipid-lipopolysaccharide (LPS)-protein bilayer with typical fluid mosaic characteristics (95). The phospholipid forms most of the inner leaflet of the bilayer, whereas the LPS forms much of the outer leaflet. The polysaccharide moiety of the LPS extends away from the outer membrane and is a prominent antigen important for serologic testing. Most LPS from wild-type bacteria have very long polysaccharide structures consisting of repeating units of tri- or tetrasaccharides. Mutations in polysaccharide synthesis may lead to truncated polysaccharide molecules, giving rise to so-called rough strains. In addition to LPS, the outer membrane contains intercalated proteins, which have one or more hydrophobic regions responsible for anchoring the molecules to the cell membrane. Most of the outer membrane proteins extend across the entire thickness of the bilayer so that they are exposed to both the periplasmic space and external environment. Many of the classic outer membrane proteins (OMPs) are porins that serve as channels to provide a selective permeability to the membrane (94, 95). Some of the other functions of the outer membrane include a selective permeability barrier and the ability to bind certain bacteriophages (the binding of bacteriophages is one means of exchanging genetic information).

The adhesins of the gram-negative bacteria are anchored on the outer membrane surface by four different mechanisms (Fig. 4.2; Table 4.1). The most common mechanism for presenting adhesins of gram-negative bacteria is via hair-like heteropolymeric structures called fimbriae or pili (Fig. 4.11). Less common mechanisms are for adhesins to be OMPs, stabilized by hydrophobic domains that span the outer membrane, glycolipids anchored in the outer membrane via fatty acid moieties, or molecules that are tethered to the surface by association with as yet undetermined surface structures.

The O-polysaccharide chain of LPS has been suggested to function as an adhesin for only a few bacterial species (e.g., *Helicobacter pylori*) (Table 4.1), probably by reacting with host cell surface lectins, which are common on macrophages (100). Although the OMPs have been less commonly identified as adhesins thus far, the OMP adhesins are usually important adhesins for the virulence of the bacteria, because in many of the cases studied they have been found to mediate intimate adhesion that is usually associated either with invasion of the bacteria into the target cell or with formation of a lesion at the site of adhesion (e.g., the attachment/effacement lesion). Examples of OMP adhesins are the invasin of yersiniae and the intimin of enteropathogenic *Escherichia coli* (see also chapter 7). The functioning of associated molecules as adhesins has thus far been

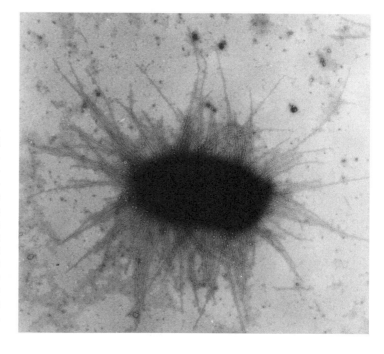

FIGURE 4.11 Electron micrograph of a negatively stained recombinant *E. coli* strain expressing type 1 fimbriae. These fimbriae and others of its class are peritrichously arranged, can number in the hundreds per cell, and are rigid, straight structures. There are 1,000 or more FimA subunits polymerized into a 7-nm-diameter helical structure. This subunit makes up the majority of the fimbrial structure. The mannose binding lectin, FimH, is located at the distal tip of the fimbriae. Magnification, ×28,615.

described in only one case, whereby an intracellular lectin of *Pseudomonas aeruginosa* is released by dying bacteria and the lectin then becomes bound to the surface of living *P. aeruginosa* cells, which subsequently can react with host cell glycoconjugates (136).

Many gram-negative bacteria, and gram-positive bacteria as well, are capable of expressing significant quantities of polysaccharide capsular material. When present, capsular material often inhibits bacterial adhesion to host cells by masking surface adhesins. Capsules are often quite thick, but they are not easily visualized morphologically. In normal electron microscopy procedures, the capsular material shrinks during processing and, even with ruthenium red staining, is seen only as a thin, electron-dense surface layer (Fig. 4.12A). Certain capsules, however, can be stabilized by pretreating bacteria with anticapsule antibodies prior to the fixation and dehydration steps. In these instances, the dramatic extent of capsular material can be clearly seen (Fig. 4.12B).

Fimbriae

Fimbriae are, by far, the most common adhesive structures expressed by gram-negative bacteria. There have been a number of attempts to classify fimbriae based on their morphological appearance and/or their hemagglutination patterns. Based in part on the most recent classification scheme presented by Ørskov and Ørskov (105), four general categories of fimbiae can be described. The first category is made up of fimbriae with rigid structures measuring approximately 7 nm in diameter with various lengths (e.g., type 1 and P fimbriae) [Fig. 4.13]). Studies of P fimbriae by quick-freeze, deep-etch electron microscopy have shown that they are composite structures composed of fibers with a distinct fibrillar tip (59, 72) (Fig. 4.14). The 7-nm fiber is composed of PapA subunits arranged in a right-handed helix with an external diameter of approximately 7 nm, an axial hole of approximately 2 nm, and 3.2 subunits per turn of the helical cylinder. The tip fibrillum is composed of PapE, PapF, PapK, and the PapG adhesin. Type 1 fimbriae, composed primarily of the FimA subunit, have a very similar structure (23), although some measurements differ slightly; however, their adhesive tip fibrillae, composed of FimF, FimG, and the FimH adhesin, are notably shorter than those of P fimbriae (62). The second category of fimbriae is com-

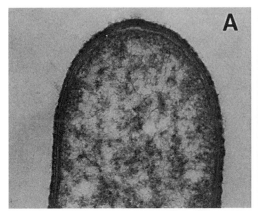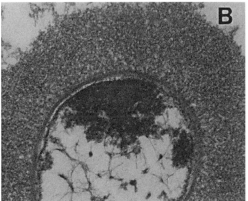

FIGURE 4.12 Electron micrographs illustrating the *Klebsiella pneumoniae* capsule. Bacteria were grown on lactose-supplemented medium, harvested, and incubated overnight at 4°C with preimmune serum or anticapsule serum. After fixation and staining with ruthenium red, the bacteria were processed for thin sectioning. (A) Without antibody stabilization, the capsular material contracts during processing and is barely visible. (B) With antibody stabilization, the extent of the capsule is clearly indicated. Magnification, ×45,000.

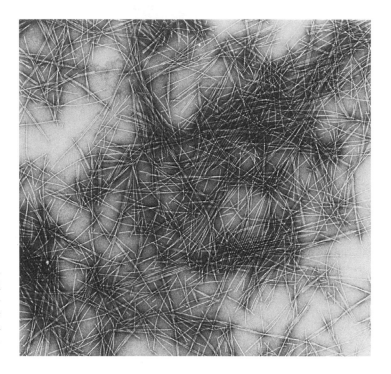

FIGURE 4.13 Electron micrograph of a negatively stained preparation of purified type 1 fimbriae. Their rigid structures are even more evident than in a micrograph of an intact bacterium. Magnification, ×39,375.

prised of thin, flexible structures, 2 to 3 nm in diameter (e.g., 987P fimbriae [Fig. 4.15]). The third category of fimbriae contains even thinner and much more flexible structures (e.g., curli [Fig. 4.16]). Curli are shared by many strains of *E. coli* and *Salmonella* and usually are highly aggregated on the surface, frequently appearing almost as a capsule under the electron microscope (111). The major subunit of *E. coli* curli is a 15-kDa protein, CsgA. A minor subunit, termed CsgB, may be distributed along the length of the curli fiber (17, 34, 103). The fourth category of fimbriae contains the type IV fimbriae expressed by a number of different bacteria including *Neisseria, Vibrio,* and enteropathogenic *E. coli*. They are usually flexible, are 4 to 6 nm in diameter, and often form into bundles (18, 126) (Fig. 4.17). Type IV fimbriae of enteropathogenic *E. coli, N. gonorrhoeae, V. cholerae* and others measure 6 nm in diameter and are up to 4 μm long, with approximately five subunits per turn (106). Type IV fimbriae typically contain a positively charged leader sequence with a modified amino acid, *N*-methylphenylalanine, at the N terminus of the mature protein. These fim-

briae also include minor components, including the tip-associated adhesin, PilC, in *N. gonorrhoeae* (115–117). The major subunit, PilE, has been crystallized and shows an α-β fold with a rather long hydrophobic N-terminal α-helical spine with an overall ladle shape (106). Unique structural features of PilE include an an O-linked GlcNAc(α1→3)Gal at Ser63 and a disulfide-containing region.

These categories of fimbriae are by no means inclusive, since there are other types of fimbriae that do not fit any of the categories listed above. Examples include the fimbria of enteropathogenic and enterotoxigenic *E. coli* termed CS1 and CS2, respectively. CS1 fimbriae are composed predominantly of one component, termed CooA, with a distally located minor component, CooD. They are morphologically similar to the type 1 and P fimbriae, although the structural proteins of the major subunits have no significant homologies (118, 119).

Fimbrial Biogenesis

The biogenesis of these categories of fimbriae follows certain common themes (131). It has been proposed that there are four dis-

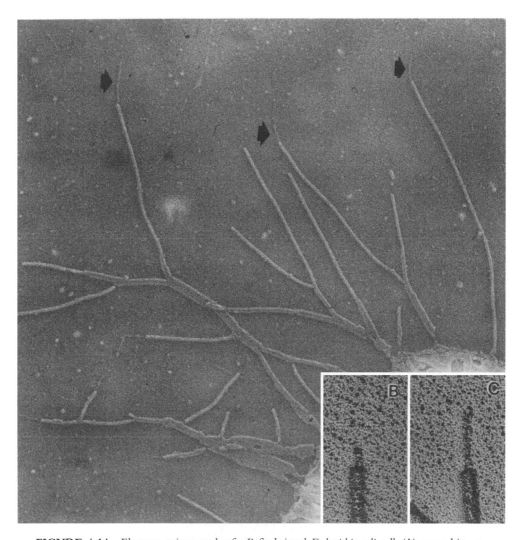

FIGURE 4.14 Electron micrograph of a P-fimbriated *Escherichia coli* cell. (A) recombinant *E. coli* cell possessing the pPAP5 plasmid, which contains the entire P fimbrial gene cluster, was fixed first with glutaraldehyde and then with osmium tetroxide and allowed to attach to freshly cleaved mica. The samples were dried, rotary shadowed with platinum and carbon, and viewed under a transmission electron microscope. The main 7-nm shafts of the heteropolymeric fimbriae extend individually from the surface of the bacterium, although sometimes they lie so closely that they appear to form bundles. Tip fibrillae typical of this class of fimbriae extend from the distal ends of several fimbriae (arrowheads). The diameter of the tip fibrillum is roughly half the diameter of the main shaft and is composed of the PapG adhesin subunit at the tips bound to several PapE subunits and finally to a PapK subunit to anchor the fibrillum to the PapA subunits that make up the primary 7-nm fimbrial shaft. The lengths of the tip fibrillae vary (see insets for two examples), but they are normally much longer on P fimbriae than on type 1 fimbriae. Magnifications: ×60,000 (A) and ×200,000 (B and C) (see also Fig. 4.20).

tinct mechanisms for the biogenesis of fimbriae (Table 4.4.), which do not completely correspond to the morphological classification given above. The most widely used mechanism for fimbrial biogenesis appears to be the chaperone-usher pathway. This pathway is shared by many enterobacteria (e.g., *Salmonella* spp., *Klebsiella pneumo-*

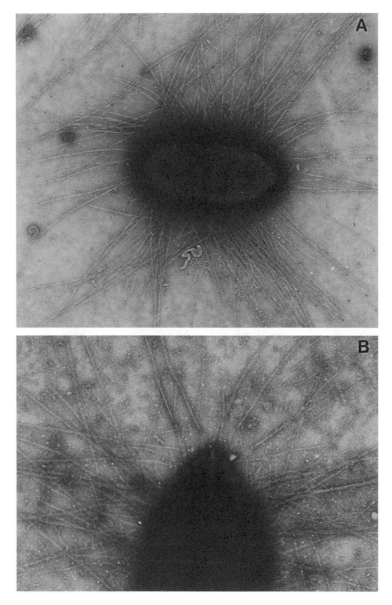

FIGURE 4.15 Electron micrographs of a negatively stained *E. coli* strain expressing 987P fimbriae. These fimbriae are also peritrichously arranged (A), but at high magnification their architecture is such that they are thinner and much more flexible than P or type 1 fimbriae (B). The *E. coli* strain was provided by Robert Edwards and Dieter Schifferli. Magnifications, ×19,125 (A) and ×25,500 (B).

niae, Yersinia spp., and *E. coli*) as well as some respiratory pathogens (e.g., *Haemophilus influenzae* and *Bordetella pertussis*). Of the many fimbriae that are built via this pathway, the most thoroughly characterized are the type 1 and P fimbriae of *E. coli*. P fimbriae are expressed on the surfaces of pyelonephritogenic *E. coli* (65, 75), while type 1 fimbriae are expressed on virtually all enterobacteria.

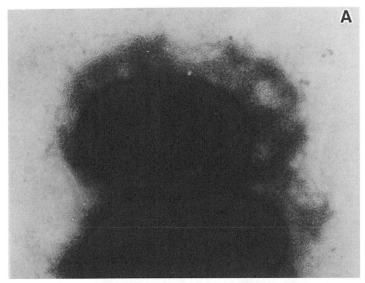

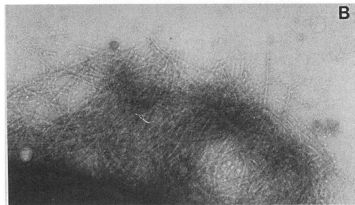

FIGURE 4.16 Electron micrographs of a negatively stained *Salmonella enterica* serovar Typhimurium strain expressing curli. These adhesins are long, thin, and much more flexible than even 987P fimbriae. (A) When cells are grown at 26°C, they produce a thick mat of curli surrounding the cell. (B) At higher magnification, individual filaments can be seen more easily. Magnifications, ×33,000 (A) and ×75,000 (B).

TYPE 1 AND P FIMBRIAL EXPRESSION AND BIOGENESIS

Although the genetic structure of the type 1 and P fimbrial operons and their regulation differ somewhat (Fig. 4.18 and 4.19) (see below), their assembly via the chaperone-usher pathway follows remarkably similar processes. Initial translocation of subunits across the cytoplasmic membrane occurs utilizing the general secretion (Sec) system (Fig. 4.20). Nascent subunit polypeptides are not immediately released into the periplasm by the Sec apparatus but are retained transiently at the cytoplasmic membrane through interactions of the membrane with their hydrophobic C termini. Periplasmic chaperones (i.e., PapD in the *pap* system and FimC in the *fim* system) rescue the subunits from their inner membrane hiatus, and they move into the periplasmic space as chaperone-subunit complexes. Interestingly, when investigators attempted to overexpress the FimH adhesin in the absence of the FimC chaperone, it was degraded in the periplasm (133). However, FimC-FimH complexes overproduced in the absence of other components of the fimbrial system are stable. Because the binding of chaperone and subunit leads to the masking of an interactive site on the subunit and prevents polymerization within the periplasm, these complexes can be overexpressed and purified from the periplasm in a soluble form, and they exhibit most of the expected binding characteristics. It is during the rescue

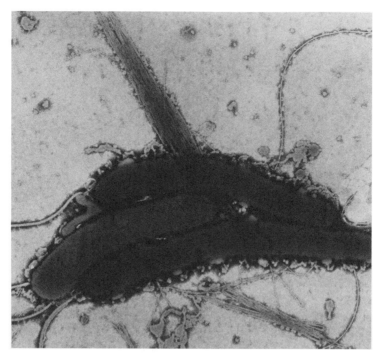

FIGURE 4.17 Electron micrograph of a negatively stained *V. cholerae* strain expressing type IV fimbriae. Type IV fimbriae are straight and rigid in appearance but form into large, easily identified bundles. In electron micrographs of *V. cholerae,* these bundles are typically seen broken off and separated from the cells, possibly due to the extreme forces applied to the large structures by surface tension of the medium or buffer as it dries during preparation of the electron microscope grid. Magnification, ×20,700.

of nascent subunits from their temporary dock in the inner membrane that initial folding of the subunits into an assembly-competent conformation occurs. The complexes are then transported through the periplasm toward the outer membrane, where an oligomeric struc-

TABLE 4.4 Mechanisms of biogenesis of gram-negative fimbriae[a]

Mechanism	Example
Chaperone-usher pathway	Thick, rigid: type 1 and P fimbriae of *E. coli*
	Thin, flexible: K99 fimbriae of *E. coli*
General secretion pathway	Type IV fimbriae of *N. gonorrhoeae*
Extracellular nucleation-precipitation pathway	Curli of *E. coli*
Alternate chaperone pathway	CS1 fimbriae of enterotoxigenic *E. coli*

[a]For more details, see reference 131.

ture composed of PapC or FimD subunits has formed. This oligomeric structure has come to be known as the outer membrane usher, although the analogy to an usher at the theater is not patently obvious. In any case, the pore of this oligomeric structure is apparently closed until a conformational change induced by the binding of chaperone-adhesin complex opens a 2- to 3-nm channel through which the polymerizing subunits are translocated. After contact with the chaperone-subunit complex, the outer membrane usher is involved in dissociating the assembly-competent subunits from the periplasmic chaperone, transporting them through the internal channel, and anchoring the developing fimbrial superstructure to the outer membrane. The order of assembly of subunits into the fimbriae is thought to be due both to the relative concentrations of subunits secreted into the

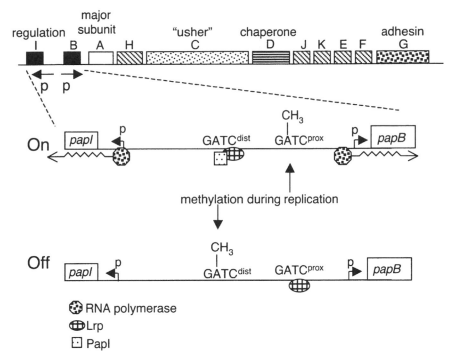

FIGURE 4.18 Schematic illustration of the locus for P-pilus expression and the mechanism of regulation. As indicated, *papI* and *papB* encode regulatory elements, *papA* encodes the major subunit (see Fig. 4.14), *papC* encodes the usher, *papD* encodes the periplasmic chaperone, *papE* and *papK* encode components of the tip fibrillum, and *papG* encodes the Gal(α1→4)Gal binding tip adhesin. Expression of fimbrial genes is regulated by methylation of proximal (GATC^{prox}) or distal (GATC^{dist}) nucleotide sequences in the region between the divergently transcribed *papI* and *papB* genes. This affects the binding of Lrp and hence gene expression, as described in the text and references cited.

periplasm and to the relative binding affinity of the individual chaperone-subunit complexes to the outer membrane usher. It is thought that the polymerizing subunits pass through the channel in the usher as approximately 2-nm linear structures and that final folding into the right-handed helix occurs as the subunits arrive on the bacterial surface. While mature type 1 fimbriae are very difficult to dissociate into monomeric subunits, they are very easily unraveled into 2-nm linear polymers by treatment with glycerol (Fig. 4.21) (2). Additional information regarding the structural details of these complicated interactions can be found elsewhere (12, 58–60, 62, 72, 121, 122, 131, 132).

REGULATION OF GRAM-NEGATIVE ADHESIN EXPRESSION

A remarkable feature of all surface adhesins is the ability of the bacteria to switch expression from one adhesin to another or to a nonadhesive phenotype. In fact, a particular adhesin may be essential only at a certain stage of the infectious process at a specific site whereas another adhesin will be essential at a different site and stage. Nonadhesive phenotypes have their own role to play, such as in avoiding recognition by phagocytes or allowing the bacteria to shed and colonize a new host. It is not surprising that complicated regulatory mechanisms have evolved to switch each adhesin "on" or "off" at relatively high fre-

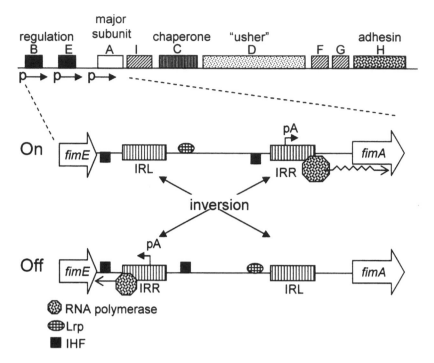

FIGURE 4.19 Schematic illustration of the locus for type 1 fimbria expression and the mechanism of regulation. As indicated, *fimB* and *fimE* encode regulatory elements, *fimA* encodes the major subunit, *fimC* encodes the periplasmic chaperone, *fimD* encodes the outer membrane usher, *fimF* and *fimG* encode components of the tip fibrillum, and *fimH* encodes the mannose binding tip adhesin. Expression of fimbrial genes is regulated by the inversion of a DNA segment located between *fimE* and *fimA* and containing the *fimA* promoter. In the "on" orientation, *fimAICDFGH* is successfully transcribed. In the "off" orientation, the transcript is aborted, as described in the text and references cited.

quencies (i.e., at least 2 orders of magnitude higher frequencies than random mutation). While such switching is commonly influenced by the environment, for a number of adhesin genes expression is restricted to a fraction of the population irrespective of the environment. In other words, expression is regulated on two different levels. One level, discussed above for gram-positive bacteria, is directed and is due to the sensing of environmental cues and the transcription of specific regulatory elements, because bacteria are exposed to inhospitable environments. The other level, described best for a number of gram-negative bacteria, is a random switching of adhesin genes on and off, because the bacterial population is also exposed to unpredictable environments and never "knows" what substrata and resources the next niche will offer. Thus,

facing such uncertainties, having cells in both on and off phases undoubtedly offers the bacterial population a survival advantage in the diverse situations it will encounter (19). These regulatory mechanisms are diverse and have been described in detail for only a relatively few adhesins; incomplete information is available for many others. The mechanisms of the regulation of type P and 1 fimbrial adhesins (Table 4.5) are described in more detail because they are the best characterized of the regulatory mechanisms for gram-negative bacterial adhesins. Variation of the PII OMP adhesin of *N. gonorrhoeae* is also described, as an unusual example of phase variation mechanisms.

Type P fimbria regulation. The mechanisms underlying the regulation of P

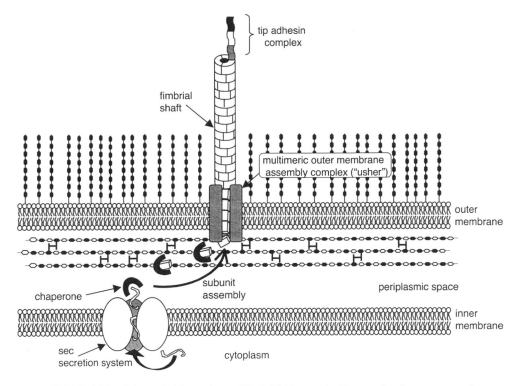

FIGURE 4.20 Schematic illustration of fimbrial biogenesis. The mechanisms are remarkably similar for type P and 1 fimbriae, the best-studied examples of fimbrial biogenesis. Nascent polypeptides of fimbrial subunits are transported across the inner membrane by the general secretion (type II) pathway. The periplasmic chaperone binds to the polypeptide as it is being transported, and the chaperone aids in folding the subunits into a polymerization-competent conformation. The chaperone-subunit complexes arrive at the outer membrane usher, where they bind to previously delivered subunits, traverse the outer membrane as a ~2-nm-diameter linear filament, and then coil into a helical form at the external face of the usher. In this pathway, the translocation of subunits is highly ordered, with translocation of tip adhesin being followed by that of other tip fibrillum subunits and, finally, the major subunit protein.

fimbriae are dependent on DNA methylation by the deoxyadenosine methylase (Dam) and the leucine response regulatory protein (Lrp). Lrp is a global regulatory protein that is involved in transcriptional activation and repression of a wide variety of genes (24). Dam mediates the transfer of a methyl group from S-adenosine-L-methionine to the adenine of the target sequence 5'-GATC-3' (78). In the *pap* operon (Fig. 4.18), transcription from the promoter of *papB* and *papA* is essential for the expression of P fimbriae. Lrp may activate or repress transcription from this promoter depending on which of two Lrp bind-ing regions is occupied. When Lrp binds at the distal binding site, it functions as an activator of transcription and, therefore, the fimbrial transcription is "on." When the Lrp binds at the proximal site that overlaps the promoter, it functions as a repressor of transcription, and so fimbrial transcription is "off" (137). Therefore, the switching between off and on phases in transcription of the *pap* fimbriae is due to specific location of Lrp binding in the regulatory region of the *pap* operon (Fig. 4.18). The methylation pattern of certain *pap* GATC sequences is essential for the phase variation because it affects Lrp binding. This

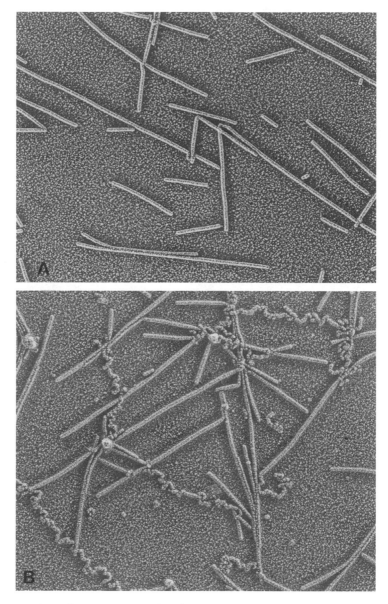

FIGURE 4.21 Electron micrographs of rotary shadowed, purified type 1 fimbriae. (A) Purified type 1 fimbriae bound to mica in water. (B) Purified type 1 fimbriae bound to mica after a brief incubation in a 50% glycerol solution. Glycerol does not dissociate subunits into monomers but instead unravels the fimbrial helix. This image illustrates an early stage in the unraveling process, but eventually the 7-nm helical structure is converted into filaments of roughly 2 nm. Magnification, ×49,500.

effect is due to the presence of the GATC sequence in the proximal and distal regions of the promoter. The GATC sequence that is not occupied by the Lrp becomes susceptible to methylation. In a strain where the distal GATC site is mutated so that it is permanently nonmethylated, *pap* expression is locked in the on phase, suggesting that the distal GATC must be nonmethylated for translocation of Lrp to the upstream binding

TABLE 4.5 Examples of phase variation of gram-negative fimbria expression[a]

Fimbrial operon	Mechanism	Target DNA
E. coli pap	DAM-dependent methylation	CH_3
		On phase GATC proximal
		CH_3
		Off phase GATC distal
E. coli fim	fimB- and fimE-mediated DNA inversion	On phase 5'-TTGGGCCA-3'
		Off phase 5'-ACCGGGTT-3'
N. gonorrhoeae pil	Recombination with partial pilS	On phase pilS in frame
		Off phase pilS out of frame
N. gonorrhoeae opa	Insertion-deletion of nucleotide repeats	On 3 CTCTT repeats in promoter
		Off 7 CTCTT repeats in promoter

[a]See the text for more details.

site to result in the on phase. Hence, the affinity of Lrp binding is decreased if the GATC distal site is methylated. Therefore, the switch to the on phase is thought to occur immediately following DNA replication, when the DNA including the pap GATC sequence is temporarily methylated. During replication, either the distal or the proximal GATC sequence is methylated. Because Lrp binds only nonmethylated GATC and because the differential methylation of the distal and proximal GATC occurs at random, switching from on to off is a random event that occurs at a frequency of approximately 1 of 100 to 1,000 daughter cells.

The adhesive OMP Ag43, recently described as being involved in adhesion and biofilm formation in E. coli, also undergoes phase variation in a Dam-dependent manner (52). In this case, the methylation of GATC is dependent on the product of the oxyR gene. OxyR in the oxidized form binds to the GATC-containing region and prevents methylation and transcription of the Ag43 protein.

Type 1 fimbria regulation. The regulation of phase variation in type 1 fimbriae is controlled primarily by a phase switch, a 314-nucleotide invertible segment of DNA immediately upstream of the fimA gene encoding the major subunit (Fig. 4.19) (1). This element includes a promoter that directs the transcription of fimAICDFGH when it is in the on orientation. Two regulatory genes at

the 5' end of the cluster, fimB and fimE, encode proteins that resemble the lambda integrase family of site-specific recombinases (38, 67) and affect the rate of DNA inversion. FimE promotes inversion to the off orientation preferentially, whereas FimB promotes inversion in both directions, with minimal preference for off-to-on inversion (47, 81). Several other accessory proteins bind to the fim switch region, resulting in reorientation of the DNA into a conformation more favorable for inversion. These accessory proteins include integration host factor (20, 38, 41), leucine-responsive regulatory protein (21, 48), and the histone-like protein (66, 104, 123). It has also been reported that wild-type E. coli possesses a variety of mutations within or adjacent to the fim switch that affect phase variation to different degrees (73).

Expression of type 1 fimbriae, as well as several other virulence factors, can also be affected in certain strains of E. coli by the excision of pathogenicity islands (PAIs) (53). The effects of the leucine-specific tRNA locus, leuX, on type 1 fimbria production were first indicated by Newman et al. (92), and PAI II of the uropathogenic E. coli strain 536 is inserted at this locus. Because excision of PAI II destroys the leuX locus, genes containing the rare leucine codon TTG, which is recognized specifically by the leuX product, $tRNA_5^{Leu}$, are not transcribed. Because fimB has five such codons (fimE has two), the interruption of leuX has a dramatic effect on type 1 fimbria production, as has been elegantly shown for the uropathogenic *E. coli* strain 536 (113).

Recently, it was also observed that the expression of type 1 fimbriae can trigger changes in the expression of other genes within the cell (124, 125). The massive disulfide bond formation associated with fimbrial expression may represent a novel signal transduction pathway that affects the oxidation state of the cellular redox sensor, OxyR, in turn affecting additional genes regulated by this protein.

It is interesting that there is also regulation via cross talk between different fimbrial gene clusters. For instance, regulatory elements of both P and S fimbrial gene clusters downregulate the expression of genes necessary for type 1 fimbrial expression (57, 139). Undoubtedly, there are other phenomena yet to be described that will add to the complexities of fimbrial regulation and expression.

PII outer membrane protein variation. Phase variation of the PII outer membrane protein of *N. gonorrhoeae* is mediated by random insertion or deletion of CTCTT nucleotide repeats within the leader sequence of the *opa* genes. There are about 12 or 13 PII gene loci, designated *opaE1, opaE2,* etc., scattered in the *N. gonorrhoeae* chromosome (see chapter 6). Translation of the transcript into mature PII protein is dependent on the number of CTCTT repeats in the 5′ regions of each of the *opaE* genes. Hence, whether mature PII protein is synthesized and expressed in the outer membrane or whether the protein is prematurely truncated depends on the number of copies. This will result in either an in-frame or an out-of-frame transcript. At a relatively high frequency, the number of CTCTT repeats can change by inserting or deleting one or more repeats. The mechanism of insertion or deletion is not known, but it is certainly responsible for the phase variation in the synthesis of this protein.

Because the ability of bacteria to regulate adhesin expression is essential for them to survive in various niches of the host, it may be possible to develop antimicrobial drugs which

target specific constituents of the complex regulatory systems. Initial studies toward this goal have recently been described for *S. aureus* (7, 8, 10, 11). Short peptides that inhibit quorum-sensing mechanisms were found to inhibit *S. aureus* adhesion to epithelial cells as well as biofilm formation on medical device plastics. This is an interesting and promising new direction in the efforts to develop antiadhesion therapies for infectious diseases. This theme is addressed in more detail in chapter 11.

REFERENCES

1. **Abraham, J. M., C. S. Freitag, J. R. Clements, and B. I. Eisenstein.** 1985. An invertible element of DNA controls phase variation of type 1 fimbriae of *Escherichia coli. Proc. Natl. Acad. Sci. USA* **82:**5724–5727.
2. **Abraham, S. N., M. Land, S. Ponniah, R. Endres, D. L. Hasty, and J. P. Babu.** 1992. Glycerol-induced unraveling of the tight helical conformation of *Escherichia coli* type 1 fimbriae. *J. Bacteriol.* **174:**5145–5148.
3. **Andersson, G., K. McIver, L. O. Hedén, and J. R. Scott.** 1996. Complementation of divergent *mga* genes in group A *Streptococcus. Gene* **175:**77–81.
4. **Arvidson, S., and K. Tegmark.** 2001. Regulation of virulence determinants in *Staphylococcus aureus. Int. J. Med. Microbiol.* **291:**159–170.
5. **Baker, N. R., V. Minor, C. Deal, M. S. Shahrabadi, D. A. Simpson, and D. E. Woods.** 1991. *Pseudomonas aeruginosa* exoenzyme S is an adhesin. *Infect. Immun.* **59:**2859–2863.
6. **Balaban, N., and R. P. Novick.** 1995. Autocrine regulation of toxin synthesis by *Staphylococcus aureus. Proc. Natl. Acad. Sci. USA* **92:**1619–1623.
7. **Balaban, N., T. Goldkorn, R. T. Nhan, L. B. Dang, S. Scott, R. M. Ridgley, A. Rasooly, S. C. Wright, J. W. Larrick, and J. R. Carlson.** 1998. Autoinducer of virulence as a target for vaccine and therapy against *Staphylococcus aureus. Science* **280:**438–440.
8. **Balaban, N., L. V. Collins, J. S. Cullor, E. B. Hume, E. Medina-Acosta, O. Vieira da Motta, R. O'Callaghan, P. V. Rossitto, M. E. Shirtliff, L. Serafim da Silveira, A. Tarkowski, and J. V. Torres.** 2000. Prevention of diseases caused by *Staphylococcus aureus* using the peptide RIP. *Peptides* **21:**1301–1311.
9. **Balaban, N., T. Goldkorn, Y. Gov, M.**

Hirshberg, N. Koyfman, H. R. Matthews, R. T. Nhan, B. Singh, and O. Uziel. 2001. Regulation of *Staphylococcus aureus* pathogenesis via target of RNAIII-activating protein (TRAP). *J. Biol. Chem.* **276:**2658–2667.

10. Balaban, N., Y. Gov, A. Bitler, and J. R. Boelaert. 2003. The quorum-sensing inhibitor RIP prevents adherence and biofilm formation of *Staphylococcus aureus* to human keratinocytes and dialysis catheters. *Kidney Int.* **63:**340–345.

11. Balaban, N., O. Cirioni, M. S. del Prete, R. Ghiselli, A. Giacometti, Y. Gov, F. Mocchegiani, V. Saba, G. Scalise, C. Viticchi, and G. Dell'Acqua. 2003. Use of the quorum-sensing inhibitor RNAIII inhibiting peptide to prevent biofilm formation *in vivo* by drug resistant *Staphylococcus aureus*. *J. Infect. Dis.* **187:**625–630.

12. Barnhart, M. M., J. S. Pinkner, G. E. Soto, F. G. Sauer, S. Langermann, G. Waksman, C. Frieden, and S. J. Hultgren. 2000. PapD-like chaperones provide the missing information for folding of pilin proteins. *Proc. Natl. Acad. Sci. USA* **97:**7709–7714.

13. Beckert, S., B. Kreikemeyer, and A. Podbielski. 2001. Group A streptococcal *rofA* gene is involved in control of several virulence genes and eukaryotic cell attachment and internalization. *Infect. Immun.* **69:**534–537.

14. Bernish, B., and I. van de Rijn. 1999. Characterization of a two-component system in *Streptococcus pyogenes* which is involved in regulation of hyaluronic acid production. *J. Biol. Chem.* **274:**4786–4793.

15. Beveridge, T. J. 2000. Ultrastructure of gram-positive cell walls, p. 3–10. *In* V. A Fischetti, R. P. Novick, J. J. Ferretti, D. A. Portnoy, and J. I. Rood (ed.), *Gram-Positive Pathogens*. ASM Press, Washington, D.C.

16. Beveridge, T. J., P. H. Pouwels, M. Sara, A. Kotiranta, K. Lounatmaa, K. Kari, E. Kerosuo, M. Haapasalo, E. M. Egelseer, I. Schocher, W. B. Sleytr, L. Morelli, M.-L. Callegari, J. F. Nomellini, W. H. Bingle, J. Smit, E. Leibovitz, M. Lemaire, I. Miras, S. Salamitou, P. Beguin, H. Ohayon, P. Gounon, M. Matuschek, K. Sahm, H. Bahl, R. Grogono-Thomas, J. Dworkin, M. J. Blaser, R. M. Woodland, D. G. Newell, M. Kessel, and S. F. Koval. 1997. Functions of S-layers. *FEMS Microbiol. Rev.* **20:**99–149.

17. Bian, Z., and S. Normark. 1997. Nucleator function of CsgB for the assembly of adhesive surface organelles in *Escherichia coli*. *EMBO J.* **16:**5827–5836.

18. Bieber, D., S. W. Ramer, C.-Y. Wu, W. J. Murray, T. Tobe, R. Fernandez, and G. K. Schoolnik. 1998. Type IV pili, transient bacterial aggregates, and virulence of enteropathogenic *Escherichia coli*. *Science* **280:**2114–2118.

19. Blomfield, I., and M. van der Woude. 2002. Regulation and function of phase variation in *Escherichia coli*, p. 89–113. *In* M. Wilson (ed.), *Bacterial Adhesion to Host Tissues. Mechanisms and Consequences*. Cambridge University Press, Cambridge, United Kingdom.

20. Blomfield, I. C., M. S. McClain, J. A. Princ, P. J. Calie, and B. I. Eisenstein. 1991. Type 1 fimbriation and *fimE* mutants of *Escherichia coli* K-12. *J. Bacteriol.* **173:**5298–5307.

21. Blomfield, I. C., P. J. Calie, K. J. Eberhardt, M. S. McClain, and B. I. Eisenstein. 1993. Lrp stimulates phase variation of type 1 fimbriation in *Escherichia coli* K-12. *J. Bacteriol.* **175:**27–36.

22. Braun, V. 1975. Covalent lipoprotein from the outer membrane of *Escherichia coli*. *Biochim. Biophys. Acta* **415:**335–377.

23. Brinton, C. C., Jr. 1965. The structure, function, synthesis and genetic control of bacterial pili and a molecular model for DNA and RNA transport in Gram-negative bacteria. *Trans. N.Y. Acad. Sci.* **27:**1003–1054.

24. Calvo, J. M., and R. G. Matthews. 1994. The leucine-responsive regulatory protein, a global regulator of metabolism in *Escherichia coli*. *Microbiol. Rev.* **58:**466–490.

25. Camara, M., G. J. Boulnois, P. W. Andrew, and T. J. Mitchell. 1994. A neuraminidase from *Streptococcus pneumoniae* has the features of a surface protein. *Infect. Immun.* **62:**3688–3695.

26. Caparon, M. G., R. T. Geist, J. Perez-Casal, and J. R. Scott. 1992. Environmental regulation of virulence in group A streptococci: transcription of the gene encoding M protein is stimulated by carbon dioxide. *J. Bacteriol.* **174:**5693–5701.

27. Chaussee, M. S., G. L. Sylva, D. E. Sturdevant, L. M. Smoot, W. R. Graham, R. O. Watson, and J. M. Musser. 2002. Rgg influences the expression of multiple regulatory loci to coregulate virulence factor expression in *Streptococcus pyogenes*. *Infect. Immun.* **70:**762–770.

28. Chaussee, M. S., R. O. Watson, J. C. Smoot, and J. M. Musser. 2001. Identification of Rgg-regulated exoproteins of *Streptococcus pyogenes*. *Infect. Immun.* **69:**822–831.

29. Chen, C., N. Bormann, and P. P. Cleary.

1993. VirR and Mry are homologous trans-acting regulators of M protein and C5a peptidase expression in group A streptococci. *Mol. Gen. Genet.* **241**:685–693.

30. **Cheung, A. L., and G. Zhang.** 2002. Global regulation of virulence determinants in *Staphylococcus aureus* by the SarA protein family. *Front. Biosci.* **7**:1825–1842.

31. **Cheung, A. L., J. M. Koomey, C. A. Butler, S. J. Projan, and V. A. Fischetti.** 1992. Regulation of exoprotein expression in *Staphylococcus aureus* by a locus (*sar*) distinct from *agr. Proc. Natl. Acad. Sci. USA* **89**:6462–6466.

32. **Cheung, A. L., K. Schmidt, B. Bateman, and A. C. Manna.** 2001. SarS, a SarA homolog repressible by *agr*, is an activator of protein A synthesis in *Staphylococcus aureus. Infect. Immun.* **69**:2448–2455.

33. **Chien, C.-T., A. C. Manna, S. J. Projan, and A. L. Cheung.** 1999. SarA, a global regulator of virulencedeterminants in *Staphylococcus aureus*, binds to a conserved motif essential for *sar* dependent gene regulation. *J. Biol. Chem.* **274**:37169–37176.

34. **Collinson, S. K., L. Emödy, K. H. Muller, T. J. Trust, and W. W. Kay.** 1991. Purification and characterization of thin, aggregative fimbriae from *Salmonella enteritidis. J. Bacteriol.* **173**:4773–4781.

35. **Courtney, H. S., D. L. Hasty, Y. Li, H. C. Chiang, J. L. Thacker, and J. B. Dale.** 1999. Serum opacity factor is a major fibronectin-binding protein and a virulence determinant of M type 2 *Streptococcus pyogenes. Mol. Microbiol.* **32**:89–98.

36. **Courtney, H. S., S. Liu, J. B. Dale, and D. L. Hasty.** 1997. Conversion of M serotype 24 of *Streptococcus pyogenes* to M serotypes 5 and 18: effect on resistance to phagocytosis and adhesion to host cells. *Infect. Immun.* **65**:2472–2474.

37. **Dale, J. B., R. G. Washburn, M. B. Marques, and M. R. Wessels.** 1996. Hyaluronate capsule and surface M protein in resistance to opsonization of group A streptococci. *Infect. Immun.* **64**:1495–1501.

38. **Dorman, C. J., and C. F. Higgins.** 1987. Fimbrial phase variation in *Escherichia coli*: dependence on integration host factor and homologies to other site-specific recombinases. *J. Bacteriol.* **169**:3840–3843.

39. **Downer, R., F. Roche, P. W. Park, R. P. Mecham, and T. J. Foster.** 2002. The elastin-binding protein of *Staphylococcus aureus* (EbpS) is expressed at the cell surface as an integral membrane protein and not as a cell wall-associated protein. *J. Biol. Chem.* **277**:243–250.

40. **Edwards, N. J., M. A. Monteiro, G. Faller,** E. J. Walsh, A. P. Moran, I. S. Roberts, and N. J. High. 2000. Lewis X structures in the O antigen side-chain promote adhesion of *Helicobacter pylori* to the gastric epithelium. *Mol. Microbiol.* **35**:1530–1539.

41. **Eisenstein, B. I., D. S. Sweet, V. Vaughn, and D. I. Freidman.** 1987. Integration host factor is required for the DNA inversion that controls phase variation in *Escherichia coli. Proc. Natl. Acad. Sci. USA* **84**:6506–6510.

41a. **Elliott, D., E. Harrison, P. S. Handley, S. K. Ford, N. Mordan, E. Jaffrey, and R. McNab.** 2003. Prevalence of Csh-like fibrillar surface proteins among mitis group oral streptococci. *Oral Microbiol. Immunol.* **18**:1–7.

42. **Federle, M. J., K. S. McIver, and J. Scott.** 1999. a response regulator that represses transcription of several virulence operons in the group A streptococcus. *J. Bacteriol.* **181**:3649–3657.

43. **Fischetti, V. A.** 1989. Streptococcal M protein: molecular design and biological behavior. *Clin. Microbiol. Rev.* **2**:285–314.

44. **Fischetti, V. A.** 2000. Surface proteins on gram-positve bacteria, p. 11–24. *In* V. A. Fischetti, R. P. Novick, J. J. Ferretti, D. A. Portnoy, and J. I. Rood (ed.), *Gram-Positive Pathogens.* ASM Press, Washington, D.C.

45. **Fogg, G. C., and M. G. Caparon.** 1997. Constitutive expression of fibronectin binding in *Streptococcus pyogenes* as a result of anaerobic activation of *rofA. J. Bacteriol.* **179**:6172–6180.

46. **Gaillard, J. L., P. Berche, C. Frehel, E. Gouin, and P. Cossart.** 1991. Entry of *L. monocytogenes* into cells is mediated by internalin, a repeat protein reminiscent of surface antigens from gram-positive cocci. *Cell* **65**:1127–1141.

47. **Gally, D. L., J. Leathart, and I. C. Blomfield.** 1996. Interaction of FimB and FimE with the *fim* switch that controls the phase variation of type 1 fimbriation in *Escherichia coli. Mol. Microbiol.* **21**:725–738.

48. **Gally, D. L., T. J. Rucker, and I. C. Blomfield.** 1994. The leucine-responsive regulatory protein binds to the *fim* switch to control phase variation of type 1 fimbrial expression in *Escherichia coli* K-12. *J. Bacteriol.* **176**:5665–5672.

49. **Gov, Y., A. Bitler, G. Dell'Acqua, J. V. Torres, and N. Balaban.** 2001. RNAIII inhibiting peptide (RIP), a global inhibitor of *Staphylococcus aureus*: structure and function analysis. *Peptides* **22**:1609–1620.

50. **Graham, M. R., L. M. Smoot, C. A. Lux Migliaccio, K. Virtaneva, D. E. Sturdevant, S. F. Porcella, M. J. Federle, G. J.**

Adams, J. R. Scott, and J. M. Musser. 2002. Virulence control in group A *Streptococcus* by a two-component gene regulatory system: global expression profiling and *in vivo* infection modeling. *Proc. Natl. Acad. Sci. USA* **99:** 13855–13860.

51. Granok, A. B., D. Parsonage, R. P. Ross, and M. G. Caparon. 2000. The RofA binding site in *Streptococcus pyogenes* is utilized in multiple transcriptional pathways. *J. Bacteriol.* **182:**1529–1540.

52. Haagmans, W., and M. van der Woude. 2000. Phase variation of Ag43 in *E. coli:* Dam-dependent methylation abrogates OxyR binding and OxyR-mediated repression of transcription. *Mol. Microbiol.* **35:**877–887.

53. Hacker, J., G. Blum-Oehler, I. Muhldorfer, and H. Tschape. 1997. Pathogenicity islands of virulent bacteria: structure, function and impact on microbial evolution. *Mol. Microbiol.* **23:**1089–1097.

54. Hanski, E., and M. Caparon. 1992. Protein F, a fibronectin-binding protein, is an adhesin of the group A streptococcus *Streptococcus pyogenes. Proc. Natl. Acad. Sci. USA* **89:**6172–6176.

55. Hedén, L. O., E. Frithz, and G. Lindahl. 1991. Molecular characterization of an IgA receptor from group B streptococci: sequence of the gene, identification of a proline-rich region with unique structure and isolation of N-terminal fragments with IgA-binding capacity. *Eur J. Immunol.* **21:**1481–1490.

56. Heinrichs, J. H., M. G. Bayer, and A. L. Cheung. 1996. Characterization of the *sar* locus and its interaction with *agr* in *Staphylococcus aureus. J. Bacteriol.* **178:**418–423.

57. Holden, N. J., B. E. Uhlin, and D. L. Gally. 2001. PapB paralogues and their effect on the phase variation of type 1 fimbriae in *Escherichia coli. Mol. Microbiol.* **42:**319–330.

58. Hung, D. L., J. S. Pinkner, S. D. Knight, and S. J. Hultgren. 1999. Structural basis of chaperone self-capping in P pilus biogenesis. *Proc. Natl. Acad. Sci. USA* **96:**8178–8183.

59. Jacob-Dubuisson, F., J. Heuser, K. Dodson, S. Normark, and S. Hultgren. 1993. Initiation of assembly and association of the structural elements of a bacterial pilus depend on two specialized tip proteins. *EMBO J.* **12:**837–847.

60. Jacob-Dubuisson, F., M. Kuehn, and S. J. Hultgren. 1993. A novel secretion apparatus for the assembly of adhesive bacterial pili. *Trends Microbiol.* **1:**50–55.

61. Jerlström, P. G., S. R. Talay, P Valentin-Weigand, K. N. Timmis, and G. S. Chhatwal. 1996. Identification of an immuno-globulin A binding motif located in the beta-antigen of the c protein complex of group B streptococci. *Infect. Immun.* **64:**2787–2793.

62. Jones, C. H., J. S. Pinkner, R. Roth, J. Heuser, A. V. Nicholes, S. N. Abraham, and S. J. Hultgren. 1995. FimH adhesin of type 1 pili is assembled into a fibrillar tip structure in the *Enterobacteriaceae. Proc. Natl. Acad. Sci. USA* **92:**2081–2085.

63. Jonquieres, R., H. Bierne, F. Fiedler, P. Gounon, and P. Cossart. 1999. Interaction between the protein InlB of *Listeria monocytogenes* and lipoteichoic acid: a novel mechanism of protein association at the surface of gram-positive bacteria. *Mol. Microbiol.* **34:**902–914.

64. Jönsson, K., C. Signas, H.-P. Muller, and M. Lindberg. 1991. Two different genes encode fibronectin binding proteins in *Staphylococcus aureus*—the complete nucleotide sequence and characterization of the 2nd gene. *Eur. J. Biochem.* **202:**1041–1048.

65. Kallenius, G., S. B. Svenson, H. Hultberg, R. Mollby, I. Helin, B. Cedergren, and J. Winberg. 1981. Occurrence of P fimbriated *Escherichia coli* in urinary tract infection. *Lancet* **ii:**1369–1372.

66. Kawula, T. H., and P. E. Orndorff. 1991. Rapid site-specific DNA inversion in *Escherichia coli* mutants lacking the histone-like protein H-NS. *J. Bacteriol.* **173:**4116–4123.

67. Klemm, P. 1986. Two regulatory *fim* genes, *fimB* and *fimE*, control the phase variation of type 1 fimbriae in *Escherichia coli. EMBO J.* **5:**1389–1393.

68. Koch, A. L., M. L. Higgins, and R. J. Doyle. 1982. The role of surface stress in the morphology of microorganisms. *J. Gen. Microbiol.* **128:**927–945.

69. Kreikemeyer, B., M. D. P. Boyle, B. A. Buttaro, M. Heinemann, and A. Podbielski. 2001. Group A streptococcal growth phase-associated virulence factor regulation by a novel operon (Fas) with homologies to two-component-type regulators requires a small RNA molecule. *Mol. Microbiol.* **39:**392–406.

70. Kreikemeyer, B., S. Beckert, A. Braun-Kiewnick, and A. Podbielski. 2002. Group A streptococcal RofA-type global regulators exhibit strain-specific genomic presence and regulation pattern. *Microbiology* **148:**1501–1511.

71. Kreikemeyer, B., S. R. Talay, and G. S. Chhatwal. 1995. Characterization of a novel fibronectin-binding surface protein in group A streptococci. *Mol. Microbiol.* **17:**137–145.

72. Kuehn, M. J., J. Heuser, S. Normark, and S. J. Hultgren. 1992. P pili in uropathogenic

E. coli are composite fibres with distinct fibrillar adhesive tips. *Nature* **356:**252–255.

73. **Leathart, J. B., and D. L. Gally.** 1998. Regulation of type 1 fimbrial expression in uropathogenic *Escherichia coli:* heterogeneity of expression through sequence changes in the *fim* switch region. *Mol. Microbiol.* **28:**371–381.

74. **Levin, J. C., and M. R. Wessels.** 1998. Identification of *csrR/csrS,* a genetic locus that regulates hyaluronic acid capsule synthesis in group A *Streptococcus. Mol. Microbiol.* **30:**209–219.

75. **Lomberg, H., L. A. Hansson, B. J. Jacobsson, U. Jodal, H. Leffler, and C. Svanborg-Eden.** 1983. Correlation of P blood group, vesicoureteral refleux and bacterial attachment in patients with recurrent pyelonephritis. *N. Engl. J. Med.* **308:**1189–1192.

76. **Lowy, F. D.** 1998. *Staphylococcus aureus* infections. *N. Engl. J. Med.* **339:**520–532.

77. **Lyon, W. R., C. M. Gibson, and M. G. Caparon.** 1998. A role for trigger factor and an rgg-like regulator in the transcription, secretion and processing of the cysteine proteinase of *Streptococcus pyogenes. EMBO J.* **17:**6263–6275.

78. **Marinus, M. G.** 1996. Methylation of DNA, p. 782–791. *In* F. C. Neidhardt, R. Curtiss III, J. L. Ingraham, E. C. C. Lin, K. B. Low, B. Magasanik, W. S. Reznikoff, M. Riley, M. Schaechter, and H. E. Umbarger (ed.), Escherichia coli *and* Salmonella: *Cellular and Molecular Biology,* 2nd ed. ASM Press, Washington, D.C.

79. **Mazmanian, S. K., G. Liu, E. R. Jensen, E. Lenoy, and O. Schneewind.** 2000. *Staphylococcus aureus* sortase mutants defective in the display of surface proteins and in the pathogenesis of animal infections. *Proc. Natl. Acad. Sci. USA* **97:**5510–5515.

80. **Mazmanian, S. K., G. Liu, H. Ton-That, and O. Schneewind.** 1999. *Staphylococcus aureus* sortase, an enxyme that anchors surface proteins to the cell wall. *Science* **285:**760–763.

81. **McClain, M. S., I. C. Blomfield, and B. I. Eisenstein.** 1991. Roles of *fimB* and *fimE* in site-specific inversion associated with phase variation of type 1 fimbriae in *Escherichia coli. J. Bacteriol.* **173:**5308–5314.

82. **McDevitt, D., P. Francois, P. Vaudaux, and T. J. Foster.** 1994. Molecular characterization of the clumping factor (fibrinogen receptor) of *Staphylococcus aureus. Mol. Microbiol.* **11:**237–248.

83. **McIver, K. S., A. S. Heath, B. D. Green, and J. R. Scott.** 1995. Specific binding of the activator Mga to promoter sequences of the *emm* and *scpA* genes in the group A streptococcus. *J. Bacteriol.* **177:**6619–6624.

84. **McIver, K. S., and J. R. Scott.** 1997. Role of *mga* in growth phase regulation of virulence genes of the group A streptococcus. *J. Bacteriol.* **179:**5178–5187.

85. **McIver, K. S., A. S. Thurman, and J. R. Scott.** 1999. Regulation of *mga* transcription in the group A streptococcus: specific binding of Mga within its own promoter and evicence for a negative regulator. *J. Bacteriol.* **181:**5373–5383.

86. **McIver, K. S., and R. L. Myles.** 2002. Two DNA-binding domains of Mga are required for virulence gene activation in the group A streptococcus. *Mol. Microbiol.* **43:**1591–1601.

86a. **McNab, R., H. Forbes, P. S. Handley, D. M. Loach, G. W. Tannock, and H. F. Jenkinson.** 1999. Cell wall-anchored CshA polypeptide (259 kilodaltons) in *Streptococcus gordonii* forms surface fibrils that confer hydrophobic and adhesive properties. *J. Bacteriol.* **181:**3087–3095.

87. **McNab, R., P. S. Handley, and H. F. Jenkinson.** 2002. Adhesive surface structures of oral streptococci, p. 59–88. *In* M. Wilson (ed.), *Bacterial Adhesion to Host Tissues. Mechanisms and Consequences.* Cambridge University Press, Cambridge, United Kingdom.

88. **Molinari, G., M. Rhode, S. R. Talay, G. S. Chhatwal, S. Beckert, and A. Podbielski.** 2001. The role played by the group A streptococcal negative regulator Nra on bacterial interactions with epithelial cells. *Mol. Microbiol.* **40:**99–114.

89. **Morfeldt, E., D. Taylor, A. von Gabain, and S. Arvidson.** 1995. Activation of alpha-toxin translation in *Staphylococcus aureus* by the trans-encoded antisense RNA, RNAIII. *EMBO J.* **14:**4569–4577.

90. **Myhre, E. B., and G. Kronvall.** 1977. Demonstration of specific binding sites for human serum albumin in group C and G streptococci. *Infect. Immun.* **17:**475–482.

91. **Navarre, W. W., and O. Schneewind.** 1999. Surface proteins of gram-positive bacteria and mechanisms of their targeting to the cell wall envelope. *Microbiol. Mol. Biol. Rev.* **63:**174–229.

92. **Newman, J. V., R. L. Burghoff, L. Pallesen, K. A. Krogfelt, C. S. Kristensen, D. C. Laux, and P. S. Cohen.** 1994. Stimulation of *Escherichia coli* F-18 Col-type 1 fimbriae synthesis by *leuX. Infect. Immun.* **22:**247–254.

93. **Ni Eidhin, D., S. Perkins, P. Freancois, P. Vaudaux, M. Höök, and T. J. Foster.** 1998.

Clumping factor B (ClfB), a new surface-located fibrinogen-binding adhesin of *Staphylococcus aureus*. *Mol. Microbiol.* **32:**245–257.

94. **Nikaido, H.** 1985. Molecular basis of bacterial outer membrane permeability. *Microbiol. Rev.* **49:**1–32.

95. **Nikaido, H.** 1996. Outer membrane, p. 29–47. In F. C. Neidhardt, R. Curtiss III, J. L. Ingraham, E. C. C. Lin, K. B. Low, B. Magasanik, W. S. Reznikoff, M. Riley, M. Schaechter, and H. E. Umbarger (ed.), Escherichia coli and Salmonella: *Cellular and Molecular Biology,* 2nd ed. ASM Press, Washington, D.C.

96. **Nilsson, M., L. Frykberg, J. I. Flock, L. Pei, M. Lindberg, and G. Guss.** 1998. A fibrinogen-binding protein of *Staphylococcus epidermidis.* *Infect. Immun.* **66:**2666–2673.

97. **Novick, R. P.** 2000. Pathogenicity factors and their regulation, p. 392–407. In V. A. Fischetti, R. P. Novick, J. J. Ferretti, D. A. Portnoy, and J. I. Rood (ed.), Gram Positive Pathogens. ASM Press, Washington, D.C.

98. **Novick, R. P., H. F. Ross, S. J. Projan, J. Kornblum, B. Kreiswirth, and S. Moghazeh.** 1993. Synthesis of staphylococcal virulence factors is controlled by a regulatory RNA molecule. *EMBO J.* **12:**3967–3975.

99. **Novick, R. P., S. Projan, J. Kornblum, H. Ross, B. Kreiswirth, and S. Moghazeh.** 1995. The agr P-2 operon: an autocatalytic sensory transduction system in *Staphyococcus aureus.* *Mol. Gen. Genet.* **248:**446–458.

100. **Ofek, I., J. Goldhar, Y. Keisari, and N. Sharon.** 1995. Nonopsonic phagocytosis of microorganisms. *Annu. Rev. Microbiol.* **49:**239–276.

101. **Ofek, I., W. A. Simpson, and E.H. Beachey.** 1982. Formation of molecular complexes between a structurally defined M-protein and acylated or deacylated lipoteichoic acid of *Streptococcus pyogenes.* *J. Bacteriol.* **149:**426–433.

102. **Olmsted, S. B., G. M. Dunny, S. I. Erlandsen, and C. L. Wells.** 1994. A plasmid-encoded surface protein on *Enterococcus faecalis* augments its internalization by cultured intestinal epithelial cells. *J. Infect. Dis.* **170:**1549–1556.

103. **Olsen, A., A. Jonsson, and S. Normark.** 1989. Fibronectin binding mediated by a novel class of surface organelles on *Escherichia coli.* *Nature* **338:**652–655.

104. **Olsen, P. B., and P. Klemm.** 1994. Localization of promoters in the *fim* gene cluster and the effect of H-NS on the transcription of *fimB* and *fimE.* *FEMS Microbiol. Lett.* **116:**95–100.

105. **Ørskov, I., and F. Ørskov.** 1990. Serologic classification of fimbriae, p. 71–90. In K. Jann and B. Jann (ed.), *Bacterial Adhesins.* Springer-Verlag KG, Berlin, Germany.

106. **Parge, H. E., K. T. Forest, M. J. Hickey, D. A. Christensen, E. D. Getzoff, and J. A. Tainer.** 1995. Structure of the fibre-forming protein pilin at 2.6 Å resolution. *Nature* **378:**32–38.

107. **Patti, J. M., H. Jonsson, B. Guss, L. M. Switalski, K. Wiberg, M. Lindberg, and M. Höök.** 1992. Molecular characterization and expression of a gene encoding a *Staphylococcus aureus* collagen adhesin. *J. Biol. Chem.* **267:**4766–4772.

108. **Perez-Casal, J., M. G. Caparon, and J. R. Scott.** 1991. Mry, a trans-acting positive regulator of the M protein gene of *Streptococcus pyogenes* with similarity to the receptor proteins of two-component regulatory systems. *J. Bacteriol.* **13:**2617–2624.

109. **Podbielski, A., A. Flosdorff, and J. Weber-Heynemann.** 1995. The group A streptococcal virR49 gene controls expression of four structural vir regulon genes. *Infect. Immun.* **63:**9–20.

110. **Podbielski, A., M. Woischnik, B. A. Leonard, and K. H. Schmidt.** 1999. Characterization of nra, a global negative regulator gene in group A streptococci. *Mol. Microbiol.* **31:**1051–1064.

111. **Provence, D. L., and R. Curtiss III.** 1992. Role of crl in avian pathogenic Escherichia coli: a knockout mutation of crl does not affect hemagglutination activity, fibronectin binding, or curli production. *Infect. Immun.* **60:**4460–4467.

112. **Rakonjac, J. V., J. C. Robbins, and V. A. Fischetti.** 1995. DNA sequence of the serum opacity factor of group A streptococci: identification of a fibronectin-binding repeat domain. *Infect. Immun.* **63:**622–631.

113. **Ritter, A., D. L. Gally, P. B. Olsen, U. Dobrindt, A. Friedrich, P. Klemm, and J. Hacker.** 1997. The Pai-associated leuX specific tRNA$_5^{Leu}$ affects type 1 fimbriation in pathogenic *Escherichia coli* by control of FimB recombinase expression. *Mol. Microbiol.* **25:**871–882.

114. **Rogers, H. J., H. R. Perkins, and J. B. Ward.** 1980. *Microbial Cell Walls and Membranes.* Chapman & Hall, London, United Kingdom.

115. **Rudel, T., D. Facius, R. Barten, I. Scheuerpflug, E. Nonnenmacher, and T. F. Meyer.** 1995. Role of pili and the phase-variable PilC protein in natural competence for transformation of *Neisseria gonorrhoeae.* *Proc. Natl. Acad. Sci. USA* **92:**7986–7990.

116. **Rudel, T., H. J. Boxberger, and T. F. Meyer.** 1995. Pilus biogenesis and epithelial cell adherence of *Neisseria gonorrhoeae pilC* double knock-out mutants. *Mol. Microbiol.* **17:** 1057–1071.

117. **Rudel, T., I. Scheuerpflug, and T. F. Meyer.** 1995. Neisseria PilC protein identified as type IV pilus tip-located adhesin. *Nature* **373:**357–359.

118. **Sakellaris, H., and J. R. Scott.** 1998. New tools in an old trade: CS1 pilus morphogenesis. *Mol. Microbiol.* **30:**681–687.

119. **Sakellaris, H., D. P. Balding, and J. R. Scott.** 1996. Assembly proteins of CS1 pili of enterotoxigenic *Escherichia coli*. *Mol. Microbiol.* **21:**529–541.

120. **Sato, Y., Y. Yamamoto, and H. Kizaki.** 1997. Cloning and sequence analysis of the *gbpC* gene encoding a novel glucan-binding protein of *Streptococcus mutans*. *Infect. Immun.* **65:**668–675.

121. **Sauer, F. G., M. Barnhart, D. Choudhury, S. D. Knight, G. Waksman, and S. J. Hultgren.** 2000. Chaperone-assisted pilus assembly and bacterial attachment. *Curr. Opin. Struct. Biol.* **10:**548–556.

122. **Saulino, E. T., E. Bullitt, and S. J. Hultgren.** 2000. Snapshots of usher-mediated protein secretion and ordered pilus assembly. *Proc. Natl. Acad. Sci. USA* **97:**9240–9245.

123. **Schembri, M. A., P. B. Olsen, and P. Klemm.** 1998. Orientation-selective enhancement by H-NS of the *Escherichia coli* type 1 fimbrial phase switch promoter. *Mol. Gen. Genet.* **259:**336–344.

124. **Schembri, M. A., and P. Klemm.** 2001. Coordinate gene regulation by fimbriae induced signal transduction. *EMBO J.* **20:**3074–3081.

125. **Schembri, M. A., M. Givskov, and P. Klemm.** 2002. An attractive surface: Gram negative bacterial biofilms. *Sci. S. T. K. E.* **132:**RE6.

126. **Shaw, C. E., and R. K. Taylor.** 1990. *Vibrio cholerae* O395 *tcpA* pilin gene sequence and comparison of predicted protein structural features to those of type 4 pilins. *Infect. Immun.* **58:**3042–3049.

127. **Shuttleworth, H. L., C. J. Duggleby, S. A. Jones, T. Atkinson, and N. P. Minton.** 1987. Nucleotide sequence analysis of the gene for protein A from *Staphylococcus aureus* Cowan 1 (NCTC8530) and its enhanced expression in *Escherichia coli*. *Gene* **58:**283–295.

128. **Signas, C., G. Raucci, K. Jönsson, P.-E. Lindgren, G. Anatharamaiah, M. Höök, and M. Lindberg.** 1989. Nucleotide sequence of the gene for a fibronectin-binding protein from *Staphylococcus aureus* and its use in the synthesis of biologically active peptides. *Proc. Natl. Acad. Sci. USA* **86:**697–703

129. **Simpson, W. J., D. LaPenta, C. Chen, and P. P. Cleary.** 1990. Coregulation of type 12 M protein and streptococcal C5a peptidase genes in group A streptococci: evidence for a virulence regulon controlled by the *virR* locus. *Infect. Immun.* **55:**2448–2455.

130. **Sokurenko, E. V., V. Chesnokova, D. E. Dykhuizen, I. Ofek, X.-R. Wu, K. A. Krogfelt, C. Struve, M. A. Schembri, and D. L. Hasty.** 1998. Pathogenic adaptation of *Escherichia coli* by natural variation of the FimH adhesin. *Proc. Natl. Acad. Sci. USA* **95:**8922–8926.

131. **Soto, G. E., and S. J. Hultgren.** 1999. Bacterial adhesins: common themes and variations in architecture and assembly. *J. Bacteriol.* **181:**1059–1071.

132. **St. Geme, J. W., III, J. S. Pinkner, G. P. Krasan, J. Heuser, E. Bullitt, A. L. Smith, and S. J. Hultgren.** 1996. *Haemophilus influenzae* pili are composite structures assembled via the HifB chaperone. *Proc. Natl. Acad. Sci. USA* **93:**11913–11918.

133. **Tewari, R., J. I. MacGregor, T. Ikeda, J. R. Little, S. J. Hultgren, and S. N. Abraham SN.** 1993. Neutrophil activation by nascent FimH subunits of type 1 fimbriae purified from the periplasm of *Escherichia coli*. *J. Biol. Chem.* **268:**3009–3015.

134. **Toleman, M., E. Aho, and M. Virji.** 2001. Expression of pathogen-like Opa adhesins in commensal *Neisseria*: genetic and functional analysis. *Cell. Microbiol.* **3:**33–44.

135. **Tuomanen, E.** 1986. Piracy of adhesins: attachment of superinfecting pathogens to respiratory cilia by secreted adhesins of *Bordetella pertussis*. *Infect. Immun.* **54:**905–908.

136. **Wentworth, J., F. E. Austin, N. Garber, N. Gilboa-Garber, C. Paterson, and R. J. Doyle.** 1991. Cytoplasmic lectins contribute to the adhesion of *Pseudomonas aeruginosa*. *Biofouling* **4:**99–104.

137. **Weyand, N., and D. Low.** 2000. Lrp is sufficient for the establishment of the phase OFF *pap* DNA methylation pattern and repression of *pap* transctiption in vitro. *J. Biol. Chem.* **275:**3192–3200.

138. **Wu, H., and P. M. Fives-Taylor.** 2001. Molecular strategies for fimbrial expression and assembly. *Crit. Rev. Oral Biol. Med.* **12:**101–115.

139. **Xia, Y., D. Gally, K. Forsman-Semb, and B. E. Uhlin.** 2000. Regulatory cross-talk between adhesin operons in *Escherichia coli*: inhibition of type 1 fimbriae expression by the PapB protein. *EMBO J.* **19:**1450–1457.

EMERGING CONCEPTS IN BACTERIAL ADHESION AND ITS CONSEQUENCES

5

There are many different conceptual areas within the overall field of bacterial adhesion. The field has grown to the extent that it is virtually impossible to discuss all topics in detail, but many of them have been discussed in excellent reviews and books published during the last 8 years (see, e.g., references 3–8, 11, and 24). A selection of some of these conceptual areas, or common themes, is listed in Table 5.1. Other areas either have not yet become sufficiently well established or have escaped our attention. Some themes are discussed briefly in this chapter, while others are covered in more detail in later chapters.

BIOLOGICAL FUNCTIONS OF ADHESION OTHER THAN MEDIATING ATTACHMENT

It is becoming more and more evident that adhesins can exhibit biological functions other than bacterial adhesion but that may nevertheless play a crucial role in the infectious process. These activities may occur either while the adhesin is still on the bacterial surface or after it is shed into the surrounding milieu. The concept that adhesins can have pathogenic activities other than in bacterial adhesion can be illustrated by the proinflammatory propterties of lipoteichoic acid (LTA).

LTAs are a family of related amphiphilic polymers consisting of a polyglycerol phosphate ester linked to a glycolipid. Structurally related LTAs are cell wall components of virtually all gram-positive bacteria (see chapter 4). LTAs are also secreted into growth media at substantial levels, which are dramatically increased in the presence of subinhibitory concentrations of β-lactam antibiotics. A number of studies have shown that although LTA is not pyrogenic, it exhibits a number of other biological properties that are reminiscent of those of endotoxin (lipopolysaccharide [LPS]). Although controversy about the proinflammatory properties of LTA existed in the field for a number of years, it has recently been shown that LTA purified under conditions that maintain D-alanine substituents induces a strong cytokine response in whole blood at concentrations of as little as 10 ng/ml, similar to the response to *Pseudomonas aeruginosa* LPS (15). It was also shown that LTAs with reduced D-alanine content (and thus more negatively charged) could induce a potent cytokine response from macrophages when the macrophage-bound LTA was cross-linked using either anti-LTA antibodies (9, 13) or a phage display-derived anti-LTA antibody fragment (9). Thus, LTA acts not only as an adhesin (10) but also as a major immuno-

TABLE 5.1 Selected common themes in the field of bacterial adhesion

Theme	Examples and comments
Biological functions of adhesins other than mediating attachment	Activation of innate immune system
Lectins as the primary type of adhesin	Oligosaccharides are infinitely more variable than proteins
Bacterial surface carbohydrates as adhesins for nonphagocytic cells	O antigen of LPS can serve as an adhesin for a number of gram-negative species
Interaction with cells of the immune system	Mechanisms for defending or defeating host systems and for inflammation dysregulation
Bacterial clones with multiple adhesins	Impediment to development of antiadhesion therapy
Diversification of receptor specificity[a]	Pathoadaptive mutations; commensals become more pathogenic (see chapter 6)
Entry into nonphagocytic cells[a]	Major mechanism of virulence in certain pathogens; escape from host defenses (see chapter 7)
Activation of nonphagocytic cells[a]	Proinflammatory properties of cytokine induction (see chapter 8)
Adhesion-dependent upregulation of bacterial genes[a]	Virulence factor expression is upregulated (see chapter 9)
Formation of biofilms[a]	Major problem for medical device-associated infections (see chapter 10)
Antiadhesion therapy[a]	Clinical trials of dietary antiadhesive components have proved successful (e.g., cranberry juice) (see chapter 11)

[a]Themes discussed in greater depth in the indicated chapters.

stimulatory component of gram-positive organisms. A few other examples of this phenomenon are listed in Table 5.2. Because some of these additional biological functions are postulated to be directly related to the development of disease, it is expected that therapies based on neutralizing these functions would protect at multiple levels.

LECTINS AS ADHESINS ON BACTERIAL SURFACES AND RECEPTORS ON ANIMAL CELLS

Lectins have been and continue to be the primary type of adhesin identified in bacteria (17, 21). Inhibition of adhesion by oligosaccharides and simple sugars continues to be a major method of identifying lectin adhesins and characterizing receptor specificities (see chapter 2). During the last decade, lectins have also been the focus for development of anti-adhesion therapies; they have been used either directly as vaccines (12) or as a target for oligosaccharide inhibitors (see chapter 11).

Although it has been known for some time that phagocytic cells, macrophages in particular, express lectins that bind specific carbohydrates on bacterial surfaces (17, 18), the

TABLE 5.2 Examples of biological functions of adhesins in infections other than mediating adhesion

Adhesin	Function	Postulated role in infection	Reference[a]
LPS	Stimulation of macrophages and other cells of the immune system	Shock	20
Staphylococcus aureus LTA	Stimulation of macrophages and other cells of immune system	Shock	15
Bordetella pertussis filamentous hemagglutinin	Induction of inflammation and apoptosis	Undefined	2
Vibrio cholerae protease-hemagglutinin	Proteolytic activity	Undefined	16

[a]References are for the adhesin's roles in infection. The roles of these molecules in adhesion can be found in chapter 12.

involvement of host lectins on nonphagocytic cells (e.g., epithelial cells) in binding bacteria has been controversial. Nevertheless, there is now compelling evidence that the O antigen of their LPS serves as a functional adhesin for a number of bacterial species (25). The lectins on the host cells, however, have not been determined. If the sole function of the putative lectins on host epithelial cells is to bind bacteria, then oligosaccharide antagonists may be effective. However, if the lectins are important for specific cell-cell interactions within the host, as is more likely to be the case, then antiadhesin therapy based on oligosaccharide antagonists should be reconsidered.

INTERACTION WITH CELLS OF THE IMMUNE SYSTEM

The interaction of bacteria with phagocytic cells has been extensively studied, and the role of this interaction in host defense has been well documented (18). In the last decade, increased interest has been focused on mast cells, immune cells that are best known for their role in allergy and that have never previously been implicated in interactions with bacterial targets. There is now compelling evidence that mast cells do interact directly with bacteria by binding to adhesins, and it is thus possible that mast cells play a role in the pathogenesis of bacterial infections. This has been best documented for type 1-fimbriated *Escherichia coli*. CD48, a receptor for the type 1 fimbrial lectin, FimH, was found to mediate internalization of the bacteria by mast cells. The entry process was found to involve caveolae and represents a distinct mechanism for phagocytosis of bacterial pathogens. Because caveolae do not fuse with endosomes, the bacteria contained within caveolar vesicles are apparently provided with a certain degree of protection from mast cell bactericidal activities. Nevertheless, mast cell-deficient mice were significantly more susceptible to infection by type 1-fimbriated *E. coli* than were wild-type mice, suggesting a protective role for mast cell-bacterium interaction (1, 14, 22).

BACTERIAL CLONES POSSESS MULTIPLE ADHESINS

The concept that bacterial clones are capable of producing multiple adhesins was established a number of years ago (for a review, see reference 19). Recent studies have not only confirmed this concept but have also shown that it is even more extensive than previously thought. This is also evident from the list of adhesins for a number of pathogens which have been studied more extensively (see chapter 12). There are still many issues that have not yet been resolved. Can bacteria utilize these multiple adhesins within each niche, or are the adhesins niche specific? This topic has major implications for the development of antiadhesion therapy. For example, in any particular niche, one adhesin phenotype may predominate because the niche is rich in receptors for that adhesin. If, however, there are a small number of receptors for a secondary phenotypic variant, a small number of organisms of this variant may survive treatment with an agent that inhibits the adhesion of the predominant variant. Phase variation would then return the primary phenotype once treatment is concluded. Thus, due to the phenomenon of phase variation among multiple adhesins, it may be necessary to produce a cocktail of antiadhesion agents that target each of the adhesins that are important in a particular niche. In one experimental system, four different adhesins had to be inactivated to prevent adhesion and the development of a symptomatic infection (23).

The remaining themes listed in Table 5.1 have been the focus of intense research during the last decade and therefore are discussed in greater depth in succeeding chapters.

REFERENCES

1. **Abraham, S. N.** 1997. Discovering the benign traits of the mast cell. *Sci. Med.* **4:**46–55.
2. **Abramson, T., H. Kedem, and D. A. Relman.** 2001. Proinflammatory and proapoptotic activities associated with *Bordetella pertussis* filamentous hemagglutinin. *Infect. Immun.* **69:**2650–2658.
3. **An, Y., and R. J. Friedman.** 2000. *Handbook*

of Bacterial Adhesion. Humana Press, Totowa, N.J.

4. **Cossart, P., P. Boquet, S. Normark, and R. Rappuoli.** 2000. *Cellular Microbiology.* ASM Press, Washington, D.C.

5. **Doyle, R. J.** 1999. Biofilms. *Methods Enzymol.* **310:**1–720.

6. **Doyle, R. J.** 2001. Microbial growth in biofilms, part A. Developmental and molecular biological aspects. *Methods Enzymol.* **336:**1–427.

7. **Doyle, R. J.** 2001. Microbial growth in biofilms, part B. Special environments and physicochemical aspects. *Methods Enzymol.* **337:**1–439.

8. **Doyle, R. J., and I. Ofek.** 1995. Adhesion of microbial pathogens. *Methods Enzymol.* **253:**1–600.

9. **Gargir, A., I. Ofek, D. Hasty, S. Meron-Sudai, H. Tsubery, Y. Keisari, and A. Nissim.** 2001. Inhibition of antibody-dependent stimulation of lipoteichoic acid-treated human monocytes and macrophages by polyglycerolphosphate-reactive peptides. *J. Leukoc. Biol.* **70:**537–542.

10. **Hasty, D. L., I. Ofek, H. S. Courtney, and R. J. Doyle.** 1992. Multiple adhesins of streptococci. *Infect. Immun.* **60:**2147–2152.

11. **Kahane, I., and I. Ofek.** 1996. *Toward Anti-Adhesion Therapy for Microbial Diseases.* Plenum Press, New York, N.Y.

12. **Langermann, S., S. Palaszynski, M. Barnhart, G. Auguste, J. S. Pinkner, J. Burlein, P. Barren, S. Koenig, S. Leath, C. H. Jones, and S. J. Hultgren.** 1997. Prevention of mucosal *Escherichia coli* infection by FimH-adhesin-based systemic vaccination. *Science* **276:**607–611.

13. **Mancuso, G., F. Tomasello, I. Ofek, and G. Teti.** 1994. Anti-lipoteichoic acid antibodies enhance release of cytokines by monocytes treated with lipoteichoic acid. *Infect. Immun.* **62:**1470–1473.

14. **McLachlan, J. B., and S. N. Abraham.** 2001. Studies of the multifaceted mast cell response to bacteria. *Curr. Opin. Microbiol.* **4:**260–266.

15. **Morath, S., A. Geyer, and T. Hartung.** 2001. Structure-function relationship of cytokine induction by lipoteichoic acid from *Staphylococcus aureus. J. Exp. Med.* **193:**393–397.

16. **Naka, A., K. Yamamoto, T. Miwatani, and T. Honda.** 1992. Characterization of two forms of hemagglutinin/protease produced by *Vibrio cholerae* non-O1. *FEMS Microbiol. Lett.* **77:**197–200.

17. **Ofek, I., and R. J. Doyle.** 1994. *Bacterial Adhesion to Cells and Tissues,* p. 94–135. Chapman & Hall, New York, N.Y.

18. **Ofek, I., and R. J. Doyle.** 1994. *Bacterial Adhesion to Cells and Tissues,* p. 171–194. Chapman & Hall, New York, N.Y.

19. **Ofek, I., and R. J. Doyle.** 1994. *Bacterial Adhesion to Cells and Tissues,* p. 513–562. Chapman & Hall, New York, N.Y.

20. **Proctor, R. A., L. C. Denlinger, and P. J. Bertis.** 1995. Lipopolysaccharides and bacterial virulence, p. 173–194. *In* J. A. Roth, C. A. Bolin, K. A. Brogden, C. Minion, and M. J. Wannemuehler (ed.), *Virulence Mechanisms of Bacterial Pathogens.* ASM Press, Washington, D.C.

21. **Sharon, N., and I. Ofek.** 2000. Safe as mother's milk: carbohydrates as future anti-adhesion drugs for bacterial diseases. *Glycoconjugates* **17:**659–664.

22. **Shin, J.-S., Z. Gao, and S. N. Abraham.** 2000. Involvement of cellular caveolae in bacterial entry into mast cells. *Science* **289:**785–788.

23. **Van der Velden, A. W. M., A. J. Bäumler, R. M. Tsolis, and F. Heffron.** 1998. Multiple fimbrial adhesins are required for full virulence of *Salmonella typhimurium* in mice. *Infect. Immun.* **66:**2803–2808.

24. **Wilson, M.** 2002. *Bacterial Adhesion to Host Tissues. Mechanisms and Consequences.* Cambridge University Press, Cambridge, United Kingdom.

25. **Zaidi, T. S., S. M. Fleiszig, M. J. Preston, J. B. Goldberg, and G. B. Pier.** 1996. Lipopolysaccharide outer core is a ligand for corneal cell binding and ingestion of *Pseudomonas aeruginosa. Investig. Ophthalmol. Visual Sci.* **37:**976–986.

DIVERSIFICATION OF RECEPTOR SPECIFICITIES
AND ITS BIOLOGICAL CONSEQUENCES

6

Because survival of microorganisms is dependent on adhesion to surfaces, adhesins must have been basic traits of the very first microbial cells emerging on Earth. As new hosts and new surfaces appeared, adhesins must have evolved to accommodate adhesion in these new niches, with a strong selective pressure toward best-fit adhesins. In the evolutionary process, a sufficient number of adhesins with distinct receptor specificity have evolved to ensure that most bacterial clones can normally colonize a range of substrata that is defined but that is sufficiently broad to promise species survival. Macroevolution occurs over a millennial time frame, but more rapid changes in receptor specificity can come about through microevolution of adhesins within a single host. The concept of antiadhesion therapy discussed in chapter 11 requires a thorough understanding of the mechanisms underlying diversification of receptor specificities. Microevolution of the adhesion process is the result of intraspecies diversification of adhesins to become complementary to new receptors (36). Genotypically, variation in receptor specificity may be achieved by horizontal transfer of foreign genes, allelic variation via mutation of a parental gene, or gene rearrangement (Table 6.1). Phenotypically, the receptor specificity variation may be directly due to conformational changes of the adhesin

itself, or its conformation may be indirectly affected by interactions with other proteins that are involved in presentation of the adhesin on the bacterial surface. For example, naturally occurring mutations in FimH affect its receptor specificity, but it has also been suggested that changes in other subunits in the fimbrial shaft (i.e., FimA, FimF, or FimG) also affect the receptor specificity of the FimH lectin (see below). Thus, microorganisms with distinct receptor specificity may emerge within a relatively short period to enable survival on new surfaces. Although the variation in receptor specificity can be subtle, it may be significant enough to allow colonization of a new niche. In the following sections, we discuss this concept and present examples of the various mechanisms for receptor specificity diversification.

PRONOUNCED AND SUBTLE CHANGES IN RECEPTOR SPECIFICITY
The bacterial lectins are good examples of adhesins that can have both pronounced and subtle changes. This family of adhesins is generally classified according to the structure of the carbohydrates they recognize. Their specificity is usually defined as the simplest carbohydrate structure (usually a monosaccharide) that best inhibits the lectin-mediated

TABLE 6.1 Major mechanisms for variation of receptor specificity of bacterial adhesins

Mechanism	Type	Example	Comments
Genotypic	PAI	PapG	Changes in receptor specifity due to acquisition or deletion of adhesion-encoding PAIs
	Allelic variation	FimH	Point mutations in the lectin domain of the FimH tip adhesin
Phenotypic	Conformational	Type 1 fimbriae	Variations in the fimbrial shaft affect the conformation of the FimH tip adhesin
		PilC	Gene rearrangement in the major pilus subunit gene affects PilC conformation indirectly

adhesion (see the discussion of lectins as adhesins in chapter 1). This is referred to as the primary sugar specificity of the lectin (Table 6.2). Within lectins possessing the same primary sugar specificity, differences in the binding of different oligosaccharides are often observed. This is referred to as the fine sugar specificity. The emergence of strains which express adhesins exhibiting a primary sugar specificity different from the parental strain would be considered a pronounced change. Changes in the fine sugar specificity of a bacterial lectin would be considered a subtle change. Monitoring of changes in fine sugar specificity in bacterial lectins is based primarily on two different assays. In one assay, the concentration of various derivatives of the primary sugar inhibitor resulting in 50% inhibition of adhesion is established for the various isolates. In the other assay, the magnitude of

adhesion to various glycoproteins that contain the primary sugar in different contexts is quantified. Table 6.2 lists examples of inter- and intraspecies variations in receptor specificity.

HORIZONTAL ACQUISITION OF NEW ADHESINS: THE PATHOGENICITY ISLAND EFFECT

Intraspecies variations of adhesin receptor specificity may be due to genes found in the chromosome of some strains but not others within a species. The best-known examples of this type of variation are the pathogenicity islands (PAIs), which carry genes that encode adhesins with distinct receptor specificity, as well as genes that confer other specific aspects of pathogenicity. In general, PAIs are selected for best fit to accomplish a variety of important survival functions, including adhesion (8). A number of PAIs have been described that encode adhesins (Table

TABLE 6.2 Inter- and intraspecies diversity in the primary and fine sugar specificities of enterobacterial fimbrial lectins[a]

Type of diversity	Group and species[b]	Primary specificity[b]	Fine specificity[b]
Interspecies	Enterobacteria		
	E. coli (T1F)	Mannoside	pNPα Man ≫ Man
	K. pneumoniae (T1F)	Mannoside	pNPα Man > Man
	S. enterica serovar Typhimurium (T1F)	Mannoside	Man > pNPα Man
Intraspecies	E. coli		
	UPEC (T1F)	Mannoside	Man = Man$_3$
	Fecal EC (T1F)	Mannoside	Man$_3$ ≫ Man
	UPEC (PF-G1)	Galα4Gal	GbO$_3$
	UPEC (PF-G2)	Galα4Gal	GbO$_4$
	UPEC (PF-G3)	Galα4Gal	GbO$_5$

[a]Data from references 11 and 20.
[b]Abbreviations: UPEC, uropathogenic E. coli; T1F, type 1 fimbriae; pNPα-Man, p-nitrophenyl α-mannoside; Man, mannose; Man$_3$, Manα3, Manα6Man; PF, P fimbriae; G1, G2, G3, PapG adhesin subclasses; GbO$_3$, Galα4, Galβ4, Glcβ Cer; GbO$_4$, GalNAc, α2Gal, α4Gal, β4Glc, βCer; GbO$_5$, GalNAc, α2GalNac, α2Gal, α4Gal, β4Glc, βCer (from reference 32); GalNAc, N-acetylgalactosamine; Gal, galactose; Glc, glucose.

6.3). The adhesin-containing PAI may be transferred from one strain to another by transduction, as has been shown for *Vibrio cholerae*. Furthermore, in other organisms such as uropathogenic *Escherichia coli*, the whole adhesin-containing PAI elements can also be deleted. The changes in receptor specificity emerging from either acquisition or deletion of PAIs are usually much more pronounced than with allelic variation (9).

GENE REARRANGEMENT

Gene rearrangement as a mechanism for diversifying receptor specificity has been best studied and characterized for the pilus genes of *Neisseria*. One characteristic of *Neisseria* species is their enormous ability to vary their adhesins (7). As a consequence, *Neisseria* strains can interact with various types of host cells, including epithelial and phagocytic cells, and thus survive in different microenvironments during the infectious process (13, 24, 25). The organisms express at least two families of adhesins, each of which can vary in its receptor specificity. One of the adhesin families includes the complex pilus adhesins, composed of major pilus subunits polymerized into a superstructure which functions to present the minor pilus protein, the PilC adhesin, on its tip at a distance from the bacterial surface (Fig. 6.1). Although there are usually two distinct copies of the gene for PilC, *pilC1* and

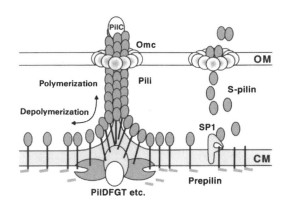

FIGURE 6.1 Hypothetical model of *N. gonorrhoea* pilus biogenesis. OM and CM indicate the outer and inner (cytoplasmic) membranes, respectively. The pre-PilE molecules initially float in the CM. They consist of a globular domain (shaded oval) on the periplasmic face of the CM, a transmembrane domain (black), and a positively charged signal sequence (gray). After processing by the pre-PilE signal peptidase (PilD), the PilE subunits associate via their hydrophobic stems to form a pilus. The core of the pilus, therefore, forms a continuous hydrophobic layer with the CM. The hydrophobic continuum facilitates polymerization and depolymerization of pili under control of the CM-associated polymerization center (PilDFGT, etc.). Alternative cleavage of pre-PilE at Ala40 leads to release of S-pilin (soluble subunits). Signal peptidase 1 (SP1) is thought to be responsible for S-pilin production. The assembled pili, and presumably also S-pilin, penetrate the OM through a pore formed by the multisubunit complex of Omc. PilC facilitates the penetration of pili through the pore and sticks to the pilus tip, where it acts as a human cell-specific adhesin. (Reprinted from reference 5 with permission from the publisher.)

TABLE 6.3 Examples of PAIs encoding adhesins[a]

Organism[b]	Description[b]	Adhesin[b]
E. coli 536 (UPEC)	PAI II$_{536}$	P fimbriae
E. coli 536 (UPEC)	PAI III$_{536}$	S fimbriae
E. coli J96 (UPEC)	PAI I$_{J96}$	P fimbriae
E. coli J96 (UPEC)	PAI II$_{J96}$	P fimbriae
E. coli CFT073 (UPEC)	PAI I$_{CFT073}$	P fimbriae
E. coli E2348/69 (EPEC)	LEE	Intimin
E. coli O157:H7 (EHEC)	LEE	Intimin
E. coli EPEC2	LEE	Intimin
E. coli ETEC	Tia PAI	?
L. ivanovii	*prf* virulence gene cluster	Internalin
V. cholerae	VPI	Type IV pilus, MSH

[a]Adapted from reference 9.
[b]UPEC, uropathogenic *E. coli*; EPEC, enteropathogenic *E. coli*; EHEC, enterohemorrhagic *E. coli*; ETEC, enterotoxigenic *E. coli*; LEE, locus of enterocyte effacement; VPI, *V. cholerae* pathogenicity island; Tia, toxigenic invasion locus A; MSH, mannose-sensitive hemagglutinin.

pilC2, which can lead to limited variation in adhesive ability, the major component of the pilus shaft, PilE is extremely variable and the protein-protein interactions that accompany polymerization can indirectly affect the conformation of the PilC tip adhesin (34). Significantly, this has been shown for PilC1 of *N. meningitidis,* where some pilin variants were capable of modifying the degree of adhesiveness through bundling of fimbriae (19).

The mechanism that controls the variation of the pilus shaft is gene rearrangement (Fig. 6.2). Genomic DNA encodes one complete pilin gene and multiple copies of incomplete pilin genes. Homologous recombination

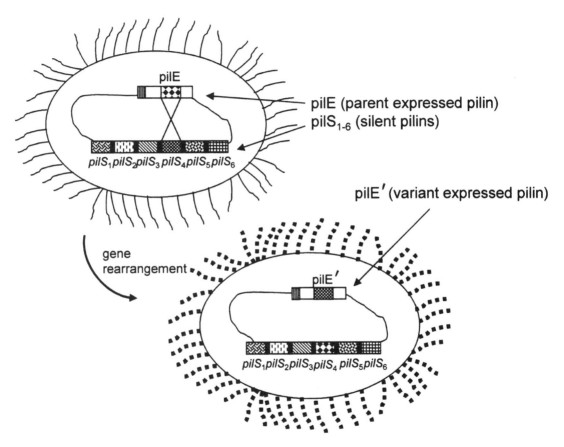

FIGURE 6.2 Schematic diagram of pilus variation in *N. gonorrhoeae.* The major subunit of the *N. gonorrhoeae* pilus shaft, PilE, is a member of the type IV pilus family. The extreme antigenic variability found in *N. gonorrhoeae,* however, appears to be unique among the type IV-expressing genera. The chromosomal locus for pilus expression contains a complete *pilE* (expressed) gene. The chromosome also contains, often in clusters, up to 15 (or more) *pilS* (silent) genes, all lacking the invariant N-terminal domain. Homologous recombination between the variable domain of *pilE* and a *pilS* partial gene leads to structural variation of the PilE protein. Silent genes may also be present due to transformation of genetic information from other *N. gonorrhoeae* strains, offering another pathway for genetic exchange with the *pilE* segment, leading to even more diversity. Because of the large number of partial sequences available and the additional possibilities for horizontal exchange, a single *Neisseria* strain could theoretically yield daughters that express thousands of different PilE proteins. *N. gonorrhoeae* also usually possess two distinct genes for the tip adhesin PilC, *pilC1* and *pilC2.* PilC appears to be essential for adhesion, but it is also clear that changes in PilE structure influence the adhesion behavior of piliated *Neisseria* strains.

results in the insertion of different lengths of one of the partial genes into the complete gene, resulting in the formation of a new pilin gene with a distinct receptor and antigenic specificity. This new pilin gene may undergo further rounds of modification, with insertion of different incomplete adhesin genes. This type of gene rearrangement can lead to the expression of thousands of antigenically different pili by a single clone. Although the genes for PilC vary, the receptor specificity of the tip adhesin may also vary due to physical constraints imposed by its association with structurally different major pilin subunits (17). For example, variations that prevent glycosylation of the major shaft pilin, PilE, gives rise to hyperadherent pili (35).

The other *Neisseria* adhesin family is that of the Opa proteins. Variation of the Opa proteins occurs by selective expression of 1 of a group of between 7 and 11 different Opa genes scattered within the *Neisseria* chromosome (21). The Opa genes are complete, but expression is regulated by the number of repeated CTCTT segments in the leader sequence. For example, the presence of three of these repeat segments results in an in-frame start site, while seven repeat segments will result in an out-of-frame start site. This slipped-strand mispairing mechanism of control does not exclude the possibility that more than one Opa can be expressed by any single cell. The variability is so high that it has been impossible in the past to study a stable culture. Only when genetic mechanisms were developed to lock the expression of a particular adhesin gene in the "on" orientation did it become possible to study stable molecules (15). It is now known that only a few of the Opa proteins, such as Opa_{50}, mediate adhesion to heparan sulfate proteoglycans expressed by certain epithelial cell lines, whereas other Opa proteins, such as Opa_{52}, mediate binding to CD66 expressed on phagocytic cells (1, 6, 15). Variations of receptor specificity within the Opa family are even more pronounced than those of the pilus family.

ALLELIC VARIATION OF ADHESINS

It is currently popular to assume that acquisition of distinct virulence genes, including adhesins of a more pathogenic nature, occurs almost exclusively by horizontal transfer of PAIs or other genetic elements. The possibility that allelic variation of a commensal trait could lead to a more pathogenic phenotype, and therefore aid in the transition from a commensal to a virulent phenotype, has not been as thoroughly investigated. Allelic variation has been shown for a number of bacterial adhesins, including the PapG adhesin of P fimbriae, the FimH adhesin of type 1 fimbriae, type IV pili of *Neisseria* (18) and Flp pili of *Actinobacillus actinomycetemcomitans* (14). Allelic variation is an important event that undoubtedly relates to the evolution of microorganisms in response to the necessity that organisms gain attachment to new substrata in order to survive. However, it is only in a few cases (e.g., PapG and FimH lectins) that the evolution of new receptor specificities has been demonstrated. It is also only in a few cases where the pathoadaptive nature of these allelic variants has been demonstrated (see the following section).

Bacterial evolution toward a more pathogenic phenotype most certainly involves a "gain-of-function" type of acquisition of extra genes that encode specific virulence factors (see "Horizontal acquisition of new adhesins" above). However, an evolutionary step toward increased pathogenicity can also occur by a "change-of-function" type of modification of existing genes. This type of change depends on the occurrence of random (i.e., not programmed and not phase-variable) genetic mutations that confer a selective advantage on the bacterial clone in a particular niche. This type of mutation has been called pathogenicity adaptive, or pathoadaptive (30), and it is well illustrated by the allelic variation of the FimH lectin. Minor sequence changes were shown to result in significant functional alterations that increased the fitness of strains for survival in a new niche (27–30).

BIOLOGICAL CONSEQUENCES

The concept that multiple adhesins can be expressed by individual clones of bacteria (10, 21) is now well established. The chemical, functional, and ultrastructural diversity of adhesins expressed by a single bacterial clone undoubtedly endows the organisms with the ability to colonize a range of substrata. To better understand the biological consequences of the rapid change in receptor specificity, this section addresses how changes in two major adhesins expressed by E. coli, type 1 and P fimbrial adhesins, affect host and tissue tropism of the organism. Essentially, receptor specificity variation is a mechanism to circumvent host defenses whereby variants of the bacteria emerge that can attach to alternative receptors.

The intraspecies diversities in the receptor specificities of FimH and PapG described in Table 6.2 are due to allelic variations in the *fimH* and *papG* genes. The allelic variants of FimH associated with changes in the fine sugar specificity became apparent when the *fimH* genes of more than a dozen normal fecal and uropathogenic E. coli strains were cloned, sequenced, and expressed in recombinant strains that varied only in the sequence of the FimH adhesin (27). Most of the *fimH* genes of urinary isolates encoded FimH lectins that bound to uroepithelial cells in dramatically increased numbers compared to the FimH lectins encoded by genes obtained from normal fecal E. coli isolates (Table 6.4). The uropathogenic FimH lectins were able to bind to either monomannose residues (Man1) or trimannose residues (Man3) of glycoproteins, while the normal fecal FimH lectins required Man3 units. For example, Man1 binding FimH variants bound to uroepithelial cell lines to a much greater degree than did Man3-requiring FimH variants (Table 6.4). However, the two FimH variants bound to buccal epithelial cells in an equivalent manner (29). FimH-mediated adhesion was equally inhibited by D-mannose in all these cases. The differences in fine sugar specificities of FimH alleles and in the binding to animal cells were observed to result in marked differences in bladder colonization by E. coli. The Man1 binding FimH variants that bound 15-fold more effectively to uroepithelial cell lines also colonized mouse bladders in 15-fold larger numbers (29).

This remarkable difference in the adhesion to uroepithelial cells between the two types of allelic variants of FimH may represent a pathoadaptation of uropathogenic E. coli through mutational change (i.e., loss or modification of function) of preexisting genes (30). E. coli is primarily a commensal inhabitant of the large intestines of humans and other animals. Almost every E. coli strain expresses type 1 fimbriae, mannose-sensitive adhesive appendages that are important for the transient oropharyngeal colonization that occurs as an essential step in the normal fecal-oral transmission cycle of E. coli (3, 4). The adhesive subunit of type 1 fimbriae, FimH, is a very highly conserved lectin, but FimH subunits from ~70% of uropathogenic strains carry minor mutations that enhance the ability of the lectin to recognize Man1 receptors. In contrast, 80% of the type 1-fimbriated E. coli isolates from feces of healthy adults are able to bind only to Man3 receptors, which is the evolutionarily ancestral phenotype. The mutant alleles confer on E. coli a significantly higher tropism for uroepithelium, and as a result, they confer a 15-fold-greater ability to colonize the bladder in a mouse model compared with isogenic strains with the ancestral allele. Therefore, these Man1-specific *fimH* mutations are pathoadaptive. It is not yet clear whether most of the mutant E. coli cells spread from the intestinal reservoir or arise during selection in the urinary tract. It has also been suggested that FimH mutations conferring a collagen binding phenotype may provide a selective advantage for meningitis strains of E. coli (23).

In fact, the differential transmission rates of FimH variants could be difficult to measure experimentally in vivo, but the population stability of the *fimH* alleles can be compared based on evolutionary history. The genealogi-

TABLE 6.4 Relationship between the fine receptor specificity and adhesion to cells from various tissues[a]

Type of diversity	Group and species	Fine specificity[b]	Relative adhesion to animal cells
Interspecies	Enterobacteria		
	E. coli (T1F)[c]	pNPα Man ≫ Man	Uro[d] > Entero
	K. pneumoniae (T1F)	pNPα Man > Man	Uro > Entero
	S. enterica serovar Typhimurium (T1F)	Man > pNPα Man	Entero > Uro
	E. coli (FimH^E/S/shaft^S)[f]	Man > pNPα Man	Entero > Uro
Intraspecies			
	E. coli		
	UPEC	Man1 = Man3	Buccal ≫ Uro[e]
	Fecal	Man3 ≫ Man	Buccal = Uro

[a]Adapted from reference 22.
[b]For designations of fine sugar specificity, see Table 6.2.
[c]T1F, type 1 fimbriae.
[d]Uro, ATCC mouse uroepithelial cell line MM45T.BL; Entero, ATCC mouse enterocyte cell line SI-H10. Buccal epithelial cells were scraped from the buccal mucosa of healthy volunteers by using cotton-tipped swabs.
[e]Uro, ATCC human uroepithelial cell line, J82.
[f]A recombinant *E. coli* strain in which the FimH of *E. coli* (E) or *Salmonella* (S) is presented on the fimbrial shaft of *Salmonella*.

cal relationships of 11 unique *fimH* alleles that confer various abilities to bind to monomannose units strongly suggest that high-monomannose binding alleles are short-lived (Fig. 6.3). All interior nodes of the phylogenetic network are represented by FimH subunits with a relatively low ability to bind to monomannose, and terminal nodes are the distinctly high-monomannose binding alleles. This strongly suggests that the low-monomannose binding alleles are ancestral. The high-monomannose binding alleles arise by different nonsynonymous mutations and do not form distinct genetic lineages. The fact that there are very few alleles in any developing lineage strongly suggests that the mutations are short-lived. This supports the notion that the pathoadaptive mutations are deleterious with regard to the fitness of FimH for commensal ecology (29, 30).

Based on the crystal structure of FimH (see chapter 1), we now know that most of the allelic variations associated with the distinct receptor specificities of uropathogenic rather than normal fecal isolates of *E. coli* are located in the lectin domain but distant from the D-mannose binding site (26) (Fig. 6.4). Interestingly, sequencing of the *fimH* genes of 200 uropathogenic *E. coli* isolates revealed that the amino acid residues of the D-mannose

binding pocket and of the immediately surrounding regions were invariant (12). Some of the natural allelic variations and *fimH* mutants from a random mutant library were located in the pilus domain (e.g., the domain that allows for incorporation of FimH into the tip of the fimbrial shaft). Some of these variations in the pilin domain exhibit distinct receptor specificities. This change in receptor specificity may be due to conformational constraints on the D-mannose binding pocket that are imposed by changes in interdomain hydrogen bonding or other interactions. In fact, changes in one or more of the other proteins of the fimbrial shaft may also result in changes in receptor specificity, as discussed below. FimH subunits found in a random mutant library having mutations near the mannose binding pocket were found to be nonadhesive (26).

The interspecies diversity described in Table 6.2 may also be due to conformational changes in the FimH tip adhesin caused by variations in one of the other fimbrial shaft components, FimA, FimF, or FimG. This was suggested in experiments in which the *fimH* gene of *Klebsiella pneumoniae* was reciprocally exchanged with the *fimH* gene of *E. coli*. When the genes encoding the fimbrial shaft and the fimbrial adhesin were swapped between these species, it became apparent that a component

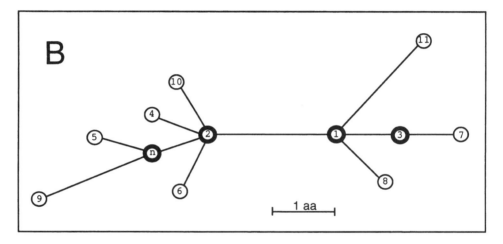

fimH allele	Residue number -++++++++++++++ 1111112 123677791111660 673603816789361 TANGNGSPGVAIVRH	M_1/M_3 ratio	wild strain (source)
1	----S-N---------	0.08	F-18 (fecal)
2	---------------	0.09	KB21 (recomb.)
3	----S-N-----A--	0.15	MJ11-2 (UTI)
4	-V-------------	0.17	K-12 (fecal)
5	N---------V----	0.20	CI3 (UTI)
6	-------------H-	0.33	1177 (UTI)
7	---DS-N-----A--	0.63	MJ2-2 (UTI)
8	----S-N------C-	0.72	CI12 (UTI)
9	N-H----L-------	0.77	CI7 (UTI)
10	--------ΔΔΔΔ---	0.91	CI10 (UTI)
11	----SEN-------D	0.93	CI4 (UTI)

FIGURE 6.3 Phylogenetic analysis of FimH alleles. (A) Amino acid sequences of FimH variants. The alleles are listed based on an increasing Man1/Man3 binding ratio. The residues listed above the 11 alleles are the amino acid residues of the new alleles that vary from the original FimH sequence. Only polymorphic residues are shown, and the positions are numbered vertically from −16 of the leader peptide to +201 of the mature FimH. Δ indicates deleted residues. (B) Inferred phylogenetic network demonstrating evolutionary relationships of the FimH alleles shown in panel A. Each node represents a distinct FimH allele, numbered as in panel A. The allele labeled n represents a hypothetical FimH that differs from allele 2 by the substitution of Asp (N) for Tyr (T) in the leader sequence (residue −16) and phenotypically should be equivalent to allele 2. Internal nodes are shown in bold. The deduced sequences of the 11 FimH proteins exhibit greater than 99% homology, and the network showing their phylogenetic relationships is fully consistent, without any homoplasty. Branch lengths are scaled to the number of amino acids that differ between alleles. The deletion of 4 amino acids in FimH allele 10 is considered to be a single event, equivalent to one amino acid substitution. (Reprinted from reference 29 with permission from the publisher.)

of the fimbrial shaft must in some way influence the fine sugar specificity of FimH. For example, the relative inhibitory activity of aromatic mannosides (e.g., *p*-nitro-*o*-chlorophenyl α-mannoside) was three- to four-fold greater, compared to that of aliphatic mannosides (e.g., methyl-αMan), for a recombinant strain expressing *E. coli* shaft and *K. pneumoniae* FimH than for a recombinant strain expressing *K. pneumoniae* shaft and *K. pneumo-*

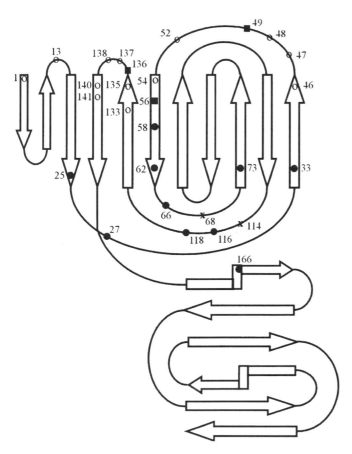

FIGURE 6.4 β-sheet topology diagram of the lectin (top) and pilin (bottom) domains of FimH (see reference 2 for more details). Indicated on this diagram are variant amino acids which affect FimH function that were found either among naturally occurring isolates or in a random mutant library. Although the diagram is not precisely to scale, the relative positions of amino acids are indicated accurately. Residues indicated are those that caused enhanced monomannose binding (solid circles), those that caused complete loss of D-mannose binding (solid squares), and those that were neutral substitutions able to act in concert with other mutations to enhance monomannose binding (X's). The residues of the mannose binding site are also indicated (open circles). Interestingly, most of the residues that resulted in increased binding to monomannose units were found at the opposite end of the barrel-like lectin domain from the actual binding site. Furthermore, it is interesting that at least one residue in the pilin domain also affected the lectin binding activity, consistent with the notion that association of FimH with the main fimbrial shaft via its pilin domain affects the fine sugar specificity of the lectin domain (see the text). (Reprinted from reference 26 with permission from the publisher.)

niae FimH. Conversely, the relative inhibitory activity of these compounds for a recombinant strain expressing the *K. pneumoniae* shaft and *E. coli* FimH was five- to six-fold lower than that for a recombinant strain expressing *E. coli* shaft and *E. coli* FimH. These results are consistent with the high degree of identity in the carboxy-terminal regions of *E. coli* and *K. pneumoniae* FimH that associate with the shaft.

In contrast to the high degree of homology between *E. coli* FimH and *K. pneumoniae* FimH, there is limited sequence homology between these proteins and FimH of *Salmonella enterica* serovar Typhimurium (33). To determine whether the marked difference in the fine sugar specificity between *S. enterica* serovar Typhimurium FimH (FimH[S]) and *E. coli* FimH (FimH[E]) in their affinity for aromatic mannosides (Table 6.1) is due to the primary structure of FimH or to the FimH-presenting shaft, a chimeric FimH was constructed (33). This fusion protein consisted of the amino terminus of *E. coli* FimH (FimH[E]), which contained the carbohydrate binding domain, and the carboxy terminus of serovar Typhimurium FimH (FimH[S]), which contained the region that associates with the fimbrial shaft. When this FimH[E/S] fusion was expressed in a recombinant strain and presented on the serovar Typhimurium type 1 fimbrial shaft, the recombinant strain bound aromatic mannosides with an affinity similar to that of FimH[S] (S. N. Abraham, unpublished observation). The data, taken together, support the concept that the affinity of FimH of enterobacteria toward aromatic mannosides is significantly influenced by the fimbrial shaft proteins. One would presume this effect to be the influence of FimF and/or FimG, which are components of the tip fibrillum, but this remains to be determined. The interspecies differences between *E. coli, K. pneumoniae* and serovar Typhimurium were reflected in the relative binding affinity of these bacteria to mouse uroepithelial and enterocyte cell lines (Table 6.4). While *E. coli* bound severalfold better to uroepithelial cells than to enterocytes, serovar Typhimurium bound in the reverse manner. The binding pattern of *K. pneumoniae* was similar to that of *E. coli,* but the differences in binding between uroepithelial cells and enterocytes were much less pronouced. Recombinant strains were used to test the effect of the fimbrial shaft on this binding. When *E. coli* FimH was expressed on the serovar Typhimurium fimbrial shaft, the resultant recombinant strain bound to cells in a serovar Typhimurium-like pattern. Furthermore, when *fimH* genes of the three enterobacterial species were fused with a maltose binding protein gene, blots of the soluble fusion proteins bound intact uroepithelial and enterocyte cells equally well (Abraham, unpublished). All of these cell binding activities were inhibited by similar concentrations of mannose, suggesting that the magnitude of differences in adhesion reflected differences in the fine sugar specificity. Collectively, these data suggest that interspecies differences in type 1 fimbria-mediated binding to various animal cell types may be influenced not only by the primary structure of FimH but also by components of the fimbrial shaft. Another example of the influence of variations in the fimbrial shaft on receptor specificity is the *N. gonorrhoeae* pilus discussed above (Fig. 6.1 and 6.2).

While the receptors for FimH are most likely to be glycoproteins on host cell membranes, receptors for the PapG lectin of P fimbriae are members of the globo series of glycosphingolipids (32) (see chapter 1). These are composed of a ceramide portion and a Gal(α1→4)Gal-containing saccharide chain that varies structurally among different tissues and different animal species. These glycolipids can be globotriosylceramide (GbO_3), globotetraosylceramide (GbO_4), or the Forssman glycosphingolipid (GbO_5) (see Fig. 1.5). Three classes of allelic variants of the PapG lectin have been identified (16, 31, 32), all of which exhibit 50 to 60% amino acid sequence homology. The three variants have distinct receptor specificities, not only for the type of glycosphingolipid they bind but also for the type of eukaryotic cells to which they attach.

The differences in PapG receptor specificity extend to distinct differences in animal tropism. *E. coli* strains carrying the GbO_4-specific PapG lectin bind to human uroepithelial cells, whereas those carrying the GbO_5-specific PapG lectin bind to canine uroepithelial cells. The former strains predominate in human pyelonephritic strains, and the latter predominate in *E. coli* strains associated with acute cystitis in humans and dogs. Thus, variation in receptor specificity resulting from different receptor architectures may represent a mechanism that allows *E. coli* PapG adhesins to recognize different animal hosts and different tissues in one individual. It has been suggested that the expression of distinct Gal(α1→4)Gal-containing globosides at different tissue sites selected allelic variants of the *papG* genes.

In conclusion, the remarkable diversity in the fine sugar specificity of the type 1 and P fimbrial lectins clearly illustrates the concept that functional diversity of fimbrial lectins plays an important role in host tropism, tissue tropism, and infectivity. It is highly likely that similar variations will also be found among the other types of fimbrial lectins. It is now recognized that most pathogens are capable of using lectins in the binding to host cells (see chapter 1). Given the enormous variability of animal cell oligosaccharide structures, it is possible that many adhesive variants will evolve. Further studies should reveal an underlying principle for such selection and its relation to bacterial pathogenicity.

REFERENCES

1. **Chen, T., R. Belland, J. Wilson, and J. Swanson.** 1995. Adherence of pilus-Opa$^+$ gonococci to epithelial cells *in vitro* involves heparan sulfate. *J. Exp. Med.* **182:**511–517.
2. **Choudhury, D., A. Thompson, V. Stojanoff, S. Langermann, J. Pinkner, S. J. Hultgren, and S. D. Knight.** 1999. X-ray structure of the FimC-FimH chaperone-adhesin complex from uropathogenic *Escherichia coli.* *Science* **285:**1061–1066.
3. **Bloch, C., and P. Orndorff.** 1990. Impaired colonization by and full invasiveness of *Escherichia coli* K1 bearing a site-directed mutation in the type 1 pilin gene. *Infect. Immun.* **58:**275–278.
4. **Bloch, C., B. A. Stocker, and P. Orndorff.** 1992. A key role for type 1 pili in enterobacterial communicability. *Mol. Microbiol.* **6:**697–701.
5. **Fussenegger, M., T. Rudel, R. Barten, R. Ryll, and T. F. Meyer.** 1997. Transformation competence and type-4 pilus biogenesis in *N. gonorrhoeae*—a review. *Gene* **192:**125–134.
6. **Gray-Owen, S. D., C. Dehio, A. Haude, F. Grunert, and T. F. Meyer.** 1997. CD66 carcinoembryonic antigens mediate interactions between Opa-expressing *Neisseria gonorrhoeae* and human polymorphonuclear phagocytes. *EMBO J.* **16:**3435–3445.
7. **Haas, R., and T. F. Meyer.** 1992. Silent pilin genes of *Neisseria gonorrhoeae* MS11 and the occurrence of related hypervariant sequences among gonococcal isolates. *Mol. Microbiol.* **6:**197–208.
8. **Hacker, J., and E. Carniel.** 2001. Ecological fitness, genomic islands and bacterial pathogenicity. A Darwinian view of the evoluation of microbes. *EMBO Rep.* **2:**376–381.
9. **Hacker, J., and J. B. Kaper.** 2000. Pathogenicity islands and the evolution of microbes. *Annu. Rev. Microbiol.* **54:**641–679.
10. **Hasty, D. L., I. Ofek, H. S. Courtney, and R. J. Doyle.** 1992. Multiple adhesins of streptococci. *Infect. Immun.* **60:**2147–2152.
11. **Hasty, D. L., and E. V. Sokurenko.** 2000. FimH, an adaptable adhesin, p. 481–515. *In* R. J. Doyle (ed.), *Glycomicrobiology.* Klewer Academic/Plenum Publishers, New York, N.Y.
12. **Hung, C.-S., J. Bouckaert, D. Hung, J. Pinkner, C. Widberg, A. DeFusco, C. G. Auguste, R. Strouse, S. Langermann, G. Waksman, and S. J. Hultgren.** 2002. Structural basis of tropism of *Escherichia coli* to the bladder during urinary tract infection. *Mol. Microbiol.* **44:**903–915.
13. **Jonsson, A.-B., D. Ilver, P. Falk, J. Pepose, and S. Normark.** 1994. Sequence changes in the pilus subunit lead to tropism variation of *Neisseria gonorrhoeae* to human tissue. *Mol. Microbiol.* **12:**403–416.
14. **Kachlany, S. C., P. J. Planet, R. Desalle, D. H. Fine, D. H. Figurski, and J. B. Kaplan.** 2001. *flp-1*, the first representative of a new pilin gene subfamily, is required for non-specific adherence of *Actinobacillus actinomycetemcomitans.* *Mol. Microbiol.* **40:**542–554.
15. **Kupsch, E.-M., B. Knepper, T. Kuroki, I. Heuer, and T. F. Meyer.** 1993. Variable opacity (Opa) outer membrane proteins account for the cell tropisms displayed by *Neisseria gonorrhoeae* for human leukocytes and epithelial cells. *EMBO J.* **12:**641–650.
16. **Lund, B., B.-I. Marklund, N. Strömberg, F. Lindberg, K.-A. Karlsson, and S. Normark.** 1988. Uropathogenic *Escherichia coli* can express serologically identical pili of different receptor binding specificities. *Mol. Microbiol.* **2:**255–263.
17. **Meyer, T. F.** 1999. Pathogenic neisseriae: complexity of pathogen-host cell interplay. *Clin. Infect. Dis.* **28:**433–441.
18. **Morand, P. C., P. Tattevin, E. Eugene, J. L. Beretti, and X. Nassif.** 2001. The adhesive property of the type IV pilus-associated component PilC1 of pathogenic *Neisseria* is supported by the conformational structure of the N-terminal part of the molecule. *Mol. Microbiol.* **40:**846–856.
19. **Nassif, X., M. Marceau, C. Pujol, B. Pron, J. L. Beretti, and M. K. Taha.** 1997. Type 4 pili and meningococcal adhesiveness. *Gene* **192:**149–153.
20. **Ofek, I., and N. Sharon.** 1990. Adhesins as lectins: specificity and role in infection. *Curr. Top. Microbiol. Immunol.* **151:**91–113.
21. **Ofek, I., and R. J. Doyle.** 1994. *Bacterial Adhesion to Cells and Tissues,* p. 239–320. Chapman & Hall, New York, N.Y.
22. **Ofek, I., D. L. Hasty, S. N. Abraham, and N. Sharon.** 2000. Role of bacterial lectins in urinary tract infections: molecular mechanisms for diversification of bacterial surface lectins. *Adv. Exp. Med. Biol.* **485:**183–192.
23. **Pouttu, R., T. Puustinen, R. Virkola, J. Hacker, P. Klemm, and T. K. Korhonen.** 1999. Amino acid residue Ala-62 in the FimH fimbrial adhesin is critical for the adhesiveness of meningitis-associated *Escherichia coli* to collagens. *Mol. Microbiol.* **31:**1747–1757.
24. **Rudel, T., H.-J. Boxberger, and T. F. Meyer.** 1995. Pilus biogenesis and epithelial cell adherence of *Neisseria gonorrhoeae pilC* double knock-out mutants. *Mol. Microbiol.* **17:**1057–1071.

25. **Rudel, T., J. P. M. van Putten, C. P. Gibbs, R. Haas, and T. F. Meyer.** 1992. Interaction of two variable proteins (PilE and PilC) required for pilus-mediated adherence of *Neisseria gonorrhoeae* to human epithelial cells. *Mol. Microbiol.* **22:**3439–3450.

26. **Schembri, M. A., E. V. Sokurenko, and P. Klemm.** 2000. Functional flexibility of the FimH adhesin: Insights from a random mutant library. *Infect. Immun.* **68:**2638–2646.

27. **Sokurenko, E. V., H. S. Courtney, J. Maslow, A. Siitonen, and D. L. Hasty.** 1995. Quantitative differences in adhesiveness of type 1 fimbriated *Escherichia coli* due to structural differences in *fimH* genes. *J. Bacteriol.* **177:**3680–3686.

28. **Sokurenko, E. V., V. Chesnokova, R. J. Doyle, and D. L. Hasty.** 1997. Diversity of the *Escherichia coli* type 1 fimbrial lectin. Differential binding to mannosides and uroepithelial cells. *J. Biol. Chem.* **272:**17880–17886.

29. **Sokurenko, E. V., V. Chesnokova, D. E. Dykhuizen, I. Ofek, X.-R. Wu, K. A. Krogfelt, C. Struve, M. A. Schembri, and D. L. Hasty.** 1998. Pathogenic adaptation of *Escherichia coli* by natural variation of the FimH adhesin. *Proc. Natl. Acad. Sci. USA* **95:**8922–8926.

30. **Sokurenko, E. V., D. L. Hasty, and D. E. Dykhuizen.** 1999. Pathoadaptive mutations: gene loss and variation in bacterial pathogens. *Trends Microbiol.* **7:**191–195.

31. **Strömberg, N., B.-I. Marklund, B. Lund, D. Ilver, A. Hamers, W. Gaastra, K.-A. Karlsson, and S. Normark.** 1990. Host-specificity of uropathogenic *Escherichia coli* depends on differences in binding specificity to Gal alpha 1–4Gal-containing isoreceptors. *EMBO J.* **9:**2001–2010.

32. **Strömberg, N., P. G. Nyholm, I. Pascher, and S. Normark.** 1991. Saccharide orientation at the cell surface affects glycolipid receptor function. *Proc. Natl. Acad. Sci. USA* **88:**9340–9344.

33. **Thankavel, K., A. H. Shah, M. S. Cohen, T. Ikeda, R.G. Lorenz, R. Curtiss III, and S.N. Abraham.** 1999. Molecular basis for the enterocyte tropism exhibited by *Salmonella typhimurium* type 1 fimbriae. *J. Biol. Chem.* **274:**5797–5809.

34. **Tonjum, T., and M. Koomey.** 1997. The pilus colonization factor of pathogenic neisserial species: organelle biogenesis and structure/function relationships—a review. *Gene* **192:**155–163.

35. **Virji, M., J. R. Saunders, G. Sims, K. Makepeace, D. Maskell, and D. J. P. Ferguson.** 1993. Pilus-facilitated adherence of *Neisseria meningitidis* to human epithelial and endothelial cells: modulation of adherence phenotype occurs concurrently with changes in primary amino acid sequence and the glycosylation status of pilin. *Mol. Microbiol.* **10:**1013–1028.

36. **Ziebuhr, W., K. Ohlsen, H. Karch, T. Korhonen, and J. Hacker.** 1999. Evolution of bacterial pathogenesis. *Cell. Mol. Life Sci.* **56:**719–728.

ENTRY OF BACTERIA INTO
NONPHAGOCYTIC CELLS

7

Certain cells are adapted for internalizing and killing pathogens via a process known as phagocytosis. Prominent examples of these types of cells are polymorphonuclear leukocytes and macrophages. Mucosal epithelial cells, which are frequently the first host cells contacted by pathogens, are considered nonphagocytic cells (NPCs). They are not professionally adapted for killing microbes and, for that matter, are not known for taking up large particles. Thus, pathogens capable of entry into NPCs must induce their own uptake into these cells via a process that has been termed invasion, internalization, or, simply, entry. Entry and internalization are preferred terms for the uptake of pathogens by NPC, because these are less confusing terms than invasion. Uptake of bacteria by NPC may or may not lead to invasive infections and, in fact, for most of the bacteria that are capable of entering NPC, there is no clear connection to true invasive infections. Entry of bacteria into NPCs has, perhaps more than any other postadhesion event, captured the interest of investigators within the field of host-pathogen interactions. Undoubtedly, the process is complicated, and understanding it will require a thorough knowledge of both animal cell biology and the complex machinery for exporting of bacterial proteins.

There are three categories of bacterial growth in the presence of NPCs: obligate intracellular, facultative intracellular, and extracellular (Table 7.1) (61). The obligate intracellular pathogens, which include members of the genera *Rickettsia* and *Chlamydia*, can replicate only within animal cells, and in this regard they resemble viruses. The facultative intracellular bacteria are readily cultivated in laboratory media, but they have a distinct intracellular phase during the infectious process. Typically they include members of the genera *Listeria, Yersinia, Shigella, Salmonella, Bartonella,* and *Brucella* (13, 23, 55). The extracellular pathogens have not been found to exhibit a well-defined stage of intracellular growth in vivo, but certain clones are able to enter NPCs in vitro, usually at relatively low levels. Since the development of reliable assays for quantitating bacterial invasion into NPCs, a growing number of extracellular species have been shown to be capable of entering NPCs (Table 7.2).

Regardless of the category to which the bacteria belong, it is assumed that bacterial entry into NPC is a consequence of initial adhesion. In some cases, the process of adhesion can be separated from that of entry, while in other cases, no clear distinction has been demonstrated. In all cases, the host cell

TABLE 7.1 Relationship between bacterial entry into and growth within NPCs

Type of relationship	Intracellular growth	Extracellular growth	Examples	Mechanisms
Obligate intracellular	+	–	*Chlamydia, Rickettsia*	Unknown
Facultative intracellular	+	+	*Yersinia, Listeria*	Zipper
			Shigella, Salmonella	Trigger
			Bartonella	Invasome
			Brucella	Unknown
Extracellular	–	+	Many (see Table 7.2)	Unknown

cytoskeletal system is exploited by the pathogen to provide motive forces for internalization. This can occur only when the binding is mediated by a receptor which has transmembrane links to intracellular contractile elements. The mechanisms responsible for entry into the cell following adhesion have been studied in detail only for some of the bacteria belonging to the facultative group.

TABLE 7.2 Examples of extracellular bacteria with the capacity to enter NPCs

Species	Reference(s)
Actinobacillus actinomycetemcomitans	59
Aeromonas salmonicida	29
Burkholderia (Pseudomonas) cepacia	5
Burkholderia pseudomallei	44
Campylobacter jejuni	71
Citrobacter freundii	71
Citrobacter rodentium	36
Clostridium difficile	16
Escherichia coli	45, 56
Fusobacterium nucleatum	35
Haemophilus ducreyi	102
Haemophilus influenzae	92
Klebsiella pneumoniae	72
Leptospira interrogans	58
Mycobacterium avium	3
Mycobacterium bovis	11
Neisseria gonorrhoeae	97
Neisseria meningitidis	93, 94
Pseudomonas aeruginosa	18, 19
Staphylococcus aureus	2, 34
Streptococcus agalactiae	39, 91
Streptococcus pyogenes	32, 50
Streptococcus pneumoniae	9, 103
Streptococcus suis	49
Porphyromonas gingivalis	98
Prevotella intermedia	14
Proteus mirabilis	99
Bacillus piliformis	22

These bacteria appear to invade NPCs primarily via three different mechanisms, termed zipper, trigger, and invasome mechanisms, which are described in more detail below (Fig. 7.1; Table 7.1). In the following discussion, we focus mainly on the initial events that lead to entry following initial adhesion. In-depth discussions about the complex signaling pathways and mechanisms of intracellular proliferation may be found in excellent reviews and books on the subject (7, 77).

OBLIGATE INTRACELLULAR PATHOGENS

Typically, the obligate intracellular pathogens that enter NPCs include members of *Rickettsia* and *Chlamydia* spp. Adhesion of representative species has been demonstrated for both genera (see chapter 12). Although all strains belonging to these species invariably are capable of penetrating into NPCs, the mechanisms that lead to the entry process are not clear. Studies utilizing primary human endometrial cells have shown that *C. trachomatis* enters through coated pits and vesicles, implying that there is a clathrin requirement (100). In contrast, more recent studies have indicated that *C. trachomatis* enters epithelial and phagocytic tissue culture cells via caveolin-containing sphingolipid- and cholesterol-enriched raft micro-domains in the host cell plasma membranes (88). It has been further suggested that glycosaminoglycan-specific adhesins on the chlamydial surface mediate attachment and entry of the organisms into NPCs (81). It is probable, therefore, that chlamydiae utilize different mechanisms of entry, depending on the host cell target.

FACULTATIVE INTRACELLULAR PATHOGENS

The mechanisms of adhesion and entry used by the facultative intracellular pathogens have been the most thoroughly studied by far. The organisms utilize all three recognized mechanisms to initiate contact-induced uptake by NPC (the zipper, trigger, and invasome mechanisms [see above]). All three mechanisms involve intracellular polymerization and reorganization of actin filaments reminiscent of the effects of a variety of host-derived stimuli (e.g., growth factors [24]).

The zipper mechanism of bacterial uptake is exemplified by *Listeria monocytogenes* and *Yersinia* spp. In both cases, a single bacterial surface molecule accomplishes both adhesion and internalization (Table 7.3). The mechanism involves direct contact between the bacterial adhesin and a complementary receptor on the NPC, and with a subsequent series of ligand-receptor interactions, the organism is gradually engulfed (Fig. 7.1A). This mechanism is very similar to the process of receptor-mediated uptake of particles by phagocytic cells (33). The *Yersinia* spp. utilize a molecule called invasin, while *L. monocytogenes* utilizes molecules called internalin A and internalin B (Table 7.3). Internalin binds to E-cadherin, and invasin binds to β_1-integrins. It is difficult to say that the adhesion process is completely separable from the entry process, yet entry can be blocked by a variety of cytoskeleton-active agents without blocking adhesion (54). Although many bacteria express adhesins, those that mediate entry via the zipper mechanism appear to bind to their receptors with a very high affinity, up to 1 or 2 orders of magnitude higher than that of the natural ligand for its receptor (e.g., the affinity of fibronectin for integrins). The relative affinity of the binding of internalin to its receptor has not been determined but is likely to be similar to that of invasin. Entry of both *L. monocytogenes* and *Yersinia* is divalent cation dependent and is inhibited by microfilament-disrupting drugs. Entry by way of the zipper mechanism exploits cellular machinery that is normally involved in the process of cell binding to a substratum.

Unlike the zipper mechanism, entry via a trigger mechanism involves dramatic cytoskeleton and membrane rearrangements (17, 24, 40, 82). The trigger mechanism is associated with ruffling of the host cell membrane over a relatively large area that continues to expand for some time following initial bacterial attachment (Fig. 7.1B). Filamentous actin accumulates around the point of contact of the bacterium with the cell and projects membrane ruffles outward in the general direction of the bacterial stimulus in a striking manner (21, 40). As a result, other bacteria in the vicinity may also be engulfed. It has been proposed that bacterial cells that are not able to induce entry on their own may be internalized if they happen to lie in the vicinity of the membrane ruffling induced by an internalization-competent bacterium.

Members of the genera *Shigella* and *Salmonella* have been the most thoroughly studied with regard to the trigger mechanism. These organisms are known to produce a variety of adhesins, but it is not clear how or even whether these adhesins trigger the cellular response described above. For example, *Salmonella* bacteria enter NPCs efficiently if the bacteria were harvested from growth under low oxygen tension (21) whereas *Salmonella* bacteria grown under high oxygen tension bind efficiently to the cells but are unable to trigger ruffling of the membrane and entry (51). In both cases, defined genetic loci required for entry have been identified (10, 89). It is not surprising that most of the genes in these loci encode proteins that are secreted by the bacterium to induce the dramatic cellular changes. In fact, both genera employ type III secretion systems in the entry process.

In *Salmonella,* the genes encoding one type III secretion system and effector proteins are contained in the *inv/spa* locus within *Salmonella* pathogenicity island 1 (SPI1) located at centisome 63 of the chromosome (25). A second type III secretion system is present within another *Salmonella* pathogenicity island, SPI2 (38). Type III secretion sys-

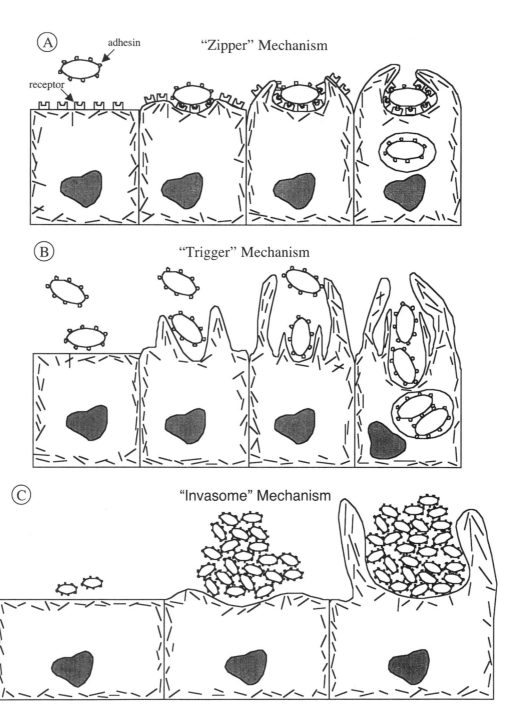

TABLE 7.3 Examples of bacterial ligands and receptors involved in adhesion to and entry into NPCs

Microorganism	Bacterial ligand	Host cell receptor	Reference
L. monocytogenes	InlA, InlB	E-cadherin	95
N. gonorrhoeae	Opa50	Heparan sulfate, vitronectin	60
Yersinia spp.	Inv	β_1-integrin	40
Salmonella spp.	InvABCD locus?	?	10
Shigella spp.	IpaBCD	$\alpha_5\beta_1$-integrin?	89

tems are activated rapidly, generally only after contact of bacteria with host cells (30, 104). It has been shown that the type III secretion loci encode a so-called needle complex that is formed between the bacterial cytoplasm and host cell cytoplasm, and it is through this channel that effector proteins required for entry and subsequent phenomena are translocated (see chapter 8) (26, 46, 47). These include SipA, SopE, SopB, and SptP (27, 28). Similar mechanisms have also been proposed for other species as well, such as *Yersinia* spp. The mechanism of action of the effector proteins on intracellular machinery is an exciting and burgeoning field that is outside the scope of this book. The reader is referred to recent reviews on this topic (6, 28, 55).

In *Shigella,* at least 27 different genes are essential for entry. Most of these genes are located in adjacent regions of a large virulence plasmid and are termed the *mxi/spa* and *ipa* operons (89). Proteins encoded by the *mxi/spa* locus appear to make up the secretion system, while the proteins encoded by the *ipa* operon are effectors. The proteins are already present in the bacterial cytoplasm when needed, so the response does not require transcription and is quite rapid. A needle complex

has also been observed for *Shigella* (Fig. 7.2), and it is necessary for insertion of IpaB and IpaC into the host cell membrane (4, 4a). Nevertheless, latex beads coated with IpaBC complex are readily internalized by NPCs (Fig. 7.3), suggesting that these proteins are sufficient to interact with putative cell surface receptors and to trigger entry (57). Indeed, a member of the β_1 family of integrins was shown to bind the IpaBD complex (96). The IpaBC complex is also capable of association with the hyaluronan receptor, CD44, during *Shigella* entry. Anti-CD44 significantly reduced the entry of *Shigella*. It is currently thought that the Ipa proteins act on the outside of the host cell to stimulate signaling mechanisms. It has been postulated that the Ipa complex acts by creating a bridge through which effector proteins are translocated into the cytoplasm of the host cell (55).

The complex machinery responsible for ruffle formation is similar to that for responses to growth factors or other normal host signals that result in ruffling (55, 89). As with *Salmonella,* the role that adhesion of the bacteria plays in entry into the host cell is not clear. It would seem that brief contact between bacteria and the host cell membrane is sufficient to trigger a

FIGURE 7.1 Schematic diagram illustrating the three primary mechanisms of bacterial internalization by NPCs. (A) Internalization of *Listeria* or *Yersinia* species is accomplished by surrounding the microorganism via a tight phagosome. High-affinity binding of bacterial cell surface components to their cognate receptors on animal cells (see Table 7.3) is required to initiate cytoskeleton-mediated zippering of the host cell plasma membrane around the bacterium. (B) In internalization of *Shigella* and *Salmonella* spp. by the trigger mechanism, bacterial effectors translocate through a type III secretion apparatus into the host cell cytosol and trigger a cascade of reactions including activation of small G proteins, which regulate the actin cytoskeleton, to induce membrane extensions. (C) Internalization by the invasome mechanism, described for *Bartonella* spp. Internalization by NPCs involves the formation of a bacterial aggregate which is engulfed and subsequently internalized.

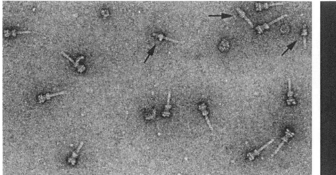
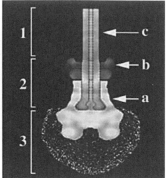

FIGURE 7.2 Structural analysis of the *Shigella flexneri* needle complex by electron microscopy. Negative staining of isolated needle complexes is shown. Arrows point to incomplete needle complexes, lacking the base. Bar, 100 nm. The model of a central axial section of the needle complex indicates the tripartite structure of a base (a), upper ring doublet (b), and needle (c). (Reprinted from reference 4a with permission from the publisher.)

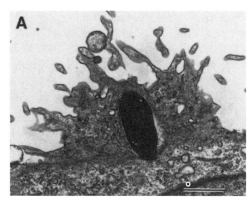
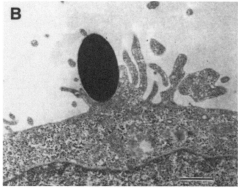

FIGURE 7.3 Electron micrographs of the trigger mechanism of bacterial entry. (A) Transmission electron micrograph of the ruffling response of the epithelial cell membrane to *S. flexneri*. (Micrograph courtesy of Philippe Sansonetti.) (B) Morphological response of HeLa cells to latex beads bearing the Ipa complex. Semiconfluent HeLa cells were incubated for 2 h at 37°C with Ipa$^+$ beads. The ultrastructural appearance of the HeLa cell apical plasma membrane in response to Ipa$^+$ beads is reminiscent of the dramatic membrane-ruffling response to *S. flexneri*. (Reprinted from reference 57 with permission from the publisher.)

bacterial response that results in the secretion of effector molecules capable of initiating the dramatic changes in the host cell. Once such changes take place, the weakly bound bacteria are trapped by the vinculin-, actin-, and ezrin-containing ruffles. It is interesting that *Shigella* cannot enter the apical surface of enterocytes but can do so only through the basolateral membranes. Thus, these cells are probably not the initial target, raising the issue of how they invade in vivo. Currently, it appears that *Shigella* is another example of the pathogenic organisms that utilize M cells to gain access to the subepithelial compartment. Once within this compartment, they invade phagocytic cells and induce apoptosis (78). In fact, interaction of bacteria with M cells resembles their interaction with phagocytic cells in many ways.

The invasome mechanism of entry has been described for *Bartonella* species, which are able to penetrate and multiply within both nucleated and nonnucleated cells. *B. bacilliformis* bacteria are able to enter erythrocytes, but, because erythrocytes have very little actin and are not endocytic, the mechanism is not completely clear (62). There appears to be little doubt, however, that entry and multiplication within erythrocytes is an important phase of the life cycle of these organisms (84). *B. henselae* bacteria are able to invade endothelial cells via a unique mechanism involving the formation of a bacterial aggregate, movement of endothelial membranes, and engulfment of the aggregate to form an "invasome" (Fig. 7.4) (13). Although this mechanism is not yet thoroughly defined, it is clearly different from the typical zipper or trigger mechanisms. The fact that a large bacterial aggregate is an apparent requisite for triggering of internalization via this mechanism suggests that expression of the effector molecules may require a quorum-sensing signaling cascade.

A unique characteristic of *Brucella abortus* is that it can enter and multiply within both phagocytic and nonphagocytic cells. *B. abortus* interacts with endocytic compartments and proliferates within the endoplasmic reticulum of host cells (80). A two-component regulatory system (*Brucella* virulence-related [*bvr*], with regulatory [BvrR] and sensory [BvrS] proteins) appears to be involved in both entry and the ability to multiply intracellularly (87).

A subgroup of the facultative pathogens which enter NPCs are bacteria that normally enter and multiply only within phagocytic cells during the infectious process, but they also are able to penetrate and multiply in epithelial cells in vitro. Typically, members of the genus *Legionella* belong to this group of intracellular bacteria (95).

EXTRACELLULAR PATHOGENS

For the two groups already discussed, the obligate and facultative intracellular organisms, it is clear that there is an intracellular phase during infections in vivo and there is a strong correlation between their infectivity and their capacity to enter and multiply within NPCs in vitro. In contrast, there is weak or nonexistent evidence for an intracellular growth phase during in vivo infections for the growing list of extracellular pathogens able to enter tissue culture cells (Table 7.3). Typically, only certain variants or phenotypes of this group of pathogens are capable of NPC entry. Thus, the role of internalization in vivo by the in vitro extracellular invaders has been challenged (52, 61). In spite of this reservation, there is circumstantial evidence that the internalization process may indeed be important during infections caused by members of the genus *Neisseria* and the species *Streptococcus pyogenes*.

The *Neisseria* species produce functionally similar, structurally diverse outer membrane proteins that mediate entry into NPCs: the gonococcal opacity-associated proteins Opa50 (also called OpaA) and the meningococcal Opc proteins (43, 60, 67, 68, 93, 97). Repulsive forces, such as those caused by the terminal sialic acid residues of neisserial lipooligosaccharide, may interfere with the interaction of the bacteria with cell surfaces. Removal of sialic acid by host neuraminidases facilitates adhesion (94). Internalization may also be facilitated by fimbrial adhesins, which may help to overcome repulsive forces and allow the intimate attachment of opacity-associated proteins and their receptors to proceed (12). The opacity proteins mediate entry by interacting with the extracellular matrix constituents heparan sulfate proteoglycans and vitronectin. The gonococcal Opa50 requires the presence of both extracellular components, while the meningococcal Opc can mediate entry when either one is present. In addition, gonococci producing Opa50 may enter NPCs (as well as phagocytic cells) by interacting with CD66 (a member of the carcinoembryonic antigen family) on the membrane of host cells. We speculate that neisserial entry does not require a type III secretion system and that it is mediated by the zipper mechanism, discussed above. This notion is supported by the findings showing that recombinant *Escherichia coli* strains producing gonococcal Opa proteins are capable

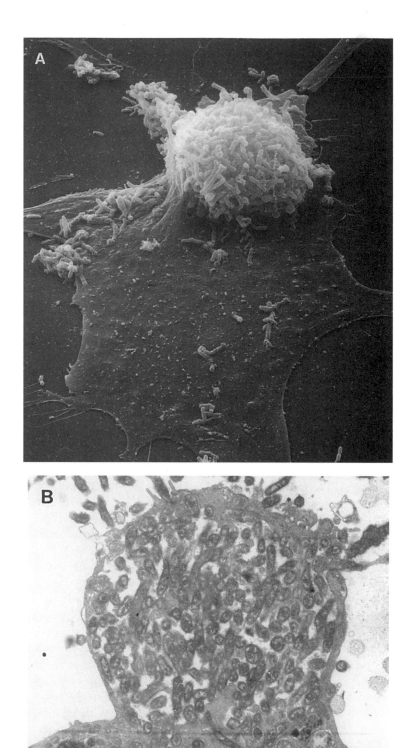

FIGURE 7.4 Electron micrographs illustrating the invasome mechanism of bacterial entry. (A) Scanning electron micrograph of a *B. henselae* invasome on the surface of a cultured endothelial cell. (Micrograph courtesy of Christoph Dehio.) (B) Transmission electron micrograph of a thin section through the invasome. (Reprinted from C. Dehio, M. Meyer, J. Berger, H. Schwarz, and C. Lanz, *J. Cell Sci.* **110:**2141–2154, 1997, with permission from the publisher.)

of invading NPCs, that Opc-mediated entry occurs only when the opacity protein is expressed at high levels, and that host cell membrane ruffling is not observed. Because Opa50 and Opc proteins are only two of many other opacity proteins encoded by one single clone and because their expression is regulated, only a small subset of the wild strains are capable of entering NPC (48, 53). Finally, an intracellular phase has been observed during natural infections, suggesting that an invader clone may be selected in vivo, although there still is no evidence of intracellular growth in vivo.

S. pyogenes is more typical of the extracellular pathogens that are internalized by NPCs than is *Neisseria*, because there is very little physical evidence to date for an intracellular phase during the natural course of infections. Nevertheless, since 1994, when it was first demonstrated that *S. pyogenes* could enter NPCs (50), there have been cumulative studies showing that the organisms can be internalized by human epithelial cells (for reviews, see references 64 and 65). Two major observations emerged from these studies. First, the entry is mediated by the fibronectin binding protein, protein F1 (also called Sfb1), and/or M protein (20, 41, 63). As in the pili of gonococci, lipoteichoic acid, which appears to be carried by both M protein and the fibronectin binding proteins, may facilitate internalization by overcoming repulsive forces and allowing the first adhesive step to occur (37). The receptors on NPC that mediate streptococcal entry are members of the integrins, with fibronectin bridging between integrins and protein F1/Sfb1 (8, 75). Latex beads coated with Sfb1 (63) or with M1 protein (15) were also ingested by NPCs, similar to the ingestion of invasin-coated beads (see above), suggesting that at the proper density Sfb1 and M1 protein bind avidly enough to cause internalization, probably by the zipper mechanism. While the receptor for M protein has not yet been defined, it is interesting that integrins have been implicated in the entry of both facultative intracellular organisms (e.g., *Yersinia*)

and extracellular organisms (e.g., group A streptococci).

Second, there is epidemiologic evidence that correlates the presence of the *prtF1* gene, the ability to be internalized by NPCs in vitro, and the ability of group A streptococci to persist in an asymptomatic state in the human oropharynx (70, 85). Moreover, the ability of the carrier strains to invade NPCs was considerably more efficient than that of isolates from the noncarriage state. Protein F1/Sfb1 is present in about half of all group A streptococcal isolates (31, 69, 90). Furthermore, internalization and expression of the adhesins/invasins M protein and protein F1/Sfb1 are under regulatory control (1, 42, 66, 76, 86). Thus, only a subset of *S. pyogenes* isolates will be able to invade NPCs via these surface molecules (15). Recently, it was suggested that the gentamicin in vitro test for measuring intracellular versus extracellular group A streptococci may underestimate the number of intracellular group A streptococci because following internalization of *nra* mutant organisms, cell membranes are sufficiently lysed to enable antibiotics to diffuse into the cells and kill the bacteria (66). At least one publication has shown streptococci within tonsillar epithelial cells by electron microscopy (74), but there is no evidence for bacterial multiplication within NPCs in vitro (32).

From the foregoing, a number of characteristic features of the extracellular pathogens that can enter NPCs appear. The organisms express adhesins/invasins which mediate entry via a zipper mechanism, they survive intracellularly but do not multiply, and they employ extracellular matrix proteins as bridges to cellular receptors (73). Perhaps most importantly, expression of the invasin/adhesin at high density is restricted to certain variants or clones and appears to provide a reservoir for the species to escape deleterious agents such as antibiotics. Indeed, in group A streptococci, strains isolated from patients with cases of penicillin treatment failure were found to be internalized by cells at a high frequency. It is possible that such variants of the species are sufficiently attenuated in their virulence to provide a persistent colonization. This notion is strengthened by the findings that nonhemolytic *Bartonella* species persist within erythrocytes, which provide a reservoir for efficient transmission by blood-sucking arthropods (84). Additionally, *Klebsiella pneumoniae,* which occupies the intestine as its common habitat and is capable of invading intestinal cells efficiently, causes symptomatic infection only in extraintestinal sites (83).

There is little clear-cut evidence to date that the entry of predominantly extracellular pathogens into NPCs in vitro truly reflects a process that is important in the pathophysiology of infections. However, the fact that this phenomenon has been described for such a large and growing list of pathogens suggests that it must play an important role in the overall survival of the species in a susceptible host. Perhaps the most likely role, at present, is as a mechanism to rescue the species from the many deleterious eukaryotic and prokaryotic agents encountered in the host. Because mucosal epithelial cells are regularly shed and eliminated, this intracellular "rescue niche" may be helpful to the pathogen for only a limited period, unless the variant can enter other cells in order to survive desquamation. For example, it has been suggested that entry of *Pseudomonas aeruginosa* into airway epithelial cells via the cystic fibrosis transmembrane regulator (79, 101) represents a mechanism of innate immunity which is missing in cystic fibrosis patients, increasing their susceptibility to infection. The process by which a variant strain becomes efficient at exploiting host cell functions in order to enter cellular compartments will undoubtedly continue to attract interest for many years, during which other roles for the invasion phenomenon may be uncovered.

REFERENCES

1. **Beckert, S., B. Kreikemeyer, and A. Podbielski.** 2001. Group A streptococcal *rofA* gene is involved in the control of several virulence genes and eukaryotic cell attachment and internalization. *Infect. Immun.* **69:**534–537.
2. **Beekhuizen, H., J. S. van de Gevel, B. Olsson, I. J. van Benten, and R. van Furth.** 1997. Infection of human vascular endothelial cells with *Staphylococcus aureus* induces hyperadhesiveness for human monocytes and granulocytes. *J. Immunol.* **158:**774–782.
3. **Bermudez, L. E., M. Petrofsky, and J. Goodman.** 1997. Exposure to low oxygen tension and increased osmolarity enhance the ability of *Mycobacterium avium* to enter intestinal epithelial (HT-29) cells. *Infect. Immun.* **65:** 3768–3773.
4. **Blocker, A., P. Gounon, E. Larquet, K. Niebuhr, V. Cabiaux, C. Parsot, and P. Sansonetti.** 1999. The tripartite type III secreton of *Shigella flexneri* inserts IpaB and IpaC into host membranes. *J. Cell Biol.* **147:**683–693.
4a. **Blocker, A., N. Jouihri, E. Larquet, P. Gounon, F. Ebel, C. Parsot, P. Sansonetti, and A. Allaoui.** 2001. Structure and composition of the *Shigella flexneri* 'needle complex,' a part of its type III secreton. *Mol. Microbiol.* **39:**652–663.
5. **Burns, J. L., M. Jonas, E. Y. Chi, D. K. Clark, A. Berger, and A. Griffith.** 1996. Invasion of respiratory epithelial cells by *Burkholderia (Pseudomonas) cepacia. Infect. Immun.* **64:**4054–4059.
6. **Cornelis, G. R.** 2002. The *Yersinia* Ysc-Yop 'type III' weaponry. *Nat. Rev. Mol. Cell Biol.* **3:**742–752.
7. **Cossart, P., P. Bouquet, S. Normark, and R. Rappuoli (ed.).** 2000. *Cellular Microbiology.* ASM Press, Washington, D.C.
8. **Cue, D. R., and P. P. Cleary.** 1997. High-frequency invasion of epithelial cells by *Streptococcus pyogenes* can be activated by fibrinogen and peptides containing the sequence RGD. *Infect. Immun.* **65:**2759–2764.
9. **Cundell, D. R., N. P. Gerard, C. Gerard, I. Idanpaan-Heikkila, and E. I. Tuomanen.** 1995. *Streptococcus pneumoniae* anchor to activated human cells by the receptor for platelet-activating factor. *Nature* **377:**435–438.
10. **Darwin, K. H., and V. L. Miller** 1999. Molecular basis of the interaction of *Salmonella* with the intestinal mucosa. *Clin. Microbiol. Rev.* **12:**405–428.
11. **de Boer, E. C., R. F. Bevers, K. H. Kurth, and D. H. Schamhart.** 1996. Double fluorescent flow cytometric assessment of bacterial internalization and binding by epithelial cells. *Cytometry* **25:**381–387.

12. **Dehio, C., S. D. Gray-Owen, and T. F. Meyer.** 1998. The role of neisserial Opa proteins in interactions with host cells. *Trends Microbiol.* **6:**489–494.

13. **Dehio, C.** 1999. Interactions of *Bartonella henselae* with vascular endothelial cells. *Curr. Opin. Microbiol.* **2:**78–82.

14. **Dogan, S., F. Gunzer, H. Guenay, G. Hillmann, and W. Geurtse.** 2000. Infection of primary human gingival fibroblasts by *Porphyromonas gingivalis* and *Prevotella intermedia*. *Clin. Oral Investig.* **4:**35–41.

15. **Dombek, P. E., D. Cue, J. Sedgewick, H. Lamb, S. Ruschkowski, B. B. Finlay, and P. P. Cleary.** 1999. High-frequency intracellular invasion of epithelial cells by serotype M1 group A streptococci: M1 protein mediated invasion and cytoskeletal rearrangements. *Mol. Microbiol.* **31:**859–870.

16. **Feltis, B. A., S. M. Wiesner, A. S. Kim, S. L. Erlandsen, D. L. Lyerly, T. D. Wilkins, and C. L. Wells.** 2000. *Clostridium difficile* toxins A and B can alter epithelial permeability and promote bacterial paracellular migration through HT-29 enterocytes. *Shock* **14:**629–634.

17. **Finlay, B. B., and P. Cossart.** 1997. Exploitation of mammalian host cell functions by bacterial pathogens. *Science* **276:**718–725.

18. **Fleiszig, S. M., T. S. Zaidi, and G. B. Pier.** 1995. *Pseudomonas aeruginosa* invasion of and multiplication within corneal cells in vitro. *Infect. Immun.* **63:**4072–4077.

19. **Fleiszig, S. M., T. S. Zaidi, M. J. Preston, M. Grout, D. J. Evans, and G. B. Pier.** 1996. Relationship between cytotoxicity and corneal epithelial cell invasion by clinical isolates of *Pseudomonas aeruginosa*. *Infect. Immun.* **64:**2288–2294.

20. **Fluckiger, U., K. F. Jones, and V. A. Fischetti.** 1998. Immunoglobulins to group A streptococcal surface molecules decrease adherence to and invasion of human pharyngeal cells. *Infect. Immun.* **66:**974–979.

21. **Francis, C., M. N. Starnback, and S. Falkow.** 1992 Morphological cytoskeletal changes in epithelial cells occur immediately upon interaction with *Salmonella typhimurium* grown under low oxygen conditions. *Mol. Microbiol.* **6:**3077–3087.

22. **Franklin, C. L., D. A. Kinden, P. L. Stogsdill, and L. K. Riley.** 1993. In vitro model of adhesion and invasion by *Bacillus piliformis*. *Infect. Immun.* **61:**876–883.

23. **Gaillard, J. L., and B. B. Finlay.** 1996. Effect of cell polarization and differentiation on entry of *Listeria monocytogenes* into the entero-cyte-like Caco-2 cell line. *Infect. Immun.* **64:**1299–1308.

24. **Galán, J. E.** 1994. Interactions of bacteria with non-phagocytic cells. *Curr. Opin. Immunol.* **6:**590–595.

25. **Galán, J. E.** 1996. Molecular and cellular bases of *Salmonella* entry into host cells. *Curr. Top. Microbiol. Immunol.* **209:**43–60.

26. **Galán, J. E.** 2000. Alternative strategies for becoming an insider: lessons from the bacterial world. *Cell* **103:**363–366.

27. **Galán, J. E., and D. Zhou.** 2000. Striking a balance: modulation of the actin cytoskeleton by *Salmonella*. *Proc. Natl. Acad. Sci. USA* **97:**8754–8761.

28. **Galán, J. E.** 2001. *Salmonella* interactions with host cells: type III secretion at work. *Annu. Rev. Cell Dev. Biol.* **17:**53–86.

29. **Garduño, R. A., A. R. Moore, G. Olivier, A. L. Lizama, E. Garduño, and W. W. Kay.** 2000. Host cell invasion and intracellular residence by *Aeromonas salmonicida*: role of the S-layer. *Can. J. Microbiol.* **46:**660–668.

30. **Ginocchio, C., S. B. Olmsted, C. L. Wells, and J. E. Galan.** 1994 Contact with epithelial cells induces the formation of surface apendages on *Salmonella typhimurium*. *Cell* **76:**717–724.

31. **Goodfellow, A. M., M. Hibble, S. R. Talay, B. Kreikemeyer, B. J. Currie, K. S. Sriprakash, and G. S. Chhatwal.** 2000. Distribution and antigenicity of fibronectin-binding proteins (SfbI and SfbII) of *Streptococcus pyogenes* clinical isolates from the Northern Territory, Australia. *J. Clin. Microbiol.* **38:**389–392.

32. **Greco, R., L. De Martino, G. Donnarumma, M. P. Conte, L. Seganti, and P. Valenti.** 1995. Invasion of cultured human cells by *Streptococcus pyogenes*. *Res. Microbiol.* **146:**551–560.

33. **Griffin, F. M., J. A. Griffin, J. E. Leider, and S. C. Silverstein.** 1975. Studies on the mechanism of phagocytosis. I. Requirements for circumferential attachment of particle-bound ligands to specific receptors on the macrophage plasma membrane. *J. Exp. Med.* **142:**1263–1282.

34. **Hamill, R. J., J. M. Vann, and R. A. Proctor.** 1986. Phagocytosis of *Staphylococcus aureus* by cultured bovine aortic endothelial cells: model for postadherence events in endovascular infections. *Infect. Immun.* **54:**833–836.

35. **Han, Y. W., W. Shi, G. T. Huang, S. Kinder Haake, N. H. Park, H. Kuramitsu, and R. J. Genco.** 2000. Interactions between periodontal bacteria and human oral epithelial

cells: *Fusobacterium nucleatum* adheres to and invades epithelial cells. *Infect. Immun.* **68:**3140–3146.

36. **Hartland, E. L., V. Huter, L. M. Higgins, N. S. Goncalves, G. Dougan, A. D. Phillips, T. T. MacDonald, and G. Frankel.** 2000. Expression of intimin gamma from enterohemorrhagic *Escherichia coli* in *Citrobacter rodentium*. *Infect. Immun.* **68:**4637–4646.

37. **Hasty, D. L., I. Ofek, H. S. Courtney, and R. J. Doyle.** 1992. Multiple adhesins of streptococci. *Infect. Immun.* **60:**2147–2152.

38. **Hensel, M.** 2000. Salmonella pathogenicity island 2. *Mol. Microbiol.* **36:**1015–1023.

39. **Hulse, M. L., S. Smith, E. Y. Chi, A. Pham, and C. E. Rubens.** 1993. Effect of type III group B streptococcal capsular polysaccharide on invasion of respiratory epithelial cells. *Infect. Immun.* **61:**4835–4841.

40. **Isberg, R. R., and G. Tranh Van Nhieu.** 1994. Two mammalian cell internalization strategies used by pathogenic bacteria. *Annu. Rev. Genet.* **28:**395–422.

41. **Jadoun, J., V. Ozeri, E. Burstein, E. Skutelsky, E. Hanski, and S. Sela.** 1998. Protein F1 is required for efficient entry of *Streptococcus pyogenes* into epithelial cells. *J. Infect. Dis.* **178:**147–158.

42. **Jadoun, J., and S. Sela.** 2000. Mutation in csrR global regulator reduces *Streptococcus pyogenes* internalization. *Microb. Pathog.* **29:**311–317.

43. **Jerse, A. E., and R. F. Rest.** 1997. Adhesion and invasion by the pathogenic neisseria. *Trends Microbiol.* **5:**217–221.

44. **Jones, A. L., D. DeShazer, and D. E. Woods.** 1997. Identification and characterization of a two-component regulatory system involved in invasion of eukaryotic cells and heavy-metal resistance in *Burkholderia pseudomallei*. *Infect. Immun.* **65:**4972–4977.

45. **Jouve, M., M. I. Garcia, P. Courcoux, A. Labigne, P. Gounon, and C. le Bouguenec.** 1997. Adhesion to and invasion of HeLa cells by pathogenic *Escherichia coli* carrying the *afa-3* gene cluster are mediated by the AfaE and AfaD proteins, respectively. *Infect. Immun.* **65:**4082–4089.

46. **Kubori, T., Y. Matsuchima, D. Nakamura, J. Uralil, M. Lara-Tejero, A. Skhan, J. E. Galán, and S. I. Aizawa.** 1998. Supramolecular structure of *Salmonella typhimurium* type III protein secretion. *Science* **280:**602–605.

47. **Kubori, T., A. Sukhan, S. I. Aizawa, and J. E. Galán.** 2000. Molecular characterization and assembly of the needle complex of the *Salmonella typhimurium* type III protein secretion system. *Proc. Natl. Acad. Sci. USA* **97:**10225–10230.

48. **Kupsch, E. M., B. Knepper, T. Kuroki, I. Heuer, and T. F. Meyer.** 1993. Variable opacity (Opa) outer membrane proteins account for the cell tropism displayed by *Neisseria gonorrhoeae* for human leukocytes and epithelial cells. *EMBO J.* **12:**641–650.

49. **Lalonde, M., M. Segura, S. Lacouture, and M. Gottschalk.** 2000. Interactions between *Streptococcus suis* serotype 2 and different epithelial cell lines. *Microbiolology* **146:**1913–1921.

50. **LaPenta, D., C. Rubens, E. Chi, and P. P. Cleary.** 1994. Group A streptococci efficiently invade human respiratory epithelial cells. *Proc. Natl. Acad. Sci. USA* **91:**12115–12119.

51. **Lee, C. A., and S. Falkow** 1990. The ability of *Salmonella* to enter mammalian cells is affected by bacterial growth state. *Proc. Natl. Acad. Sci. USA* **87:**4304–4308.

52. **Lowy, F. D.** 2000. Is *Staphylococcus aureus* an intracellular pathogen? *Trends Microbiol.* **8:**341–343.

53. **Makino, S., J. P. van Putten, and T. F. Meyer.** 1991. Phase variation of the opacity outer membrane protein controls invasion by *Neisseria gonorrhoea* into human epithelial cells. *EMBO J.* **10:**1307–1315.

54. **Marra, A., and R. R. Isberg.** 1996. Bacterial pathogenesis: common entry mechanisms. *Curr. Biol.* **6:**1084–1086.

55. **McCallum, S. J., and J. A. Theriot.** 2000. Bacterial manipulation of the host cell cytoskeleton, p. 171–191. *In* P. Cossart, P. Boquet, S. Normark, and R. Rappuoli (ed.), *Cellular Microbiology*. ASM Press, Washington, D.C.

56. **Meier, C., T. A. Oelschlaeger, H. Merkert, T. K. Korhonen, and J. Hacker.** 1996. Ability of *Escherichia coli* isolates that cause meningitis in newborns to invade epithelial and endothelial cells. *Infect. Immun.* **64:**2391–2399.

57. **Ménard, R., M. C. Prevost, P. Gounon, P. Sansonetti, and C. Dehio.** 1996. The secreted Ipa complex of Shigella flexneri promotes entry into mammalian cells. *Proc. Natl. Acad..Sci. USA* **93:**1254–1258.

58. **Merien, F., G. Baranton, and P. Perolat.** 1997. Invasion of Vero cells and induction of apoptosis in macrophages by pathogenic *Leptospira interrogans* are correlated with virulence. *Infect. Immun.* **65:**729–738.

59. **Meyer, D. H., J. E. Lippmann, and P. M. Fives-Taylor.** 1996. Invasion of epithelial cells by *Actinobacillus actinomycetemcomitans*: a dynamic, multistep process. *Infect. Immun.* **64:**2988–2997.

60. **Meyer, T. F.** 1999. Pathogenic *Neisseriae:* complexity of pathogen-host cell interplay. *Clin. Infect. Dis.* **28:**433–441.

61. **Miller, V. L.** 1995. Tissue-culture invasion: fact or artefact? *Trends Microbiol.* **3:**69–71.

62. **Minnick, M. F., S. J. Mitchell, and S. J. McAllister.** 1996. Cell entry and the pathogenesis of *Bartonella* infections. *Trends Microbiol.* **4:**343–347.

63. **Molinari, G., S. R. Talay, P. Valentin-Weigand, M. Rohde, and G. S. Chhatwal.** 1997. The fibronectin-binding protein of *Streptococcus pyogenes,* SfbI, is involved in the internalization of group A streptococci by epithelial cells. *Infect. Immun.* **65:**1357–1363.

64. **Molinari, G., and G. S. Chhatwal.** 1999. Streptococcal invasion. *Curr. Opin. Microbiol.* **2:**56–61.

65. **Molinari, G., and G. S. Chhatwal.** 1999. Role played by the fibronectin-binding protein SfbI (protein F1) of *Streptococcus pyogenes* in bacterial internalization by epithelial cells. *J. Infect. Dis.* **179:**1049–1050.

66. **Molinari, G., M. Rhode, S. R. Talay, G. S. Chhatwal, S. Beckert, and A. Podbielski.** 2001. The role played by the group A streptococcal negative regulator Nra on bacterial interactions with epithelial cells. *Mol. Microbiol.* **40:**99–114.

67. **Nassif, X.** 1999. Interaction mechanisms of encapsulated meningococci with eukaryotic cells: what does this tell us about the crossing of the blood-brain barrier by *Neisseria meningitidis? Curr. Opin. Microbiol.* **2:**71–77.

68. **Nassif, X., and M. So.** 1995. Interaction of pathogenic neisseriae with nonphagocytic cells. *Clin. Microbiol. Rev.* **8:**376–388.

69. **Natanson, S., S. Sela, A. E. Moses, J. M. Musser, M. G. Caparon, and E. Hanski.** 1995. Distribution of fibronectin-binding proteins among group A streptococci among different M types. *J. Infect. Dis.* **171:**871–878.

70. **Neeman, R., N. Keller, A. Barzilai, Z. Korenman, and S. Sela.** 1998. Prevalence of internalization-associated gene, *prtF1,* among persisting group-A streptococcus strains isolated from asymptomatic carriers. *Lancet* **352:**1974–1977.

71. **Oelschlaeger, T. A., P. Guerry, and D. J. Kopecko.** 1993. Unusual microtubult-dependent endocytosis mechanisms triggered by *Campylobacter jejuni* and *Citrobacter freundii. Proc. Natl. Acad. Sci. USA* **90:**6884–6888.

72. **Oelschlaeger, T. A., and B. Tall.** 1997. Invasion of cultured human epithelial cells by *Klebsiella pneumoniae* isolated from the urinary tract. *Infect. Immun.* **65:**2950–2958.

73. **Oelschlaeger, T. A.** 2001. Adhesins as invasins. *Int. J. Med. Microbiol.* **291:**7–14.

74. **Osterlund, A., R. Popa, T. Nikkila, A. Scheynius, and L. Engstrand.** 1997. Intracellular reservoir of *Streptococcus pyogenes in vivo:* a possible explanation for recurrent pharyngotonsillitis. *Laryngoscope* **107:**640–647.

75. **Ozeri, V., I. Rosenshine, D. F. Mosher, R. Fässler, and E. Hanski.** 1998. Roles of integrins and fibronectin in the entry of *Streptococcus pyogenes* into cells via protein F1. *Mol. Microbiol.* **30:**625–637

76. **Perez-Casal, J., M. G. Caparon, and J. R. Scott.** 1991. Mry, a *trans-*acting positive regulator of the M protein gene of *Streptococcus pyogenes* with similarities to the receptor proteins of two-component regulatory systems. *J. Bacteriol.* **173:**2617–2624.

77. **Perraud, A. L., V. Weiss, and R. Gross.** 1999. Signalling pathways in two-component phosphorelay systems. *Trends Microbiol.* **7:**115–120.

78. **Phalipon, A., and P. J. Sansonetti.** 1999. Microbial-host interactions at mucosal sites. Host response to pathogenic bacteria at mucosal sites. *Curr. Top. Microbiol. Immunol.* **236:**163–189.

79. **Pier, G. B., M. Grout, and T. S. Zaidi.** 1997. Cystic fibrosis transmembrane conductance regulator is an epithelial cell receptor for clearance of *Pseudomonas aeruginosa* from the lung. *Proc. Natl. Acad. Sci. USA* **94:**12088–12093.

80. **Pizaro-Cerda, J., E. Moreno, and J. P. Gorvel.** 2000. Invasion and intracellular trafficking of *Brucella abortus* in nonphagocytic cells. *Microbes Infect.* **2:**829–835.

81. **Rasmussen-Lathrop, S. J., K. Koshiyama, N. Phillips, and R. S. Stephens.** 2000. *Chlamydia*-dependent biosynthesis of a heparan sulphate-like compound in eukaryotic cells. *Cell. Microbiol.* **2:**137–144.

82. **Raupach, B., J. Mecsas, U. Heczko, S. Falkow, and B. B. Finlay.** 1999. Bacterial epithelial cell cross talk. *Curr. Top. Microbiol. Immunol.* **236:**137–161.

83. **Sahly, H., R. Podschun, T. A. Oelschlaeger, M. Greiwe, H. Parolis, D. Hasty, J. Kekow, U. Ullmann, I. Ofek, and S. Sela.** 2000. Capsule impedes adhesion to and invasion of epithelial cells by *Klebsiella pneumoniae. Infect. Immun.* **68:**6744–6749.

84. **Schülein R., A. Seubert, C. Gille, C. Lanz, Y. Hansmann, Y. Piemont, and C. Dehio.** 2001. Invasion and persistence intracellular colonization of erythrocytes: a unique parasitic strategy of the emerging pathogen *Bartonella. J. Exp. Med.* **193:**1077–1086.

85. **Sela, S., R. Neeman, N. Keller, and A. Barzilai.** 2000. Relationship between asymptomatic carriage of *Streptococcus pyogenes* and ability of the strains to adhere and internalise cultured epithelial cells. *J. Med. Microbiol.* **49:**499–502.

86. **Simpson, W. J., D. LaPenta, C. Chen, and P. P. Cleary.** 1990. Coregulation of type 12 M protein and streptococcal C5a peptidase genes in group A streptococci: evidence for a virulence regulon controlled by the VirR locus. *J. Bacteriol.* **172:**696–700.

87. **Sola-Landa, A., J. Pizarro-Cerda, M. J. Grillo, E. Moreno, I. Moriyon, J. M. Blasco, J. P. Gorvel, and I. Lopez-Goni.** 1998. A two-component regulatory system playing a critical role in plant pathogens and endosymbionts is present in *Brucella abortus* and controls cell invasion and virulence. *Mol. Microbiol.* **29:**125–138.

88. **Stephens, R. S., F. S. Fawaz, K. A. Kennedy, K. Koshiyama, B. Nichols, C. van Ooij, and J. N. Engel.** 2000. Eukaryotic cell uptake of heparin-coated microspheres: a model of host cell invasion by *Chlamydia trachomatis. Infect. Immun.* **68:**1080–1085.

89. **Tran Van Nhieu, G., and P. J. Sansonetti.** 1999. Mechanism of *Shigella* entry into epithelial cells. *Curr. Opin. Microbiol.* **2:**51–55.

90. **Valentin-Wiegand, P., S. R. Talay, A. Kaufhold, K. N. Timmis, and G. S. Chhatwal.** 1994. The fibronectin-binding domain of Sfb protein of *Streptococcus pyogenes* occurs in many group A streptococci and does not cross-react with heart myosin. *Microb. Pathog.* **17:**111–120.

91. **Valentin-Wiegand, P., P. Benkel, M. Rohde, and G. S. Chhatwal.** 1996. Entry and intracellular survival of group B streptococci in J774 macrophages. *Infect. Immun.* **64:**2467–2473.

92. **van Schilfgaarde, M., P. van Ulsen, W. van Der Steeg, V. Winter, P. Eijk, V. Everts, J. Dankert, and L. van Alphen.** 2000. Cloning of genes of nontypeable *Haemophilus influenzae* involved in penetration between human lung epithelial cells. *Infect. Immun.* **68:**4616–4623.

93. **Virji, M., K. Makepeace, D. G. P. Ferguson, M. Achtman, J. Sarkari, and E. R. Moxon.** 1992. Expression of the Opc protein correlates with invasion of epithelial and endothelial cells by *Neisseria meningitidis. Mol. Microbiol.* **6:**2785–2795.

94. **Virji, M., K. Makepeace, D. G. P. Ferguson, M. Achtman, and E. R. Moxon.** 1993.

Meningococcal Opa and Opc proteins: their role in colonization and invasion of human epithelial and endothelial cells. *Mol. Microbiol.* **10:**499–510.

95. **Vogel, P. J., and R. R. Isberg.** 1999. Cell biology of *Legionella pneumophila. Curr. Opin. Microbiol.* **2:**30–34.

96. **Watarai, M., S. Funato, and C. Sasakawa.** 1996. Interaction of Ipa proteins of *Shigella flexneri* with alpha5-beta1 integrin promotes entry of the bacteria into mammalian cells. *J. Exp. Med.* **183:**991–999.

97. **Weel, J. F. L., C. T. P. Hopman, and J. P. M. van Putten.** 1991. *In situ* expression and localization of *Neisseria gonorrhoeae* opacity proteins in infected epithelial cells: apparant role of Opa proteins in cellular invasion. *J. Exp. Med.* **173:**1395–1405.

98. **Weinberg, A., C. M. Belton, Y. Park, and R. J. Lamont.** 1997. Role of fimbriae in *Porphyromonas gingivalis* invasion of gingival epithelial cells. *Infect. Immun.* **65:**313–316.

99. **Wells, C. L., E. M. A. van de Westerlo, R. P. Jechorek, H. M. Haines, and S. L. Erlandsen.** 1998. Cytochalasin-induced actin disruption of polarized enterocytes can augment internalization of bacteria. *Infect. Immun.* **66:**2410–2419.

100. **Wyrick, P. B., J. Choong, C. H. Davis, S. T. Knight, M. O. Royal, A. S. Maslow, and C. R. Bagnell.** 1989. Entry of genital *Chlamydia trachomatis* into polarized human epithelial cells. *Infect. Immun.* **57:** 2378–2389.

101. **Zaidi, T. S., J. Lyczak, M. Preston, and G. B. Pier.** 1999. Cystic fibrosis transmembrane conductance regulator-mediated corneal epithelial cell ingestion of *Pseudomonas aeruginosa* is a key component in the pathogenesis of experimental murine keratitis. *Infect. Immun.* **67:**1481–1492.

102. **Zaretzky, F. R., and T. H. Kawula.** 1999. Examination of early interactions between *Haemophilus ducreyi* and host cells by using cocultured HaCaT keratinocytes and foreskin fibroblasts. *Infect. Immun.* **67:**5352–5360.

103. **Zhang, J. R., K. E. Mostov, M. E. Lamm, M. Nanno, S. Shimida, M. Ohwaki, and E. Tuomanen.** 2000. The polymeric immunoglobulin receptor translocates pneumococci across human nasopharyngeal epithelial cells. *Cell* **102:**827–837.

104. **Zierler, M. K., and J. E Galán.** 1995. Contact with cultured epithelial cells stimulates secretion of *Salmonella* invasion protein InvJ. *Infect. Immun.* **63:**4024–4028.

POSTADHESION EVENTS INDUCED IN NONPHAGOCYTIC CELLS

8

Logic states that when a bacterium makes contact with a host cell, it is likely that biochemical changes in host cell metabolism will ensue. For only a few cases, however, have these events been studied in detail. In even fewer cases has the role of these events in the infectious process been elucidated. In this chapter, we focus on contact-induced signaling of nonphagocytic cells as an immediate postadhesion event. We discuss two major events, one involving an immediate activation of cytoskeletal responses and the other involving activation of transcriptional responses. The former is needed for completion of internalization, while the latter results in the production of inflammatory mediators that can have a great systemic influence on the course of infection.

CYTOSKELETAL RESPONSES

While activation of the cytoskeletal system is likely to occur during the entry of virtually all categories of invaders discussed in chapter 7, the molecular events leading to cytoskeletal activation is different for the various pathogens (48). Three major mechanisms have been described for the activation of the cytoskeletal system by pathogens (Table 8.1). One is the receptor-mediated mechanism whereby the bacteria bind to a transmembrane receptor either directly or indirectly through a bridging molecule (Fig. 8.1). The direct interaction is illustrated by the invasin of *Yersinia* and the internalin of *Listeria,* which bind with high affinity to β-integrin and E-cadherin, respectively. The bridging mechanism is illustrated by a fibronectin binding protein adhesin, protein F1, of *Streptococcus pyogenes* (38, 52, 56), and by YadA of *Yersinia* (37, 59), which utilize the extracellular matrix component fibronectin as a bridge to an integrin receptor. In both cases, ligand-receptor interactions lead to clustering of integrins in the host cell. Focal adhesion kinase (FAK) binds to the clustered cytoplasmic tails of integrin, and phosphorylation cascades lead to accumulation of actin binding proteins that are components of focal adhesion sites (e.g., α-actinin, vinculin, and talin) (37).

The second mechanism of cytoskeletal rearrangement is the type III secretion-dependent ruffling of host cell membranes, whereby bacterial contact initiates formation or completion of a secretory apparatus that "injects" effector molecules into the host cell (Fig. 8.2). At the site of bacterial contact, an arrangement of cytoskeletal molecules reminiscent of a focal adhesion site forms in response to the activation of the small GTPase, Rho, by the translocated effector

TABLE 8.1 Some mechanisms of cytoskeleton rearrangements caused by bacteria encountering host cells

Mechanism	Organism	Receptor[a]	Adhesin	Adhesin/invasin	Effector molecules[a]	Reference(s)
Receptor-dependent zipper-like						
Direct						
	Yersinia spp.	Integrin	?	Invasin	None	37
	Listeria spp.	E-cadherin	?	Internalin	None	43, 44
Bridging						
	Yersinia spp.	Fn/integrin	?	YadA	None	37
	S. pyogenes	Integrin	Lipoteichoic acid	F protein	None	50, 51
Secretion-dependent ruffles						
	Shigella spp. (type III secretion)	?	?	?	IpaB, IpaC	67
	Salmonella spp.	?	?	?	SipA, SipB	10
Secretion-dependent pedestal						
	EPEC (type III secretion)	Tir	Bundle-forming pili	Intimin	Tir	11, 18
	Helicobacter spp. (type IV secretion)	Sialyl	Sialic or Lewis acid specific		CagA	53

[a]Receptors or effector molecules implicated in cytoskeleton rearrangement.

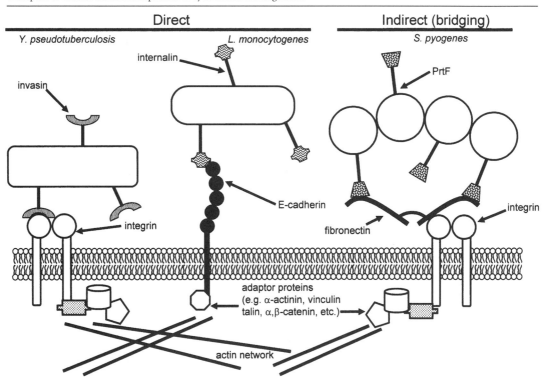

FIGURE 8.1 Schematic presentation of direct and indirect ("bridging") mechanisms of cytoskeleton activation by invading bacteria. Initiation of signal transduction that activates adaptor proteins linking the receptor to the actin cytoskeleton is shared by the two mechanisms. The direct mechanism is illustrated by binding of *Yersinia pseudotuberculosis* to integrin via invasin and the binding of *Listeria monocytogenes* to cadherin via internalin. The indirect mechanism is illustrated by the binding of *S. pyogenes* to integrin via binding of fibronectin binding proteins (e.g., protein F1) to a fibronectin bridge.

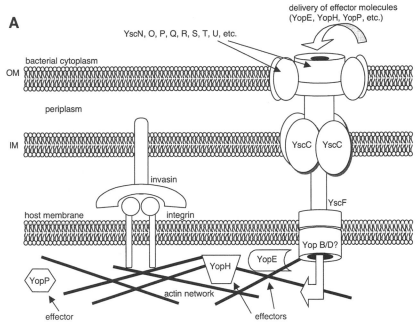

FIGURE 8.2 Schematic presentation of type III secretion apparatus-mediated activation of the cytoskeletal system. Effector molecules are translocated from the bacterial cytosol into the animal cell cytosol via a complex type III apparatus formed by more than a dozen proteins. This apparatus creates a channel, or translocation pore, that connects across both inner and outer bacterial membranes as well as the animal cell membrane, enabling molecular transport. (A) In *Yersinia,* the effector molecules (e.g., YopE and YopH proteins) cause rearrangement of the actin network connected to the *Yersinia* invasin molecule via integrin, leading to internalization. In phagocytic cells, these effector molecules may, instead, inhibit phagocytosis. (B) In EPEC, one of the effector molecules tranlocated via the type III apparatus becomes inserted into the animal cell membrane and there acts as a receptor for intimin, called the translocated intimin receptor (Tir). In each case, there is rearrangement of cytoskeletal elements and animal cell membranes at the site of bacterial contact. (C) In *Salmonella,* the type III secretion system also involves the formation of a tubular structure by which effector molecules are translocated into the host cell. These effectors, numbering more than a dozen, exert dramatic effects on host cell membranes, resulting in remarkable membrane ruffling and macropinocytic ingestion of bacteria.

molecules. This mechanism is exemplified by members of the genera *Salmonella* and *Shigella,* which employ type III secretion apparatuses to translocate effector proteins, such as SipB and SipC or IpaB and IpaC, respectively. In *Salmonella,* more than 15 proteins are translocated into host cells (21). These bacterial effectors bypass the typical ligand-induced activation of integrins. Although they are assumed to exist, neither an adhesin nor receptor has yet been defined. It has been suggested that the needle-like secretory apparatus (Fig. 8.2) itself may stabilize the bacterium-host cell contact (10, 20, 22).

The third mechanism is secretion-dependent pedestal formation. In this mechanism, initial contact between the bacteria and the host cell is mediated by an adhesin, and this is followed by more intimate adhesion mediated by a specific outer membrane protein. This mechanism is exemplified by enteropathogenic *Escherichia coli* (EPEC) (Fig. 8.3) and by *Helicobacter pylori.* Bundle-forming pili of EPEC initiate the initial contact of the bacteria with the host cell, and this is followed by intimin-mediated avid adhesion supported by the concomitant translocation of the intimin receptor, Tir (for "translocated intimin recep-

B

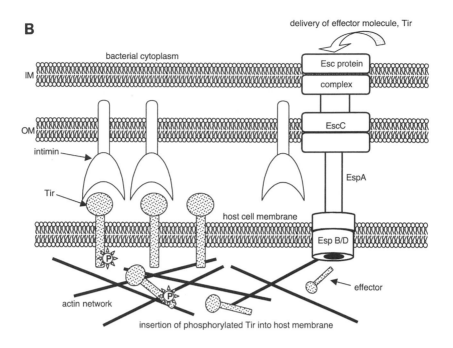

delivery of effector molecule, Tir

bacterial cytoplasm

IM

Esc protein

complex

OM

EscC

intimin

EspA

Tir

host cell membrane

Esp B/D

effector

actin network

insertion of phosphorylated Tir into host membrane

C

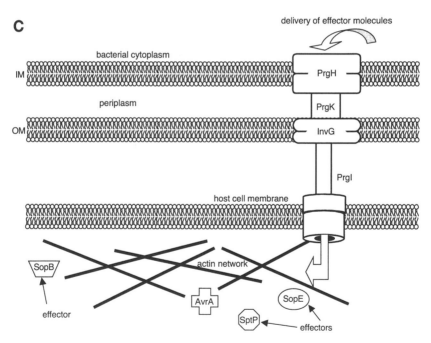

delivery of effector molecules

bacterial cytoplasm

IM

PrgH

periplasm

PrgK

OM

InvG

PrgI

host cell membrane

SopB

actin network

SopE

effector

AvrA

SptP

effectors

FIGURE 8.2 *Continued*

tor"), into the host cell membrane via a type III secretory apparatus. Tir apparently becomes embedded in the host cell membrane as a transmembrane protein whose cytoplasmic portion is phosphorylated and interacts with filamentous actin and other cytoskeletal proteins to activate major cytoskeletal rearrangements that result in the formation of a structure that has been called a pedestal (18).

For *H. pylori*, the initial attachment is mediated by one of its various adhesins (e.g., sialic acid-specific or Lewis[B]-specific adhesins [see chapter 12]). The organism then translocates CagA into the host cell via a type IV secretion system to induce alterations of the cytoskeleton in the formation of a pedestal. In both EPEC and *H. pylori,* cytoskeletal rearrangements result in pedestal formation. While in each of the previous examples bacteria are completely engulfed by the nonphagocytic cells (NPCs), in these examples there appear to be only abortive attempts at invasion, because the bacteria remain externally attached at the tip of the NPC pedestal (Fig. 8.4). This is also called attachment/effacement adhesion, which is characterized by the effacement of microvilli following attachment and also by the formation of a dense concentration of microfilaments beneath the adherent bacteria (40). Confocal microscopy beautifully illustrates the linkage of EPEC, Tir, and actin accumulation (Fig. 8.4) (18).

SECRETORY SYSTEM-DEPENDENT TRANSCRIPTIONAL RESPONSES

While the cytoskeletal response is associated mainly with actual or abortive entry of bacteria into the cell, the transcriptional response may result either from adherent or from internalized bacteria. An adherent bacterium can transmit a transcriptional signal to NPCs in two ways. One way is by using a specific secretory system to create a channel through which effector molecules can be passed into the host cell cytosol, and the second is by secreting effector molecules into the space between the bacterium and the cell, resulting in a relatively high concentration at the cell surface (72). Cells of the immune system, however, are much more reactive, and signals can sometimes be stimulated by binding of isolated adhesins.

Translocation of Effector Molecules via Type III or Type IV Secretion Apparatus

Examples of effector molecules translocated by secretion systems to induce transcriptional activation include the CagA of *H. pylori*, the IpaA of *Shigella,* the Sip proteins and SopE of *Salmonella,* and the Yop proteins of *Yersinia* (35). These effector molecules can activate or inhibit transcription either by the NF-κβ pathway or via mitogen-activated protein kinase kinase pathways (MAPKK) (Table 8.2). The end result of transcriptional activation could be either stimulation or inhibition of cytokine production or, in phagocytic cells only, stimulation of apoptosis. Some of these effects are more pronounced in phagocytic cells.

The effector molecules of the Yop family are encoded by a large plasmid which is essential for *Yersinia* virulence. The mechanisms at the molecular level have been elucidated for only a few of these proteins. Their activity in the host cell cytoplasm is dependent on both adhesion and the formation of the secretory apparatus, which directs the effector proteins into the host cell cytoplasm via the attached bacteria. It has been suggested that the activity of these proteins during the infectious process is important mainly when *Yersinia* encounters phagocytes or specialized epithelial cells (e.g., M cells), although similar activities have been seen in NPCs in vitro. YopJ/YopP (homologues from different species of *Yersinia*) is translocated via a type III secretion system into the host cell, where it becomes responsible for downregulating NF-κβ and MAPKK activities, resulting in a lowered cytokine response to lipopolysaccharide (LPS) or other immunostimulatory molecules. YopH also inhibits cellular responses by dephosphorylating FAK and targets the focal adhesion-associated protein, p130[cas], resulting in reduced

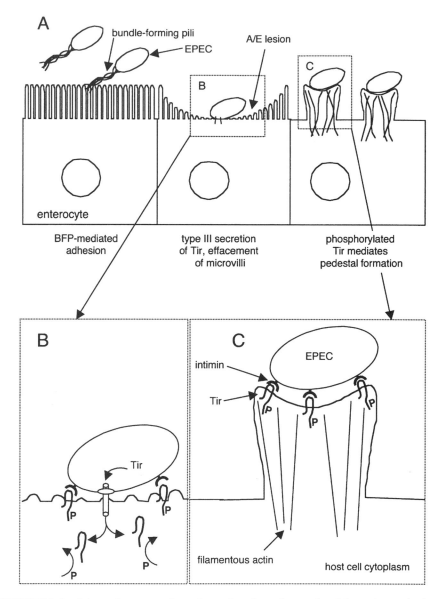

FIGURE 8.3 Schematic presentation of secretion-dependent pedestal formation and subsequent cytoskeletal rearrangement. (A) EPEC adheres to enterocytes via bundle-forming pili. Attachment enables the EPEC type III secretion apparatus to contact epithelial cells and initiates the local destruction of microvilli. (B) Tir is the major effector molecule and is thought to be translocated into the host cells through a channel formed by an EspA- and an EspB/D-generated pore in the host cell plasmalemma (see also Fig. 8.2B). Tir integrates into the host cell membrane and becomes a receptor for the bacterial adhesin, intimin. (C) Phosphorylation of Tir triggers cytoskeletal rearrangements that result in the formation of a structure that has been called a pedestal. Pedestals have a core filled with filamentous actin and other cytoskeletal proteins, such as α-actinin, ezrin, and talin, and can extend up to 10 μm outward from the surface of the cells.

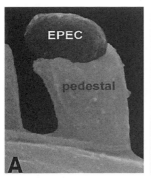
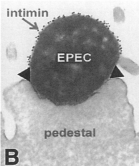
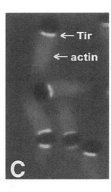

FIGURE 8.4 Microscopic examination of pedestal formation by EPEC. (A) Scanning electron micrograph of a pedestal formed in response to EPEC by an epithelial cell in tissue culture. The plasmalemma is dramatically deformed during the formation of this attaching/effacing lesion, with pedestals extending up to 10 μm from the surface of the epithelial cell. However, the bacteria are not engulfed and remain at the tips of the pedestals. (B) Transmission electron micrograph of an immunogold-labeled EPEC cell on a pedestal. Colloidal gold marks the uniform expression of intimin on the bacterial surface (arrow). The marker is excluded from the bacterium/pedestal interface only because antibody cannot penetrate into the area of intimate attachment (i.e., the area demarcated by arrowheads). (C) Immunolocalization of Tir by confocal immunofluorescence microscopy. The "body" of the pedestal is indicated by actin filament immunostaining, and Tir is localized to the bacterium/pedestal interfaces. Bacteria appear as dark ovals. (Reprinted from reference 18 with permission from the publisher.)

uptake of the bacteria by NPCs. Nevertheless, *Yersinia* can be found within NPCs, at least in vitro.

In *Salmonella,* a large number of proteins are translocated into target cells by adherent bacteria via a type III secretion system. They are encoded by the SPI1 pathogenicity island (see chapter 7). A few of these proteins are analogs to other translocated proteins from *Shigella* or *Yersinia.* For example, SptP and AvrA are structural homologs of YopE/YopH and YopJ/YopP, respectively, as well of to exoenzyme S of *Pseudomonas aeruginosa* (10). Functional homology is less apparent. In contrast, SopE appears to be unique. SopE stimulates both activation of the Rho subfamily of small GTPases to generate cytoskeleton changes required for membrane ruffling and NF-κβ pathways which activate cytokine gene transcription. Mutations in SopE eliminate membrane ruffling, while microinjection of SopE into mammalian cells causes membrane ruffling and nuclear responses (25). Unlike the above effector molecules, SopB is translocated into NPCs in an invasion-independent manner, but it still requires a type III secretion system (10). In contrast, mutations in AvrA do not affect virulence and mutations in SptP have only a slight effect on virulence (26).

Although the effector molecules produced by *Shigella* and translocated into NPCs may affect caspase-1 activity, there is little evidence that events similar to those that occur in phagocytic cells also occur in epithelia. Colonic epithelial cells produce interleukin-8 (IL-8) in response to *Shigella* infections, presumably through activity of IpaB (67).

The only protein that is clearly translocated into host cells via the type IV secretion system of *H. pylori* is the CagA protein, although other *H. pylori* proteins are likely to be translocated in this manner (53). This effector molecule is phosphorylated and probably binds to Src homology region 2 domain-containing proteins. It causes reorganization of cytoskeletal elements in the pedestal formation and, in addition, causes the release of

TABLE 8.2 Examples of secretion system-dependent transcriptional responses induced by bacteria interacting with NPC

Micro-organism	Effector molecule	Secretory system	Proposed mechanisms	Type of response	Reference(s)
Yersinia[a]	YopJ/YopP	Type III secretion	Blocks MAPKKs and NF-κβ pathways	Inhibition of cytokine production; blocking of focal adhesion site	55
	YopH	Type III secretion	Dephosphorylates FAK kinase and p130^{cas}	Inhibition of bacterial uptake	4, 37, 48, 58, 62
	YopE	Type III secretion	?	Actin disruption	60
Salmonella	SopE	Type III secretion	Activates Rho GTPases and MAPKKs and NF-κβ pathways	Membrane ruffling, IL-8 secretion	20, 23
	SptP	Type III secretion	Protein tyrosine phosphorylase (PTPase)	?	19
	SigD/SopB	Type III secretion	Inositol phosphate phosphatase	Transepithelial signaling, reduction in inflammation	20, 23
	AvrA	Type III secretion	?	Prostaglandin production	26
	?	?			15
Shigella	IpaB	Type III secretion	Activates NF-κβ	Cytokine production	67
Helicobacter	CagA	Type IV secretion	Activates NF-κβ	Cytokine production	53

[a]Very little is known about the Yops in NPCs.

proinflammatory cytokines, such as IL-8, via activation of the NF-κβ system (53). Thus far, there is also only a single molecule known to be translocated into host cells by EPEC, Tir (see above).

Other Adhesion- and Invasion-Dependent Mechanisms for Activation of NPCs

A number of studies have described adhesion- and invasion-dependent activation of NPCs, but in many cases it is not known whether the effector molecules are directly translocated into the host cell via a type III or type IV secretion system. In some cases, the effector molecule has been identified and shown to be independent of either of these specialized secretory systems. Table 8.3 lists examples only of studies where

activation is thought to be adhesion or invasion dependent, and the underlying mechanisms are listed where they are known. These mechanisms have not been thoroughly investigated in terms of the identity of the effector molecule or the strict dependence on secretory systems. Bacteria produce many immunostimulatory molecules, also termed modulins (32), that could be effectors.

There are four possible mechanisms for these types of effects (Fig. 8.5). One of the mechanisms that affects the host cells (the high concentration mechanism) is one in which a relatively high concentration of one or more of the modulins may be formed at the interface between the adherent bacteria and the epithelial cells. This mechanism was shown for E. coli type 1 fimbria-mediated adhesion

TABLE 8.3 Examples of adhesion/invasion-dependent activation of NPC by bacteria

Microorganism	Type of cell	Type of response	Mechanism	Reference(s)
Staphylococcus aureus	Human endothelial cells	Adhesion molecule expression; tissue factor activity; IL-1β and IL-6 production; *S. aureus* may also produce a factor that inhibits cytokine production in an adhesion-dependent manner	Invasion dependent mechanism; unknown	3, 68, 70, 71
Haemophilus influenzae	Human tracheal epithelial cells	IL-8 and IL-6	Association mechanism; high-concentration mechanism	5, 24
Pseudomonas aeruginosa	Human respiratory epithelial cells	IL-8; interferon regulatory factor 1	Adhesin-dependent mechanism; high-concentration or association mechanisms	12, 14, 36
Streptococcus pneumoniae	Human respiratory epithelial cells	Cytokine production; apoptosis	Association mechanism; high-concentration mechanism	24
Treponema denticola	Human gingival fibroblasts and KB epithelial cells	Cytoskeletal rearrangements	High-concentration mechanism	17
Escherichia coli	Human bladder epithelial cells; cortical tubular epithelial cells; human kidney epithelial cells	IL-6 and IL-8 production	Adhesin-dependent mechanism; P fimbria-mediated mechanism; association mechanism, type 1 fimbria mediated.	1, 28–31, 41, 42, 45, 63
	Uroepithelial cells	Chloride response; bacterial killing		47
		Cytoskeleton rearrangement		57
	Intestinal cells		Not known; adhesin-dependent mechanism	
Bordetella pertussis	Endothelial cells	Integrin upregulation and bacterial uptake	Cooperativity mechanism	33
Streptococcus pyogenes	Epithelial cells	IL-6 secretion	Not known	9
Actinobacillus actinomycetem-comitans	Keratinocytes	IL-8 secretion	Not known	34
Yersinia enterocolitica	Epithelial cells	IL-8 and IL-1 secretion	Not known	8, 39, 64, 65
Haemophilus ducreyi	Keratinocytes	IL-8 secretion	Association mechanism	73
Porphyromonas gingivalis	KB cells	IL-6 and IL-8 secretion	Invasion-dependent mechanism	61
Salmonella enterica serovar Typhi	Intestinal epithelial cells	IL-6	Invasion-dependent mechanism	69
Salmonella enterica serovar Typhimurium	Intestinal epithelial cells	Neutrophil migration	Not known	49
Helicobacter pylori	Gastric cells	IL-8 secretion	Not known	2, 6

and delivery of labile toxin and cyclic AMP (cAMP) as effector molecules (72). It may also be responsible for the cytopathic damage induced by *Treponema denticola* adhesion to epithelial cells, where proteolytic enzymes are the effector molecules delivered (16).

In another mechanism (the association mechanism), a modulin molecule is tightly associated with the adhesin and thus may be presented to its receptors on the host cell in a cooperative manner. This mechanism has been described for a number of bacterial

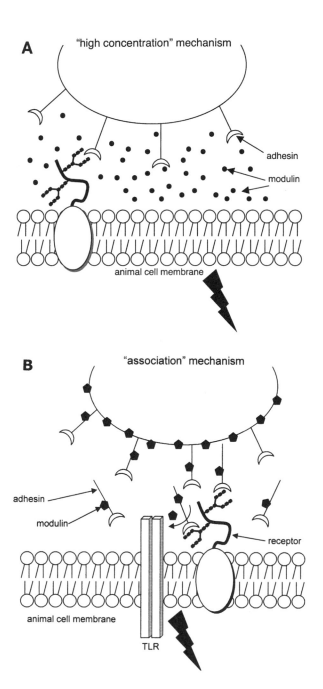

FIGURE 8.5 Diagram describing four mechanisms of adhesion-dependent activation of nonphagocytic cells by bacterial modulins. (A) In the high-concentration mechanism, accumulation of modulins at the interface between the bacterial and animal cell reach a threshold concentration necessary to affect host cell metabolism, as exemplified by the increase in the concentration of cAMP caused by labile-toxin delivery by type 1 fimbria-mediated attachment of ETEC. (B) In the association mechanism, the modulin is presented to its receptor on animal cells in an adhesin-bound form, which enables it to trigger a cascade of biological events in the animal cell. This is exemplified by association of an LPS modulin with P-fimbrial adhesin. (C) In the cooperativity mechanism, the modulin released by loosely adherent bacteria upregulates the expression of receptors for the bacterial adhesins, leading to firm adhesion, further stimulation of cells, and uptake of bacteria. This is exemplified by the upregulation of integrin receptors for the filamentous hemagglutinin adhesin by pertussis toxin. The high density of integrins induces binding of additional bacteria via the filamentous hemagglutinin and also transduces a signal that results in the invasion of the bacteria. (D) In the adhesin-mediated mechanism, the adhesin itself acts as a modulin, triggering a cascade of events in the animal cells on binding to its cognate receptor. This mechanism is exemplified by *P. aeruginosa*. For further details, see the text.

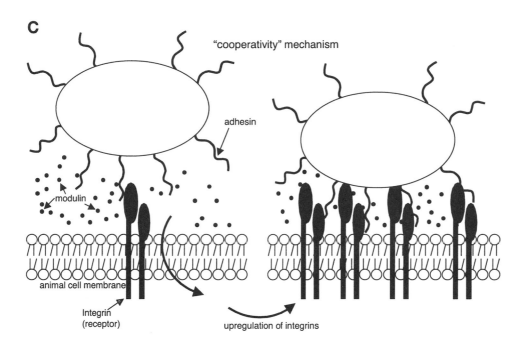

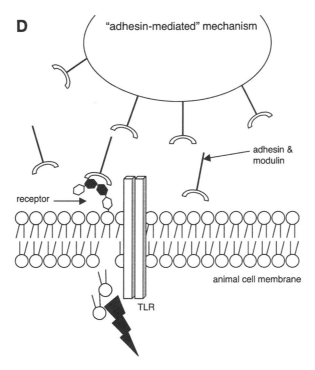

pathogens (Table 8.3). Interestingly, LPS, when presented via adherent type 1 fimbriae to epithelial cells which do not express CD14, can trigger the formation of cytokines (29, 45, 63). Normally, LPS alone would not stimulate these CD14-negative cells. It has been shown that some fraction of *E. coli* endotoxin molecules become tightly associated with fimbrial adhesins and remain associated even when the fimbriae are purified (54).

Another mechanism (the cooperativity mechanism) involves cooperativity between adhesins and effector molecules. For example, in *Bordetella pertussis,* pertussis toxin upregulates integrin, which, in turn, binds the filamentous hemagglutinin adhesin, leading to uptake of the bacteria and stimulation of the cells (33).

Yet another mechanism (the adhesin-dependent mechanism) may be mediated by the adhesin itself. This was the case for *P. aeruginosa,* where intact bacteria as well as isolated pili were major stimuli for the translocation of transcription factor NF-κβ and initiation of IL-8 expression by epithelial cells (13). This mechanism was also proposed for the P-fimbrial adhesins of uropathogenic *E. coli.* The P fimbriae bind to ceramide-anchored glycolipids containing Gal-Gal epitopes and trigger the intracellular release of ceramide residues to activate serine/threonine kinases (30, 31). The process of degrading the glycolipid into ceramide and oligosaccharide may be a host defense mechanism, because the released carbohydrate may act as a receptor analog and inhibit the adhesion of other bacteria and the intracellular ceramide release may trigger the formation of cytokines and induce the recruitment of inflammatory cells (66).

It has been shown that adhesion follows multiple steps (see chapter 1). The first of these is weak, to allow repulsive forces at the approaching cell surfaces to be overcome, while the second is to provide for firm adhesion. In one study, the relative importance of the two steps in allowing group A streptococci to induce IL-6 in HEp-2 cells was investigated (9). It was found that the initial, weak adhesion was insufficient to stimulate IL-6 secretion and that firm adhesion was required. It is possible that for initial, weak adhesion to result in cell activation, the organism must have a secretory system that is capable of rapidly secreting effector molecules into the host cell, as is the case for *Shigella* and *Salmonella* (discussed above). For group A streptococci, it has been suggested that streptolysin O, a potent cytolytic toxin, may create channels in the host cell membrane through which effector molecules may be translocated (46). It remains to be seen what circumstances are required for the initial, weak adhesion to be sufficient to deliver streptolysin O and an effector molecule to host cells. In summary, considering the four types of activation mechanisms discussed above and the vast number of modulins produced by pathogenic bacteria (32), it appears that adhesion per se is only one prerequisite for a wide variety of potential subsequent reactions between the bacterium and target cell that influence the infectious process.

REFERENCES

1. **Agace, W. W., S. R. Hedges, M. Ceska, and C. Svanborg.** 1993. Interleukin-8 and the neutrophil response to Gram-negative infection. *J. Clin. Investig.* **92:**780–785
2. **Aihara, M., D. Tsuchimoto, H. Takizawa, A. Azuma, H. Wakebe, Y. Ohmoto, K. Imagawa, M. Kikuchi, N. Mukaida, and K. Matsuchima.** 1997. Mechanisms involved in *Helicobacter pylori*-induced interleukin-8 production by a gastric cancer cell line, MKN45. *Infect. Immun.* **65:**3218–3224.
3. **Beekhuizen, H., J. S. van de Gevel, B. Olsson, I. J. van Benten, and R. van Furth.** 1997. Infection of human vascular endothelial cells with *Staphylococcus aureus* induces hyperadhesiveness for human monocytes and granulocytes. *J. Immunol.* **158:**774–782.
4. **Black, D. S., and J. B. Bliska.** 1997. Identification of p130cas as a substrate of *Yersinia* YopH (Yop51), a bacterial protein tyrosine phosphatase that translocates into mammalian cells and targets focal adhesion. *EMBO J.* **16:**2730–2744.

5. Clemans, D. L., R. J. Bauer, J. A. Hanson, M. V. Hobbs, J. W. St. Geme III, C. F. Marrs, and J. R. Gilsdorf. 2000. Induction of proinflammatory cytokines from human respiratory epithelial cells after stimulation by nontypeable *Haemophilus influenzae*. *Infect. Immun.* **68:**4430–4440.

6. Cole, S. P., D. Cirillo, M. F. Kagnoff, D. G. Guiney, and L. Eckmann. 1997. Coccoid and spiral *Helicobacter pylori* differ in their abilities to adhere to gastric epithelial cells and induce interleukin-8 secretion. *Infect. Immun.* **65:**843–846.

7. Collington, G. K., I. W. Booth, and S. Knutton. 1998. Rapid modulation of electrolyte transport in Caco-2 cell monolayers by enteropathogenic *Escherichia coli* (EPEC) infection. *Gut* **42:**200–207.

8. Cornelis, G. R. 2002. The *Yersinia* Ysc-Yop 'type III' weaponry. *Nat. Rev. Mol. Cell Biol.* **3:**742–752.

9. Courtney, H. S., I. Ofek, and D. L. Hasty. 1997. M protein mediated adhesion of M type 24 *Streptococcus pyogenes* stimulates release of interleukin-6 by HEp-2 tissue culture cells. *FEMS Microbiol. Lett.* **151:**65–70.

10. Darwin, K. H., and V. L. Miller. 1999. Molecular basis of the interaction of *Salmonella* with the intestinal mucosa. *Clin. Microbiol. Rev.* **12:**405–428.

11. DeVinney, R., A. Gauthier, A. Abe, and B. B. Finlay. 1999. Enteropathogenic *Escherichia coli*: a pathogen that inserts its own receptor into host cells. *Cell. Mol. Life Sci.* **55:**961–976.

12. DiMango, E., A. J. Ratner, R. Bryan, S. Tabibi, and A. Prince. 1996. Activation of NF-kappaB by adherent *Pseudomonas aeruginosa* in normal and cystic fibrosis respiratory epithelial cells. *Am. J. Respir. Crit. Care Med.* **154:**S187-S191.

13. DiMango, E., A. J. Ratner, R. Bryan, S. Tabibi, and A. Prince. 1998. Activation of NF-kappaB by adherent *Pseudomonas aeruginosa* in normal and cystic fibrosis respiratory epithelial cells. *J. Clin. Investig.* **101:**2598–2605.

14. DiMango, E., H. J. Zar, and A. P. R. Bryan. 1995. Diverse *Pseudomonas aeruginosa* gene products stimulate respiratory epithelial cells to produce interleukin-8. *J. Clin. Investig.* **96:**2204–2210.

15. Eckmann, L., W. F. Stenson, T. C. Savidge, D. C. Lowe, K. E. Barrett, J. Fierer, J. R. Smith, and M. F. Kagnoff. 1997. Role of epithelial cells in the host secretory response to infection by invasive bacteria. *J. Clin. Investig.* **100:**296–309.

16. Ellen, R. P. 2002. Adhesion of oral spirochaetes to host cells and its cytopathogenic consequences, p. 247–276. *In* M. Wilson (ed.), *Bacterial Adhesion to Host Cells. Mechanisms and Consequences.* Cambridge University Press, Cambridge, United Kingdom.

17. Ellen, R. P., J. R. Dawson, and P. F. Yang. 1994. *Treponema denticola* as a model for polar adhesion and cytopathogenicity of spirochetes. *Trends Microbiol.* **2:**114–119.

18. Frankel, G., A. D. Phillips, L. R. Trabulsi, S. Knutton, G. Dougan, and S. Matthews. 2001. Intimin and the host cell—is it bound to end in Tir(s)? *Trends Microbiol.* **9:**214–218.

19. Fu, Y., and J. E. Galán. 1998. The *Salmonella typhimurium* tyrosine phosphatase, SptP, is translocated into host cells and disrupts the actin cytoskeleton. *Mol. Microbiol.* **27:**359–368.

20. Galán, J. E. 1999. Interaction of *Salmonella* with host cells through the centisome 63 type III secretion system. *Curr. Opin. Microbiol.* **2:**46–50.

21. Galán, J. E. 2001. Salmonella interactions with host cells: type III secretion at work. *Annu. Rev. Cell Dev. Biol.* **17:**53–86.

22. Galán, J. E., and A. Collmer. 1999. Type III secretion machines: bacterial devices for protein delivery into host cells. *Science* **284:**1322–1328.

23. Galyov, E. E., M. W. Wood, R. Roskvist, P. B. Mullan, P. R. Watson, S. Hedges, and T. S. Wallis. 1997. A secreted effector protein of *Salmonella dublin* is translocated into eukaryotic cells and mediates inflammation and fluid secretion in infected ileal mucosa. *Mol. Microbiol.* **25:**903–912.

24. Hakansson, A., I. Carlstedt, J. Davies, A. K. Mossberg, H. Sabharwal, and C. Svanborg. 1995. Aspects on the interaction of *Streptococcus pneumoniae* and *Haemophilus influenzae* with human respiratory tract mucosa. *J. Clin. Investig.* **96:**2204–2210

25. Hardt, W.-D., L.-M. Chen, K. E. Schuebel, X. R. Bustelo, and J. E. Galán. 1998. *Salmonella typhimurium* encodes an activator of Rho GTPases that induces membrane ruffling and nuclear responses in host cells. *Cell* **93:**815–826.

26. Hardt, Y. W.-D., and J. E. Galán. 1997. A secreted *Salmonella* protein with homology to an avirulence determinant of plant pathogenic bacteria. *Proc. Natl. Acad. Sci. USA* **94:**9887–9892.

27. Hardt, Y. W.-D., H. Urlaub, and J. E. Galán. 1998. A substrate of the centisome 63 type III protein secretion system of *Salmonella typhimurium* is encoded by a cryptic prophage. *Proc. Natl. Acad. Sci. USA* **95:**2574–2579.

28. Hedges, S., W. W. Agace, M. Svensson, A. C. Sjogren, M. Ceska, and C. Svanborg. 1994. Uroepithelial cells are part of a mucosal cytokine network. *Infect. Immun.* **62:**2315–2321.

29. Hedlund, M., B. Freundèus, W. Wachtler,

L. Hang, H. Fischer, and C. Svanborg. 2001. Type 1 fimbriae deliver an LPS and TLR4-dependent activation signal to CD14-negative cells. *Mol. Microbiol.* **39**:542–552.

30. Hedlund, M., C. Wachtler, E. Johansson, L. Hang, J. E. Somerville, R. P. Darveau, and C. Svanborg. 1999. P fimbriae-dependent lipopolysaccharide-independent activation of epithelial cytokine response. *Mol. Microbiol.* **33**:693–703.

31. Hedlund, M., M. Svensson, A. Nilsson, R.-D. Duan, and C. Svanborg. 1996. Role of the ceramide-signalling pathway in cytokine responses to P fimbriated *Escherichia coli. J. Exp. Med.* **183**:1037–1044.

32. Henderson, B., S. Poole, and M. Wilson. 1996. Bacterial modulins: a novel class of virulence factors which cause host tissue pathology by inducing cytokine synthesis. *Microbiol. Rev.* **60**:316–341.

33. Hoepelman, A. I. M., and E. I. Tuomanen. 1992. Consequences of bacterial attachment: directing host cells functions with adhesins. *Infect. Immun.* **60**:1729–1733.

34. Huang, G. T., S. K. Haake, and N. H. Park. 1998. Gingival epithelial cells increase interleukin-8 secretion in response to *Actinobacillus actinomycetemcomitans* challenge. *J. Periodontol.* **69**:1105–1110.

35. Hueck, C. J. 1998. Type III protein secretion systems in bacterial pathogens of animals and plants. *Microbiol. Mol. Biol. Rev.* **62**:379–433.

36. Ichikawa, J. K., A. Norris, M. G. Bangera, G. K. Geiss, A. B. van't Wout, R. E. Bumgarner, and S. Lory. 2000. Interaction of *Pseudomonas aeruginosa* with epithelial cells: identification of differentially regulated genes by expression microarray analysis of human cDNAs. *Proc. Natl. Acad. Sci. USA* **97**:9659–9664.

37. Isberg, R. R., Z. Hamburger, and P. Dersch. 2000. Signaling and invasin-promoted uptake via integrin receptors. *Microbes Infect.* **2**:793–801.

38. Jadoun, J., V. Ozeri, E. Burstein, E. Skutelsky, E. Hanski, and S. Sela. 1998. Protein F1 is required for efficient entry of *Streptococcus pyogenes* into epithelial cells. *J. Infect. Dis.* **178**:147–158.

39. Kampik, D., R. Schulte, and I. B. Autenrieth. 2000. *Yersinia enterocolitica* invasin protein triggers differential production of interleukin-1, interleukin-8, monocyte chemoattractant protein 1, granulocyte-macrophage colony-stimulating factor, and tumor necrosis factor alpha in epithelial cells: implications for understanding the early cytokine network in *Yersinia* infections. *Infect. Immun.* **68**:2484–2492.

40. Knutton, S., T. Baldwin, P. H. Williams, and A. S. McNeish. 1989. Actin accumulation at sites of bacterial adhesion to tissue culture cells: basis of a new diagnostic test for enteropathogenic and enterohemorrhagic *Escherichia coli. Infect. Immun.* **57**:1290–1298.

41. Kreft, B., S. Bohnet, O. Carstensen, J. Hacker, and R. Marre. 1993. Differential expression of interleukin-6, intracellular adhesion molecule 1, and major histocompatibility complex class II molecules in renal carcinoma cells stimulated with S fimbriae of uropathogenic *Escherichia coli. Infect. Immun.* **61**:3060–3063.

42. Kruger, S., E. Brandt, M. Klinger, S. Kruger, and B. Kreft. 2000. Interleukin-8 secretion of cortical tubular epithelial cells is directed to the basolateral environment and is not enhanced by apical exposure to *Escherichia coli. Infect. Immun.* **68**:328–334.

43. Kuhn, M., and W. Goebel. 1998. Host cell signalling during *Listeria monocytogenes* infection. *Trends Microbiol.* **6**:11–15.

44. Kuhn, M., T. Pfeuffer, L. Greiffenberg, and W. Goebel. 1999. Host cell signal transduction during *Listeria monocytogenes* infection. *Arch. Biochem. Biophys.* **372**:166–172.

45. Linder, H., I. Engberg, I. M. Baltzer, K. Jann, and C. Svanborg-Edén. 1988. Induction of inflammation by *Escherichia coli* on the mucosal level: requirement for adherence and endotoxin. *Infect. Immun.* **56**:1309–1313.

46. Madden, J. C., N. Ruiz, and M. Caparon. 2001. Cytolysin-mediated translocation (CMT): a functional equivalent of type III secretion in Gram-positive bacteria. *Cell* **104**:143–152.

47. Mannhardt, W., M. Putzer, F. Zepp, and H. Schulte-Wissermann. 1996. Host defense within the urinary tract. II. Signal transducing events activate uroepithelial defense. *Pediatr. Nephrol.* **10**:573–577.

48. McCallum, S. J., and J. A. Theriot. 2000. Bacterial manipulation of the host cell cytoskeleton, p. 171–192. *In* P. Cossart, P. Boquet, S. Normark, and R. Rappuoli (ed.), *Cellular Microbiology.* ASM Press, Washington, D.C.

49. McCormick, B. A., S. P. Colgan, C. Delp-Archer, S. I. Miller, and J. L. Madara. 1993. *Salmonella typhimurium* attachment to human intestinal epithelial monolayers: transcellular signalling to subepithelial neutrophils. *J. Cell Biol.* **123**:895–907.

50. Molinari, G., and G. S. Chhatwal. 1999. Streptococcal invasion. *Curr. Opin. Microbiol.* **2**:56–61.

51. Molinari, G., and G. S. Chhatwal. 1999. Role played by the fibronectin-binding protein

SfbI (protein F1) of *Streptococcus pyogenes* in bacterial internalization by epithelial cells. *J. Infect. Dis.* **179:**1049–1050.

52. **Molinari, G., S. R. Talay, P. Valentin-Weigand, M. Rohde, and G. S. Chhatwal.** 1997. The fibronectin-binding protein of *Streptococcus pyogenes*, SfbI, is involved in the internalization of group A streptococci by epithelial cells. *Infect. Immun.* **65:**1357–1363.

53. **Montecucco, C., and R. Rappuoli.** 2001. Living dangerously: how *Helicobacter pylori* survives in the human stomach. *Nat. Rev. Mol. Cell Biol.* **2:**457–466.

54. **Morschhausser, J., H. Hoschutzky, K. Jann, and J. Hacker.** 1990. Functional analysis of the sialic acid-binding adhesin SfaS of pathogenic *Escherichia coli* by site-specific mutagenesis. *Infect. Immun.* **38:**2133–2138.

55. **Orth, K., L. E. Palmer, Z. Q. Bao, S. Stewart, A. E. Rudolph, J. Bliska, and J. E. Dixon.** 1999. Inhibition of the mitogen-activated protein kinase kinase superfamily by a *Yersinia* effector. *Science* **285:**1920–1923.

56. **Ozeri, V., I. Rosenshine, D. F. Mosher, R. Fässler, and E. Hanski.** 1998. Roles of integrins and fibronectin in the entry of *Streptococcus pyogenes* into cells via protein F1. *Mol. Microbiol.* **30:**625–637.

57. **Peiffer, I., A. L. Servin, and M.-F. Bernet-Camard.** 1998. Piracy of decay-accelerating factor (CD55) signal transduction by the diffusely adhering strain *Escherichia coli* C1845 promotes cytoskeletal F-actin rearrangements in cultured human intestinal INT407 cells. *Infect. Immun.* **66:**4036–4042.

58. **Persson, C., N. Carballeira, H. Wolf-Watz, and M. Fallman.** 1997. The PTPase, YopH, inhibits uptake of *Yersinia*, tyrosine phosphorylation of p130cas and FAK and associated accumulation of these proteins in peripheral focal adhesions. *EMBO J.* **16:**2307–2318.

59. **Raupach, B., J. Mecsas, U. Heczko, S. Falkow, and B. B. Finlay.** 1999. Bacterial epithelial cell cross talk. *Curr. Top. Microbiol. Immunol.* **236:**137–161.

60. **Roskvist, R., A. Forsberg, and H. Wolf-Watz.** 1991. Intracellular targeting of the *Yersinia* YopE cytotoxin in mammalian cells induces actin microfilament disruption. *Infect. Immun.* **59:**4562–4569.

61. **Sandros, J., C. Karlsson, D. F. Lappin, P. N. Madianos, D. F. Kinane, and P. N. Papapanou.** 2000. Cytokine responses of oral epithelial cells to *Porphyromonas gingivalis. J. Dent. Res.* **79:**1808–1814.

62. **Schesser, K., M. S. Francis, Å. Forsberg, and H. Wolf-Watz.** 2000. Type III secretion

systems in animal- and plant-interacting bacteria, p. 239–263. *In* P. Cossart, P. Boquet, S. Normark, and R. Rappuoli (ed.), *Cellular Microbiology.* ASM Press, Washington, D.C.

63. **Schilling, J. D., M. A. Mulvey, C. D. Vincent, R. G. Lorenz, and S. J. Hultgren.** 2001. Bacterial invasion augments epithelial cytokine responses to *Escherichia coli* through a lipopolysaccharide-dependent mechanism. *J. Immunol.* **166:**1148–1155.

64. **Schulte, R., and I. B. Autenreith.** 1998. *Yersinia enterocolitica*-induced interleukin-8 secretion by human intestinal epithelial cells depends on cell differentiation. *Infect. Immun.* **66:**1216–1224.

65. **Schulte, R., R. Zumbihl, D. Kampik, A. Fauconnier, and I. B. Autenriety.** 1998. Wortmannin blocks *Yersinia* invasin-triggered internalization, but not interleukin-8 production by epithelial cells. *Med. Microbiol. Immunol.* **187:**53–60.

66. **Svanborg, C., G. Bergsten, H. Fischer, B. Freudéus, G. Godaly, E. Gustafsson, L. Hang, M. Hedlund, A.-C. Lundstedt, M. Samuelsson, P. Samuelsson, M. Svensson, and B. Wullt.** 2002. Adhesion, signal transduction and mucosal inflammation, p. 223–246. *In* M. Wilson (ed.), *Bacterial Adhesion to Host Cells. Mechanisms and Consequences.* Cambridge University Press, Cambridge, United Kingdom.

67. **Tran van Nhieu, G., and P. J. Sansonetti.** 1999. Mechanism of *Shigella* entry into epithelial cells. *Curr. Opin. Microbiol.* **2:**51–55.

68. **Veltrop, M. H. A. M., H. Beekhuizen, and J. Thompson.** 1999. Bacterial species- and strain-dependent induction of tissue factor in human vascular endothelial cells. *Infect. Immun.* **67:**6130–6138.

69. **Weinstein, D. L., B. L. O'Neill, and E. S. Metcalf.** 1997. *Salmonella typhi* stimulation of human intestinal epithelial cells induces secretion of epithelial cell-derived interleukin-6. *Infect. Immun.* **65:**395–404.

70. **Yao, L., V. Bengualid, F. D. Lowy, J. Gibbons, V. B. Hatcher, and J. W. Berman.** 1995. Internalization of *Staphylococcus aureus* by endothelial cells induces cytokine gene expression. *Infect. Immun.* **63:**1835–1839.

71. **Yao, L., V. Bengualid, J. W. Berman, and F. D. Lowy.** 2000. Prevention of endothelial cell cytokine induction by a *Staphylococcus aureus* lipoprotein. *FEMS Immunol. Med. Microbiol.* **28:**301–305.

72. **Zafriri, D., Y. Oron, B. I. Eisenstein, and I. Ofek.** 1987. Growth advantage and enhanced toxicity of *Escherichia coli* adherent to tissue culture cells due to restricted diffusion of products

secreted by the cells. *J. Clin. Investig.* **79:**1210–1216.

73. **Zaretzky, F. R., and T. H. Kawula.** 1999. Examination of early interactions between *Haemo-philus ducreyi* and host cells by using cocultured HaCaT keratinocytes and foreskin fibroblasts. *Infect. Immun.* **67:**5352–5360.

ADHESION-DEPENDENT UPREGULATION
OF BACTERIAL GENES

9

The concept that bacteria are likely to upregulate certain genes on attachment to host cell surfaces is not new (2). That the mere tethering of an adhesin results in signaling for upregulating virulence factors is, however, rather new, and even though it is quite logical, it has not yet been thoroughly investigated in many systems.

There are many studies showing upregulation of genes in vivo (5). Whether adhesion is required for the activation of many these genes is not known. In a few cases (discussed below), however, it has been shown that adhesion itself activates gene expression.

The first evidence directly implicating adhesion in gene activation came from studies of uropathogenic *Escherichia coli*. The urinary tract is generally nutritionally poor for sustaining bacterial growth, especially its low levels of free iron. It was shown that the intrinsic iron acquisition machinery of uropathogenic *E. coli* is activated on the coupling of the PapG fimbrial adhesin with its globoseries glycolipid receptor (12). When the fimbriae of the uropathogen bind to the immobilized receptor, the transcription of the gene for an outer membrane sensor protein, AirS, is activated (Fig. 9.1). AirS (for "attachment and iron regulation sensor;" also called *barA*) belongs to a two-component signal transduc-

tion family that is thought to activate the bacterial iron acquisition system, composed of high-affinity iron binding proteins and iron-regulated membrane proteins. The evidence that binding of the PapG adhesin to its receptor is required for transcription of the *airS* gene comes from the observation that *airS* transcription did not occur in an isogenic PapG mutant exposed to Gal-Gal receptors immobilized on erythrocytes or plastic. Only bacteria coupled to the receptor were able to grow in urine. The data suggest the fimbria-adhesin complex acts as a sensor to elicit a response that could facilitate bacterial growth in urine. This sensory activity is dependent on the interaction of the tip-fimbrial adhesin, PapG, with its complementary receptor.

A second case is illustrated by experiments performed with *Neisseria meningitidis,* which binds to epithelial cells by a multistep mechanism (Fig. 9.2) (10). In this case, it was shown that the initial stage of adhesion is mediated by the contact of a relatively few PilC1 adhesin molecules, presumably in the outer membrane (8), with their receptor. Although the PilC1 adhesin may be found on the bacterial surface as a minor, pilus tip-associated adhesin, it is also thought to be found as an outer membrane protein in this initial phase of attachment. This basal level of PilC1 is suffi-

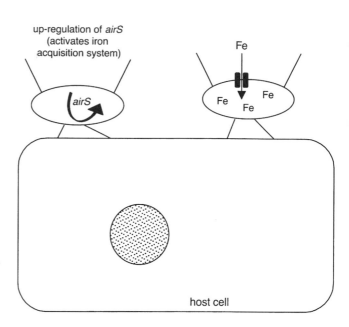

FIGURE 9.1 Schematic presentation of the effect of PapG-mediated adhesion on the expression of bacterial genes, exemplified by uropathogenic *E. coli* bound to host cells. On contact with epithelial cells, a number of bacterial genes become upregulated; this, in turn, may either downregulate or upregulate the expression of other genes. It has been shown that PapG-mediated adhesion of uropathogenic *E. coli* causes upregulation of the *airS* gene, which is involved in the regulation of iron acquisition.

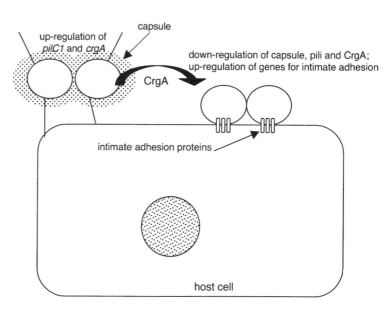

FIGURE 9.2 Schematic presentation of the effect of PilC-mediated adhesion on the expression of bacterial genes, exemplified by *N. meningitidis* bound to host cells. As with uropathogenic *E. coli,* on contact of *N. meningitidis* with epithelial cells, a number of genes become upregulated; this may either downregulate or upregulate the expression of other genes. In this example, PilC1-mediated initial adhesion of *N. meningitidis* is a signal that induces the expression of *crgA* and *pil* genes. Once induced, CrgA downregulates capsule, pili, and itself and upregulates the expression of as yet unidentified genes involved in intimate adhesion.

cient to initiate contact. The PilC1-mediated contact then activates the CREN (for "contact regulatory element of *Neisseria*") site within the promoter, which induces expression of bacterial genes for a second stage of the process. In this step, more PilC1 adhesin molecules are expressed, eventually leading to the presentation of PilC1 adhesin at the pilus tip.

This step is followed by upregulation of the machinery for intimate attachment associated with attaching/effacing lesions and concomitant downregulation of pilus expression (11). In this step, capsule production is also downregulated. A CREN-like element also participates in the intimate-adhesion step of the process. The CREN element is under the control of the product of the *crgA* (for "contact-regulated gene A") gene. It appears that the Opc and Opa proteins may also be involved in some aspects of the adhesion process. The target cell receptor involved in PilC1-mediated adhesion is the CD46 glycoprotein (4). That the outer membrane form of PilC1 could mediate adhesion in vitro was shown by centrifuging nonpiliated bacteria onto epithelial cells (8). Pili are required to bridge the repulsive charges separating prokaryotic and eukaryotic cells; therefore, in the absence of pili the centrifugal force is required to overcome the charge repulsion.

In *Yersinia pseudotuberculosis,* bacterium-host contact activates the expression of several genes. In particular, the genes of the type III secretion system are affected by contact with host cells (3, 6, 7). *Yersinia* organisms do not secrete Yops into culture medium via their type III secretion apparatus in the absence of cells, but they do so when they contact cells (9). In fact, it appears that the type of mammalian cell is not important, but contact with cells is critical (1). It is assumed that adhesins are important, at least when *Yersinia* organisms contact nonphagocytic cells, but the specific

bacterial and host molecules are not known at present. However, the needle apparatus of the type III secretion system is not sufficient. At least some parts of the type III apparatus and effector molecules are expressed once bacteria enter the 37°C host, but there is no secretion because the LcrG protein blocks the secretion channel from within the cytosol and the YopN protein blocks the translocation pore on the outside of the bacterial surface. Mutant strains incapable of expressing these proteins secrete Yops into the medium even in the absence of mammalian cells. The expression of these molecules at 37°C is controlled by VirF. Contact with epithelial cells appears to be sensed by the YopN protein present on the external surface of the translocation apparatus (Fig. 9.3). Contact then induces opening of this pore and leads to escape of a negative regulator, YscM, to a level that allows induction of *yop* and *lcrV* gene expression and removal of the LcrG internal block of the secretion apparatus. YopB and YopD expression, also induced in this contact-dependent cascade, results in the formation of a pore in the host cell membrane, aiding the delivery of effector molecules by the type III apparatus. The precise mechanisms whereby

YadA-mediated adhesion

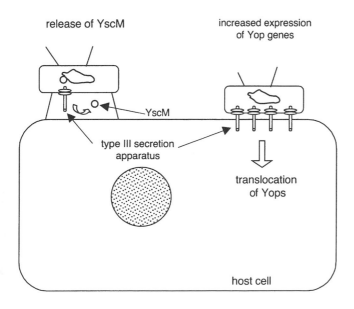

FIGURE 9.3 Schematic presentation of the effect of YadA-mediated adhesion on the expression of bacterial genes, exemplified by *Y. enterocolitica* bound to host cells. As with uropathogenic *E. coli* and *N. meningitidis,* on contact with epithelial cells, a number of genes become upregulated; this may either upregulate or downregulate the expression of other genes. YadA/invasin-mediated adhesion of *Y. enterocolitica* causes YopN to open the translocation pore and release the negative regulator YscM, which leads to increased expression of *yop* genes and other genes of the type III secretion system and translocation of Yop effector proteins into the host cell.

adhesion leads to the formation of a pore by the needle apparatus or associated molecules are not known. Nevertheless, tight regulatory coupling of these events to contact is essential for the efficient utilization of bacterial resources only under conditions where the mammalian cell is in a position where it can be effectively targeted.

It is only logical to assume that most adherent bacteria require the expression of genes that were not essential while the microorganisms existed in a planktonic phase. It stands to reason that at least some of these genes are activated only as the bacteria become attached to a target host cell. The few examples provided here show that this phenomenon does occur, but the true extent to which the process of adhesion activates bacterial gene expression remains to be more fully explored.

REFERENCES

1. **Boyd, A. P., N. Grosdent, S. Totemeyer, C. Geuijen, S. Bleves, M. Iriarte, I. Lambermont, J. N. Octave, and G. R. Cornelis.** 2000. *Yersinia enterocolitica* can deliver Yop proteins into a wide range of cell types: development of a delivery system for heterologous proteins. *Eur. J. Cell Biol.* **79:**659–671.
2. **Cornelis, G.** 1997. Contact with eukaryotic cells: a new signal triggering bacterial gene expression. *Trends Microbiol.* **5:**43–45.
3. **Cornelis, G.** 2002. The *Yersinia* Ysc-YOP 'type III' weaponry. *Nat. Rev. Mol. Cell Biol.* **3:** 742–752.
4. **Kallström, H., M. K. Liszewski, J. P. Atkinson, and A. B. Jonsson.** 1997. Membrane cofactor protein (MCP or CD46) is a cel-

lular pilus receptor for pathogenic *Neisseria*. *Mol. Microbiol.* **25:**639–647.
5. **Mahan, M. J., D. M. Heithoff, R. L. Sinsheimer, and D. A. Low.** 2000. Assessment of bacterial pathogenesis by analysis of gene expression in the host. *Annu. Rev. Genet.* **34:** 139–164.
6. **Pettersson, J., R. Nordfelth, E. Dubinina, T. Bergman, M. Gustafsson, K. E. Magnusson, and H. Wolf-Watz.** 1996. Modulation of virulence factor expression by pathogen target cell contact. *Science* **273:**1231–1233.
7. **Pierson, D. E.** 2001. Induction of protein secretion by *Yersinia enterocolitica* through contact with eukaryotic cells, p. 183–1202. *In* M. Wilson (ed.), *Bacterial Adhesion to Host Cells.* Cambridge University Press, Cambridge, United Kingdom.
8. **Rahman, M., H. Källström, S. Normark, and A.-B. Jonsson.** 1997. PilC of pathogenic *Neisseria* is associated with the bacterial cell surface. *Mol. Microbiol.* **25:**11–25.
9. **Rosqvist, R., K. E. Magnusson, and H. Wolf-Watz.** 1994. Target cell contact triggers expression and polarized transfer of *Yersinia* YopE cytotoxin into mammalian cells. *EMBO J.* **13:**964–972.
10. **Taha, M.-K.** 2002. Transcriptional regulation of meningococcal gene expression upon adhesion to target cells, p. 165–182. *In* M. Wilson, (ed.), *Bacterial Adhesion to Host Cells.* Cambridge University Press, Cambridge, United Kingdom.
11. **Taha, M.-K., P. C. Morand, Y. Pereira, E. Eugène, D. Giorgini, M. Larribe, and X. Nassif.** 1998. Pilus-mediated adhesion of *Neisseria meningitidis*: the essential role of cell contact-dependent transcriptional upregulation of the PilC1 protein. *Mol. Microbiol.* **28:**1153–1163.
12. **Zhang, J. P., and S. Normark.** 1996. Induction of gene expression in *Escherichia coli* after pilus-mediated adherence. *Science* **273:** 1234–1236.

ROLE OF ADHESION
IN BIOFILM FORMATION

10

INTRODUCTION

Because bacterial adhesion is an indispensable component of biofilm formation, the factors involved in bacterium-substratum interactions are an important basis for understanding biofilm physiology (8, 10, 11, 36). In fact, biofilm formation may be regarded as one of the most important sequelae of bacterial adhesion. Biofilms can be divided into those of medical relevance and those of environmental importance. This chapter focuses on biofilms of medical significance. However, many of the physicochemical characteristics of biofilms in the environment are shared by those of the medical biofilms. In fact, many of the principles evolved from studies of environmental biofilms and were applied to the characterization of medical biofilms. Biofilm, whether environmental or medical, constitutes a surface-bound community of microorganisms, often of different species, embedded in a matrix consisting of high- and low-molecular-mass organic compounds derived either from the microorganisms or from the surrounding milieu. A biofilm is dynamic, such that the microorganisms divide or are released and shed into the medium and its organic constituents are turned over. Medically relevant biofilms may form either on tissues (hard or soft tissues) or on implants. The former are

more complex and difficult to study, whereas the latter are more amenable to experimentation and manipulations, to the eventual benefit of the host.

Although it is accepted that biofilms on implants such as catheters and artificial joints are associated with infections, there is a growing body of evidence suggesting that formation of biofilms on tissues is a prerequisite for the development of symptomatic infection (2). In this chapter, we discuss the advantages gained by bacteria from participation in a biofilm community, the mechanisms of biofilm formation, and attempts to block biofilm development in an effort to prevent infections. Finally we focus on two types of medically important biofilms which have been extensively studied: hard tissue biofilms on tooth surfaces (i.e., dental plaque) and biomedical implant-associated biofilms.

MECHANISM OF BIOFILM FORMATION

Adhesion of one or more bacterial cells, usually of a single species, to a substratum is the first event in biofilm formation (Fig. 10.1). This initial event occurs either on unconditioned substratum or on substratum coated with organic materials from the surrounding environment. It is not known whether this

Development of biofilms

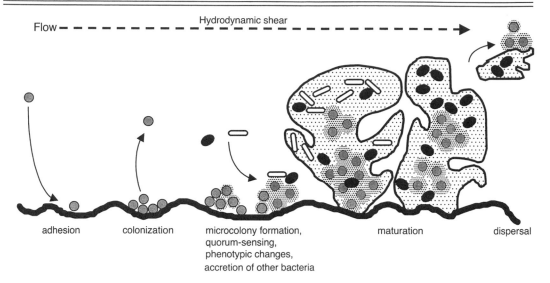

adhesion   colonization   microcolony formation,   maturation   dispersal
                                    quorum-sensing,
                                    phenotypic changes,
                                    accretion of other bacteria

FIGURE 10.1 Schematic representation of biofilm formation on a substratum. The major steps involved in this process are illustrated in a sequential manner, whereby initial adhesion is followed by colonization, accretion, maturation, and dispersal phases. The mature biofilm can be a pure population or, more probably, a mixture of various genera entrapped in a protein-polysaccharide gel. Note that the channels and concavities formed within the biofilm increase the surface area. Over time, dispersal of the mature biofilm may eventually occur whereby individual cells or fragments of the protein-polysaccharide gel become detached and then reattach at a new region or new substratum. Several different species of bacteria are shown within the developing biofilm, one of which is outcompeted for nutrients and eliminated in the mature biofilm.

event requires a minimal number of bacterial cells or whether a single bacterium will suffice. It is well known that some bacteria adhere more avidly to surfaces than do other bacteria and even that the same bacteria may undergo time-dependent changes in avidity of adhesion. Regardless of the strength of adhesion, the bacterium must remain surface localized in order to become a community. It is likely that there is an inverse relationship between the initial number of adherent bacteria required for biofilm formation and the avidity of the adhesion. For example, a biofilm community may rise from few avidly adherent bacteria on the target substratum or from a large number of weakly adherent bacteria. This notion has yet to be proven experimentally.

Once the initial adhesion is successful, accretion of bacteria can proceed with con-

comitant proliferation. Daughter cells may also become bound to the surface. The composite population consisting of the initially adherent cells and their progeny may now secrete various polymers, especially acidic polysaccharides. The bacteria and the extracellular components produced or acquired may then serve as a substratum for additional progeny or for the adhesion of additional bacterial species. For example, *Escherichia coli* expresses Ag43 protein, which induces autoaggregation of nonfimbriated bacteria (18). In fact, intergeneric bacterium-bacterium interactions are well known for oral bacteria (23, 24, 35) and undoubtedly occur in other, nonoral, bacteria.

The secreted polysaccharides of the biofilm matrix are capable of trapping ions and various organic materials in addition to their ability to trap additional bacteria. Perhaps the most compelling evidence that a secreted product of bacteria colonizing a medical device promotes the accretion of bacteria to form biofilm is the polysaccharide intracellular adhesin of *Staphylococcus aureus* and *S. epidermidis* consisting of linear (β1→6)-glucosamine (16). The presence of this polysaccharide seems to be required for biofilm formation by strains of these organisms. This material appears to be the major constituent of the so-called slime, previously shown to be essential for biofilm formation by *S. epidermidis* (5). The organisms produce another polysaccharide rich in galactose (PS/A), which also contributes to biofilm formation. It has been postulated that this polysaccharide promotes adhesion of the bacteria to medical devices (15, 40). However, the distinction between initial adhesion and subsequent accretion and eventual biofilm formation has not been well defined. In contrast, a clear distinction between the molecular species involved in the initial adhesion event and those involved in the subsequent accretion and biofilm formation was made for *Pseudomonas aeruginosa*. The pathogen, on initial adhesion mediated by hydrophobins and/or by a surface adhesin/lectin, begins to secrete a polyuronic acid, alginate, which promotes accretion and biofilm development (6).

In summary, the accretion phase involves initial colonization, whereby bacteria attach, undergo cell division, and form a primitive biofilm that involves the secretion of polysaccharides which entrap nonadherent bacteria at the surface (6). Finally, a microcolony community is formed, consisting of bacteria enmeshed in a complex hydrophilic matrix that contains solvent channels, valleys, and crevices (Fig. 10.1). Channel formation in biofilms is dictated by the genetics of the inhabiting bacteria. Their role is probably to increase the surface area of the biofilm, making it possible to more efficiently trap nutrients from the bulk medium. At the bottom of the biofilm, the oxygen tension is relatively low, creating a microaerophilic environment, although in an undefined manner. Oxygen levels at the very periphery of the biofilm are equivalent to those in the medium. Growth of the biofilm, combined with flow of the surrounding liquid, will eventually lead to dispersal of small fragments, thereby seeding other areas.

REGULATION: QUORUM SENSING

The density of the developing bacterial community in the various stages of biofilm formation varies and depends on many factors, such as the rate of multiplication; a complete description is beyond the scope of this chapter. Some of the bacteria in the biofilm community adhere to each other and to the underlying substratum, while others are entrapped within the gel matrix; the metabolic activity of each of the individual species depends on bacterial cell densities. Cell density-dependent regulation of gene expression, a phenomenon known as quorum sensing, contributes significantly to the size and development of the biofilm. A type of quorum sensing used by many gram-negative bacteria is the system involving acyl homoserine lactones. At a critical population density, the lactone compounds produced reach a concentration that affects the synthesis of a number of products, some of which may be the adhesins required for biofilm development and integrity. For example, it was recently shown that production of the *P. aeruginosa* adhesins/lectins PA-I and PA-II and of *Yersinia pseudotuberculosis* clumping factor is influenced by the quorum sensing of acyl homoserine lactone (3, 45).

Another mechanism for density-related regulation of biofilm development is the ability to sense the proportion of dead bacteria in the community. Studies which employed differential staining revealed that a significant fraction of bacteria are dead (12, 29). Dead bacteria may influence the biofilm population in a number of ways, some of which may be related to adhesion. For example, dying cells of *P. aeruginosa* may release intracellular lectins which may tether the intact bacteria to each other or to other species or to the underlying surfaces via carbohydrate-lectin interactions (43). An intracellular protein secreted or released by staphylococci binds to the bacterial surface and to components of the extracellular matrix to enhance the adhesion of the microorganisms to target substrata (37).

It is thought that most bacteria are found in biofilms, including pathogens. It follows that formation of biofilms confers a number of crucial advantages unavailable to adherent bacteria that are not participating in a biofilm (Table 10.1). It should be noted that bacteria inhabiting a biofilm may undergo a reduction in their rate of growth and may also display minimal metabolic activity. This leads to these organisms being less susceptible to antimicrobial agents. Biofilm on medical devices may represent a source of constant seeding of pathogens into the host tissues.

DENTAL PLAQUE BIOFILM: A HOME FOR MANY BACTERIAL SPECIES

Dental plaque is an excellent example of a complex, multispecies biofilm. The human mouth is thought to contain 300 to 500 distinct species and strains of bacteria (32). Many of these bacteria are found in biofilms at specific sites in the oral cavity (30, 38). The choice of sites is dictated by the physiology of the bacteria as as well as the physicochemical and nutritional characteristics of the sites. For example, some bacteria (e.g., anaerobes and microaerophiles) are found exclusively in

oxygen-limited sites, such as subgingival spaces or root surfaces. Others (e.g., streptococci) tend to be found in sites with higher oxygen tensions, such as tooth surfaces. The mechanism for the formation of biofilms on these surfaces follows the common pathways discussed above but may have additional requirements imposed by the environment and by the genetic background of the bacteria. Subgingival biofilms may involve numerous intergeneric interactions, whereas the coronal (i.e., tooth surface) plaque may involve relatively fewer species (44).

Development of the biofilm into mature dental plaque is unique to each site. On teeth, biofilm formation is dependent on the so-called early colonizers (20, 21). Some bacteria, such as *Streptococcus sanguis,* are the early colonizers and adhere tenaciously to saliva-coated teeth (Fig. 10.2). High densities of *S. sanguis* generally reflect good oral health. In the presence of sucrose, however, *S. sanguis* and other adherent oral streptococci may produce extracellular glucans (Fig. 10.3). These glucans, sometimes called dextrans, serve as substrata for the adhesion of other streptococci, especially the cariogenic mutans streptococci. The mutans streptococci produce several glucan binding proteins which mediate the binding of the organisms to the glucan matrix (28). Thus, the sucrose-dependent formation of biofilms on the tooth surface is characterized by the fact that the bacteria produce their own adhesin, the glucan binding protein, and their

TABLE 10.1 Advantages gained by bacteria as a result of participating in biofilms

Advantage	Explanation
Exchange of genetic elements	Conjugation and transformation is more efficient due to the close apposition of bacteria
Quorum sensing	Signal transduction induces the expression of virulence factors
Resistance to phagocytosis	Phagocytes cannot penetrate the biofilm matrix, enabling bacteria to escape immune surveillance and death
Nutrient enhancement	Certain species inhabiting multiple-species biofilms may produce metabolites required by the other species
Resistance to desiccation	Bacteria on medical devices before insertion are at risk of desiccation unless they are in biofilms
Resistance to antimicrobial agents	Because bacteria in biofilms reproduce at a reduced rate, they are less susceptible to many bioactive agents

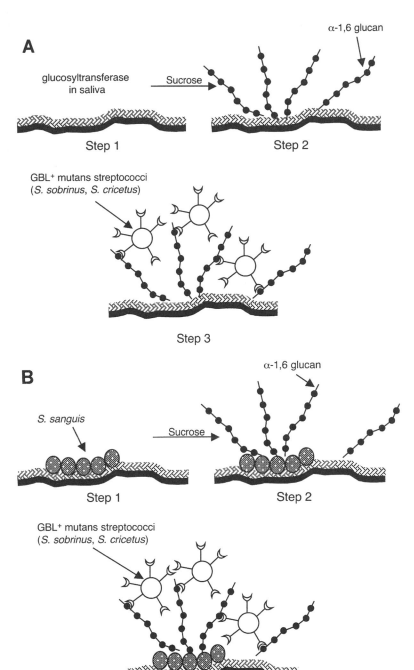

A

glucosyltransferase in saliva

Sucrose

α-1,6 glucan

Step 1

Step 2

GBL⁺ mutans streptococci
(*S. sobrinus, S. cricetus*)

Step 3

B

S. sanguis

Sucrose

α-1,6 glucan

Step 1

Step 2

GBL⁺ mutans streptococci
(*S. sobrinus, S. cricetus*)

FIGURE 10.2 Schematic diagram showing glucosyltransferase promotion of adhesion of oral streptococci on tooth surfaces. There are two mechanisms by which this occurs. (A) In one mechanism, the bacteria bind to glucan, and the glucosyltransferases secreted by oral bacteria into saliva synthesize glucan on the pellicle of tooth surfaces by using dietary sucrose. This enables mutans streptococci that express glucan binding lectins (GBL$^+$ bacteria, most notably *S. sobrinus* and *S. cricetus*) to adhere to the growing plaque. (B) In the second mechanism, the mutans streptococci adhere to the adsorbed pellicle on tooth surfaces via adhesin-specific pellicle constituents. Glucosyltransferases of the bound streptococci synthesize α1→6-glucan in the presence of sucrose, some of which remains bound and some of which is released. The bound glucan promotes the adhesion of other GBL$^+$ mutans streptococci, which express receptors for other genera of oral bacteria (see Fig. 10.3).

own receptor, the glucan. Another interesting feature of this type of biofilm formation is that the sucrose-dependent production of both adhesin and receptor by the streptococci is constitutive and density independent (i.e., not influenced by a quorum-sensing component). When oral hygeine is poor, the constitutive, density-independent nature of this biofilm formation can result in the formation of a very thick biofilm, such that it can be readily

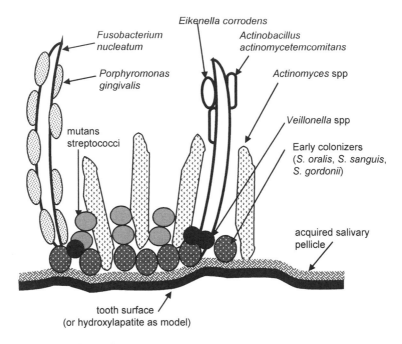

FIGURE 10.3   Ecology of the microbiota of the tooth biofilm. Salivary components bind to the tooth surface after cleaning, forming the acquired salivary pellicle. Subsequent bacterial attachment tends to occur in a somewhat ordered sequence, obviously varying among individuals. The bacteria that attach initially are usually called primary or early colonizers. This group includes primarily *Streptococcus* spp. and *Actinomyces* spp. There are many other bacterial species (e.g., *Porphyromonas gingivalis* and *Fusobacterium nucleatum*) that attach to the early colonizers or their products.

observed by the naked eye after proper staining. This biofilm can reach a thickness of 100 to 200 μm when the diet is relatively low in sucrose and can reach even greater thicknesses following a sucrose-rich diet (30). Also, only facultative aerobes, i.e., species that either produce glucan binding proteins or are capable of binding the glucan-producing streptococci, are able to participate in the formation of the biofilm on coronal tooth surfaces. The number of species that fulfill these requirements is relatively limited; they typically include *Actinomyces* strains which produce a lectin capable of binding to complementary sugars on the streptococcal surfaces (4).

In subgingival tissues, a pellicle can also be formed onto which various facultative anerobes or anerobes can reside, either on the hard tissue or on the adjacent soft tissue (Fig. 10.3). Accretion of these bacteria does not depend on sucrose but, rather, on intergeneric inter-

actions that involve lectin-sugar and protein-protein interactions, as well as the hydrophobic effect. Such intra- and intergeneric interactions can be monitored in vitro by the so-called coaggregation reactions, whereby a member of one genus is mixed with another genus (24, 35).

In summary, the dental biofilm is based on many bacterium-bacterium interactions, especially metal ion-requiring lectin-carbohydrate interactions. Saliva bathing the oral cavity is supersaturated with Ca^{2+}, as well as other trace metals, allowing the interactions to proceed even in the presence of chelators such as lactate or citrate (27). Some bacteria, including *S. aureus, P. aeruginosa, Escherichia coli,* and *Helicobacter pylori,* are regarded as transient members of the oral microbiota. Even though these bacteria elaborate adhesins, including lectins that allow them to participate in the biofilm community of the tooth/gingival tissue

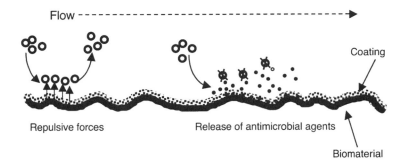

FIGURE 10.4 Illustration of two methods for altering adhesion to smooth inert surfaces. One mechanism involves coating the target surfaces with various materials which alter the surface properties and provide repulsive forces that keep bacteria from adhering. The other involves impregnating the material with antimicrobial agents which slowly leach from the surface and kill any bacteria that may adhere.

interface or the coronal biofilm, their transient nature may be due to lack of essential nutrients, predation from other bacteria, or susceptibility to salivary components. The role of transient colonization of *E. coli* in the oral cavity is discussed in more detail in chapter 6.

ADHESION TO BIOMATERIALS AND ASSOCIATED, HOST-DERIVED MACROMOLECULES

Infectious diseases associated with indwelling medical devices are now considered a major problem in health care (42). Formation of bacterial biofilms on these devices is a prerequisite for the development of infection. Initial adhesion to indwelling medical devices may occur either to unconditioned material or to materials coated with host-derived macromolecules. The mechanisms of adhesion to coated surfaces are no different from those of adhesion to tissues with respect to specificity of the adhesion. For example, adhesion of staphylococci to indwelling material is often mediated by fibronectin binding protein expressed by the staphylococci and fibronectin bound to the indwelling material (Table 10.2). In contrast, some bacteria bind best to uncoated biomedical devices, and coating with various compounds can prevent bacterial adhesion. For example, adhesion of *S. epidermidis* to catheters is inhibited by coating the biomaterial with serum-derived macromolecules (13,

22). Coating of restorative material with saliva reduces the sucrose-dependent adhesion of *S. mutans* (39). While major forces driving the initial adhesion of the bacteria to conditioned devices consist essentially of avid interaction between bacterial adhesins and their cognate receptors bound to the device, the forces driving adhesion to uncoated devices consist primarily of hydrophobic interactions (9, 34).

INFECTIOUS DISEASES AS CONSEQUENCES OF BIOFILM DEVELOPMENT ON BIOMATERIALS

Biomaterials inserted into the host quickly become coated with host-derived proteins and glycoproteins, forming a pellicle or "conditioned surface." The primary source of so-called biomaterial associated infection biofilms is usually the bacteria that are found as indigenous inhabitants near the area the biomaterial is inserted. For example, commensal bacteria of the skin often cause biofilm infections of catheters inserted through the skin (31). The various kinds of biomaterials that can become infected include indwelling vascular and intracardiac prostheses, vascular and urinary catheters, dialysis membranes, extended-wear contact lenses, orthopedic prostheses, and prosthetic joints. The indwelling devices associated with bloodstream infections are the most dangerous and costly in terms of hospital care. While a variety of bacterial species have

TABLE 10.2 Examples of bacterial adhesion to indwelling medical devices mediated by adhesins specific for devices coated with host-derived macromolecules

Bacterial species	Adhesin	Medical device	Coating compound	Reference
S. aureus	Fibronectin binding protein	Catheters	Fibronectin	13
	Fibrinogen binding protein	Vascular prostheses	Fibrinogen	46
E. coli	Undefined	Vascular prostheses	Fibrinogen	46

been implicated in these blood infections, staphylococci account for the majority of infections of intravascular devices (7, 41).

The biofilm formation discussed above is typical for biofilms formed by staphyloccoci on biomaterial. The ability of staphylococci to colonize the skin cannot explain the remarkable ability of the staphylococci to form persistent biofilms on conditioned biomaterials. For example, *Propionibacterium* spp. or coryneform bacteria are skin colonizers, but they rarely participate in the formation of biofilms on biomaterials and therefore are rare in biomaterial-associated infections. There must therefore be particular traits in the staphylococcal genome that favor the formation of persistent biofilms on biomaterials. These traits includes three types of constituents. One trait is the expression of adhesins specific for serum and extracellular matrix macromolecules, such as fibronectin binding proteins, fibrinogen binding proteins, and collagen binding proteins. A second trait is the production of a secreted polysaccharide which can form a hydrated gel in which the bacteria can become enveloped. The third type of trait is the presence of a hydrophobic bacterial surface with reduced surface charge (17).

EXOGENOUS AGENTS OR FACTORS MODULATING BIOFILM FORMATION

Because bacterial adhesion is critical to the development of biofilms and because biofilms on medical devices lead to serious diseases, attempts have been made to block bacterial adhesion in order to reduce morbidity and mortality of patients requiring foreign-body insertion. Antiadhesion strategies required to prevent the colonization of medical devices are different from those discussed in chapter 11 to prevent tissue colonization. First, and perhaps most important, is the fact that the target substrata are inanimate and metabolically inactive and can be selectively modified. It is not surprising that management of biofilms by impregnation of antibiotics into the target device has been the most extensively studied and applied antiadhesion mechanism (Fig. 10.4). This approach proved very efficient in cases where an antimicrobial agent was introduced during the process of device construction, such as antibiotic impregnation of bone cements (14, 26, 33). Otherwise, uniform coating of surfaces is difficult to achieve. In addition, increasing instances of resistance of bacterial strains to antibiotics has been a major drawback to this approach. An alternative approach is aimed at preventing the initial adhesion phase of bilofilm formation by a number of means, including coating the biomaterial with metals such as silver (1, 19, 25**).** Uniform coating of the devices remains a major problem, and coating of both sides of devices such as the inside and outside of catheters has not yet been achieved.

REFERENCES

1. **Ahearn, D. G., D. T. Grace, M. J. Jennings, R.N. Borazjani, K. J. Boles, L. J. Rose, R. B. Simmons, and E. N. Ahanotu.** 2000. Effects of hydrogel/silver coatings on *in vitro* adhesion to catheters of bacteria associated with urinary tract infections. *Curr. Microbiol.* **41:**120–125.
2. **An, Y. H., and R. J. Friedman.** 2000. *Handbook of Bacterial Adhesion. Principles, Methods and Applications.* Humana Press, Totowa, N.J.
3. **Atkinson, S., J. P. Throup, G. S. Stewart, and P. Williams.** 1999. A hierarchical quorum-sensing system in *Yersinia pseudotuberculosis* is involved in the regulation of motility and clumping. *Mol. Microbiol.* **33:**1267–1277.
4. **Cassels, F. J., C. V. Hughes, and J. L.**

Nauss. 1995. Adhesin receptors of human oral bacteria and modeling of putative adhesin-binding domains. *J. Ind. Microbiol.* **15:**176–185.

5. **Christensen, G. D., L. M. Baddour, and W. A. Simpson.** 1990. Colonial morphology of staphyloccoci on Memphis agar: phase production in vitro and in vivo. *Infect. Immun.* **55:**2870–2877.

6. **Costerton, J. W., Z. Lewandowski, D. E. Caldwell, D. R. Korber, and H. M. Lappin-Scott.** 1995. Microbial biofilms. *Annu. Rev. Microbiol.* **49:**711–745.

7. **Costerton, J. W., P. S. Stewart, and E. P. Greenberg.** 1999. Bacterial biofilms: a common cause of persistent infections. *Science* **284:**1318–1322.

8. **Doyle, R. J.** 1999. Biofilms. *Methods Enzymol.* **310:**1–720.

9. **Doyle, R. J.** 2000. Contribution of the hydrophobic effect to microbial infection. *Microbes Infect.* **2:**391–400.

10. **Doyle, R. J.** 2001. Microbial growth in biofilms, part A. Developmental and molecular biological aspects. *Methods Enzymol.* **336:**1–500.

11. **Doyle, R. J.** 2001. Microbial growth in biofilms, part B. Special environments and physicochemical aspects. *Methods Enzymol.* **337:**1–439.

12. **Dunne, W. M., Jr.** 2002. Bacterial adhesion: seen any good biofilms lately? *Clin. Microbiol. Rev.* **15:**155–166.

13. **Francois, P., P. Vaudaux, and P. D. Lew.** 1998. Role of plasma and extracellualr matrix proteins in the pathophysiology of foreign body infections. *Ann. Vasc. Surg.* **12:**34–40.

14. **Garvin, K. L., B. G. Evans, E. A. Salvati, and B. D. Brause.** 1994. Palacos Gentamicin for the treatment of deep periprosthetic hip infections. *Clin. Orthop.* **298:**97–105.

15. **Goldmann D. A., and G. B. Pier.** 1993. Pathogenesis of infections related to intravascular catheterization. *Clin. Microbiol. Rev.* **6:**176–192.

16. **Götz, F., and G. Peters.** 2000. Colonization of medical devices by coagulase-negative staphylococci, p. 55–88. *In* F. A. Waldvogel and A. L. Bisno (ed.), *Infections Associated with Indwelling Medical Devices.* ASM Press, Washington, D.C.

17. **Gross, M., S. E. Cramton, F. Götz, and A. Peschel.** 2001. Key role of teichoic acid net charge in *Staphylococcus aureus* colonization of artificial surfaces. *Infect. Immun.* **69:**3423–3426.

18. **Hasman, H., T. Chakraborty, and P. Klemm.** 1999. Antigen-43-mediated autoaggregation of *Escherichia coli* is blocked by fimbriation. *J. Bacteriol.* **181:**4834–4841.

19. **Jansen, B., M. Rinck, P. Wolbring, A. Strohmeier, and T. Jahns.** 1994. *In vitro* eval-

uation of the antimicrobial efficacy and biocompatibility of a silver-coated central venous catheter. *J. Biomater. Appl.* **9:**55–70

20. **Jenkinson, H. F.** 1994. Adherence and accumulation of oral streptococci. *Trends Microbiol.* **2:**209–212.

21. **Jenkinson, H. M., and H. F. Lappin-Scott.** 2001. Biofilms adhere to stay. *Trends Microbiol.* **9:**9–10.

22. **John, S. F., M. R. Derrick, A. E. Jacob, and P. S. Handley.** 1996. The combined effects of plasma and hydrogel coating on adhesion of *Staphylococcus epidermidis* and *Staphylococcus aureus* to polyurethane catheters. *FEMS Microbiol. Lett.* **144:**241–247.

23. **Kolenbrander, P.** 2000. Oral microbial communities: biofilms, interactions, and genetic systems. *Annu. Rev. Microbiol.* **54:**413–437.

24. **Kolenbrander, P. E., and J. London.** 1993. Adhere today, here tomorrow: oral bacterial adherence. *J. Bacteriol.* **175:**3247–3252.

25. **Leung, J. W., G. T. Lau, J. J. Sung, and J. W. Costerton.** 1992. Decreased bacterial adherence to silver-coated stent material: an *in vitro* study. *Gastrointest. Endosc.* **38:**338–340.

26. **Leung, J. W., Y. Liu, S. Cheung, R. C. Chan, J. F. Inciardi, and A. F. Cheng.** 2001. Effect of antibiotic-loaded hydrophilic stent in the prevention of bacterial adherence: a study of the charge, discharge, and recharge concept using ciprofloxacin. *Gastrointest. Endosc.* **53:**431–437.

27. **Lu-Lu, J. S. Singh, M. Y. Galperin, D. Drake, K. G. Taylor, and R. J. Doyle.** 1992. Chelating agents inhibit activity and prevent expression of streptococcal glucan-binding lectins. *Infect. Immun.* **60:**3807–3813.

28. **Ma, Y., M. Y. Galperin, K. G. Taylor, and R. J. Doyle.** 1994. Glucan-binding proteins of *Streptococcus sobrinus*, p. 275–286. *In* E. van Driessche, J. Fischer, S. Beekmans, and T. C. Bog-Hansen (ed.), *Lectins, Biology, Chemistry, Clinical Chemistry*, vol. 1. TEXTOP, Hellerup, Denmark.

29. **Mah, T. F., and G. A. O'Toole.** 2001. Mechanisms of biofilm resistance to antimicrobial agents. *Trends Microbiol.* **9:**34–39.

30. **Marsh, P. D.** 1995. Dental plaque, p. 282–300. *In* H. M. Lappin-Scott and J. W. Costerton (ed.), *Microbial Biofilm.* Cambridge University Press, Cambridge, United Kingdom.

31. **Montdargent, B., and D. Letourneur.** 2000. Toward new biomaterials. *Infect. Control Hosp. Epidemiol.* **21:**404–410.

32. **Moore, N., and S. Moore.** 1994. Microbiology and immunology of periodontal diseases. *Periodontology* **6:**78–111.

33. **Multanen, M., M. Talja, S. Hallanvuo, A.**

Siitonen, T. Valimaa, T. L. Tammela, J. Seppala, and P. Tormala. 2000. Bacterial adherence to silver nitrate coated poly-L-lactic acid urological stents *in vitro*. *Urol. Res.* **28:** 327–331.

34. Nomura, S., F. Lundberg, M. Stollenwerk, K. Nakamura, and Å. Ljungh. 1997. Adhesion of staphylococci to polymers with and without immobilized heparin in cerebrospinal fluid. *J. Biomed. Mater. Res.* **38:**35–42.

35. Ofek, I., and R. J. Doyle. 1994. *Bacterial Adhesion to Cells and Tissues,* p. 195–238. Chapman & Hall, New York, N.Y.

36. O'Toole, G., H. B. Kaplan, and R. Kolter. 2000. Biofilm formation as microbial development. *Annu. Rev. Microbiol.* **54:**49–79.

37. Palma, M., A. Haggar, and J. I. Flock. 1999. Adherence of *Staphylococcus aureus* is enhanced by an endogenous secreted protein with broad binding activity. *J. Bacteriol.* **181:**2840–2845.

38. Rosan, B., and R. J. Lamont. 2000. Dental plaque formation. *Microbes Infect.* **2:**1599–1607.

39. Steinberg, D., C. Mor, H. Dogan, B. Zacks, and I. Rotstein. 1999. Effect of salivary biofilm on the adherence of oral bacteria to bleached and non-bleached restorative material. *Dent. Mater.* **15:**14–20.

40. Tojo, M., N. Yamashita, D. A. Goldmann, and G. B. Pier. 1988. Isolation and characterization of a capsular polysaccharide adhesin from *Staphylococcus epidermidis*. *J. Infect. Dis.* **157:** 713–722.

41. von Eiff, C., C. Heilmann, and G. Peters. 1999. New aspects in the molecular basis of polymer-associated infections due to staphylococci. *Eur. J. Clin. Microbiol. Infect. Dis.* **18:**843–846.

42. Waldvogel, F. A., and A. L. Bisno (ed.). 2000. *Infections Associated with Indwelling Medical Devices.* ASM Press, Washington, D.C.

43. Wentworth, J., F. E. Austin, N. Garber, N. Gilboa-Garber, C. Paterson, and R. J. Doyle. 1991. Cytoplasmic lectins contribute to the adhesion of *Pseudomonas aeruginosa*. *Biofouling* **4:**99–104.

44. Whittaker, C. J., C. M. Klier, and P. E. Kolenbrander. 1996. Mechanisms of adhesion by oral bacteria. *Annu. Rev. Microbiol.* **50:**513–552.

45. Winzer, K., C. Falconer, N. C. Garber, S. P. Diggle, M. Camara, and P. Williams. 2000. The *Pseudomonas aeruginosa* lectins PA-IL and PA-IIL are controlled by quorum sensing and by RpoS. *J. Bacteriol.* **182:**6401–6411.

46. Zdanowski, Z., E. Ribbe, and C. Schalen. 1993. Influence of some plasma proteins on *in vitro* bacterial adherence to PTFE and Dacron vascular prostheses. *APMIS* **101:**926–932.

ANTIADHESION THERAPY

11

Antiadhesion therapy and immunity are meant to reduce contact between host tissues and pathogens, either by preventing adhesion or by dissociating the infectious agent. There is now compelling evidence showing that adhesion of enteric, oral, and respiratory bacteria is required for colonization and that colonization is required for subsequent development of symptoms of diseases (see chapter 1). Moreover, when adherent to animal cells or tissues, bacteria assume greater resistance to normal immune factors, to bacteriolytic enzymes, and to antibiotics. In addition, adherent bacteria are better able to acquire nutrients (71). Thus, the adherent state better enables bacterial survival; in theory, prevention of adhesion at an early time following exposure to the pathogen would prevent the disease.

Bacteria seemingly can adapt to many noxious or deleterious agents, either by mutation, by acquisition of genetic material, or by phenotypic variations. Bacteria resistant to antiadhesion agents will be expected to emerge. It is likely that there is some proportionality between the fraction of resistant survivors and the nature of the antibacterial agents used. Because antiadhesion agents act not by killing or arresting the growth of the pathogen, as antibiotics do, it is reasonable to expect that spread of resistant strains to antiadhesion agents will occur at significantly lower frequencies than will resistance to antibiotics (Fig. 11.1 and 11.2). Neither antibiotics, vaccines, nor antiadhesion agents prevent the spread of resistant strains in the population. However, antiadhesion agents allow sensitive pathogens to propagate and be transmitted in a manner that permits them to cause new infections that can still be treated by the agent because the bacteria remain sensitive. It would appear, therefore, that time-dependent spread of strains resistant to the antiadhesion agent would be significantly delayed compared to that of strains resistant to antibiotics. The major drawback of antiadhesion therapy is that most pathogens possess genes encoding more than one type of adhesin (see chapter 1), a phenomenon that requires the use of either multiple agents for multiple adhesins of a particular pathogen or an agent that exhibits a broad spectrum of antiadhesion activity. The following sections review various traditional approaches of antiadhesion therapy and immunity, including the use of adhesin-based vaccines, receptor and adhesin analogs, sublethal concentrations of antibiotics, dietary constituents, and innate host-derived antiadhesion factors. Early discussions on antiadhesion therapy and immunity may be found in a number of review articles (41, 61, 71).

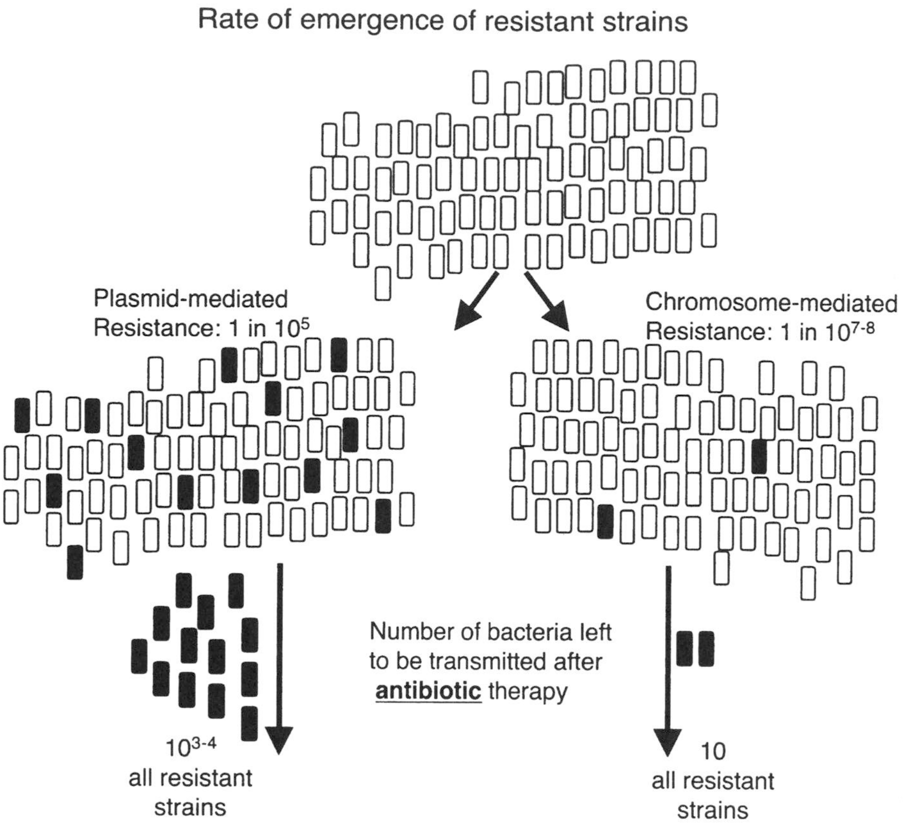

Rate of emergence of resistant strains

FIGURE 11.1 Schematic diagram illustrating a proposed effect of antibiotic usage on the survival and spread of resistant bacteria. The proposed effect assumes that if the resistance is plasmid mediated, the rate of spread of resistance within the population will be equal to the rate of plasmid transfer within the population. The rate of conjugative plasmid transfer observed in vitro is approximately 1 in 10^5 bacterial cells. If the resistance is chromosomally mediated, mostly arising from point mutations, the rate of spread will be much lower, approximately 1 in 10^7 to 10^8 bacteria. Thus, assuming that an infection caused by 10^7 to 10^8 bacteria is treated with antibiotics, 1 to 10 would remain viable in chromosomal resistance population while 10^3 to 10^4 would remain viable in the plasmid-mediated resistance population. The probability that the resistant population will grow and spread to other individuals would therefore be much higher for plasmid-mediated resistance than for chromosome-mediated resistance. Continued use of the same antibiotics will inevitably result, in a relatively short period (e.g., 5 to 6 years), in infections caused mostly by resistant strains, ultimately requiring alternative antibiotic treatment.

ADHESIN-BASED VACCINE

Prevention of symptomatic infections by blocking adhesion with adhesin-based vaccines can be achieved by either active or passive means. Although the active antiadhesin immunity is expected to prevent infection by stimulation of secretory immunoglobulin A (IgA) on mucosal surfaces, significant amounts of serum IgG appear to reach mucosal surfaces, such as the gut, the oral cavity, and even the urinary tract (84). In passive immunity, the target host acquires antiadhesin antibodies made in another host. The best example of the latter is where the K88 fimbriae and related adhesins of farm animal pathogens were used to vaccinate pregnant females so that the suckling newborns

Rate of emergence of resistant strains

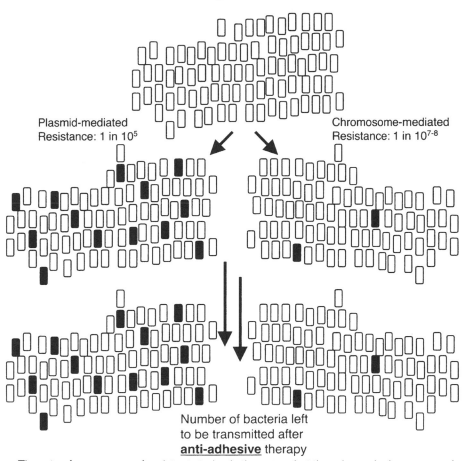

Plasmid-mediated
Resistance: 1 in 10^5

Chromosome-mediated
Resistance: 1 in 10^{7-8}

Number of bacteria left
to be transmitted after
anti-adhesive therapy

The rate of emergence of resistant strains is the same, but there is much slower spread
due to the presence of competing, sensitive strains

FIGURE 11.2 Schematic diagram illustrating a proposed effect of the use of antiadhesion
agents on the survival and spread of resistant bacteria. The rate of spread of resistance to antiadhesion agents within the population for either plasmid- or chromosome-mediated resistance
is expected to be essentially the same as that described for antibiotic resistance in Fig. 11.1.
Unlike antibiotic treatments, however, treatment with antiadhesion agents does not have a differential effect on the viability of resistant and sensitive strains. Thus, the rate of spread of
resistant strains will remain low, due to the continued viability of sensitive strains. It is predicted, therefore, that over a given period, antiadhesion therapy would result in a much slower
spread of resistant strains within the population than would antibiotic therapy.

acquired the milk-secreted antibodies, which
functioned to prevent the development of
infection by the pathogen (60). It has been
assumed that these antibodies prevent infection
by inhibiting adhesion.

Antigenic variability, especially in proteinaceous adhesins, may compromise the

efficacy of the antiadhesin vaccine, if one con-
siders that each of the multiple adhesins produced by a particular pathogen undergo allelic
variations giving rise to so-called isoadhesins
(e.g., adhesins that maintain their receptor
specificity but otherwise differ antigenically).
For example, there are numerous allelic vari-

ants of the FimH adhesin of type 1 fimbriae of *Escherichia coli* (102). Recently, it has been suggested that a conserved region of the adhesin may serve as a vaccine, especially when this region contains the receptor binding domain (48, 119). However, convincing data for the efficacy of such a vaccine in humans have not yet been published.

Although it has been known for more than three decades that antiadhesin antibodies effectively inhibit adhesion, there are no reports of herd immunizations with purified adhesins or with recombinant strains containing the DNA encoding the adhesin. It has been suggested that mucosal immunity, as opposed to systemic, IgG-mediated immunity, can improve the protective effect of adhesin-based vaccines (58, 119). In some cases, whole attenuated or inactivated bacteria which carry the adhesin have been used. In these cases, the antiadhesin antibodies may function in concert with antibodies against other virulence factors to protect the vaccinated host from symptomatic infection. For example, the inactivated *Bordetella* vaccine contains the hemagglutinin adhesin, and vaccinated individuals develop antihemagglutinin antibodies, which have been shown to inhibit adhesion of *Bordetella* to animal cells (83). In animal models of experimental infections, there seems to be ample evidence for active immunization with adhesins in providing IgG- or secretory IgA-mediated protection against infection (71). The choice of animal models for mucosal infection is of paramount importance and may be a major drawback (119). Active immunization with a complex of the FimH adhesin and its periplasmic chaperone, FimC, was shown to protect against urinary tract infection caused by *E. coli* in both mice and nonhuman primates (47, 48). It is very important in this latter case that an adhesin-chaperone complex could be used, rather than a whole organism or purified fimbriae, since in those cases the FimH adhesin would be an exceedingly minor part of the antigenic complement in the immunization mixture. Even when using purified fimbriae,

most of the antibodies are directed against the major subunit, FimA.

In contrast to active immunization, the passive application of antibodies against adhesins may be more attractive. The success of milk-containing antibodies in protecting against diarrheal infections in suckling piglets and calves is well known (60). In another study, anti-*Streptococcus mutans* monoclonal antibodies directed against the SA I/II adhesins were applied to the tooth surfaces of human volunteers (54). The latter were treated with chlorhexadine to eliminate the *S. mutans* microbiota prior to the passive application of the antibodies. Reacquisition of *S. mutans* in the placebo group receiving nonrelevant monoclonal antibodies occurred within 2 to 3 months, whereas the passively immunized subjects remained virtually free of *S. mutans* for at least a year. Because antiadhesin antibodies were no longer detectable a day after application, it is difficult to explain these findings. It has been speculated that other nonpathogenic oral bacteria colonized the tooth surfaces, preventing *S. mutans* from recolonizing the occupied surfaces (42). Nevertheless, the studies are promising, and in the future, consumption of milk from immunized cows might be especially useful in the prevention of infections in targeted populations. While the beneficial effects of acquired immunity that follows vaccinations have been known for a long time, the possible efficacy of antiadhesion approaches began to become clear only after the essentiality of adhesion in the infectious process was realized. Therefore, antiadhesion approaches to the treatment of infectious diseases are comparatively recent.

RECEPTOR ANALOGS AS ANTIADHESIVE AGENTS

The use of the receptor analog as an agent for antiadhesion therapy would be practical primarily against pathogens that bind to animal cells via carbohydrate-specific adhesins (e.g., lectin). In this case, the receptor analog is a saccharide that is structurally similar to that of the glycoprotein or glycolipid receptors and therefore acts by competitively inhibiting the adhesin (Fig. 11.3A). It was less than three decades ago that mannose was first shown to be a receptor for enterobacteria (72). Since then, the sugar specificities of many bacteria have been determined, allowing the targeting of carbohydrate-containing receptors for specific pathogens (see chapter 12). In vitro, the molar concentrations of the carbohydrate required for effective inhibition of adhesion are usually high (in the millimolar range),

probably because the affinity of the saccharides for the bacterial lectins is low. The affinity can be increased 10- to 600-fold by covalently linking the saccharide to a hydrophobic residue such as a phenyl or methylumbelliferyl group (30). It can also be increased by several orders of magnitude by ligating the saccharide to a suitable carrier, yielding multidentate adhesin-inhibitors, as demonstrated with *Helicobacter pylori* (63, 98) and with the type 1-fimbriated *E. coli* (51).

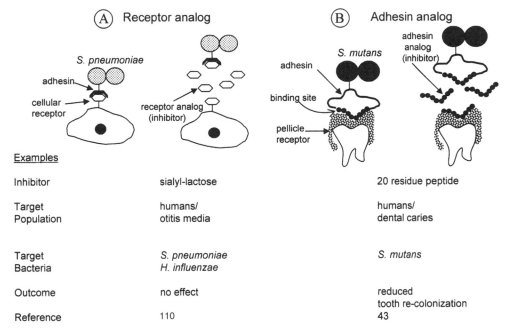

FIGURE 11.3 Schematic diagram and examples of receptor analog (A) and adhesin analog (B) inhibition of adhesion. (A) Adhesion mediated by a lectin on the surface of *S. pneumoniae* binding to sialic acid residues on host cells. Excess of sialyllactose in the reaction mixture competitively inhibits adhesion by occupying the sugar-combining sites of the bacterial lectin. A clinical study based on this type of therapy was conducted by Ukkonen et al. (110), in which children were given synthetic sialyllactose, administered intranasally, to prevent otitis media caused mainly by *S. pneumoniae* and *H. influenzae* which express adhesins specific for this sugar. There was no effect on the incidence of otitis media in the experimental group compared to the placebo group. The lack of effect was postulated to be due to the presence of multiple adhesins expressed by these pathogens for distinct receptors. (B) Adhesion mediated by a protein adhesin of *S. mutans* to its cognate receptor on the acquired pellicle on teeth. In vitro, a synthetic oligopeptide representing the binding site of the streptococcal adhesin competitively inhibited adhesion by occupying the adhesin binding site in the acquired pellicle. In a clinical study, the oligopeptide mimicking the adhesin binding site was applied directly to the tooth surfaces of four human volunteers (43). The rate of recolonization of mutans streptococci was significantly reduced compared to that in patients given placebo, consisting of a control peptide administered in a similar fashion. The effect was specific for mutans streptococci and not for other oral bacterial genera.

The feasibility of using saccharides to protect against experimental infections by bacteria expressing adhesive lectins was shown more than two decades ago (6). Administration of methyl α-mannoside and *E. coli* expressing the mannose-specific type 1 fimbrial lectin into the bladders of mice reduced the extent of bladder colonization by uropathogenic *E. coli* by about two-thirds, compared to animals receiving methyl α-glucoside, which does not inhibit the mannose-specific bacterial lectin. Subsequently, many studies have revealed the ability of saccharides to prevent experimental infections caused by different pathogenic bacteria (Table 11.1). Perhaps the most impressive studies are those that employed saccharides to treat *H. pylori* infections of the stomachs of monkeys (64). In these studies, six naturally infected rhesus monkeys were treated with sialyllactose [N-acetyl-neuraminyl-(α2→3)-galactosyl-(β1→4)-glucose), a potent inhibitor of *H. pylori* adhesion to human gastric cells. Two of the animals were cured of the *H. pylori* infection, and one animal was transiently cleared. The three other animals remained colonized. Six additional animals received sialyllactose in combination with conventional antiulcer drugs. Decreases in colony counts were found in three of these animals, while two animals

cleared the *H. pylori* and there was no effect in the sixth. The results suggested that sialyllactose could be a safe and effective antiadhesive therapeutic agent for *H. pylori* infections.

Clinical trials employing saccharides or sugars in antiadhesion therapy of humans have been limited. In a recent study, 254 children were given a sialyllactose spray twice daily intranasally for three months to prevent otitis media (Fig. 11.3A). This infection is caused primarily by *Streptococcus pneumoniae* and *Haemophilus influenzae* and to a lesser extent by *Moraxella catarrhalis* (110). These bacteria are known to produce multiple adhesins (see chapter 12), only some of which are specific for sialyllactose. The efficacy of treatment was negative compared to the placebo group in preventing otitis media, as well as in preventing the colonization of the upper respiratory tract by the bacteria. It is possible, therefore, that, as stated above, targeting only one of several adhesins of the pathogens will frequently be insufficient to prevent colonization and symptomatic infection.

In a study of 58 patients with acute otitis externa, 36 were culture positive for *Pseudomonas aeruginosa* (12, 106). The patients were treated by local administration of gentamicin or a mixture of galactose, mannose, and N-acetylneuraminic acid. The gentamicin-

TABLE 11.1 Carbohydrates prevent bacterial infection in vivo

Organism	Animal and site of infection[a]	Inhibitor	Reference(s)
E. coli (type 1 fimbriated)	Mouse UT	Methyl α-mannoside	6
	Mouse GT	Mannose	36
K. pneumoniae (type 1 fimbriated)	Rat UT	Methyl α-mannoside	28
E. coli (P fimbriated)	Mouse UT	Globotetraose	108
		P1 antigen [Gal(α1→4)Gal] containing Gp	40
	Monkey UT	Gal(α1→4)GalOMe	86
Shigella flexneri (type 1 fimbriated)	Guinea pig eye	Mannose	5
E. coli K99	Calf GT	Glycopeptides	61, 62
H. pylori	Monkey GT	Sialyl 3'Lac	64
S. pneumoniae	Rabbit and rat lungs	Sialyl-Gal(β1→4)GlcNAc	38
S. sobrinus	Rat oral cavity	Oxidized (α1→6)glucan	115

[a]UT, urinary tract; GT, gastrointestinal tract; Gp, glycoprotein found in dove and pigeon egg white; OMe, O-linked methyl; Lac, lactose.

treated patients recovered more rapidly than did the patients treated with the sugars, although the recovery of both groups was more rapid than that of the untreated control patients. The solution contained sugars specific for two *Pseudomonas* lectins, PA-I and PA-II, which may act as adhesins following their release from the bacterial cytoplasm (118). It is possible that the failure of sialyl-3′-lactose neo-tetraose to prevent *S. pneumoniae* and *H. influenzae* infections, as opposed to the positive effect obtained with the sugar mixture in preventing *P. aeruginosa* infections, is due to the fact that in the former case only one adhesin of the bacterium was inhibited whereas in the latter at least two types of adhesins were inhibited.

ADHESIN ANALOGS IN ANTIADHESION THERAPY

The adhesin analog strategy is based on the assumption that the isolated adhesin molecule or its synthetic or recombinant fragment binds to the receptor and thereby competitively blocks adhesion of the bacteria (Fig. 11.3B). It has so far been impractical to use adhesin analogs in antiadhesion therapy because they are almost always macromolecules that must be employed in relatively high molar concentrations and they are available only in limited supply. In addition, careful consideration must be given to their toxicity and immunogenicity. Nevertheless, modern proteomics and recombinant biotechnology have permitted the development of unique types of relatively small peptides for antiadhesion therapy, as reported by Kelly et al. (43). In these studies, a synthetic peptide of 20 amino acids copied from the sequence of an *S. mutans* cell surface adhesin which binds a salivary protein was constructed (Fig. 11.3B). In vitro, this peptide inhibited the binding of the streptococci to the immobilized salivary receptor (i.e., an artificial tooth pellicle). Application of the peptide to teeth pretreated with chlorhexidine gluconate (to reduce the normal microbiota and eliminate *S. mutans*) significantly retarded the recolonization of the teeth with *S. mutans*

but had no effect on recolonization by *Actinomyces naeslundii*. Recolonization by *S. mutans* was observed in control volunteers receiving saline or placebo peptide. These results are promising, especially since the investigators employed small peptides. However, while conceptually encouraging, these specific results must be interpreted with caution because the adhesion of *S. mutans* in the presence of sucrose is mediated by adhesins other than the one used to derive the peptide (see chapter 12). A mixture of peptides corresponding to the multiple adhesins known for the streptococci may prove to be even more effective. The premise that peptides derived from a proteinaceous adhesin can serve as potent inhibitors of adhesion supports this notion. For example, synthetic peptides copying a fragment of the fimbrillin adhesin of *Porphyromonas gingivalis* were found to inhibit the adhesion of the organisms to hydroxylapatite (49, 50).

Nonproteinaceous adhesins may also be effective as ligand analogs for antiadhesion therapy. This was the case in lipoteichoic acid (LTA)-mediated adhesion of groups A and B streptococci. In one study, LTA was applied to the oral cavity, perineum, and nape of 5-day-old mice (23). The mother was painted with a suspension of group B streptococci in her oral cavity and vagina and on the nipples. None of the experimental pups were culture positive after 3 days, whereas 47% of the control pups were culture positive. In the second study, LTA was applied into the nasal cavity of mice and a group A streptococcal suspension was then administered (26). The incidence of colonization and death of the LTA-treated mice was significantly reduced compared to that of control mice. Although this illustrates a principle, the use of LTA in antiadhesive therapy would not be advised, due to its toxic properties. Group A streptococci may also bind via its hyaluronic acid capsule to the CD44 receptor on epithelial cells (25). The receptor normally binds hyaluronic acid of the host extracellular matrix (111). Application of hyaluronic acid in the oral cavity of mice prevented adhesion and colonization of the challenge group A strepto-cocci (25), but the precise mechanisms for this are not known. Under some circumstances, the hyaluronic acid capsule of *Streptococcus pyogenes* acts indirectly to block adhesion, by masking adhesins and blocking their access to cellular receptors by steric hindrance (21), and it also acts as a chelator of cations, which may be required for proper adhesin function (35).

SUBLETHAL CONCENTRATIONS OF ANTIBIOTICS

Subinhibitory concentrations of antibiotics reduce the ability of pathogens to adhere to various substrata (Fig. 11.4A). There is little evidence, however, that the sublethal concentrations of antibiotics are beneficial in the therapy of infections. This apparent contradiction may be due to a number of factors, including the paucity of pilot clinical trails and a lack of well-designed in vivo animal experiments. In a study of the effects of sublethal concentrations of penicillin on urinary tract infections, it was reported that 16 of 20 patients with more than 10^5 *E. coli* cells per ml of urine were 3 to 7 days later (Fig. 11.4A) (11). Cure, in this instance, was a reduction in the number of *E. coli* cells in urine to less than 10^4 per ml. None of 18 age- and sex-matched

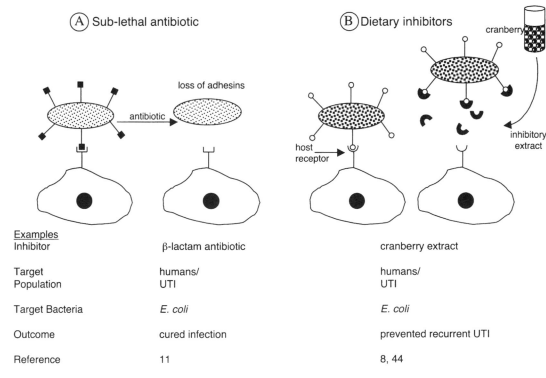

	A) Sub-lethal antibiotic	B) Dietary inhibitors
Examples Inhibitor	β-lactam antibiotic	cranberry extract
Target Population	humans/ UTI	humans/ UTI
Target Bacteria	*E. coli*	*E. coli*
Outcome	cured infection	prevented recurrent UTI
Reference	11	8, 44

FIGURE 11.4 Schematic diagram and examples of sublethal antibiotic (A) and dietary product (B) inhibition of adhesion. (A) Treatment of *E. coli* with subinhibitory concentrations of β-lactam antibiotics in vitro leads to lack of fimbrial expression and reduced adhesion. In a study of the effects of sublethal concentrations of penicillin on urinary tract infections, the number of *E. coli* CFU in urine was significantly reduced (11). (B) Example of the inhibitory effects of dietary compounds on the adhesion of various bacteria. With respect to cranberry extract, these effects can be observed on bacteria ranging from strains of *E. coli* causing urinary tract infections, diarrhea, and meningitis to oral bacteria and *H. pylori*. In two independent clinical studies, cranberry juice consumption was found to reduce significantly the incidence of urinary tract infections both in elderly (8) and young (44) women.

control patients were considered cured. It must also be pointed out that there are a limited number of reports showing, instead, an increase in the adhesion of pathogens grown in the presence of sublethal concentrations of antibiotics (71).

It is likely that a significant fraction of recipients of antibiotics experience a period when the concentration of antibiotics in their system is sublethal, at least transiently, due to poor patient compliance. For this reason alone, the effects of antibiotics on adhesion is a subject worthy of additional study. Interestingly, the advertisement of a new antibiotic (Monurol; Rafa Laboratories, Ltd.) contained a reference to a study describing the loss of adhesion of treated pathogens to uroepithelial cells (79), although the advertisement did not contain any details about the antibiotic concentration necessary to cause this effect in vitro. A number of publications dealing with the effect of sublethal concentration of antibiotics on adhesion have continued to appear over the years, but no major breakthroughs with regard to mechanism or clear clinical effect have been found.

PROBIOTICS AS ANTIADHESIN THERAPEUTIC AGENTS

An emerging subdiscipline in the prevention of infectious diseases is that of probiotics. Probiotics involves the study of bacteria that can prevent pathogens from achieving the critical density required to cause disease. Although this principle has been known for some time, the rapid spread of antibiotic-resistant strains has only now stimulated a reexamination of the use of probiotics in infectious-disease therapy. Beneficial bacteria may be derived from the normal microbiota. In theory, they can be also be genetically engineered, but present regulations prohibit the commercial use of such bacteria. A priori, the beneficial bacteria must possess the ability to adhere to and colonize specific sites on hard or soft tissues or biomaterials. Several members of the normal microbiota are being tested for probiotic use, but because of the propri-etary considerations, only the genus being tested is public knowledge in many cases. In general, most probiotic bacteria belong to the genera *Lactobacillus* and *Bifidobacterium* and in some cases to the species *E. coli* (29, 39, 82). The identities of adhesins of probiotic bacteria have not been determined in most cases. In one case, it has been shown that a proteinaceous constituent of the surface of *Lactobacillus* functions as an adhesin (20, 57).

Research has revealed that beneficial bacteria may either modify the metabolic activity of the pathogens and/or neutralize their virulence factors (10). For example, probiotic bacteria produce bacteriocins that kill bacteria and factors that neutralize toxins of the pathogen, and they deprive the pathogen of essential nutrients. Recently, attention has been paid to the antiadhesion activity of probiotic bacteria. The adhesion of pathogenic bacteria can be modulated by the probiotic bacteria in a number ways, including steric exclusion of the pathogen from a required niche, changing the local pH to a point that pathogen adhesins do not function efficiently, detachment of the pathogen, and production of bacteriolytic or otherwise deleterious enzymes capable of altering adhesion of the pathogen. In some studies, a specific inhibitor of pathogen adhesion was identified in the culture supernatants of probiotic bacteria (Table 11.2).

Beneficial bacteria may also coaggregate with pathogens to enhance the clearance of the latter. This activity could be exhibited even by dead bacteria if the adhesin is an unusually stable molecule. Coaggregation-mediated clearance was shown in a study where *Lactobacillus* strains isolated from the genitourinary tract of weanling piglets coaggregated with enterotoxigenic K88 fimbriated, but not with K88-negative, *E. coli*, suggesting that a putative *Lactobacillus* adhesin binds to the K88 fimbria (103). It should be noted that in spite of the promising in vitro studies and commercialization of probiotics, there have not been any carefully controlled clinical studies of humans to test whether the

TABLE 11.2 Examples of probiotic bacteria inhibiting the adhesion of pathogens

Probiotic	Pathogen	Substrata for adhesion assay	Probiotic inhibitor[a]	Reference
Lacobacillus spp.	*E. coli* K88	Ileum mucus	>250 kDa	16
Lactobacillus fermentum	*E. coli* K88	Ileum mucus	Carbohydrate?, 1,700 kDa	77
Lactobacillus acidophilus	*Enterococcus faecalis, E. coli, S. epidermidis, Candida* spp.	Silicone rubber	Biosurfactant	113
Lactobacillus spp.[b]	*E. faecalis*	Glass	Biosurfactant, >6,000 mol wt	114
Lactobacillus spp.	*E. coli* K88	Coaggregation		103
Bifidobacterium spp.	ETEC[c]	Immobilized asialo-GM$_1$, GA1	Protein, 10^5 kDa	32

[a]Found in culture supernatants of the probiotic bacteria.
[b]Fifteen isolates.
[c]ETEC, Enterotoxigenic *E. coli.*

claimed beneficial effects are due to antiadhesive effects.

DIETARY INHIBITORS OF ADHESION

The most successful antiadhesins may be present in foodstuffs, especially those which contain either a mixture of inhibitors or an inhibitor with a broad spectrum of activity (Fig. 11.4B). While it may be possible to find the appropriate inhibitors for some pathogens, it is likely to be impossible to match every individual or group of pathogens with specific diets that contain complementary adhesin inhibitors. Empirical observations over the years suggest that certain dietary constituents may have beneficial effects for the therapy of infections. These dietary constituents represent good candidates for antiadhesion studies. However, caution should be used, because diets may contain bactericidal and bacteriostatic compounds (22), and selective pressures such as these should be avoided. Another caution is that there are also dietary substances that promote adhesion, such as lectins, so that critical evaluation of each of these materials will be important. Two examples serve to highlight the importance of modulating adhesin activities by dietary constituents. One involves milk, which is rich in several different types of adhesin-inhibiting glycoconjugates, and the other involves plant-derived constituents.

Milk has long been known to be beneficial in preventing certain bacterial infections (7, 46). It was not until the adhesion of many pathogens was found to be specifically mediated by lectin-carbohydrate interactions that researchers began to explore the possibility that milk oligosaccharides and glycoproteins might contribute considerably to its therapeutic effect by inhibiting the adhesion of diverse pathogens. In this regard, it is interesting that human milk is much richer in glycoconjugates than is either cow milk or formula (46, 68). Table 11.3 lists some types of glycoconjugates and the pathogens that are known to express corresponding adhesins. Milk may also contain other constituents which bind pathogenic bacteria and inhibit adhesion. For example, lactoferrin, a common constituent of milk, inhibits the adhesion of *Actinobacillus actinomycetemcomitans, Prevotella intermedia,* and *P. nigrescens* to fibroblast monolayers and reconstituted basement membranes via its proteinaceous moiety (2, 3). In other studies, lactoferrin inhibited the binding of *E. coli* to HeLa cells and to red blood cells (34, 52). Secretory component, often found in human milk in adequate concentrations, also was found to inhibit CFA-I-carrying enterotoxigenic *E. coli* (34). Mucins in human milk were shown to

TABLE 11.3 Glycoconjugates and saccharides from milk capable of interacting with bacteria

Glycoconjugate	Bacterium	Reference(s)
Caseinoglycopeptides	*H. pylori*	107
Glycoprotein (40 kDa)	*S. mutans*	112
Glycoprotein (periodate sensitive)	*S. aureus*	56
Caseinoglycopeptides	*S. sobrinus* and *S. mutans*	95
Fucosylated saccharides	*E. coli*	24, 69
Gal(β1→4)GlcnNAc and Gal(β1→3)GlcNAc	*P. aeruginosa*	81
Mannosylated proteins	*E. coli* (type 1 fimbriae)	27
Sialylated glycoproteins	*Mycoplasma pneumoniae*	85
Sialylated poly(*N*-acetyllactosamine)	*M. pneumoniae*	53
Neutral oligosaccharides	*S. pneumoniae* and *H. influenzae*	4
Sialyl lactose and sialylated proteins	*E. coli* (S fimbriae)	45
Sialylgalactosides	*E. coli* (S fimbriae)	94

inhibit the adhesion of S-fimbriated *E. coli* bacteria to buccal epithelial cells (94).

Milk also contains fat globules which can carry glycoconjugates specific for bacterial adhesins. For example, the adhesion of S-fimbriated *E. coli* to buccal epithelial cells was inhibited by sialylated fat globules (93). Colostrum lipids could inhibit the adhesion of *H. pylori* and *H. mustelae* to immobilized glycolipids (14); β-lactoglobulin, which was shown to bind to *Listeria monocytogenes*, is another potential inhibitor present in milk (1). It must be acknowledged, however, that milk contains other antimicrobial agents such as lysozyme, lactoperoxidase, and immunoglobulins (70). A well-designed study is needed to define the individual contributions of these various antimicrobial agents, including the antiadhesive components, to the overall beneficial effect of milk in preventing infections.

While human milk has a restricted availability, plant materials are generally abundant or can be engineered to become available. For this reason, plant materials with antiadhesin activities are attractive candidates for antiadhesion therapeutic agents. Nevertheless, during the last decade there has been a relative paucity of information about the antiadhesive effects of plant materials (Table 11.4). Although plant lectins are well represented in the human diet and many of these are well characterized (22, 65), their application to antiadhesion therapy is very limited.

Theoretically, these lectins could interact with animal cell surface saccharides to block adhesion mediated by lectin-carrying bacteria (see chapter 1). They also could interact with glycoconjugates on bacterial surfaces to enhance the clearance of bacteria (99).

Because lectins are abundant in food (65), there is a strong possibility that dietary lectins may become associated with mucosal cells and thus function as receptors for bacterial glycans. In fact, this was examined with regard to the oral microbiota (33). Buccal epithelial cells scraped from individuals who recently had eaten raw wheat germ had high concentrations of the lectin bound to their surfaces. As a result, and presumably due the presence of this GlucNAc-specific lectin, the buccal cells bound an increased number of *Streptococcus sanguis* bacteria. Certain dietary lectins may also reach the alimentary tract in a functional form (17), and so it is possible that here, too, adhesion could be affected (66).

A practical advantage of searching dietary plant extracts for the ability to treat bacterial infections is that clinical trials are sometimes easier to perform because toxicity is usually not an issue. Among the various plant extracts listed in Table 11.4, the extract obtained from cranberry is probably the most thoroughly studied and, thus far, appears to be the most effective in clinical trials. In the following, we describe the studies performed to investigate the antiadhesion activities of cranberry mate-

TABLE 11.4 Plant extracts containing bacterial antiadhesin activities

Plant	Constituent	Bacterium affected	Clinical experience	Reference
Camellia sinensis (green tea)	(−)-Epicatechin gallate, (−)-gallocatechin gallate	*Porphyromonas gingivalis*	Reduction of caries risk	90
Camellia sinensis (oolong tea)	Polyphenol	*S. mutans* and *S. sobrinus*	NT[b]	76
Gloipeltis furcata and *Gigartina teldi* (seaweeds)	Sulfated polysaccharides	*S. sobrinus*	NT	88
Azadirachta indica (neem stick)	ND[a]	*S. sanguis*	NT	120
Melaphis chinensis	Gallotannin	*S. sanguis*	NT	120
Hop bract	Polyphenols (36–40 kDa)	*S. mutans*	NT	109
Persea americana (avocado)	Tannins	*S. mutans*	NT	104
Various legume storage proteins	Glycoprotein	*E. coli*	NT	67
Gilanthus nivalis (snowdrop)	Mannose-sensitive lectin	*E. coli*	NT	80

[a]ND, not determined.
[b]NT, not tried.

rials (Table 11.5). It should be noted that one important antiadhesive substance in cranberry juice cocktail, fructose, is added artificially and that its effect, therefore, is not unique to the overall effects of cranberry extract.

There are two lines of evidence implicating different components of cranberry extracts as the active antiadhesive agent. One study found that extracts containing proto-anthocyanin in its condensed form inhibited the adhesion of P-fimbriated *E. coli* (31). Another study found that the active material in cranberry juice cocktail was in the form of high-molecular-weight nondialyzable materials (i.e., material retained when using >15,000-Da-cutoff membranes). This nondialyzable material contained oxygen, hydrogen, and carbon in the ratio of 2:41.4:56.6 but lacked detectable nitrogen (74). Nuclear magnetic resonance spectroscopy analyses have, thus far, been unable to resolve the structural details of the high-molecular-mass material (73, 121).

A number of conclusions have emerged from the studies of cranberry extracts. First, the materials exhibit a broad spectrum of activity but not a universal one. For example the extract acts on adhesins of uropathogenic *E. coli*, including P fimbriae, S fimbriae, and

nonfimbrial adhesin I, but does not act on adhesins of diarrheal *E. coli* or on the type 1 fimbrial lectin (73). The cranberry extract inhibited the coaggregation between pairs of gram-negative oral bacteria more often than it did those between pairs of gram-positive bacteria. The target of the antiadhesive activity is the bacterial adhesin, not the animal cell receptors or human mucus.

While the initial in vitro experiments were guided by the long-known observations about the beneficial effects of cranberry juice in therapy of urinary tract infections (59, 78, 105), they have now been substantiated by careful clinical trials (8, 44). In one trial, elderly women were asked to drink 300 ml of cranberry cocktail or placebo per day for 6 months. In the other study, young women were asked to drink 50 ml of cranberry-lingonberry juice concentrate (or diluted to 200 ml with water) per day for 6 months. In both studies, regular drinking of cranberry or cranberry-lingonberry cocktail reduced the incidence of bacteriuria (e.g., the presence of $>10^5$ bacteria per ml of urine) considerably compared to placebo (Fig. 11.4B). Cranberry juice is usually consumed as a cocktail that contains a number of additives, including 20%

TABLE 11.5 Examples of antiadhesion effects of juice or extracts of *Vaccinium* spp. (cranberry)

Bacterium	Disease	Adhesion assay	Effect[a]	Clinical trial	Reference(s)
E. coli	Urinary tract infection and pyelonephritis	HA[f], uroepithelial adhesion	I	Prevents urinary tract infection in women	8, 44, 73, 91, 101, 121
E. coli	Diarrhea	HA	NI	ND[b]	73
E. coli	Meningitis	HA	I	ND	73
Oral bacteria	Dental decay and periodontitis	Coaggregation and buccal epithelial cell adhesion	I[c]	Decreases salivary S. mutans[d]	116, 117
H. pylori	Gastric ulcer	Human gastric mucus, TC cells[e]	I	ND	19

[a] I or NI, inhibition or no inhibition of adhesion by either cranberry juice or a high-molecular-weight nondialyzable material.
[b] ND, not done.
[c] Nondialyzable material caused inhibition most frequently when at least one of the interacting partners was a gram-negative bacteria.
[d] Results are from reference 116.
[e] Results for adhesion to gastric carcinoma tissue culture cells are from references 18 and 19. In all assays, nondialyzable material was assayed and found to inhibit the sialic acid-specific adhesion and the heparin-specific adhesion. TC, tissue culture.
[f] HA, hemagglutination.

fructose. High concentrations of fructose inhibit the *E. coli* type 1 fimbrial lectin, FimH, so that cranberry cocktail also inhibits type 1-fimbriated *E. coli* adhesion, while the cranberry extract does not (121).

In conclusion, there is little doubt that despite the clear hurdles remaining, the use of antiadhesion agents for the therapy of bacterial and perhaps other microbial infections has a bright future. As information continues to accumulate on the complexities of the adhesive interactions involved, this therapeutic approach may soon move from the realm of theory to one of reality.

HOST-DERIVED ANTIADHESINS IN INNATE IMMUNITY

In recent years, increased scientific attention has been given to immediate defense mechanisms based on nonclonal recognition of microbial components (i.e., innate immunity). Inhibition of adhesion of pathogens, and subsequent reduction of colonization, by components of various body fluids may be considered an important component of the innate immune system. In fact, there is an abundance of potential inhibitors of adhesion in various body fluids, but only for some of them is there evidence to suggest that they act to provide immunity to pathogens. For example, the hydrophobic molecule sphinganin, a component of sphingolipids, decreases the adhesion of *Streptococcus mitis* to buccal epithelial cells and the adhesion of *Staphylococcus aureus* to nasal mucosal cells (13). The most important host-derived components which potentially may provide innate immunity by inhibiting adhesion are those found associated with mucus on mucosal surfaces. It is assumed that these components act by undergoing specific binding to the pathogen, thus preventing its access to the underlying epithelial cells. In this case, reduction in colonization and infectivity would be dependent on the flow rate of the mucus. It follows that conditions which retard the mucus flow rate may lead to the promotion of colonization (97). Mucus may also prevent access to the underlying epithelial cells by forming a passive protective barrier. In either case, it is important to identify the adhesin-reactive components in mucus in order to resolve their role in innate immunity. Mucus is a complex mixture containing acidic polysaccharides, glycoproteins, glycolipids, and ions (see chapter 3). Glycoconjugates containing oligosaccharide residues specific for lectins expressed by the pathogen are most likely to modulate adhesion. Early studies have identified two major mucus-associated glycoproteins, Tamm Horsfall glycoprotein

TABLE 11.6 Example of studies describing the interaction with mucus and effect on adhesion

Source of mucus	Pathogen	Adhesion assay and effect[a]	Reference
Intestinal (rabbit)	EPEC[b]	Tissue culture, I	100
Meconium (human)	E. coli, S fimbriae	Buccal cells, I	92
Saliva (human) (lysozyme)	S. mutans	ND	96
Gastric (human)	H. pylori	Gastric cells, I	37
Bladder (rabbit)	Mycobacterium (BCG)	Uroepithelial, I[c]	9
Colon (rabbit)	E. coli RDEC-1	Tissue culture, I	55
Colon (rabbit)[d]	E. coli RDEC-1	Tissue culture, NI	55
Bladder (bovine)	E. coli	Bladder cells[e], I	87

[a]I, inhibition of adhesion; NI, no inhibition of adhesion: ND, not detected.
[b]EPEC, enteropathogenic E. coli.
[c]Adhesion to mucin-depleted bladder was low compared to adhesion to nondepleted bladder.
[d]Mucus from colons of rabbits treated with dinitrochlorobenzine.
[e]Mucin-depleted rabbit bladder.

and IgA, which interact with type 1 fimbrial lectin in a mannose-specific fashion (71). More recently, there have been a number of studies of the effect of mucus on the adhesion of pathogenic bacteria (Table 11.6). In only one case was lysozyme the mucus component identified. In most cases, soluble constituents of the mucus inhibited adhesion. In one case, mucus acted as a passive barrier, preventing access of the pathogen to the underlying cells (9). Interestingly, mucus constituents may change during inflammation, resulting in an inability to inhibit adhesion (55).

The secretor or nonsecretor status of individuals is of particular interest in relation to their susceptibility to infection and the possible role of soluble adhesin receptors. It has long been known that there are polymorphisms with respect to the presence or absence of blood group determinants in the fluids bathing mucosal surfaces (i.e., secretors or nonsecretors, respectively [15]). Because most blood group determinants are glycoconjugates, some of which act as receptors for bacterial adhesins, it has been postulated that the decreased susceptibility of secretors to infections may be due to the inhibition of adhesion of the pathogens by soluble blood group determinants in the fluid bathing mucosal surfaces. This is exemplified by showing that carriage of Neisseria meningitidis was significantly higher among nonsecretors during a school outbreak of meningitis (122). Saliva pooled from secretors inhibited the adhesion of meningococci to buccal epithelial cells more strongly than did saliva of nonsecretors.

The future of antiadhesion therapy depends on developing a combination of powerful, nontoxic inhibitory agents (e.g., carbohydrates), each targeted to a distinct bacterial surface adhesin (e.g., lectin), and/or on finding inhibitors present in dietary products that target a number of adhesins at the same time. Once such compounds become available, they could become the drugs of choice for the management of at least some infectious diseases. Finally, any host-derived protein of nonimmune origin that is capable of agglutinating bacteria can potentially act as an antiadhesion agent and provide innate immunity to infection. This notion was postulated for the lung collectin SP-D, which agglutinates a number of gram-negative bacteria and is capable of inhibiting their adhesion to nonphagocytic cells (75, 89). Further studies are needed, however, to determine precisely how these inhibitors, constitutively produced in body fluids, provide innate immunity to infections.

REFERENCES

1. **al-Makhlafi, H., J. McGuire, and M. Daeschel.** 1994. Influence of preadsorbed milk proteins on adhesion of Listeria monocytogenes to hydrophobic and hydrophilic silica surfaces. Appl. Environ. Microbiol. **60:**3560–3565.

2. **Alugupalli, K. R., and S. Kalfas.** 1995. Inhibitory effect of lactoferrin on the adhesion of *Actinobacillus actinomycetemcomitans* and *Prevotella intermedia* to fibroblasts and epithelial cells. *APMIS* **103:**154–160.
3. **Alugupalli, K. R., and S. Kalfas.** 1997. Characterization of the lactoferrin-dependent inhibition of the adhesion of *Actinobacillus actinomycetemcomitans, Prevotella intermedia* and *Prevotella nigrescens* to fibroblasts and to a reconstituted basement membrane. *APMIS* **105:** 680–688.
4. **Andersson, B., O. Porras, L. A. Hanson, T. Lagergard, and C. Svanborg-Edén.** 1986. Inhibition of attachment of *Streptococcus pneumoniae* and *Haemophilus influenzae* by human milk and receptor oligosaccharides. *J. Infect. Dis.* **153:**232–237.
5. **Andrade, J. R. C.** 1980. Role of fimbrial adhesiveness in guinea pig keratoconjunctivitis by *Shigella flexneri. Rev. Microbiol.* **11:**117–125.
6. **Aronson, M., O. Medalia, L. Schori, D. Mirelman, N. Sharon, and I. Ofek.** 1979. Prevention of colonization of the urinary tract of mice with *Escherichia coli* by blocking of bacterial adherence with methyl alpha-D-mannopyranoside. *J. Infect. Dis.* **139:**329–332.
7. **Ashkenazi, S.** 1994. A review of the effect of human milk fractions on the adherence of diarrheogenic *Escherichia coli* to the gut in an animal model. *Isr. J. Med. Sci.* **30:**335–338.
8. **Avorn, J., M. Monane, J. H. Gurwitz, R. J. Glynn, I. Choodnovskiy, and L. A. Lipsitz.** 1994. Reduction of bacteriuria and pyuria after ingestion of cranberry juice. *JAMA* **271:**751–754.
9. **Badalament, R. A., G. L. Franklin, C. M. Page, B. M. Dasani, M. G. Wientjes, and J. R. Drago.** 1992. Enhancement of bacillus Calmette-Guerin attachment to the urothelium by removal of the rabbit bladder mucin layer. *J. Urol.* **147:**482–485.
10. **Bengmark, S.** 1998. Ecological control of the gastrointestinal tract. The role of probiotic flora. *Gut* **42:**2–7.
11. **Ben-Redjeb, S., A. Slim, A. Horchani, S. Zmerilli, A. Boujnah, and V. Lorian.** 1982. Treatment of urinary tract infections with ten milligrams of ampicillin per day. *Antimicrob. Agents Chemother.* **22:**1084–1086.
12. **Beuth, J., B. Stoffel, and G. Pulverer.** 1996. Inhibition of bacterial adhesion and infections by lectin blocking. *Adv. Exp. Med. Biol.* **408:**51–56.
13. **Bibel, D. J., R. Aly, and H. R. Shinefield.** 1992. Inhibition of microbial adherence by sphinganine. *Can. J. Microbiol.* **38:**983–985.
14. **Bitzan, M. M., B. D. Gold, D. J. Philpott, M. Huesca, P. M. Sherman, H. Karch, R. Lissner, C. A. Lingwood, and M. A. Karmali.** 1998. Inhibition of *Helicobacter pylori* and *Helicobacter mustelae* binding to lipid receptors by bovine colostrum. *J. Infect. Dis.* **177:**955–961.
15. **Blackwell, C. C.** 1989. Genetic susceptibility to infectious agents. *Proc. R. Coll. Physicians Edinburgh* **19:**129–138.
16. **Blomberg, L., A. Henriksson, and P. L. Conway.** 1993. Inhibition of adhesion of *Escherichia coli* K88 to piglet ileal mucus by *Lactobacillus* spp. *Appl. Environ. Microbiol.* **59:**34–39.
17. **Brady, P. G., A. M. Vannier, and J. G. Banwell.** 1978. Identification of the dietary lectin, wheat germ agglutin, in inhuman intestinal contents. *Gastroenterology* **75:**236–239.
18. **Burger O., E. Weiss, N. Sharon, M. Tabak, I. Neeman, and I. Ofek.** 2002 Inhibition of *Helicobacter pylori* adhesion to human gastric mucus by a high-molecular-weight constituent of cranberry juice. *Crit. Rev. Food Sci. Nutr.* **42:**279–284.
19. **Burger, O., I. Ofek, M. Tabak, E. I. Weiss, N. Sharon, and I. Neeman.** 2000. A high molecular mass constituent of cranberry juice inhibits *Helicobacter pylori* adhesion to human gastric mucus. *FEMS Immunol. Med. Microbiol.* **29:**295–301.
20. **Coconnier, M. H., T. R. Klaenhammer, S. Kerneis, M. F. Bernet, and A. L. Servin.** 1992. Protein-mediated adhesion of *Lactobacillus acidophilus* BG2FO4 on human enterocyte and mucus-secreting cell lines in culture. *Appl. Environ. Microbiol.* **58:**2034–2039.
21. **Courtney, H. S., I. Ofek, and D. L. Hasty.** 1997. M protein mediated adhesion of M type 24 *Streptococcus pyogenes* stimulates release of interleukin-6 by HEp-2 tissue culture cells. *FEMS Microbiol. Lett.* **151:**65–70.
22. **Cowan, M. M.** 1999. Plant products as antimicrobial agents. *Clin. Microbiol. Rev.* **2:**564–582.
23. **Cox, F.** 1982. Prevention of group B streptococcal colonization with topically applied lipoteichoic acid in a maternal-newborn mouse model. *Pediatr. Res.* **16:**816–819.
24. **Cravioto, A., A. Tello, H. Villafan, J. Ruiz, S. del Vedovo, and J.-R. Neeser.** 1991. Inhibition of localized adhesion of enteropathogenic *Escherichia coli* to HEp-2 cells by immunoglobulin and oligosaccharide fractions of human colostrum and breast milk. *J. Infect. Dis.* **163:**1247–1255.
25. **Cywes, C., I. Stamenkovic, and M. R.**

Wessels. 2000. CD44 as a receptor for colonization of the pharynx by group A streptococcus. *J. Clin. Investig.* **106:**995–1002.

26. **Dale, J. B., R. W. Baird, H. S. Courtney, D. L. Hasty, and M. S. Bronze.** 1994. Passive protection of mice against group A streptococcal pharyngeal infection by lipoteichoic acid. *J. Infect. Dis.* **169:**319–323.

27. **Duguid, J. P., and R. R. Gillies.** 1957. Fimbriae and adhesive properties of dysentery bacilli. *J. Pathol. Bacteriol.* **74:**397–411.

28. **Fader, R. C., and C. P. Davis.** 1980. Effect of piliation on *Klebsiella pneumoniae* infection in rat bladders. *Infect. Immun.* **30:**554–561.

29. **Faubion, W. A., and W. J. Sandborn.** 2000. Probiotic therapy with *E. coli* for ulcerative colitis: take the good with the bad. *Gastroenterology* **118:**630–631.

30. **Firon, N., S. Ashkenazi, D. Mirelman, I. Ofek, and N. Sharon.** 1987. Aromatic alphaglycosides of mannose are powerful inhibitors of the adherence of type 1 fimbriated *Escherichia coli* to yeast and intestinal epithelial cells. *Infect. Immun.* **55:**472–476.

31. **Foo, L. Y., Y. Lu, A. B. Howell, and N. Vorsa.** 2000. A-type proanthocyanidin trimers from cranberry that inhibit adherence of uropathogenic P-fimbriated *Escherichia coli*. *J. Nat. Prod.* **63:**1225–1228.

32. **Fujiwara, S., H. Hashiba, T. Hirota, and J. F. Forstner.** 1997. Proteinaceous factor(s) in culture supernatant fluids of bifidobacteria which prevents the binding of enterotoxigenic *Escherichia coli* to gangliotetraosylceramide. *Appl. Environ. Microbiol.* **63:**506–512.

33. **Gibbons, R. J., and I. Denkers.** 1983. Association of food lectins with humanoral epithelial cells *in vivo*. *Arch. Oral Biol.* **28:**561–566.

34. **Giugliano, L. G., S. T. Ribeiro, M. H. Vainstein, and C. J. Ulhoa.** 1995. Free secretory component and lactoferrin of human milk inhibit the adhesion of enterotoxigenic *Escherichia coli*. *J. Med. Microbiol.* **42:**3–9.

35. **Goh, C. T., S. Taweechaisupapong, K. G. Taylor, and R. J. Doyle.** 2000. Polycarboxylates inhibit the glucan-binding lectin of *Streptococcus sobrinus*. *Biochim. Biophys. Acta* **1523:**111–116.

36. **Goldhar, J., A. Zilberberg, and I. Ofek.** 1986. Infant mouse model of adherence and colonization of intestinal tissues by enterotoxigenic strains of *Escherichia coli* isolated from humans. *Infect. Immun.* **52:**205–208.

37. **Hayashi, S., T. Sugiyama, M. Asaka, K. Yokota, K. Oguma, and Y. Hirai.** 1998. Modification of *Helicobacter pylori* adhesion to human gastric epithelial cells by antiadhesion agents. *Dig. Dis. Sci.* **43:**56S–60S.

38. **Idänpään-Heikkilä, I., P. M. Simon, D. Zopf, T. Vullo, P. Cahill, K. Sokol, and E. Tuomanen.** 1997. Oligosaccharides interfere with the establishment and progression of experimental pneumococcal pneumonia. *J. Infect. Dis.* **176:**704–712.

39. **Isolauri, E.** 2001. Probiotics in human disease. *Am. J. Clin. Nutr.* **73:**1142S–1146S.

40. **Johnson, J. R., and T. Berggren.** 1994. Pigeon and dove eggwhite protect mice against renal infection due to P fimbriated *Escherichia coli*. *Am. J. Med. Sci.* **307:**335–339.

41. **Kahane, I., and I. Ofek.** 1996. *Toward Anti-Adhesion Therapy of Microbial Infectious Diseases*. Plenum Publishers, New York, N.Y.

42. **Kelly, C. G., and J. S. Younson.** 2000. Anti-adhesive strategies in the prevention of infectious disease at mucosal surfaces. *Expert Opin. Investig. Drugs* **9:**1711–1712.

43. **Kelly, C. G., J. S. Younson, B. Y. Hikmat, S. M. Todryk, M. Czisch, P. I. Haris, I. R. Flindall, C. Newby, A. I. Mallet, J. K. Ma, and T. Lehner.** 1999. A synthetic peptide adhesion epitope as a novel antimicrobial agent. *Nat. Biotechnol.* **17:**42–47.

44. **Kontiokari, T., K. Sundqvist, M. Nuutinen, T. Pokka, M. Koskela, and M. Uhari.** 2001. Randomized trial of cranberry-lingonberry juice and Lactobacillus GG drink for the prevention of urinary tract infections in women. *Br. Med. J.* **322:**1571.

45. **Korhonen, T. K., M. V. Valtonen, J. Parkkinen, V. Vaisanen-Rhen, J. Finne, F. Orskov, I. Orskov, S. B. Svenson, and P. H. Makela.** 1985. Serotypes, hemolysin production, and receptor recognition of *Escherichia coli* strains associated with neonatal sepsis and meningitis. *Infect. Immun.* **48:**486–491.

46. **Kunz, C., and S. Rudloff.** 1993. Biological functions of oligosaccharides in human milk. *Acta Paediatr.* **82:**903–912.

47. **Langermann, S., R. Mollby, J. E. Burlein, S. R. Palaszynski, C. G. Auguste, A. DeFusco, R. Strouse, M. A. Schenerman, S. J. Hultgren, J. S. Pinkner, J. Winberg, L. Guldevall, M. Soderhall, K. Ishikawa, S. Normark, and S. Koenig.** 2000. Vaccination with FimH adhesin protects cynomolgus monkeys from colonization and infection by uropathogenic *Escherichia coli*. *J. Infect. Dis.* **181:**774–778.

48. **Langermann, S., S. Palaszynski, M. Barnhart, G. Auguste, J. S. Pinkner, J. Burlein, P. Barren, S. Koenig, S. Leath, C. H. Jones, and S. J. Hultgren.** 1997.

Prevention of mucosal *Escherichia coli* infection by FimH-adhesin-based systemic vaccination. *Science* **276:**607–611.

49. **Lee, J. Y., H. T. Sojar, G. S. Bedi, and R. J. Genco.** 1992. Synthetic peptides analogous to the fimbrillin sequence inhibit adherence of *Porphyromonas gingivalis. Infect. Immun.* **60:** 1662–1670.

50. **Lee, K. K., W. Y. Wong, H. B. Sheth, R. S. Hodges, W. Paranchych, and R. T. Irvin.** 1995. Use of synthetic peptides in characterization of microbial adhesins. *Methods Enzymol.* **253:**115–131.

51. **Lindhorst, T. K., C. Kieburg, and U. Krallmann-Wenzel.** 1998. Inhibition of the type 1 fimbriae-mediated adhesion of *Escherichia coli* to erythrocytes by multiantennary alpha-mannosyl clusters: the effect of multivalency. *Glycoconj. J.* **15:**605–613.

52. **Longhi, C., M. P. Conte, L. Seganti, M. Polidoro, A. Alfsen, and P. Valenti.** 1993. Influence of lactoferrin on the entry process of *Escherichia coli* HB101 (pRI203) in HeLa cells. *Med. Microbiol. Immunol.* **182:**25–35.

53. **Loveless, R. W., and T. Feizi.** 1989. Sialo-oligosaccharide receptors for *Mycoplasma pneumoniae* and related oligosaccharides of poly-N-acetyllactosamine series are polarized at the cilia and apical-microvillar domains of the ciliated cells in human bronchial epithelium. *Infect. Immun.* **57:**1285–1289.

54. **Ma, J. K., B. Y. Hikmat, K. Wycoff, N. D. Vine, D. Chargelegue, L. Yu, M. B. Hein, and T. Lehner.** 1998. Characterization of a recombinant plant monoclonal secretory antibody and preventive immunotherapy in humans. *Nat. Med.* **4:**601–606.

55. **Mack, D. R., T. S. Gaginella, and P. M. Sherman.** 1992. Effect of colonic inflammation on mucin inhibition of *Escherichia coli* RDEC-1 binding *in vitro. Gastroenterology* **102:**1199–1211.

56. **Mamo, W., and G. Froman.** 1994. Adhesion of *Staphylococcus aureus* to bovine mammary epithelial cells induced by growth in milk whey. *Microbiol. Immunol.* **38:**305–308.

57. **McGroarty, J. A.** 1994. Cell surface appendages of lactobacilli. *FEMS Microbiol. Lett.* **124:**405–410.

58. **Mestecky, J., S. M. Michalek, Z. Moldoveanu, and M. W. Russell.** 1997. Routes of immunization and antigen delivery systems for optimal mucosal immune responses in humans. *Behring Inst. Mitt.* **98:**33–43.

59. **Moen, D. V.** 1962. Observations on the effectiveness of cranberry juice in urinary infections. *Wis. Med. J.* **61:**282–283.

60. **Moon, H. W., and T. O. Bunn.** 1993. Vaccines for preventing enterotoxigenic *Escherichia coli* infections in farm animals. *Vaccine* **11:**200–213.

61. **Mouricout, M.** 1991. Swine and cattle enterotoxigenic *Escherichia coli*-mediated diarrhea. Development of therapies based on inhibition of bacteria-host interactions. *Eur. J. Epidemiol.* **7:**588–604.

62. **Mouricout, M., J. M. Petit, J. R. Carias, and R. Julien.** 1990. Glycoprotein glycans that inhibit adhesion of *Escherichia coli* mediated by K99 fimbriae: treatment of experimental colibacillosis. *Infect. Immun.* **58:**98–106.

63. **Mulvey, G., P. I. Kitov, P. Marcato, D. R. Bundle, and G. D. Armstrong.** 2001. Glycan mimicry as a basis for novel anti-infective drugs. *Biochimie* **83:**841–847.

64. **Mysore, J. V., T. Wigginton, P. M. Simon, D. Zopf, L. M. Heman-Ackah, and A. Dubois.** 1999. Treatment of *Helicobacter pylori* infection in rhesus monkeys using a novel antiadhesion compound. *Gastroenterology* **117:**1316–1325.

65. **Nachbar, M. S., and J. D. Oppenheim.** 1980. Lectins in the United States diet: a survey of lectins in commonly consumed foods and review of the literature. *Am. J. Clin. Nutr.* **33:**2338–2345.

66. **Naughton, P. J., G. Grant, S. Bardocz, and A. Pusztai.** 2000. Modulation of *Salmonella* infection by the lectins of *Canavalia ensiformis* (Con A) and *Galanthus nivalis* (GNA) in a rat model *in vivo. J. Appl. Microbiol.* **88:**720–727.

67. **Neeser, J. R., B. Koellreutter, and P. Wuersch.** 1986. Oligomannoside-type glycopeptides inhibiting adhesion of *Escherichia coli* strains mediated by type 1 pili: preparation of potent inhibitors from plant glycoproteins. *Infect. Immun.* **52:**428–436.

68. **Newburg, D. S.** 2000. Oligosaccharides in human milk and bacterial colonization. *J. Pediatr. Gastroenterol. Nutr.* **30:**S8–S17.

69. **Newburg, D. S., L. K. Pickering, R. H. McCluer, and T. G. Cleary.** 1990. Fucosylated oligosaccharides of human milk protect suckling mice from heat-stabile enterotoxin of *Escherichia coli. J. Infect. Dis.* **162:** 1075–1080.

70. **Newman, J.** 1995. How breast milk protects newborns. *Sci. Am.* **273:**76–79.

71. **Ofek, I., and R. J. Doyle.** 1994. *Bacterial Adhesion to Cells and Tissues,* p. 513–562. Chapman & Hall, New York, N.Y.

72. **Ofek, I., D. Mirelman, and N. Sharon.** 1977. Adherence of *Escherichia coli* to human

mucosal cells mediated by mannose receptors. *Nature* **265**:623–625.

73. **Ofek, I., J. Goldhar, D. Zafriri, H. Lis, R. Adar, and N. Sharon.** 1991. Anti-*Escherichia coli* adhesin activity of cranberry and blueberry juices. *N. Engl. J. Med.* **324**:1599.

74. **Ofek, I., J. Goldhar, and N. Sharon.** 1996. Anti-*Escherichia coli* adhesin activity of cranberry and blueberry juices. *Adv. Exp. Med. Biol.* **408**:179–184.

75. **Ofek, I., and E. Crouch.** 2000. Interactions of microbial glycoconjugates with collectins, p. 517–537. *In* R. J. Doyle (ed.), *Glycobiology.* Kluwer Academic/Plenum Publishers, New York, N.Y.

76. **Ooshima, T., T. Minami, W. Aono, A. Izumitani, S. Sobue, T. Fujiwara, S. Kawabata, and S. Hamada.** 1993. Oolong tea polyphenols inhibit experimental dental caries in SPF rats infected with mutans streptococci. *Caries Res.* **27**:124–129.

77. **Ouwehand, A. C., and P. L. Conway.** 1996. Purification and characterization of a component produced by *Lactobacillus fermentum* that inhibits the adhesion of K88 expressing *Escherichia coli* to porcine ileal mucus. *J. Appl. Bacteriol.* **80**:311–318.

78. **Papas, P. N., C. A. Brusch, and G. C. Cresia.** 1968. Cranberry juice in the treatment of urinary tract infections. *Southwest. Med.* **47**:17–20.

79. **Patel, S. S., J. A. Balfour, and H. M. Bryson.** 1997. Fosfomycin tromethamine. A review of its antibacterial activity, pharmacokinetic properties and therapeutic efficacy as a single-dose oral treatment for acute uncomplicated lower urinary tract infections. *Drugs* **53**:637–656.

80. **Pusztai, A., G. Grant, R. J. Spencer, T. J. Duguid, D. S. Brown, S. W. B. Ewen, W. J. Peumans, E. J. M. van Damme, and S. Bardocz.** 1993. Kidney bean lectin-induced *Escherichia coli* overgrowth in the small intestine is blocked by GNA, a mannose specific lectin. *J. Appl. Bacteriol.* **75**:360–368.

81. **Ramphal, R., C. Carnoy, S. Fievre, J. C. Michalski, N. Houdret, G. Lamblin, G. Strecker, and P. Roussel.** 1991. *Pseudomonas aeruginosa* recognizes carbohydrate chains containing type 1 (Gal beta 1–3GlcNAc) or type 2 (Gal beta 1–4GlcNAc) disaccharide units. *Infect. Immun.* **59**:700–704.

82. **Reid, G., J. Howard, and B. S. Gan.** 2001. Can bacterial interference prevent infection? *Trends Microbiol.* **9**:424–428.

83. **Relman, D. A., M. Domenighini, E. Tuomanen, R. Rappuoli, and S. Falkow.** 1989. Protein filamentous hemagglutinin of *Bordetella pertussis:* nucleotide sequence and crucial role in adherence. *Proc. Natl. Acad. Sci. USA* **86**:2637–2641.

84. **Robbins, J. B., R. Schneerson, and S. C. Szu.** 1995. Perspective: hypothesis: serum IgG antibody is sufficient to confer protection against infectious diseases by inactivating the inoculum. *J. Infect. Dis.* **171**:1387–1398.

85. **Roberts, D. D., L. D. Olson, M. F. Barile, V. Ginsburg, and H. C. Krivan.** 1989. Sialic acid-dependent adhesion of *Mycoplasma pneumoniae* to purified glycoproteins. *J. Biol. Chem.* **264**:9289–9293.

86. **Roberts, J. A., B. Kaack, G. Kallenius, R. Møllby, J. Winberg, and S. B. Svenson.** 1984. Receptors for pyelonephritogenic *Escherichia coli* in primates. *J. Urol.* **131**:163–168.

87. **Ruggieri, M. R., R. K. Balagani, J. J. Rajter, and P. M. Hanno.** 1992. Characterization of bovine bladder mucin fractions that inhibit *Escherichia coli* adherence to the mucin deficient rabbit bladder. *J. Urol.* **148**:173–178.

88. **Saeki, Y.** Effect of seaweed extracts on *Streptococcus sobrinus* adsorption to saliva-coated hydroxyapatite. *Bull. Tokyo Dent. Coll.* **35**:9–15.

89. **Sahly, H., I. Ofek, R. Podschun, H. Brade, Y. He, U. Ullmann, and E. Crouch.** 2002. Surfactant protein D binds selectively to *Klebsiella pneumoniae* lipopolysaccharides containing mannose-rich O-antigens. *J. Immunol.* **169**:3267–3274.

90. **Sakanaka, S., M. Aizawa, M. Kim, and T. Yamamoto.** 1996. Inhibitory effects of green tea polyphenols on growth and cellular adherence of an oral bacterium, *Porphyromonas gingivalis. Biosci. Biotechnol. Biochem.* **60**:745–749.

91. **Schmidt, D. R., and A. E. Sobota.** 1988. An examination of the anti-adherence activity of cranberry juice on urinary and nonurinary bacterial isolates. *Microbios* **55**:173–181.

92. **Schröten, H., A. Lethen, F. G. Hanisch, R. Plogmann, J. Hacker, R. Nobis-Bosch, and V. Wahn.** 1992. Inhibition of adhesion of S-fimbriated *Escherichia coli* to epithelial cells by meconium and feces of breast-fed and formula-fed newborns: mucins are the major inhibitory component. *J. Pediatr. Gastroenterol. Nutr.* **15**:150–158.

93. **Schröten, H., F. G. Hanisch, R. Plogmann, J. Hacker, G. Uhlenbruck, R. Nobis-Bosch, and V. Wahn.** 1992. Inhibition of adhesion of S-fimbriated *Escherichia coli* to buccal epithelial cells by human milk fat globule membrane components: a novel aspect of the protective function of mucins in the nonimmunoglobulin fraction. *Infect. Immun.* **60**:2893–2899.

94. Schröten, H., R. Plogmann, F. G. Hanisch, J. Hacker, R. Nobis-Bosch, and V. Wahn. 1993. Inhibition of adhesion of S-fimbriated *E. coli* to buccal epithelial cells by human skim milk is predominantly mediated by mucins and depends on the period of lactation. *Acta Paediatr.* **82:**6–11.

95. Schupbach, P., J. R. Neeser, M. Golliard, M. Rouvet, and B. Guggenheim. 1996. Incorporation of caseinoglycomacropeptide and caseinophosphopeptide into the salivary pellicle inhibits adherence of mutans streptococci. *J. Dent. Res.* **75:**1779–1788.

96. Senpuku, H., H. Kato, M. Todoroki, N. Hanada, and T. Nisizawa. 1996. Interaction of lysozyme with a surface protein antigen of *Streptococcus mutans. FEMS Microbiol. Lett.* **139:**195–201.

97. Sherman, P. M., and E. C. Boedeker. 1987. Pilus-mediated interactions of the *Escherichia coli* strain RDEC-1 with mucosal glycoproteins in the small intestine of rabbits. *Gastroenterology* **93:**734–743.

98. Simon, P. M., P. L. Goode, A. Mobasseri, and D. Zopf. 1997. Inhibition of *Helicobacter pylori* binding to gastrointestinal epithelial cells by sialic acid-containing oligosaccharides. *Infect. Immun.* **65:**750–757.

99. Slifkin, M., and R. J. Doyle. 1990. Lectins and their application to clinical microbiology. *Clin. Microbiol. Rev.* **3:**197–218.

100. Smith, C. J., J. B. Kaper, and D. R. Mack. 1995. Intestinal mucin inhibits adhesion of human enteropathogenic *Escherichia coli* to HEp-2 cells. *J. Pediatr. Gastroenterol. Nutr.* **21:**269–276.

101. Sobota, A. E. 1984. Inhibition of bacterial adherence by cranberry juice: Potential use for treatment of urinary tract infection. *J. Urol.* **131:**1031–1016.

102. Sokurenko, E. V., V. Chesnokova, R. J. Doyle, and D. L. Hasty. 1997. Diversity of the *Escherichia coli* type 1 fimbrial lectin. Differential binding to mannosides and uroepithelial cells. *J. Biol. Chem.* **272:**17880–17886.

103. Spencer, R. J., and A. Chesson. 1994. The effect of *Lactobacillus* spp. on the attachment of enterotoxigenic *Escherichia coli* to isolated porcine enterocytes. *J. Appl. Bacteriol.* **77:**215–220.

104. Staat, R. H., R. J. Doyle, S. D. Langley, and R. P. Suddick. 1978. Modification of *in vitro* adherence of *Streptococcus mutans* by plant lectins. *Adv. Exp. Med. Biol.* **107:**639–647.

105. Sternlieb, P. 1963. Cranberry juice in renal disease. *N. Engl. J. Med.* **268:**57.

106. Steuer, M. K., H. Herbst, J. Beuth, M. Steuer, G. Pulverer, and R. Matthias. 1993. Inhibition of lectin mediated bacterial adhesion by receptor blocking carbohydrates in patients with *Pseudomonas aeruginosa* induced otitis externa: a prospective phase II study, *Otorhinolaryngol. Nova* **3:**19.

107. Stromqvist, M., P. Falk, S. Bergstrom, L. Hansson, B. Lonnerdal, S. Normark, and O. Hernell. 1995. Human milk kappa-casein and inhibition of *Helicobacter pylori* adhesion to human gastric mucosa. *J. Pediatr. Gastroenterol. Nutr.* **21:**288–296.

108. Svanborg-Edén, C., R. Freter, L. Hagberg, R. Hull, S. Hull, H. Leffler, and G. Schoolnik. 1982. Inhibition of experimental ascending urinary tract infection by an epithelial cell-surface receptor analogue. *Nature* **298:**560–562.

109. Tagashira, M., K. Uchiyama, T. Yoshimura, M. Shirota, and N. Uemitsu. 1997. Inhibition by hop bract polyphenols of cellular adherence and water-insoluble glucan synthesis of mutans streptococci. *Biosci. Biotechnol. Biochem.* **61:**332–335.

110. Ukkonen, P., K. Varis, M. Jernfors, E. Herva, J. Jokinen, E. Ruokokoski, D. Zopf, and T. Kilpi. 2000. Treatment of acute otitis media with an antiadhesive oligosaccharide: a randomised, double-blind, placebo-controlled trial. *Lancet* **356:**1398–1402.

111. Underhill C. 1992. CD44: the hyaluronan receptor. *J. Cell Sci.* **103:**293–298.

112. Vacca-Smith, A. M., B. C. Van Wuyckhuyse, L. A. Tabak, and W. H. Bowen. 1994. The effect of milk and casein proteins on the adherence of *Streptococcus mutans* to saliva-coated hydroxyapatite. *Arch. Oral Biol.* **39:**1063–1069.

113. Velraeds, M. M. B. van de Belt-Gritter, H. C. van der Mei, G. Reid, and H. J. Busscher. 1998. Interference in initial adhesion of uropathogenic bacteria and yeasts to silicone rubber by a *Lactobacillus acidophilus* biosurfactant. *J. Med. Microbiol.* **47:**1081–1085.

114. Velraeds, M. M., H. C. van der Mei, G. Reid, and H. J. Busscher. 1996. Inhibition of initial adhesion of uropathogenic *Enterococcus faecalis* by biosurfactants from *Lactobacillus* isolates. *Appl. Environ. Microbiol.* **62:**1958–1963.

115. Wang, Q., S. Singh, K. G. Taylor, and R. J. Doyle. 1996. Anti-adhesins of *Streptococcus sobrinus. Adv. Exp. Med. Biol.* **408:**249–262.

116. Weiss, E. I., R. Lev-Dor, N. Sharon, and I. Ofek. 2002. Inhibitory effect of a high-molecular-weight constituent of cranberry on adhesion of oral bacteria. *Crit. Rev. Food Sci. Nutr.* **42:**285–293.

117. **Weiss, E. I., R. Lev-Dor, Y. Kashman, J. Goldhar, N. Sharon, and I. Ofek.** 1998. Inhibiting interspecies coaggregation of plaque bacteria with cranberry juice constituent. *J. Am. Dent. Assoc.* **129:**1719–1723.

118. **Wentworth, J. S., F. E. Austin, N. Garber, N. Gilboa-Garber, C. A. Paterson, and R. J. Doyle.** 1991. Cytoplasmic lectins contribute to the adhesion of *Pseudomonas aeruginosa*. *Biofouling* **4:**99–104.

119. **Wizemann, T. M., J. E. Adamou, and S. Langermann.** 1999. Adhesins as targets for vaccine development. *Emerg. Infect. Dis.* **5:**395–403.

120. **Wolinsky, L. E., S. Mania, S. Nachnani, and S. Ling.** 1996. The inhibiting effect of aqueous *Azadirachta indica* (Neem) extract upon bacterial properties influencing *in vitro* plaque formation. *J. Dent. Res.* **75:**816–822.

121. **Zafriri, D., I. Ofek, R. Adar, M. Pocino, and N. Sharon.** 1989. Inhibitory activity of cranberry juice on adherence of type 1 and type P fimbriated *Escherichia coli* to eucaryotic cells. *Antimicrob. Agents Chemother.* **33:**92–98.

122. **Zorgani, A. A., J. Stewart, C. C. Blackwell, R. A. Elton, and D. M. Weir.** 1994. Inhibitory effect of saliva from secretors and non-secretors on binding of meningococci to epithelial cells. *FEMS Immunol. Med. Microbiol.* **9:**135–142.

ADHESINS, RECEPTORS, AND TARGET SUBSTRATA INVOLVED IN THE ADHESION OF PATHOGENIC BACTERIA TO HOST CELLS AND TISSUES

12

The study of adhesion during the 1960s, 1970s, and 1980s focused in large part on establishing the importance of this field of biological research. During the 1990s, studies focused more and more on determining the detailed molecular characteristics of the critical molecules involved. However, there remain many organisms for which the adhesins and receptors are still not known and many others for which critical adhesin-receptor information is still too limited to be useful in the design of clinically effective treatments. The molecules involved in adhesion have become clearly linked to infections, and information regarding their structure and molecular function is constantly growing. Thus, it is increasingly possible to design more knowledgeable approaches to inhibit pathogen adhesion in order to prevent infection (see chapter 11). For some of the more thoroughly studied adhesin-receptor systems, antiadhesive therapies (e.g., antiadhesin vaccines or receptor analog therapy) are under development. There are, of course, many other pathogenic organisms for which much less is known.

This section contains a tabulation of all of the studies between January 1992 and December 2002 in which an adhesion experiment involving one or more of the various pathogenic species was performed. The species studied number almost 300, but it remains true that studies of the adhesion of *Escherichia coli* outnumber those of many of the other species combined. For this reason, adhesion studies of *E. coli* have been placed in a separate table. Although some studies undoubtedly escaped our attention, the other table will quickly reveal which pathogenic organisms are receiving the most attention and which have received the least attention. It is interesting that the number of adhesion studies performed for some of the pathogens does not reflect the clinical significance of the infections in each case. For many of those organisms which are being studied extensively, the numbers of putative adhesins and receptors identified have increased dramatically over the last 10 years compared to the previous decade. As the number of substratum types that have been tested for any particular species increases, it can be reasonably assumed that more types of adhesins with distinct receptor specificities will be identified. For instance, for *Staphylococcus aureus*, the numbers of putative adhesins, receptors, and substrata tested over the last 10 years has increased four- to eightfold over the previous decade.

The studies cited here were essentially limited to those which involved adhesion tests of host cells or tissue components, materials of prosthetic devices, or other indwelling medical equipment. They do not include those which focused only on biochemical or genetic characterization of adhesion apparatus components without performing adhesion assays. Many of those publications are included in earlier sections. The table also includes only a very few studies of adhesion to foods (e.g., lettuce leaves) and essentially no studies of adhesion to food-processing components (e.g., stainless steel tables) or to environmental materials (e.g., pipes in water-processing plants).

ABBREVIATIONS USED IN THIS CHAPTER

APEC, avian pathogenic *E. coli*; BC, buccal cells; Bfp, bundle-forming pili; Biomat., biomaterial; BLD, blood source; BSA, bovine serum albumin; Bvg, *Bordetella* virulence gene; CEACAM, carcinoembryonic antigen cell adhesion molelcule; CF, cystic fibrosis; CF mucin, cystic fibrosis mucin; CFA, colonization factor antigen; CNF I, cytotoxic necrotizing factor I; CoA, coenzyme A; Coagg, coaggregation (partner provided within parentheses); Coll, collagen; CR3, complement receptor 3; DAEC, diffuse-adhering *E. coli*; EAEC, enteroadherent *E. coli*; EAggEC, enteroaggregative *E. coli*; ECM, extracellular matrix; EHEC, enterohemorrhagic *E. coli*; EIEC, enteroinvasive *E. coli*; ENV, environmental source; EPEC, enteropathogenic *E. coli*; ERT, erythrocytes; ETEC, enterotoxigenic *E. coli*; EXC, excised; EXT, excised tissue; Fbgn, fibrinogen; FHA, filamentous hemagglutinin; fim, fimbriae; Fimb, fimbrillae; Fn, fibronectin; Fr.Tis.Sect., frozen tissue sections; FSH, follicle-stimulating hormone; Gal, galactose; GD1a, ganglioside GD1a; GD1b, ganglioside GD1b; GI, gastrointestinal; GL, glycolipid; Glc, glucose; Glp, glycoprotein; GM_1, ganglioside GM_1; GM_2, ganglioside GM_2; GT1b, ganglioside GT1b; GTF, glucosyltransferase; GU, genitourinary tract source; HA, hydroxylapatite; HC, undefined hydrocarbon; Hxd, hexadecane; IC, immobilized carbohydrate; Ig, immunoglobulin; IGL, immobilized glycolipid; IGlp, immobilized glycoprotein; IL, immobilized lipid; IM, immobilized mucus; immob, immobilized; InlA, internalin A; InlB, internalin B; Lac, lactose; LAEC, localized adhesion *E. coli*; LF, lactoferrin; LH, luteinizing hormone; Lm, laminin; LOS, lipooligosaccharide; LPS, lipopolysaccharide; LTA, lipoteichoic acid; Man, mannose, mannoside; MC, mast cells; MHC, major histocompatibility complex; Mø, macrophages; MR, mannose receptor of macrophages; MR fim, mannose-resistant fimbriae; MSCRAMM, microbial surface component reactive against matrix molecules; MUC5AC, type of mucin; MUC5B, type of mucin; NANA, *N*-acetylmuraminic acid; NTEC, necrotoxigenic *E. coli*; OC, organ culture; OMP, outer membrane proteins; ORL, oral source; PAF, platelet-activating factor; PC, primary culture; PEG, polyethylene glycol partitioning; PMN, polymorphonuclear leukocyte; prot, protein; Rabbit intest., rabbit intestine; RDEC, rabbit diarrheagenic *E. coli*; Resp. tract cells, respiratory tract cells; sC, soluble carbohydrate; SGL, soluble glycolipid; sGlp, soluble glycoprotein; SHA, saliva-coated hydroxylapatite; sIgA, secretory immunoglobulin A; SKN, skin source; Spa, protein A; SR-A, class A scavenger receptor; STEC, shigatoxigenic *E. coli*; TC, tissue culture cells; Tir, transported intimin receptor; Tis. Sect., tissue section; UC, urinary cells; UPEC, uropathogenic *E. coli*; URG, urogenital; URT, upper respiratory tract; UT, urinary tract; Vn, vitronectin; VTEC, verotoxigenic *E. coli*.

TABLE 12.1 Studies of adhesion from 1992 to 2002: *E. coli* pathotypes, adhesins, receptors, and target substrata used

Bacterium	Clinical source	Adhesins	Receptor specificity	Substrata (target cells)	References
Escherichia coli (VTEC)	GI	Fimbriae		Intestinal cells, TC	151, 422, 560, 905, 907, 1654, 1690, 2575, 2654, 3107, 4035
Escherichia coli (ETEC)	GI	CFA I, CFA II, CFA III 25 kDa, prot CS19, Tia, CS15	Glycosphingolipid GM1, mucus	SKN tissue, ERT, TC brush border, IGL, EXT, IGlp	276, 319, 537, 782, 945, 1134, 1300, 1613, 1688, 1864, 1882, 1891, 2168, 2473, 2576, 2823, 2876, 2823, 2876, 2956, 3213, 3226, 3880
Escherichia coli (EPEC, LAEC)	GI	Fim, intimin, Bfp, type 1 fim	Integrin, phosphatidyl choline, phosphatidyl serine, Tir, N-acetyllactosamine	ERT, TC, IGP, biopsy specimen, brush border, intestinal cells, Tis.Sect, PEG, EXT, OC, IGL, Mφ	13, 187, 199, 214, 320, 373, 480, 506, 716, 797, 906,961, 1039, 1040, 1079, 1223, 1341, 1424, 1509, 1688, 1715, 1830, 1947, 2000, 2028, 2180, 2239, 2295, 2385, 2778, 2839, 2870, 3031, 3096, 3123, 3267, 3268, 3385, 3732, 3840, 3841, 3889, 4175
Escherichia coli (EIEC, STEC)	GI	LPS		ERT, TC	910, 1688, 1760, 2148, 2807, 2808
Escherichia coli (EHEC, O157)	UT, GI	Type 1 fim, intimin	MR, mannose	TC, EXT, Hxd, Coagg (*S. boulardii*), M cell OC, lettuce	181, 187, 213, 660, 747, 748, 797, 809, 1148, 1177, 1690, 2142, 2180, 2392, 2636, 2700, 2871, 3684, 3685, 3935, 4032
Escherichia coli	Not defined	Type 1 fim, curli	MHC class I	BC, UC, OC, Biomat., Hxd, sGlp mesentric lymph nodes, lymphoma cells, ERT, PMN	130, 305, 389, 390, 400, 401, 439, 474, 1649, 2723, 3489, 3779, 3988
Escherichia coli (DAEC)	GI	F1845, afa-Dr Dr-II	Decay-accelerating factor (CD55), CD66E	ERT, TC	253, 1320, 1321, 1322, 1688, 1747, 2301, 2352, 2834, 2835, 2836, 2837, 3364
Escherichia coli (NTEC)	GI			TC	2735

(Continued)

TABLE 12.1 Studies of adhesion from 1992 to 2002: *E. coli* pathotypes, adhesins, receptors, and target substrata used (*Continued*)

Bacterium	Clinical source	Adhesins	Receptor specificity	Substrata (target cells)	References
Escherichia coli	GI, UT	Type 1 fim, LPS	Integrin, Lm, Man, LF, LH, FSH, lysosomal membranes, bladder mucus, mucus, uroplakins, CD48, 82- and 55-kDa glycoproteins (sIgA), 35- to 170-kDa lectin, IgA, IgG, SR–A	Yeast cells, EXT, ERT, Mφ, Man-BSA, IGlp, mucus, TC, IC, PMN, MC, lymphocytes bladder EXT, polyethylene oxide–coated glass, amoeba, endometrial cells, glass, stents	73, 141, 202, 288–290, 445, 693, 1274, 1391, 1747, 1844, 1911, 1980, 2024, 2050, 2071, 2135, 2165, 2243, 2265, 2266, 2270, 2355, 2544, 2629, 2754, 2769, 2838, 2910, 2943, 2960, 2984, 3199, 3233, 3285, 3286, 3484, 3609, 3706, 3712, 3779, 3869, 4058
Escherichia coli (EAggEC)		Bfp, Fim II	Sialic acid	TC, ERT, plastic, OC, EXT, glass	4, 198, 703, 1309, 1508, 1688, 1943, 2610, 2611, 2963, 3386, 4084
Escherichia coli (RDEC, EPEC)	Rabbit	32-kDa fim, AF/R2 fim, AF/R1 fim (K88– and CS31-like fim)		Tiss. Sect., TC, enterocytes, Peyer's patch EXT, M cells	16, 481, 673, 788, 1012, 2388, 2674, 3184, 3915
Escherichia coli	Porcine and bovine GI	Type 1 fim, 987P fim, F107 fim, F165 fim, K88 fim, K99 fim, MR fim, 31A fim, Prs, F4, F5 fim, F6 fim, Fas fim, F17 fim, Pap fim, F41 fim	GL, Man, NANA, mucus, fucose, Fbgn, 210- and 240-kDa Glp, GM$_3$, transferrin, aggregation-promoting factor, mucin, neolacotetraosylceramide; 26-, 41-, and 80-kDa mucus Glp	Brush border, IGlp, IGL, ERT, intestinal villi, gastric cells, brush border, TC, IGlp, PMN, Coagg (*L. acidophilus*)	3, 195, 196, 292, 320, 336, 337, 570, 732a, 746, 854, 855, 955, 988, 1066, 1279, 1281, 1418, 1419, 1622, 1719, 1726, 1727, 1811, 1895, 1937, 1983, 2039a, 2177, 2247, 2633, 2747, 2755, 2842, 3358, 3580, 3599, 3767, 3817, 4019
Escherichia coli	GI	Type 1 fim, MR fim, Ag43	Caveolae, Ag43	BC, uroepithelial cells, TC, ERT, Coagg (*E. coli*), plastic, stainless steel, Mφ, silicone rubber, intestinal cells, Biomat., MC	211, 228, 387, 399, 401, 714, 1181, 1392, 1433, 1657, 1927, 2395, 2543, 2982, 3113, 3284, 3404, 3419, 3941, 4060
Escherichia coli (EAEC)	GI	CS31a fim, CS1fim (CSOA), 20-kDa fimbrial prot.		ERT, TC, EXT, brush border	1983, 2388, 3992
Escherichia coli	Meninges, UT	S fim, type 1 fim	NANA, GM1, Lm, albumin, collagen, 65- and 130-kDa sialo–Glp, sIgA, Fn, sialyllactose	TC, brain endothelial cells, ECM, IGlp, BC, ERT	1402, 1980, 1995, 1996, 2256, 2761, 2930, 2940, 3240, 3317, 3546, 3609, 3903

Organism	Site	Adhesins	Receptors	Substrata	References
Escherichia coli (APEC)	Avian	P fim, curli, type 1 fim, Tsh protein (106 kDa), Tsh-β (33 kDa)	Man	ERT, EXT, tracheal cells, pharyngeal cells, Mφ	137, 873, 1261, 1410, 2074, 2075, 2115, 2952, 3104, 3516, 3815
Escherichia coli (UPEC)	UT	P fim (pyelonephritis), F1C fim, type 1 fim, NFA, G fim, S fim (meningitis), Afa, Dr family fim (DrI, Dr-II, F1845, Afa-I, Afa-III), S fim	Gal(α1→4)Gal, Lm, globoseries GL, Man, CD66, plasminogen, heparin, N and Dr blood groups, 80- to 200-kDa whey Glp, asialo-GM_1	ERT, IGL, IGP, silicone, EXT, TC, IC, SGlp, urinary bladder, PMN, glycosphingolipid P-coated beads, Hxd, endometrial cells, catheter, UC, bladder and kidney in vivo, OC, PMN, BC, colonocyte membranes, primary tubular epithelial cells	23, 24, 43, 321–323, 340, 382, 394, 448, 451, 633, 753, 1047, 1117, 1130, 1236, 1246, 1276, 1324, 1355, 1471, 1493, 1566, 1588, 1747–1749, 1758, 1766, 1808, 1857, 1858, 1882, 1892, 1929, 2009, 2025, 2073, 2085, 2207, 2253, 2282, 2313, 2437, 2586, 2631, 2771, 2780, 2787, 2867, 2868, 3042, 3209, 3253, 3304, 3311, 3341, 3450, 3477–3479, 3482, 3573, 3609, 3707, 3767, 4030, 4061

TABLE 12.2 Studies of adhesion from 1992 to 2002: pathogens, adhesins, receptors, and target substrata used

Bacterium	Clinical source	Adhesins	Receptor specificity	Substrata (target cells)	References
Acinetobacter baumanii	UT		Coll, Fn, Vn, Fbgn	ERT, bladder cells, Hxd, coated latex beads	1270, 1563, 1564, 1961, 3372
Acetobacterium spp.	ENV			Hxd	845
Actinobacillus actinomyce temcomitans	ORL	29-kDa prot, fim, Flp pili	Coll, Lm, Fn, saliva	PMN, TC, galactan, Hxd, IGP, Teflon, SHA, BC, saliva–coated titanium, ECM, coated plastic, reconstituted basement membrane, tissue regeneration membrane, titanium implants, HA	90, 91, 92, 300, 333, 459, 664, 1015, 1016, 1413, 1545, 1802, 1803, 2265, 2423, 2424, 2452, 2535, 2717, 2892, 3360, 3509, 3523, 3943, 4117
Actinobacillus pleuropneumoniae	URT, SKN	10- to 11-kDa prot, fim, LPS porcine	Mucus, respiratory Glp	EXT, TC, Fr.Tis.Sect.	254, 255, 256, 839, 1030, 1359, 2790, 2791, 3078, 3079, 3956
Actinomyces israelii				Coagg (oral bacteria)	459
Actinomyces naeslundii	ORL	Type 1 fim, type 2 fim, 200-kDa GP	Saliva, Coll, Fbgn, enamel, proline-rich Glp, statherin, 95-kDa prot, GalNAcβ Gal(β1→3)GalNAc	IGlp, GalNAcβ, ERT, tongue cells, BC, Coagg (*P. gingivalis*, *A. odontolyticus*, streptococci) SHA, PMN, Hxd, Sglp	251, 362, 405, 492, 924, 937, 1374, 1375, 1376, 1442, 1933, 2143, 2202, 2622, 2985, 3223, 3227, 3435, 3569, 3570, 4075, 4109
Actinomyces odontolyticus	ORL	Fim	Proline-rich Glp, Nacβ-D-galactosamine	GalNAcβ, tongue cells, BC, Coagg (*A. naeslundii*)	1374, 1375, 1376
Actinomyces pleuropneu- moniae	ORL	LPS	10- and 11-kDa mucus proteins	Resp. tract cells and IM, organ in vivo, IGlp, ERT	254, 255, 839, 2791
Actinomyces pyogenes	SKN (bovine)		Fbgn	Hxd, TC, sGL	823, 824, 2058
Actinomyces viscosus	ORL	Dextran binding protein, type 1 fim	Proline-rich Glp, Coll, saliva, amylase, enamel, galactan, salivary proteins, fetuin, statherin	Coagg (streptococci, *Porphy- romonas*), plastic, tooth, IGlp, dentures, sGlp, dental alloy, Teflon, amalgam, dentine, saliva–coated titanium, SHA, tissue regeneration membrane, dental restorative material	935, 1518, 1807, 1912, 1977, 2143, 2371, 2372, 2624, 2682, 2774, 2932, 3143, 3296, 3435, 3520, 3530, 3595, 3626, 3626, 3943, 4062, 4107, 4108
Actinomyces serovar	ORL	95-kDa prot	Lac	Coagg (*S. oralis*), SHA	491, 1934

Aeromonas caviae	GI	Type IV fim, Bfp	Mucin	Mucus–coated plastic, TC	148, 1301, 1921, 1922, 2012, 2975, 3099, 3722, 4011
Aeromonas hydrophila	GI, ENV	OMPs, LPS, type IV fim, Bfp and Tap fim	Mucin	ERT, TC, IM	148, 221, 1301, 1401, 1917, 2008, 2411, 2413, 2627, 4009
Aeromonas salmonicida		50.7-kDa prot, LPS	Coll IV, Lm, Fn	sGlp, IGlp, TC	835, 1164, 3756
Aeromononas sobria	ENV, GI	LPS, Ae24 pili, TAP13 pili (23 kDa)	Mucin	TC, IM, EXC rabbit intestine, ERT	148, 221, 1069, 1069, 1301, 1401, 2627, 4009
Aeromonas trota		20-kDa fimbriae		EXC rabbit intestine, ERT	2588
Aeromonas veronii		Type IV fim, Bfp and Tap fim		TC, PMN	1068, 1301, 1347, 1826, 1917, 1918, 1922
Aeromonas spp.				TC	1311, 2653
Agrobacterium spp.				Silicone	86
Alcaligenes spp.	ENV			Glass	2505
Alteromonas spp.	ENV			Glass	2505
Anaplasma marginale					769
Arcobacter spp.		20-kDa lectin	Galactose	ERT	3760
Arcanobacterium pyogenes					1789
Arthrobacter spp.	SKN			Plastic, glass	258
Bacillus cereus	ENV, GI	S layer (97-kDa prot)	Coll, Lm, Fn	PMN, IGlp, TC	114, 1986, 2841, 3511
Bacillus piliformis	Rat			TC	1083
Bacillus subtilis			Amniotic fluid, LPS–binding protein	Luminal surface of amnion, sGlp	33, 733, 886, 987
Bacteroides distasonis	GI		Lm	IGlp	928
Bacteroides forsythus	ORL	98-kDa protein		ECM, coated plastic, ERT, Coagg (streptococci), PMN, TC	1395, 2546, 3382, 4099
Bacteroides fragilis	GI	Capsular polysaccharide	Lm, Fn, Coll, Vn, CR3b, NANA	IGlp, ERT, mesothelium, TC, BC, PMN, cecal mucosa, Lm-, Vn-, Fn-, and Coll-coated plastic	252, 842, 928, 1009, 1049, 1346, 2595, 2810, 3632, 3831, 3990

(Continued)

TABLE 12.2 Studies of adhesion from 1992 to 2002: pathogens, adhesins, receptors, and target substrata used (*Continued*)

Bacterium	Clinical source	Adhesins	Receptor specificity	Substrata (target cells)	References
Bacteroides merdae	GI		Lm	IGlp	928
Bacteroides multiciacidus	GI		Coll	IGlp	3632
Bacteroides ovatus	GI		Lm, Fn, Coll, Vn	IGlp	928, 3632
Bacteroides spp.	GI		Fn, Coll, Vn	IGlp, ERT	3632
Bacteroides stercoris	GI		Lm,	IGlp	928
Bacteroides thetaiotaomicron	GI		Lm, Coll, CR3b	IGlp, PMN	928, 1049, 3632
Bacteroides uniformis	GI		Lm	IGlp	928
Bacteroides vulgatus	GI		NANA, Fn	ERT, sGlp, IGlp, Lm-coated plastic	928, 2004, 3632
Bartonella bacilliformis			44-kDa Glp (glycophorin B)	TC, ERT	434, 621, 1515
Bartonella henselae	BLD, GI	43-kDa prot, OMP		ERT, TC, endothelial cells, lymphocytes	229, 443, 444, 762, 2394, 2828
Bartonella quintana				TC, in vivo endothelial cells	229, 424
Bifidobacterium adolescentis	GI	36- and 52-kDa prot	Galactose-containing Coll V	IGlp	2534
Bifidobacterium breve	GI			TC	273
Bifidobacterium infantis	GI	Prot		TC	273
Bifidobacterium lactis	GI		Mucus	IM, IGlp	443, 1913, 2759
Bifidobacterium suis				Intestinal biopsy	775
Bifidobacterium spp.	GI		Mucus	TC, glass, ERT, IGlp	681, 1462, 1463, 2758, 2850

Organism	Source	Adhesin	Receptor	Substrata/Target	Reference
Bordetella avium	Avian URT	41-kDa protein	NANA, GD1a, GD1b	ERT, ciliated tracheal cells, turkey tracheal mucosa	145, 841, 2492, 2493, 3608, 3696
Bordetella bronchiseptica	Porcine URT	Fim 2, fim 3, flagella, FHA (RGD), 200-kDa prot, pertactin, Bvg	CD11/CD18, NANA	TC, porcine nasal cells, PMN, ERT, organ in vivo, Mφ	420, 421, 654, 1048, 1651, 2313, 2895, 3035, 3036, 3215, 3258, 3811
Bordetella parapertussis	URT	FHA	FHA	TC, ERT	1465, 2218, 2919, 3812, 3813
Bordetella pertussis	URT	FHA, P69 fim, minor fimbrial subunit, fim 2, pertactin (RGD), pertussis toxin	GL, heparin, CR3 VLA-5, Lewis^B, Lewis^X, Lewis^A	ERT, respiratory, IGL, monocytes, OC, TC, IS, EXT, factor H, C4B binding protein, BC, m:onocytes, Mφ, PMN, nasal mucosa	10, 134, 227, 268, 930, 975, 1118, 1147, 1191, 1404, 1451, 1452, 1453, 1454, 1647, 1648, 2112, 2113, 2191, 2404, 2405, 2937, 3187, 3264, 3463, 3549, 3813, 3814, 3849, 3979, 3980
Borrelia burgdorferi	BLD	39-kDa OMP; 70-, 67-, and 62-kDa Hsp60; 41- and 31-kDa OspA; P66 prot, DbpA, Bgp	Proteoglycan, integrin, GD1a, GT1b, Coll, plasminogen, CR3, heparin, heparan sulfate, Gal-ceramide, galactocerebroside, integrin $\alpha_{IIb}\beta_3$, decorin	TC (fibroblast, neural, endothelial), platelets, decorin, IGlp, ERT, IC, PMN, lymphocytes, IGL, platelets, sGlp, peripheral blood, fibrocytes, tick gut	175, 425, 580, 581, 582, 607, 608, 610, 620, 630, 758, 862, 863, 1156, 1287, 1327, 1392, 1467, 1572, 1640, 1827, 2123, 2124, 2125, 2424, 2773, 2803, 2858, 3224, 3409, 359
Borrelia garinii	BLD	147-kDa prot	Integrin, Fn	TC, IGlp, sGlp	1979
Borrelia hermsii		VspA	Plasminogen, sulfated glycosaminoglycan	IGlp, sGlp, endothelial TC, other TC, IC	620, 2249, 3224
Borrelia japonica			Galactosylceramide	IGL	1828
Borrelia turicatae		VspA	Sulfated glycosaminoglycan	TC, IC	2249
Brucella abortus	Bovine		CD11/CD18	Mononuclear phagocytes	478
Burkholderia (Pseudomonas) cepacia		22-kDa pilus protein (cable pili)	Mucin, CF mucin, 55-kDa prot	TC, BC, IGlp, EXT	516, 561, 3202, 3203, 3205, 3207, 3208
Burkholderia pseudomallei	RT, ENV		GM_1 and GM_2 gangliosides	TC, Mφ	40, 1266, 1764
Campylobacter coli		Pili	Fn	IGlp, PMN	838, 974, 1924, 2519
Campylobacter curvus				TC	1395

(Continued)

TABLE 12.2 Studies of adhesion from 1992 to 2002: pathogens, adhesins, receptors, and target substrata used *(Continued)*

Bacterium	Clinical source	Adhesins	Receptor specificity	Substrata (target cells)	References
Campylobacter fetus		Pili		TC	838, 1277
Campylobacter jejuni	GI	CBF1, FlaA (flagellin), PEB1 pili, 59- and 27-kDa prot, 42- and 43-kDa OMP, flagella	Unsaturated fatty acids	ERT, TC, IL, coated plastic, Hxd, IGlp, Coagg (*C. jejuni*)	177, 484, 551, 974, 1285, 1506, 1729, 1885, 1966, 1967, 2167, 2318, 2442, 2454, 2518, 2519, 2559, 2833, 2935, 2936, 3084, 3314, 3474, 3633, 3634, 3688, 4102
Campylobacter rectus		150-kDa prot (S layer)		TC	3942
Campylobacter upsaliensis	GI	Pili	Fn, mucins, ethanolamine, ganglioteraocylceramide	Immob Fn, IGlp, Hxd	2519, 3629
Capnocytophaga sputigena				Coagg (oral bacteria)	459
Chlamydia pneumoniae					2166
Chlamydia psittaci		16- to 18-kDa OMP		TC	182
Chlamydia trachomatis	URT, GU	40-kDa glycoprotein high-Man surface, oligosaccharides, 32-kDa Glp	Heparin, heparan sulfate, β₁-integrin, Lm, Vn, Fn, Coll I, Coll IV	TC, endothelial cells	739, 1380, 1496, 1788, 1902, 2258, 3531, 3584, 3585, 3616, 3617, 3618, 4065, 4148, 4157
Citrobacter diversus				TC	4049
Citrobacter freundii	GI			Intestinal cells, MC	141, 1078, 3280
Clostridium difficile	GI	12- 27-, and 40-kDa OMP; 70- and 50-kDa OMP	Mucus, Glc, Gal, Lac, Fn, Coll, Vn	TC, ERT, IGLp, Hxd	882, 973, 1485, 1841, 2006, 2563, 3632, 3683, 3937
Clostridium pefringens				Intestinal cells, biomat., TC	170, 1265, 1955
Corynebacterium diphtheriae				ERT, Hxd	2344
Corynebacterium jeikeium				Biomat.	3490

Organism	Source	Adhesin	Receptor	Substrata	Reference(s)
Corynebacterium matruchotii	ORL			Plastic, glass	258
Corynebacterium parvum			Scavenger receptor	Mφ	1360
Corynebacterium spp.				Biomat.	1870
Corynebacterium urealyticum				Biomat.	3490
Coryneforms	ENV			Glass, Teflon	258
Coxiella burnetii			CR3 integrin	Mφ	483, 770
Desulfomonile tiedjei	ENV			Glass	3847
Desulfovibrio spp.	ENV			Glass	3847
Dichelobacter nodosus	GI			ERT	2973
Edwardsiella ictaluri	Fish	Type 1 fim	Man	ERT, PMN	45
Edwardsiella tarda				TC	3564
Ehrlichia chaffeensis	Horse	120-kDa prot		TC	2916
Ehrlichia risticii	Horse			Mφ, TC	2420
Eikenella corrodens	ORL	31-kDa prot, 17-kDa lectin	NAc-galactosamine	ERT, Coagg (*S. milleri*, *Actinomyces*, Streptococci), sGlp	166, 913, 1395, 2581, 2997, 3743, 4119, 4136, 4137
Endogenous bacteria				Rat intestine	3506
Enterobacter aerogenes				TC, Biomat.	992, 2182
Enterobacter cloacae		Type 1 fim, 35-kDa subunit, MR fim	Man	ERT, TC, HC	1329, 1871, 2182, 2260, 2784, 2856, 2980, 3349, 3778
Enterobacter spp.				Coagg (*Fusobacterium*)	1184
Enterococcus spp.	GI, URT			Biliary stent	4131

(Continued)

TABLE 12.2 Studies of adhesion from 1992 to 2002: pathogens, adhesins, receptors, and target substrata used (*Continued*)

Bacterium	Clinical source	Adhesins	Receptor specificity	Substrata (target cells)	References
Enterococcus (Streptococcus) faecalis	GI, UT	RGD protein (Asc10), sex phermone aggregation factor, Asa373, Ace (75-kDa prot) AceA (105-kDa prot)	Vn, LF, thrombospondin, Coll, Fn, Lm, sulfated ECM, CD11b/CD18, fibrin	TC, Biomat., sGlp, ERT, catheter, EXT, glass, MC, *Lactobacillus* biosurfactant on glass, silicone rubber, vitreous, plasma-coated Biomat., IGlp,PMN, Mφ, stent	141, 502, 647, 1126, 1138, 1355, 1718, 1792, 1994, 2135, 2445, 2543, 2558, 2594, 2987, 3038, 3042, 3044, 3069, 3243, 3407, 3578, 3586, 3607, 3610, 3737, 3832, 3866, 3867, 3868, 3989, 4130, 4147
Enterococcus faccium				PMN, rubber	1497, 1681, 1705, 1727, 1792
Erwinia carotovora	ENV	Type 1 fim	MR	ERT	3939
Erwinia chrysanthemi	GI			TC	885
Flavobacterium columnare	Fish			Gills in vivo	408, 757
Flavobacterium meningosepticum				Contact lens material, Hxd	408, 667
Fusarium oxysporum					947
Fusobacterium necrophorum	ORL		Coll	ERT, TC, BC, IGlp, rumen cells	1834, 2714, 3652, 4074
Fusobacterium nucleatum	ORL	42-kDa prot, 30-kDa prot	Arginine, Coll, NANA, NAc-galactosamine, Lac, Fn, galactose, raffinose	ERT, saliva, IGLp, lymphocytes, Coagg (*P. gingivalis*, streptococci, *Treponema* spp.), Mφ, PMN, Teflon, galactan, TC, buccal cells, glass, tissue regeneration membrane	167, 173, 384, 1128, 1328, 1395, 1910, 1960, 3121, 3378, 3379, 3653, 3654, 3776, 3943, 3986
Gardnerella vaginalis				Vaginal epithelial cells, ERT	513, 781
Gordona spp.	Skin			Plastic, glass	258
Haemophilus aphrophilus	ORL			PMN	1545
Haemophilus ducreyi	URG	58.5-kDa prot, LOS, 260-kDa-prot	Heparin, Fn, ECM, Lm, Coll.	Foreskin fibroblasts, EXT, primary keratinocytes, IGlp, ERT, TC	5, 63, 64, 65, 231, 232, 236, 412, 793, 1098, 1099, 1202, 1544, 2045, 2054, 2261, 2421, 2801, 3748, 3955, 4046, 4149

Organism	Site	Adhesin	Receptor	Substrata	References
Haemophilus influenzae	URT	Fim, OMP A, capsule?, Hsp60, Hif fim, HifA, HifD, Hap, serine protease, 114-kDa OMP (Hia protein), 37-kDa prot, P2 and P5 prot, OapA, LPS	Sialyl-Glp, GM1, glycosaminoglycan, mucin, Glc, bronchoepithelial Glp, galactosamine, LF, glycosphingolipid, ganglioside, mucus, ECM, Coll I, Coll III, Fn, Lm, Lewis[a], sialylated Glp, PAF	Nasopharyngeal cells, IGlp, ERT, tracheal cells, BC, EXT, TC, IC, synthetic fibers, IGL, genital cells, PMN, adenoid OC, organ in vivo, primary fetal lung cells, Mdb, mucus-coated plastic, bronchoepithelial cells, Coagg (*H. influenzae*)	190, 191, 193, 218, 338, 415, 460, 520, 735, 740, 749, 867, 930, 960, 978, 980, 996, 1055, 1214, 1217, 1268, 1400, 1421, 1430, 1483, 1548, 1585, 1668, 1687, 1721, 1887, 1972, 2013, 2014, 2033, 2214, 2296, 2339, 2359, 2360, 2361, 2458, 2459, 2461, 2615, 2659, 2660, 2806, 2875, 2938, 2967, 3013, 3021, 3026, 3071, 3126, 3433, 3524, 3525, 3526, 3535, 3537, 3538, 3559, 3627, 3689, 3809, 3836, 3837, 3848, 3849, 3904, 3905, 3909, 3973
Haemophilus paragallinarum				ERT	403, 3646
Haemophilus parainfluenzae	URT		Mucin Glp	Coagg (streptococci), saliva, ERT	2159, 3861
Haemophilus parasuis	Porcine URT		MR	ERT	2545
Haemophilus somnus				TC	2040
Hafnia alvei				TC	60, 1078, 1656
Halomonas marina	ENV			Poly-N-isopropylacrylamide	1661
Helicobacter acinonyx	Cheetah GI	HxaA		ERT	968
Helicobacter felis					3691
Helicobacter mustelae	Ferret GI (stomach)	gangliotetra-O-ceramide	Phosphatidylethanolamine, stomach biopsy, Hxd, TC gangliosides, lysophosphatidylethanolamine	IGL, ERT, immob. glycolipid, 2329, 2739, 3691, IGP, primary gastric cells, organ	316, 606, 968, 1052, 1238, 1239, 1240
Helicobacter nemestrinae	GI (primate)	HnaA	NANA	ERT	968

(Continued)

TABLE 12.2 Studies of adhesion from 1992 to 2002: pathogens, adhesins, receptors, and target substrata used (*Continued*)

Bacterium	Clinical source	Adhesins	Receptor specificity	Substrata (target cells)	References
Helicobacter pylori	GI	20-kDa and 63-kDa prot, LPS, Lewis^X, 25-kDa prot, HpaA (29-kDa lipoprotein), 70-kDa prot, 16-kDa prot, BabA, Hsp60, Hsp70, 24-kDa sialic acid binding prot, Hopz, AlpA, AlpB	NANA, fetuin, phosphatidylethanolamine, Lewis^b, Lm, plasminogen, sialyl-Lac, heparin, heparan sulfate, lactosylceramide, Lewis^a, sulfo-Lewis^a, sulfogalactose, Lewis^X, sulfated mucin, mucus, lyosphosphatidylethanolamine, sulfated glycolipid, sulfated glycoconjugates, gangliosides, LF, N-acetylneuraminyllactose, mucin, sialyl glycoconjugate, MUC5AC, MUC5B (sulfo-Lewis)	TC, IGL, sGlp, ERT, BC, gastric biopsy, IGLp, IC, PMN, EXT, Mφ, yeast cells, sGL, Tis.Sect., mucus–coated plastic, organ in vivo, Coagg (*Fusobacterium*), gastric mucin, gastric secretions, primary gastric cells	44, 112, 122, 123, 147, 154, 263, 316, 354, 365, 409, 441, 563, 564, 565, 566, 567, 600, 602, 603, 605, 606, 619, 652, 702, 803, 837, 906, 918, 942, 960, 966, 968, 984, 985, 1014, 1183, 1185, 1231, 1238, 1239, 1348, 1446, 1447, 1448, 1449, 1450, 1470, 1522, 1523, 1524, 1525, 1526, 1534, 1555, 1574, 1580, 1581, 1582, 1619, 1693, 1736, 1801, 1823, 1824, 1825, 1863, 1908, 1924, 1948, 2076, 2088, 2103, 2117, 2118, 2119, 2120, 2175, 2195, 2315, 2393, 2429, 2443, 2596, 2644, 2651, 2656, 2658, 2668, 2683, 2741, 2742, 2752, 2763, 2783, 2840, 2860, 2861, 2933, 3005, 3050, 3070, 3076, 3076, 3100, 3156, 3352, 3396, 3418, 3423, 3459, 3572, 3581, 3582, 3628, 3650, 3691, 3804, 3805, 3806, 3816, 3862, 3890, 3916, 3928, 3930, 4072, 4073, 4088
Klebsiella oxytoca Klebsiella ozaenae	UT, GI ENV			UC, ERT, pectin, TC, Biomat. Glass	1899, 1697, 2182, 2956, 3146, 3675 3558
Klebsiella pneumoniae	UT, GI	LPS, CF29K fim, 28-kDa fim (KPF-28), type 3 fim, MR fim, (MrkD), type 1 fim, capsular polysaccharide, Man(α1→2)Man	Galectin-3, GlcNAc, ECM, Coll, Man	UC, EXT, IGLp, ERT, buccal cells, Hxd, TC, endothelial cells, bladder cells, yeast, Mφ, plastic, Biomat, PMN, catheter material	720, 235, 818, 819, 820, 992, 998, 999, 1000, 1057, 1115, 1126, 1295, 1559, 1560, 1694, 1903, 2181, 2182, 2243, 2296, 2331, 2412, 2422, 2437, 2554, 2903, 2956, 3089, 3196, 3331, 3346, 3347, 3349, 3675, 3676, 3765, 3909
Klebsiella spp.	GI	Type 3 fim	Collagen	IGLp	3331
Lactobacillus acidophilus	GI, vaginal	S-layer, prot?, fim	GL, Coll I, Coll V, mucus, Fn	TC, Hxd, avian intestinal cells, vaginal cells, Coagg (C. albicans, E. coli, G. vaginalis), platelets, IM, enterocytes, rubber, glass, Ieal cells, iIGlp, Tr Sect.	274, 356, 443, 513, 536, 537, 613, 614, 694, 1061, 1293, 1431, 1725, 1837, 2370, 2532, 3040, 3306, 3648, 3770, 3771, 3773, 4071, 4135

Organism	Site	Adhesin	Receptor	Substrata	Reference(s)
Lactobacillus brevis	GI, vaginal		Mucus		426
Lactobacillus casei			GM$_1$ ganglioside, Coll, mucus	Hxd, tooth, dentine, IGL, sGlp, epithelial TC, intestinal cells, IGlp	443, 482, 694, 1053, 1988, 2369, 2499, 2760, 2774, 3040, 3041, 3626, 3770, 4076
Lactobacillus crispatus	GI		Coll, mucus	sGlp, IM, Tis.Sect.	1913, 3731, 3846, 4135
Lactobacillus delbrueckii				TC	3239
Lactobacillus fermentum		S layer, 50- to 67-kDa prot	Coll I, Coll V, mucus (30- to 70-kDa Glp), Man	Ileal cells, rubber, glass, porcine squamous cells, sGlp, mucus-coated plastic, platelets, Coagg (streptococci), rabbit intest. cells, platelets, ERT	628, 1339, 1431, 1490, 1491, 1725, 3043, 3114, 3254, 3846, 4015, 4124
Lactobacillus GG	ENV		Mucus	IM, TC	482, 1913, 3770
Lactobacillus gasseri	Vaginal	Prot	GI, colonic mucin	Vaginal cells, TC, Coagg (*C. albicans*, *E. coli*, *G. vaginalis*), IGlp	356, 1293, 2338, 3771
Lactobacillus jensenii	Vaginal	Prot	GL	Vaginal, Coagg (*C. albicans*, *E. coli*, *G. vaginalis*)	356
Lactobacillus johnsonii	Vaginal	LTA		TC	1278, 2617
Lactobacillus leichmanii				TC	3239
Lactobacillus oris			Coll I, Coll V	Platelets	1431
Lactobacillus paracasei			Mucus, Coll I, Coll V	IM, TC, platelets	42, 443, 1431, 1913
Lactobacillus plantarum			Man, Coll I, Coll V	TC, platelets, Coagg (*Lactobacillus* spp.), ERT	18, 42, 628, 1431, 3830
Lactobacillus reuteri	ENV, GI, vaginal	29- and 56-kDa prot	Coll, Fn, β-galactosyl residues, mucus	sGlp, IGlp, Coagg (*E. coli*), ERT, IGL, Tis.Sect.	42, 62, 426, 1783, 1936, 2172, 2533, 3119, 3120, 4135
Lactobacillus rhamnosus	ORL, GI		Coll I, Coll V, mucus	TC, platelets, IGlp, ERT	426, 443, 628, 1053, 1431, 1832, 2107, 3770, 3771
Lactobacillus salivarius	ORL, GI		Coll I, Coll V	Coagg (streptococci), platelets, ERT	628, 1413, 4015

(Continued)

TABLE 12.2 Studies of adhesion from 1992 to 2002: pathogens, adhesins, receptors, and target substrata used (*Continued*)

Bacterium	Clinical source	Adhesins	Receptor specificity	Substrata (target cells)	References
Lactobacillus spp.	GI, ORL		Fn	Catheter, SGP, ERT, TC, Coagg (streptococci)	117, 2577, 2748, 2893, 3040, 3042, 3306, 3576, 3734
Lactococcus lactis	GI			TC	1531, 1547, 1906, 2996
Legionella haemophilae	ENV, URT		Hemin	TC, SGP	2678
Legionella micdadei	ENV, URT	170-kDa lectin	Gal/GalNAc-specific lectin	Amoeba	2039
Legionella pneumophila	ENV, RT	170-kDa lectin, Hsp60, pili, 25-kDa OMP	Gal/GalNAc-specific lectin	PMN, Mφ, TC, *Acanthamoeba polyphaga*, *Hartmanella verniformis*, brass, copper	299, 381, 815, 1013, 1149, 1161, 1162, 1163, 1205, 1514, 1601, 2005, 3102, 3621, 3871, 3872, 4086, 4087
Leptospira interrogans	Arthopod Fbp BLD		Fn	TC, Mφ, IGlp	2409, 2410
Leptotrichia buccalis				ERT	3458
Listeria innocua				TC	3300
Listeria ivanovii				TC	3300, 3863
Listeria monocytogenes	GI	ActA, InlA, InlB, LAP, p104, 100-kDa heat shock prot	Heparin, NANA, β-lacto–globulin	Mφ, TC, polycarbonate membrane, EXT, Mφ, biomat. (contact lenses) silica, silicone, stainless steel, IGlp	79, 80, 93, 211, 328, 375, 376, 408, 416, 417, 423, 595, 636, 890, 997, 1235, 1296, 1505, 1714, 1774, 1818, 1849, 2086, 2133, 2245, 2430, 2448, 2449, 2541, 2578, 2670, 2785, 2880, 3237, 3300, 3670, 3671, 3988, 3989, 3991
Mobiluncus curtsii				Vaginal epithelial cells	513
Moraxella bovis	Eye, bovine			Cornea	309, 2128, 2129, 2130, 3161, 3162
Moraxella (Branhamella) catarrhalis	URT	Type 4 fim, 57-kDa OMP, 200-kDa prot, UspA1, UspA2	Mucin Glp, glycosphingolipid, glucosamine, galactosamine	Endothelium, TC, bronchial and oropharyngeal cells, IM, ERT	22, 34, 35, 36, 37, 39, 412, 930, 1027, 1028, 1029, 1870, 1972, 2044, 2296, 2556, 3026, 3073, 3909, 4172
Mycobacterium avium (intracellulare)	URT, ENV	68- and 70-kDa heat shock prot, β1-integrin (?), Fn attachment prot (Fap)	αVβ3-integrin, Lm, Fn, Coll, mucus, CD43	Amoeba, monocyte-derived Mφ, TC, IGlp	270, 584, 1089, 2435, 2994, 2995, 3004, 3348
Mycobacterium bovis (BCG)	ND	Lipoarabinomannan, 30- to 32-kDa adhesin (85B prot), FapA	Fn, Man	IGlp, TC, ERT, heparin, bladder cells, phagocytes, Mφ	178, 753, 1097, 2032, 2206, 2406, 2826, 3003, 3274, 3305, 3550, 3699, 3870

Organism	Site	Adhesin	Receptor	Substrata/Cells	References
Mycobacterium fortetum	Skin			Plastic, glass	258
Mycobacterium leprae	Skin	29.5-kDa prot ML-LBP21	α-Dystroglycan Lm, Fn	Schwann cells, TC, IGlp	718, 2303, 2988, 3312, 3403
Mycobacterium smegmatis	RT		Phosphatidylinositol, Man	Man-binding prot, TC, sGlp	271, 1556, 2864
Mycobacterium tuberculosis	URT	Heparin binding agglutinin, micotin, lipoarabinomannan, 28-kDa heparin binding hemagglutinin	Heparin, Man, CR3 complement receptor, CD11, Man receptor, CD43	ERT, Mφ, phagocytes	155, 411, 773, 1004, 1089, 1271, 2100, 2302, 2312, 2407, 2863, 3550, 3551, 3726, 3870, 4139
Mycobacterium tuberculosis	RT	Phosphatidylinositol, Man, capsular polysaccharide	Man binding prot, CR4, CD11b/CD18 (CR3), glucan receptor	TC, sGlp, monocyte-derived Mφ	699, 1556, 4141
Mycoplasma arthritidis	URT			TC	3958, 3959
Mycoplasma bovis		P26 antigen (32-kDa prot)		TC	3191, 3192
Mycoplasma bovoculi		P94 antigen		ERT	3219
Mycoplasma fermentans				Mφ	3061
Mycoplasma floccale	Swine			Nasal epithelium	3060
Mycoplasma gallisepticum	Avian URT	Membrane prot p30, p48, p50, and p80; 64- and 67-kDa OMP, GapA, MGC2 (32.6-kDa OMP), 64-kDa lipoprotein (LP64), 150-kDa prot		ERT, TC, tracheal rings	157, 1056, 1237, 1532, 1867, 2298, 2299, 2789
Mycoplasma genitalium		153-kDa adhesin, 140-kDa prot (P140)		ERT, TC	804, 805, 2416, 2734, 3030
Mycoplasma hominis		P50 prot, 100-kDa membrane prot	Sulfated sugar	TC, IGL	1486, 1489, 1926, 2726
Mycoplasma hyopneumoniae	Pig trachea	97-kDa prot, P97, 124.9-kDa OMP, 100-kDa multimeric prot	GL, heparan sulfate, dextran sulfate, chondroitin sulfate, Lm, mucin	OC, IGL, TC, swine cilia, tracheal epithelial cells, plastic, respiratory cells	541, 1570, 1571, 1680, 2450, 4161, 4163, 4164, 4178

(Continued)

TABLE 12.2 Studies of adhesion from 1992 to 2002: pathogens, adhesins, receptors, and target substrata used *(Continued)*

Bacterium	Clinical source	Adhesins	Receptor specificity	Substrata (target cells)	References
Mycoplasma imitans	Avian			OC	2
Mycoplasma iowae	Avian (lymphoma)	65-kDa prot		TC	1023
Mycoplasma mobile				Glass	2460
Mycoplasma ovipneumoniae		Capsule		Organ in vivo	2635
Mycoplasma penetrans		65-kDa Fn binding prot		Coated plastic, TC, ERT	1225, 2186, 2187
Mycoplasma pirum	URT	P1-like prot		ND	3711
Mycoplasma pneumoniae	LRT	P1 prot, HMW3, P30	Sulfated Glp, NANA	ERT, TC, Mφ	200, 711, 828, 1026, 1363, 1472, 1677, 1784, 1991, 2084, 2915, 3029, 3030, 3116, 3308, 3373
Mycoplasma pulmonis	URT	46-kDa prot		OC, ERT	298, 1704, 3534
Mycoplasma synoviae	Avian	RT 45- to 50-kDa ULB925 hemagglutinin		ERT, TC	257, 2192, 2600, 2666
Neisseria gonorrhoeae	URG	Opa, 97-kDa prot, pili, type IV fim, PilE, Opa50, Opa52, PilC, LOS (lacto-N-neotetraose)	LOS (autoaggregative), heparin, hemoglobin, Opa-like sphingomyelinase, heparan sulfate, CD66, Vn, lutropin, 30-kDa prot, GM1a, CEACAM	Endothelial cells, EXT, autoaggregation, SC, TC, S GP, ERT, PMN, monocytes, urethral cells, OC (fallopian), IGP, Mφ, primary cervical cells	318, 357, 485, 538, 546, 547, 548, 763, 917, 1066, 1076, 1091, 1200, 1251, 1262, 1264, 1288, 1302, 1438, 1439, 1440, 1441, 1471, 1496, 1504, 1557, 1620, 1780, 1782, 1815, 1816, 1817, 2090, 2198, 2368, 2418, 2419, 2522, 2917, 2918, 3057, 3148, 3149, 3151, 3289, 3421, 3488, 3497, 3498, 3620, 3842, 3843, 3902, 4033, 4036
Neisseria lactamica	URT			BC	930

Organism	Source	Adhesin	Receptor/Substratum	Target cells/substrata	References
Neisseria meningitidis	URT	Fim (pili), Opc, fim bundle, Opa, PilC	Fn, Glp glycosphingolipid, ganglioside, CD46, CD66, Lewis[a], CD14, CD18	TC, BC, PMN, nasopharyngeal OC, IGlp, Hec-1B endothelial cells, virus-infected cells, Tis. Sect.	487, 760, 912, 930, 931, 960, 963, 1390, 1416, 1539, 1585, 1782, 1814, 1815, 1930, 2294, 2386, 2387, 2419, 2498, 2529, 2530, 2602, 2603, 2950, 2958, 2959, 3010, 3012, 3013, 3014, 3015, 3019, 3151, 3168, 3183, 3289, 3530, 3642, 3738, 3891, 3892, 3893, 3894, 3895, 3896, 3897, 3898, 3902, 4184
Nocardia asteroides	Mice	Filament (43 kDa)		Brain, glial cells, Mφ, EXT, TC, pulmonary cells	242, 243, 244, 246, 247, 2688
Ochrobactrum anthropi				Silicone	86
Pasteurella haemolytica			GlcNAc	ERT, tracheal cells	1703
Pasteurella multocida	Rabbit	Capsular polysaccharide (hyaluronic acid), pili	CD44	Mφ, ERT, TC, mucus, porcine and rabbit respiratory tract cells, tracheal rings and explants	66, 351, 892, 1136, 1642, 1681, 1682, 2955
Pasteuria penetrans	Nematodes			Nematodes	738
Peptostreptococcus micros	ORL			Hxd, glass, TC, Coagg (*F. nucleatum, P. gingivalis*)	664, 665, 2001, 2002
Peptostreptococcus productis	ORL			Hxd	845
Porphyromonas gingivalis	ORL	Protease 2, fim, 43-kDa fim subunit, 40- and 41-kDa OMP, HA-Ag2 (fim), 35-kDa prot, HagB, HagC, 132-kDa prot (45PrtR) (15-, 17-, 27-, and 45-kDa complex), ArgI protease, 130-kDa hemagglutinin, FimA	Coll, Fn, Lm, proline-rich prot, Fbgn, statherin, 200-kDa Glp, LF, N-acetyllactosamine, fucose, CD11/CD18, neuroglobin, saliva, Coll I, N-acetylgalactosamine, ERT band III, thrombospondin	ERT, saliva, IGlp, TC, BC, Coagg (streptococci, *A. naeslundii, S. oralis, T. denticola* and *T. medium*), bone, gingival cells, sGlp, immob. LF, SHA, titanium, embryonic calvarial cells, tissue regeneration membrane, galactan, Teflon, IGP, platelets	6, 7, 26, 99, 100, 101, 102, 104, 435, 459, 521, 522, 550, 573, 697, 756, 833, 861, 883, 884, 893, 894, 911, 937, 1297, 1298, 1383, 1395, 1398, 1445, 1521, 1530, 1562, 1578, 1666, 1821, 1850, 1851, 1866, 1872, 1969, 1970, 1971, 2015, 2031, 2060, 2061, 2062, 2064, 2065, 2092, 2093, 2104, 2105, 2126, 2271, 2421, 2552, 2553, 2567, 2568, 2569, 2579, 2585, 2592,

(Continued)

TABLE 12.2 Studies of adhesion from 1992 to 2002: pathogens, adhesins, receptors, and target substrata used *(Continued)*

Bacterium	Clinical source	Adhesins	Receptor specificity	Substrata (target cells)	References
					2652, 2657, 2729, 2796, 2847, 2883, 2884, 2947, 2970, 3200, 3228, 3230, 3231, 3322, 3361, 3380, 3381, 3383, 3394, 3398, 3439, 3469, 3470, 3472, 3473, 3493, 3523, 3548, 3651, 3655, 3735, 3736, 3787, 3943, 3963, 3978, 4062, 4075, 4099, 4113, 4114, 4117
Prevotella denticola				Coagg (oral bacteria)	459
Prevotella intermedius (nigrescens)	ORL	Prot/type C fimbriae, 3.8-kDa fimbrial prot, LPS	Lm, LF, ECM	Coagg (*Actinomyces*), TC, dentine, HA, sGlp, ECM (reconstituted basement membrane), ERT, *S. mutans*-coated glass	89, 90, 92, 468, 662, 706, 860, 395, 1519, 1521, 1812, 2137, 2138, 2623, 2716
Prevotella loeschii			Gelatin, fibrin, casein	Coagg (streptococci), coated plastic	459, 515
Prevotella melaninogenica		Fim		ERT	1412
Propionibacterium acnes				Biomaterials	868
Propionibacterium freundenreichii			Mucus	IGlp	2760, 3771
Proteus mirabilis	UT	MR/P fim, 18- and 16-kDa prot, ATF fim	MR	ERT, plastic, UC, TC, endothelium, EXT, Biomat., biliary stent	72, 183, 235, 524, 642, 1109, 1126, 1697, 1743, 2078, 2144, 2145, 2146, 2181, 2325, 2327, 2543, 3038, 3089, 3687, 3739, 3764, 3988, 3989, 3991, 4131, 4187
Proteus rutgeri				TC	2856
Proteus vulgaris				Uroepithelial cells	235, 3764
Providencia alcalifaciens				Organ in vivo	1896, 2332
Providencia rettgeri				TC	235, 2980

Organism	Source	Adhesin	Receptor	Substrata	References
Pseudomonas aeruginosa	Multiple ENV	OMP, alginate, type IV fim, LPS, flagella (FliC), 15-, 16-, 18-kDa nonfimbrial prot, PAI/PAII lectins, PAK-P, 57- and 59-kDa nonpilus adhesins, 50-kDa OMP	57-kDa Glp, heparin, phosphatidylinositol, phosphatidylglycerol, asialo-GM1, perlecan, GM1, GD1, fucose, GalNAc(β1→4)Gal, sialic acid, cytokeratin, phosphatidylserine, respiratory mucin, mucus, Gal, Man, NANA, GlcNAc, galectin-3, 21- to 97-kDa Glp, Lm, Fn, 29-kDa urine protein, α_1-microglobulin, saliva, cystatin, Gal(β1→4)Gluc, Gal(α1→4)Gal, human blood group antigen, Coll I, Coll II, cholesterol, cholesterol esters, sialyl-Lewis^x, $\alpha_v\beta_1$-integrin	ERT, cornea, corneal cells, corneal proteins, damaged cornea, saliva, gastric mucus cells, Biomat., UC, synthetic polymer, OC, IGLp, IGL, TC, Mϕ, rubber, EXT, IGlp, contact lenses, glass, Coll, Hxd, nasal cells, tracheo-bronchial mucin, urine–coated glass, bronchial epithelial cells, BC, glass, respiratory cells, gallbladder epithelial cells, titanium, bone, silicone rubber, Fr.Tis.Sect., exfoliated corneal cells, CF nasoepithelial cells, polystyrene, endometrial cells, alginate beads	9, 30, 32, 90, 130, 133, 136, 142, 143, 144, 169, 188, 208, 209, 220, 233, 236, 348, 427, 429, 495, 496, 516, 539, 542, 543, 544, 611, 629, 640, 667, 718, 736, 737, 742, 751, 761, 795, 802, 817, 822, 866, 915, 923, 992, 995, 1002, 1005, 1034, 1037, 1038, 1109, 1125, 1127, 1151, 1209, 1222, 1273, 1295, 1334, 1335, 1336, 1355, 1356, 1457, 1458, 1460, 1517, 1536, 1610, 1621, 1699, 1737, 1911, 1915, 2066, 2082, 2089, 2097, 2098, 2162, 2212, 2250, 2251, 2254, 2263, 2301, 2316, 2321, 2324, 2481, 2512, 2513, 2543, 2725, 2730, 2787, 2798, 2875, 2885, 2891, 2896, 2898, 2899, 2900, 2901, 2905, 2923, 2969, 3023, 3024, 3025, 3027, 3045, 3047, 3051, 3052, 3053, 3081, 3108, 3197, 3198, 3203, 3278, 3279, 3315, 3316, 3340, 3374, 3393, 3427, 3489, 3495, 3532, 3605, 3692, 3727, 3749, 3750, 3752, 3874, 3918, 3961, 3962, 4041, 4043, 4057, 4089, 4133, 4142, 4143, 4144, 4171
Pseudomonas (Burkholderia) cepacia				ERT	561, 1244, 2016, 3202, 3204, 3205, 3206, 3207, 3208
Pseudomonas fluorescens	Marine, plant			Glass, catheter material, TC	449, 503, 812, 1691, 2877, 3089

(Continued)

TABLE 12.2 Studies of adhesion from 1992 to 2002: pathogens, adhesins, receptors, and target substrata used *(Continued)*

Bacterium	Clinical source	Adhesins	Receptor specificity	Substrata (target cells)	References
Pseudomonas fragi				Stainless steel	211
Pseudomonas pseudomallei	BLD		Insulin	sGlp	4050
Pseudomonas spp.	ENV			Glass, Teflon	3072
Rhodococcus equi			Mac1 (CD11/CD18?)	ERT, TC, Mφ	272, 1106, 1552, 1553
Rhodococcus spp.	Skin			Plastic, glass	258
Rothia denticariosa				Coagg (oral bacteria)	459
Salmonella enterica serovar Abortusovis				TC	3144
Salmonella enterica serovar Braenderup	GI	Type 1 fim		ERT, TC	1558
Salmonella enterica serovar California	Avian	Type 1 fim		ERT	674, 675
Salmonella enterica serovar Choleraesuis		35-kDa protein	Man	TC, intestinal EXT	88, 2641, 3118
Salmonella enterica serovar Dublin				EXT (M cells)	529, 3982, 4047
Salmonella enterica	GI	SEF 14 fim, flagella, type 1 fim, SEF 17 fim	Man	TC, EXT, plastic, Mφ	70, 328, 922, 1686, 2612, 2705, 2986, 3063, 3093, 3810, 4025, 4053, 4166
Salmonella enterica serovar Enteritidis	GI	Type 1 fim, type 3 fim (aggregated fim), SEF 17 fim, SEF 14, SEF 21	Man, plasminogen, Fn, Lm, Coll	BC, intestinal cells, ERT, SGP, HC, TC, Teflon, steel, glass, Fn-coated plastic, granulosa cells, organ in vivo, polycarbonate membrane	152, 162, 595, 596, 806, 807, 808, 977, 1722, 2259, 2846, 3434, 3474, 3485, 3590, 3717, 3718, 3725
Salmonella enterica serovar Gallinarum				TC	217, 4025
Salmonella enterica serovar Pullorum				EXT, TC	1722, 4173

Organism	Site	Adhesin	Receptor	Substrata	References
Salmonella enterica serovar Typhi	GI	37-kDa OMP	Fucose, galactose	TC, EXT (M cells), yeast, ERT	1277, 1575, 2444, 2457, 3682, 3981, 3982, 4166
Salmonella enterica serovar Typhimurium	GI (mice)	Aggregative fim, type 3 fim, D OMP C, type 1 fim, FimH, flagella, LPS, 44-kDa prot, *fim, lpf, pef* operons, Rck, PagC, Agf	Coll, Lm, Man, mucus, CR3, Man receptor?, plasminogen, 60-kDa GP Gal(β1→3)GalNAc	Intestinal cells, ERT, TC (Caco2), IGP, urinary cells, IGlp, Mϕ, EXT (M cells), sGlp, skin cells, OC, cecal mucosa, muscle tissue, coated plastic, Tis.Sect., organ in vivo, brush border, poultry skin/feather follicle, poly-ethyleneoxide-coated glass, Coagg (*S. boulardii*)	57, 70, 88, 171, 236, 237, 238, 240, 286, 328, 493, 589, 670, 675, 676, 677, 693, 951, 952, 977, 1065, 1104, 1177, 1196, 1199, 1310, 1399, 1543, 1558, 1595, 1714, 1722, 1905, 1981, 2024, 2025, 2035, 2049, 2052, 2099, 2110, 2228, 2356, 2434, 2474, 2614, 2620, 2621, 2638, 3032, 3033, 3066, 3291, 3430, 3489, 3554, 3594, 3682, 3713, 3772, 3924, 3982, 3988, 3989, 3999, 4066, 4087, 4179
Salmonella spp.				TC	825
Selenomonas sputigena	ORL		Coll	Teflon, galactan, IGlp, tissue regeneration membrane	3943
Serpulina pilosicoli				Colonic mucosa	1426
Serratia liquefaciens				Plastic	2798
Serratia marcescens	SKN, UT, GI	Type 1 fim, LPS, MR–T fimbriae	Man, MR	ERT, Hxd, EXT, contact lens, UC, plastic, glass, hydro-carbon, stainless steel	90, 211, 667, 1109, 1247, 1594, 2132, 2348, 2625, 2782, 2798, 2800, 3324
Serratia spp.				TC (Caco-2)	2182
Shewanella algae	ENV			Hydrous ferric oxide	462
Shigella boydii	GI	Type 1 fim	Man	ERT	3792
Shigella dysenteriae	GI		NANA, fucose, NAc–mannosamine	TC, HXD, ERT, guinea pig colon cells	1315, 1317, 1319, 1411, 2914, 2978, 3587
Shigella flexneri	GI	Type 1 and 3 fim	NANA, Man fucose, NAc–mannosamine	TC, HXD, EXT, ERT, PMN, virus-infected TC, *Entamoeba histolytica*, guinea pig colon cells	1176, 1315, 1316, 1411, 1714, 2469, 2470, 2914, 3705, 3791, 3873

(Continued)

TABLE 12.2 Studies of adhesion from 1992 to 2002: pathogens, adhesins, receptors, and target substrata used (*Continued*)

Bacterium	Clinical source	Adhesins	Receptor specificity	Substrata (target cells)	References
Shigella sonnei			NANA	ERT	1411
Staphylococcus aureus	URT, UT, BLD, SKN, peritoneal dialysis	FgBP, ExoPS, clumping facto, (ClfA), capsular polysaccharide, teichoic acid, 50-kDa prot, slime, FnbpA, 150- and 170-kDa prot, MSCRAMM, Spa, CoA, FnbA, FnbB, 14.4- and 16.5-kDa proteins, Can, PIA (β-1,6-glucosamino-glycans), Eap	Fn, clusterin, heparin, Vn, Coll I, Fbgn, milk, fibrin, Lewis[a], LF, Lm, nasal mucin, ECM, CD44, kininogen, GM$_1$ ganglioside, mucus, heparan sulfate, plasma components (Fbgn, thrombin), fibrin, von Willebrand factor, rabbit plasma, Mφ receptors with collagenous structure, integrins, gClqR/p33 serum constituent	BC, UC, ECM, coated plastic, catheter, endothelium, HXD, glass, mammary cells, biomat., EXT, PMN, IGLp, TC, synthetic polymers, hair, respiratory cells, intestinal cells, sGlp, platelets, fibrin clots, lung cells, MC, silicone, diaper fibers, plasma–coated silicone oxide, oropharyngeal cells, epidermal cells (horny layer), titanium, bone, steel, teat cells/ductal cells/secretory mammary gland cells, meso-thelial cells, plasma–coated plastic, primary renal cells, ERT, valvular vegetation, biliary stent, amniotic mem-brane, polymethyl meth-acrylatc– and polyethylene-imine–coated glass slides, Coagg (*S. aureus*), polymethyl methacrylate beads, ortho-dontic brackets, polyethylene oxide–coated glass	1, 9, 27, 28, 29, 41, 47, 48, 49, 50, 51, 87, 121, 127, 129, 130, 133, 141, 211, 224, 234, 236, 241, 282, 295, 302, 314, 315, 330, 338, 340, 354, 364, 387, 390, 391, 393, 394, 395, 397, 399, 466, 494, 499, 500, 552, 553, 569, 572, 575, 576, 577, 578, 587, 665, 671, 672, 678, 693, 726, 733, 771, 790, 794, 811, 812, 843, 908, 930, 934, 942, 948, 960, 1025, 1036, 1041, 1057, 1063, 1071, 1072, 1074, 1107, 1212, 1234, 1291, 1292, 1295, 1307, 1423, 1427, 1492, 1498, 1499, 1537, 1551, 1605, 1629, 1663, 1731, 1734, 1739, 1742, 1771, 1794, 1835, 1836, 1877, 1925, 1954, 1998, 2056, 2082, 2181, 2225, 2246, 2273, 2275, 2301, 2319, 2320, 2329, 2339, 2363, 2364, 2365, 2398, 2410, 2456, 2471, 2472, 2473, 2485, 2486, 2500, 2548, 2560, 2566, 2573, 2634, 2663, 2671, 2720, 2776, 2777, 2802, 2815, 2824, 2825, 2852, 2875, 2901, 2906, 2911, 2922, 2923, 2925, 2928, 2941, 2972, 2977, 3022, 3039, 3047, 3048, 3062, 3068, 3074, 3089, 3172, 3186, 3188, 3262, 3288, 3332, 3388, 3389, 3406, 3412, 3462, 3466, 3489, 3525, 3552, 3575, 3596, 3600, 3686, 3693, 3709, 3720, 3721, 3740, 3783, 3789, 3790, 3800, 3850, 3853, 3854, 3855, 3856, 3875, 3906, 3913, 3921, 3954, 3957, 3961, 3996, 4020, 4021, 4042, 4101, 4105, 4123, 4131, 4150, 4151

Organism	Site/source	Adhesin components	Receptor	Substrata	References
Staphylococcus aureus			LF, Vn, heparan sulfate	IGlp, TC	2055, 3579
Staphylococcus caprae	Animal strains	Ica	Fn, Fbgn, Coll	IGlp	71
Staphylococcus epidermidis (coagulase-negative staphylococci)	SKN, BLD (endocarditis, peritoneal dialysis)	Capsular polysaccharide adhesin, (β(1→6)-linked GlcNac), 119-kDa prot, slime, β-lactose-containing polysaccharide, SspI, PS/A, ICA (β(1→6)-linked glucosaminoglycan), PIA, Fbgn binding prot	Heparin, fibrin, albumin, Lm, Fbgn (β chain), Fn, Vn, mucin, von Willebrand factor	Plastic, glass, TC, ERT, EXT, catheters, bone, synthetic polymers, IG1p, Biomat, hair, SG1p, polyvinyl chloride, PMN, plasma-coated plastic, gold film, titanium/aluminum/vanadium, contact lenses, vascular graft, fibrin clot, silicone, Fn-coated plastic Hxd, steel, intravascular catheter, vancomycin-treated silicone, rat bladder, urine, poly(N-isopropylacrylamide), Coagg (*S. epidermidis*)	11, 12, 30, 31, 50, 85, 90, 120, 121, 125, 128, 130, 131, 132, 133, 135, 174, 197, 236, 265, 266, 267, 343, 345, 418, 419, 429, 432, 486, 499, 501, 524, 525, 528, 641, 665, 672, 680, 694, 730, 731, 765, 766, 771, 778, 790, 801, 868, 896, 923, 959, 972, 993, 1011, 1035, 1042, 1071, 1124, 1135, 1140, 1141, 1159, 1221, 1267, 1273, 1295, 1355, 1422, 1474, 1476, 1495, 1511, 1602, 1606, 1660, 1661, 1695, 1696, 1699, 1738, 1739, 1740, 1773, 1806, 1877, 1881, 1949, 1963, 1964, 1965, 2140, 2141, 2176, 2181, 2219, 2220, 2221, 2229, 2230, 2231, 2232, 2233, 2234, 2264, 2301, 2320, 2365, 2377, 2417, 2482, 2510, 2528, 2540, 2560, 2584, 2630, 2647, 2664, 2686, 2687, 2798, 2805, 2815, 2816, 2830, 2831, 2832, 2888, 2889, 2891, 2922, 2923, 2951, 2977, 3001, 3038, 3065, 3089, 3137, 3170, 3173, 3174, 3175, 3176, 3177, 3238, 3292, 3303, 3329, 3335, 3350, 3399, 3406, 3408, 3517, 3528, 3529, 3692, 3694, 3695, 3721, 3757, 3798, 3799, 3800, 3860, 3868, 3874, 3875, 3945, 3946, 3947, 3974, 4010, 4011, 4123, 4126, 4150, 4151, 4177
Staphylococcus haemolyticus	Skin, blood		Vn	sGlp, Biomat, plastic, ERT, ureteral epithelial cells	1107, 2221, 2482, 3961
Staphylococcus hyicus			Fn, Fbgn, Coll	Teat canal orifice, organ in vivo, Hxd	1294, 2560

(Continued)

TABLE 12.2 Studies of adhesion from 1992 to 2002: pathogens, adhesins, receptors, and target substrata used *(Continued)*

Bacterium	Clinical source	Adhesins	Receptor specificity	Substrata (target cells)	References
Staphylococcus intermedius			Fn, Fbgn	Contact lenses, Hxd, EXT, canine keratinocytes	667, 679, 1294, 2366
Staphylococcus lugdunensis				sGlp, IGlp	2818
Staphylococcus saprophyticus		Ssp, 160-kDa prot	Lm, Fbgn, Fn, Coll, I, Coll IV, Vn	Plastic, TC, ERT, IGP	1107, 1168, 1170, 1171, 2427, 2428, 2482, 3008
Stentrophomonas maltophilia			Fn, 60-kDa Glp	Biomat, glass, Teflon contact lenses, Hxd	667, 796, 947, 1793,
Streptococcus agalactiae	UG, bovine	Surface protein, 21-kDa pilus prot, Lral prot	Fn, Lm, thrombin, NAc-glucosamine, cytokeratin 8	Vaginal cells, Hxd, ERT, HC, glass, TC, IGlp, Mφ, PMN, amnion cells, chorion cells, BC, IGlp, endothelial cells, plastic, coated plastic	340, 439, 1282, 1516, 1591, 1813, 2565, 2843, 2981, 3443, 3457, 3489, 3496, 3661, 3662, 3663, 3720, 3803, 4002, 4031, 4140
Streptococcus anginosus	ORL	14-kDa protein	Fn, saliva, proline-rich protein	SHA, HA, IGP, ERT, Coagg (*S. mitis*), titanium	916, 2475, 3679, 4016
Streptococcus bovis			Sulfated ECM	Scraped rumen epithelial cells, IGlp, sGlp	3577, 3576, 3578
Streptococcus constellatus	ORL		Saliva	SHA, HA, titanium, IGLp	1611, 2475, 3679
Streptococcus cricetus	ORL		Saliva	IGlp, SC (glucan)	452, 2215, 2475
Streptococcus crista	ORL		Amylase	sGlp, Coagg (*Fusobacterium*)	650, 651, 3272
Streptococcus defectivus (*Abiotrophia adiacens, defectiva,* and *elegans*)	ORL	Emb	ECM	Coated plastic	2279, 2711, 3681
Streptococcus downei		Glucan binding protein		Plastic	617
Streptococcus dysgalactiae	SKN	MSCRAMM, FnbA, FnbB, 140-kDa Fn binding prot	Fn	TC, bovine mammary epithelial cells, EXT, ECM, Mφ, IGlp, Hxd, coated plastic, sGlp	472, 473, 1731, 3339, 3505, 3720

Organism	Host	Adhesins	Receptors	Substrata	References
Streptococcus equi (*zooepidemicus*)	Horses, pigs, monkeys	Hyaluronic acid	Man	TC, Mφ, endometrium	1010, 1911, 4004
Streptococcus gordonii	ORL	SspA, SspB >200-kDa prot, hemagglutinin, 45-, 62-kDa prot, glucosyltransferase (150-kDa prot, AP153), LTA, CshA (290-kDa prot), SpaP (antigen I/II)	Saliva, amylase, sialic(α2→3)glucan, proline-rich prot, lysozyme, lactose, LF, Coll, Fn, mucus, sialic acid, Gal-GalNAc (CD43 and CD45)	Coagg (*Actinomyces, P. gingivalis, S. sanguis, S. oralis*), *H. parainfluenzae*), dentine, IGlp, ERT, Fn-coated plastic, enamel, sGlp, Hxd, endothelial cells, HA, saliva-coated HA, immob, glucan, IGlp, PMN, Mφ, saliva-coated glass	111, 153, 349, 364, 423, 459, 591, 779, 924, 1016, 1527, 1547, 1706, 1711, 1772, 2063, 2154, 2155, 2156, 2209, 2210, 2211, 2379, 2380, 2382, 2383, 2446, 2475, 3110, 3143, 3159, 3269, 3272, 3273, 3296, 3435, 3650, 3679, 3794, 3884, 3885, 3886
Streptococcus intermedius	ORL		Fn, saliva, thrombin	SHA, IGLp	2475, 3679, 3800
Streptococcus milleri	ORL		Saliva, Fn	IGLp, HA, sGlp, platelets, fibrin clots	2475, 4013, 4016
Streptococcus mitior				Thrombin-activated platelets, heart tissue	719
Streptococcus mitis	ORL		Fn (RGD), saliva, amylase, proline-rich prot	HC, IGLp, SGP, ERT, Coagg (*S. oralis*), coated plastic, glass, Hxd, BC, nasal cells	261, 302, 665, 2475, 2774, 3223, 3272, 3589, 3665, 3821
Streptococcus mutans	ORL	P1 (185-kDa), glucan binding prot, SpaP (AgI/II), SSp-5, WapA, GTF, PAc (190 kDa), FBP130	(α1→6)-glucan, LF, Coll, saliva, albumin, glucan, GalNH₂, sulfo-mucin, Fn, Lm, 300-kDa salivary agglutinin, *P. gingivalis* vesicle, sulfated polysaccharide, sIgA-salivary Glp complex	Tooth, Hxd, glass, SHA, Teflon, galactan-derivatized plastic, IGlp, IC, sGlp, enamel, dentine, tissue regeneration membrane, Coagg (*S. sobrinus*), polyethylene oxide–coated glass, dental restorative material, dextrans on SHA, mucins, dextran, epithelium in vivo, E-glass fibers	86, 113, 296, 385, 456, 465, 489, 490, 491, 492, 554, 555, 618, 684, 1062, 1112, 1337, 1367, 1369, 1455, 1546, 1575, 1822, 1859, 1874, 1975, 1988, 2069, 2077, 2101, 2196, 2210, 2335, 2502, 2547, 2582, 2583, 2628, 2643, 2708, 2727, 2733, 2774, 2933, 2964, 3009, 3109, 3169, 3182, 3194, 3249, 3250, 3295, 3296, 3342, 3369, 3370, 3375, 3446, 3493, 3520, 3521, 3626, 3638, 3673, 3674, 3678, 3679, 3761, 3795, 3796, 3839, 3852, 3944, 4125

(Continued)

TABLE 12.2 Studies of adhesion from 1992 to 2002: pathogens, adhesins, receptors, and target substrata used (*Continued*)

Bacterium	Clinical source	Adhesins	Receptor specificity	Substrata (target cells)	References
Streptococcus oralis	ORL	Fimbriae	Saliva, Fbgn, proline-rich prot, lysozyme, LF, sulfated polysaccharide	Enamel, ERT, sGlp, IGlp, Coagg (*S. sangius, A. naeslundii, P. gingivalis, S. gordonii, A. viscosus, H. parainfluenzae, S. salivarius*), Hxd, SHA, plastic, teeth, glass, dentures	124, 296, 358, 361, 362, 459, 916, 937, 1491, 1807, 1934, 1988, 2156, 2568, 3159, 3194, 3435, 3679, 3886, 4077
Streptococcus parasanguis		Fim A (36-kDa prot)	Fibrin, saliva	sGlp, IGlp, Biomat., platelets	446, 1103, 2718, 3201
Streptococcus pneumoniae	URT blood	Peptidoglycan, 37-kDa prot (PsaA), pneumolysin, LTA, phosphocholine, choline binding prot (Cbp, 75 kDa) choline-containing C-polysaccharide, CbpE, CbpG, CbpA, ClpC	GalNAc(β1→3)Gal, GalNAc(β1→4)Gal, asialo-gangliosides, NeuAc(α2→3)/6GalβI lacto-N-neotetraose, sialyllactose, Lm, Coll IV, Vn, asialo GM_1, galactosylceramide, lactosylceramide, PAF receptor, hemin, plasminogen, mucin, C3	OC, PC, ERT, IC, TC, nasopharyngeal cells, IGlp, bronchopharyngeal cells, BC, immobFn, tracheal cells, nasopharyngeal cells from children, leukocytes, type II pneumocytes, platelets, endothelial cells, IGL, cytokine-activated TC, endometrium, sGlp virus-infected cells, liposomes	15, 146, 219, 279, 280, 293, 294, 338, 528, 687, 689, 691, 717, 749, 911, 930, 931, 1001, 1010, 1157, 1178, 1179, 1269, 1371, 1461, 1612, 1653, 1690, 1804, 1985, 2033, 2046, 2182, 2354, 2731, 2737, 2875, 3011, 3071, 3127, 3145, 3452, 3524, 3525, 3526, 3559, 3601, 3602, 3644, 3658, 3741, 3742, 3818, 4036, 4159
Streptococcus pyogenes	URT, SKN	Prot F (PrtF, SfbI), M prot, LTA, FBP 54, R 28 prot, 127-kDa prot, hyaluronic acid	Fn, Fbgn, C4B binding prot, thrombin, Vn, CD44	IGP, TC, EXT, coated plastic, BC, sGlp, Biomat., phage-coated plastic, phage-displayed Vn, OC, sGlp, glass	51, 249, 260, 268, 269, 534, 655, 656, 657, 658, 659, 700, 701, 710, 722, 723, 724, 725, 1042, 1153, 1361, 1407, 1409, 1607, 1684, 1853, 1860, 1901, 1970, 1999, 2027, 2034, 2095, 2150, 2158, 2240, 2391, 2478, 2479, 2597, 2598, 2606, 2710, 2713, 2766, 2851, 3007, 3034, 3098, 3167, 3181, 3313, 3361, 3362, 3363, 3510, 3657, 3800, 3802, 3838, 3943, 3950, 4123
Streptococcus rattus	ORL		Saliva	IGLp	452, 2475, 3201

Organism	Source	Adhesin	Receptor	Substrata	References
Streptococcus salivarius	ORL		Saliva, amylase	Hxd, sGlp, BC, IGlp, amalgam, titanium, SHA, Coagg (*S. gordonii*)	465, 528, 665, 916, 3272, 3287, 3420, 3595, 3679
Streptococcus sanguinis	ORL			Glass	440, 1978, 3975
Streptococcus sanguis	ORL	Lipoprotein, 150-kDa (PAAP) glucan binding prot	Saliva, milk, LF, proline–rich prot, 170-kDa Glp, 23-kDa Glp, NeuNac($\alpha2\rightarrow3$)Gal, sulfated polysaccharide, sulfomucin, mucins, glucan, sIgA	Hxd, platelets, IGL, catheter, IGlp, ERT, Biomat., Coagg (*S. oralis, A. viscosus*), buccal cells, amalgam, titanium, dentine, SGlp, Goretex dental material, soft tissue patch, enamel, saliva–coated plastic, platelets, SHA, mucin–dependent aggregation, thrombin–activated platelets, PMN, heart tissue, medical devices, Biomat, dental alloy	9, 86, 192, 339, 358, 360, 467, 546, 665, 719, 877, 949, 957, 958, 1254, 1255, 1256, 1257, 1312, 1530, 1608, 1977, 2287, 2619, 2680, 2774, 2890, 2932, 3194, 3201, 3446, 3493, 3595, 3664, 3666, 3679, 3797, 3886, 4062
Streptococcus sobrinus	ORL	210-kDa prot, glucan-binding prot (16- to 145-kDa Glp)	$\alpha(1\rightarrow6)$Glucan, saliva, milk, dextran, amylase, proline-rich prot	IGlp, IGL, IC, Hxd, sGlp, Teflon, SHA, glass, Coagg (*S. mutans*), saliva-coated enamel disk, glass, tooth, soluble glucan, dextran-coated glass, SHA ± GTF, dentral restorative material	113, 296, 440, 663, 669, 1977, 2038, 2196, 2215, 2227, 2582, 2618, 3195, 3330, 3377, 3521, 3522, 3844, 3845, 3948, 4002, 4014, 4063, 4064
Streptococcus spp. (group G)		GfbA (64.9 kDa)	Fn	Coated plastic, TC	1935
Streptococcus suis	Porcine	18-kDa	Gal($\alpha1\rightarrow4$)Gal	ERT, IGL, TC, Mϕ	527, 1352, 1353, 1354, 2053, 3217, 3218, 3354, 3730, 4003
Streptococcus thermophilus	Milk			Hxd	3820
Streptococcus uberis	Bovine	LTA	ECM, Lm, Fn, Coll, glycosaminoglycans	TC, EXT	81, 82, 83, 84, 340, 830, 989, 3720
Streptococcus vestibularis	ORL			SHA	3679
Syntrophobacter wolinii	ENV			Glass	3847
Syntrophomonas wolfei	ENV			Glass	3842

(Continued)

TABLE 12.2 Studies of adhesion from 1992 to 2002: pathogens, adhesins, receptors, and target substrata used *(Continued)*

Bacterium	Clinical source	Adhesins	Receptor specificity	Substrata (target cells)	References
Treponema denticola	ORL	27-, 38-, 51-, 53-, and 72-kDa prot, flagella	NANA, Coll I, Coll IV, Coll V, Fn, hyaluronan, albumin, Lm, Fbgn, NAc-galactosamine	ERT, IGlp, palate epithelial cells, Coagg (*F. nucleatum*, *P. gingivalis*, streptococci), coated latex beads, coated plastic, tissue regeneration membrane	497, 742, 936, 1297, 1349, 1350, 1889, 1890, 1960, 2439, 3361, 3784, 3785, 3786, 4099
Treponema medium		37- and 95-kDa prot	Coll I, Coll IV, Coll V		3786, 3787
Treponema pallidum				EXT	497
Treponema pectinovorum	ORL	LPS		Coagg (*F. nucleatum*), TC	1960, 3938
Treponema phagedenis		51- and 53-kDa protein	Coll, Lm	IGlp	3785
Treponema scoliodontum				TC	497
Treponema socranskii	ORL	51-, 53-, 95-, and 110-kDa prot	Coll I, IV, V, Lm	Coagg (*F. nucleatum*), TC, coated plastic	497, 1960, 3785, 3786
Treponema vincentii	ORL	51- and 53-kDa prot	Coll, Lm	Coagg (*F. nucleatum*), IGlp	1960, 3785
Ureaplasma urealyticum			NANA	TC	3454
Veillonella atypica		45-kDa prot	Lactose	Coagg (streptococci)	1583
Vibrio alginolyticus	ENV			Glass, ERT, TC	1952, 2505, 4145
Vibrio cholerae	GI	53-kDa OMP, fim, LPS, 38-kDa OMP U. 20-kDa pilus prot, 47-kDa lectin, protease, 32- and 34-kDa Glp, 20-kDa OMP, MshA, pilus, PilD, 40-kDa prot	MR, Man, fucose, RGD, Coll, Lm, Fbgn, Fn, mucus, NAc-D-glucosamine, glucosamine, *N,N'*-deacetyl-chitobiose, Man-Glu	ERT, IC, IGlp, lymphocytes, TC, EXT, mucus-coated plastic, scraped intestinal cells, homogenized GI mucosa, M cells, organ in vivo, rabbit intestinal cells, autoagglutination, mouse intestine, glass abiotic surface	55, 159, 160, 259, 518, 556, 559, 713, 926, 927, 1017, 1085, 1114, 1318, 1472, 1669, 2026, 2475, 2536, 2537, 2580, 2589, 2630, 2746, 2772, 2924, 2962, 3245, 3246, 3247, 3248, 3365, 3367, 3431, 3503, 3639, 3640, 3659, 3680, 3968, 3969, 4081, 4088
Vibrio hollisae				TC	2440
Vibrio mimicus	GI	39- and 31-kDa OMP, LPS	Peptide of Glp	ERT, EXT, intestinal tissue	52, 53, 54, 56, 3780

Vibrio parahaemolyticus	GI, marine	Non-fim prot, OMP	Man	TC, ERT, intestinal cells, HC	707, 929, 2570, 2571, 2590
Vibrio vulnificus	ENV	Type IV fim, 45-kDa prot		TC, ERT	2462, 2792
Yersinia enterocolitica	GI	YadA, YopH	Fn, nidogen/entactin, Gal, GalNAc, Coll 1, Coll II, Coll IV, Lm, NAc-galactosamine, mucin, basal lamina	TC, Fn plates, PMN, intestinal mucin, sGlp, intestinal cells, IGlp, polycarbonate membrane	595, 696, 810, 891, 933, 1043, 1304, 2289, 2829, 2865, 3111, 3112, 3322, 3323, 3326, 3327, 3328, 3390, 3438, 3643, 3702, 3703, 3908, 3996, 4088, 4121
Yersinia pestis	URT, GI	pH6 antigen, fim, 36-kDa OMP	ECM, Coll, Lm, GM1A, GM2A, lactosylceramide	TC, ECM-coated plastic, IGL, IGlp	661, 1900, 2048, 2174, 2822, 3132
Yersinia pseudotuberculosis	GI	YadA, invasin (103 kDa), RGD-containing prot	β-Integrin, Fn	TC, platelets, IGlp, PMN, T lymphocytes, Mφ, ERT, suspended cells binding to immobilized bacteria	115, 332, 472, 634, 784, 810, 950, 2011, 2121, 2122, 2866, 3132, 3222, 3390, 3424, 3637, 3656, 3703, 4095
Yersinia ruckeri				TC	3115

REFERENCES

1. **Aathithan, S., R. Dybowski, and G. L. French.** 2001. Highly epidemic strains of methicillin-resistant *Staphylococcus aureus* not distinguished by capsule formation, protein A content or adherence to HEp-2 cells. *Eur. J. Clin. Microbiol. Infect. Dis.* **20**:27–32.
2. **Abdul-Wahab, O. M. S., G. Ross, and J. M. Bradbury.** 1995. Pathogenicity and cytadherence of *Mycoplasma imitans* in chicken and duck embryo tracheal organ cultures. *Infect. Immun.* **64**:563–568.
3. **Abe, A., B. Kenny, M. Stein, and B. B. Finlay.** 1997. Characterization of two virulence proteins secreted by rabbit enteropathogenic *Escherichia coli*, EspA and EspB, whose maximal expression is sensitive to host body temperature. *Infect. Immun.* **65**:3547–3555.
4. **Abe, C. M., S. Knutton, M. Z. Pedroso, E. Freymuller, and T. A. Gomes.** 2001. An enteroaggregative *Escherichia coli* strain of serotype O111:H12 damages and invades cultured T84 cells and human colonic mucosa. *FEMS Microbiol. Lett.* **203**:199–205.
5. **Abeck, D., A. P. Johnson, and H. Mensing.** 1992. Binding of *Haemophilus ducreyi* to extracellular matrix proteins. *Microb. Pathog.* **13**:81–84.
6. **Abiko, Y., K. Kaizuka, and Y. Hosogi.** 2001. Inhibition of hemolysis by antibody against the *Porphyromonas gingivalis* 130-kDa hemagglutinin domain. *J. Oral Sci.* **43**:159–163.
7. **Abiko, Y., N. Ogura, U. Matsuda, K. Yanagi, and H. Takiguchi.** 1997. A human monoclonal antibody which inhibits the coaggregation activity of *Porphyromonas gingivalis*. *Infect. Immun.* **65**:3966–3969.
8. **Abraham, S. N., M. Land, S. Ponniah, R. Endres, D. L. Hasty, and J. P. Babu.** 1992. Glycerol-induced unraveling of the tight helical conformation of *Escherichia coli* type 1 fimbriae. *J. Bacteriol.* **174**:5145–5148.
9. **Abramson, A. L., E. Gilberto, V. Mullooly, K. France, P. Alperstein, and H. D. Isenberg.** 1993. Microbial adherence to and disinfection of laryngoscopes used in office practice. *Laryngoscope* **103**:503–508.
10. **Abramson, T., H. Kedemm, and D. A. Relman.** 2001. Proinflammatory and proapoptotic activities associated with *Bordetella pertussis* filamentous hemagglutinin. *Infect. Immun.* **69**:2650–2658.
11. **Abu el-Asrar, A. M., A. M. Shibl, K. F. Tabbara, and S. A. al-Kharashi.** 1997. Heparin and heparin-surface-modification reduce *Staphylococcus epidermidis* adhesion to intraocular lenses. *Int. Ophthalmol.* **21**:71–74.
12. **Abu el-Asrar, A. M., A. A. Kadry, A. M. Shibl, S. A. al-Kharashi, and A. A. al-Mosallam.** 2000. Antibiotics in the irrigating solutions reduce *Staphylococcus epidermidis* adherence to intraocular lenses. *Eye* **14**:225–230.
13. **Abul-Milh, M., Y. Wu, B. Lau, C. A. Lingwood, and D. B. Foster.** 2001. Induction of epithelial cell death including apoptosis by enteropathogenic *Escherichia coli* expressing bundle-forming pili. *Infect. Immun.* **69**:7356–7364.
14. **Adam, E. C., B. S. Mitchell, D. U. Schumacher, G. Grant, and U. Schumacher.** 1997. *Pseudomonas aeruginosa* II lectin stops human ciliary beating: therapeutic implications of fucose. *Am. Respir. Crit. Care Med.* **155**:2102–2104.
15. **Adamou, J. E., T. M. Wizemann, P. Barren, and S. Langermann.** 1998. Adherence of *Streptococcus pneumoniae* to human bronchial epithelial cells (BEAS-2B). *Infect. Immun.* **66**:820–822.
16. **Adams, L. M., C. P. Simmons, L. Rezmann, R. A. Strugnell, and R. M. Robins-Browne.** 1997. Identification and characterization of a K88- and CS31A-like operon of a rabbit enteropathogenic *Escherichia coli* strain which encodes fimbriae involved in the colonization of rabbit intestine. *Infect. Immun.* **65**:5222–5230.
17. **Adlerberth, I., L. A. Hanson, C. Svanborg, A.-M. Svennerholm, S. Nordgren, and A. E. Wold.** 1995. Adhesins of *Escherichia coli* associated with extra-intestinal pathogenicity confer binding to colonic epithelial cells. *Microb. Pathog.* **18**:373–385.
18. **Adlerberth, I., S. Ahrne, M. L. Johansson, G. Molin, L. A. Hanson, and A. E. Wold.** 1996. A mannose-specific adherence mechanism in *Lactobacillus plantarum* conferring binding to the human colonic cell line HT-29. *Appl. Environ. Microbiol.* **62**:2244–2251.
19. **Adlerberth, I., C. Svanborg, B. Carlsson, L. Mellander, L. A. Hanson, F. Jalil, K. Khalil, and A. E. Wold.** 1998. P fimbriae and other adhesins enhance intestinal persistence of *Escherichia coli* in early infancy. *Epidemiol. Infect.* **121**:599–608.
20. **Adu-Bobie, J., G. Frankel, C. Bain, A. G. Goncalves, L. R. Trabulsi, G. Douce, S. Knutton, and G. Dougan.** 1988. Detection of intimins alpha, beta, gamma, and delta, four intimin derivatives expressed by attaching and effacing microbial pathogens. *J. Clin. Microbiol.* **36**:662–668.

21. **Adu-Bobie, J., L. R. Trabulsi, M. M. S. Carneiro-Sampaio, G. Dougan, and G. Frankel.** 1998. Identification of immunodominant regions within the C-terminal cell binding domain of intimin α and intimin β from enteropathogenic *Escherichia coli. Infect. Immun.* **66:**5643–5649.

22. **Aebi, C., E. R. Lafontaine, L. D. Cope, J. L. Latimer, S. L. Lumbley, G. H. McCracken, Jr., and E. J. Hansen.** 1998. Phenotypic effect of isogenic *uspA1* and *uspA2* mutations on *Moraxella catarrhalis* 035E. *Infect. Immun.* **66:**3113–3119.

23. **Agace, W. W., S. R. Hedges, M. Ceska, and C. Svanborg.** 1993. Interleukin-8 and the neutrophil response to mucosal gram-negative infection. *J. Clin. Investig.* **92:**780–785.

24. **Agace, W., S. Hedges, U. Andersson, J. Andersson, M. Ceska, and C. Svanborg.** 1993. Selective cytokine production by epithelial cells following exposure to *Escherichia coli. Infect. Immun.* **61:**602–609.

25. **Agin, T. S., and M. K. Wolf.** 1997. Identification of a family of intimins common to *Escherichia coli* causing attaching-effacing lesions in rabbits, humans, and swine. *Infect. Immun.* **65:**320–326.

26. **Agnani, G., S. Tricot-Doleux, L. Du, and M. Bonnaure-Mallet.** 2000. Adherence of *Porphyromonas gingivalis* to gingival epithelial cells: modulation of bacterial protein expression. *Oral Microbiol. Immunol.* **15:**48–52.

27. **Aguilar, B., M. Iturralde, R. Baselga, and B. Amorena.** 1992. An efficient microtest to study adherence of bacteria to mammalian cells. *FEMS Microbiol. Lett.* **69:**161–164.

28. **Aguilar, B., B. Amorena, and M. Iturralde.** 2001. Effect of slime on adherence of *Staphylococcus aureus* isolated from bovine and ovine mastitis. *Vet. Microbiol.* **78:**183–191.

29. **Aguilar, B., and M. Iturralde.** 2001. Binding of a surface protein of *Staphylococcus aureus* to cultured ovine mammary gland epithelial cells. *Vet. Microbiol.* **82:**165–175.

30. **Ahanotu, E. N., M. D. Hyatt, M. J. Graham, and D. G. Ahearn.** 2001. Comparative radiolabel and ATP analyses of adhesion of *Pseudomonas aeruginosa* and *Staphylococcus epidermidis* to hydrogel lenses. *CLAO J.* **27:**89–93.

31. **Ahanotu, E. N., J. H. Stone, S. K. McAllister, J. M. Miller, and D. G. Ahearn.** 2001. Vancomycin resistance among strains of *Staphylococcus epidermidis:* effects on adherence to silicone. *Curr. Microbiol.* **43:**124–128.

32. **Ahearn, D. G., D. T. Grace, M. J. Jennings, R. N. Borazjani, K. J. Boles, L. J. Rose, R. B. Simmons, and E. N. Ahanotu.** 2000. Effects of hydrogel/silver coatings on *in vitro* adhesion to catheters of bacteria associated with urinary tract infections. *Curr. Microbiol.* **41:**120–125.

33. **Ahimou, F., M. Paquot, P. Jacques, P. Thonart, and P. G. Rouxhet.** 2001. Influence of electrical properties on the evaluation of the surface hydrophobicity of *Bacillus subtilis. J. Microbiol Methods* **45:**119–126.

34. **Ahmed, K., T. Nakagawa, Y. Nakano, G. Martinez, A. Ichinose, C. H. Zheng, M. Akaki, M. Aikawa, and T. Nagatake.** 2000. Attachment of *Moraxella catarrhalis* occurs to the positively charged domains of pharyngeal epithelial cells. *Microb. Pathog.* **28:**203–209.

35. **Ahmed, K.** 1992. Fimbriae of *Branhamella catarrhalis* as possible mediators of adherence to pharyngeal epithelial cells. *APMIS* **100:**1066–1072.

36. **Ahmed, K., N. Rikitomi, and K. Matsumoto.** 1992. Fimbriation, hemagglutination and adherence properties of fresh clinical isolates of *Branhamella catarrhalis. Microbiol. Immunol.* **36:**1009–1017.

37. **Ahmed, K., N. Rikitomi, T. Nagatake, and K. Matsumoto.** 1992. Ultrastructural study on the adherence of *Branhamella catarrhalis* to oropharyngeal epithelial cell. *Microbiol. Immunol.* **36:**563–573.

38. **Ahmed, K., H. Masaki, T. C. Dai, A. Ichinose, Y. Utsunomiya, M. Tao, T. Nagatake, and K. Matsumoto.** 1994. Expression of fimbriae and host response in *Branhamella catarrhalis* respiratory infections. *Microbiol. Immunol.* **38:**767–771.

39. **Ahmed, K., K. Matsumoto, N. Rikitomi, and T. Nagatake.** 1996. Attachment of *Moraxella catarrhalis* to pharyngeal epithelial cells is mediated by a glycosphingolipid receptor. *FEMS Microbiol. Lett.* **135:**305–309.

40. **Ahmed, K., H. D. Enciso, H. Masaki, M. Tao, A. Omori, P. Tharavichikul, and T. Nagatake.** 1999. Attachment of *Burkholderia pseudomallei* to pharyngeal epithelial cells: a highly pathogenic bacteria with low attachment ability. *Am. J. Trop. Med. Hyg.* **60:**90–93.

41. **Ahmed, S., S. Meghji, R. J. Williams, B. Henderson, J. H. Brock, and S. P. Nair.** 2001. *Staphylococcus aureus* fibronectin binding proteins are essential for internalization by osteoblasts but do not account for differences in intracellular levels of bacteria. *Infect. Immun.* **69:**2872–2877.

42. **Ahrne, S., S. Nobaek, B. Jeppsson, I. Adlerberth, A. E. Wold, and G. Molin.** 1998. The normal *Lactobacillus* flora of healthy human rectal and oral mucosa. *J. Appl. Microbiol.* **85**:88–94.

43. **Ahuja, S., B. Kaack, and J. Roberts.** 1998. Loss of fimbrial adhesion with the addition of *Vaccinum macrocarpon* to the growth medium of P-fimbriated *Escherichia coli. J. Urol.* **159**:559–562.

44. **Aihara, M., D. Tsuchimoto, H. Takizawa, A. Azuma, H. Wakebe, Y. Ohmoto, K. Imagawa, M. Kikuchi, N. Mukaida, and K. Matsuchima.** 1997. Mechanisms involved in *Helicobacter pylori*-induced interleukin-8 production by a gastric cancer cell line, MKN45. *Infect. Immun.* **65**:3218–3224.

45. **Ainsworth, A. J.** 1993. Carbohydrate and lectin interactions with *Edwardsiella ictaluri* and channel catfish, *Ictalurus punctatus* (Rafinesque), anterior kidney leucocytes and hepatocytes. *J. Fish Dis.* **16**:449–459.

46. **Akaza, H., A. Iwasaki, M. Ohtani, N. Ikeda, K. Niijima, I. Toida, and K. Koiso.** 1993. Expression of antitumor response. Role of attachment and viability of bacillus Calmette-Guerin to bladder cancer cells. *Cancer* **72**:558–563.

47. **Akiyama, H., R. Torigoe, and J. Arata.** 1993. Interaction of *Staphylococcus aureus* cells and silk threads in vitro and in mouse skin. *J. Dermatol. Sci.* **6**:247–257.

48. **Akiyama, H., O. Yamasaki, H. Kanzaki, J. Tada, and J. Arata.** 1998. Effects of zinc oxide on the attachment of *Staphylococcus aureus* strains. *J. Dermatol. Sci.* **17**:67–74.

49. **Akiyama, H., O. Yamasaki, H. Kanzaki, J. Tada, and J. Arata.** 1998. Effects of various salts and irradiation with UV light on the attachment of *Staphylococcus aureus* strains. *J. Dermatol. Sci.* **16**:216–225.

50. **Akiyama, H., O. Yamasaki, H. Kanzaki, J. Tada, and J. Arata.** 1998. Adherence characteristics of *Staphylococcus aureus* and coagulase-negative staphylococci isolated from various skin lesions. *J. Dermatol. Sci.* **18**:132–136.

51. **Akiyama, H., O. Yamasaki, J. Tada, and J. Arata.** 1999. Characteristics in adherence of streptococci and *Staphylococcus aureus* isolated from various infective skin lesions: serum IgA decreases adherence of *Streptococcus pyogenes* but not *Staphylococcus aureus. J. Dermatol. Sci.* **21**:165–169.

52. **Alam, M., S. Miyoshi, I. Maruo, C. Ogawa, and S. Shinoda.** 1994. Existence of a novel hemagglutinin having no protease activity in *Vibrio mimicus. Microbiol. Immunol.* **38**:467–470.

53. **Alam, M., S. Miyoshi, S. Yamamoto, K. Tomochika, and S. Shinoda.** 1996. Expression of virulence-related properties by, and intestinal adhesiveness of, *Vibrio mimicus* strains isolated from aquatic environments. *Appl. Environ. Microbiol.* **62**:3871–3874.

54. **Alam, M., S.-I. Miyoshi, K.-I. Tomochika, and S. Shinoda.** 1996. Purification and characterization of novel hemagglutinins from *Vibrio mimicus*: a 39-kilodalton major outer membrane protein and lipopolysaccharide. *Infect. Immun.* **64**:4035–4041.

55. **Alam, M., S. Miyoshi, K. Tomochika, and S. Shinoda.** 1997. Hemagglutination is a novel biological function of lipopolysaccharide (LPS), as seen with the *Vibrio cholerae* O139 LPS. *Clin. Diagn. Lab. Immunol.* **4**:604–606.

56. **Alam, M., S.-I. Miyoshi, K.-I. Tomochika, and S. Shinoda.** 1997. *Vibrio mimicus* attaches to the intestinal mucosa by outer membrane hemagglutinins specific to polypeptide moieties of glycoproteins. *Infect. Immun.* **65**:3662–3665.

57. **Al-Bahry, S. N., and T. G. Pistole.** 1997. Adherence of *Salmonella typhimurium* to murine peritoneal macrophages is mediated by lipopolysaccharide and complement receptors. *Zentbl. Bakteriol.* **286**:83–92.

58. **Albanese, C. T., M. Cardona, S. D. Smith, S. Watkins, A. G. Kurkchubasche, I. Ulman, R. L. Simmons, and M. I. Rowe.** 1994. Role of intestinal mucus in transepithelial passage of bacteria across the intact ileum *in vitro. Surgery* **116**:76–82.

59. **Albanyan, E. A., J. G. Vallejo, C. W. Smith, and M. S. Edwards.** 2000. Nonopsonic binding of type III group B streptococci to human neutrophils induces interleukin-8 release mediated by the p38 mitogen-activated protein kinase pathway. *Infect. Immun.* **68**:2053–2060.

60. **Albert, M. J., S. M. Faruque, M. Ansaruzzaman, M. M. Islam, K. Haider, K. Alam, I. Kabir, and R. Robins-Browne.** 1992. Sharing of virulence-associated properties at the phenotypic and genetic levels between enteropathogenic *Escherichia coli* and *Hafnia alvei. J. Med. Microbiol.* **37**:310–314.

61. **Alderete, J. F., R. Arroyo, and M. W. Lehker.** 1994. Identification of fibronectin as a receptor for bacterial cytoadherence. *Methods Enzymol.* **236**:318–333.

62. **Aleljung, P., W. Shen, B. Rozalska, U. Hellman, Å. Ljungh, and T. Wadström.**

1994. Purification of collagen binding proteins of *Lactobacillus reuteri* NCIB 11951. *Curr. Microbiol.* **28**:231–236.

63. **Alfa, M. J., P. DeGagne, and T. Hollyer.** 1993. *Haemophilus ducreyi* adheres to but does not invade cultured human foreskin cells. *Infect. Immun.* **61**:1735–1742.

64. **Alfa, M. J., and P. DeGagne.** 1994. A quantitative chemiluminescent ribosomal probe method for monitoring adherence of *Haemophilus ducreyi* to eukaryotic cells. *Microb. Pathog.* **17**:167–174.

65. **Alfa, M. J., and P. DeGagne.** 1997. Attachment of *Haemophilus ducreyi* to human foreskin fibroblasts involves LOS and fibronectin. *Microb. Pathog.* **22**:39–46.

66. **Al-Haddawi, M. H., S. Jasni, M. Zamri-Saad, A. R. Mutalib, I. Zulkifli, R. Son, and A. R. Sheikh-Omar.** 2000. *In vitro* study of *Pasteurella multocida* adhesion to trachea, lung and aorta of rabbits. *Vet. J.* **159**:274–281.

67. **Allen-Vercoe, E., M. Dibb-Fuller, C. J. Thorns, and M. J. Woodward.** 1997. SEF17 fimbriae are essential for the convoluted colonial morphology of *Salmonella enteritidis*. *FEMS Microbiol. Lett.* **153**:33–42.

68. **Allen-Vercoe, E., R. Collighan, and M. J. Woodward.** 1998. The variant *rpoS* allele of *S. enteritidis* strain 27655R does not affect virulence in a chick model nor constitutive curliation but does generate a cold-sensitive phenotype. *FEMS Microbiol. Lett.* **167**:245–253.

69. **Allen-Vercoe, E., and M. J. Woodward.** 1999. Colonisation of the chicken caecum by afimbriate and aflagellate derivatives of *Salmonella enterica* serotype *Enteritidis*. *Vet. Microbiol.* **69**:265–275.

70. **Allen-Vercoe, E., and M. J. Woodward.** 1999. The role of flagella, but not fimbriae, in the adherence of *Salmonella enterica* serotype *Enteritidis* to chick gut explant. *J. Med. Microbiol.* **48**:771–780.

71. **Allignet, J., J. O. Galdbart, A. Morvan, K. G. Dyke, P. Vaudaux, S. Aubert, N. Desplaces, and N. el Solh.** 1999. Tracking adhesion factors in *Staphylococcus caprae* strains responsible for human bone infections following implantation of orthopaedic material. *Microbiology* **145**:2033–2042.

72. **Allison, C., N. Coleman, P. L. Jones, and C. Hughes.** 1992. Ability of *Proteus mirabilis* to invade human urothelial cells is coupled to motility and swarming differentiation. *Infect. Immun.* **60**:4740–4746.

73. **Allison, D. G., M. A. Cronin, J. Hawker,** and **S. Freeman.** 2000. Influence of cranberry juice on attachment of *Escherichia coli* to glass. *J. Basic Microbiol.* **40**:3–6.

74. **Alm, R. A., and J. S. Mattick.** 1995. Identification of a gene, *pilV*, required for type 4 fimbrial biogenesis in *Pseudomonas aeruginosa*, whose product possesses a pre-pilin-like leader sequence. *Mol. Microbiol.* **16**:485–496.

75. **Alm, R. A., A. J. Bodero, P. D. Free, and J. S. Mattick.** 1996. Identification of a novel gene, *pilZ*, essential for type 4 fimbrial biogenesis in *Pseudomonas aeruginosa*. *J. Bacteriol.* **178**:46–53.

76. **Alm, R. A., and J. S. Mattick.** 1996. Identification of two genes with prepilin-like leader sequences involved in type 4 fimbrial biogenesis in *Pseudomonas aeruginosa*. *J. Bacteriol.* **178**:3809–3817.

77. **Alm, R. A., J. P. Hallinan, A. A. Watson, and J. S. Mattick.** 1996. Fimbrial biogenesis genes of *Pseudomonas aeruginosa*: *pilW* and *pilX* increase the similarity of type 4 fimbriae to the GSP protein-secretion systems and *pilY1* encodes a gonococcal PilC homologue. *Mol. Microbiol.* **22**:161–173.

78. **Alm, R. A., and J. S. Mattick.** 1997. Genes involved in the biogenesis and function of type-4 fimbriae in *Pseudomonas aeruginosa*. *Gene* **192**:89–98.

79. **al-Makhlafi, H., J. McGuire, and M. Daeschel.** 1994. Influence of preadsorbed milk proteins on adhesion of *Listeria monocytogenes* to hydrophobic and hydrophilic silica surfaces. *Appl. Environ. Microbiol.* **60**:3560–3565.

80. **al-Makhlafi, H., A. Nasir, J. McGuire, and M. Daeschel.** 1995. Adhesion of *Listeria monocytogenes* to silica surfaces after sequential and competitive adsorption of bovine serum albumin and beta-lactoglobulin. *Appl. Environ. Microbiol.* **61**:2013–2015.

81. **Almeida, R. A., D. A. Luther, S. J. Kumar, L. F. Calvinho, M. S. Bronze, and S. P. Oliver.** 1996. Adherence of *Streptococcus uberis* to bovine mammary epithelial cells and to extracellular matrix proteins. *Zentbl. Veterinaermed. Reihe B* **43**:385–392.

82. **Almeida, R. A., D. A. Luther, and S. P. Oliver.** 1999. Incubation of *Streptococcus uberis* with extracellular matrix proteins enhances adherence to and internalization into bovine mammary epithelial cells. *FEMS Microbiol. Lett.* **178**:81–85.

83. **Almeida, R. A., W. Fang, and S. P. Oliver.** 1999. Adherence and internalization of *Streptococcus uberis* to bovine mammary

epithelial cells are mediated by host cell proteoglycans. *FEMS Microbiol. Lett.* **177:**
313–317.

84. **Almeida, R. A., and S. P. Oliver.** 2001.
Role of collagen in adherence of *Streptococcus uberis* to bovine mammary epithelial cells. *J. Vet. Med. Ser. B* **48:**759–763.

85. **Almeida, R. A., and S. P. Oliver.** 2001.
Interaction of coagulase-negative *Staphylococcus* species with bovine mammary epithelial cells. *Microb. Pathog.* **31:**205–212.

86. **Alnor, D., N. Frimodt-Møller, F. Espersen, and W. Frederiksen.** 1994.
Infections with the unusual human pathogens *Agrobacterium* species and *Ochrobactrum anthropi*. *Clin. Infect. Dis.* **18:**914–920.

87. **Alston, W. K., D. A. Elliott, M. E. Epstein, V. B. Hatcher, M. Tang, and F. D. Lowy.** 1997. Extracellular matrix heparan sulfate modulates endothelial cell susceptibility to *Staphylococcus aureus*. *J. Cell. Physiol.* **173:**102–109.

88. **Altmeyer, R. M., J. K. McNern, J. C. Bossio, I. Rosenshine, B. B. Finlay, and J. E. Galán.** 1993. Cloning and molecular characterization of a gene involved in *Salmonella* adherence and invasion of cultured epithelial cells. *Mol. Microbiol.* **7:**89–98.

89. **Alugupalli, K. R., S. Kalfas, S. Edwardsson, A. Forsgren, R. R. Arnold, and A. S. Naidu.** 1994. Effect of lactoferrin on interaction of *Prevotella intermedia* with plasma and subepithelial matrix proteins. *Oral Microbiol. Immunol.* **9:**174–179.

90. **Alugupalli, K. R., and S. Kalfas.** 1995.
Inhibitory effect of lactoferrin on the adhesion of *Actinobacillus actinomycetemcomitans* and *Prevotella intermedia* to fibroblasts and epithelial cells. *APMIS* **103:**154–160.

91. **Alugupalli, K. R., S. Kalfas, and A. Forsgren.** 1996. Laminin binding to a heat-modifiable outer membrane protein of *Actinobacillus actinomycetemcomitans*. *Oral Microbiol. Immunol.* **11:**326–331.

92. **Alugupalli, K. R., and S. Kalfas.** 1997.
Characterization of the lactoferrin-dependent inhibition of the adhesion of *Actinobacillus actinomycetemcomitans, Prevotella intermedia* and *Prevotella nigrescens* to fibroblasts and to a reconstituted basement membrane. *APMIS* **105:**680–688.

93. **Alvarez-Dominguez, C., J. A. Vazquez-Boland, E. Carrasco-Marin, P. Lopez-Mato, and F. Leyva-Cobian.** 1997. Host cell heparan sulfate proteoglycans mediate attachment and entry of *Listeria monocytogenes,* and the listerial surface protein ActA is involved in heparan sulfate receptor recognition. *Infect. Immun.* **65:**78–88.

94. **Alverdy, J. C., and E. Aoys.** 1992. The effect of dexamethasone and endotoxin administration on biliary IgA and bacterial adherence. *J. Surg. Res.* **53:**450–454.

95. **Alverdy, J. C., J. Spitz, G. Hecht, and S. Ghandi.** 1994. Causes and consequences of bacterial adherence to mucosal epithelia during critical illness. *New Horiz.* **2:**264–272.

96. **Alverdy, J. C., B. Hendrickson, S. S. Guandalini, R. J. Laughlin, K. Kent, and R. Banerjee.** 1999. Perturbed bioelectrical properties of the mouse cecum following hepatectomy and starvation: the role of bacterial adherence. *Shock* **12:**235–241.

97. **Alves, A. M., M. O. Lasaro, D. F. Almeida, and L. C. Ferreira.** 1998.
Epitope specificity of antibodies raised against enterotoxigenic *Escherichia coli* CFA/I fimbriae in mice immunized with naked DNA. *Vaccine* **16:**9–15.

98. **Amano, A., A. Sharma, H. T. Sojar, H. K. Kuramitsu, and R. J. Genco.** 1994.
Effects of temperature stress on expression of fimbriae and superoxide dismutase by *Porphyromonas gingivalis*. *Infect. Immun.* **62:**4682–4685.

99. **Amano, A., H. T. Sojar, J.-Y. Lee, A. Harma, M. J. Levine, and R. J. Genco.** 1994. Salivary receptors for recombinant fimbrillin of *Porphyromonas gingivalis*. *Infect. Immun.* **62:**3372–3380.

100. **Amano, A., A. Sharma, J. Y. Lee, H. T. Sojar, P. A. Raj, and R. J. Genco.** 1996.
Structural domains of *Porphyromonas gingivalis* recombinant fimbrillin that mediate binding to salivary proline-rich protein and statherin. *Infect. Immun.* **64:**1631–1637.

101. **Amano, A., T. Fujiwara, H. Nagata, M. Kuboniwa, A. Sharma, H. T. Sojar, R. J. Genco, S. Hamada, and S. Shizukuishi.** 1997. *Porphyromonas gingivalis* fimbriae mediate coaggregation with *Streptococcus oralis* through specific domains. *J. Dent. Res.* **76:**852–857.

102. **Amano, A., S. Shizukuishi, H. Horie, S. Kimura, I. Morisaki, and S. Hamada.** 1998. Binding of *Porphyromonas gingivalis* fimbriae to proline-rich glycoproteins in parotid saliva via a domain shared by major salivary components. *Infect. Immun.* **66:**2072–2077.

103. **Amano, A., T. Nakamura, S. Kimura, I. Morisaki, I. Nakagawa, S. Kawabata, and S. Hamada.** 1999. Molecular interactions of *Porphyromonas gingivalis* fimbriae with host proteins: kinetic analyses based on surface plasmon resonance. *Infect. Immun.* **67:**2399–2405.

104. **Amano, A., T. Premaraj, M. Kuboniwa, I. Nakagawa, S. Shizukuishi, I. Morisaki, and S. Hamada.** 2001. Altered antigenicity in periodontitis patients and decreased adhesion of *Porphyromonas gingivalis* by environmental temperature stress. *Oral Microbiol. Immunol.* **16:**124–128.

105. **Amara, A., S. E. Had, T. Jirrari, and K. Bouzoubaa.** 1996. Lethality, hemagglutination and adhesion of *Escherichia coli* strains (serotype 01) isolated in Morocco from chickens with colibacillosis. *Avian Dis.* **40:**540–545.

106. **Ambrozic, J., A. Ostroversnik, M. Starcic, I. Kuhar, M. Grabnar, and D. Zgur-Bertok.** 1998. *Escherichia coli* ColV plasmid pRK100: genetic organization, stability and conjugal transfer. *Microbiology* **144:**343–352.

107. **Amsbaugh, D. F., Z. M. Li, and R. D. Shahin.** 1993. Long-lived respiratory immune response to filamentous hemagglutinin following *Bordetella pertussis* infection. *Infect. Immun.* **61:**1447–1452.

108. **An, Y. H., and R. J. Friedman.** 1996. Prevention of sepsis in total joint arthroplasty. *J. Hosp. Infect.* **33:**93–108.

109. **Anantha, R. P., K. D. Stone, and M. S. Donnenberg.** 1998. Role of BfpF, a member of the PilT family of putative nucleotide-binding proteins, in type IV pilus biogenesis and in interactions between enteropathogenic *Escherichia coli* and host cells. *Infect. Immun.* **66:**122–131.

110. **Anantha, R. P., K. D. Stone, and M. S. Donnenberg.** 2000. Effects of *bfp* mutations on biogenesis of functional enteropathogenic *Escherichia coli* type IV pili. *J. Bacteriol.* **182:**2498–2506.

111. **Andersen, R. N., N. Ganeshkumar, and P. E. Kolenbrander.** 1993. Cloning of the *Streptococcus gordonii* PK488 gene, encoding an adhesin which mediates coaggregation with *Actinomyces naeslundii* PK606. *Infect. Immun.* **61:**981–987.

112. **Andersen, R. N., N. Ganeshkumar, and P. E. Kolenbrander.** 1998. *Helicobacter pylori* adheres selectively to *Fusobacterium* spp. *Oral Microbiol. Immunol.* **13:**51–54.

113. **Anderson, L. C., S. C. Yang, H. Xie, and R. J. Lamont.** 1994. The effects of streptozotocin diabetes on salivary-mediated bacterial aggregation and adherence. *Arch. Oral Biol.* **39:**261–269.

114. **Andersson, A., P. E. Granum, and U. Ronner.** 1998. The adhesion of *Bacillus cereus* spores to epithelial cells might be an additional virulence mechanism. *Int. J. Food Microbiol.* **39:**93–99.

115. **Andersson, K., K. E. Magnusson, M. Majeed, O. Stendahl, and M. Fallman.** 1999. *Yersinia pseudotuberculosis*-induced calcium signaling in neutrophils is blocked by the virulence effector YopH. *Infect. Immun.* **67:**2567–2574.

116. **Andreev, J., Z. Borovsky, I. Rosenshine, and S. Rottem.** 1995. Invasion of HeLa cells by *Mycoplasma penetrans* and the induction of tyrosine phosphorylation of a 145-kDa host cell protein. *FEMS Microbiol. Lett.* **132:**189–194.

117. **Andreu, A., A. E. Stapleton, C. L. Fennell, S. L. Hillier, and W. E. Stamm.** 1995. Hemagglutination, adherence, and surface properties of vaginal *Lactobacillus* species. *J. Infect. Dis.* **171:**1237–1243.

118. **Andreu, A., A. E. Stapleton, C. Fennell, H. A. Lockman, M. Xercavins, F. Fernandez, and W. E. Stamm.** 1997. Urovirulence determinants in *Escherichia coli* strains causing prostatitis. *J. Infect. Dis.* **176:**464–469.

119. **Andrews, C. S., S. P. Denyer, B. Hall, G. W. Hanlon, and A. W. Lloyd.** 2001. A comparison of the use of an ATP-based bioluminescent assay and image analysis for the assessment of bacterial adhesion to standard HEMA and biomimetic soft contact lenses. *Biomaterials* **22:**3225–3233.

120. **Anglen, J. O., S. Apostoles, G. Christensen, and B. Gainor.** 1994. The efficacy of various irrigation solutions in removing slime-producing *Staphylococcus*. *J. Orthopaed. Trauma* **8:**390–396.

121. **Anglen, J., P. S. Apostoles, G. Christensen, B. Gainor, and J. Lane.** 1996. Removal of surface bacteria by irrigation. *J. Orthopaed. Res.* **14:**251–254.

122. **Ångstrom, J., S. Tenebeg, M. A. Milh, T. Larsson, I. Leonardsson, B.-M. Olsson, M. Ö. Halvarsson, D. Danielsson, I. Näslund, Å. Ljungh, T. Wadström, and K.-A. Karlsson.** 1998. The lactosylceramide binding specificity of *Helicobacter pylori*. *Glycobiology* **8:**297–309.

123. **Ansorg, R., and E. N. Schmid.** 1998. Adhesion of *Helicobacter pylori* to yeast cells. *Zentbl. Bakteriol.* **288:**501–508.

124. **Apodaca, G., M. Bomsel, R. Lindstedt, J. Engel, D. Frank, K. E. Mostov, and J. Wiener-Kronish.** 1995. Characterization of *Pseudomonas aeruginosa*-induced MDCK cell injury: glycosylation-defective host cells are resistant to bacterial killing. *Infect. Immun.* **63:**1541–1551.

125. **Appelgren, P., U. Ransjo, L. Bindslev, F.**

Espersen, and O. Larm. 1996. Surface heparinization of central venous catheters reduces microbial colonization *in vitro* and *in vivo:* results from a prospective, randomized trial. *Crit. Care Med.* **24:**1482–1489.

126. Appelmelk, B. J., B. Shiberu, C. Trinks, N. Tapsi, P. Y. Zheng, T. Verboom, J. Maaskant, C. H. Hokke, W. E. C. M. Schiphorst, D. Blanchard, I. M. Simoons-Smit, D. H. van den Eijnden, and C. M. J. E. Vandenbroucke-Grauls. 1998. Phase variation in *Helicobacter pylori* lipopolysaccharide. *Infect. Immun.* **66:**70–76.

127. Arciola, C. R., L. Radin, P. Alvergna, E. Cenni, and A. Pizzoferrato. 1993. Heparin surface treatment of poly(methylmethacrylate) alters adhesion of a *Staphylococcus aureus* strain: utility of bacterial fatty acid analysis. *Biomaterials* **14:**1161–1164.

128. Arciola, C. R., R. Caramazza, and A. Pizzoferrato. 1994. *In vitro* adhesion of *Staphylococcus epidermidis* on heparin-surface-modified intraocular lenses. *J. Cataract Refract. Surg.* **20:**158–161.

129. Arciola, C. R., M. C. Maltarello, E. Cenni, and A. Pizzoferrato. 1995. Disposable contact lenses and bacterial adhesion. *In vitro* comparison between ionic/high-water-content and non-ionic/low-water-content lenses. *Biomaterials* **16:**685–690.

130. Arciola, C. R., L. Montanaro, R. Caramazza, V. Sassoli, and D. Cavedagna. 1998. Inhibition of bacterial adherence to a high-water-content polymer by a water-soluble, nonsteroidal, anti-inflammatory drug. *J. Biomed. Mater. Res.* **42:**1–5.

131. Arciola, C. R., L. Montanaro, A. Moroni, M. Giordano, A. Pizzoferrato, and M. E. Donati. 1999. Hydroxyapatite-coated orthopaedic screws as infection resistant materials: *in vitro* study. *Biomaterials* **20:**323–327.

132. Arciola, C. R., M. E. Donati, and L. Montanaro. 2001. Adhesion to a polymeric biomaterial affects the antibiotic resistance of *Staphylococcus epidermidis. New Microbiol.* **24:**63–68.

133. Ardehali, R., and S. F. Mohammad. 1993. [111]Indium labeling of microorganisms to facilitate the investigation of bacterial adhesion. *J. Biomed. Mater. Res.* **27:**269–275.

134. Arico, B., S. Nuti, V. Scarlato, and R. Rappuoli. 1993. Adhesion of *Bordetella pertussis* to eukaryotic cells requires a time-dependent export and maturation of filamentous hemagglutinin. *Proc. Natl. Acad. Sci. USA* **90:**9204–9208.

135. Arizono, T., M. Oga, and Y. Sugioka. 1992. Increased resistance of bacteria after adherence to polymethyl methacrylate. An *in vitro* study. *Acta Orthopaed. Scand.* **63:**661–664.

136. Armstrong, E. A., B. Ziola, B. F. Habbick, and K. Komiyama. 1993. Role of cations and IgA in saliva-mediated aggregation of *Pseudomonas aeruginosa* in cystic fibrosis patients. *J. Oral Pathol. Med.* **22:**207–213.

137. Arne, P., D. Marc, A. Bree, C. Schouler, and M. Dho-Moulin. 2000. Increased tracheal colonization in chickens without impairing pathogenic properties of avian pathogenic *Escherichia coli* MT78 with a *fimH* deletion. *Avian Dis.* **44:**343–355.

138. Arnold, A. J., D. Sunderland, A. M. Rickwood, and C. A. Hart. 1993. Bacterial factors in the formation of renal scars. An experimental study on the role of *Escherichia coli* P-fimbriation and hydrophobicity. *Br. J. Urol.* **72:**549–553.

139. Arnold, R., J. Scheffer, B. Konig, and W. Konig. 1993. Effects of *Listeria monocytogenes* and *Yersinia enterocolitica* on cytokine gene expression and release from human polymorphonuclear granulocytes and epithelial (HEp-2) cells. *Infect. Immun.* **61:**2545–2552.

140. Arnold, R., and W. Konig. 1998. Interleukin-8 release from human neutrophils after phagocytosis of *Listeria monocytogenes* and *Yersinia enterocolitica. J. Med. Microbiol.* **47:**55–62.

141. Arock, M., E. Ross, R. Lai-Kuen, G. Averlant, Z. Gao, and S. N. Abraham. 1998. Phagocytic and tumor necrosis factor alpha response of human mast cells following exposure to gram-negative and gram-positive bacteria. *Infect. Immun.* **66:**6030–6034.

142. Arora, S. K., B. W. Ritchings, E. C. Almira, S. Lory, and R. Ramphal. 1996. Cloning and characterization of *Pseudomonas aeruginosa fliF,* necessary for flagellar assembly and bacterial adherence to mucin. *Infect. Immun.* **64:**2130–2136.

143. Arora, S. K., B. W. Ritchings, E. C. Almira, S. Lory, and R. Ramphal. 1997. A transcriptional activator, FleQ, regulates mucin adhesion and flagellar gene expression in *Pseudomonas aeruginosa* in a cascade manner. *J. Bacteriol.* **179:**5574–5581.

144. Arora, S. K., B. W. Ritchings, E. C. Almira, S. Lory, and R. Ramphal. 1998. The *Pseudomonas aeruginosa* flagellar cap protein, FliD, is responsible for mucin adhesion. *Infect. Immun.* **66:**1000–1007.

145. Arp, L. H., E. L. Huffman, and D. H. Hellwig. 1993. Glycoconjugates as compo-

nents of receptors for *Bordetella avium* on the tracheal mucosa of turkeys. *Am. J. Vet. Res.* **54:**2027–2030.

146. **Arva, E., U. Dahlgren, R. Lock, and B. Andersson.** 1996. Antibody response in bronchoalveolar lavage and serum of rats after aerosol immunization of the airways with a well-adhering and a poorly adhering strain of *Streptococcus pneumoniae*. *Int. Arch. Allergy Immunol.* **109:**35–43.

147. **Ascencio, F., L. Å. Fransson, and T. Wadström.** 1998. Affinity of the gastric pathogen *Helicobacter pylori* for the N-sulphated glycosaminoglycan heparan sulfate. *J. Med. Microbiol.* **38:**240–244.

148. **Ascencio, F., W. Martinez-Arias, M. J. Romero, and T. Wadström.** 1998. Analysis of the interaction of *Aeromonas caviae, A. hydrophila* and *A. sobria* with mucins. *FEMS Immunol. Med. Microbiol.* **20:**219–229.

149. **Ascon, M. A., D. M. Hone, N. Walters, and D. W. Pascual.** 1998. Oral immunization with a *Salmonella typhimurium* vaccine vector expressing recombinant enterotoxigenic *Escherichia coli* K99 fimbriae elicits elevated antibody titers for protective immunity. *Infect. Immun.* **66:**5470–5476.

150. **Ashkenazi, S.** 1994. A review of the effect of human milk fractions on the adherence of diarrheogenic *Escherichia coli* to the gut in an animal model. *Isr. J. Med. Sci.* **30:**335–338.

151. **Ashkenazi, S., M. Larocco, B. E. Murray, and T. G. Cleary.** 1992. The adherence of verocytotoxin-producing *Escherichia coli* to rabbit intestinal cells. *J. Med. Microbiol.* **37:**304–309.

152. **Aslanzadeh, J., and L. J. Paulissen.** 1992. Role of type 1 and type 3 fimbriae on the adherence and pathogenesis of *Salmonella enteritidis* in mice. *Microbiol. Immunol.* **36:**351–359.

153. **Aspiras, M. B., K. M. Kazmerzak, P. E. Kolenbrander, R. McNab, N. Hardegen, and H. F. Jenkinson.** 2000. Expression of green fluorescent protein in *Streptococcus gordonii* DL1 and its use as a species-specific marker in coadhesion with *Streptococcus oralis* 34 in saliva-conditioned biofilms in vitro. *Appl. Environ. Microbiol.* **66:**4074–4083.

154. **Assmann, I. A., G. A. Enders, J. Puls, G. Rieder, R. Haas, and R. A. Hatz.** 2001. Role of virulence factors, cell components and adhesion in *Helicobacter pylori*-mediated iNOS induction in murine macrophages. *FEMS Immunol. Med. Microbiol.* **30:**133–138.

155. **Astarie-Dequeker, C., E. N. N'Diaye, V. Le Cabec, M. G. Rittig, J. Prandi, and I. Maridonneau-Parini.** 1999. The mannose receptor mediates uptake of pathogenic and nonpathogenic mycobacteria and bypasses bactericidal responses in human macrophages. *Infect. Immun.* **67:**469–477.

156. **Athamna, A., M. R. Kramer, and I. Kahane.** 1996. Adherence of *Mycoplasma pneumoniae* to human alveolar macrophages. *FEMS Immunol. Med. Microbiol.* **15:**135–141.

157. **Athamna, A., R. Rosengarten, S. Levisohn, I. Kahane, and D. Yogev.** 1997. Adherence of *Mycoplasma gallisepticum* involves variable surface membrane proteins. *Infect. Immun.* **65:**2468–2471.

158. **Atkinson, S., J. P. Throup, G. S. Stewart, and P. Williams.** 1999. A hierarchical quorum-sensing system in *Yersinia pseudotuberculosis* is involved in the regulation of motility and clumping. *Mol. Microbiol.* **33:**1267–1277.

159. **Attridge, S. R., E. Voss, and P. A. Manning.** 1993. The role of toxin-coregulated pili in the pathogenesis of *Vibrio cholerae* O1 El Tor. *Microb. Pathog.* **15:**421–431.

160. **Attridge, S. R., P. A. Manning, J. Holmgren, and G. Jonson.** 1996. Relative significance of mannose-sensitive hemagglutinin and toxin-coregulated pili in colonization of infant mice by *Vibrio cholerae* El Tor. *Infect. Immun.* **64:**3369–3373.

161. **Aubel, D., A. Darfeuille-Michaud, C. Martin, and B. Joly.** 1992. Nucleotide sequence of the *nfaA* gene encoding the antigen 8786 adhesive factor of enterotoxigenic *Escherichia coli*. *FEMS Microbiol. Lett.* **77:**277–284.

162. **Austin, J. W., G. Sanders, W. W. Kay, and S. K. Collinson.** 1998. Thin aggregative fimbriae enhance *Salmonella enteritidis* biofilm formation. *FEMS Microbiol. Lett.* **162:**295–301.

163. **Autenrieth, I. B., V. Kempf, T. Sprinz, S. Preger, and A. Schnell.** 1996. Defense mechanisms in Peyer's patches and mesenteric lymph nodes against *Yersinia enterocolitica* involve integrins and cytokines. *Infect. Immun.* **64:**1357–1368.

164. **Avakian, A. P., and D. H. Ley.** 1993. Inhibition of *Mycoplasma gallisepticum* growth and attachment to chick tracheal rings by antibodies to a 64-kilodalton membrane protein of *M. gallisepticum*. *Avian Dis.* **37:**706–714.

165. **Avakian, A. P., and D. H. Ley.** 1993. Protective immune response to *Mycoplasma gallisepticum* demonstrated in respiratory-tract washings from *M. gallisepticum*-infected chickens. *Avian Dis.* **37:**697–705.

166. **Azakami, H., H. Yumoto, T. Nakae, T.**

Matsuo, and S. Ebisu. 1996. Molecular analysis of the gene encoding a protein component of the *Eikenella corrodens* adhesin complex that is close to the carbohydrate recognition domain. *Gene* **180:**207–212.

167. **Azen, E., A. Prakobphol, and S. J. Fisher.** 1993. PRB3 null mutations result in absence of the proline-rich glycoprotein Gl and abolish *Fusobacterium nucleatum* interactions with saliva in vitro. *Infect. Immun.* **61:**4434–4439.

168. **Azghani, A. O., A. Y. Kondepudi, and A. R. Johnson.** 1992. Interaction of *Pseudomonas aeruginosa* with human lung fibroblasts: role of bacterial elastase. *Am. J. Respir. Cell Mol. Biol.* **6:**652–657.

169. **Azghani, A. O., I. Williams, D. B. Holiday, and A. R. Johnson.** 1995. A beta-linked mannan inhibits adherence of *Pseudomonas aeruginosa* to human lung epithelial cells. *Glycobiology* **5:**39–44.

170. **Baba, E., H. Wakeshima, K. Fufui, T. Fukata, and A. Arakawa.** 1992. Adhesion of bacteria to the cecal mucosal surface of conventional and germ-free chickens infected with *Eimeria tenella*. *Am. J. Vet. Res.* **53:**194–197.

171. **Baba, E., Y. Tsukamoto, T. Fukata, K. Sasai, and A. Arakawa.** 1993. Increase of mannose residues as *Salmonella typhimurium*-adhering factor on the cecal mucosa of germ-free chickens infected with *Eimeria tenella*. *Am. J. Vet. Res.* **54:**1471–1475.

172. **Babai, R., G. Blum-Oehler, B. E. Stern, J. Hacker, and E. Z. Ron.** 1997. Virulence patterns from septicemic *Escherichia coli* O78 strains. *FEMS Microbiol. Lett.* **149:**99–105.

173. **Babu, J. P., J. W. Dean, and M. J. Pabst.** 1995. Attachment of *Fusobacterium nucleatum* to fibronectin immobilized on gingival epithelial cells or glass coverslips. *J. Periodontol.* **66:**285–290.

174. **Bach, A., H. Bohrer, J. Motsch, E. Martin, H. K. Geiss, and H. G. Sonntag.** 1994. Prevention of bacterial colonization of intravenous catheters by antiseptic impregnation of polyurethane polymers. *J. Antimicrob. Chemother.* **33:**969–978.

175. **Backenson, P. B., J. L. Coleman, and J. L. Benach.** 1995. *Borrelia burgdorferi* shows specificity of binding to glycosphingolipids. *Infect. Immun.* **63:**2811–2817.

176. **Backman, M., H. Kallström, and A. B. Jonsson.** 1998. The phase-variable pilus-associated protein PilC is commonly expressed in clinical isolates of *Neisseria gonorrhoeae*, and shows sequence variability among strains. *Microbiology* **144:**149–156.

177. **Bacon, D. J., R. A. Alm, D. H. Burr, L. Hu, D. J. Kopecko, C. P. Ewing, T. J. Trust, and P. Guerry.** 2000. Involvement of a plasmid in virulence of *Campylobacter jejuni* 81–176. *Infect. Immun.* **68:**4384–4390.

178. **Badalament, R. A., G. L. Franklin, C. M. Page, B. M. Dasani, M. G. Wientjes, and J. R. Drago.** 1992. Enhancement of bacillus Calmette-Guerin attachment to the urothelium by removal of the rabbit bladder mucin layer. *J. Urol.* **147:**482–485.

179. **Baddour, L. M.** 1994. Virulence factors among gram-positive bacteria in experimental carditis. *Infect. Immun.* **62:**2143–2148.

180. **Badger, J. L., and V. L. Miller.** 1998. Expression of invasin and motility are coordinately regulated in *Yersinia enterocolitica*. *J. Bacteriol.* **180:**793–800.

181. **Baehler, A. A., and R. A. Moxley.** 2000. *Escherichia coli* O157:H7 induces attaching-effacing lesions in large intestinal mucosal explants from adult cattle. *FEMS Microbiol. Lett.* **185:**239–242.

182. **Baghian, A., and K. L. Schnorr.** 1992. Detection and antigenicity of chlamydial proteins that bind eukaryotic cell membrane proteins. *Am. J. Vet. Res.* **53:**980–986.

183. **Bahrani, F. K., and H. L. T. Mobley.** 1993. *Proteus mirabilis* MR/P fimbriae: Molecular cloning, expression, and nucleotide sequence of the major fimbrial subunit gene. *J. Bacteriol.* **175:**457–464.

184. **Bahrani, F. K., S. Cook, R. A. Hull, G. Massad, and H. L. Mobley.** 1993. *Proteus mirabilis* fimbriae: N-terminal amino acid sequence of a major fimbrial subunit and nucleotide sequences of the genes from two strains. *Infect. Immun.* **61:**884–891.

185. **Bahrani, F. K., and H. L. Mobley.** 1994. *Proteus mirabilis* MR/P fimbrial operon: genetic organization, nucleotide sequence, and conditions for expression. *J. Bacteriol.* **176:**3412–3419.

186. **Bahrani, F. K., G. Massad, C. V. Lockatell, D. E. Johnson, R. G. Russell, J. W. Warren, and H. L. Mobley.** 1994. Construction of an MR/P fimbrial mutant of *Proteus mirabilis*: role in virulence in a mouse model of ascending urinary tract infection. *Infect. Immun.* **62:**3363–3371.

187. **Bain, C., R. Keller, G. K. Collington, L. R. Trabulsi, and S. Knutton.** 1998. Increased levels of intracellular calcium are not required for the formation of attaching and effacing lesions by enteropathogenic and enterohemorrhagic *Escherichia coli*. *Infect. Immun.* **66:**3900–3908.

188. Bajolet-Laudinat, O., S. Girod-de-Bentzmann, J. M. Tournier, C. Madoulet, M. C. Plotkowski, C. Chippaux, and E. Puchelle. 1994. Cytotoxicity of *Pseudomonas aeruginosa* internal lectin PA-I to respiratory epithelial cells in primary culture. *Infect. Immun.* **62**:4481–4487.

189. Bakaletz, L. O. 1995. Viral potentiation of bacterial superinfection of the respiratory tract. *Trends Microbiol.* **3**:110–114.

190. Bakaletz, L. O., M. A. Ahmed, P. E. Kolattukudy, D. J. Lim, and L. J. Forney. 1992. Cloning and sequence analysis of a pilin-like gene from an otitis media isolate of nontypeable *Haemophilus influenzae*. *J. Infect. Dis.* **165**:S201–S203.

191. Bakaletz, L. O., and S. J. Barenkamp. 1994. Localization of high-molecular-weight adhesion proteins of nontypeable *Haemophilus influenzae* by immunoelectron microscopy. *Infect. Immun.* **62**:4460–4468.

192. Bakaletz, L. O., E. R. Leake, J. M. Billy, and P. T. Kaumaya. 1997. Relative immunogenicity and efficacy of two synthetic chimeric peptides of fimbrin as vaccinogens against nasopharyngeal colonization by nontypeable *Haemophilus influenzae* in the chinchilla. *Vaccine* **15**:955–961.

193. Bakaletz, L. O., B. M. Tallan, T. Hoepf, T. F. DeMaria, H. G. Birck, and D. J. Lim. 1988. Frequency of fimbriation of nonetypeable *Haemophilus influenzae* and its ability to adhere to chinchilla and human respiratory epithelium. *Infect. Immun.* **56**:331–335.

194. Baker, D. R., L. O. Billey, and D. H. Francis. 1997. Distribution of K88 *Escherichia coli*-adhesive and nonadhesive phenotypes among pigs of four breeds. *Vet. Microbiol.* **54**:123–132.

195. Bakker, D., P. T. J. Willemsen, L. H. Simons, F. G. van Zijderveld, and F. K. de Graaf. 1992. Characterization of the antigenic and adhesive properties of FaeG, the major subunit of K88 fimbriae. *Mol. Microbiol.* **6**:247–255.

196. Bakker, D., P. T. J. Willemsen, R. H. Willems, T. T. Huisman, F. R. Mooi, B. Oudega, F. Stegehuis, and F. K. de Graaf. 1992. Identification of minor fimbrial subunits involved in biosynthesis of K88 fimbriae. *J. Bacteriol.* **174**:6350–6358.

197. Baldassarri, L., G. Donelli, A. Gelosia, W. A. Simpson, and G. D. Christensen. 1997. Expression of slime interferes with in vitro detection of host protein receptors of *Staphylococcus epidermidis*. *Infect. Immun.* **65**:1522–1526.

198. Baldwin, T. J., S. Knutton, L. Sellers, H. A. Manjarrez Hernandez, A. Aitken, and P. H. Williams. 1992. Enteroaggregative *Escherichia coli* strains secrete a heat-labile toxin antigenically related to *E. coli* hemolysin. *Infect. Immun.* **60**:2092–2095.

199. Baldwin, T. J., M. B. Lee-Delaunay, S. Knutton, and P. H. Williams. 1993. Calcium-calmodulin dependence of actin accretion and lethality in cultured HEp-2 cells infected with enteropathogenic *Escherichia coli*. *Infect. Immun.* **61**:760–763.

200. Balish, M. F., T. W. Hahn, P. L. Popham, and D. C. Krause. 2001. Stability of *Mycoplasma pneumoniae* cytadherence-accessory protein HMW1 correlates with its association with the triton shell. *J. Bacteriol.* **183**:3680–3688.

201. Ballyk, M., and H. Smith. 1999. A model of microbial growth in a plug flow reactor with wall attachment. *Math. Biosci.* **158**:95–126.

202. Baorto, D. M., Z. Gao, R. Malaviya, M. L. Dustin, A. van der Merwe, D. M. Lublin, and S. N. Abraham. 1997. Survival of FimH-expressing enterobacteria in macrophages relies on glycolipid traffic. *Nature* **389**:636–639.

203. Barenkamp, S. J., and E. Leininger. 1992. Cloning, expression, and DNA sequence analysis of genes encoding nontypeable *Haemophilus influenzae* high-molecular-weight surface-exposed proteins related to filamentous hemagglutinin of *Bordetella pertussis*. *Infect. Immun.* **60**:1302–1313.

204. Barenkamp, S. J., and J. W. St. Geme III. 1994. Genes encoding high-molecular-weight adhesion proteins of nontypeable *Haemophilus influenzae* are part of gene clusters. *Infect. Immun.* **62**:3320–3328.

205. Barenkamp, S. J. 1996. Immunization with high-molecular-weight adhesion proteins of nontypeable *Haemophilus influenzae* modifies experimental otitis media in chinchillas. *Infect. Immun.* **64**:1246–1251.

206. Barenkamp, S. J., and J. W. St. Geme III. 1996. Identification of surface-exposed B-cell epitopes on high molecular-weight adhesion proteins of nontypeable *Haemophilus influenzae*. *Infect. Immun.* **64**:3032–3037.

207. Barenkamp, S. J., and J. W. St. Geme III. 1996. Identification of a second family of high-molecular-weight adhesion proteins expressed by non-typable *Haemophilus influenzae*. *Mol. Microbiol.* **19**:1215–1223.

208. Barghouthi, S., K. D. Everett, and D. P. Speert. 1995. Nonopsonic phagocytosis of

Pseudomonas aeruginosa requires facilitated transport of D-glucose by macrophages. *J. Immunol.* **154:**3420–3428.

209. **Barghouthi, S., L. M. Guerdoud, and D. P. Speert.** 1996. Inhibition by dextran of *Pseudomonas aeruginosa* adherence to epithelial cells. *Am. J. Respir. Crit. Care Med.* **154:**1788–1793.

210. **Barker, A., C. A. Clark, and P. A. Manning.** 1994. Identification of VCR, a repeated sequence associated with a locus encoding a hemagglutinin in *Vibrio cholerae* O1. *J. Bacteriol.* **176:**5450–5458.

211. **Barnes, L. M., M. F. Lo, M. R. Adams, and A. H. Chamberlain.** 1999. Effect of milk proteins on adhesion of bacteria to stainless steel surfaces. *Appl. Environ. Microbiol.* **65:**4543–4548.

212. **Barnett, T. C., S. M. Kirov, M. S. Strom, and K. Sanderson.** 1997. *Aeromonas* spp. possess at least two distinct type IV pilus families. *Microb. Pathog.* **23:**241–247.

213. **Barnett-Foster, D., M. Abul-Milh, M. Huesca, and C. A. Lingwood.** 2000. Enterohemorrhagic *Escherichia coli* induces apoptosis which augments bacterial binding and phosphatidylethanolamine exposure on the plasma membrane outer leaflet. *Infect. Immun.* **68:**3108–3115.

214. **Barnett-Foster, D., D. Philpott, M. Abul-Milh, M. Huesca, P. M. Sherman, and C. A. Lingwood.** 1999. Phosphatidylethanolamine recognition promotes enteropathogenic *E. coli* and enterohemorrhagic *E. coli* host cell attachment. *Microb. Pathog.* **27:**289–301.

215. **Barros, H. C., M. L. Silva, M. Z. Laporta, R. M. Silva, and L. R. Trabulsi.** 1992. Inhibition of enteropathogenic *Escherichia coli* adherence to HeLa cells by immune rabbit sera. *Braz. J. Med. Biol. Res.* **25:**809–812.

216. **Barros, H. C., S. R. Ramos, L. R. Trabulsi, and M. L. Silva.** 1995. Inhibition of enteropathogenic *Escherichia coli* adhesion to HeLa cells by serum of infants with diarrhea and by cord serum. *Braz. J. Med. Biol. Res.* **28:**83–87.

217. **Barrow, P. A., M. A. Lovell, and D. C. Old.** 1992. *In-vitro* and *in-vivo* characteristics of TnphoA mutant strains of *Salmonella* serotype *Gallinarum* not invasive for tissue culture cells. *J. Med. Microbiol.* **36:**389–397.

218. **Barsum, W., R. Wilson, R. C. Read, A. Rutman, H. C. Todd, N. Houdret, P. Roussel, and P. J. Cole.** 1995. Interaction of fimbriated and nonfimbriated strains of

unencapsulated *Haemophilus influenzae* with human respiratory tract mucus *in vitro*. *Eur. Respir. J.* **8:**709–714.

219. **Barthelson, R., A. Mobasseri, D. Zopf, and P. Simon.** 1998. Adherence of *Streptococcus pneumoniae* to respiratory epithelial cells is inhibited by sialylated oligosaccharides. *Infect. Immun.* **66:**1439–1444.

220. **Bartkova, G., and I. Ciznar.** 1992. Adherence of intestinal and extraintestinal *Pseudomonas aeruginosa* to tissue culture cells. *Folia Microbiol.* **37:**140–145.

221. **Bartkova, G., and I. Ciznar.** 1994. Adherence pattern of non-pilated *Aeromonas hydrophila* strains to tissue cultures. *Microbios* **77:**47–55.

222. **Bartkova, G., I. Ciznar, V. Lehotska, and T. Kernova.** 1994. Characterization of adhesion associated surface properties of uropathogenic *Escherichia coli*. *Folia Microbiol.* **39:**373–377.

223. **Barzu, S., Z. Benjelloun-Touimi, A. Phalipon, P. Sansonetti, and C. Parsot.** 1997. Functional analysis of the *Shigella flexneri* IpaC invasin by insertional mutagenesis. *Infect. Immun.* **65:**1599–1605.

224. **Baselga, R., I. Albizu, M. de la Cruz, E. del Cacho, M. Barberan, and B. Amorena.** 1993. Phase variation of slime production in *Staphylococcus aureus*: implications in colonization and virulence. *Infect. Immun.* **61:**4857–4862.

225. **Baseman, J. B.** 1993. The cytadhesins of *Mycoplasma pneumoniae* and *M. genitalium*. *SubCell. Biochem.* **20:**243–259.

226. **Baseman, J. B., S. P. Reddy, and D. F. Dallo.** 1996. Interplay between mycoplasma surface proteins, airway cells, and the protean manifestations of mycoplasma-mediated human infections. *Am. J. Respir. Crit. Care Med.* **154:**S137–S144.

227. **Bassinet, L., P. Gueirard, B. Maitre, B. Housset, P. Gounon, and N. Guiso.** 2000. Role of adhesins and toxins in invasion of human tracheal epithelial cells by *Bordetella pertussis*. *Infect. Immun.* **68:**1934–1941.

228. **Batai, I., and M. Kerenyi.** 1999. Halothane decreases bacterial adherence *in vitro*. *Acta Anaesthesiol. Scand.* **43:**760–763.

229. **Batterman, H. J., J. A. Peek, J. S. Loutit, S. Falkow, and L. S. Tompkins.** 1995. *Bartonella henselae* and *Bartonella quintana* adherence to and entry into cultured human epithelial cells. *Infect. Immun.* **63:**4553–4556.

230. **Battikhi, T., W. Lee, C. A. McCulloch, and R. P. Ellen.** 1999. *Treponema denticola* outer membrane enhances the phagocytosis of

collagen-coated beads by gingival fibroblasts. *Infect. Immun.* **67:**1220–1226.

231. **Bauer, M. E., and S. M. Spinola.** 1999. Binding of *Haemophilus ducreyi* to extracellular matrix proteins. *Infect. Immun.* **67:**2649–2652.

232. **Bauer, M. E., M. P. Goheen, C. A. Townsend, and S. M. Spinola.** 2001. *Haemophilus ducreyi* associates with phagocytes, collagen, and fibrin and remains extracellular throughout infection of human volunteers. *Infect. Immun.* **69:**2549–2557.

233. **Baumann, U., J. J. Fischer, P. Gudowius, M. Lingner, S. Herrmann, B. Tummler, and H. von der Hardt.** 2001. Buccal adherence of *Pseudomonas aeruginosa* in patients with cystic fibrosis under long-term therapy with azithromycin. *Infection* **29:**7–11.

234. **Baumgartner, J. N., and S. L. Cooper.** 1998. Influence of thrombus components in mediating *Staphylococcus aureus* adhesion to polyurethane surfaces. *J. Biomed. Mater. Res.* **40:**660–670.

235. **Baumler, A. J., and F. Heffron.** 1995. Identification and sequence analysis of *lpfABCDE,* a putative fimbrial operon of *Salmonella typhimurium. J. Bacteriol.* **177:**2087–2097.

236. **Baumler, A. J., R. M. Tsolis, and F. Heffron.** 1996. Contribution of fimbrial operons to attachment to and invasion of epithelial cell lines by *Salmonella typhimurium. Infect. Immun.* **64:**1862–1865.

237. **Baumler, A. J., R. M. Tsolis, and F. Heffron.** 1996. The *lpf* fimbrial operon mediates adhesion of *Salmonella typhimurium* to murine Peyer's patches. *Proc. Natl. Acad. Sci. USA* **93:**279–283.

238. **Baumler, A. J., R. M. Tsolis, F. A. Bowe, J. G. Kusters, S. Hoffmann, and F. Heffron.** 1996. The *pef* fimbrial operon of *Salmonella typhimurium* mediates adhesion to murine small intestine and is necessary for fluid accumulation in the infant mouse. *Infect. Immun.* **64:**61–68.

239. **Baumler, A. J., A. J. Gilde, R. M. Tsolis, A. W. van der Velden, B. M. Ahmer, and F. Heffron.** 1997. Contribution of horizontal gene transfer and deletion events to development of distinctive patterns of fimbrial operons during evolution of *Salmonella* serotypes. *J. Bacteriol.* **179:**317–322.

240. **Baumler, A. J., R. M. Tsolis, and F. Heffron.** 1997. Fimbrial adhesins of *Salmonella typhimurium.* Role in bacterial interactions with epithelial cells. *Adv. Exp. Med. Biol.* **412:**149–158.

241. **Bayer, A. S., P. M. Sullam, M. Ramos, C. Li, A. L. Cheung, and M. R. Yeaman.** 1995. *Staphylococcus aureus* induces platelet aggregation via a fibrinogen-dependent mechanism which is independent of principal platelet glycoprotein IIb/IIIa fibrinogen-binding domains. *Infect. Immun.* **63:**3634–3641.

242. **Beaman, B. L.** 1993. Ultrastructural analysis of growth of *Nocardia asteroides* during invasion of the murine brain. *Infect. Immun.* **61:**274–283.

243. **Beaman, B. L.** 1996. Differential binding of *Nocardia asteroides* in the murine lung and brain suggests multiple ligands on the nocardial surface. *Infect. Immun.* **64:**4859–4862.

244. **Beaman, B. L., and S. A. Ogata.** 1993. Ultrastructural analysis of attachment to and penetration of capillaries in the murine pons, midbrain, thalamus, and hypothalamus by *Nocardia asteroides. Infect. Immun.* **61:**955–965.

245. **Beaman, B. L., and L. Beaman.** 1994. *Nocardia* species: host-parasite relationships. *Clin. Microbiol. Rev.* **7:**213–264.

246. **Beaman, B. L., and L. Beaman.** 1998. Filament tip-associated antigens involved in adherence to and invasion of murine pulmonary epithelial cells in vivo and HeLa cells in vitro by *Nocardia asteroides. Infect. Immun.* **66:**4676–4689.

247. **Beaman, L., and B. L. Beaman.** 1994. Differences in the interactions of *Nocardia asteroides* with macrophage, endothelial, and astrocytoma cell lines. *Infect. Immun.* **62:**1787–1798.

248. **Beaudry, M., C. Zhu, J. M. Fairbrother, and J. Harel.** 1996. Genotypic and phenotypic characterization of *Escherichia coli* isolates from dogs manifesting attaching and effacing lesions. *J. Clin. Microbiol.* **34:**144–148.

249. **Beckert, S., B. Kreikemeyer, and A. Podbielski.** 2001. Group A streptococcal *rofA* gene is involved in the control of several virulence genes and eukaryotic cell attachment and internalization. *Infect. Immun.* **69:**534–537.

250. **Beekhuizen, H., J. S. van de Gevel, B. Olsson, J. J. van Benten, and R. van Furth.** 1997. Infection of human vascular endothelial cells with *Staphylococcus aureus* induces hyperadhesiveness for human monocytes and granulocytes. *J. Immunol.* **158:**774–782.

251. **Beem, J. E., C. G. Hurley, W. W. Nesbitt, D. F. Croft, R. G. Marks, J. O. Cisar, and W. B. Clark.** 1996. Fimbrial-mediated colonization of murine teeth by *Actinomyces naeslundii. Oral Microbiol. Immunol.* **11:**259–265.

252. **Beena, V. K., and P. G. Shivananda.** 1997. *In vitro* adhesiveness of *Bacteroides fragilis*

group in relation to encapsulation. *Indian J. Med. Res.* **105**:258–261.

253. **Beinke, C., S. Laarmann, C. Wachter, H. Karch, L. Greune, and M. A. Schmidt.** 1998. Diffusely adhering *Escherichia coli* strains induce attaching and effacing phenotypes and secrete homologs of Esp proteins. *Infect. Immun.* **66**:528–539.

254. **Bélanger, M., S. Rioux, B. Foiry, and M. Jacques.** 1992. Affinity for porcine respiratory tract mucus is found in some isolates of *Actinobacillus pleuropneumoniae. FEMS Microbiol. Lett.* **76**:119–125.

255. **Bélanger, M., D. Dubreuil, and M. Jacques.** 1993. Proteins found within porcine respiratory tract secretions bind lipopolysaccharides of *Actinobacillus pleuropneumoniae. Infect. Immun.* **62**:868–873.

256. **Bélanger, M., D. Dubreuil, and M. Jacques.** 1994. Identification of porcine respiratory tract mucus proteins binding lipopolysaccharides of *Actinobacillus pleuropneumoniae. Ann. N. Y. Acad. Sci.* **730**:249–251.

257. **Bencina, D., M. Narat, P. Dovc, M. Drobnic-Valic, F. Habe, and S. H. Kleven.** 1999. The characterization of *Mycoplasma synoviae* EF-Tu protein and proteins involved in hemadherence and their N-terminal amino acid sequences. *FEMS Microbiol. Lett.* **173**:85–94.

258. **Bendinger, B., H. H. M. Rijnaarts, K. Altendorf, and A. J. B. Zehnder.** 1993. Physicochemical cell surface and adhesive properties of coryneform bacteria related to the presence and chain length of mycolic acids. *Appl. Environm. Microbiol.* **59**:3973–3977.

259. **Benitez, J. A., R. G. Spelbrink, A. Silva, T. E. Phillips, C. M. Stanley, M. Boesman-Finkelstein, and R. A. Finkelstein.** 1997. Adherence of *Vibrio cholerae* to cultured differentiated human intestinal cells: an in vitro colonization model. *Infect. Immun.* **65**:3474–3477.

260. **Bennett-Wood, V. R., J. R. Carapetis, and R. M. Robins-Browne.** 1998. Ability of clinical isolates of group A streptococci to adhere to and invade HEp-2 epithelial cells. *J. Med. Microbiol.* **47**:899–906.

261. **Bensing, B. A., C. E. Rubens, and P. M. Sullam.** 2001. Genetic loci of *Streptococcus mitis* that mediate binding to human platelets. *Infect. Immun.* **69**:1373–1380.

262. **Benson, M., U. Jodal, A. Andreasson, A. Karlsson, J. Rydberg, and C. Svanborg.** 1994. Interleukin 6 response to urinary tract infection in childhood. *Pediatr. Infect. Dis. J.* **13**:612–616.

263. **Benz, I., and M. A. Schmidt.** 1992. Isolation and serologic charactrerization of AIDA-I, the adhesin mediating the diffuse adherence phenotype of the diarrhea-associated *Escherichia coli* strain 2787 (O126:H27). *Infect. Immun.* **60**:13–18.

264. **Benz, I., and M. A. Schmidt.** 1993. Diffuse adherence of enteropathogenic *Escherichia coli* strains—processing of AIDA-I. *Zentbl. Bakteriol.* **278**:197–208.

265. **Bergamini, T. M., R. A. Corpus, Jr., K. R. Brittian, J. C. Peyton, and W. G. Cheadle.** 1994. The natural history of bacterial biofilm graft infection. *J. Surg. Res.* **56**:393–396.

266. **Bergamini, T. M., R. A. Corpus, Jr., T. M. McCurry, J. C. Peyton, K. R. Brittian, and W. G. Cheadle.** 1995. Immunosuppression augments growth of graft-adherent *Staphylococcus epidermidis. Arch. Surg.* **130**:1345–1350.

267. **Bergamini, T. M., T. M. McCurry, J. D. Bernard, K. L. Hoeg, R. A. Corpus, B. E. James, J. C. Peyton, K. R. Brittian, and W. G. Cheadle.** 1996. Antibiotic efficacy against *Staphylococcus epidermidis* adherent to vascular grafts. *J. Surg. Res.* **60**:3–6.

268. **Berggard, K., E. Johnsson, F. R. Mooi, and G. Lindahl.** 1997. *Bordetella pertussis* binds the human complement regulator C4BP: role of filamentous hemagglutinin. *Infect. Immun.* **65**:3638–3643.

269. **Berkower, C., M. Ravins, A. E. Moses, and E. Hanski.** 1999. Expression of different group A streptococcal M proteins in an isogenic background demonstrates diversity in adherence to and invasion of eukaryotic cells. *Mol. Microbiol.* **31**:1463–1475.

270. **Bermudez, L. E., L. S. Young, and C. B. Inderlied.** 1994. Rifabutin and sparfloxacin but not azithromycin inhibit binding of *Mycobacterium avium* complex to HT-29 intestinal mucosal cells. *Antimicrob. Agents Chemother.* **38**:1200–1202.

271. **Bermudez, L. E., K. Shelton, and L. S. Young.** 1995. Comparison of the ability of *Mycobacterium avium, M. smegmatis* and *M. tuberculosis* to invade and replicate within HEp-2 epithelial cells. *Tubercle Lung Dis.* **76**:240–247.

272. **Bern, D., and C. Lammler.** 1996. Relationship between haemagglutination and HeLa-cell adherence of *Rhodococcus equi. Zentbl. Veterinarmed. Reihe B.* **43**:147–153.

273. **Bernet, M. F., D. Brassart, J. R. Neeser, and A. L. Servin.** 1993. Adhesion of human bifidobacterial strains to cultured human intes-

tinal epithelial cells and inhibition of enteropathogen-cell interactions. *Appl. Environ. Microbiol.* **59:**4121–4128.

274. **Bernet, M. F., D. Brassart, J. R. Neeser, and A. L. Servin.** 1994. *Lactobacillus acidophilus* LA 1 binds to cultured human intestinal cell lines and inhibits cell attachment and cell invasion by enterovirulent bacteria. *Gut* **35:**483–489.

275. **Bernet-Camard, M. F., M. H. Coconnier, S. Hudault, and A. L. Servin.** 1996. Pathogenicity of the diffusely adhering strain *Escherichia coli* C1845: F1845 adhesin-decay accelerating factor interaction, brush border microvillus injury and actin disassembly in cultured human intestinal epithelial cells. *Infect. Immun.* **64:**1918–1928.

276. **Bernet-Camard, F. Duigou, S. Kernéis, M.-H. Coconnier, and A. L. Servin.** 1997. Glucose up-regulates expression of the differentiation-associated brush border binding site for enterotoxigenic *Escherichia coli* colonization factor antigen I in cultured human erythrocyte-like cells. *Infect. Immun.* **65:**1299–1306.

277. **Bernhard, W., A. Gbarah, and N. Sharon.** 1992. Lectinophagocytosis of type 1 fimbriated (mannose-specific) *Escherichia coli* in the mouse peritoneum. *J. Leukoc. Biol.* **52:**343–348.

278. **Bernkop-Schnurch, A., F. Gabor, and P. Spiegl.** 1997. Bacterial adhesins as a drug carrier: covalent attachment of K99 fimbriae to 6-methylprednisolone. *Pharmazie* **52:**41–44.

279. **Bernstein, J. M., and M. Reddy.** 2000. Bacteria-mucin interaction in the upper aerodigestive tract shows striking heterogeneity: implications in otitis media, rhinosinusitis and pneumonia. *Otolaryngol. Head Neck Surg.* **122:**514–520.

280. **Berry, A. M., and J. C. Paton.** 1996. Sequence heterogeneity of PsaA, a 37-kilodalton putative adhesin essential for virulence of *Streptococcus pneumoniae. Infect. Immun.* **64:**5255–5262.

281. **Berry, E. D., and G. R. Siragusa.** 1997. Hydroxyapatite adherence as a means to concentrate bacteria. *Appl. Environ. Microbiol.* **63:**4069–4074.

282. **Berthaud, N., and J. F. Desnottes.** 1997. In-vitro bactericidal activity of quinupristin/dalfopristin against adherent *Staphylococcus aureus. J. Antimicrob. Chemother.* **39:**99–102.

283. **Berti, M., G. Candiani, A. Kaufhold, A. Muscholl, and R. Wirth.** 1998. Does aggregation substance of *Enterococcus faecalis* contribute to development of endocarditis? *Infection* **26:**48–53.

284. **Bertin, A.** 1998. Phenotypic expression of K88 adhesion alone or simultaneously with K99 and/or F41 adhesins in the bovine enterotoxigenic *Escherichia coli* strain B41. *Vet. Microbiol.* **59:**283–294.

285. **Bertin, Y., C. Martin, E. Oswald, and J. P. Girardeau.** 1996. Rapid and specific detection of F17-related pilin and adhesin genes in diarrheic and septicemic *Escherichia coli* strains by multiplex PCR. *J. Clin. Microbiol.* **34:** 2921–2928.

286. **Bertin, Y., J.-P. Girardeau, A. Darfeuille-Michaud, and M. Contrepois.** 1996. Characterization of 20K fimbria, a new adhesin of septicemic and diarrhea-associated *Escherichia coli* strains, that belongs to a family of adhesins with N-acetyl-D-glucosamine recognition. *Infect. Immun.* **64:**332–342.

287. **Bertin, Y., C. Martin, J. P. Girardeau, P. Pohl, and M. Contrepois.** 1998. Association of genes encoding P fimbriae, CS31A antigen and EAST 1 toxin among CNF1-producing *Escherichia coli* strains from cattle with septicemia and diarrhea. *FEMS Microbiol. Lett.* **162:**235–239.

288. **Bertozzi, C., and M. Bednarski.** 1992. C-Glycosyl compounds bind to receptors on the surface of *Escherichia coli* and can target proteins to the organism. *Carbohydr. Res.* **223:**243–253.

289. **Bertozzi, C. R., and M. D. Bednarski.** 1992. Antibody targeting to bacterial cells using receptor specific ligands. *J. Am. Chem. Soc.* **114:**2242–2245.

290. **Bertozzi, C. R., and M. D. Bednarski.** 1992. A receptor-mediated immune response using synthetic glycoconjugates. *J. Am. Chem. Soc.* **114:**5543–5546.

291. **Bertram, E. M., S. R. Attridge, and I. Kotlarski.** 1994. Immunogenicity of the *Escherichia coli* fimbrial antigen K99 when expressed by *Salmonella enteritidis* 11RX. *Vaccine* **12:**1372–1378.

292. **Bertschinger, H. U., M. Stamm, and P. Vögeli.** 1993. Inheritance of resistance to oedema disease in the pig: experiments with an *Escherichia coli* strain expressing fimbriae107. *Vet. Microbiol.* **35:**79–89.

293. **Berube, L. R., H. Jouishomme, and H. C. Jarrell.** 1998. The nonrandom binding distribution of *Streptococcus pneumoniae* to type II pneumocytes in culture is dependent on the relative distribution of cells among the phases of the cell cycle. *Can. J. Microbiol.* **44:**448–455.

294. **Berube, L. R., M. K. Schur, R. K. Latta, T. Hirama, C. R. McKenzie, and H. C. Jarrell.** 1999. Phosphatidyl choline-mediated inhibition of *Streptococcus pneumoniae* adherence

to type II pneumocytes *in vitro*. *Microb. Pathog.* **26**:65–75.

295. **Betjes, M. G., C. W. Tuk, D. G. Struijk, R. T. Krediet, L. Arisz, and R. H. Beelen.** 1992. Adherence of *Staphylococci* to plastic, mesothelial cells and mesothelial extracellular matrix. *Adv. Periton. Dial.* **8**:215–218.

296. **Bevenius, J., L. Linder, and K. Hultenby.** 1994. Site-related streptococcal attachment to buccocervical tooth surfaces. A correlative micromorphologic and microbiologic study. *Acta Odontol. Scand.* **52**:294–302.

297. **Beveridge, T. J., P. H. Pouwels, M. Sara, A. Kotiranta, K. Lounatmaa, K. Kari, E. Kerosuo, M. Haapasalo, E. M. Egelseer, I. Schocher, U. B. Sleytr, L. Morelli, M. L. Callegari, J. F. Nomellini, W. H. Bingle, J. Smit, E. Leibovitz, M. Lemaire, I. Miras, S. Salamitou, P. Beguin, H. Ohayon, P. Gounon, M. Matuschek, and S. F. Koval.** 1997. Functions of S-layers. *FEMS Microbiol. Rev.* **20**:99–149.

298. **Beyers, T. M., W. C. Lai, R. W. Read, and S. P. Pakes.** 1994. *Mycoplasma pulmonis* 46-kDa trypsin-resistant protein adheres to rat tracheal epithelial cells. *Lab. Anim. Sci.* **44**:573–578.

299. **Bezanson, G., S. Burbridge, D. Haldane, and T. Marrie.** 1992. *In situ* colonization of polyvinyl chloride, brass and copper by *Legionella pneumophila*. *Can. J. Microbiol.* **38**:328–330.

300. **Bhattacharjee, M. K., S. C. Kachlany, D. H. Fine, and D. H. Figurski.** 2001. Nonspecific adherence and fibril biogenesis by *Actinobacillus actinomycetemcomitans*:TadA protein is an ATPase. *J. Bacteriol.* **183**:5927–5936.

301. **Bian, Z., and S. Normark.** 1997. Nucleator function of CsgB for the assembly of adhesive surface organelles in *Escherichia coli*. *EMBO J.* **16**:5827–5836.

302. **Bibel, D. J., R. Aly, and H. R. Shinefield.** 1992. Inhibition of microbial adherence by sphinganine. *Can. J. Microbiol.* **38**:983–985.

303. **Biedlingmaier, J. F., R. Samaranayake, and P. Whelan.** 1998. Resistance to biofilm formation on otologic implant materials. *Otolaryngol. Head Neck Surg.* **118**:444–451.

304. **Bijlsma, I. G., L. van Dijk, J. G. Kusters, and W. Gaastra.** 1995. Nucleotide sequences of two fimbrial major subunit genes, *pmpA* and *ucaA*, from canine-uropathogenic *Proteus mirabilis* strains. *Microbiology* **141**:1349–1357.

305. **Bilbruck, J., G. W. Hanlon, and G. P. Martin.** 1993. The effects of polyHEMA

coating on the adhesion of bacteria to polymer monofilaments. *Int. J. Pharm.* **99**:293–301.

306. **Bilge, S. S., J. M. Apostol, Jr., K. J. Fullner, and S. L. Moseley.** 1993. Transcriptional organization of the F1845 fimbrial adhesin determinant of *Escherichia coli*. *Mol. Microbiol.* **7**:993–1006.

307. **Bilge, S. S., J. C. Vary, Jr., S. F. Dowell, and P. I. Tarr.** 1996. Role of the *Escherichia coli* O157:H7 O side chain in adherence and analysis of an *rfb* locus. *Infect. Immun.* **64**:4795–4801.

308. **Billey, L. O., A. K. Erickson, and D. H. Francis.** 1998. Multiple receptors on porcine intestinal epithelial cells for the three variants of *Escherichia coli* K88 fimbrial adhesin. *Vet. Microbiol.* **59**:203–212.

309. **Billson, F. M., J. L. Hodgson, J. R. Egerton, A. W. Lepper, W. P. Michalski, and C. L. Schwartzkoff.** 1994. A haemolytic cell-free preparation of *Moraxella bovis* confers protection against infectious bovine keratoconjunctivitis. *FEMS Microbiol. Lett.* **124**:69–73.

310. **Birkness, K. A., V. G. George, D. S. Stephens, E. Ribot, and F. D. Quinn.** 1994. Use of tissue culture invasion assays to compare strains of *Neisseria meningitidis*. *Ann. N. Y. Acad. Sci.* **730**:257–259.

311. **Birkness, K. A., B. L. Swisher, E. H. White, E. G. Long, E. P. Ewing, Jr., and F. D. Quinn.** 1995. A tissue culture bilayer model to study the passage of *Neisseria meningitidis*. *Infect. Immun.* **63**:402–409.

312. **Birkness, K. A., B. D. Gold, E. H. White, J. H. Bartlett, and F. D. Quinn.** 1996. *In vitro* models to study attachment and invasion of *Helicobacter pylori*. *Ann. N. Y. Acad. Sci.* **797**:293–295.

313. **Bisno, A. L.** 1995. Molecular aspects of bacterial colonization. *Infect. Control Hosp. Epidemiol.* **16**:648–657.

314. **Bisognano, C., P. E. Vaudaux, D. P. Lew, E. Y. Ng, and D. C. Hooper.** 1997. Increased expression of fibronectin-binding proteins by fluoroquinolone-resistant *Staphylococcus aureus* exposed to subinhibitory levels of ciprofloxacin. *Antimicrob. Agents Chemother.* **41**:906–913.

315. **Bisognano, C., P. Vaudaux, P. Rohner, D. P. Lew, and D. C. Hooper.** 2000. Induction of fibronectin-binding proteins and increased adhesion of quinolone-resistant *Staphylococcus aureus* by subinhibitory levels of ciprofloxacin. *Antimicrob. Agents Chemother.* **44**:1428–1437.

316. **Bitzan, M. M., B. D. Gold, D. J. Philpott, M. Huesca, P. M. Sherman, H.**

Karch, R. Lissner, C. A. Lingwood, and M. A. Karmali. 1998. Inhibition of *Helicobacter pylori* and *Helicobacter mustelae* binding to lipid receptors by bovine colostrum. *J. Infect. Dis.* **177:**955–961.

317. Bjerkli, I. H., R. Myklebust, S. Raisanen, S. Telimaa, and L. E. Stenfors. 1996. Bacterial attachment to oropharyngeal epithelial cells in breastfed newborns. *Int. J. Pediatr. Otorhinolaryngol.* **36:**205–213.

318. Blake, M. S., C. M. Blake, M. A. Apicella, and R. E. Mandrell. 1995. Gonococcal opacity: lectin-like interactions between Opa proteins and lipooligosaccharide. *Infect. Immun.* **63:**1434–1439.

319. Blanco, J., E. A. González, P. Espinosa, M. Blanco, J. I. Garabal, and M. P. Alonso. 1992. Enterotoxigenic and necrotizing *Escherichia coli* in human diarrhoea in Spain. *Eur. J. Epidemiol.* **8:**548–552.

320. Blanco, J., M. Blanco, E. A. González, J. I. Garabal, J. E. Blanco, and W. H. Jansen. 1992. Bovine *Escherichia coli* of serotypes O55:H4 and O55:H21 which produce CNF2 express P fimbriae and Vir surface antigen, respectively. *FEMS Microbiol. Lett.* **94:**149–154.

321. Blanco, J., M. Blanco, M. P. Alonso, J. E. Blanco, E. A. González, and J. I. Garabal. 1992. Characteristics of haemolytic *Escherichia coli* with particular reference to production of cytotoxic necrotizing factor type 1 (CNF1). *Res. Microbiol.* **143:**869–878.

322. Blanco, J., M. Blanco, M. P. Alonso, J. E. Blanco, J. I. Garabal, and E. A. González. 1992. Serogroups of *Escherichia coli* strains producing cytotoxic necrotizing factors CNF1 and CNF2. *FEMS Microbiol. Lett.* **96:**155–160.

323. Blanco, J., M. P. Alonso, M. Blanco, J. E. Blanco, E. A. González, and J. I. Garabal. 1992. Establishment of three categories of P-fimbriated *Escherichia coli* strains that show different toxic phenotypes and belong to particular O serogroups. *FEMS Microbiol. Lett.* **99:**131–136.

324. Blanco, J., J. E. Blanco, M. Blanco, E. A. Gonzalez, J. I. Garabal, and M. P. Alonso. 1993. Toxic and adhesive properties of *Escherichia coli* strains belonging to classic enteropathogenic serogroups. *Acta Microbiol. Hung.* **40:**335–341.

325. Blanco, J. E., J. Blanco, M. Blanco, M. P. Alonso, and W. H. Jansen. 1994. Serotypes of CNF1-producing *Escherichia coli* strains that cause extraintestinal infections in humans. *Eur. J. Epidemiol.* **10:**707–711.

326. Blanco, M., J. E. Blanco, M. P. Alonso, and J. Blanco. 1994. Virulence factors and O groups of *Escherichia coli* strains isolated from cultures of blood specimens from urosepsis and non-urosepsis patients. *Microbiologia* **10:**249–256.

327. Blanco, M., J. E. Blanco, M. P. Alonso, A. Mora, C. Balsalobre, F. Munoa, A. Juarez, and J. Blanco. 1997. Detection of *pap*, *sfa* and *afa* adhesin-encoding operons in uropathogenic *Escherichia coli* strains: relationship with expression of adhesins and production of toxins. *Res. Microbiol.* **148:**745–755.

328. Bleiweis, A. S., P. C. Oyston, and L. J. Brady. 1992. Molecular, immunological and functional characterization of the major surface adhesin of *Streptococcus mutans*. *Adv. Exp. Med. Biol.* **327:**229–241.

329. Bleumink-Pluym, N. M., E. A. ter Laak, D. J. Houwers, and B. A. van der Zeijst. 1996. Differences between *Taylorella equigenitalis* strains in their invasion of and replication in cultured cells. *Clin. Diagn. Lab. Immunol.* **3:**47–50.

330. Blevins, J. S., A. F. Gillaspy, T. M. Rechtin, B. K. Hurlburt, and M. S. Smeltzer. 1999. The staphylococcal accessory regulator (*sar*) represses transcription of the *Staphylococcus aureus* collagen adhesin gene (*cna*) in an *agr*-independent manner. *Mol. Microbiol.* **33:** 317–326.

331. Bliska, J. B., J. E. Galán, and S. Falkow. 1993. Signal transduction in the mammalian cell during bacterial attachment and entry. *Cell* **73:**903–920. .

332. Bliska, J. B., M. C. Copass, and S. Falkow. 1993. The *Yersinia pseudotuberculosis* adhesin YadA mediates intimate bacterial attachment to and entry into HEp-2 cells. *Infect. Immun.* **61:**3914–3921.

333. Blix, I. J., R. Hars, H. R. Preus, and K. Helgeland. 1992. Entrance of *Actinobacillus actinomycetemcomitans* into HEp-2 cells *in vitro*. *J. Periodontol.* **63:**723–728.

334. Blom, J., I. Heron, and J. O. Hendley. 1994. Immunoelectron microscopy of antigens of *Bordetella pertussis* using monoclonal antibodies to agglutinogens 2 and 3, filamentous haemagglutinin, pertussis toxin, pertactin and adenylate cyclase toxin. *APMIS* **102:**681–968.

335. Blomberg, L., A. Henriksson, and P. L. Conway. 1993. Inhibition of adhesion of *Escherichia coli* K88 to piglet ileal mucus by *Lactobacillus* spp. *Appl. Environ. Microbiol.* **59:**34–39.

336. Blomberg, L., H. C. Krivan, P. S. Cohen, and P. L. Conway. 1993. Piglet

ileal mucus contains protein and glycolipid (galactosylceramide) receptors specific for *Escherichia coli* K88 fimbriae. *Infect. Immun.* **61:**2526–2531.

337. **Blomberg, L., P. S. Cohen, and P. L. Conway.** 1993. A study of the adhesive capacity of *Escherichia coli* strain Bd 1107/7508 (K88ac) in relation to growth phase. *Microb. Pathog.* **14:**67–74.

338. **Blomfield, I. C., P. J. Calie, K. J. Eberhardt, M. S. McClain, and B. I. Eisenstein.** 1993. Lrp stimulates phase variation of type 1 fimbriation in *Escherichia coli* K-12. *J. Bacteriol.* **175:**27–36.

339. **Bloomquist, C. G., B. E. Reilly, and W. F. Liljemark.** 1996. Adherence, accumulation and cell division of a natural adherent bacterial population. *J. Bacteriol.* **178:**1172–1177.

340. **Blum, G., V. Falbo, A. Caprioli, and J. Hacker.** 1995. Gene clusters encoding the cytotoxic necrotizing factor type 1, Prs-fimbriae and α-hemolysin form the pathogenicity island II of the uropathogenic *Escherichia coli* strain J96. *FEMS Microbiol. Lett.* **126:**189–196.

341. **Blunden, R. E., R. G. Oliver, and C. O. O'Kane.** 1994. Microbial growth on the surfaces of various orthodontic bonding cements. *Br. J. Orthodont.* **21:**125–132.

342. **Boedeker, E. C.** 1994. Adherent bacteria: breaching the mucosal barrier? *Gastroenterology* **106:**255–257.

343. **Boelens, J. J., J. Dankert, J. L. Murk, J. J. Weening, T. van der Poll, K. P. Dingemans, L. Koole, J. D. Laman, and S. A. Zaat.** 2000. Biomaterial-associated persistence of *Staphylococcus epidermidis* in pericatheter macrophages. *J. Infect. Dis.* **181:**1337–1349.

344. **Boelens, J. J., W. F. Tan, J. Dankert, and S. A. Zaat.** 2000. Antibacterial activity of antibiotic-soaked polyvinylpyrrolidone-grafted silicon elastomer hydrocephalus shunts. *J. Antimicrob. Chemother.* **45:**221–224.

345. **Boelens, J. J., S. A. Zaat, J. Meeldijk, and J. Dankert.** 2000. Subcutaneous abscess formation around catheters induced by viable, and nonviable *Staphylococcus epidermidis* as well as by small amounts of bacterial cell wall components. *J. Biomed. Mater. Res.* **50:**546–556.

346. **Bokete, T. N., C. M. O'Callahan, C. R. Clausen, N. M. Tang, N. Tran, S. L. Moseley, T. R. Fritsche, and P. I. Tarr.** 1993. Shiga-like toxin-producing *Escherichia coli* in Seattle children: a prospective study. *Gastroenterology* **105:**1724–1731.

347. **Bokete, T. N., T. S. Whittam, R. A. Wilson, C. R. Clausen, C. M.** O'Callahan, S. L. Moseley, T. R. Fritsche, and P. I. Tarr. 1997. Genetic and phenotypic analysis of *Escherichia coli* with enteropathogenic characteristics isolated from Seattle children. *J. Infect. Dis.* **175:**1382–1389.

348. **Boles, S. F., M. F. Refojo, and F. L. Leong.** 1992. Attachment of *Pseudomonas* to human-worn, disposable etafilcon A contact lenses. *Cornea* **11:**47–52.

349. **Bolken, T. C., C. A. Franke, K. F. Jones, G. O. Zeller, C. H. Jones, E. K. Dutton, and D. E. Hruby.** 2001. Inactivation of the *srtA* gene in *Streptococcus gordonii* inhibits cell wall anchoring of surface proteins and decreases *in vitro* and *in vivo* adhesion. *Infect. Immun.* **69:**75–80.

350. **Bonci, A., A. Chiesurin, P. Muscas, and G. M. Rossolini.** 1997. Relatedness and phylogeny within the family of periplasmic chaperones involved in the assembly of pili or capsule-like structures of gram-negative bacteria. *J. Mol. Evol.* **44:**299–309.

351. **Bonilla-Ruz, L. F., and G. A. Garcia-Delgado.** 1993. Adherence of *Pasteurella multocida* to rabbit respiratory epithelial cells *in vitro. Rev. Latinoam. Microbiol.* **35:**361–369.

352. **Bonten, M. J., and R. A. Weinstein.** 1996. The role of colonization in the pathogenesis of nosocomial infections. *Infect. Control Hosp. Epidemiol.* **17:** 193–200.

353. **Booth, V., and T. Lehner.** 1997. Characterization of the *Porphyromonas gingivalis* antigen recognized by a monoclonal antibody which prevents colonization by the organism. *J. Periodont. Res.* **32:**54–60.

354. **Borén, T., P. Falk, K. A. Roth, G. Larson, and S. Normark.** 1993. Attachment of *Helicobacter pylori* to human gastric epithelium mediated by blood group antigens. *Science* **262:**1892–1895.

355. **Borén, T., S. Normark, and P. Falk.** 1994. *Helicobacter pylori*:molecular basis for host recognition and bacterial adherence. *Trends Microbiol.* **2:**221–228.

356. **Boris, S., J. E. Suárez, F. Vásquez, and C. Barbés.** 1998. Adherence of human vaginal lactobacilli to vaginal epithelial cells and interaction with uropathogens. *Infect. Immun.* **66:**1985–1989.

357. **Bos, M. P., F. Grunert, and R. J. Belland.** 1997. Differential recognition of members of the carcinoembryonic antigen family by Opa variants of *Neisseria gonorrhoeae.Infect. Immun.* **65:**2353–2361.

358. **Bos, R., H. C. van der Mei, J. M. Meinders, and H. J. Busscher.** 1994. A quantitative method to study co-adhesion of

microorganisms in a parallel plate flow chamber: basic principles of the analysis. *J. Microbiol. Methods* **20**:289–305.

359. **Bos, R., H. C. van der Mei, and H. J. Busscher.** 1995. A quantitative method to study co-adhesion of microorganisms in a parallel plate flow chamber. II. Analysis of the kinetics of co-adhesion. *J. Microbiol. Methods* **23**:169–182.

360. **Bos, R., H. C. van der Mei, and H. J. Busscher.** 1996. Co-adhesion of oral microbial pairs under flow in the presence of saliva and lactose. *J. Dent. Res.* **75**:809–815.

361. **Bos, R., H. C. van der Mei, and H. J. Busscher.** 1996. Influence of ionic strength and substratum hydrophobicity on the co-adhesion of oral microbial pairs. *Microbiology* **142**:2355–2361.

362. **Bos, R., H. C. van der Mei, and H. J. Busscher.** 1996. Influence of temperature on the co-adhesion of oral microbial pairs in saliva. *Eur. J. Oral Sci.* **104**:372–377.

363. **Bos, R., H. C. van der Mei, and H. J. Busscher.** 1999. Physico-chemistry of initial microbial adhesive interactions—its mechanisms and methods for study. *FEMS Microbiol. Rev.* **23**:179–230.

364. **Bosch, J. A., H. S. Brand, T. J. M. Ligtenberg, B. Bermond, J. Hoogstraten, and A. V. N. Amerongen.** 1996. Psychological stress as a determinant of protein levels and salivary-induced aggregation of *Streptococcus gordonii* in human whole saliva. *Psychosom. Med.* **58**:374–382.

365. **Bosch, J. A., E. J. de Geus, T. J. Ligtenberg, K. Nazmi, E. C. Veerman, J. Hoogstraten, and A. V. Amerongen.** 2000. Salivary MUC5B-mediated adherence (*ex vivo*) of *Helicobacter pylori* during acute stress. *Psychosom. Med.* **62**:40–49.

366. **Boschwitz, J. S., H. G. van der Heide, F. R. Mooi, and D. A. Relman.** 1997. *Bordetella bronchiseptica* expresses the fimbrial structural subunit gene *fimA.J. Bacteriol.* **179**:7882–7885.

367. **Bosworth, B. T., E. A. Dean-Nystrom, T. A. Casey, and H. L. Neibergs.** 1998. Differentiation of F18ab⁺ from F18ac⁺ Escherichia coli by single-strand conformational polymorphism analysis of the major fimbrial subunit gene (*fedA*). *Clin. Diagn. Lab. Immunol.* **5**:299–302.

368. **Botta, G. A., A. Arzese, R. Minisini, and G. Trani.** 1994. Role of structural and extracellular virulence factors in gram-negative anaerobic bacteria. *Clin. Infect. Dis.* **18**:S260–S264.

369. **Boucher, P. E., K. Murakami, A.

Ishihama, and S. Stibitz.** 1997. Nature of DNA binding and RNA polymerase interaction of the *Bordetella pertussis* BvgA transcriptional activator at the *fha* promoter. *J. Bacteriol.* **179**:1755–1763.

370. **Boudeau, J., N. Barnich, and A. Darfeuille-Michaud.** 2001. Type 1 pili-mediated adherence of *Escherichia coli* strain LF82 isolated from Crohn's disease is involved in bacterial invasion of intestinal epithelial cells. *Mol. Microbiol.* **39**:1272–1284.

371. **Bouzari, S., B. R. Vatsala, and A. Varghese.** 1992. *In vitro* adherence property of cytolethal distending toxin (CLDT) producing EPEC strains and effect of the toxin on rabbit intestine. *Microb. Pathog.* **12**:153–157.

372. **Bouzari, S., A. Jafari, A. A. Farhoudi-Moghaddam, F. Shokouhi, and M. Parsi.** 1994. Adherence of non-enteropathogenic *Escherichia coli* to HeLa cells. *J. Med. Microbiol.* **40**:95–97.

373. **Bouzari, S., M. N. Jafari, F. Shokouhi, M. Parsi, and A. Jafari.** 2000. Virulence-related DNA sequences and adherence patterns in strains of enteropathogenic *Escherichia coli.* *FEMS Microbiol. Lett.* **185**:89–93.

374. **Bowden, G. H., and I. R. Hamilton.** 1998. Survival of oral bacteria. *Crit. Rev. Oral Biol. Med.* **9**:54–85.

375. **Bower, C. K., J. McGuire, and M. A. Daeschel.** 1995. Influences on the antimicrobial activity of surface-adsorbed nisin. *J. Ind. Microbiol.* **15**:227–233.

376. **Bower, C. K., J. McGuire, and M. A. Daeschel.** 1995. Suppression of *Listeria monocytogenes* colonization following adsorption of nisin onto silica surfaces. *Appl. Environ. Microbiol.* **61**:992–997.

377. **Bower, C. K., M. A. Daeschel, and J. McGuire.** 1998. Protein antimicrobial barriers to bacterial adhesion. *J. Dairy Sci.* **81**:2771–2778.

378. **Boyd, E. F., and D. L. Hartl.** 1998. Chromosomal regions specific to pathogenic isolates of *Escherichia coli* have a phylogenetically clustered distribution. *J. Bacteriol.* **180**:1159–1165.

379. **Boyd, E. F., and D. L. Hartl.** 1998. Diversifying selection governs sequence polymorphism in the major adhesin proteins *fimA, papA,* and *sfaA* of *Escherichia coli.* *J. Mol. Evol.* **47**:258–267.

380. **Boyd, J. M., T. Koga, and S. Lory.** 1994. Identification and characterization of PilS, an essential regulator of pilin expression in *Pseudomonas aeruginosa.* *Mol. Gen. Genet.* **243**:565–574.

381. **Bozue, J. A., and W. Johnson.** 1996. Interaction of *Legionella pneumophila* with *Acanthamoeba castellanii:* uptake by coiling phagocytosis and inhibition of phagosome-lysosome fusion. *Infect. Immun.* **64:**668–673.

382. **Braaten, B. A., J. V. Platko, M. W. van der Woude, B. H. Simons, F. K. de Graaf, J. M. Calvo, and D. A. Low.** 1992. Leucine-responsive regulatory protein controls the expression of both the *pap* and *fan* pili operons in *Escherichia coli. Proc. Natl. Acad. Sci. USA* **89:**4250–4254.

383. **Bradley, D. E., and C. R. Thompson.** 1992. Synthesis of unusual thick pili by *Escherichia coli* of EPEC serogroup O119. *FEMS Microbiol. Lett.* **73:**31–36.

384. **Bradshaw, D. J., P. D. Marsh, G. K. Watson, and C. Allison.** 1998. Role of *Fusobacterium nucleatum* and coaggregation in anaerobe survival in planktonic and biofilm oral microbial communities during aeration. *Infect. Immun.* **66:**4729–4732.

385. **Brady, L. J., D. A. Piacentini, P. J. Crowley, P. C. Oyston, and A. S. Bleiweis.** 1992. Differentiation of salivary agglutinin-mediated adherence and aggregation of mutans streptococci by use of monoclonal antibodies against the major surface adhesin P1. *Infect. Immun.* **60:**1008–1017.

386. **Brady, L. J., D. G. Cvitkovitch, C. M. Geric, M. N. Addison, J. C. Joyce, P. J. Crowley, and A. S. Bleiweis.** 1998. Deletion of the central proline-rich repeat domain results in altered antigenicity and lack of surface expression of the *Streptococcus mutans* P1 adhesin molecule. *Infect. Immun.* **66:**4274–4282.

387. **Braga, P. C., and G. Piatti.** 1992. Influence of enoxacin Sub-MICs on the adherence of *Staphylococcus aureus* and *Escherichia coli* to human buccal and urinary epithelial cells. *Chemotherapy* **38:**261–266.

388. **Braga, P. C., G. Piatti, and S. Reggio.** 1992. Sub-inhibitory concentrations of brodimoprim inhibit adhesion of *E. coli* to human uroepithelial cells. *Drugs Exp. Clin. Res.* **18:**233–237.

389. **Braga, P. C., and G. Piatti.** 1993. Favourable effects of sub-MIC rufloxacin concentrations in decreasing the pathogen-host cell adhesion. *Pharmacol. Res.* **28:**11–19.

390. **Braga, P. C., and G. Piatti.** 1993. Interference by subinhibitory concentrations of azithromycin with the mechanism of bacterial adhesion to human epithelial cells. *Chemotherapy* **39:**432–437.

391. **Braga, P. C., and G. Piatti.** 1993. Sub-lethal concentrations of clarithromycin interfere with the expression of *Staphylococcus aureus* adhesiveness to human cells. *J. Chemother.* **5:**159–163.

392. **Braga, P. C., and G. Piatti.** 1993. Subminimum inhibitory concentrations of ceftibuten reduce adherence of *Escherichia coli* to human cells and induces formation of long filaments. *Microbiol. Immunol.* **37:**175–179.

393. **Braga, P. C.** 1994. Effects of subinhibitory concentrations of seven macrolides and four fluoroquinolones on adhesion of *Staphylococcus aureus* to human mucosal cells. *Chemotherapy* **40:**304–310.

394. **Braga, P. C., G. Piatti, A. Limoli, M. Santoro, and T. Gazzola.** 1994. Inhibition of bacterial adhesion by sub-inhibitory concentrations: brodimoprim vs trimethoprim. *J. Chemother.* **5:**447–452.

395. **Braga, P. C., and S. Reggio.** 1995. Correlation between reduction of surface hydrophobicity of *S. aureus* and the decrease in its adhesiveness induced by subinhibitory concentrations of brodimoprim. *Pharmacol. Res.* **32:**315–319.

396. **Braga, P. C., M. Dal Sasso, S. Maci, S. Reggio, and G. Piatti.** 1995. Influence of subinhibitory concentrations of brodimoprim and trimethoprim on the adhesiveness, hydrophobicity, hemagglutination and motility of *Escherichia coli. Chemotherapy* **41:**50–58.

397. **Braga, P. C., M. Dal Sasso, and S. Maci.** 1997. Cefodizime: effects of sub-inhibitory concentrations on adhesiveness and bacterial morphology of *Staphylococcus aureus* and *Escherichia coli:* comparison with cefotaxime and ceftriaxone. *J. Antimicrob. Chemother.* **39:**79–84.

398. **Braga, P. C., M. Dal Sasso, L. Mancini, and M. T. Sala.** 1999. Influence of sub-minimum inhibitory concentrations of cefodizime on the phagocytosis, intracellular killing and oxidative bursts of human polymorphonuclear leukocytes. *Chemotherapy* **45:**166–174.

399. **Braga, P. C., M. Dal Sasso, M. T. Sala, and V. Gianelle.** 1999. Effects of erdosteine and its metabolites on bacterial adhesiveness. *Arzneim. Forsch.* **49:**344–350.

400. **Braga, P. C., M. T. Sala, and M. Dal Sasso.** 1999. Pharmacodynamic effects of subinhibitory concentrations of rufloxacin on bacterial virulence factors. *Antimicrob. Agents Chemother.* **43:**1013–1019.

401. **Braga, P. C., M. D. Sasso, and M. T. Sala.** 2000. Sub-MIC concentrations of cefodizime interfere with various factors affecting bacterial virulence. *J. Antimicrob. Chemother.* **45:**15–25.

402. **Braga, P. C., T. Zuccotti, and M. Dal Sasso.** 2001. Bacterial adhesiveness: effects of the SH metabolite of erdosteine (mucoactive drug) plus clarithromycin versus clarithromycin alone. *Chemotherapy* **47:**208–214.

403. **Bragg, R. R., G. Purdan, L. Coetzee, and J. A. Verschoor.** 1995. Effects of transformation on the hemagglutinins of *Haemophilus paragallinarum*. *Onderstepoort J. Vet. Res.* **62:**261–270.

404. **Brant, E. E., H. T. Sojar, A. Sharma, G. S. Bedi, R. J. Genco, and E. De Nardin.** 1995. Identification of linear antigenic sites on the *Porphyromonas gingivalis* 43-kDa fimbrillin subunit. *Oral Microbiol. Immunol.* **10:**146–150.

405. **Bratt, P., T. Boren, and N. Strömberg.** 1999. Secretory immunoglobulin A heavy chain presents Galbeta1-3GalNAc binding structures for *Actinomyces naeslundii* genospecies 1. *J. Dent. Res.* **78:**1238–1244.

406. **Breines, D. M., and J. C. Burnham.** 1994. Modulation of *Escherichia coli* type 1 fimbrial expression and adherence to uroepithelial cells following exposure of logarithmic phase cells to quinolones at subinhibitory concentrations. *J. Antimicrob. Chemother.* **34:**205–221.

407. **Breines, D. M., and J. C. Burnham.** 1995. The effects of quinolones on the adherence of type-1 fimbriated *Escherichia coli* to mannosylated agarose beads. *J. Antimicrob. Chemother.* **36:**911–925.

408. **Bremer, P. J., I. Monk, and C. M. Osborne.** 2001. Survival of *Listeria monocytogenes* attached to stainless steel surfaces in the presence or absence of *Flavobacterium* spp. *J. Food Prot.* **64:**1369–1376.

409. **Brenciaglia, M. I., A. M. Fornara, M. M. Scaltrito, and F. Dubini.** 1995. Influence of amoxicillin, erythromycin and metronidazole on adherence of *Helicobacter pylori*. *New Microbiol.* **18:**283–288.

410. **Brennan, M. J., and R. D. Shahin.** 1996. Pertussis antigens that abrogate bacterial adherence and elicit immunity. *Am. J. Respir. Crit. Care Med.* **154:**S145–S149.

411. **Brennan, M. J., G. Delogu, Y. Chen, S. Bardarov, J. Kriakov, M. Alavi, and W. R. Jacobs, Jr.** 2001. Evidence that mycobacterial PE_PGRS proteins are cell surface constituents that influence interactions with other cells. *Infect. Immun.* **69:**7326–7333.

412. **Brentjens, R. J., S. M. Spinola, and A. A. Campagnari.** 1994. *Haemophilus ducreyi* adheres to human keratinocytes. *Microb. Pathog.* **16:**243–247.

413. **Brentjens, R. J., M. Ketterer, M. A. Apicella, and S. M. Spinola.** 1996. Fine tangled pili expressed by *Haemophilus ducreyi* are a novel class of pili. *J. Bacteriol.* **178:**808–816.

414. **Bresser, P., L. van Alphen, F. J. Habets, A. A. Hart, J. Dankert, H. M. Jansen, and R. Lutter.** 1997. Persisting *Haemophilus influenzae* strains induce lower levels of interleukin-6 and interleukin-8 in H292 lung epithelial cells than nonpersisting strains. *Eur. Respir. J.* **10:**2319–2326.

415. **Bresser, P., R. Virkola, M. Jonsson-Vihanne, H. M. Jansen, T. K. Korhonen, and L. Van Alphen.** 2000. Interaction of clinical isolates of nonencapsulated *Haemophilus influenzae* with mammalian extracellular matrix proteins. *FEMS Immunol. Med. Microbiol.* **28:**129–132.

416. **Briandet, R., T. Meylheuc, C. Maher, and M. N. Bellon-Fontaine.** 1999. *Listeria monocytogenes* Scott A: cell surface charge, hydrophobicity, and electron donor and acceptor characteristics under different environmental growth conditions. *Appl. Environ. Microbiol.* **65:**5328–5333.

417. **Briandet, R., V. Leriche, B. Carpentier, and M. N. Bellon-Fontaine.** 1999. Effects of the growth procedure on the surface hydrophobicity of *Listeria monocytogenes* cells and their adhesion to stainless steel. *J. Food Prot.* **62:**994–998.

418. **Bridgett, M. J., M. C. Davies, and S. P. Denyer.** 1992. Control of staphylococcal adhesion to polystyrene surfaces by polymer surface modification with surfactants. *Biomaterials* **13:**411–416.

419. **Bridgett, M. J., M. C. Davies, S. P. Denyer, and P. R. Eldridge.** 1993. *In vitro* assessment of bacterial adhesion to Hydromer-coated cerebrospinal fluid shunts. *Biomaterials* **14:**184–188.

420. **Brockmeier, S. L.** 1999. Early colonization of the rat upper respiratory tract by temperature modulated *Bordetella bronchiseptica*. *FEMS Microbiol. Lett.* **174:**225–229.

421. **Brockmeier, S. L., and K. B. Register.** 2000. Effect of temperature modulation and bvg mutation of *Bordetella bronchiseptica* on adhesion, intracellular survival and cytotoxicity for swine alveolar macrophages. *Vet. Microbiol.* **73:**1–12.

422. **Brook, M. G., H. R. Smith, B. A. Bannister, M. McConnell, H. Chart, S. M. Scotland, A. Sawyer, M. Smith, and B. Rowe.** 1994. Prospective study of verocytotoxin-producing, enteroaggregative and diffusely adherent *Escherichia coli* in different diarrhoeal states. *Epidemiol. Infect.* **112:**63–67.

423. **Brooks, W., D. R. Demuth, S. Gil, and R. J. Lamont.** 1997. Identification of a *Streptococcus gordonii* SspB domain that mediates adhesion to *Porphyromonas gingivalis*. *Infect. Immun.* **65:**3753–3758.

424. **Brouqui, P., and D. Raoult.** 1996. *Bartonella quintana* invades and multiplies within endothelial cells *in vitro* and *in vivo* and forms intracellular blebs. *Res. Microbiol.* **147:**719–731.

425. **Brown, E. L., B. P. Guo, P. O'Neal, and M. Höök.** 1999. Adherence of *Borrelia burgdorferi*. Identification of critical lysine residues in DbpA required for decorin binding. *J. Biol. Chem.* **274:**26272–26278.

426. **Brown, R. C., and R. K. Taylor.** 1995. Organization of *tcp, acf,* and *toxT* genes within a ToxT-dependent operon. *Mol. Microbiol.* **16:**425–439.

427. **Bruinsma, G. M., M. Rustema-Abbing, H. C. van der Mei, and H. J. Busscher.** 2001. Effects of cell surface damage on surface properties and adhesion of *Pseudomonas aeruginosa*. *J Microbiol. Methods* **45:**95–101.

428. **Bruinsma, G. M., H. C. van der Mei, and H. J. Busscher.** 2001. Bacterial adhesion to surface hydrophilic and hydrophobic contact lenses. *Biomaterials* **22:**3217–3224.

429. **Brunstedt, M. R., S. Sapatnekar, K. R. Rubin, K. M. Kieswetter, N. P. Ziats, K. Merritt, and J. M. Anderson.** 1995. Bacteria/blood/material interactions. I. Injected and preseeded slime-forming *Staphylococcus epidermidis* in flowing blood with biomaterials. *J. Biomed. Mater. Res.* **29:**455–466.

430. **Brunton, J. L.** 1992. Characterization of EPEC adhesion. *J. Pediatr. Gastroenterol. Nutr.* **15:**214–215.

431. **Bryan, R., D. Kube, A. Perez, P. Davis, and A. Prince.** 1998. Overproduction of the CFTR R domain leads to increased levels of asialoGM1 and increased *Pseudomonas aeruginosa* binding by epithelial cells. *Am. J. Respir. Cell Mol. Biol.* **19:**269–277.

432. **Brydon, H. L., R. Bayston, R. Hayward, and W. Harkness.** 1996. Reduced bacterial adhesion to hydrocephalus shunt catheters mediated by cerebrospinal fluid proteins. *J. Neurol. Neurosurg. Psychiatry* **60:**671–675.

433. **Bubert, A., M. Kuhn, W. Goebel, and S. Kohler.** 1992. Structural and functional properties of the p60 proteins from different *Listeria* species. *J. Bacteriol.* **174:**8166–8171.

434. **Buckles, E. L., and E. M. Hill.** 2000. Interaction of *Bartonella bacilliformis* with human erythrocyte membrane proteins. *Microb. Pathog.* **29:**165–174.

435. **Bullitt, E., and L. Makowski.** 1995. Structural polymorphism of bacterial adhesion pili. *Nature* **373:**164–167.

436. **Bullitt, E., C. H. Jones, R. Striker, G. Soto, F. Jacob-Dubuisson, J. Pinkner, M. J. Wick, L. Makowski, and S. J. Hultgren.** 1996. Development of pilus organelle subassemblies *in vitro* depends on chaperone uncapping of a beta zipper. *Proc. Natl. Acad. Sci. USA* **93:**12890–12895.

437. **Bundy, K. J., L. G. Harris, B. A. Rahn, and R. G. Richards.** 2001. Measurement of fibroblast and bacterial detachment from biomaterials using jet impingement. *Cell Biol. Int.* **25:**289–307.

438. **Bunning, V. K., and R. B. Raybourne.** 1994. Fluorescent labeling of *Listeria* and *Salmonella*. Analysis of bacteria-host cell interactions by flow microfluorimetry. *Ann. N. Y. Acad. Sci.* **730:**273–275.

439. **Bunt, C. R., D. S. Jones, and I. G. Tucker.** 1993. The effects of pH, ionic strength and organic phase on the bacterial adhesion to hycrocarbons (BATH) test. *Int. J. Pharm.* **99:**93–98.

440. **Burgemeister, S., E. M. Decker, R. Weiger, and M. Brecx.** 2001. Bactericidal effect of delmopinol on attached and planktonic *Streptococcus sanguinis* cells. *Eur. J. Oral Sci.* **109:**425–427.

441. **Burger, O., I. Ofek, M. Tabak, E. I. Weiss, N. Sharon, and I. Neeman.** 2000. A high molecular mass constituent of cranberry juice inhibits *Helicobacter pylori* adhesion to human gastric mucus. *FEMS Immunol. Med. Microbiol.* **29:**295–301.

442. **Burgess, A. W., L. J. Paradise, D. Hilbelink, and H. Friedman.** 1994. Adherence of *Treponema pallidum* subsp. *pallidum* in the rabbit placenta. *Proc. Soc. Exp. Biol. Med.* **207:**180–185.

443. **Burgess, A. W., and B. E. Anderson.** 1998. Outer membrane proteins of *Bartonella henselae* and their interaction with human endothelial cells. *Microb. Pathog.* **25:**157–164.

444. **Burgess, A. W., J. Y. Paquet, J. J. Letesson, and B. E. Anderson.** 2000. Isolation, sequencing and expression of *Bartonella henselae omp43* and predicted membrane topology of the deduced protein. *Microb. Pathog.* **29:**73–80.

445. **Burghoff, R. L., L. Pallesen, K. A. Krogfelt, J. V. Newman, M. Richardson, J. L. Bliss, D. C. Laux, and P. S. Cohen.** 1993. Utilization of the mouse large intestine to select an *Escherichia coli* F-18 DNA sequence that enhances colonizing ability and stimulates

synthesis of type 1 fimbriae. *Infect. Immun.* **61:**1293–1300.

446. **Burnette-Curley, D., V. Wells, H. Viscount, C. L. Munro, J. C. Fenno, P. Fives-Taylor, and F. L. Macrina.** 1995. FimA, a major virulence factor associated with *Streptococcus parasanguis* endocarditis. *Infect. Immun.* **63:**4669–4674.

447. **Burns, E. H., Jr., J. M. Norman, M. D. Hatcher, and D. A. Bemis.** 1993. Fimbriae and determination of host species specificity of *Bordetella bronchiseptica. J. Clin. Microbiol.* **31:**1838–1844.

448. **Burns, S. M., and S. I. Hull.** 1999. Loss of resistance to ingestion and phagocytic killing by O(−) and K(−) mutants of a uropathogenic *Escherichia coli* O75:K5 strain. *Infect. Immun.* **67:**3757–3762.

449. **Busalmen, J. P., and S. R. de Sanchez.** 2001. Adhesion of *Pseudomonas fluorescens* (ATCC 17552) to nonpolarized and polarized thin films of gold. *Appl. Environ. Microbiol.* **67:**3188–3194.

450. **Busch, S., C. Rosenplanter, and B. Averhoff.** 1999. Identification and characterization of ComE and ComF, two novel pilin-like competence factors involved in natural transformation of *Acinetobacter* sp. strain BD413. *Appl. Environ. Microbiol.* **65:**4568–4574.

451. **Busolo, F., L. Francescon, F. Aragona, and F. Pagano.** 1995. Solid-phase binding of clinical isolates of *Escherichia coli* expressing different piliation phenotypes. Effect of glycosaminoglycans. *Urol. Res.* **22:**399–402.

452. **Busscher, H. J., G. I. Doornbusch, and H. C. van der Mei.** 1992. Adhesion of mutans streptococci to glass with and without a salivary coating as studied in a parallel-plate flow chamber. *J. Dent. Res.* **71:**491–500.

453. **Busscher, H. J., M. M. Cowan, and H. C. van der Mei.** 1992. On the relative importance of specific and non-specific approaches to oral microbial adhesion. *FEMS Microbiol. Rev.* **8:**199–209.

454. **Busscher, H. J., G. I. Geertsema-Doornbusch, and H. C. van der Mei.** 1993. On mechanisms of oral microbial adhesion. *J. Appl. Bacteriol.* **74:**136S–142S.

455. **Busscher, H. J., and H. C. van der Mei.** 1995. Use of flow chamber devices and image analysis methods to study microbial adhesion. *Methods Enzymol.* **253:**455–477.

456. **Busscher, H. J., R. Bos, and H. C. van der Mei.** 1995. Initial microbial adhesion is a determinant for the strength of biofilm adhesion. *FEMS Microbiol. Lett.* **128:**229–234.

457. **Busscher, H. J., and H. C. van der Mei.** 1997. Physico-chemical interactions in initial microbial adhesion and relevance for biofilm formation. *Adv. Dent. Res.* **11:**24–32.

458. **Busscher, H. J., G. I. Geertsema-Doornbusch, and H. C. van der Mei.** 1997. Adhesion to silicone rubber of yeasts and bacteria isolated from voice prostheses: influence of salivary conditioning films. *J. Biomed. Mater. Res.* **34:**201–209.

459. **Busscher, H. J., A. T. Poortinga, and R. Bos.** 1998. Lateral and perpendicular interaction forces involved in mobile and immobile adhesion of microorganisms on model solid surfaces. *Curr. Microbiol.* **37:**319–323.

460. **Busse, J., E. Hartmann, and C. A. Lingwood.** 1997. Receptor affinity purification of a lipid-binding adhesin from *Haemophilus influenzae. J. Infect. Dis.* **175:**77–83.

461. **Bustos, C., L. Zurita, P. Smith, P. Vallejos, J. C. Aguillon, M. Lopez, and A. Ferreira.** 1995. Humoral immune response anti K99 pilus from enterotoxigenic *Escherichia coli* in experimentally inoculated calves. *Biol. Res.* **28:**277–282.

462. **Caccavo, F., Jr.** 1999. Protein-mediated adhesion of the dissimilatory Fe(III)-reducing bacterium *Shewanella alga* BrY to hydrous ferric oxide. *Appl. Environ. Microbiol.* **65:**5017–5022.

463. **Cahill, E. S., D. T. O'Hagan, L. Illum, and K. Redhead.** 1993. Mice are protected against *Bordetella pertussis* infection by intranasal immunization with filamentous haemagglutinin. *FEMS Microbiol. Lett.* **107:**211–216.

464. **Cahill, E. S., D. T. O'Hagan, L. Illum, A. Barnard, K. H. Mills, and K. Redhead.** 1995. Immune responses and protection against *Bordetella pertussis* infection after intranasal immunization of mice with filamentous haemagglutinin in solution or incorporated in biodegradable microparticles. *Vaccine* **13:**455–462.

465. **Cai, S., M. R. L. Simionato, M. P. A. Mayer, N. F. Novo, and F. Zelante.** 1994. Effects of subinhibitory concentrations of chemical agents on hydrophobicity and *in vitro* adherence of *Streptococcus mutans* and *Streptococcus sanguis. Caries Res.* **28:**335–341.

466. **Calame, W., C. Afram, N. Blijleven, R. J. Hendrickx, F. Namavar, and R. H. Beelen.** 1995. Phagocytosis and killing of suspended and adhered bacteria by peritoneal cells after dialysis. *Periton. Dial. Int.* **15:**320–327.

467. **Calas, P., T. Rochd, and G. Michel.**

1994. *In vitro* attachment of *Streptococcus sanguis* to the dentin of the root canal. *J. Endodont.* **20:**71–74.

468. **Calas, P., T. Rochd, P. Druilhet, and J. M. Azais.** 1998. *In vitro* adhesion of two strains of *Prevotella nigrescens* to the dentin of the root canal: the part played by different irrigation solutions. *J. Endodont.* **24:**112–115.

469. **Caloca, M. J., J. Soler, and S. Suarez.** 1996. Adhesion of K88ab to guinea pig erythrocytes: effect on membrane enzyme activities. *Infect. Immun.* **64:**3416–3418.

470. **Caloca, M. J., S. Suarez, and J. Soler.** 1997. Metabolic changes in red cells in response to adhesion of porcine K88 fimbriated *Escherichia coli.* *Vet. Microbiol.* **58:**45–52.

471. **Caloca, M. J., S. Suarez, and J. Soler.** 1998. Binding characteristics of purified *Escherichia coli* K88ab fimbriae to guinea pig erythrocyte membrane. *Vet. Microbiol.* **61:**51–58.

472. **Calvinho, L. F., R. A. Almeida, and S. P. Oliver.** 1996. Influence of *Streptococcus dysgalactiae* surface hydrophobicity on adherence to mammary epithelial cells and phagocytosis by mammary macrophages. *Zentbl. Veterinaermed. Reihe B* **43:**257–266.

473. **Calvinho, L. F., and S. P. Oliver.** 1998. Factors influencing adherence of *Streptococcus dysgalactiae* to bovine mammary epithelial cell monolayers. *Zentbl. Veterinaermed. Reihe B* **45:**161–170.

474. **Camara, L. M., S. B. Carbonare, M. L. Silva, and M. M. Carneiro-Sampaio.** 1994. Inhibition of enteropathogenic *Escherichia coli* (EPEC) adhesion to HeLa cells by human colostrum: detection of specific sIgA related to EPEC outer-membrane proteins. *Int. Arch. Allergy Immunol.* **103:**307–310.

475. **Camara, L. M., S. B. Carbonare, I. C. Scaletsky, M. L. da Silva, and M. M. Carneiro-Sampaio.** 1995. Inhibition of enteropathogenic *Escherichia coli* (EPEC) adherence to HeLa cells by human colostrum. Detection of specific sIgA related to EPEC outer-membrane proteins. *Adv. Exp. Med. Biol.* **371A:**673–676.

476. **Campbell, A. P., C. McInnes, R. S. Hodges, and B. D. Sykes.** 1995. Comparison of NMR solution structures of the receptor binding domains of *Pseudomonas aeruginosa* pili strains PAO, KB7, and PAK: implications for receptor binding and synthetic vaccine design. *Biochemistry* **34:**16255–16268.

477. **Campbell, A. P., H. Sheth, R. S. Hodges, and B. D. Sykes.** 1996. NMR solution structure of the receptor binding domain of *Pseudomonas aeruginosa* pilin strain P1. Identification of a beta-turn. *Int. J. Pept. Protein Res.* **48:**539–552.

478. **Campbell, G. A., L. G. Adams, and B. A. Sowa.** 1994. Mechanisms of binding of *Brucella abortus* to mononuclear phagocytes from cows naturally resistant or susceptible to brucellosis. *Vet. Immunol. Immunopathol.* **41:**295–306.

479. **Campos, L. C., T. S. Whittam, T. A. Gomes, J. R. Andrade, and L. R. Trabulsi.** 1994. *Escherichia coli* serogroup O111 includes several clones of diarrheagenic strains with different virulence properties. *Infect. Immun.* **62:**3282–3288.

480. **Canil, C., I. Rosenshine, S. Ruschkowski, M. S. Donnenberg, J. B. Kaper, and B. B. Finlay.** 1993. Enteropathogenic *Escherichia coli* decreases the transepithelial electrical resistance of polarized epithelial monolayers. *Infect. Immun.* **61:**2755–2762.

481. **Cantey, J. R., R. K. Blake, J. R. Williford, and S. L. Moseley.** 1999. Characterization of the *Escherichia coli* AF/R1 pilus operon: novel genes necessary for transcriptional regulation and for pilus-mediated adherence. *Infect. Immun.* **67:**2292–2298.

482. **Cao, J., A. S. Khan, M. E. Bayer, and D. M. Schifferli.** 1995. Ordered translocation of 987P fimbrial subunits through the outer membrane of *Escherichia coli.* *J. Bacteriol.* **177:**3704–3713.

483. **Capo, C., F. P. Lindberg, S. Meconim, Y. Zaffran, G. Tardei, E. J. Brown, D. Raoult, and J. L. Mege.** 1999. Subversion of monocyte functions by *Coxiella burnetii*: impairment of the cross-talk between alphav-beta3 integrin and CR3. *J. Immunol.* **163:**6078–6085.

484. **Cappelier, J. M., J. Minet, C. Magras, R. R. Colwell, and M. Federighi.** 1999. Recovery in embryonated eggs of viable but nonculturable *Campylobacter jejuni* cells and maintenance of ability to adhere to HeLa cells after resuscitation. *Appl. Environ. Microbiol.* **65:**5154–5157.

485. **Caprilli, F., F. Mongio, G. Palamara, G. Prignano, and L. Baldassarri.** 1993. *In vitro* adherence of *Neisseria gonorrhoeae* to human epithelial cells. *New Microbiol.* **16:**51–56.

486. **Carballo, J., C. M. Ferreiros, and M. T. Criado.** 1992. Factor analysis in the evaluation of the relationship between bacterial adherence to biomaterials and changes in free energy. *J. Biomater. Appl.* **7:**130–141.

487. **Carbonare, S. B., C. Arslanian, M. L.**

Silva, C. K. Farhat, and M. M. Carneiro-Sampaio. 1995. The antimeningococcal vaccine VAMENGOC B-C induced poor serum and salivary antibody response in young Brazilian children. *Pediatr. Infect. Dis. J.* **14:**797–803.

488. Carbonare, S. B., P. Palmeira, M. L. Silva, and M. M. Carneiro-Sampaio. 1996. Effect of microwave radiation, pasteurization and lyophilization on the ability of human milk to inhibit *Escherichia coli* adherence to HEp-2 cells. *J. Diarrhoeal Dis. Res.* **14:**90–94.

489. Carlen, A., and J. Olsson. 1995. Monoclonal antibodies against a high-molecular-weight agglutinin block adherence to experimental pellicles on hydroxyapatite and aggregation of *Streptococcus mutans. J. Dent. Res.* **74:**1040–1047.

490. Carlen, A., J. Olsson, and A. C. Borjesson. 1996. Saliva-mediated binding *in vitro* and prevalence *in vivo* of *Streptococcus mutans. Arch. Oral Biol.* **41:**35–39.

491. Carlen, A., J. Olsson, and P. Ramberg. 1996. Saliva mediated adherence, aggregation and prevalence in dental plaque of *Streptococcus mutans, Streptococcus sanguis* and *Actinomyces* spp, in young and elderly humans. *Arch. Oral Biol.* **41:**1133–1140.

492. Carlen, A., P. Bratt, C. Stenudd, J. Olsson, and N. Strömberg. 1998. Agglutinin and acidic proline-rich protein receptor patterns may modulate bacterial adherence and colonization on tooth surfaces. *J. Dent. Res.* **77:**81–90.

493. Carlson, S. A., M. Browning, K. E. Ferris, and B. D. Jones. 2000. Identification of diminished tissue culture invasiveness among multiple antibiotic resistant *Salmonella typhimurium* DT104. *Microb. Pathog.* **28:**37–44.

494. Carneiro, C. R., E. Postol, C. Boilesen, and R. R. Brentani. 1993. Participation of glycosylation sites in the binding of *Staphylococcus aureus* to laminin. *Braz. J. Med. Biol. Res.* **26:**689–697.

495. Carnoy, C., R. Ramphal, A. Scharfman, J. M. Lo-Guidice, N. Houdret, A. Klein, C. Galabert, G. Lamblin, and P. Roussel. 1993. Altered carbohydrate composition of salivary mucins from patients with cystic fibrosis and the adhesion of *Pseudomonas aeruginosa. Am. J. Respir. Cell Mol. Biol.* **9:**323–334.

496. Carnoy, C., A. Scharfman, E. van Brussel, G. Lamblin, R. Ramphal, and P. Roussel. 1994. *Pseudomonas aeruginosa* outer membrane adhesins for human respiratory mucus glycoproteins. *Infect. Immun.* **62:**1896–1900.

497. Carranza, N., Jr., G. R. Riviere, K. S. Smith, D. F. Adams, and T. Maier. 1997. Differential attachment of oral treponemes to monolayers of epithelial cells. *J. Periodontol.* **68:**1010–1018.

498. Carrick, C. S., J. A. Fyfe, and J. K. Davies. 1997. The normally silent sigma54 promoters upstream of the *pilE* genes of both *Neisseria gonorrhoeae* and *Neisseria meningitidis* are functional when transferred to *Pseudomonas aeruginosa. Gene* **198:**89–97.

499. Carsenti-Etesse, H., J. Durant, E. Bernard, V. Mondain, J. Entenza, and P. Dellamonica. 1992. Effect of subinhibitory concentrations of cefamandole and cefuroxime on adherence of *Staphylococcus aureus* and *Staphylococcus epidermidis* to polystyrene culture plates. *Eur. J. Clin. Microbiol. Infect. Dis.* **11:**732–737.

500. Carsenti-Etesse, H., J. Entenza, J. Durant, C. Pradier, V. Mondain, E. Bernard, and P. Dellamonica. 1992. Efficacy of subinhibitory concentration of pefloxacin in preventing experimental *Staphylococcus aureus* foreign body infection in mice. *Drugs Exp. Clin. Res.* **18:**415–422.

501. Carsenti-Etesse, H., J. Durant, J. Entenza, V. Mondain, C. Pradier, E. Bernard, and P. Dellamonica. 1993. Effects of subinhibitory concentrations of vancomycin and teicoplanin on adherence of staphylococci to tissue culture plates. *Antimicrob. Agents Chemother.* **37:**921–923.

502. Carvalho, M. G., and L. M. Teixeira. 1995. Hemagglutination properties of *Enterococcus. Curr. Microbiol.* **30:**265–168.

503. Casaz, P., A. Happel, J. Keithan, D. L. Read, S. R. Strain, and S. B. Levy. 2001. The *Pseudomonas fluorescens* transcription activator AdnA is required for adhesion and motility. *Microbiology* **147:**355–361.

504. Casey, T. A., B. Nagy, and H. W. Moon. 1992. Pathogenicity of porcine enterotoxigenic *Escherichia coli* that do not express K88, K99, F41, or 987P adhesins. *Am. J. Vet. Res.* **53:**1488–1492.

505. Casey, T. A., R. A. Schneider, and E. A. Dean-Nystrom. 1993. Identification of plasmid and chromosomal copies of 987P pilus genes in enterotoxigenic *Escherichia coli* 987. *Infect. Immun.* **61:**2249–2252.

506. Casey, T. A., C. J. Herring, R. A. Schneider, B. T. Bosworth, and S. C. Whipp. 1998. Expression of heat-stable enterotoxin STb by adherent *Escherichia coli* is not sufficient to cause severe diarrhea in neonatal pigs. *Infect. Immun.* **66:**1270–1272.

507. **Cassels, F. J., C. D. Deal, R. H. Reid, D. L. Jarboe, J. L. Nauss, J. M. Carter, and E. C. Boedeker.** 1992. Analysis of *Escherichia coli* colonization factor antigen I linear B-cell epitopes, as determined by primate responses, following protein sequence verification. *Infect. Immun.* **60:**2174–2181.

508. **Cassels, F. J., C. V. Hughes, and J. L. Nauss.** 1995. Adhesin receptors of human oral bacteria and modeling of putative adhesin-binding domains. *J. Ind. Microbiol.* **15:**176–185.

509. **Cassels, F. J., and H. van Halbeek.** 1995. Isolation and structural characterization of adhesin polysaccharide receptors. *Methods Enzymol.* **253:**69–91.

510. **Cassels, F. J., and M. K. Wolf.** 1995. Colonization factors of diarrheagenic *E. coli* and their intestinal receptors. *J. Ind. Microbiol.* **15:**214–226.

511. **Castellazzo, A., M. Shero, M. A. Apicella, and S. M. Spinola.** 1992. Expression of pili by *Haemophilus ducreyi. J. Infect. Dis.* **165:**S198–S199.

512. **Castric, P. A., and C. D. Deal.** 1994. Differentiation of *Pseudomonas aeruginosa* pili based on sequence and B-cell epitope analyses. *Infect. Immun.* **62:**371–376.

513. **Catalanotti, P., F. Rossano, P. de Paolis, A. Baroni, G. Buttini, and M. A. Tufano.** 1994. Effects of cetyltrimethylammonium naproxenate on the adherence of *Gardnerella vaginalis, Mobiluncus curtisii,* and *Lactobacillus acidophilus* to vaginal epithelial cells. *Sex. Transm. Dis.* **21:**338–344.

514. **Catani, C. F., A. T. Yamada, M. C. Vidotto, and T. Yano.** 1996. Adhesion of bovine enterotoxigenic *Escherichia coli* (ETEC) by type 1-like fimbriae. *FEMS Microbiol. Lett.* **137:**241–245.

515. **Cavedon, K., and J. London.** 1993. Adhesin degradation: a possible function for a *Prevotella loescheii* protease? *Oral Microbiol. Immunol.* **8:**283–287.

516. **Cervin, M. A., D. A. Simpson, A. L. Smith, and S. Lory.** 1994. Differences in eucaryotic cell binding of *Pseudomonas. Microb. Pathog.* **17:**291–299.

517. **Chakrabarti, M. K., J. Bhattacharya, A. K. Sinha, and S. P. De.** 1998. Role of outer membrane proteins on the adherence of *Vibrio parahaemolyticus* to rabbit intestinal epithelial cell *in vitro. Zentbl. Bakteriol.* **282:**436–441.

518. **Chakrabarti, S. R., K. Chaudhuri, K. Sen, and J. Das.** 1996. Porins of *Vibrio cholerae:* purification and characterization of OmpU. *J. Bacteriol.* **178:**524–530.

519. **Chan, K. N., A. D. Phillips, S. Knutton, H. R. Smith, and J. A. Walker-Smith.** 1994. Enteroaggregative *Escherichia coli:*another cause of acute and chronic diarrhoea in England? *J. Pediatr. Gastroenterol. Nutr.* **18:**87–91.

520. **Chance, D. L., T. J. Reilly, and A. L. Smith.** 1999. Acid phosphatase activity as a measure of *Haemophilus influenzae* adherence to mucin. *J. Microbiol. Methods* **39:**49–58.

521. **Chandad, F., and C. Mouton.** 1995. Antigenic, structural, and functional relationships between fimbriae and the hemagglutinating adhesin HA-Ag2 of *Porphyromonas gingivalis. Infect. Immun.* **63:**4755–4763.

522. **Chandad, F., D. Mayrand, D. Grenier, D. Hinode, and C. Mouton.** 1996. Selection and phenotypic characterization of nonhemagglutinating mutants of *Porphyromonas gingivalis. Infect. Immun.* **64:**952–958.

523. **Chandler, D. S., T. L. Mynott, R. K. Luke, and J. A. Craven.** 1994. The distribution and stability of *Escherichia coli* K88 receptor in the gastrointestinal tract of the pig. *Vet. Microbiol.* **38:**203–215.

524. **Chang, C. C., and K. Merritt.** 1992. Microbial adherence on poly(methyl methacrylate) (PMMA) surfaces. *J. Biomed. Mater. Res.* **26:**197–207.

525. **Chang, C. C., and K. Merritt.** 1994. Infection at the site of implanted materials with and without preadhered bacteria. *J. Orthop. Res.* **12:**526–531.

526. **Charland, N., J. Harel, M. Kobisch, S. Lacasse, and M. Gottschalk.** 1998. *Streptococcus suis* serotype 2 mutants deficient in capsular expression. *Microbiology* **144:**325–332.

527. **Charland, N., V. Nizet, C. E. Rubens, K. S. Kim, S. Lacouture, and M. Gottschalk.** 2000. *Streptococcus suis* serotype 2 interactions with human brain microvascular endothelial cells. *Infect. Immun.* **68:**637–643.

528. **Charpentier, E., R. Novak, and E. Tuomanen.** 2000. Regulation of growth inhibition at high temperature, autolysis, transformation and adherence in *Streptococcus pneumoniae* by *clpC. Mol. Microbiol.* **37:**717–726.

529. **Chart, H., H. R. Smith, and B. Rowe.** 1994. Antigenic cross-reactions between fimbriae expressed by *Salmonella enteritidis, S. dublin* and an 18 kDa outer membrane associated protein expressed by *Escherichia coli* O126:H27. *FEMS Microbiol. Lett.* **121:**19–23.

530. **Chart, H., H. R. Smith, and B. Rowe.** 1995. Enteroaggregative strains of *Escherichia coli* belonging to serotypes O126:H27 and

O44:H18 express antigenically similar 18 kDa outer membrane-associated proteins. *FEMS Microbiol. Lett.* **132:**17–22.

531. **Chart, H., J. Spencer, H. R. Smith, and B. Rowe.** 1997. Magnesium ions are required for HEp-2 cell adhesion by enteroaggregative strains of *Escherichia coli* O126:H27 and O44:H18. *FEMS Microbiol. Lett.* **148:**49–52.

532. **Chart, H., J. Spencer, H. R. Smith, and B. Rowe.** 1997. Identification of enteroaggregative *Escherichia coli* based on surface properties. *J. Appl. Microbiol.* **83:**712–717.

533. **Chart, H., H. R. Smith, R. M. La Ragione, and M. J. Woodward.** 2000. An investigation into the pathogenic properties of *Escherichia coli* strains BLR, BL21, DH5alpha and EQ1. *J. Appl. Microbiol.* **89:**1048–1058.

534. **Chaussee, M. S., R. L. Cole, and J. P. van Putten.** 2000. Streptococcal erythrogenic toxin B abrogates fibronectin-dependent internalization of *Streptococcus pyogenes* by cultured mammalian cells. *Infect. Immun.* **68:**3226–3232.

535. **Chauvière, G., M. H. Coconnier, S. Kerneis, A. Darfeuille-Michaud, B. Joly, and Å. L. Servin.** 1992. Competitive exclusion of diarrheagenic *Escherichia coli* (ETEC) from human enterocyte-like Caco-2 cells by heat-killed *Lactobacillus*. *FEMS Microbiol. Lett.* **70:**213–217.

536. **Chauvière, G., M. H. Coconnier, S. Kerneis, J. Fourniat, and Å. L. Servin.** 1992. Adhesion of human *Lactobacillus acidophilus* strain LB to human enterocyte-like Caco-2 cells. *J. Gen. Microbiol.* **138:**1689–1696.

537. **Chauvière, G., M.-H. Coconnier, S. Kerneis, A. Darfeuille-Michaud, B. Joly, and Å. L. Servin.** 1992. Competitive exclusion of diarrheagenic *Escherichia coli* (ETEC) from human enterocyte-like Caco-2 cells by heat-killed *Lactobacillus*. *FEMS Microbiol. Lett.* **91:**213–218.

538. **Chen, C.-J., P. F. Sparling, L. A. Lewis, D. W. Dyer, and C. Elkins.** 1996. Identification and purification of a hemoglobin-binding outer membrane protein from *Neisseria gonorrhoeae*. *Infect. Immun.* **64:**5008–5014.

539. **Chen, C. P., S. C. Song, N. Gilboa-Garber, K. S. Chang, and A. M. Wu.** 1998. Studies on the binding site of the galactose-specific agglutinin PA-IL from *Pseudomonas aeruginosa*. *Glycobiology* **8:**7–16.

540. **Chen, J. C., and R. S. Stephens.** 1997. *Chlamydia trachomatis* glycosaminoglycan-dependent and independent attachment to eukaryotic cells. *Microb. Pathog.* **22:**23–30.

541. **Chen, J. R., J. H. Lin, C. N. Weng, and S. S. Lai.** 1998. Identification of a novel adhesin-like glycoprotein from *Mycoplasma hyopneumoniae*. *Vet. Microbiol.* **62:**97–110.

542. **Chen, L., J. A. Hobden, S. A. Masinick, and L. D. Hazlett.** 1998. Environmental factors influence *P. aeruginosa* binding to the wounded mouse cornea. *Curr. Eye Res.* **17:**231–237.

543. **Chen, L., and L. D. Hazlett.** 2001. Human corneal epithelial extracellular matrix perlecan serves as a site for *Pseudomonas aeruginosa* binding. *Curr Eye Res.* **22:**19–27.

544. **Chen, L. D., and L. D. Hazlett.** 2000. Perlecan in the basement membrane of corneal epithelium serves as a site for *P. aeruginosa* binding. *Curr. Eye Res.* **20:**260–267.

545. **Chen, Q., A. Barragan, V. Fernandez, A. Sundström, M. Schlichtherle, A. Sahlén, J. Carlson, S. Datta, and M. Wahlgren.** 1998. Identification of *Plasmodium falciparum* erythrocyte membrane protein 1 (PfEMP1) as the rosetting ligand of the malaria parasite *P. falciparum*. *J. Exp. Med.* **187:**15–23.

546. **Chen, T., R. J. Belland, J. Wilson, and J. Swanson.** 1995. Adherence of pilus- Opa+ gonococci to epithelial cells *in vitro* involves heparan sulfate. *J. Exp. Med.* **182:**511–517.

547. **Chen, T., and E. C. Gotschlich.** 1996. CGM1a antigen of neutrophils, a receptor of gonococcal opacity proteins. *Proc. Natl Acad Sci. USA* **93:**14851–14856.

548. **Chen, T., F. Grunert, A. Medina-Marino, and E. C. Gotschlich.** 1997. Several carcinoembryonic antigens (CD66) serve as receptors for gonococcal opacity proteins. *J. Exp. Med.* **185:**1557–1564.

549. **Chen, T., H. Dong, R. Yong, and M. J. Duncan.** 2000. Pleiotropic pigmentation mutants of *Porphyromonas gingivalis*. *Microb. Pathog.* **28:**235–247.

550. **Chen, T., K. Nakayama, L. Belliveau, and M. J. Duncan.** 2001. *Porphyromonas gingivalis* gingipains and adhesion to epithelial cells. *Infect. Immun.* **69:**3048–3056.

551. **Cheng, X., J. D. Cirillo, and G. E. Duhamel.** 1999. Coiling phagocytosis is the predominant mechanism for uptake of the colonic spirochetosis bacterium *Serpulina pilosicoli* by human monocytes. *Adv. Exp. Med. Biol.* **473:**207–214.

552. **Cheung, A. L., K. J. Eberhardt, E. Chung, M. R. Yeaman, P. M. Sullam, M. Ramos, and A. S. Bayer.** 1994. Diminished virulence of a *sar+agr+* mutant of *Staphylococcus aureus* in the rabbit model of endocarditis. *J. Clin. Investig.* **94:**1815–1822.

553. Cheung, A. L., M. R. Yeaman, P. M. Sullam, M. D. Witt, and A. S. Bayer. 1994. Role of the *sar* locus of *Staphylococcus aureus* in induction of endocarditis in rabbits. *Infect. Immun.* **62**:1719–1725.

554. Chia, J. S., R. H. Lin, S. W. Lin, J. Y. Chen, and C. S. Yang. 1993. Inhibition of glucosyltransferase activities of *Streptococcus mutans* by a monoclonal antibody to a subsequence peptide. *Infect. Immun.* **61**:4689–4695.

555. Chia, J. S., C. Y. Yeh, and J. Y. Chen. 2000. Identification of a fibronectin binding protein from *Streptococcus mutans*. *Infect. Immun.* **68**:1864–1870.

556. Chiang, S. L., R. K. Taylor, M. Koomey, and J. J. Mekalanos. 1995. Single amino acid substitutions in the N-terminus of *Vibrio cholerae* TcpA affect colonization, autoagglutination, and serum resistance. *Mol. Microbiol.* **17**:1133–1142.

557. Chiang, S. L., and J. J. Mekalanos. 1998. Use of signature-tagged transposon mutagenesis to identify *Vibrio cholerae* genes critical for colonization. *Mol. Microbiol.* **27**:797–805.

558. Chiang, S. L., and J. J. Mekalanos. 1999. *rfb* mutations in *Vibrio cholerae* do not affect surface production of toxin-coregulated pili but still inhibit intestinal colonization. *Infect. Immun.* **67**:976–980.

559. Chiavelli, D. A., J. W. Marsh, and R. K. Taylor. 2001 The mannose-sensitive hemagglutinin of *Vibrio cholerae* promotes adherence to zooplankton. *Appl. Environ. Microbiol.* **67**:3220–3225.

560. China, B., V. Pirson, E. Jacquemin, P. Pohl, and J. G. Mainil. 1997. Pathotypes of bovine verotoxigenic *Escherichia coli* isolates producing attaching/effacing (AE) lesions in the ligated intestinal loop assay in rabbits. *Adv. Exp. Med. Biol.* **412**:311–316.

561. Chiu, C. H., S. Wong, R. E. Hancock, and D. P. Speert. 2001. Adherence of *Burkholderia cepacia* to respiratory tract epithelial cells and inhibition with dextrans. *Microbiology* **147**:2651–2658.

562. Chmiela, M., J. Lelwala-Guruge, and T. Wadström. 1994. Interaction of cells of *Helicobacter pylori* with human polymorphonuclear leucocytes: possible role of haemagglutinins. *FEMS Immunol. Med. Microbiol.* **9**:41–48.

563. Chmiela, M., B. Paziak-Domanska, and T. Wadström. 1995. Attachment, ingestion and intracellular killing of *Helicobacter pylori* by human peripheral blood mononuclear leukocytes and mouse peritoneal inflammatory macrophages. *FEMS Immunol. Med. Microbiol.* **10**:307–316.

564. Chmiela, M., B. Paziak-Domanska, W. Rudnicka, and T. Wadström. 1995. The role of heparan sulphate-binding activity of *Helicobacter pylori* bacteria in their adhesion to murine macrophages. *APMIS* **103**:469–474.

565. Chmiela, M., A. Ljungh, W. Rudnicka, and T. Wadström. 1996. Phagocytosis of *Helicobacter pylori* bacteria differing in the heparan sulfate binding by human polymorphonuclear leukocytes. *Zentbl. Bakteriol.* **283**:346–350.

566. Chmiela, M., E. Czkwianianc, T. Wadström, and W. Rudkicka. 1997. Role of *Helicobacter pylori* surface structures in bacterial interaction with macrophages. *Gut* **40**:20–24.

567. Chmiela, M., M. Lawnik, E. Czkwianianc, T. Rechcinski, I. Planeta-Malecka, T. Wadström, and W. Rudnicka. 1997. Attachment of *Helicobacter pylori* strains to human epithelial cells. *J. Physiol. Pharmacol.* **48**:393–404.

568. Chmiela, M., T. Wadström, H. Folkesson, I. P. Malecka, E. Czkwianianc, T. Rechcinski, and W. Rudnicka. 1998. Anti-Lewis X antibody and Lewis X-anti-Lewis X immune complexes in *Helicobacter pylori* infection. *Immunol. Lett.* **61**:119–125.

569. Cho, S. H., I. Strickland, A. Tomkinson, A. P. Fehringer, E. W. Gelfand, and D. Y. Leung. Preferential binding of *Staphylococcus aureus* to skin sites of Th2-mediated inflammation in a murine model. *J Investig. Dermatol.* **116**:658–663.

570. Choi, B. K., and D. M. Schifferli. 1999. Lysine residue 117 of the FasG adhesin of enterotoxigenic *Escherichia coli* is essential for binding of 987P fimbriae to sulfatide. *Infect. Immun.* **67**:5755–5761.

571. Choi, J. I., R. E. Schifferle, F. Yoshimura, and B. W. Kim. 1998. Capsular polysaccharide-fimbrial protein conjugate vaccine protects against *Porphyromonas gingivalis* infection in SCID mice reconstituted with human peripheral blood lymphocytes. *Infect. Immun.* **66**:391–393.

572. Chuard, C., P. Vaudaux, F. A. Waldvogel, and D. P. Lew. 1993. Susceptibility of *Staphylococcus aureus* growing on fibronectin-coated surfaces to bactericidal antibiotics. *Antimicrob. Agents Chemother.* **37**:625–632.

573. Chung, W. O., D. R. Demuth, and R. J. Lamont. 2000. Identification of a *Porphyromonas gingivalis* receptor for the *Streptococcus gordonii* SspB protein. *Infect. Immun.* **68**:6758–6762.

574. Cid, D., J. A. Ruiz Santa Quiteria, and R. de la Fuente. 1993. F17 fimbriae in *Escherichia coli* from lambs and kids. *Vet. Rec.* 132:251.

575. Cifrian, E., A. J. Guidry, C. N. O'Brien, S. C. Nickerson, and W. W. Marquardt. 1994. Adherence of *Staphylococcus aureus* to cultured bovine mammary epithelial cells. *J. Dairy Sci.* 77:970–983.

576. Cifrian, E., A. J. Guidry, C. N. O'Brien, and W. W. Marquardt. 1995. Effect of alpha-toxin and capsular exopolysaccharide on the adherence of *Staphylococcus aureus* to cultured teat, ductal and secretory mammary epithelial cells. *Res. Vet. Sci.* 58:20–25.

577. Cifrian, E., A. J. Guidry, A. J. Bramley, N. L. Norcross, F. D. Bastida-Corcuera, and W. W. Marquardt. 1996. Effect of staphylococcal beta toxin on the cytotoxicity, proliferation and adherence of *Staphylococcus aureus* to bovine mammary epithelial cells. *Vet. Microbiol.* 48:187–198.

578. Cifrian, E., A. J. Guidry, C. N. O'Brien, and W. W. Marquardt. 1996. Effect of antibodies to staphylococcal alpha and beta toxins and *Staphylococcus aureus* on the cytotoxicity for and adherence of the organism to bovine mammary epithelial cells. *Am. J. Vet. Res.* 57:1308–1311.

579. Cimolai, N., A. C. Cheong, and C. Trombley. 1997. Markers of virulence among prospectively acquired putative enteropathogenic *Escherichia coli* serogroups. *Pediatr. Pathol. Lab. Med.* 17:267–274.

580. Cinco, M., R. Murgia, G. Presani, and S. Perticarari. 1997. Integrin CR3 mediates the binding of nonspecifically opsonized *Borrelia burgdorferi* to human phagocytes and mammalian cells. *Infect. Immun.* 65:4784–4789.

581. Cinco, M., E. Panfili, G. Presani, and S. Perticarari. 2000. Interaction with *Borrelia burgdorferi* causes increased expression of the CR3 integrin and increased binding affinity to fibronectin via CR3. *J. Mol. Microbiol. Biotechnol.* 2:575–579.

582. Cinco, M., B. Cini, R. Murgia, G. Presani, M. Prodan, and S. Perticarari. 2001. Evidence of involvement of the mannose receptor in adhesion of *Borrelia burgdorferi* to monocyte/macrophages. *Infect. Immun.* 69:2743–2747.

583. Cirillo, J. D., S. Falkow, and L. S. Tompkins. 1994. Growth of *Legionella pneumophila* in *Acanthamoeba castellanii* enhances invasion. *Infect. Immun.* 62:3254–3261.

584. Cirillo, J. D., S. Falkow, L. S. Tomkins, and L. E. Bermudez. 1997. Interaction of *Mycobacterium avium* with environmental amoebae enhznces virulence. *Infect. Immun.* 65:3759–3767.

585. Cisar, J. O., A. L. Sandberg, C. Abeygunawardana, G. P. Reddy, and C. A. Bush. 1995. Lectin recognition of host-like saccharide motifs in streptococcal cell wall polysaccharides. *Glycobiology* 5:655–662.

586. Cisar, J. O., Y. Takahashi, S. Ruhl, J. A. Donkersloot, and A. L. Sandberg. 1997. Specific inhibitors of bacterial adhesion: observations from the study of gram-positive bacteria that initiate biofilm formation on the tooth surface. *Adv. Dent. Res.* 11:168–175.

587. Clark, B. A., J. P. Rissing, T. B. Buxton, N. H. Best, and G. K. Best. 1994. The effect of growth temperature on *Staphylococcus aureus* binding to type I collagen. *Microb. Pathog.* 17:239–251.

588. Clark, C. A., M. W. Heuzenroeder, and P. A. Manning. 1992. Colonization factor antigen CFA/IV (PCF8775) of human enterotoxigenic *Escherichia coli*: nucleotide sequence of the CS5 determinant. *Infect. Immun.* 60:1254–1257.

589. Clark, M. A., M. A. Jepson, N. L. Simmons, and B. H. Hirst. 1994. Preferential interaction of *Salmonella typhimurium* with mouse Peyer's patch M cells. *Res. Microbiol.* 145:543–552.

590. Cleary, P. P., L. McLandsborough, L. Ikeda, D. Cue, J. Krawczak, and H. Lam. 1998. High-frequency intracellular infection and erythrogenic toxin A expression undergo phase variation in M1 group A streptococci. *Mol. Microbiol.* 28:157–167.

591. Clemans, D. L., and P. E. Kolenbrander. 1995. Isolation and characterization of coaggregation-defective (Cog) mutants of *Streptococcus gordonii* DL1 (Challis). *J. Ind. Microbiol.* 15:193–197.

592. Clemans, D. L., and P. E. Kolenbrander. 1995. Identification of a 100-kilodalton putative coaggregation-mediating adhesin of *Streptococcus gordonii* DL1 (Challis). *Infect. Immun.* 63:4890–4893.

593. Clemans, D. L., C. F. Marrs, M. Patel, M. Duncan, and J. R. Gilsdorf. 1998. Comparative analysis of *Haemophilus influenzae* hifA (pilin) genes. *Infect. Immun.* 66:656–663.

594. Clemans, D. L., P. E. Kolenbrander, D. V. Debabov, Q. Zhang, R. D. Lunsford, H. Sakone, C. J. Whittaker, M. P. Heaton, and F. C. Neuhaus. 1999. Insertional inactivation of genes responsible for the D-alanylation of lipoteichoic acid in *Streptococcus gordonii* DL1 (Challis) affects intra-

generic coaggregations. *Infect. Immun.* **67:**2464–2474.

595. **Cloak, O. M., G. Duffy, J. J. Sheridan, I. S. Blair, and D. A. McDowell.** 1999. Isolation and detection of *Listeria* spp, *Salmonella* spp and *Yersinia* spp using a simultaneous enrichment step followed by a surface adhesion immunofluorescent technique. *J. Microbiol. Methods* **39:**33–43.

596. **Cloak, O. M., G. Duffy, J. J. Sheridan, D. A. McDowell, and I. S. Blair.** 1999. Development of a surface adhesion immunofluorescent technique for the rapid detection of *Salmonella* spp. from meat and poultry. *J. Appl. Microbiol.* **86:**583–590.

597. **Clouthier, S. C., K. H. Muller, J. L. Doran, S. K. Collinson, and W. W. Kay.** 1993. Characterization of three fimbrial genes, *sefABC,* of *Salmonella enteritidis. J. Bacteriol.* **175:**2523–2533.

598. **Clouthier, S. C., S. K. Collinson, and W. W. Kay.** 1994. Unique fimbriae-like structures encoded by *sefD* of the SEF14 fimbrial gene cluster of *Salmonella enteritidis. Mol. Microbiol.* **12:**893–901.

599. **Clouthier, S. C., S. K. Collinson, A. P. White, P. A. Banser, and W. W. Kay.** 1998. tRNA(Arg) *(fimU)* and expression of SEF14 and SEF21 in *Salmonella enteritidis. J. Bacteriol.* **180:**840–845.

600. **Clyne, M., and B. Drumm.** 1993. Adherence of *Helicobacter pylori* to primary human gastrointestinal cells. *Infect. Immun.* **61:**4051–4057.

601. **Clyne, M., and B. Drumm.** 1996. Cell envelope characteristics of *Helicobacter pylori:* their role in adherence to mucosal surfaces and virulence. *FEMS Immunol. Med. Microbiol.* **16:**141–155.

602. **Clyne, M., and B. Drumm.** 1996. The urease enzyme of *Helicobacter pylori* does not function as an adhesin. *Infect. Immun.* **64:**2817–2820.

603. **Clyne, M., and B. Drumm.** 1997. Absence of effect of Lewis A and Lewis B expression on adherence of *Helicobacter pylori* to human gastric cells. *Gastroenterology* **113:**72–80.

604. **Clyne, M., and B. Drumm.** 1997. Adherence of *Helicobacter pylori* to the gastric mucosa. *Can. J. Gastroenterol.* **11:**243–248.

605. **Clyne, M., J. Thomas, L. Weaver, and B. Drumm.** 1997. *In vitro* evaluation of the role of antibodies against *Helicobacter pylori* in inhibiting adherence of the organism to gastric cells. *Gut* **40:**731–738.

606. **Clyne, M., T. Ocroinin, S. Suerbaum, C. Josenhans, and B. Drumm.** 2000.

Adherence of isogenic flagellum-negative mutants of *Helicobacter pylori* and *Helicobacter mustelae* to human and ferret gastric epithelial cells. *Infect. Immun.* **68:**4335–4339.

607. **Coburn, J., J. M. Leong, and J. K. Erban.** 1993. Integrin alpha IIb beta 3 mediates binding of the Lyme disease agent *Borrelia burgdorferi* to human platelets. *Proc. Natl. Acad. Sci. USA* **90:**7059–7063.

608. **Coburn, J., S. W. Barthold, and J. M. Leong.** 1994. Diverse Lyme disease spirochetes bind integrin $\alpha_{IIb}\beta_3$ on human platelets. *Infect. Immun.* **62:**5559–5567.

609. **Coburn, J., L. Magoun, S. C. Bodary, and J. M. Leong.** 1998. Integrins $\alpha_v\beta_3$ and $\alpha_5\beta_1$ mediate attachment of Lyme disease spirochetes to human cells. *Infect. Immun.* **66:**1946–1952.

610. **Coburn, J., W. Chege, L. Magoun, S. C. Bodary, and J. M. Leong.** 1999. Characterization of a candidate *Borrelia burgdorferi* beta3-chain integrin ligand identified using a phage display library. *Mol. Microbiol.* **34:**926–940.

611. **Cochran, W. L., G. A. McFeters, and P. S. Stewart.** 2000. Reduced susceptibility of thin *Pseudomonas aeruginosa* biofilms to hydrogen peroxide and monochloramine. *J. Appl. Microbiol.* **88:**22–30.

612. **Cockerill, F., III, G. Beebakhee, R. Soni, and P. Sherman.** 1996. Polysaccharide side chains are not required for attaching and effacing adhesion of *Escherichia coli* O157:H7. *Infect. Immun.* **64:**3196–3200.

613. **Coconnier, M. H., T. R. Klaenhammer, S. Kerneis, M. R. Bernet, and A. L. Servin.** 1992. Protein-mediated adhesion of *Lactobacillus acidophilus* BG2FO4 on human enterocyte and mucus-secreting cell lines in culture. *Appl. Environ. Microbiol.* **58:** 2034–2039.

614. **Coconnier, M. H., M. F. Bernet, G. Chauviere, and A. L. Servin.** 1993. Adhering heat-killed human *Lactobacillus acidophilus,* strain LB, inhibits the process of pathogenicity of diarrhoeagenic bacteria in cultured human intestinal cells. *J. Diarroheal Dis. Res.* **11:**235–242.

615. **Coconnier, M. H., M. F. Bernet, S. Kerneis, G. Chauviere, J. Fourniat, and A. L. Servin.** 1993. Inhibition of adhesion of enteroinvasive pathogens to human intestinal Caco-2 cells by *Lactobacillus acidophilus* strain LB decreases bacterial invasion. *FEMS Microbiol. Lett.* **110:**299–305.

616. **Cohen, P. S., and D. C. Laux.** 1995. Bacterial adhesion to and penetration of intes-

tinal mucus *in vitro*. *Methods Enzymol.* **253:**309–314.

617. **Colby, S. M., R. E. McLaughlin, J. J. Ferretti, and R. R. Russell.** 1999. Effect of inactivation of *gtf* genes on adherence of *Streptococcus downei*. *Oral Microbiol. Immunol.* **14:**27–32.

618. **Colby, S. M., G. C. Whiting, L. Tao, and R. R. Russell.** 1995. Insertional inactivation of the *Streptococcus mutans dexA* (dextranase) gene results in altered adherence and dextran catabolism. *Microbiology* **141:**2929–2936.

619. **Cole, S. P., D. Cirillo, M. F. Kagnoff, D. G. Guiney, and L. Eckmann.** 1997. Coccoid and spiral *Helicobacter pylori* differ in their abilities to adhere to gastric epithelial cells and induce interleukin-8 secretion. *Infect. Immun.* **65:**843–846.

620. **Coleman, J. L., T. J. Sellati, J. E. Testa, R. R. Kew, M. B. Furie, and J. L. Benach.** 1995. *Borrelia burgdorferi* binds plasminogen, resulting in enhanced penetration of endothelial monolayers. *Infect. Immun.* **63:**2478–2484.

621. **Coleman, S. A., and M. F. Minnick.** 2001. Establishing a direct role for the *Bartonella bacilliformis* invasion-associated locus B (IalB) protein in human erythrocyte parasitism. *Infect. Immun.* **69:**4373–4381.

622. **Collazo, C. M., and J. E. Galán.** 1996. Requirement for exported proteins in secretion through the invasion-associated type III system of *Salmonella typhimurium*. *Infect. Immun.* **64:**3524–3531.

623. **Collings, S., and D. N. Love.** 1992. Further studies on some physical and biochemical characteristics of asaccharolytic pigmented *Bacteroides* of feline origin. *J. Appl. Bacteriol.* **72:**529–535.

624. **Collington, G. K., I. W. Booth, and S. Knutton.** 1998. Rapid modulation of electrolyte transport in Caco-2 cell monolayers by enteropathogenic *Escherichia coli* (EPEC) infection. *Gut* **42:**200–207.

625. **Collinson, S. K., S. C. Clouthier, J. L. Doran, P. A. Banser, and W. W. Kay.** 1996. *Salmonella enteritidis agfBAC* operon encoding thin, aggregative fimbriae. *J. Bacteriol.* **178:**662–667.

626. **Collinson, S. K., S. L. Liu, S. C. Clouthier, P. A. Banser, J. L. Doran, K. E. Sanderson, and W. W. Kay.** 1996. The location of four fimbrin-encoding genes, *agfA, fimA, sefA* and *sefD*, on the *Salmonella enteritidis* and/or *S. typhimurium* XbaI-BlnI genomic restriction maps. *Gene* **169:**75–80.

627. **Colliot, G., S. de Bentzmann, M. C.**
Plotkowski, S. Lebonvallet, E. Puchelle, and N. Bonnet.** 1993. Quantitative analysis and cartography in scanning electron microscopy: application to the study of bacterial adhesion to respiratory epithelium. *Microsc. Res. Tech.* **24:**527–536.

628. **Colloca, M. E., M. C. Ahumada, M. E. Lopez, and M. E. Nader-Macias.** 2000. Surface properties of lactobacilli isolated from healthy subjects. *Oral Dis.* **6:**227–233.

629. **Comolli, J. C., L. L. Waite, K. E. Mostov, and J. N. Engel.** 1999. Pili binding to asialo-GM1 on epithelial cells can mediate cytotoxicity or bacterial internalization by *Pseudomonas aeruginosa*. *Infect. Immun.* **67:**3207–3214.

630. **Comstock, L. E., E. Fikrig, R. J. Shoberg, R. A. Flavell, and D. D. Thomas.** 1993. A monoclonal antibody to OspA inhibits association of *Borrelia burgdorferi* with human endothelial cells. *Infect. Immun.* **61:**423–431.

631. **Condorelli, F., G. Scalia, G. Cali, B. Rossetti, G. Nicoletti, and A. M. Lo Bue.** 1998. Isolation of *Porphyromonas gingivalis* and detection of immunoglobulin A specific to fimbrial antigen in gingival crevicular fluid. *J. Clin. Microbiol.* **36:**2322–2325.

632. **Connell, H., M. Hedlund, W. Agace, and C. Svanborg.** 1997. Bacterial attachment to uroepithelial cells: mechanisms and consequences. *Adv. Dent. Res.* **11:**50–58.

633. **Connell, H., L. K. Poulsen, and P. Klemm.** 2000. Expression of type 1 and P fimbriae *in situ* and localisation of a uropathogenic *Escherichia coli* strain in the murine bladder and kidney. *Int. J. Med. Microbiol.* **290:**587–597.

634. **Conte, M. P., C. Longhi, V. Buonfiglio, M. Polidoro, L. Seganti, and P. Valenti.** 1994. The effect of iron on the invasiveness of *Escherichia coli* carrying the *inv* gene of *Yersinia pseudotuberculosis*. *J. Med. Microbiol.* **40:**236–240.

635. **Conte, M. P., C. Longhi, M. Polidoro, G. Petrone, V. Buonfiglio, S. Di Santo, E. Papi, L. Seganti, P. Visca, and P. Valenti.** 1996. Iron availability affects entry of *Listeria monocytogenes* into the enterocytelike cell line Caco-2. *Infect. Immun.* **64:**3925–3929.

636. **Conte, M. P., G. Petrone, A. M. Di Biase, M. G. Ammendolia, F. Superti, and L. Seganti.** 2000. Acid tolerance in *Listeria monocytogenes* influences invasiveness of enterocyte-like cells and macrophage-like cells. *Microb. Pathog.* **29:**137–144.

637. **Contrepois, M., Y. Bertin, J. P.**

Girardeau, B. Picard, and P. Goullet. 1993. Clonal relationships among bovine pathogenic *Escherichia coli* producing surface antigen CS31A. *FEMS Microbiol. Lett.* **106**:217–222.

638. **Contrepois, M., Y. Bertin, P. Pohl, B. Picard, and J. P. Girardeau.** 1998. A study of relationships among F17 a producing enterotoxigenic and non-enterotoxigenic *Escherichia coli* strains isolated from diarrheic calves. *Vet. Microbiol.* **64**:75–81.

639. **Cook, A. D., R. D. Sagers, and W. G. Pitt.** 1993. Bacterial adhesion to poly(HEMA)-based hydrogels. *J. Biomed. Mater. Res.* **27**:119–126.

640. **Cook, A. D., R. D. Sagers, and W. G. Pitt.** 1993. Bacterial adhesion to protein-coated hydrogels. *J. Biomater. Appl.* **8**:72–89.

641. **Cook, G., J. W. Costerton, and R. O. Darouiche.** 2000. Direct confocal microscopy studies of the bacterial colonization *in vitro* of a silver-coated heart valve sewing cuff. *Int. J. Antimicrob. Agents* **13**:169–173.

642. **Cook, S. W., N. Mody, J. Valle, and R. Hull.** 1995. Molecular cloning of *Proteus mirabilis* uroepithelial cell adherence (*uca*) genes. *Infect. Immun.* **63**:2082–2086.

643. **Cookson, S. T., and J. P. Nataro.** 1996. Characterization of HEp-2 cell projection formation induced by diffusely adherent *Escherichia coli*. *Microb. Pathog.* **21**:421–434.

644. **Cooper, G. L., and C. J. Thorns.** 1996. Evaluation of SEF14 fimbrial dot blot and flagellar western blot tests as indicators of *Salmonella enteritidis* infection in chickens. *Vet. Rec.* **138**:149–153.

645. **Cormio, L., J. Vuopio-Varkila, A. Siitonen, M. Talja, and M. Ruutu.** 1996. Bacterial adhesion and biofilm formation on various double-J stents *in vivo* and *in vitro*. *Scand. J. Urol. Nephrol.* **30**:19–24.

646. **Cormio, L., P. La Forgia, D. La Forgia, A. Siitonen, and M. Ruutu.** 1997. Is it possible to prevent bacterial adhesion onto ureteric stents? *Urol. Res.* **25**:213–216.

647. **Cormio, L., P. La Forgia, A. Siitonen, M. Ruutu, P. Tormala, and M. Talja.** 1997. Immersion in antibiotic solution prevents bacterial adhesion onto biodegradable prostatic stents. *Br. J. Urol.* **79**:409–143.

648. **Cormio, L., P. La Forgia, D. La Forgia, A. Siitonen, and M. Ruutu.** 2001. Bacterial adhesion to urethral catheters: role of coating materials and immersion in antibiotic solution. *Eur. Urol.* **40**:354–358.

649. **Cornacchione, P., L. Scaringi, K.**

Fettucciari, E. Rosati, R. Sabatini, G. Orefici, C. von Hunolstein, A. Modesti. A. Modica, F. Minelli, and P. Marconi. 1998. Group B streptococci persist inside macrophages. *Immunology* **93**:86–95.

650. **Correia, F. F., J. M. DiRienzo, R. J. Lamont, C. Anderman, T. L. McKay, and B. Rosan.** 1995. Insertional inactivation of binding determinants of *Streptococcus crista* CC5A using Tn*916*. *Oral Microbiol. Immunol.* **10**:220–226.

651. **Correia, F. F., J. M. DiRienzo, T. L. McKay, and B. Rosan.** 1996. *scbA* from *Streptococcus crista* CC5A: an atypical member of the *lraI* gene family. *Infect. Immun.* **64**:2114–2121.

652. **Corthesy-Theulaz, I., N. Porta, E. Pringault, L. Racine, A. Bogdanova, J. P. Kraehenbuhl, A. L. Blum, and P. Michetti.** 1996. Adhesion of *Helicobacter pylori* to polarized T84 human intestinal cell monolayers is pH dependent. *Infect. Immun.* **64**:3827–3832.

653. **Costerton, J. W., Z. Lewandowski, D. E. Caldwell, D. R. Korber, and H. M. Lappin-Scott.** 1995. Microbial biofilms. *Annu. Rev. Microbiol.* **49**:711–745.

654. **Cotter, P. A., M. H. Yuk, S. Mattoo, B. J. Akerley, J. Boschwitz, D. A. Relman, and J. F. Miller.** 1998. Filamentous hemagglutinin of *Bordetella bronchiseptica* is required for efficient establishment of tracheal colonization. *Infect. Immun.* **66**:5921–5929.

655. **Courtney, H. S., C. von Hunolstein, J. B. Dale, M. S. Bronze, E. H. Beachey, and D. L. Hasty.** 1992. Lipoteichoic acid and M protein: dual adhesins of group A streptococci. *Microb. Pathog.* **12**:199–208.

656. **Courtney, H. S., M. S. Bronze, J. B. Dale, and D. L. Hasty.** 1994. Analysis of the role of M24 protein in group A streptococcal adhesion and colonization by use of Ω-interposon mutagenesis. *Infect. Immun.* **62**:4868–4873.

657. **Courtney, H. S., J. B. Dale, and D. L. Hasty.** 1996. Differential effects of the streptococcal fibronectin-binding protein, FBP54, on adhesion of group A streptococci to human buccal cells and HEp-2 tissue culture cells. *Infect. Immun.* **64**:2415–2419.

658. **Courtney, H. S., S. Liu, J. B. Dale, and D. L. Hasty.** 1997. Conversion of M serotype 24 of *Streptococcus pyogenes* to M serotypes 5 and 18: effect on resistance to phagocytosis and adhesion to host cells. *Infect. Immun.* **65**:2472–2474.

659. **Courtney, H. S., I. Ofek, and D. L.**

Hasty. 1997. M protein mediated adhesion of M type 24 *Streptococcus pyogenes* stimulates release of interleukin-6 by HEp-2 tissue culture cells. *FEMS Microbiol. Lett.* **151**:65–70.

660. **Coutino-Rodriguez, R., P. Hernandez-Cruz, and H. Giles-Rios.** 2001. Lectins in fruits having gastrointestinal activity: their participation in the hemagglutinating property of *Escherichia coli* O157:H7. *Arch. Med. Res.* **32**:251–257.

661. **Cowan, C., H. A. Jones, Y. H. Kaya, R. D. Perry, and S. C. Straley.** 2000. Invasion of epithelial cells by *Yersinia pestis:* evidence for a *Y. pestis*-specific invasin. *Infect. Immun.* **68**:4523–4530.

662. **Cowan, M., and H. J. Busscher.** 1993. Flow chamber study of the adhesion of *Prevotella intermedia* to glass after preconditioning with mutans streptococcal species: kinetics and spatial arrangement. *Microbios* **73**:135–144.

663. **Cowan, M. M., E. A. Horst, S. Luengpailin, and R. J. Doyle.** 2000. Inhibitory effects of plant polyphenoloxidase on colonization factors of *Streptococcus sobrinus* 6715. *Antimicrob. Agents Chemother.* **44**:2578–2580.

664. **Cowan, M. M., H. C. van der Mei, I. Stokroos, and H. H. Busscher.** 1992. Heterogeneity of surfaces of subgingival bacteria as detected by zeta potential measurements. *J. Dent. Res.* **71**:1803–1806.

665. **Cowan, M. M., H. C. van der Mei, P. G. Rouxhet, and H. J. Busscher.** 1992. Physico-chemical and structural properties of the surfaces of *Peptostreptococcus micros* and *Streptococcus mitis* as compared to those of mutans streptococci, *Streptococcus sanguis* and *Streptococcus salivarius*. *J. Gen. Microbiol.* **138**:2702–2714.

666. **Cowan, M. M.** 1995. Kinetic analysis of microbial adhesion. *Methods Enzymol.* **253**:179–189.

667. **Cowell, B. A., M. D. Willcox, and R. P. Schneider.** 1998. A relatively small change in sodium chloride concentration has a strong effect on adhesion of ocular bacteria to contact lenses. *J. Appl. Microbiol.* **84**:950–958.

668. **Cox, E., and A. Houvenaghel.** 1993. Comparison of the *in vitro* adhesion of K88, K99, F41 and P987 positive *Escherichia coli* to intestinal villi of 4- to 5-week-old pigs. *Vet. Microbiol.* **34**:7–18.

669. **Cox, S. D., M. O. Lassiter, K. G. Taylor and R. J. Doyle.** 1994. Fluoride inhibits the glucan-binding lectin of *Streptococcus sobrinus*. *FEMS Microbiol. Lett.* **123**:331–334.

670. **Crago, A. M., and V. Koronakis.** 1999. Binding of extracellular matrix laminin to *Escherichia coli* expressing the *Salmonella* outer membrane proteins Rck and PagC. *FEMS Microbiol. Lett.* **176**:495–501.

671. **Cramton, S. E., C. Gerke, N. F. Schnell, W. W. Nichols and F. Götz.** 1999. The intercellular adhesion (*ica*) locus is present in *Staphylococcus aureus* and is required for biofilm formation. *Infect. Immun.* **67**:5427–5433.

672. **Cramton, S. E., M. Ulrich, F. Götz, and G. Doring.** 2001. Anaerobic conditions induce expression of polysaccharide intercellular adhesin in *Staphylococcus aureus* and *Staphylococcus epidermidis*. *Infect. Immun.* **69**:4079–4085.

673. **Crane, J. K., S. Majumdar, and D. F. Pickhardt, III.** 1999. Host cell death due to enteropathogenic *Escherichia coli* has features of apoptosis. *Infect. Immun.* **67**:2575–2584.

674. **Craven, S. E., N. A. Cox, J. S. Bailey, and L. C. Blankenship.** 1992. Binding of *Salmonella* strains to immobilized intestinal mucosal preparations from broiler chickens. *Avian Dis.* **36**:296–303.

675. **Craven, S. E., N. A. Cox, J. S. Bailey, N. J. Stern, R. J. Meinersmann, and L. C. Blankenship.** 1993. Characterization of *Salmonella california* and *S. typhimurium* strains with reduced ability to colonize the intestinal tract of broiler chicks. *Avian Dis.* **37**:339–348.

676. **Craven, S. E., and D. D. Williams.** 1997. Inhibition of *Salmonella typhimurium* attachment to chicken cecal mucus by intestinal isolates of *Enterobacteriaceae* and lactobacilli. *Avian Dis.* **41**:548–558.

677. **Craven, S. E., and D. D. Williams.** 1998. *In vitro* attachment of *Salmonella typhimurium* to chicken cecal mucus: effect of cations and pretreatment with *Lactobacillus* spp. isolated from the intestinal tracts of chickens. *J. Food Prot.* **61**:265–271.

678. **Cree, R. G., P. Aleljung, M. Paulsson, W. Witte, W. C. Noble, A. Ljungh, and T. Wadström.** 1994. Cell surface hydrophobicity and adherence to extra-cellular matrix proteins in two collections of methicillin-resistant *Staphylococcus aureus*. *Epidemiol. Infect.* **112**:307–314.

679. **Cree, R. G., and W. C. Noble.** 1995. *In vitro* indices of tissue adherence in *Staphylococcus intermedius*. *Lett. Appl. Microbiol.* **20**:168–170.

680. **Cree, R. G., I. Phillips, and W. C. Noble.** 1995. Adherence characteristics of coagulase-negative staphylococci isolated from patients with infective endocarditis. *J. Med. Microbiol.* **43**:161–168.

681. **Crociani, J., J. P. Grill, M. Huppert, and J. Ballongue.** 1995. Adhesion of different bifidobacteria strains to human enterocyte-like Caco-2 cells and comparison with *in vivo* study. *Lett. Appl. Microbiol.* **21:**146–148.

682. **Crocker, I. C., W. K. Liu, P. O. Byrne, and T. S. Elliott.** 1992. A novel electrical method for the prevention of microbial colonization of intravascular cannulae. *J. Hosp. Infect.* **22:**7–17.

683. **Cross, A., L. Asher, M. Seguin, L. Yuan, N. Kelly, C. Hammack, J. Sadoff, and P. Gemski, Jr.** 1995. The importance of a lipopolysaccharide-initiated, cytokine-mediated host defense mechanism in mice against extraintestinally invasive *Escherichia coli*. *J. Clin. Investig.* **96:**676–686.

684. **Crowley, P. J., L. J. Brady, D. A. Piecentini, and A. S. Bleiweiss.** 1993. Identification of a salivary agglutinin-binding domain within cell surface adhesin P1 of *Streptococcus mutans*. *Infect. Immun.* **61:**1547–1552.

685. **Cue, D., P. E. Dombek, H. Lam, and P. P. Cleary.** 1998. *Streptococcus pyogenes* serotype M1 encodes multiple pathways for entry into human epithelial cells. *Infect. Immun.* **66:**4593–4601.

686. **Culham, D. E., C. Dalgado, C. L. Gyles, D. Mamelak, S. MacLellan, and J. M. Wood.** 1998. Osmoregulatory transporter ProP influences colonization of the urinary tract by *Escherichia coli*. *Microbiology* **144:**91–102.

687. **Cundell, D. R., and E. I. Tuomanen.** 1994. Receptor specificity of adherence of *Streptococcus pneumoniae* to human type-II pneumocytes and vascular endothelial cells *in vitro*. *Microb. Pathog.* **17:**361–374.

688. **Cundell, D., H. R. Masure, and E. I. Tuomanen.** 1995. The molecular basis of pneumococcal infection: a hypothesis. *Clin. Infect. Dis.* **21:**S204–S211.

689. **Cundell, D. R., N. P. Gerard, C. Gerard, I. Idanpaan-Heikkila, and E. I. Tuomanen.** 1995. *Streptococcus pneumoniae* anchor to activated human cells by the receptor for platelet-activating factor. *Nature* **377:**435–438.

690. **Cundell, D. R., B. J. Pearce, J. Sandros, A. M. Naughton, and H. R. Masure.** 1995. Peptide permeases from *Streptococcus pneumoniae* affect adherence to eucaryotic cells. *Infect. Immun.* **63:**2493–2498.

691. **Cundell, D. R., J. N. Weiser, J. Shen, A. Young, and E. I. Tuomanen.** 1995. Relationship between colonial morphology and adherence of *Streptococcus pneumoniae*. *Infect. Immun.* **63:**757–761.

692. **Cundell, D. R., C. Gerard, I. Idänpään-Heikkilä, E. I. Tuomanen, and N. P. Gerard.** 1996 PAf receptor anchors *Streptococcus pneumoniae* to activated human endothelial cells. *Adv. Exp. Med. Biol.* **416:**89–94.

693. **Cunliffe, D., C. A. Smart, C. Alexander, and E. N. Vulfson.** 1999. Bacterial adhesion at synthetic surfaces. *Appl. Environ. Microbiol.* **65:**4995–5000.

694. **Cuperus, P. L., H. C. van der Mei, G. Reid, A. W. Bruce, A. E. Khoury, M. van der Kuijl-Booij, J. Noordmans, and H. J. Busscher.** 1995. Effects of ciprofloxacin and vancomycin on physicochemical surface properties of *Staphylococcus epidermidis*, *Escherichia coli*, *Lactobacillus casei* and *Lactobacillus acidophilus*. *Microbios* **82:**49–67.

695. **Curfs, J., J. Meis, H. van der Lee, W. Kraak, and J. Hoogkamp-Korstanje.** 1995. Invasion of Caco-2 cells by *Yersinia enterocolitica* O:8 WA. *Contrib. Microbiol. Immunol.* **13:**251–253.

696. **Curfs, J. H., J. F. Meis, J. A. Fransen, H. A. van der Lee, and J. A. Hoogkamp-Korstanje.** 1995. Interactions of *Yersinia enterocolitica* with polarized human intestinal Caco-2 cells. *Med. Microbiol. Immunol.* **184:**123–127.

697. **Curtis, M. A., J. Aduse-Opoku, J. M. Slaney, M. Rangarajan, V. Booth, J. Cridland, and P. Shepherd.** 1996. Characterization of an adherence and antigenic determinant of the ArgI protease of *Porphyromonas gingivalis* which is present on multiple gene products. *Infect. Immun.* **64:**2532–2539.

698. **Cywes, C., N. L. Godenir, H. C. Hope, R. R. Scholle, L. M. Steyn, R. E. Kirsch, and M. R. W. Ehlers.** 1996. Nonopsonic binding of *Mycobacterium tuberculosis* to human complement receptor type 3 expressed in Chinese hamster ovary cells. *Infect. Immun.* **64:**5373–5383.

699. **Cywes, C., H. C. Hoppe, M. Daffé, and M. R. W. Ehlers.** 1997. Nonopsonic binding of *Mycobacterium tuberculosis* to complement receptor type 3 is mediated by capsular polysaccharides and is strain dependent. *Infect. Immun.* **65:**4258–4266.

700. **Cywes, C., I. Stamenkovic, and M. R. Wessels.** 2000. CD44 as a receptor for colonization of the pharynx by group A streptococcus. *J. Clin. Investig.* **106:**995–1002.

701. **Cywes, C., and M. R. Wessels.** 2001. Group A streptococcus tissue invasion by

CD44-mediated cell signalling. *Nature* **414:**648–652.

702. **Czajkowski, A., J. Piotrowski, F. Yotsumoto, A. Slomiany, and B. L. Slomiany.** 1993. Inhibition of *Helicobacter pylori* colonization by an antiulcer agent, sulglycotide. *Biochem. Mol. Biol. Int.* **29:**965–971.

703. **Czeczulin, J. R., S. Balepur, S. Hicks, A. Phillips, R. Hall, M. H. Kothary, F. Navarro-Garcia, and J. P. Nataro.** 1997. Aggregative adherence fimbriae II, a second fimbrial antigen mediating aggregative adherence in enteroaggregative *Escherichia coli. Infect. Immun.* **65:**4135–4145.

704. Reference deleted.

705. **Daefler, S.** 1999. Type III secretion by *Salmonella typhimurium* does not require contact with a eukaryotic host. *Mol. Microbiol.* **31:**45–51.

706. **Dahlen, G. G., J. R. Johnson, and R. Gmur.** 1996. *Prevotella intermedia* and *Prevotella nigrescens* serotypes, ribotypes and binding characteristics. *FEMS Microbiol. Lett.* **138:**89–95.

707. **Dai, J.-H., Y.-S. Lee, and H.-C. Wong.** 1992. Effects of iron limitation on production of a siderophore, outer membrane proteins, and hemolysin and on hydrophobicity, cell adherence, and lethality for mice of *Vibrio parahaemolyticus. Infect. Immun.* **60:**2952–2959.

708. **Daigle, F., C. M. Dozois, M. Jacques, and J. Harel.** 1997. Mutations in the f165(1)A and f165(1)E fimbrial genes and regulation of their expression in an *Escherichia coli* strain causing septicemia in pigs. *Microb. Pathog.* **22:**247–252.

709. **Dale, J. B., R. G. Washburn, M. B. Marques, and M. R. Wessels.** 1996. Hyaluronate capsule and surface M protein in resistance to opsonization of group A streptococci. *Infect. Immun.* **64:**1495–1501.

710. **Dale, J. B., R. W. Baird, H. S. Courtney, D. L. Hasty, and M. S. Bronze.** 1994. Passive protection of mice against group A streptococcal pharyngeal infection by lipoteichoic acid. *J. Infect. Dis.* **169:**319–323.

711. **Dallo, S. F., A. L. Lazzell, A. Chavoya, S. P. Reddy, and J. B. Baseman.** 1996. Biofunctional domains of the *Mycoplasma pneumoniae* P30 adhesin. *Infect. Immun.* **64:**2595–2601.

712. **Dal Nogare, A. R., and S. Q. Azizi.** 1992. Chinese hamster ovarian cell glycoproteins that mediate type 1 piliated gram-negative bacterial adherence. *Am. J. Respir. Cell Mol. Biol.* **7:**399–405.

713. **Dalsgaard, A., M. J. Albert, D. N. Taylor, T. Shimada, R. Meza, O. Serichantalergs, and P. Echeverria.** 1995. Characterization of *Vibrio cholerae* non-O1 serogroups obtained from an outbreak of diarrhea in Lima, Peru. *J. Clin. Microbiol.* **33:**2715–2722.

714. **Danese, P. N., L. A. Pratt, S. L. Dove, and R. Kolter.** 2000. The outer membrane protein, antigen 43, mediates cell-to-cell interactions within *Escherichia coli* biofilms. *Mol. Microbiol.* **37:**424–432.

715. **Danese, P. N., L. A. Pratt, and R. Kolter.** 2000. Exopolysaccharide production is required for development of *Escherichia coli* K-12 biofilm architecture. *J. Bacteriol.* **182:**3593–3596.

716. **Daniell, S. J., R. M. Delahay, R. K. Shaw, E. L. Hartland, M. J. Pallen, F. Booy, F. Ebel, S. Knutton, and G. Frankel.** 2001. Coiled-coil domain of enteropathogenic *Escherichia coli* type III secreted protein EspD is involved in EspA filament-mediated cell attachment and hemolysis. *Infect. Immun.* **69:**4055–4064.

717. **Danino, J., H. Z. Joachims, and M. Barak.** 1998. Predictive value of an adherence test for acute otitis media. *Otolaryngol. Head Neck Surg.* **118:**400–403.

718. **Danjo, Y., L. D. Hazlett, and I. K. Gipson.** 2000. C57BL/6 mice lacking Muc1 show no ocular surface phenotype. *Investig. Ophthalmol. Visual Sci.* **41:**4080–4084.

719. **Dankert, J., J. van der Werff, S. A. Zaat, W. Joldersma, D. Klein, and J. Hess.** 1995. Involvement of bactericidal factors from thrombin-stimulated platelets in clearance of adherent viridans streptococci in experimental infective endocarditis. *Infect. Immun.* **63:**663–671.

720. **Darfeuille-Michaud, A., C. Jallat, D. Aubel, D. Sirot, C. Rich, J. Sirot, and B. Joly.** 1992. R-plasmid-encoded adhesive factor in *Klebsiella pneumoniae* strains responsible for human nosocomial infections. *Infect. Immun.* **60:**44–55.

721. **Darfeuille-Michaud, A., C. Neut, N. Barnich, E. Lederman, P. Di Martino, P. Desreumaux, L. Gambiez, B. Joly, A. Cortot, and J. F. Colombel.** 1998. Presence of adherent *Escherichia coli* strains in ileal mucosa of patients with Crohn's disease. *Gastroenterology* **115:**1405–1413.

722. **Darmstadt, G. L., P. Fleckman, M. Jonas, E. Chi, and C. E. Rubens.** 1998. Differentiation of cultured keratinocytes promotes the adherence of *Streptococcus pyogenes. J. Clin. Investig.* **101:**128–136.

723. **Darmstadt, G. L., P. Fleckman, and C. E. Rubens.** 1999. Tumor necrosis factor-alpha and interleukin-1alpha decrease the adherence of *Streptococcus pyogenes* to cultured keratinocytes. *J. Infect. Dis.* **180:**1718–1721.

724. **Darmstadt, G. L., K, Mentele, P. Fleckman, and C. E. Rubens.** 1999. Role of keratinocyte injury in adherence of *Streptococcus pyogenes. Infect. Immun.* **67:**6707–6709.

725. **Darmstadt, G. L., L. Mentele, A. Podbielski, and C. E. Rubens.** 2000. Role of group A streptococcal virulence factors in adherence to keratinocytes. *Infect. Immun.* **68:**1215–1521.

726. **Darouiche, R. O., G. C. Landon, J. M. Patti, L. L. Nguyen, R. C. Fernau, D. McDevitt, C. Greene, T. Foster, and M. Klima.** 1997. Role of *Staphylococcus aureus* surface adhesins in orthopaedic device infections: are results model-dependent? *J. Med. Microbiol.* **46:**75–79.

727. **Darzins, A.** 1993, The *pilG* gene product, required for *Pseudomonas aeruginosa* pilus production and twitching motility, is homologous to the enteric, single-domain response regulator CheY. *J. Bacteriol.* **175:**5934–5944.

728. **Darzins, A.** 1994. Characterization of a *Pseudomonas aeruginosa* gene cluster involved in pilus biosynthesis and twitching motility: sequence similarity to the chemotaxis proteins of enterics and the gliding bacterium *Myxococcus xanthus. Mol. Microbiol.* **11:**137–153.

729. **Darzins, A., and M. A. Russell.** 1997. Molecular genetic analysis of type-4 pilus biogenesis and twitching motility using *Pseudomonas aeruginosa* as a model system—a review. *Gene* **192:**109–115.

730. **Das, J. R., M. Bhakoo, M. V. Jones, and P. Gilbert.** 1998. Changes in the biocide susceptibility of *Staphylococcus epidermidis* and *Escherichia coli* cells associated with rapid attachment to plastic surfaces. *J. Appl. Microbiol.* **84:**852–858.

731. **Dasgupta, M. K., K. Ward, P. A. Noble, M. Larabie, and J. W. Costerton.** 1994. Development of bacterial biofilms on silastic catheter materials in peritoneal dialysis fluid. *Am. J. Kidney Dis.* **23:**709–716.

732. **Dashper, S. G., N. M. O'Brien-Simpson, P. S. Bhogal, A. D. Franzmann, and E. C. Reynolds.** 1998. Purification and characterization of a putative fimbrial protein/receptor of *Porphyromonas gingivalis. Aust. Dent. J.* **43:**99–104.

732a. **da Silvaria, W. D., F. Fantinatti, G. P. Manfio, and A. F. P. de Castro.** 1994. Studies of the genetic expression of 31A fimbriae by two bovine septicaemic *Escherichia coli* strains. *Rev. Brasil. Genet.* **17:**365–370.

733. **Daunter, B., K. L. Forbes, B. M. Sanderson, J. Morrison, and G. Wright.** 1992. Inhibition of binding of bacteria to amniochorionic membranes by amniotic fluid. *Eur. J. Obstet. Gynecol. Reprod. Biol.* **47:**95–102.

734. **Davies, D. G., A. M. Chakrabarty, and G. G. Geesey.** 1993. Exopolysaccharide production in biofilms: substratum activation of alginate gene expression by *Pseudomonas aeruginosa. Appl. Environ. Microbiol.* **59:**1181–1186.

735. **Davies, J., I. Carlstedt, A.-K. Nilsson, A. Håkansson, H. Sabharwal, L. van Alphen, M. van Ham, and C. Svanborg.** 1995. Binding of *Haemophilus influenzae* to purified mucins from the human respiratory tract. *Infect. Immun.* **63:**2485–2492.

736. **Davies, J., A. Dewar, A. Bush, T. Pitt, D. Gruenert, D. M. Geddes, and E. W. Alton.** 1999. Reduction in the adherence of *Pseudomonas aeruginosa* to native cystic fibrosis epithelium with anti-asialoGM1 antibody and neuraminidase inhibition. *Eur. Respir. J.* **13:**565–570.

737. **Davies, J. C., M. Stern, A. Dewar, N. J. Caplen, F. M. Munkonge, T. Pitt, F. Sorgi, L. Huang, A. Bush, D. M. Geddes, and E. W. Alton.** 1997. CFTR gene transfer reduces the binding of *Pseudomonas aeruginosa* to cystic fibrosis respiratory epithelium. *Am. J. Respir. Cell Mol. Biol.* **16:**657–663.

738. **Davies, K. G., P. Afolabi, and P. O'Shea.** 1996. Adhesion of *Pasteuria penetrans* to the cuticle of root-knot nematodes (*Meloidogyne* spp.) inhibited by fibronectin: a study of electrostatic and hydrophobic interactions. *Parasitology* **112:**553–559.

739. **Davis, C. H., and P. B. Wyrick.** 1997. Differences in the association of *Chlamydia trachomatis* serovar E and serovar L2 with epithelial cells in vitro may reflect biological differences in vivo. *Infect. Immun.* **65:**2914–2924.

740. **Dawid, S., S. Grass, and J. W. St. Geme III.** 2001. Mapping of binding domains of nontypeable *Haemophilus influenzae* HMW1 and HMW2 adhesins. *Infect. Immun.* **69:**307–314.

741. **Dawid, S., S. J. Barenkamp, and J. W. St. Geme III.** 1999. Variation in expression of the *Haemophilus influenzae* HMW adhesins: a prokaryotic system reminiscent of eukaryotes. *Proc. Natl. Acad. Sci. USA* **96:**1077–1082.

742. **Dawson, J. R., and R. P. Ellen.** 1994. Clustering of fibronectin adhesins toward *Treponema denticola* tips upon contact with immobilized fibronectin. *Infect. Immun.* **62:**2214–2221.

743. **Deal, C. D., and H. C. Krivan.** 1994. Solid-phase binding of microorganisms to glycolipids and phospholipids. *Methods Enzymol.* **236:**346–353.

744. **Dean-Nystrom, E. A.** 1995. Identification of intestinal receptors for enterotoxigenic *Escherichia coli*. *Methods Enzymol.* **253:**315–324.

745. **Dean-Nystrom, E. A., T. A. Casey, R. A. Schneider, and B. Nagy.** 1993. A monoclonal antibody identifies 2134P fimbriae as adhesins on enterotoxigenic *Escherichia coli* isolated from postweaning pigs. *Vet. Microbiol.* **37:**101–114.

746. **Dean-Nystrom, E. A., and J. E. Samuel.** 1994. Age-related resistance to 987P fimbriae-mediated colonization correlates with specific glycolipid receptors in intestinal mucus in swine. *Infect. Immun.* **62:**4789–4794.

747. **Dean-Nystrom, E. A., B. T. Bosworth, W. C. Cray, Jr., and H. W. Moon.** 1997. Pathogenicity of *Escherichia coli* O157:H7 in the intestines of neonatal calves. *Infect. Immun.* **65:**1842–1848.

748. **Dean-Nystrom, E. A., B. T. Bosworth, H. W. Moon, and A. D. O'Brien.** 1998. *Escherichia coli* O157:H7 requires intimin for enteropathogenicity in calves. *Infect. Immun.* **66:**4560–4563.

749. **de Bentzmann, S., C. Plotkowski, and E. Puchelle.** 1996. Receptors in the *Pseudomonas aeruginosa* adherence to injured and repairing airway epithelium. *Am. J. Respir. Crit. Care Med.* **154:**S155–S162.

750. **de Bentzmann, S., P. Roger, and E. Puchelle.** 1996. *Pseudomonas aeruginosa* adherence to remodelling respiratory epithelium. *Eur. Respir. J.* **9:**2145–2150.

751. **de Bentzmann, S. G., O. Bajolet-Laudinat, F. Dupuit, D. Pierrot, C. Fuchey, M. C. Plotkowski, and E. Puchelle.** 1993. Protection of human respiratory epithelium from *Pseudomonas aeruginosa* adherence by phosphatidylglycerol liposomes. *Infect. Immun.* **62:**704–708.

752. **de Bentzmann, S., P. Roger, F. Dupuit, O. Bajolet-Laudinat, C. Fuchey, M. C. Plotkowski, and E. Puchelle.** 1996. Asialo GM1 is a receptor for *Pseudomonas aeruginosa* adherence to regenerating respiratory epithelial cells. *Infect. Immun.* **64:**1582–1588.

753. **de Boer, E. C., R. F. M. Bevers, K.-H. Kurth, and D. H. J. Schamhart.** 1996. Double fluorescent flow cytometric assessment of bacterial internalization and binding by epithelial cells. *Cytometry* **25:**381–387.

754. **Debroy, C., B. D. Bright, R. A. Wilson, J. Yealy, R. Kumar, and M. K. Bhan.** 1994. Plasmid-coded DNA fragment developed as a specific gene probe for the identification of enteroaggregative *Escherichia coli*. *J. Med. Microbiol.* **41:**393–398.

755. **Debroy, C., J. Yealy, R. A. Wilson, M. K. Bhan, and R. Kumar.** 1995. Antibodies raised against the outer membrane protein interrupt adherence of enteroaggregative *Escherichia coli*. *Infect. Immun.* **63:**2873–2879.

756. **DeCarlo, A. A., and G. J. Harber.** 1997. Hemagglutinin activity and heterogeneity of related *Porphyromonas gingivalis* proteinases. *Oral Microbiol. Immunol.* **12:**47–56.

757. **Decostere, A., F. Haesebrouck, G. Charlier, and R. Ducatelle.** 1999. The association of *Flavobacterium columnare* strains of high and low virulence with gill tissue of black mollies (*Poecilia sphenops*). *Vet. Microbiol.* **67:**287–298.

758. **Defoe, G., and J. Coburn.** 2001. Delineation of *Borrelia burgdorferi* p.66 sequences required for integrin alpha(IIb)beta(3) recognition. *Infect. Immun.* **69:**3455–3459.

759. **de Geus, B., M. Harmsen, and F. van Zijderveld.** 1998. Prevention of diarrhoea using pathogen specific monoclonal antibodies in an experimental enterotoxigenic *E. coli* infection in germfree piglets. *Vet. Q.* **20:**S87–S89.

760. **Deghmane, A. E., S. Petit, A. Topilko, Y. Pereira, D. Giorgini, M. Larribe, and M. K. Taha.** 2000. Intimate adhesion of *Neisseria meningitidis* to human epithelial cells is under the control of the *crgA* gene, a novel LysR-type transcriptional regulator. *EMBO J.* **19:**1068–1078.

761. **de Graaf-Miltenburg, L. A., K. E. van Vliet, T. L. Ten Hagen, J. Verhoef, and J. A. van Strijp.** 1994. The role of HSV-induced Fc- and C3b (i)-receptors in bacterial adherence. *J. Med. Microbiol.* **40:**48–54.

762. **Dehio, C., M. Meyer, J. Berger, H. Schwarz, and C. Lanz.** 1997. Interaction of *Bartonella henselae* with endothelial cells results in bacterial aggregation on the cell surface and the subsequent engulfment and internalisation of the bacterial aggregate by a unique structure, the invasome. *J. Cell Sci.* **110:**2141–2154.

763. **Dehio, C., E. Freissler, C. Lanz, O. G.**

Gomez-Duarte, G. David, and T. F. Meyer. 1998. Ligation of cell surface heparan sulfate proteoglycans by antibody-coated beads stimulates phagocytic uptake into epithelial cells: a model for cellular invasion by *Neisseria gonorrhoeae*. *Exp. Cell Res.* **242**:528–539.

764. **Dehio, M., O. D. Gomez-Duarte, C. Dehio, and T. F. Meyer.** 1998. Vitronectin-dependent invasion of epithelial cells by *Neisseria gonorrhoeae* involves alpha(v) integrin receptors. *FEBS Lett.* **424**:84–88.

765. **Deighton, M. A., J. /Capstick, and R. Borland.** 1992. A study of phenotypic variation of *Staphylococcus epidermidis* using Congo red agar. *Epidemiol. Infect.* **109**:423–432.

766. **Deighton, M., S. Pearson, J. Capstick, D. Spelman, and R. Borland.** 1992. Phenotypic variation of *Staphylococcus epidermidis* isolated from a patient with native valve endocarditis. *J. Clin. Microbiol.* **30**:2385–2390.

767. **Deighton, M. A., R. Borland, and J. A. Capstick.** 1996. Virulence of *Staphylococcus epidermidis* in a mouse model: significance of extracellular slime. *Epidemiol. Infect.* **117**:267–280.

768. **de Kort, G., A. Bolton, G. Martin, J. Stephen, and J. A. van de Klundert.** 1994. Invasion of rabbit ileal tissue by *Enterobacter cloacae* varies with the concentration of OmpX in the outer membrane. *Infect. Immun.* **62**:4722–4726.

769. **de la Fuente, J., J. C. Garcia-Garcia, E. F. Blouin, and K. M. Kocan.** 2001. Differential adhesion of major surface proteins 1a and 1b of the ehrlichial cattle pathogen *Anaplasma marginale* to bovine erythrocytes and tick cells. *Int. J. Parasitol.* **31**:145–153.

770. **Dellacasagrande, J., E. Ghigo, S. M. Hammami, R. Toman, D. Raoult, C. Capo, and J. L. Mege.** 2000 Alpha(v)beta(3) integrin and bacterial lipopolysaccharide are involved in *Coxiella burnetii*-stimulated production of tumor necrosis factor by human monocytes. *Infect. Immun.* **68**:5673–5678.

771. **Delmi, M., P. Vaudaux, D. P. Lew, and H. Vasey.** 1994. Role of fibronectin in staphylococcal adhesion to metallic surfaces used as models of orthopaedic devices. *J. Orthop. Res.* **12**:432–438.

772. **Delneri, M. T., S. B. Carbonare, M. L. Silva, P. Palmeira, and M. M. Carneiro-Sampaio.** 1997. Inhibition of enteropathogenic *Escherichia coli* adhesion to HEp-2 cells by colostrum and milk from mothers delivering low-birth-weight neonates. *Eur. J. Pediatr.* **156**:493–498.

773. **Delogu, G., and M. J. Brennan.** 1999. Functional domains present in the mycobacterial hemagglutinin, HBHA. *J. Bacteriol.* **181**:7464–7469.

774. **Del Re, B., B. Sgorbati, M. Miglioli, and D. Palenzona.** 2000. Adhesion, autoaggregation and hydrophobicity of 13 strains of *Bifidobacterium longum*. *Lett. Appl. Microbiol.* **31**:438–442.

775. **Del Re, B., A. Busetto, G. Vignola, B. Sgorbati, and D. L. Palenzona.** 1998. Autoaggregation and adhesion ability in a *Bifidobacterium suis* strain. *Lett. Appl. Microbiol.* **27**:307–310.

776. **de Luna, M. G., A. Rudin, S. A. Vinhas, D. F. de Almeida, and L. C. de Souza Ferreira.** 1998. Epitope specificity of a monoclonal antibody generated against the dissociated CFA/I fimbriae of enterotoxigenic *Escherichia coli*. *Microbiol. Immunol.* **42**:341–346.

777. **de Melo Marques, M. A., S. Mahapatra, D. Nandan, T. Dick, E. N. Sarno, P. J. Brennan, and M. C. Vidal Pessolani.** 2000. Bacterial and host-derived cationic proteins bind alpha2-laminins and enhance *Mycobacterium leprae* attachment to human Schwann cells. *Microbes Infect.* **2**:1407–1417.

778. **Demirer, S., I. E. Gecim, K. Aydinuraz, H. Ataoglu, M. A. Yerdel, and E. Kuterdem.** 2001. Affinity of *Staphylococcus epidermidis* to various prosthetic graft materials. *J. Surg. Res.* **99**:70–74.

779. **Demuth, D. R., Y. Duan, W. Brooks, A. R. Holmes, R. McNab, and H. F. Jenkinson.** 1996. Tandem genes encode cell-surface polypeptides SspA and SspB which mediate adhesion of the oral bacterium *Streptococcus gordonii* to human and bacterial receptors. *Mol. Microbiol.* **20**:403–413.

780. **Demuth, D. R., Y. Duan, H. F. Jenkinson, R. McNab, S. Gil, and R. J. Lamont.** 1997. Interruption of the *Streptococcus gordonii* M5 *sspA*/*sspB* intergenic region by an insertion sequence related to IS1167 of *Streptococcus pneumoniae*. *Microbiol.* **143**: 2047–2055.

781. **Deodhar, L., and J. Karnad.** 1994. *In vitro* adhesiveness of *Gardnerella vaginalis* strains in relation to the occurrence of clue cells in vaginal discharge. *Indian J. Med. Res.* **100**:59–61.

782. **de Oliveira, I. R., A. N. de Araujo, S. N. Bao, and L. G. Giugliano.** 2001. Binding of lactoferrin and free secretory component to enterotoxigenic *Escherichia coli*. *FEMS Microbiol. Lett.* **203**:29–33.

783. **de Reuse, H., and M. K. Taha.** 1997. RegF, an SspA homologue, regulates the expression of the *Neisseria gonorrhoeae pilE* gene. *Res. Microbiol.* **148**:289–303.

784. **Dersch, P., and R. R. Isberg.** 1999. A region of the *Yersinia pseudotuberculosis* invasin protein enhances integrin-mediated uptake into mammalian cells and promotes self-association. *EMBO J.* **18:**1199–1213.

785. **Der Vartanian, M., J. P. Girardeau, C. Martin, E. Rousset, M. Chavarot, H. Laude, and M. Contrepois.** 1997. An *Escherichia coli* CS31A fibrillum chimera capable of inducing memory antibodies in outbred mice following booster immunization with the entero-pathogenic coronavirus transmissible gastroenteritis virus. *Vaccine* **15:**111–120.

786. **Der Vartanian, M., M. C. Mechin, B. Jaffeux, Y. Bertin, I. Felix, and B. Gaillard-Martinie.** 1994. Permissible peptide insertions surrounding the signal peptide-mature protein junction of the ClpG prepilin: CS31A fimbriae of *Escherichia coli* as carriers of foreign sequences. *Gene* **148:**23–32.

787. **De Rycke, J., P. Mazars, J. P. Nougayrede, C. Tasca, M. Boury, F. Herault, A. Valette, and E. Oswald.** 1996. Mitotic block and delayed lethality in HeLa epithelial cells exposed to *Escherichia coli* BM2–1 producing cytotoxic necrotizing factor type 1. *Infect. Immun.* **64:**1694–1705.

788. **De Rycke, J., E. Comtet, C. Chalareng, M. Boury, C. Tasca, and A. Milon.** 1997. Enteropathogenic *Escherichia coli* O103 from rabbit elicits actin stress fibers and focal adhesions in HeLa epithelial cells, cytopathic effects that are linked to an analog of the locus of enterocyte effacement. *Infect. Immun.* **65:**2555–2563.

789. **De Rycke, J., J. P. Nougayrede, E. Oswald, and P. Mazars.** 1997. Interaction of *Escherichia coli* producing cytotoxic necrotizing factor with HeLa epithelial cells. *Adv. Exp. Med. Biol.* **412:**363–366.

790. **Desai, N. P., S. F. Hossainy, and J. A. Hubbell.** 1992. Surface-immobilized polyethylene oxide for bacterial repellence. *Biomaterials* **13:**417–420.

791. **Desjardins, M., L. G. Filion, S. Robertson, and D. W. Cameron.** 1995. Inducible immunity with a pilus preparation booster vaccination in an animal model of *Haemophilus ducreyi* infection and disease. *Infect. Immun.* **63:**2012–2020.

792. **Desjardins, M., L. G. Filion, S. Robertson, L. Kobylinski, and D. W. Cameron.** 1996. Evaluation of humoral and cell-mediated inducible immunity to *Haemophilus ducreyi* in an animal model of chancroid. *Infect. Immun.* **64:**1778–1788.

793. **Deslauriers, M., and C. Mouton.** 1992. Epitope mapping of hemagglutinating adhesin HA-Ag2 of *Bacteroides* (*Porphyromonas*) *gingivalis. Infect. Immun.* **60:**2791–2799.

794. **Desnottes, J. F., and N. Diallo.** 1994. Effect of sparfloxacin on *Staphylococcus aureus* adhesiveness and phagocytosis. *J. Antimicrob. Chemother.* **33:**737–746.

795. **Devaraj, N., M. Sheykhnazari, W. S. Warren, and V. P. Bhavanandan.** 1994. Differential binding of *Pseudomonas aeruginosa* to normal and cystic fibrosis tracheobronchial mucins. *Glycobiology* **4:**307–316.

796. **De Vidipo, L. A., E. A. De Marques, E. Puchelle, and M. C. Plotkowski.** 2001. *Stenotrophomonas maltophilia* interaction with human epithelial respiratory cells *in vitro. Microbiol. Immunol.* **45:**563–569.

797. **DeVinney, R., J. L. Puente, A. Gauthier, D. Goosney, and B. B. Finlay.** 2001. Enterohaemorrhagic and enteropathogenic *Escherichia coli* use a different Tir-based mechanism for pedestal formation. *Mol. Microbiol.* **41:**1445–1458.

798. **DeVinney, R., M. Stein, D. Reinscheid, A. Abe, S. Ruschkowski, and B. B. Finlay.** 1999. Enterohemorrhagic *Escherichia coli* O157:H7 produces Tir, which is translocated to the host cell membrane but is not tyrosine phosphorylated. *Infect. Immun.* **67:**2389–2398.

799. **De Vries, F. P., A. van Der Ende, J. P. van Putten, and J. Dankert.** 1996. Invasion of primary nasopharyngeal epithelial cells by *Neisseria meningitidis* is controlled by phase variation of multiple surface antigens. *Infect. Immun.* **64:**2998–3006.

800. **De Vries, F. P., R. Cole, J. Dankert, M. Frosch, and J. P. M. van Putten.** 1998. *Neisseria meningitidis* producing the Opc adhesin binds epithelial cell proteoglycan receptors. *Mol. Microbiol.* **27:**1203–1212.

801. **Dexter, S. J., R. G. Pearson, M. C. Davies, M. Camara, and K. M. Shakesheff.** 2001. A comparison of the adhesion of mammalian cells and *Staphylococcus epidermidis* on fibronectin-modified polymer surfaces. *J. Biomed. Mater. Res.* **56:**222–227.

802. **Deziel, E., Y. Comeau, and R. Villemur.** 2001. Initiation of biofilm formation by *Pseudomonas aeruginosa* 57RP correlates with emergence of hyperpiliated and highly adherent phenotypic variants deficient in swimming, swarming, and twitching motilities. *J. Bacteriol.* **183:**1195–1204.

803. **Dhaenens, L., F. Szczebara, and M. O. Husson.** 1997. Identification, characterization, and immunogenicity of the lactoferrin-

binding protein from *Helicobacter pylori*. *Infect. Immun.* **65:**514–518.

804. **Dhandayuthapani, S., W. G. Rasmussen, and J. B. Baseman.** 1999. Disruption of gene *mg218* of *Mycoplasma genitalium* through homologous recombination leads to an adherence-deficient phenotype. *Proc. Natl. Acad. Sci. USA* **96:**5227–5232.

805. **Dhandayuthapani, S., M. W. Blaylock, C. M. Bebear, W. G. Rasmussen, and J. B. Baseman.** 2001. Peptide methionine sulfoxide reductase (MsrA) is a virulence determinant in *Mycoplasma genitalium*. *J. Bacteriol.* **183:**5645–5650.

806. **Dhir, V. K., and C. E. Dodd.** 1995. Susceptibility of suspended and surface-attached *Salmonella enteritidis* to biocides and elevated temperatures. *Appl. Environ. Microbiol.* **61:**1731–1738.

807. **Dibb-Fuller, M., E. Allen-Vercoe, M. J. Woodward, and C. J. Thorns.** 1997. Expression of SEF17 fimbriae by *Salmonella enteritidis*. *Lett. Appl. Microbiol.* **25:**447–452.

808. **Dibb-Fuller, M. P., E. Allen-Vercoe, C. J. Thorns, and M. J. Woodward.** 1999. Fimbriae- and flagella-mediated association with and invasion of cultured epithelial cells by *Salmonella enteritidis*. *Microbiology* **145:**1023–1031.

809. **Dibb-Fuller, M. P., A. Best, D. A. Stagg, W. A. Cooley, and M. J. Woodward.** 2001. An *in-vitro* model for studying the interaction of *Escherichia coli* O157:H7 and other enteropathogens with bovine primary cell cultures. *J. Med. Microbiol.* **50:**759–769.

810. **Di Biase, A. M., G. Petrone, M. P. Conte, L. Seganti, M. G. Ammendolia, A. Tinari, F. Iosi, M. Marchetti, and F. Superti.** 2000. Infection of human enterocyte-like cells with rotavirus enhances invasiveness of *Yersinia enterocolitica* and *Y. pseudotuberculosis*. *J. Med. Microbiol.* **49:**897–904.

811. **Dickinson, R. B., J. A. Nagel, D. McDevitt, T. J. Foster, R. A. Proctor, and S. L. Cooper.** 1995. Quantitative comparison of clumping factor- and coagulase-mediated *Staphylococcus aureus* adhesion to surface-bound fibrinogen under flow. *Infect. Immun.* **63:**3143–3150.

812. **Dickinson, R. B., J. A. Nagel, R. A. Proctor, and S. L. Cooper.** 1997. Quantitative comparison of shear-dependent *Staphylococcus aureus* adhesion to three polyurethane ionomer analogs with distinct surface properties. *J. Biomed. Mater. Res.* **36:**152–162.

813. **Didenko, L. W., I. B. Buchwalow, W. Schulze, K. Augsten, M. Susa, and E.** Unger. 1996. Localization of G-proteins in macrophages and *E. coli* during phagocytosis. *Acta Histochem.* **98:**399–409.

814. **Diemer, T., W. Weidner, H. W. Michelmann, H. G. Schiefer, E. Rovan, and F. Mayer.** 1996. Influence of *Escherichia coli* on motility parameters of human spermatozoa *in vitro*. *Int. J. Androl.* **19:**271–277.

815. **Dietrich, C., K. Heuner, B. C. Brand, J. Hacker, and M. Steinert.** 2001. Flagellum of *Legionella pneumophila* positively affects the early phase of infection of eukaryotic host cells. *Infect. Immun.* **69:**2116–2122.

816. **DiMango, E., H. J. Zar, R. Bryan, and A. Prince.** 1995. Diverse *Pseudomonas aeruginosa* gene products stimulate respiratory epithelial cells to produce interleukin-8. *J. Clin. Investig.* **96:**2204–2210.

817. **DiMango, E., A. J. Ratner, R. Bryan, S. Tabibi, and A. Prince.** 1998. Activation of NF-kappaB by adherent *Pseudomonas aeruginosa* in normal and cystic fibrosis respiratory epithelial cells. *J. Clin. Investig.* **101:**2598–2605.

818. **Di Martino, P., Y. Bertin, J. P. Girardeau, V. Livrelli, B. Joly, and A. Darfeuille-Michaud.** 1995. Molecular characterization and adhesive properties of CF29K, an adhesin of *Klebsiella pneumoniae* strains involved in nosocomial infections. *Infect. Immun.* **63:**4336–4344.

819. **Di Martino, P., V. Livrelli, D. Sirot, B. Joly, and A. Darfeuille-Michaud.** 1996. A new fimbrial antigen harbored by CAZ-5/SHV-4-producing *Klebsiella pneumoniae* strains involved in nosocomial infections. *Infect. Immun.* **64:**2266–2273.

820. **Di Martino, P., D. Sirot, B. Joly, C. Rich, and A. Darfeuille-Michaud.** 1997. Relationship between adhesion to intestinal Caco-2 cells and multidrug resistance in *Klebsiella pneumoniae* clinical isolates. *J. Clin. Microbiol.* **35:**1499–1503.

821. **Di Martino, P., J. P. Girardeau, M. Der Vartanian, B. Joly, and A. Darfeuille-Michaud.** 1997. The central variable V2 region of the CS31A major subunit is involved in the receptor-binding domain. *Infect. Immun.* **65:**609–616.

822. **Di Martino, P., J. Rebiere-Huet, and C. Hulen.** 2000. Effects of antibiotics on adherence of *Pseudomonas aeruginosa* and *Pseudomonas fluorescens* to A549 pneumocyte cells. *Chemotherapy* **46:**129–134.

823. **Ding, H., and C. Lämmler.** 1992. Cell surface hydrophobicity of *Actinomyces pyogenes* determined by hexadecane adherence- and salt

aggregation studies. *J. Vet. Med. Ser. B* **39:**132–138.

824. **Ding, H., C. Lammler, and R. S. Seleim.** 1993. Adherence of *Actinomyces pyogenes* to HeLa cells mediated by hydrophobic surface proteins. *Zentbl. Bakteriol.* **279:**299–306.

825. **Dinjus, U., I. Hanel, W. Rabsch, and R. Helmuth.** 1998. Studies of the presence of the virulence factors, adhesion, invasion, intracellular multiplication and toxin formation in salmonellas of different origin. *Zentbl. Bakteriol.* **287:**387–398.

826. **DiNovo, B. B., R. Doan, R. B. Dyer, S. Baron, N. K. Herzog, and D. W. Niesel.** 1996. Treatment of HeLa cells with bacterial water extracts inhibits *Shigella flexneri* invasion. *FEMS Immunol. Med. Microbiol.* **15:**149–158.

827. **Dintilhac, A., G. Alloing, C. Granadel, and J. P. Claverys.** 1997. Competence and virulence of *Streptococcus pneumoniae:* Adc and PsaA mutants exhibit a requirement for Zn and Mn resulting from inactivation of putative ABC metal permeases. *Mol. Microbiol.* **25:**727–739.

828. **Dirksen, L. B., K. A. Krebes, and D. C. Krause.** 1994. Phosphorylation of cytadherence-accessory proteins in *Mycoplasma pneumoniae. J. Bacteriol.* **176:**7499–7505.

829. **Dirksen, L. B., T. Proft, H. Hilbert, H. Plagens, R. Herrmann, and D. C. Krause.** 1996. Sequence analysis and characterization of the *hmw* gene cluster of *Mycoplasma pneumoniae. Gene* **171:**19–25.

830. **Ditcham, W. G., J. A. Leigh, A. P. Bland, and A. W. Hill.** 1996. Adhesion of *Streptococcus uberis* to monolayers of cultured epithelial cells derived from the bovine mammary gland. *FEMS Immunol. Med. Microbiol.* **14:**145–150.

831. **DiTizio, V., G. W. Ferguson, M. W. Mittelman, A. E. Khoury, A. W. Bruce, and F. DiCosmo.** 1998. A liposomal hydrogel for the prevention of bacterial adhesion to catheters. *Biomaterials* **19:**1877–1884.

832. **Dodson, K. W., F. Jacob-Dubuisson, R. T. Striker, and S. J. Hultgren.** 1993. Outer-membrane PapC molecular usher discriminately recognizes periplasmic chaperone-pilus subunit complexes. *Proc. Natl. Acad. Sci. USA* **90:**3670–3674.

833. **Dogan, S., F. Gunzer, H. Guenay, G. Hillmann, and W. Geurtse.** 2000. Infection of primary human gingival fibroblasts by *Porphyromonas gingivalis* and *Prevotella intermedia. Clin. Oral Investig.* **4:**35–41.

834. **Dohi, H., Y. Nishida, M. Mizuno, M. Shinkai, T. Kobayashi, T. Takeda, H.** Uzawa, and K. Kobayashi. 1999. Synthesis of an artificial glycoconjugate polymer carrying Pk-antigenic trisaccharide and its potent neutralization activity against Shiga-like toxin. *Bioorgan. Med. Chem.* **7:**2053–2062.

835. **Doig, P., L. Emody, and T. J. Trust.** 1992. Binding of laminin and fibronectin by the trypsin-resistant major structural domain of the crystalline virulence surface array protein of *Aeromonas salmonicida. J. Biol. Chem.* **267:**43–49.

836. **Doig, P., J. W. Austin, and T. J. Trust.** 1992. The *Helicobacter pylori* 19.6-kilodalton protein is an iron-containing protein resembling ferritin. *J. Bacteriol.* **175:**557–560.

837. **Doig, P., J. W. Austin, M. Kostryzynska, and T. J. Trust.** 1992. Production of a conserved adhesin by the human gastroduodenal pathogen *Helicobacter pylori. J. Bacteriol.* **174:**2539–2547.

838. **Doig, P., R. Yao, D. H. Burr, P. Guerry, and T. J. Trust.** 1996. An environmentally regulated pilus-like appendage involved in *Campylobacter* pathogenesis. *Mol. Microbiol.* **20:**885–894.

839. **Dom, P., F. Haesebrouck, R. Ducatelle, and G. Charlier.** 1994. In vivo association of *Actinobacillus pleuropneumoniae* serotype 2 with the respiratory epithelium of pigs. *Infect. Immun.* **62:**1262–1267.

840. **Domenico, P., R. J. Salo, A. S. Cross, and B. A. Cunha.** 1994. Polysaccharide capsule-mediated resistance to opsonophagocytosis in *Klebsiella pneumoniae. Infect. Immun.* **62:**4495–4499.

841. **Domingo, D. T., M. W. Jackwood, and T. P. Brown.** 1992. Filamentous forms of *Bordetella avium:*culture conditions and pathogenicity. *Avian Dis.* **36:**707–713.

842. **Domingues, R. M., S. M. Cavalcanti, A. F. Andrade, and M. C. Ferreira.** 1992. Sialic acid as receptor of *Bacteroides fragilis* lectin-like adhesin. *Zentbl. Bakteriol.* **277:**340–344.

843. **Dominiecki, M. E., and J. Weiss.** 1999. Antibacterial action of extracellular mammalian group IIA phospholipase A2 against grossly clumped *Staphylococcus aureus. Infect. Immun.* **67:**2299–2305.

844. **Donabedian, H., E. O'Donnell, C. Drill, L. M. Lipton, S. A. Khuder, and J. C. Burnham.** 1995. Prevention of subsequent urinary tract infections in women by the use of anti-adherence antimicrobial agents: a double-blind comparison of enoxacin with co-trimoxazole. *J. Antimicrob. Chemother.* **35:**409–420.

845. **Donlon, B., and E. Colleran.** 1993. A

comparison of different methods to determine the hydrophobicity of acetogenic bacteria. *J. Microbiol. Methods* **17:**27–37.

846. **Donnenberg, M. S., J. A. Giron, J. P. Nataro and J. B. Kaper.** 1992. A plasmid-encoded type IV fimbrial gene of enteropathogenic *Escherichia coli* associated with localized adherence. *Mol. Microbiol.* **6:**3427–3437.

847. **Donnenberg, M. S., J. B. Kaper, and B. B. Finlay.** 1997. Interactions between enteropathogenic *Escherichia coli* and host epithelial cells. *Trends Microbiol.* **5:**109–114.

848. **Donnenberg, M. S., and J. B. Kaper.** 1992. Enteropathogenic *Escherichia coli*. *Infect. Immun.* **60:**3953–3961.

849. **Donnenberg, M. S., C. O. Tacket, S. P. James, G. Losonsky, J. P. Nataro, S. S. Wasserman, J. B. Kaper, and M. M. Levine.** 1993. Role of the *eaeA* gene in experimental enteropathogenic *Escherichia coli* infection *J. Clin. Investig.* **92:**1412–1427.

850. **Donnenberg, M. S., T. Tzipori, M. L. McKee, A. D. O'Brien, J. Alroy, and J. B. Kaper.** 1993. The role of the *eae* gene of enterohemorrhagic *Escherichia coli* in intimate attachment *in vitro* and in a porcine model. *J. Clin. Invest.* **92:**1418–1424.

851. **Donnenberg, M. S., J. Yu, and J. B. Kaper.** 1993. A second chromosomal gene necessary for intimate attachment of enteropathogenic *Escherichia coli* to epithelial cells. *J. Bacteriol.* **175:**4670–4680.

852. **Donnenberg, M. S., and J. P. Nataro.** 1995. Methods for studying adhesion of diarrheagenic *Escherichia coli*. *Methods Enzymol.* **253:**324–336.

853. **Donnenberg, M. S., H. Z. Zhang, and K. D. Stone.** 1997. Biogenesis of the bundle-forming pilus of enteropathogenic *Escherichia coli*: reconstitution of fimbriae in recombinant *E. coli* and role of DsbA in pilin stability—a review. *Gene* **192:**33–38.

854. **Dopfer, D., R. A. Almeida, T. J. Lam, H. Nederbragt, S. P. Oliver, and W. Gaastra.** 2000. Adhesion and invasion of *Escherichia coli* from single and recurrent clinical cases of bovine mastitis *in vitro*. *Vet. Microbiol.* **74:**331–343.

855. **Dopfer, D., H. Nederbragt, R. A. Almeida, and W. Gaastra.** 2001. Studies about the mechanism of internalization by mammary epithelial cells of *Escherichia coli* isolated from persistent bovine mastitis. *Vet. Microbiol.* **80:**285–296.

856. **Doran, J. L., S. K. Collinson, J. Burian, G. Sarlos, E. C. Todd, C. K. Munro, C. M. Kay, P. A. Banser, P. I. Peterkin, and W. W. Kay.** 1993. DNA-based diagnostic tests for *Salmonella* species targeting *agfA*, the structural gene for thin, aggregative fimbriae. *J. Clin. Microbiol.* **31:**2263–2273.

857. **Doran, J. L., S. K. Collinson, C. M. Kay, P. A. Banser, J. Burian, C. K. Munro, S. H. Lee, J. M. Somers, E. C. Todd, and W. W. Kay.** 1994. *fimA* and *tctC* based DNA diagnostics for *Salmonella*. *Mol. Cell. Probes* **8:**291–310.

858. **Dorman, C. J.** 1995. DNA topology and the global control of bacterial gene expression: implications for the regulation of virulence gene expression. *Microbiology* **141:**1271–1280.

859. **Dorman, C. J., and N. N. Bhriain.** 1992. Thermal regulation of *fimA*, the *Escherichia coli* gene coding for the type 1 fimbrial subunit protein. *FEMS Microbiol. Lett.* **78:**125–130.

860. **Dorn, B. R., K.-P. Jeung, and A. Progulske-Fox.** 1998. Invasion of human oral epithelial cells by *Prevotella intermedia*. *Infect. Immun.* **66:**6054–6057.

861. **Dorn, B. R., J. N. Burks, K. N. Seifert, and A. Progulske-Fox.** 2000. Invasion of endothelial and epithelial cells by strains of *Porphyromonas gingivalis*. *FEMS Microbiol. Lett.* **187:**139–144.

862. **Dorward, D. W., E. R. Fischer, and D. M. Brooks.** 1997. Invasion and cytopathic killing of human lymphocytes by spirochetes causing Lyme disease. *Clin. Infect. Dis.* **25:**S2–S8.

863. **Dorward, D. W., and R. S. Larson.** 2001. Murine model for lymphocytic tropism by *Borrelia burgdorferi*. *Infect. Immun.* **69:**1428–1432.

864. **Douglas, C. W.** 1994. Bacterial-protein interactions in the oral cavity. *Adv. Dent. Res.* **8:**254–262.

865. **Dove, S. L., and C. J. Dorman.** 1994. The site-specific recombination system regulating expression of the type 1 fimbrial subunit gene of *Escherichia coli* is sensitive to changes in DNA supercoiling. *Mol. Microbiol.* **14:**975–988.

866. **Dowling, R. B., M. Johnson, P. J. Cole, and R. Wilson.** 1999. Effect of fluticasone propionate and salmeterol on *Pseudomonas aeruginosa* infection of the respiratory mucosa *in vitro*. *Eur. Respir. J.* **14:**363–369.

867. **Dowling, R. B., M. Johnson, P. J. Cole, and R. Wilson.** 1998. Effect of salmeterol on *Haemophilus influenzae* infection of respiratory mucosa in vitro. *Eur. Respir. J.* **11:**86–90.

868. **Doyle, A., B. Beigi, A. Early, A. Blake, P. Eustace, and R. Hone.** 1995. Adherence of bacteria to intraocular lenses: a prospective study. *Br. J. Ophthalmol.* **79:**347–349.

869. **Doyle, R. J.** 1998. Prevention of mucosal *Escherichia coli* infection by Fim H-adhesin-based systemic vaccination. *Chemtracts* **11:**212–216.

870. **Doyle, R. J., and M. Rosenberg.** 1995. Measurement of microbial adhesion to hydrophobic substrata. *Methods Enzymol.* **253:**542–550.

871. **Dozois, C. M., J. M. Fairbrother, J. Harel, and M. Bossé.** 1992. *pap*- and *pil*-related DNA sequences and other virulence determinants associated with *Escherichia coli* isolated from septicemic chickens and turkeys. *Infect. Immun.* **60:**2648–2656.

872. **Dozois, C. M., N. Chanteloup, M. Dho-Moulin, A. Bree, C. Desautels, and J. M. Fairbrother.** 1994. Bacterial colonization and *in vivo* expression of F1 (type 1) fimbrial antigens in chickens experimentally infected with pathogenic *Escherichia coli*. *Avian Dis.* **38:**231–239.

873. **Dozois, C. M., S. A. Pourbakhsh, and J. M. Fairbrother.** 1995. Expression of P and type 1 (F1) fimbriae in pathogenic *Escherichia coli* from poultry. *Vet. Microbiol.* **45:**297–309.

874. **Dozois, C. M., J. Harel, and J. M. Fairbrother.** 1996. P-fimbriae-producing septicaemic *Escherichia coli* from poultry possess *fel*-related gene clusters whereas *pap*-hybridizing P-fimbriae-negative strains have partial or divergent P fimbrial gene clusters. *Microbiology* **142:**2759–2766.

875. **Dozois, C. M., S. Clement, C. Desautels, E. Oswald, and J. M. Fairbrother.** 1997. Expression of P, S, and F1C adhesins by cytotoxic necrotizing factor 1-producing *Escherichia coli* from septicemic and diarrheic pigs. *FEMS Microbiol. Lett.* **152:**307–312.

876. **Drago, L., E. de Vecchi, B. Mombelli, L. Nicola, M. Valli, and M. R. Gismondo.** 2001. Activity of levofloxacin and ciprofloxacin against urinary pathogens. *J. Antimicrob. Chemother.* **48:**37–45.

877. **Drake, D. R., J. Pau, and J. C. Keller.** 1999. Primary bacterial colonization of implant surfaces. *Int. J. Oral Maxillofac. Implants* **14:**226–232.

878. **Dramsi, S., M. Lebrun, and P. Cossart.** 1996. Molecular and genetic determinants involved in invasion of mammalian cells by *Listeria monocytogenes*. *Curr. Top. Microbiol. Immunol.* **209:**61–77.

879. **Drevets, D. A., R. T. Sawyer, T. A. Potter, and P. A. Campbell.** 1995. *Listeria monocytogenes* infects human endothelial cells by two distinct mechanisms. *Infect. Immun.* **63:**4268–4276.

880. **Drolet, R., J. M. Fairbrother, J. Harel, and P. Helie.** 1994. Attaching and effacing and enterotoxigenic *Escherichia coli* associated with enteric colibacillosis in the dog. *Can. J. Vet. Res.* **58:**87–92.

881. **Drolet, R., J. M. Fairbrother, and D. Vaillancourt.** 1994. Attaching and effacing *Escherichia coli* in a goat with diarrhea. *Can. Vet. J.* **35:**122–123.

882. **Drudy, D., D. P. O'Donoghue, A. Baird, L. Fenelon, and C. O'Farrelly.** 2001. Flow cytometric analysis of *Clostridium difficile* adherence to human intestinal epithelial cells. *J. Med. Microbiol.* **50:**526–534.

883. **Du, L., P. Pellen-Mussi, F. Chandad, C. Mouton, and M. Bonnaure-Mallet.** 1997. Fimbriae and the hemagglutinating adhesin HA-Ag2 mediate adhesion of *Porphyromonas gingivalis* to epithelial cells. *Infect. Immun.* **65:**3875–3881.

884. **Du, L., P. Pellen-Mussi, F. Chandad, C. Mouton, and M. Bonnaure-Mallet.** 1997. Conservation of fimbriae and the hemagglutinating adhesin HA-Ag2 among *Porphyromonas gingivalis* strains and other anaerobic bacteria studied by epitope mapping analysis. *Clin. Diagn. Lab. Immunol.* **4:**711–714.

885. **Duarte, X., C. T. Anderson, M. Grimson, R. D. Barabote, R. E. Strauss, L. S. Gollahon, and M. J. San Francisco.** 2000. *Erwinia chrysanthemi* strains cause death of human gastrointestinal cells in culture and express an intimin-like protein. *FEMS Microbiol. Lett.* **190:**81–86.

886. **Dubnau, D.** 1997 Binding and transport of transforming DNA by *Bacillus subtilis:* the role of type-IV pilin-like proteins—a review. *Gene* **192:**191–198.

887. **Dubreuil, J. D., and J. M. Fairbrother.** 1992. Biochemical and serological characterization of *Escherichia coli* fimbrial antigen F165(2). *FEMS Microbiol. Lett.* **74:**219–224.

888. **Duchet-Suchaux, M., P. Menanteau, and F. G. van Zijderveld.** 1992. Passive protection of suckling infant mice against F41-positive enterotoxigenic *Escherichia coli* strains by intravenous inoculation of the dams with monoclonal antibodies against F41. *Infect. Immun.* **60:**2828–2834.

889. **Duerden, B. I.** 1994. Virulence factors in anaerobes. *Clin. Infect. Dis.* **18:**S253–S259.

890. **Duffy, G., and J. J. Sheridan.** 1997. The effect of temperature, pH and medium in a surface adhesion immunofluorescent technique for detection of *Listeria monocytogenes*. *J. Appl. Microbiol.* **83:**95–101.

891. **Duffy, G., and J. J. Sheridan.** 1999. The

effect of pH and culture system on the attachment of plasmid-bearing and plasmid-cured *Yersinia enterocolitica* to a polycarbonate membrane in a surface adhesion immunofluorescent technique. *J. Appl. Microbiol.* **86:**867–873.

892. **Dugal, F., M. Belanger, and M. Jacques.** 1992. Enhanced adherence of *Pasteurella multocida* to porcine tracheal rings preinfected with *Bordetella bronchiseptica*. *Can. J. Vet. Res.* **56:**260–264.

893. **Duncan, M. J., S. A. Emory, and E. C. Almira.** 1996. *Porphyromonas gingivalis* genes isolated by screening for epithelial cell attachment. *Infect. Immun.* **64:**3624–3631.

894. **Duncan, M. J., S. Nakao, Z. Skobe, and H. Xie.** 1993. Interactions of *Porphyromonas gingivalis* with epithelial cells. *Infect. Immun.* **61:**2260–2265.

895. **Dunn, K. L., M. Virji, and E. R. Moxon.** 1995. Investigations into the molecular basis of meningococcal toxicity for human endothelial and epithelial cells: the synergistic effect of LPS and pili. *Microb. Pathog.* **18:**81–96.

896. **Dunne, W. M., Jr., and E. M. Burd.** 1992. The effects of magnesium, calcium, EDTA, and pH on the *in vitro* adhesion of *Staphylococcus epidermidis* to plastic. *Microbiol. Immunol.* **36:**1019–1027.

897. **Dunne, W. M., and E. M. Burd.** 1993. Fibronectin and proteolytic fragments of fibronectin interfere with the adhesion of *Staphylococcus epidermidis* to plastic. *J. Appl. Bacteriol.* **74:**411–416.

898. **Dupuy, B., and A. P. Pugsley.** 1994. Type IV prepilin peptidase gene of *Neisseria gonorrhoeae* MS11: presence of a related gene in other piliated and nonpiliated *Neisseria* strains. *J. Bacteriol.* **176:**1323–1331.

899. **Dupuy, B., M. K. Taha, O. Possot, C. Marchal, and A. P. Pugsley.** 1992. PulO, a component of the pullulanase secretion pathway of *Klebsiella oxytoca,* correctly and efficiently processes gonococcal type IV prepilin in *Escherichia coli*. *Mol. Microbiol.* **6:**1887–1894.

900. **Duran, J. A., A. Malvar, M. T. Rodriguez-Ares, and C. Garcia-Riestra.** 1993. Heparin inhibits *Pseudomonas* adherence to soft contact lenses. *Eye* **7:**152–154.

901. **Dusek, D. M., A. Progulske-Fox, J. Whitlock, and T. A. Brown.** 1993. Isolation and characterization of a cloned *Porphyromonas gingivalis* hemagglutinin from an avirulent strain of *Salmonella typhimurium*. *Infect. Immun.* **61:**940–946.

902. **Dusek, D. M., A. Progulske-Fox, and T. A. Brown.** 1994. Systemic and mucosal immune responses in mice orally immunized with avirulent *Salmonella typhimurium* expressing a cloned *Porphyromonas gingivalis* hemagglutinin. *Infect. Immun.* **62:**1652–1657.

903. **Dutta, S., S. Pal, S. Chakrabarti, P. Dutta, and B. Manna.** 1999. Use of PCR to identify enteroaggregative *Escherichia coli* as an important cause of acute diarrhoea among children living in Calcutta, India. *J. Med. Microbiol.* **48:**1011–1016.

904. **Dytoc, M., L. Fedorko, and P. M. Sherman.** 1994. Signal transduction in human epithelial cells infected with attaching and effacing *Escherichia coli in vitro*. *Gastroenterology*. **106:**1150–1161.

905. **Dytoc, M., R. Soni, F. Cockerill III, J. De Azavedo, M. Louie, J. Brunton, and P. Sherman.** 1993. Multiple determinants of verotoxin-producing *Escherichia coli* O157:H7 attachment-effacement. *Infect. Immun.* **61:**3382–3391.

906. **Dytoc, M., B. Gold, M. Louie, M. Huesca, L. Fedorko, S. Crowe, C. Lingwood, J. Brunton, and P. Sherman.** 1993. Comparison of *Helicobacter pylori* and attaching-effacing *Escherichia coli* adhesion to eukaryotic cells. *Infect. Immun.* **61:**448–456.

907. **Dytoc, M. T., A. Ismaili, D. J. Philpott, R. Soni, J. L. Brunton, and P. M. Sherman.** 1994. Distinct binding properties of *eaeA*-negative verocytotoxin-producing *Escherichia coli* of serotype O113:H21. *Infect. Immun.* **62:**3494–3505.

908. **Dziewanowska, K., J. M. Patti, C. F. Deobald, K. W. Bayles, W. R. Trumble, and G. A. Bohach.** 1999. Fibronectin binding protein and host cell tyrosine kinase are required for internalization of *Staphylococcus aureus* by epithelial cells. *Infect. Immun.* **67:**4673–4678.

909. **Dzik, W.** 1995. Use of leukodepletion filters for the removal of bacteria. *Immunol. Investig.* **24:**95–115.

910. **Ebel, F., T. Podzadel, M. Rohde, A. U. Kresse, S. Krämer, C. Deibel, C. A. Guzmán, and T. Chakraborty.** 1998. Initial binding of Shiga toxin-producing *Escherichia coli* to host cells and subsequent induction of actin rearrangements depend on filamentous EspA-containing surface appendages. *Mol. Microbiol.* **30:**147–161.

911. **Eberhard, T., G. Kronvall, and M. Ullberg.** 1999. Surface bound plasmin promotes migration of *Streptococcus pneumoniae* through reconstituted basement membranes. *Microb. Pathog.* **26:**175–181.

912. **Eberhard, T., R. Virkola, T. Korhonen, G. Kronvall, and M. Ullberg.** 1998.

Binding to human extracellular matrix by *Neisseria meningitidis*. *Infect. Immun.* **66**:1791–1794.

913. **Ebisu, S., H. Nakae, H. Fukuhara, and H. Okada.** 1992. The mechanisms of *Eikenella corrodens* aggregation by salivary glycoprotein and the effect of the glycoprotein on oral bacterial aggregation. *J. Periodontal Res.* **27**:615–622.

914. **Echeverria, P., O. Serichantalerg, S. Changchawalit, B. Baudry, M. M. Levine, F. Ørskov, and I. Ørskov.** 1992. Tissue culture-adherent *Escherichia coli* in infantile diarrhea. *J. Infect. Dis.* **165**:141–143.

915. **Edfors-Lilja, I., U. Gustafsson, Y. Duval-Iflah, H. Ellergren, M. Johansson, R. K. Juneja, L. Marklund, and L. Andersson.** 1995. The porcine intestinal receptor for *Escherichia coli* K88ab, K88ac: regional localization on chromosome 13 and influence of IgG response to the K88 antigen. *Anim. Genet.* **26**:237–242.

916. **Edgerton, M., S. E. Lo, and F. A. Scannapieco.** 1996. Experimental salivary pellicles formed on titanium surfaces mediate adhesion of streptococci. *Int. J. Oral Maxillofac. Implants.* **11**:443–449.

917. **Edwards, J. L., J. Q. Shao, K. A. Ault, and M. A. Apicella.** 2000. *Neisseria gonorrhoeae* elicits membrane ruffling and cytoskeletal rearrangements upon infection of primary human endocervical and ectocervical cells. *Infect. Immun.* **68**:5354–5363.

918. **Edwards, N. J., M. A. Monteiro, G. Faller, E. J. Walsh, A. P. Moran, I. S. Roberts, and N. J. High.** 2000. Lewis X structures in the O antigen side-chain promote adhesion of *Helicobacter pylori* to the gastric epithelium. *Mol. Microbiol.* **35**:1530–1539.

919. **Edwards, R. A., J. Cao, and D. M. Schifferli.** 1996. Identification of major and minor chaperone proteins involved in the export of 987P fimbriae. *J. Bacteriol.* **178**:3426–3433.

920. **Edwards, R. A., and D. M. Schifferli.** 1997. Differential regulation of *fasA* and *fasH* expression of *Escherichia coli* 987P fimbriae by environmental cues. *Mol. Microbiol.* **25**:797–809.

921. **Edwards, R. A., L. H. Keller, and D. M. Schifferli.** 1998. Improved allelic exchange vectors and their use to analyze 987P fimbria gene expression. *Gene* **207**:149–157.

922. **Edwards, R. A., D. M. Schifferli, and S. R. Maloy.** 2000. A role for *Salmonella* fimbriae in intraperitoneal infections. *Proc. Natl. Acad. Sci. USA* **97**:1258–1262.

923. **Eginton, P. J., J. Holah, D. G. Allison, P. S. Handley, and P. Gilbert.** 1998. Changes in the strength of attachment of microorganisms to surfaces following treatment with disinfectants and cleansing agents. *Lett. Appl. Microbiol.* **27**:101–105.

924. **Egland, P. G., L. D. Du, and P. E. Kolenbrander.** 2001. Identification of independent *Streptococcus gordonii* SspA and SspB functions in coaggregation with *Actinomyces naeslundii*. *Infect. Immun.* **69**:7512–7516.

925. **Ehara, M., Y. Ichinose, M. Iwami, A. Utsunomiya, S. Shimodori, S. K. Kangethe, B. C. Neves, K. Supawat, and S. Nakamura.** 1993. Immunogenicity of *Vibrio cholerae* O1 fimbriae in animal and human cholera. *Microbiol. Immunol.* **37**:679–688.

926. **Ehara, M., M. Iwami, Y. Ichinose, and T. Hirayama.** 2000. Development of hyperfimbriated strains of *Vibrio cholerae* O1. *Microbiol. Immunol.* **44**:439–446.

927. **Ehara, M., M. Iwami, Y. Ichinose, T. Hirayama, M. J. Albert, S. B. Sack, and S. Shimodori.** 1998. Induction of fimbriated *Vibrio cholerae* O139. *Clin. Diagn. Lab. Immunol.* **5**:65–69.

928. **Eiring, P., B. Manncke, K. Gerbracht, and H. Werner.** 1995. *Bacteroides fragilis* adheres to laminin significantly stronger than *Bacteroides thetaiotaomicron* and other species of the genus. *Zentbl. Bakteriol.* **282**:279–286.

929. **Eko, F. O., V. O. Rotimi, and A. O. Coker.** 1992. Haemagglutinating and buccal epithelial cell adherence activities of *Vibrio parahaemolyticus*: correlation with virulence. *J. Trop. Med. Hyg.* **95**:202–205.

930. **el Ahmer, O. R., S. D. Essery, A. T. Saadi, M. W. Raza, M. M. Ogilvie, D. M. Weir, and C. C. Blackwell.** 1999. The effect of cigarette smoke on adherence of respiratory pathogens to buccal epithelial cells. *FEMS Immunol. Med. Microbiol.* **23**:27–36.

931. **el Ahmer, O. R., M. W. Raza, M. M. Ogilvie, D. M. Weir, and C. C. Blackwell.** 1999. Binding of bacteria to HEp-2 cells infected with influenza A virus. *FEMS Immunol. Med. Microbiol.* **23**:331–341.

932. **Eliades, T., G. Eliades, and W. A. Brantley.** 1995. Microbial attachment on orthodontic appliances: I. Wettability and early pellicle formation on bracket materials. *Am. J. Orthodont. Dentofacial Orthop.* **108**:351–360.

933. **Elias, W. P., Jr., J. R. Czeczulin, I. R. Henderson, L. R. Trabulsi, and J. P. Nataro.** 1999. Organization of biogenesis

genes for aggregative adherence fimbria II defines a virulence gene cluster in enteroaggregative *Escherichia coli*. *J. Bacteriol.* **181:**1779–1785.

934. **Ellen, R. P., J. R. Dawson, and P. F. Yang.** 1994. *Treponema denticola* as a model for polar adhesion and cytopathogenicity of spirochetes. *Trends Microbiol.* **2:**114–119.

935. **Ellen, R. P., M. Song, and I. A. Buivids.** 1992. Inhibition of *Actinomyces viscosus–Porphyromonas gingivalis* coadhesion by trypsin and other proteins. *Oral Microbiol. Immunol.* **7:**198–203.

936. **Ellen, R. P., M. Song, and C. A. McCulloch.** 1994. Degradation of endogenous plasma membrane fibronectin concomitant with *Treponema denticola* 35405 adhesion to gingival fibroblasts. *Infect. Immun.* **62:**3033–3037.

937. **Ellen, R. P., H. Veisman, I. A. Buivids, and M. Rosenberg.** 1994. Kinetics of lactose-reversible coadhesion of *Actinomyces naeslundii* WVU 398A and *Streptococcus oralis* 34 on the surface of hexadecane droplets. *Oral Microbiol. Immunol.* **9:**364–371.

938. **Ellen, R. P., G. Lepine, and P. M. Nghiem.** 1997. *In vitro* models that support adhesion specificity in biofilms of oral bacteria. *Adv. Dent. Res.* **11:**33–42.

939. **Elliott, S. J., and J. B. Kaper.** 1997. Role of type 1 fimbriae in EPEC infections. *Microb. Pathog.* **23:**113–118.

940. **el Mazouari, K., E. Oswald, J. P. Hernalsteens, P. Lintermans, and H. De Greve.** 1994. F17-like fimbriae from an invasive *Escherichia coli* strain producing cytotoxic necrotizing factor type 2 toxin. *Infect. Immun.* **62:**2633–2638.

941. **Elomaa, O., M. Sankala, T. Pikkarainen, U. Bergmann, A. Tuuttila, A. Raatikainen-Ahokas, H. Sariola, and K. Tryggvason.** 1998. Structure of the human macrophage MARCO receptor and characterization of its bacteria-binding region. *J. Biol. Chem.* **273:**4530–4538.

942. **el-Shoura, S. M.** 1995. *Helicobacter pylori*. I. Ultrastructural sequences of adherence, attachment, and penetration into the gastric mucosa. *Ultrastruct. Pathol.* **19:**323–333.

943. **Elshourbagy, N. A., X. Li, J. Terrett, S. Vanhorn, M. S. Gross, J. E. Adamou, K. M. Anderson, C. L. Webb, and P. G. Lysko.** 2000. Molecular characterization of a human scavenger receptor, human MARCO. *Eur. J. Biochem.* **267:**919–926.

944. **Elsinghorst, E. A., and J. A. Weitz.** 1994. Epithelial cell invasion and adherence directed by the enterotoxigenic *Escherichia coli tib* locus is associated with a 104-kilodalton outer membrane protein. *Infect. Immun.* **62:**3463–3471.

945. **Elsinghorst, E. A., and D. J. Kopecko.** 1992. Molecular cloning of epithelial cell invasion determinants from enterotoxigenic *Escherichia coli*. *Infect. Immun.* **60:**2409–2417.

946. **el Tahir, Y., P. Toivanen, and M. Skurnik.** 1997. Application of an enzyme immunoassay to monitor bacterial binding and to measure inhibition of binding to different types of solid surfaces. *J. Immunoassay* **18:**165–183.

947. **Elvers, K. T., K. Leeming, and H. M. Lappin-Scott.** 2001. Binary culture biofilm formation by *Stenotrophomonas maltophilia* and *Fusarium oxysporum*. *J. Ind. Microbiol. Biotechnol.* **26:**178–183.

948. **Elwing, H., and A. Askendal.** 1994. Lens-on-surface method for investigating adhesion of *Staphylococcus aureus* to solid surfaces incubated in blood plasma. *J. Biomed. Mater. Res.* **28:**775–782.

949. **Embleton, J. V., H. N. Newman, and M. Wilson.** 2001. Amine and tin fluoride inhibition of *Streptococcus sanguis* adhesion under continuous flow. *Oral Microbiol. Immunol.* **16:**182–184.

950. **Ennis, E., R. R. Isberg, and Y. Shimizu.** 1993. Very late antigen 4-dependent adhesion and costimulation of resting human T cells by the bacterial β1 integrin ligand invasin. *J. Exp. Med.* **177:**207–212.

951. **Ensgraber, M., and M. Loos.** 1992. A 66-kilodalton heat shock protein of *Salmonella typhimurium* is responsible for binding of the bacterium to intestinal mucus. *Infect. Immun.* **60:**3072–3078.

952. **Ensgraber, M., R. Genitsariotis, S. Störkel, and M. Loos.** 1992. Purification and characterization of a *Salmonella typhimurium* agglutinin from gut mucus secretions. *Microb. Pathog.* **12:**255–266.

953. **Enss, M. L., H. Muller, U. Schmidt-Wittig, R. Kownatzki, M. Coenen, and H. J. Hedrich.** 1996. Effects of perorally applied endotoxin on colonic mucins of germfree rats. *Scand. J. Gastroenterol.* **31:**868–874.

954. **Erickson, A. K., J. A. Willgohs, S. Y. McFarland, D. A. Benfield, and D. H. Francis.** 1992. Identification of two porcine brush border glycoproteins that bind the K88ac adhesin of *Escherichia coli* and correlation of these glycoproteins with the adhesive phenotype. *Infect. Immun.* **60:**983–988.

955. **Erickson, A. K., D. R. Baker, B. T.**

Bosworth, T. A. Casey, D. A. Benfield, and D. H. Francis. 1994. Characterization of porcine intestinal receptors for the K88ac fimbrial adhesin of *Escherichia coli* as mucin-type sialoglycoproteins. *Infect. Immun.* 62:5404–5410.

956. Erickson, A. K., L. O. Billey, G. Srinivas, D. R. Baker, and D. H. Francis. 1997. A three-receptor model for the interaction of the K88 fimbrial adhesin variants of *Escherichia coli* with porcine intestinal epithelial cells. *Adv. Exp. Med. Biol.* 412:167–173.

957. Erickson, P. R., and M. C. Herzberg. 1993. Evidence for the covalent linkage of carbohydrate polymers to a glycoprotein from *Streptococcus sanguis*. *J. Biol. Chem.* 268:23780–23783.

958. Erickson, P. R., and M. C. Herzberg. 1995. Altered expression of the platelet aggregation-associated protein from *Streptococcus sanguis* after growth in the presence of collagen. *Infect. Immun.* 63:1084–1088.

959. Espersen, F., M. Wurr, L. Corneliussen, A. L. Hog, V. T. Rosdahl, N. Frimodt-Møller, and P. Skinhoj. 1994. Attachment of staphylococci to different plastic tubes *in vitro*. *J. Med. Microbiol.* 40:37–42.

960. Essery, S. D., D. M. Weir, V. S. James, C. C. Blackwell, A. T. Saadi, A. Busuttil, and G. Tzanakaki. 1994. Detection of microbial surface antigens that bind Lewis(a) antigen. *FEMS Immunol. Med. Microbiol.* 9:15–21.

961. Esslinger, J., M. Boury, R. S. Seleim, and A. Milon. 1993. Regulation of expression of the adhesin AF/R2 of *Escherichia coli* O103 by Bacto-Peptone™. Preliminary results. *Rev. Med Vet.* 145:125–131.

962. Estabrook, M. M., N. C. Christopher, J. M. Griffiss, C. J. Baker, and R. E. Mandrell. 1992. Sialylation and human neutrophil killing of group C *Neisseria meningitidis*. *J. Infect. Dis.* 166:1079–1088.

963. Estabrook, M. M., D. Zhou, and M. A. Apicella. 1998. Nonopsonic phagocytosis of group C *Neisseria meningitidis* by human neutrophils. *Infect. Immun.* 66:1028–1036.

964. Estes, R. J., and G. U. Meduri. 1995. The pathogenesis of ventilator-associated pneumonia: I. Mechanisms of bacterial transcolonization and airway inoculation. *Intensive Care Med.* 21:365–383.

965. Evans, D. G., D. J. Evans, Jr., and D. Y. Graham. 1992. Adherence and internalization of *Helicobacter pylori* by HEp-2 cells. *Gastroenterology* 102:1557–1567.

966. Evans, D. G., T. K. Karjalainen, D. J.

Evans, Jr., D. Y. Graham, and C.-H. Lee. 1993. Cloning, nucleotide sequence, and expression of a gene encoding an adhesin subunit protein of *Helicobacteri pylori*. *J. Bacteriol.* 175:674–683.

967. Evans, D. G., D. J. Evans, Jr., H. C. Lampert, and D. Y. Graham. 1995. Restriction fragment length polymorphism in the adhesin gene *hpaA* of *Helicobacter pylori*. *Am. J. Gastroenterol.* 90:1282–1288.

968. Evans, D. G., H. C. Lampert, H. Nakano, K. A. Eaton, A. P. Burnens, M. A. Bronsdon, and D. J. Evans, Jr. 1995. Genetic evidence for host specificity in the adhesin-encoding genes *hxaA* of *Helicobacter acinonyx*, *hnaA* of *H. nemestrinae* and *hpaA* of *H. pylori*. *Gene* 163:97–102.

969. Evans, D. G., and D. J. Evans, Jr. 1995. Adhesion properties of *Helicobacter pylori*. *Methods Enzymol.* 253:336–360.

970. Evans, R. T., B. Klausen, and R. J. Genco. 1992. Immunization with fimbrial protein and peptide protects against *Porphyromonas gingivalis*-induced periodontal tissue destruction. *Adv. Exp. Med. Biol.* 327:255–262.

971. Evans, R. T., B. Klausen, H. T. Sojar, G. S. Bedi, C. Sfintescu, N. S. Ramamurthy, L. M. Golub, and Genco. 1992. Immunization with *Porphyromonas* (*Bacteroides*) *gingivalis* fimbriae protects against periodontal destruction. *Infect. Immun.* 60:2926–2935.

972. Evans-Hurrell, J. A., J. Adler, S. Denyer, T. G. Rogers, and P. Williams. 1993. A method for the enumeration of bacterial adhesion to epithelial cells using image analysis. *FEMS Microbiol. Lett.* 107:77–82.

973. Eveillard, M., V. Fourel, M. C. Barc, S. Kerneis, M. H. Coconnier, T. Karjalainen, P. Bourlioux, and A. L. Servin. 1993. Identification and characterization of adhesive factors of *Clostridium difficile* involved in adhesion to human colonic enterocyte-like Caco-2 and mucus-secreting HT29 cells in culture. *Mol. Microbiol.* 7:371–381.

974. Everest, P. H., H. Goossens, J. P. Butzler, D. Lloyd, S. Knutton, J. M. Ketley, and P. H. Williams. 1992. Differentiated Caco-2 cells as a model for enteric invasion by *Campylobacter jejuni* and *C. coli*. *J. Med. Microbiol.* 37:319–325.

975. Everest, P., J. Li, G. Douce, I. Charles, J. De Azavedo, S. Chatfield, G. Dougan, and M. Roberts. 1996. Role of the *Bordetella pertussis* P.69/pertactin protein and the

P.69/pertactin RGD motif in the adherence to and invasion of mammalian cells. *Microbiology* **142:**3261–3268.

976. **Everiss, K. D., K. J. Hughes, and K. M. Peterson.** 1994. The accessory colonization factor and toxin-coregulated pilus gene clusters are physically linked on the *Vibrio cholerae* O395 chromosome. *DNA Seq.* **5:**51–55.

977. **Ewen, S. W., P. J. Naughton, G. Grant, M. Sojka, E. Allen-Vercoe, S. Bardocz, C. J. Thorns, and A. Pusztai.** 1997. *Salmonella enterica var Typhimurium* and *Salmonella enterica var Enteritidis* express type 1 fimbriae in the rat *in vivo. FEMS Immunol. Med. Microbiol.* **18:**185–192.

978. **Faden, H., L. Duffy, T. Foels, and J. J. Hong.** 1996. Adherence of nontypeable *Haemophilus influenzae* to respiratory epithelium of otitis-prone and normal children. *Ann. Otol. Rhinol. Laryngol.* **105:**367–370.

979. **Fairbrother, J., J. Harel, C. Forget, C. Desautels, and J. Moore.** 1993. Receptor binding specificity and pathogenicity of *Escherichia coli* F165-positive strains isolated from piglets and calves and possessing *pap* related sequences. *Can. J. Vet. Res.* **57:**53–55.

980. **Fakih, M. G., T. F. Murphy, M. A. Pattoli, and C. S. Berenson.** 1997. Specific binding of *Haemophilus influenzae* to minor gangliosides of human respiratory epithelial cells. *Infect. Immun.* **65:**1695–1700.

981. **Falero, G., B. L. Rodriguez, T. Valmaseda, M. E. Perez, J. L. Perez, R. Fando, A. Robert, J. Campos, A. Silva, G. Sierra, and J. A. Benitez.** 1998. Production and characterization of a monoclonal antibody against mannose-sensitive hemagglutinin of *Vibrio cholerae. Hybridoma* **17:**63–67.

982. **Falk, P., T. Boren, D. Haslam, and M. Caparon.** 1994. Bacterial adhesion and colonization assays. *Methods Cell Biol.* **45:**165–192.

983. **Falk, P., T. Boren, and S. Normark.** 1994. Characterization of microbial host receptors. *Methods Enzymol.* **236:**353–374.

984. **Falk, P., K. A. Roth, T. Boren, T. U. Westblom, J. I. Gordon, and S. Normark.** 1993. An *in vitro* adherence assay reveals that *Helicobacter pylori* exhibits cell lineage-specific tropism in the human gastric epithelium. *Proc. Natl. Acad. Sci. USA* **90:**2035–2039.

985. **Falk, P. G., L. Bry, J. Holgersson, and J. I. Gordon.** 1995. Expression of a human α-1,3/4-fucosyltransferase in the pit cell lineage of FVB/N mouse stomach results in the production of Le^b-containing glycoconjugates: A potential transgenic mouse model for studying *Helicobacter pylori* infection. *Proc. Natl. Acad. Sci. USA* **92:**1515–1519.

986. **Falkow, S., R. R. Isberg, and D. A. Portnoy.** 1992. The interaction of bacteria with mammalian cells. *Ann. Rev. Cell Biol.* **8:**333–363.

987. **Fan, X., F. Stelter, R. Menzel, R. Jack, I. Spreitzer, T. Hartung, and C. Schutt.** 1999. Structures in *Bacillus subtilis* are recognized by CD14 in a lipopolysaccharide binding protein-dependent reaction. *Infect. Immun.* **67:**2964–2968.

988. **Fang, L., Z. Gan, and R. R. Marquardt.** 2000. Isolation, affinity purification, and identification of piglet small intestine mucosa receptor for enterotoxigenic *Escherichia coli* k88ac+ fimbriae. *Infect. Immun.* **68:**564–569.

989. **Fang, W., R. A. Almeida, and S. P. Oliver.** 2000. Effects of lactoferrin and milk on adherence of *Streptococcus uberis* to bovine mammary epithelial cells. *Am. J. Vet. Res.* **61:**275–279.

990. **Fantinatti, F., W. D. Silveira, and A. F. Castro.** 1994. Characteristics associated with pathogenicity of avian septicaemic *Escherichia coli* strains. *Vet. Microbiol.* **41:**75–86.

991. **Farber, B. F., H. C. Hsieh, E. D. Donnenfeld, H. D. Perry, A. Epstein, and A. Wolff.** 1995. A novel antibiofilm technology for contact lens solutions. *Ophthalmology* **102:**831–836.

992. **Farber, B. F., and A. G. Wolff.** 1993. Salicylic acid prevents the adherence of bacteria and yeast to silastic catheters. *J. Biomed. Mater. Res.* **27:**599–602.

993. **Farber, B. F., and A. G. Wolff.** 1992. The use of nonsteroidal antiinflammatory drugs to prevent adherence of *Staphylococcus epidermidis* to medical polymers. *J. Infect. Dis.* **166:**861–865.

994. **Farber, B. F., and A. G. Wolff.** 1993. The use of salicylic acid to prevent the adherence of *Escherichia coli* to silastic catheters. *J. Urol.* **149:**667–670.

995. **Farinha, M. A., B. D. Conway, L. M. G. Glasier, N. W. Ellert, R. T. Irvin, R. Sherburne, and W. Paranchych.** 1994. Alteration of the pilin adhesin of *Pseudomonas aeruginosa* PAO results in normal pilus biogenesis but a loss of adherence to human pneumocyte cells and decreased virulence in mice. *Infect. Immun.* **62:**4118–4123.

996. **Farley, M. M., A. M. Whitney, P. Spellman, F. D. Quinn, R. S. Weyant, L. Mayer, and D. S. Stephens.** 1992. Analysis of the attachment and invasion of human

epithelial cells by *Haemophilus influenzae* biogroup aegyptius. *J. Infect. Dis.* **165:**S111–S114.

997. **Fatemi, P., and J. F. Frank.** 1999. Inactivation of *Listeria monocytogenes/Pseudomonas* biofilms by peracid sanitizers. *J. Food Prot.* **62:**761–765.

998. **Favre-Bonte, S., A. Darfeuille-Michaud, and C. Forestier.** 1995. Aggregative adherence of *Klebsiella pneumoniae* to human intestine-407 cells. *Infect. Immun.* **63:**1318–1328.

999. **Favre-Bonte, S., B. Joly, and C. Forestier.** 1999. Consequences of reduction of *Klebsiella pneumoniae* capsule expression on interactions of this bacterium with epithelial cells. *Infect. Immun.* **67:**554–561.

1000. **Favre-Bonte, S., C. Forestier, and B. Joly.** 1998. Inhibitory effect of roxithromycin on adhesion of *Klebsiella pneumoniae* strains 3051, CF504 and LM21. *J. Antimicrob. Chemother.* **41:**S51–S55.

1001. **Feldman, C., R. Read, A. Rutman, P. K. Jeffery, A. Brain, V. Lund, T. J. Mitchell, P. W. Andres, G. J. Boulnois, H. C. Todd, P. J. Cole, and R. Wilson.** 1992. The interaction of *Streptococcus pneumoniae* with intact human respiratory mucosa. *Eur. Respir. J.* **5:**576–583.

1002. **Feldman, M., R. Bryan, S. Rajan, L. Scheffler, S. Brunnert, H. Tang, and A. Prince.** 1998. Role of flagella in pathogenesis of *Pseudomonas aeruginosa* pulmonary infection. *Infect. Immun.* **66:**43–51.

1003. **Fenno, J. C., A. Shaikh, G. Spatafora, and P. Fives-Taylor.** 1995. The *fimA* locus of *Streptococcus parasanguis* encodes an ATP-binding membrane transport system. *Mol. Microbiol.* **15:**849–863.

1004. **Ferguson, J. S., D. R. Voelker, F. X. McCormack, and L. S. Schlesinger.** 1999. Surfactant protein D binds to *Mycobacterium tuberculosis* bacilli and lipoarabinomannan via carbohydrate-lectin interactions resulting in reduced phagocytosis of the bacteria by macrophages. *J. Immunol.* **163:**312–321.

1005. **Ferguson, M. I., E. M. Scott, and P. S. Collier.** 1992. Factors affecting quantitative assessment of *Pseudomonas aeruginosa* adherence to buccal epithelial cells. *APMIS* **100:**876–882.

1006. **Fernandez, R., P. Nelson, J. Delgado, J. Aguilera, R. Massai, L. Velasquez, M. Imarai, H. B. Croxatto, and H. Cardenas.** 2001. Increased adhesiveness and internalization of *Neisseria gonorrhoeae* and changes in the expression of epithelial gonococcal receptors in the Fallopian tube of cop-

per T and Norplant users. *Hum. Reprod.* **16:**463–468.

1007. **Fernandez, V., C. J. Treutiger, G. B. Nash, and M. Wahlgren.** 1998. Multiple adhesive phenotypes linked to rosetting binding of erythrocytes in *Plasmodium falciparum* malaria. *Infect. Immun.* **66:**2969–2975.

1008. **Ferrara, A., C. Dos Santos, and A. Lupi.** 2001. Effect of different antibacterial agents and surfactant protein-A (SP-A) on adherence of some respiratory pathogens to bronchial epithelial cells. *Int. J. Antimicrob. Agents* **17:**401–405.

1009. **Ferreira, R., M. C. Alexandre, E. N. Antunes, A. T. Pinhao, S. R. Moraes, M. C. Ferreira, and R. M. Domingues.** 1999. Expression of *Bacteroides fragilis* virulence markers *in vitro.* *J. Med. Microbiol.* **48:**999–1004.

1010. **Ferreira-Dias, G., L. G. Nequin, and S. S. King.** 1994. Influence of estrous cycle stage on adhesion of *Streptococcus zooepidemicus* to equine endometrium. *Am. J. Vet. Res.* **55:**1028–1031.

1011. **Fey, P. D., J. S. Ulphani, F. Götz, C. Heilmann, D. Mack, and M. E. Rupp.** 1999. Characterization of the relationship between polysaccharide intercellular adhesin and hemagglutination in *Staphylococcus epidermidis. J. Infect. Dis.* **179:**1561–1564.

1012. **Fiederling, F., M. Boury, C. Petit, and A. Milon.** 1997. Adhesive factor/rabbit 2, a new fimbrial adhesin and a virulence factor from *Escherichia coli* O103, a serogroup enteropathogenic for rabbits. *Infect. Immun.* **65:**847–851.

1013. **Fields, B. S., S. R. Fields, J. N. Loy, E. H. White, W. L. Steffens, and E. B. Shotts.** 1993. Attachment and entry of *Legionella pneumophila* in *Hartmannella vermiformis. J. Infect. Dis.* **167:**1146–1150.

1014. **Figueroa, G., D. P. Portell, V. Soto, and M. Troncoso.** 1992. Adherence of *Helicobacter pylori* to HEp-2 cells. *J. Infect.* **24:**263–267.

1015. **Fine, D. H., P. Goncharoff, H. Schreiner, K. M. Chang, D. Furgang, and D. Figurski.** 2001. Colonization and persistence of rough and smooth colony variants of *Actinobacillus actinomycetemcomitans* in the mouths of rats. *Arch. Oral Biol.* **46:**1065–1078.

1016. **Fine, D. H., D. Furgang, J. Kaplan, J. Charlesworth, and D. H. Figurski.** 1999. Tenacious adhesion of *Actinobacillus actinomycetemcomitans* strain CU1000 to salivary-coated hydroxyapatite. *Arch. Oral Biol.* **44:**1063–1076.

1017. Finkelstein, R. A., M. Boesman-Finkelstein, Y. Chang, and C. C. Hase. 1992. *Vibrio cholerae* hemagglutinin/protease, colonial variation, virulence, and detachment. *Infect. Immun.* **60**:472–478.

1018. Finlay, B. B., and S. Falkow. 1997. Common themes in microbial pathogenicity revisited. *Microbiol. Mol. Biol. Rev.* **61**:136–169.

1019. Finlay, B. B., I. Rosenshine, M. S. Donnenberg, and J. B. Kaper. 1992. Cytoskeletal composition of attaching and effacing lesions associated with enteropathogenic *Escherichia coli* adherence to HeLa cells. *Infect. Immun.* **60**:2541–2543.

1020. Finlay, B. B. 1995. Interactions between *Salmonella typhimurium*, enteropathogenic *Escherichia coli* (EPEC), and host epithelial cells. *Adv. Dent. Res.* **9**:31–36.

1021. Finlay, B. B., and P. Cossart. 1997. Exploitation of mammalian host cell functions by bacterial pathogens. *Science* **276**:718–725.

1022. Finlay, B. B., S. Ruschkowski, B. Kenny, M. Stein, D. J. Reinscheid, M. A. Stein, and I. Rosenshine. 1996. Enteropathogenic *E. coli* exploitation of host epithelial cells. *Ann. N. Y. Acad. Sci.* **797**:26–31.

1023. Fiorentin, L., V. S. Panangala, Y. Zhang, and M. Toivio-Kinnucan. 1998. Adhesion inhibition of *Mycoplasma iowae* to chicken lymphoma DT40 cells by monoclonal antibodies reacting with a 65-kD polypeptide. *Avian Dis.* **42**:721–731.

1024. Fischer, J., C. Maddox, R. Moxley, D. Kinden, and M. Miller. 1994. Pathogenicity of a bovine attaching effacing *Escherichia coli* isolate lacking Shiga-like toxins. *Am. J. Vet. Res.* **55**:991–999.

1025. Fischer, B., P. Vaudaux, M. Magnin, Y. el Mestikawy, R. A Proctor, D. P. Lew, and H. Vasey. 1996. Novel animal model for studying the molecular mechanisms of bacterial adherion to bone-implanted metallic devices: role of fibronectin in *Staphylococcus aureus* adhesion. *J. Orthopaed. Res.* **14**:914–920.

1026. Fisseha, M., H. W. Gohlmann, R. Herrmann, and D. C. Krause. 1999. Identification and complementation of frameshift mutations associated with loss of cytadherence in *Mycoplasma pneumoniae*. *J. Bacteriol.* **181**:4404–4410.

1027. Fitzgerald, M., R. Mulcahy, S. Murphy, C. Keane, D. Coakley, and T. Scott. 1997. A 200 kDa protein is associated with haemagglutinating isolates of *Moraxella (Branhamella) catarrhalis*. *FEMS Immunol. Med. Microbiol.* **18**:209–216.

1028. Fitzgerald, M., S. Murphy, R. Mulcahy, C. Keane, D. Coakley, and T. Scott. 1999. Tissue culture adherence and haemagglutination characteristics of *Moraxella (Branhamella) catarrhalis*. *FEMS Immunol. Med. Microbiol.* **24**:105–114.

1029. Fitzgerald, M., S. Murphy, R. Mulcahy, C. Keane, D. Coakley, and T. Scott. 1996. Haemagglutination properties of *Moraxella (Branhamella) catarrhalis*. *Br. J. Biomed. Sci.* **53**:257–262.

1030. Fives-Taylor, P., D. Meyer, and K. Mintz. 1995. Characteristics of *Actinobacillus actinomycetemcomitans* invasion of and adhesion to cultured epithelial cells. *Adv. Dent. Res.* **9**:55–62.

1031. Fleckenstein, J. M., L. E. Lindler, E. A. Elsinghorst, and J. B. Dale. 2000. Identification of a gene within a pathogenicity island of enterotoxigenic *Escherichia coli* H10407 required for maximal secretion of the heat-labile enterotoxin. *Infect. Immun.* **68**:2766–2774.

1032. Fleiszig, S. M., N. Efron, and G. B. Pier. 1992. Extended contact lens wear enhances *Pseudomonas aeruginosa* adherence to human corneal epithelium. *Investig. Ophthalmol. Visual. Sci.* **33**:2908–2916.

1033. Fleiszig, S. M., T. S. Zaidi, E. L. Fletcher, M. J. Preston, and G. B. Pier. 1994. *Pseudomonas aeruginosa* invades corneal epithelial cells during experimental infection. *Infect. Immun.* **62**:3485–3493.

1034. Fleiszig, S. M., T. S. Zaidi, R. Ramphal, and G. B. Pier. 1994. Modulation of *Pseudomonas aeruginosa* adherence to the corneal surface by mucus. *Infect. Immun.* **62**:1799–1804.

1035. Fleiszig, S. M., D. J. Evans, M. F. Mowrey-McKee, R. Payor, T. S. Zaidi, V. Vallas, E. Muller, and G. B. Pier. 1996. Factors affecting *Staphylococcus epidermidis* adhesion to contact lenses. *Optom. Vision Sci.* **73**:590–594.

1036. Flemming, R. G., R. A. Proctor, and S. L. Cooper. 1999. Bacterial adhesion to functionalized polyurethanes. *J. Biomater. Sci. Polym. Edn.* **10**:679–697.

1037. Fletcher, E. L., S. M. Fleiszig, and N. A. Brennan. 1993. Lipopolysaccharide in adherence of *Pseudomonas aeruginosa* to the cornea and contact lenses. *Investig. Ophthalmol. Visual Sci.* **34**:1930–1936.

1038. Fletcher, E. L., B. A. Weissman, N. Efron, S. M. Fleiszig, A. J. Curcio, and N. A. Brennan. 1993. The role of pili in the attachment of *Pseudomonas aeruginosa* to unworn hydrogel contact lenses. *Curr. Eye Res.* **12:**1067–1071.

1039. **Fletcher, J. N., H. E. Embaye, B. Getty, R. M. Batt, C. A. Hart, and J. R. Saunders.** 1992. Novel invasion determinant of enteropathogenic *Escherichia coli* plasmid pLV501 encodes the ability to invade intestinal epithelial cells and HEp-2 cells. *Infect. Immun.* **60:**2229–2236.

1040. **Fletcher, J. N., J. R. Saunders, H. Embaye, R. M. Odedra, R. M. Batt, and C. A. Hart.** 1997. Surface properties of diarrhoeagenic *Escherichia coli* isolates. *J. Med. Microbiol.* **46:**67–74.

1041. **Flock, J. I., S. A. Hienz, A. Heimdahl, and T. Schennings.** 1996. Reconsideration of the role of fibronectin binding in endocarditis caused by *Staphylococcus aureus*. *Infect. Immun.* **64:**1876–1878.

1042. **Fluckiger, U., K. F. Jones, and V. A. Fischetti.** 1998. Immunoglobulins to group A streptococcal surface molecules decrease adherence to and invasion of human pharyngeal cells. *Infect. Immun.* **66:**974–979.

1043. **Flugel, A., H. Schulze-Koops, J. Heesemann, K. Kuhn, L. Sorokin, H. Burkhardt, K. von der Mark, and F. Emmrich.** 1994. Interaction of enteropathogenic *Yersinia enterocolitica* with complex basement membranes and the extracellular matrix proteins collagen type IV, laminin-1 and -2, and nidogen/entactin. *J. Biol. Chem.* **269:**29732–29738.

1044. **Fogg, G. C., and M. G. Caparon.** 1997. Constitutive expression of fibronectin binding in *Streptococcus pyogenes* as a result of anaerobic activation of *rofA. J. Bacteriol.* **179:**6172–6180.

1045. **Folkesson, A., A. Advani, S. Sukupolvi, J. D. Pfeifer, S. Normark, and S. Lofdahl.** 1999. Multiple insertions of fimbrial operons correlate with the evolution of *Salmonella* serovars responsible for human disease. *Mol. Microbiol.* **33:**612–622.

1046. **Fontana, M., L. E. Gfell, and R. L. Gregory.** 1995. Characterization of preparations enriched for *Streptococcus mutans* fimbriae: salivary immunoglobulin A antibodies in caries-free and caries-active subjects. *Clin. Diagn. Lab. Immunol.* **2:**719–725.

1047. **Foo, L. Y., Y. Lu, A. B. Howell, and N. Vorsa.** 2000. A-Type proanthocyanidin trimers from cranberry that inhibit adherence of uropathogenic P-fimbriated *Escherichia coli*. *J. Nat. Prod.* **63:**1225–1228.

1048. **Forde, C. B., X. Shi, J. Li, and M. Roberts.** 1999. *Bordetella bronchiseptica*-mediated cytotoxicity to macrophages is dependent on *bvg*-regulated factors, including pertactin. *Infect. Immun.* **67:**5972–5978.

1049. **Foreman, K. E., and A. B. Bjornson.** 1994. The alternative complement pathway promotes IgM antibody-dependent and -independent adherence of *Bacteroides* to polymorphonuclear leukocytes through CR3 and CR1. *J. Leukoc. Biol.* **55:**603–611.

1050. **Forest, K. T., S. L. Bernstein, E. D. Getzoff, M. So, G. Tribbick, H. M. Geysen, C. D. Deal, and J. A. Tainer.** 1997. Assembly and antigenicity of the *Neisseria gonorrhoeae* pilus mapped with antibodies. *Infect. Immun.* **64:**644–652.

1051. **Forest, K. T., and J. A. Tainer.** 1997. Type-4 pilus-structure: outside to inside and top to bottom—a minireview. *Gene* **192:**165–169.

1052. **Forester, N., J. S. Lumsden, T. O'Croinin, and P. W. O'Toole.** 2001. Sequence and antigenic variability of the *Helicobacter mustelae* surface ring protein Hsr. *Infect. Immun.* **69:**3447–3450.

1053. **Forestier, C., C. De Champs, C. Vatoux, and B. Joly.** 2001. Probiotic activities of *Lactobacillus casei rhamnosus*: in vitro adherence to intestinal cells and antimicrobial properties. *Res. Microbiol.* **152:**167–173.

1054. **Forney, L. J., J. R. Gilsdorf, C. F. Marrs, K. McCrea, and S. Bektesh.** 1992. Comparison of the alpha-pilins of *Haemophilus influenzae* type b strains Eagan (p+), M43 (p+), and 770235bof+. *J. Infect. Dis.* **165:**S106–S107.

1055. **Forney, L. J., J. R. Gilsdorf, and D. C. Wong.** 1992. Effect of pili-specific antibodies on the adherence of *Haemophilus influenzae* type b to human buccal cells. *J. Infect. Dis.* **165:**464–470.

1056. **Forsyth, M. H., M. E. Tourtellotte, and S. J. Geary.** 1992. Localization of an immunodominant 64 kDa lipoprotein (LP 64) in the membrane of *Mycoplasma gallisepticum* and its role in cytadherence. *Mol. Microbiol.* **6:**2099–2106.

1057. **Fortis, A. A., P. E. Lianou, and J. T. Papavassilliou.** 1998. Adherence of *Staphylococcus aureus, Klebsiella pneumoniae* and *Candida albicans* to human buccal epithelial cells, from healthy persons and HIV carriers, under the influence of *Broncho Vaxom in vitro* and ascorbic acid *in vivo*. *APMIS* **106:**441–448.

1058. **Foster, T. J., and D. McDevitt.** 1994. Surface-associated proteins of *Staphylococcus aureus*: their possible roles in virulence. *FEMS Microbiol. Lett.* **118:**199–205.

1059. **Foubister, V., I. Rosenshine, and B. B. Finlay.** 1994. A diarrheal pathogen, enteropathogenic *Escherichia coli* (EPEC), triggers a flux of inositol phosphates in infected epithelial cells. *J. Exp. Med.* **179:**993–998.

1060. Reference deleted.

1061. **Fourniat, J., C. Colomban, C. Linxe, and D. Karam.** 1992. Heat-killed *Lactobacillus acidophilus* inhibits adhesion of *Escherichia coli* B41 to HeLa cells. *Ann. Rech. Vet.* **23:**361–370.

1062. **Fournier, A., L. Payant, and R. Bouclin.** 1998. Adherence of *Streptococcus mutans* to orthodontic brackets. *Am. J. Orthodont. Dentofac. Orthop.* **114:**414–417.

1063. **Fournier, B., and D. C. Hooper.** 2000. A new two-component regulatory system involved in adhesion, autolysis, and extracellular proteolytic activity of *Staphylococcus aureus*. *J. Bacteriol.* **182:**3955–3964.

1064. **Francetic, O., S. Lory, and A. P. Pugsley.** 1998. A second prepilin peptidase gene in *Escherichia coli* K-12. *Mol. Microbiol.* **27:**763–775.

1065. **Francis, C. L., M. S. Starnbach, and S. Falkow.** 1992. Morphological and cytoskeletal changes in epithelial cells occur immediately upon interaction with *Salmonella typhimurium* grown under low-oxygen conditions. *Mol. Microbiol.* **6:**3077–3087.

1066. **Francis, D. H., P. A. Grange, D. H. Zeman, D. R. Baker, R. Sun, and A. K. Erickson.** 1998. Expression of mucin-type glycoprotein K88 receptors strongly correlates with piglet susceptibility to K88$^+$ enterotoxigenic *Escherichia coli*, but adhesion of this bacterium to brush borders does not. *Infect. Immun.* **66:**4050–4055.

1067. **Franck, S. M., B. T. Bosworth, and H. W. Moon.** 1998. Multiplex PCR for enterotoxigenic, attaching and effacing, and Shiga toxin-producing *Escherichia coli* strains from calves. *J. Clin. Microbiol.* **36:**1795–1797.

1068. **Francki, K. T., B. J. Chang, B. J. Mee, P. J. Collignon, V. Susai, and P. K. Keese.** 2000. Identification of genes associated with copper tolerance in an adhesion-defective mutant of *Aeromonas veronii* biovar *sobria*. *FEMS Immunol. Med. Microbiol.* **29:**115–121.

1069. **Francki, K. T., and B. J. Chang.** 1994. Variable expression of O-antigen and the role of lipopolysaccharide as an adhesin in *Aeromonas sobria*. *FEMS Microbiol. Lett.* **122:**97–102.

1070. **Francois, P., P. Vaudaux, T. J. Foster, and D. P. Lew.** 1996. Host-bacteria interactions in foreign body infections. *Infect. Control Hosp. Epidemiol.* **17:**514–520.

1071. **Francois, P., P. Vaudaux, N. Nurdin, H. J. Mathieu, P. Descouts, and D. P. Lew.** 1996. Physical and biological effects of a surface coating procedure on polyurethane catheters. *Biomaterials* **17:**667–678.

1072. **Francois, P., P. Vaudaux, M. Taborelli, M. Tonetti, D. P. Lew, and P. Descouts.** 1997. Influence of surface treatments developed for oral implants on the physical and biological properties of titanium. II. Adsorption isotherms and biological activity of immobilized fibronectin. *Clin. Oral Implant Res.* **8:**217–225.

1073. **Francois, P., P. Vaudaux, and P. D. Lew.** 1998. Role of plasma and extracellular matrix proteins in the physiopathology of foreign body infections. *Ann. Vasc. Surg.* **12:**34–40.

1074. **Francois, P., J. Schrenzel, C. Stoerman-Chopard, H. Favre, M. Herrmann, T. J. Foster, D. P. Lew, and P. Vaudaux.** 2000. Identification of plasma proteins adsorbed on hemodialysis tubing that promote *Staphylococcus aureus* adhesion. *J. Lab. Clin. Med.* **135:**32–42.

1075. **Franek, J., J. Malina, and H. Kratka.** 1992. Bacterial infection modulated by glucan: a search for the host defense potentiation mechanisms. *Folia Microbiol.* **37:**146–152.

1076. **Frangipane, J. V., and R. F. Rest.** 1992. Anaerobic growth of gonococci does not alter their Opa-mediated interactions with human neutrophils. *Infect. Immun.* **60:**1793–1799.

1077. **Franke, J., S. Franke, H. Schmidt, A. Schwarzkopf, L. H. Wieler, G. Baljer, L. Beutin, and H. Karch.** 1994. Nucleotide sequence analysis of enteropathogenic *Escherichia coli* (EPEC) adherence factor probe and development of PCR for rapid detection of EPEC harboring virulence plasmids. *J. Clin. Microbiol.* **32:**2460–2463.

1078. **Frankel, G., D. C. Candy, P. Everest, and G. Dougan.** 1994. Characterization of the C-terminal domains of intimin-like proteins of enteropathogenic and enterohemorrhagic *Escherichia coli*, *Citrobacter freundii*, and *Hafnia alvei*. *Infect. Immun.* **62:**1835–1842.

1079. **Frankel, G., D. C. A. Candy, E. Fabiani, J. Adu-Bobie, S. Gil, M. Novakova, A. D. Phillips, and G. Dougan.** 1995. Molecular characterization of a carboxy-terminal eukaryotic-cell-binding domain of intimin from enteropathogenic *Escherichia coli*. *Infect. Immun.* **63:**4323–4328.

1080. **Frankel, G., O. Lider, R. Hershkoviz, A. P. Mould, S. G. Kachalsky, D. C. A. Candy, L. Cahalon, M. J. Humphries,** and G. Dougan. 1996. The cell-binding domain of intimin from enteropathogenic *Escherichia coli* binds to beta1 integrins. *J. Biol. Chem.* **271:**20359–10364.

1081. Frankel, G., A. D. Phillips, S. Hicks, and G. Dougan. 1996. Enteropathogenic *Escherichia coli*—mucosal infection models. *Trans. R. Soc. Trop. Med. Hyg.* **90:**347–352.

1082. Frankel, G., A. D. Phillips, I. Rosenshine, G. Dougan, J. B. Kaper, and S. Knutton. 1998. Enteropathogenic and enterohaemorrhagic *Escherichia coli:* more subversive elements. *Mol. Microbiol.* **30:**911–921.

1083. Franklin, C. L., D. A. Kinden, P. L. Stogsdill, and L. K. Riley. 1993. In vitro model of adhesion and invasion by *Bacillus piliformis. Infect. Immun.* **61:**876–883.

1084. Franklin, M. A., D. H. Francis, D. Baker, and A. G. Mathew. 1996. A PCR-based method of detection and differentiation of K88⁺ adhesive *Escherichia coli. J. Vet. Diagn. Investig.* **8:**460–463.

1085. Franzon, V. L., A. Barker, and P. A. Manning. 1993. Nucleotide sequence encoding the mannose-fucose-resistant hemagglutinin ov *Vibrio cholerae* O1 and construction of a mutant. *Infect. Immun.* **61:**3032–3037.

1086. Franzoso, G., P. C. Hu, G. A. Meloni, and M. F. Barile. 1993. The immunodominant 90-kilodalton protein is localized on the terminal tip structure of *Mycoplasma pneumoniae. Infect. Immun.* **61:**1523–1530.

1087. Frasa, H., B. Benaissa-Trouw, L. Tavares, K. van Kessel, M. Poppelier, K. Kraaijeveld, and J. Verhoef. 1996. Enhanced protection by use of a combination of anticapsule and antilipopolysaccharide monoclonal antibodies against lethal *Escherichia coli* O18:K5 infection of mice. *Infect. Immun.* **64:**775–781.

1088. Fratamico, P. M., S. Bhaduri, and R. L. Buchanan. 1993. Studies on *Escherichia coli* serotype O157:H7 strains containing a 60-MDa plasmid and on 60-MDa plasmid-cured derivatives. *J. Med. Microbiol.* **39:**371–381.

1089. Fratazzi, C., N. Manjunath, R. D. Arbeit, C. Carini, T. A. Gerken, B. Ardman, E. Remold-O'Donnell, and H. G. Remold. 2000. A macrophage invasion mechanism for mycobacteria implicating the extracellular domain of CD43. *J. Exp. Med.* **192:**183–192.

1090. Fredriksen, F., S. Raisanen, R. Myklebust, and L. E. Stenfors. 1996. Bacterial adherence to the surface and isolated cell epithelium of the palatine tonsils. *Acta Otoaryngol.* **116:**620–626.

1091. Freissler, E., A. Meyer auf der Heyde, G. David, T. F. Meyer, and C. Dehio. 2000. Syndecan-1 and syndecan-4 can mediate the invasion of OpaHSPG-expressing *Neisseria gonorrhoeae* into epithelial cells. *Cell. Microbiol.* **2:**69–82.

1092. Freitag, N. E., H. S. Seifert, and M. Koomey. 1995. Characterization of the *pilF-pilD* pilus-assembly locus of *Neisseria gonorrhoeae. Mol. Microbiol.* **16:**575–586.

1093. Frey, A., K. T. Giannasca, R. Weltzin, P. J. Giannasca, H. Reggio, W. I. Lencer, and M. R. Neutra. 1996. Role of the glycocalyx in regulating access of microparticles to apical plasma membranes of intestinal epithelial cells: implications for microbial attachment and oral vaccine targeting. *J. Exp. Med.* **184:**1045–1059.

1094. Frick, I.-M., K. L. Crossin, G. M. Edelman, and L. Björck. 1995. Protein H-a bacterial protein with affinity for both immunoglobulin and fibronectin type III domains. *EMBO J.* **14:**1674–1679.

1095. Friedman, R. L., K. Nordensson, L. Wilson, E. T. Akporiaye, and D. E. Yocum. 1992. Uptake and intracellular survival of *Bordetella pertussis* in human macrophages. *Infect. Immun.* **60:**4578–4585.

1096. Friedrich, M. J., N. E. Kinsey, J. Vila, and R. J. Kadner. 1993. Nucleotide sequence of a 13.9 kb segment of the 90 kb virulence plasmid of *Salmonella typhimurium:* the presence of fimbrial biosynthetic genes. *Mol. Microbiol.* **8:**543–558.

1097. Friman, V., I. Adlerberth, H. Connell, C. Svanborg, L. A. Hanson, and A. E. Wold. 1996. Decreased expression of mannose-specific adhesins by *Escherichia coli* in the colonic microflora of immunoglobulin A-deficient individuals. *Infect. Immun.* **64:**2794–2798.

1098. Frisk, A., and T. Lagergård. 1998. Characterization of mechanisms involved in adherence of *Haemophilus ducreyi* to eukaryotic cells. *APMIS* **106:**539–546.

1099. Frisk, A., C. A. Ison, and T. Lagergård. 1998. GroEL heat shock protein of *Haemophilus ducreyi:* association with cell surface and capacity to bind to eukaryotic cells. *Infect. Immun.* **66:**1252–1257.

1100. Frithz-Lindsten, E., Y. Du, R. Rosqvist, and A. Forsberg. 1997. Intracellular targeting of exoenzyme S of *Pseudomonas aeruginosa* via type III-dependent translocation induces phagocytosis resistance, cytotoxicity and disruption of actin microfilaments. *Mol. Microbiol.* **25:**1125–1139.

1101. **Froehlich, B., L. Husmann, J. Caron, and J. R. Scott.** 1994. Regulation of *rns,* a positive regulatory factor for pili of enterotoxigenic *Escherichia coli. J. Bacteriol.* **176:**5385–5392.

1102. **Froehlich, B. J., A. Karakashian, H. Sakellaris, and J. R. Scott.** 1995. Genes for CS2 pili of enterotoxigenic *Escherichia coli* and their interchangeability with those for CS1 pili. *Infect. Immun.* **63:**4849–4856.

1103. **Froeliger, E. H., M. Tomich, and P. Fives-Taylor.** 1999. Construction and analysis of a *Streptococcus parasanguis recA* mutant: homologous recombination is not required for adhesion in an in vitro tooth surface model. *J. Bacteriol.* **181:**63–67.

1104. **Frost, A. J., A. P. Bland, and T. S. Wallis.** 1997. The early dynamic response of the calf ileal epithelium to *Salmonella typhimurium. Vet. Pathol.* **34:**369–386.

1105. **Fu, Y., G. Rieg, W. A. Fonzi, P. H. Belanger, J. E. Edwards, Jr., and S. G. Filler.** 1998. Expression of the *Candida albicans* gene *ALS1* in *Saccharomyces cerevisiae* induces adherence to endothelial and epithelial cells. *Infect. Immun.* **66:**1783–1786.

1106. **Fuhrmann, C., L. Soedarmanto, and C. Lammler.** 1997. Studies on the rod-coccus life cycle of *Rhodococcus equi. Zentbl. Veterinaermed.* **44:**287–294.

1107. **Fujita, K., T. Yokota, T. Oguri, M. Fujime, and R. Kitagawa.** 1992. *In vitro* adherence of *Staphylococcus saprophyticus, Staphylococcus epidermidis, Staphylococcus haemolyticus,* and *Staphylococcus aureus* to human ureter. *Urol. Res.* **20:**399–402.

1108. **Fujiwara, S., H. Hashiba, T. Hirota, and J. F. Forstner.** 1997. Proteinaceous factor(s) in culture supernatant fluids of bifidobacteria which prevents the binding of enterotoxigenic *Escherichia coli* to gangliotetraosylceramide. *Appl. Environ. Microbiol.* **63:**506–512.

1109. **Fujiwara, S., Y. Miyake, T. Usui, and H. Suginaka.** 1998. Effect of adherence on antimicrobial susceptibility of *Pseudomonas aeruginosa, Serratia marcescens,* and *Proteus mirabilis. Hiroshima J. Med. Sci.* **47:**1–5.

1110. **Fujiwara,T., S. Morishima, I. Takahashi, and S. Hamada.** 1993. Molecular cloning and sequencing of the fimbrilin gene of *Porphyromonas gingivalis* strains and characterization of recombinant proteins. *Biochem. Biophys. Res. Commun.* **197:**241–247.

1111. **Fujiwara, T., I. Nakagawa, S. Morishima, I. Takahashi, and S. Hamada.** 1994. Inconsistency between the fimbrilin gene and the antigenicity of lipopolysaccharides in selected strains of *Porphyromonas gingivalis. FEMS Microbiol. Lett.* **124:**333–341.

1112. **Fujiwara, T., M. Tamesada, Z. Bian, S. Kawabata, S. Kimura, and S. Hamada.** 1996. Deletion and reintroduction of glucosyltransferase genes of *Streptococcus mutans* and role of their gene products in sucrose dependent cellular adherence. *Microb. Pathog.* **20:**225–233.

1113. **Fukui, H., M. Sueyoshi, M. Haritani, M. Nakazawa, S. Naitoh, H. Tani, and Y. Uda.** 1995. Natural infection with attaching and effacing *Escherichia coli* (O 103:H−) in chicks. *Avian Dis.* **39:**912–918.

1114. **Fullner, K. J.,, and J. J. Mekalanos.** 1999. Genetic characterization of a new type IV-A pilus gene cluster found in both classical and El Tor biotypes of *Vibrio cholerae. Infect. Immun.* **67:**1393–1404.

1115. **Fumagalli, O., B. D. Tall, C. Schipper, and T. A. Oelschlaeger.** 1997. N-glycosylated proteins are involved in efficient internalization of *Klebsiella pneumoniae* by cultured human epithelial cells. *Infect. Immun.* **65:**4445–4451.

1116. **Funfstuck, R., J. W. Smith, H. Tschape, and G. Stein.** 1997. Pathogenetic aspects of uncomplicated urinary tract infection: recent advances. *Clin. Nephrol.* **47:**13–18.

1117. **Funfstuck, R., M. Wolfram, J. Gerth, K. Schubert, E. Straube, and G. Stein.** 1999. The influence of ofloxacin (Tarivid) on the parasite-host inter-relationship in patients with chronic urinary tract infection. *Int. J. Antimicrob. Agents* **11:**297–303.

1118. **Funnell, S. G., and A. Robinson.** 1993. A novel adherence assay for *Bordetella pertussis* using tracheal organ cultures. *FEMS Microbiol. Lett.* **110:**197–203.

1119. **Fussenegger, M., A. F. Kahrs, D. Facius, and T. F. Meyer.** 1996. Tetrapac (*tpc*), a novel genotype of *Neisseria gonorrhoeae* affecting epithelial cell invasion, natural transformation competence and cell separation. *Mol. Microbiol.* **19:**1357–1372.

1120. **Fussenegger, M., T. Rudel, R. Barten, R. Ryll, and T. F. Meyer.** 1997. Transformation competence and type-4 pilus biogenesis in *Neisseria gonorrhoeae*—a review. *Gene* **192:**125–134.

1121. **Gaastra, W., R. A. van Oosterom, E. W. Pieters, H. E. Bergmans, L. van Dijk, A. Agnes, and H. M. ter Huurne.** 1996. Isolation and characterisation of dog uropathogenic *Proteus mirabilis* strains. *Vet. Microbiol.* **48:**57–71.

1122. **Gaastra, W., and A. M. Svennerholm.** 1996. Colonization factors of human enterotoxigenic *Escherichia coli* (ETEC). *Trends Microbiol.* **4:**444–452.

1123. **Gabastou, J. M., S. Kerneis, M. F. Bernet-Camard, A. Barbat, M. H. Coconnier, J. B. Kaper, and A. L. Servin.** 1995. Two stages of enteropathogenic *Escherichia coli* intestinal pathogenicity are up and down-regulated by the epithelial cell differentiation. *Differentiation* **59:**127–134.

1124. **Gabriel, B. L., J. Gold, A. G. Gristina, B. Kasemo, J. Lausmaa, C. Harrer, and Q. N. Myrvik.** 1994. Site-specific adhesion of *Staphylococcus epidermidis* (RP12) in Ti-Al-V metal systems. *Biomaterials* **15:**628–634.

1125. **Gabriel, M. M., M. S. Mayo, L. L. May, R. B. Simmons, and D. G. Ahearn.** 1996. *In vitro* evaluation of the efficacy of a silver-coated catheter. *Curr. Microbiol.* **33:**1–5.

1126. **Gabriel, M. M., A. D. Sawant, R. B. Simmons, and D. G. Ahearn.** 1995. Effects of silver on adherence of bacteria to urinary catheters: *in vitro* studies. *Curr. Microbiol.* **30:**17–22.

1127. **Gabriel, M. M., D. G. Ahearn, K. Y. Chan, and A. S. Patel.** 1998. *In vitro* adherence of *Pseudomonas aeruginosa* to four intraocular lenses. *J. Cataract Refract. Surg.* **24:**124–129.

1128. **Gaetti-Jardim Junior, E., and M. J. Avila-Campos.** 1999. Haemagglutination and haemolysis by oral *Fusobacterium nucleatum.* *New Microbiol.* **22:**63–67.

1129. **Gaffney, R. A., A. J. Schaeffer, B. E. Anderson, and J. L. Duncan.** 1994. Effect of Lewis blood group antigen expression on bacterial adherence to COS-1 cells. *Infect. Immun.* **62:**3022–3026.

1130. **Gaffney, R. A., M. F. Venegas, C. Kdanerva, E. L. Navas, B. E. Anderson, J. L. Duncan, and A. J. Schaeffer.** 1995. Effect of vaginal fluid on adherence of type 1 piliated *Escherichia coli* to epithelial cells. *J. Infect. Dis.* **172:**1528–1535.

1131. **Gaillard, J. L., and B. B. Finlay.** 1996. Effect of cell polarization and differentiation on entry of *Listeria monocytogenes* into the enterocyte-like Caco-2 cell line. *Infect. Immun.* **64:**1299–1308.

1132. **Gaillard, J. L., F. Jaubert, and P. Berche.** 1996. The *inlAB* locus mediates the entry of *Listeria monocytogenes* into hepatocytes *in vivo.* *J. Exp. Med.* **183:**359–369.

1133. **Galán, J. E.** 1994. Interactions of bacteria with non-phagocytic cells. *Curr. Opin. Immunol.* **6:**590–595.

1134. **Galane, P. M., and M. Le Roux.** 2001. Molecular epidemiology of *Escherichia coli* isolated from young South African children with diarrhoeal diseases. *J. Health Popul. Nutr.* **19:**31–38.

1135. **Galdbart, J. O., J. Allignet, H. S. Tung, C. Ryden, and N. el Solh.** 2000. Screening for *Staphylococcus epidermidis* markers discriminating between skin-flora strains and those responsible for infections of joint prostheses. *J. Infect. Dis.* **182:**351–355.

1136. **Galdiero, M., L. De Martino, U. Pagnini, M. G. Pisciotta, and E. Galdiero.** 2001. Interactions between bovine endothelial cells and *Pasteurella multocida:* association and invasion. *Res. Microbiol.* **152:**57–65.

1137. **Galfi, P., S. Neogrady, G. Semjen, S. Bardocz, and A. Pusztai.** 1998. Attachment of different *Escherichia coli* strains to cultured rumen epithelial cells. *Vet. Microbiol.* **61:**191–197.

1138. **Gallardo-Moreno, A. M., H. C. van der Mei, H. J. Busscher, M. L. Gonzalez-Martin, J. M. Bruque, and C. Perez-Giraldo.** 2001. Adhesion of *Enterococcus faecalis* 1131 grown under subinhibitory concentrations of ampicillin and vancomycin to a hydrophilic and a hydrophobic substratum. *FEMS Microbiol. Lett.* **203:**75–79.

1139. **Galli, D., A. Friesenegger, and R. Wirth.** 1992. Transcriptional control of sex-pheromone-inducible genes on plasmid pAD1 of *Enterococcus faecalis* and sequence analysis of a third structural gene for (pPD1-encoded) aggregation substance. *Mol. Microbiol.* **6:**1297–1308.

1140. **Galliani, S., A. Cremieux, P. van der Auwera, and M. Viot.** 1996. Influence of strain, biomaterial, proteins, and oncostatic chemotherapy on *Staphylococcus epidermidis* adhesion to intravascular catheters *in vitro.* *J. Lab. Clin. Med.* **127:**71–80.

1141. **Galliani, S., M. Viot, A. Cremieux, and P. van der Auwera.** 1994. Early adhesion of bacteremic strains of *Staphylococcus epidermidis* to polystyrene: influence of hydrophobicity, slime production, plasma, albumin, fibrinogen, and fibronectin. *J. Lab. Clin. Med.* **123:**685–692.

1142. **Gally, D. L., J. A. Bogan, B. I. Eisenstein, and I. C. Blomfield.** 1993. Environmental regulation of the *fim* switch controlling type 1 fimbrial phase variation in *Escherichia coli* K-12: effects of temperature and media. *J. Bacteriol.* **175:**6186–6193.

1143. **Galvez, J., F. Lajarin, and P. Garcia-Penarrubia.** 1997. Penetration of host cell lines by bacteria. Characteristics of the process of intracellular bacterial infection. *Bull. Math. Biol.* **59**:857–879.

1144. **Galvez, J., F. Lajarin, and P. Garcia-Penarubia.** 1997. Mathematical modeling of adhesion of bacteria to host cell lines. *Bull. Math. Biol.* **59**:833–856.

1145. **Ganaba, R., M. Bigras-Poulin, J. M. Fairbrother, and D. Belanger.** 1995. Importance of *Escherichia coli* in young beef calves from northwestern Quebec. *Can. J. Vet. Res.* **59**:20–25.

1146. **Ganderton, L., J. Chawla, C. Winters, J. Wimpenny, and D. Stickler.** 1992. Scanning electron microscopy of bacterial biofilms on indwelling bladder catheters. *Eur. J. Clin. Microbiol. Infect. Dis.* **11**:789–796.

1147. **Ganeshkumar, N., N. Arora, and P. E. Kolenbrander.** 1993. Saliva-binding protein (SsaB) from *Streptococcus sanguis* 12 is a lipoprotein. *J. Bacteriol.* **175**:572–574.

1148. **Gansheroff, L. J., M. R. Wachtel, and A. D. O'Brien.** 1999. Decreased adherence of enterohemorrhagic *Escherichia coli* to HEp-2 cells in the presence of antibodies that recognize the C-terminal region of intimin. *Infect. Immun.* **67**:6409–6417.

1149. **Gao, L.-Y., O. S. Harb, and Y. A. Kwaik.** 1997. Utilization of similar mechanisms by *Legionella pneumophila* to parasitize two evolutionarily distant host cells, mammalian macrophages and protozoa. *Infect. Immun.* **65**:4738–4746.

1150. **Garabal, J. I., F. Vazquez, J. Blanco, M. Blanco, and E. A. Gonzalez.** 1997. Colonization antigens of enterotoxigenic *Escherichia coli* strains isolated from piglets in Spain. *Vet. Microbiol.* **54**:321–328.

1151. **Garber, N., U. Guempel, A. Belz, N. Gilboa-Garber, and R. J. Doyle.** 1992. On the specificity of the D-galactose-binding lectin (PA-I) of *Pseudomonas aeruginosa* and its strong binding to hydrophobic derivatives of D-galactose and thiogalactose. *Biochim. Biophys. Acta* **1116**:331–333.

1152. **Garcia, E., H. E. Bergmans, B. A. van der Zeijst, and W. Gaastra.** 1992. Nucleotide sequences of the major subunits of F9 and F12 fimbriae of uropathogenic *Escherichia coli*. *Microb. Pathog.* **13**:161–166.

1153. **Garcia, L. B., L. C. Benchetrit, and L. Barrucand.** 1995. Penicillin post-antibiotic effects on the biology of group A streptococci. *J. Antimicrob. Chemother.* **36**:475–482.

1154. **Garcia, M. I., P. Gounon, P. Courcoux, A. Labigne, and C. Le Bouguenec.** 1996. The afimbrial adhesive sheath encoded by the afa-3 gene cluster of pathogenic *Escherichia coli* is composed of two adhesins. *Mol. Microbiol.* **19**:683–693.

1155. **Garcia, M. I., A. Labigne, and C. Le Bouguenec.** 1994. Nucleotide sequence of the afimbrial-adhesin-encoding afa-3 gene cluster and its translocation via flanking IS1 insertion sequences. *J. Bacteriol.* **176**:7601–7613.

1156. **Garcia Monco, J. C., B. Fernandez Villar, R. C. Rogers, A. Szczepanski, C. M. Wheeler, and J. L. Benach.** 1992. *Borrelia burgdorferi* and other related spirochetes bind to galactocerebroside. *Neurology* **42**:1341–1348.

1157. **Garcia Rodriguez, C., D. R. Cundell, E. I. Tuomanen, L. F. Kolakowski, Jr., C. Gerard, and N. P. Gerard.** 1995. The role of N-glycosylation for functional expression of the human platelet-activating factor receptor. Glycosylation is required for efficient membrane trafficking. *J. Biol. Chem.* **270**:25178–25184.

1158. **Garcia-Rivera, G., M. A. Rodriguez, R. Ocadiz, M. C. Martinez-Lopez, R. Arroyo, A. Gonzalez-Robles, and E. Orozco.** 1999. *Entamoeba histolytica*: a novel cysteine protease and an adhesin form the 112 kDa surface protein. *Mol. Microbiol.* **33**:556–568.

1159. **Garcia-Saenz, M. C., A. Arias-Puente, M. J. Fresnadillo-Martinez, and A. Matilla-Rodriguez.** 2000. *In vitro* adhesion of *Staphylococcus epidermidis* to intraocular lenses. *J. Cataract Refract. Surg.* **26**:1673–1679.

1160. **Gardel, C. L., and J. J. Mekalanos.** 1996. Alterations in *Vibrio cholerae* motility phenotypes correlate with changes in virulence factor expression. *Infect. Immun.* **64**:2246–2655.

1161. **Garduño, R. A., E. Garduño, and P. S. Hoffman.** 1998. Surface-associated Hsp60 chaperonin of *Legionella pneumophila* mediates invasion in a HeLa cell model. *Infect. Immun.* **66**:4602–4610.

1162. **Garduño, R. A., F. D. Quinn, and P. S. Hoffman.** 1998. HeLa cells as a model to study the invasiveness and biology of *Legionella pneumophila*. *Can. J. Microbiol.* **44**:430–440.

1163. **Garduño, R. A., G. Faulkner, M. A. Trevors, N. Vats, and P. S. Hoffman.** 1998. Immunolocalization of Hsp60 in *Legionella pneumophila*. *J. Bacteriol.* **180**:505–513.

1164. **Garduño, R. A., A. R. Moore, G. Olivier, A. L. Lizama, E. Garduno, and W. W. Kay.** 2000. Host cell invasion and intracellular residence by *Aeromonas salmoni-*

cida: role of the S-layer. *Can. J. Microbiol.* **46:**660–668.

1165. **Garner, A. M., and K. S. Kim.** 1996. The effects of *Escherichia coli* S-fimbriae and outer membrane protein A on rat pial arterioles. *Pediatr. Res.* **39:**604–608.

1166. **Garvis, S. G., G. J. Puzon, and M. E. Konkel.** 1996. Molecular characterization of a *Campylobacter jejuni* 29-kilodalton periplasmic binding protein. *Infect. Immun.* **64:**3537–3543.

1167. **Gasser, T. C., and P. O. Madsen.** 1993. Influence of urological irrigation fluids on urothelial bacterial adherence. *Urol. Res.* **21:**401–405.

1168. **Gatermann, S., H. G. Meyer, and G. Wanner.** 1992. *Staphylococcus saprophyticus* hemagglutinin is a 160-kilodalton surface polypeptide. *Infect. Immun.* **60:**4127–4132.

1169. **Gatermann, S., H. G. Meyer, R. Marre, and G. Wanner.** 1993. Identification and characterization of surface proteins from *Staphylococcus saprophyticus. Zentbl. Bakteriol.* **278:**258–274.

1170. **Gatermann, S., and H. G. Meyer.** 1994. *Staphylococcus saprophyticus* hemagglutinin binds fibronectin. *Infect. Immun.* **62:**4556–4563.

1171. **Gatermann, S. G., and H. G. Meyer.** 1995. Expression of *Staphylococcus saprophyticus* surface properties is modulated by composition of the atmosphere. *Med. Microbiol. Immunol.* **184:**81–85.

1172. **Gatter, N., W. Kohnen, and B. Jansen.** 1998. *In vitro* efficacy of a hydrophilic central venous catheter loaded with silver to prevent microbial colonization. *Zentbl. Bakteriol.* **287:**157–169.

1173. **Gauthier, V., S. Redercher, and J. C. Block.** 1999. Chlorine inactivation of *Sphingomonas* cells attached to goethite particles in drinking water. *Appl. Environ. Microbiol.* **65:**355–357.

1174. **Gaynor, C. D., F. X. McCormack, D. R. Voelker, S. E. McGowan, and L. S. Schlesinger.** 1995. Pulmonary surfactant protein A mediates enhanced phagocytosis of *Mycobacterium tuberculosis* by a direct interaction with human macrophages. *J. Immunol.* **155:**5343–5351.

1175. **Gbarah, A., C. G. Gahmberg, G. Boner, and N. Sharon.** 1993. The leukocyte surface antigens CD11b and CD18 mediate the oxidative burst activation of human peritoneal macrophages induced by type 1 fimbriated *Escherichia coli. J. Leukoc. Biol.* **54:**111–113.

1176. **Gbarah, A., D. Mirelman, P. J. Sansonetti, R. Verdon, W. Bernhard, and N. Sharon.** 1993. *Shigella flexneri* trans-

formants expressing type 1 (mannose-specific) fimbriae bind to, activate, and are killed by phagocytic cells. *Infect. Immun.* **61:**1687–1693.

1177. **Gedek, B. R.** 1999. Adherence of *Escherichia coli* serogroup O157 and the *Salmonella typhimurium* mutant DT 104 to the surface of *Saccharomyces boulardii. Mycoses* **42:**261–264.

1178. **Geelen, S., C. Bhattacharyya, and E. Tuomanen.** 1992. Induction of procoagulant activity on human endothelial cells by *Streptococcus pneumoniae. Infect. Immun.* **60:**4179–4183.

1179. **Geelen, S., C. Bhattacharyya, and E. Tuomanen.** 1993. The cell wall mediates pneumococcal attachment to and cytopathology in human endothelial cells. *Infect. Immun.* **61:**1538–1543.

1180. **Geluk, F., P. P. Eijk, S. M. van Ham, H. M. Jansen, and L. van Alphen.** 1998. The fimbria gene cluster of nonencapsulated *Haemophilus influenzae. Infect. Immun.* **66:**406–417.

1181. **Genevaux, P., P. Bauda, M. S. DuBow, and B. Oudega.** 1999. Identification of Tn*10* insertions in the *dsbA* gene affecting *Escherichia coli* biofilm formation. *FEMS Microbiol. Lett.* **173:**403–409.

1182. **Genevaux, P., S. Muller, and P. Bauda.** 1996. A rapid screening procedure to identify mini-Tn*10* insertion mutants of *Escherichia coli* K-12 with altered adhesion properties. *FEMS Microbiol. Lett.* **142:**27–30.

1183. **Genta, R. M., I. E. Gurer, D. Y. Graham, B. Krishnan, A. M. Segura, O. Gutierrez, J. G. Kim, and J. L. Burchette, Jr.** 1996. Adherence of *Helicobacter pylori* to areas of incomplete intestinal metaplasia in the gastric mucosa. *Gastroenterology* **111:**1206–1211.

1184. **George, K. S., and W. A. Falkler, Jr.** 1992. Coaggregation studies of the *Eubacterium* species. *Oral Microbiol. Immunol.* **7:**285–290.

1185. **Gerhard, M., N. Lehn, N. Neumayer, T. Boren, R. Rad, W. Schepp, S. Miehlke, M. Classen, and C. Prinz.** 1999. Clinical relevance of the *Helicobacter pylori* gene for blood-group antigen-binding adhesin. *Proc. Natl. Acad. Sci. USA* **96:**12778–12783.

1186. **Gerke, C., A. Kraft, R. Sussmuth, O. Schweitzer, and F. Götz.** 1998. Characterization of the N-acetylglucosaminyl-transferase activity involved in the biosynthesis of the *Staphylococcus epidermidis* polysaccharide intercellular adhesin. *J. Biol. Chem.* **273:**18586–18593.

1187. **Germani, Y., E. Begaud, P. Duval, and C. Le Bouguenec.** 1996. Prevalence of enteropathogenic, enteroaggregative, and diffusely adherent *Escherichia coli* among isolates from children with diarrhea in new Caledonia. *J. Infect. Dis.* **174:**1124–1126.

1188. **Germani, Y., E. Begaud, P. Duval, and C. Le Bouguenec.** 1997. An *Escherichia coli* clone carrying the adhesin-encoding *afa* operon is involved in both diarrhoea and cystitis in twins. *Trans. R. Soc. Trop. Med. Hyg.* **91:**573.

1189. **Geuijen, C. A. W., R. J. L. Willems, and F. R. Mooi.** 1996. The major fimbrial subunit of *Bordetella pertussis* binds to sulfated sugars. *Infect. Immun.* **64:**2657–2665.

1190. **Geuijen, C. A., R. J. Willems, M. Bongaerts, J. Top, H. Gielen, and F. R. Mooi.** 1997. Role of the *Bordetella pertussis* minor fimbrial subunit, FimD, in colonization of the mouse respiratory tract. *Infect. Immun.* **65:**4222–4228.

1191. **Geuijen, C. A. W., R. J. L. Willems, P. Hoogerhout, W. C. Puijk, R. H. Meloen, and F. R. Mooi.** 1998. Identification and characterization of heparin binding regions of the Fim2 subunit of *Bordetella pertussis*. *Infect. Immun.* **66:**2256–2263.

1192. **Geyid, A., J. Fletcher, B. A. Gashe, and Å. Ljungh.** 1996. Invasion of tissue culture cells by diarrhoeagenic strains of *Escherichia coli* which lack the enteroinvasive *inv* gene. *FEMS Immunol. Med. Microbiol.* **14:**15–24.

1193. **Ghose, A. C.** 1996. Adherence & colonization properties of *Vibrio cholerae* & diarrhoeagenic *Escherichia coli*. *Indian J. Med. Res.* **104:**38–51.

1194. **Ghosh, A. R., G. B. Nair, T. N. Naik, M. Paul, S. C. Pal, and D. Sen.** 1992. Enteroadherent *Escherichia coli* is an important diarrhoeagenic agent in infants aged below 6 months in Calcutta, India. *J. Med. Microbiol.* **36:**264–268.

1195. **Ghosh, S., A. Mittal, and N. K. Ganguly.** 1994. Purification and characterization of distinct type of mannose-sensitive fimbriae from *Salmonella typhimurium*. *FEMS Microbiol. Lett.* **115:**229–234.

1196. **Ghosh, S., A. Mittal, H. Vohra, and N. K. Ganguly.** 1996. Interaction of a rat intestinal brush border membrane glycoprotein with type-1 fimbriae of *Salmonella typhimurium*. *Mol. Cell. Biochem.* **158:**125–131.

1197. **Ghosh, C., R. K. Nandy, S. K. Dasgupta, G. B. Nair, R. H. Hall, and A. C. Ghose.** 1997. A search for cholera toxin (CT), toxin coregulated pilus (TCP), the regulatory element ToxR and other virulence factors in non-O1/non-O139 *Vibrio cholerae*. *Microb. Pathog.* **22:**199–208.

1198. **Giaffer, M. H., C. D. Holdsworth, and B. I. Duerden.** 1992. Virulence properties of *Escherichia coli* strains isolated from patients with inflammatory bowel disease. *Gut* **33:**646–650.

1199. **Giannasca, K. T., P. J. Giannasca, and M. R. Neutra.** 1996. Adherence of *Salmonella typhimurium* to Caco-2 cells: identification of a glycoconjugate receptor. *Infect. Immun.* **64:**135–145.

1200. **Giardina, P. C., R. Williams, D. Lubaroff, and M. A. Apicella.** 1998. *Neisseria gonorrhoeae* induces focal polymerization of actin in primary human urethral epithelium. *Infect. Immun.* **66:**3416–3419.

1201. **Gibbons, R. J.** 1996. Role of adhesion in microbial colonization of host tissues: a contribution of oral microbiology. *J. Dent. Res.* **75:**866–870.

1202. **Gibson, B. W., A. A. Campagnari, W. Melaugh, N. J. Phillips, M. A. Apicella, S. Grass, J. Wang, K. L. Palmer, and R. S. Munson, Jr.** 1997. Characterization of a transposon Tn916-generated mutant of *Haemophilus ducreyi* 35000 defective in lipooligosaccharide biosynthesis. *J. Bacteriol.* **179:**5062–5071.

1203. **Gibson, C., G. Fogg, N. Okada, R. T. Geist, E. Hanski, and M. Caparon.** 1995. Regulation of host cell recognition in *Streptococcus pyogenes*. *Dev. Biol. Stand.* **85:**137–144.

1204. **Gibson, C. M., and M. G. Caparon.** 1996. Insertional inactivation of *Streptococcus pyogenes sod* suggests that *prtF* is regulated in response to a superoxide signal. *J. Bacteriol.* **178:**4688–4695.

1205. **Gibson, F. C., III, A. O. Tzianabos, and F. G. Rodgers.** 1994. Adherence of *Legionella pneumophila* to U-937 cells, guinea-pig alveolar macrophages, and MRC-5 cells by a novel, complement-independent binding mechanism. *Can. J. Microbiol.* **40:**865–872.

1206. **Gibson, R. L., M. K. Lee, C. Soderland, E. Y. Chi, and C. E. Rubens.** 1993. Group B streptococci invade endothelial cells: type III capsular polysaccharide attenuates invasion. *Infect. Immun.* **61:**478–485.

1207. **Gilbert, P., J. Das, and I. Foley.** 1997. Biofilm susceptibility to antimicrobials. *Adv. Dent. Res.* **11:**160–167.

1208. **Gilboa-Garber, N., D. Sudakevitz, M. Sheffi, R. Sela, and C. Levene.** 1994. PA-I and PA-II lectin interactions with the ABO(H) and P blood group glycosphingolipid antigens may contribute to the broad spectrum adherence of *Pseudomonas aeruginosa* to human tissues in secondary infections. *Glycoconj. J.* **11**:414–417.

1209. Gilboa-Garber, N., and D. Sudakevitz. 1999. The hemagglutinating activities of *Pseudomonas aeruginosa* lectins PA-IL and PA-IIL exhibit opposite temperature profiles due to different receptor types. *FEMS Immunol. Med. Microbiol.* **25**:365–369.

1210. Gilchrist, C. A., B. J. Mann, and W. A. Petri, Jr. 1998. Control of ferridoxin and Gal/GalNAc lectin gene expression in *Entamoeba histolytica* by a *cis*-acting DNA sequence. *Infect. Immun.* **66**:2383–2386.

1211. Gillaspy, A. F., J. M. Patti, and M. S. Smeltzer. 1997. Transcriptional regulation of the *Staphylococcus aureus* collagen adhesin gene, *cna. Infect. Immun.* **65**:1536–1540.

1212. Gillaspy, A. F., C. Y. Lee, S. Sau, A. L. Cheung, and M. S. Smeltzer. 1998. Factors affecting the collagen binding capacity of *Staphylococcus aureus*. *Infect. Immun.* **66**:3170–3178.

1213. Gilsdorf, J. R., H. Y. Chang, K. W. McCrea, L. J. Forney, and C. F. Marrs. 1992. Comparison of hemagglutinating pili of type b and nontypeable *Haemophilus influenzae*. *J. Infect. Dis.* **165**:S105–S106.

1214. Gilsdorf, J. R., H. Y. Chang, K. W. McCrea, and L. O. Bakaletz. 1992. Comparison of hemagglutinating pili of *Haemophilus influenzae* type b with similar structures of nontypeable *H. influenzae*. *Infect. Immun.* **60**:374–379.

1215. Gilsdorf, J. R., L. J. Forney, and K. W. McCrea. 1993. Reactivity of antibodies against conserved regions of pilins of *Haemophilus influenzae* type b. *J. Infect. Dis.* **167**:962–965.

1216. Gilsdorf, J. R., M. Tucci, L. J. Forney, W. Watson, C. F. Marrs, and E. J. Hansen. 1993. Paradoxical effect of pilus expression on binding of antibodies by *Haemophilus influenzae*. *Infect. Immun.* **61**:3375–3381.

1217. Gilsdorf, J. R., M. Tucci, and C. F. Marrs. 1996. Role of pili in *Haemophilus influenzae* adherence to, and internalization by, respiratory cells. *Pediatr. Res.* **39**:343–348.

1218. Gilsdorf, J. R., K. W. McCrea, and C. F. Marrs. 1997. Role of pili in *Haemophilus influenzae* adherence and colonization. *Infect. Immun.* **65**:2997–3002.

1219. Ginocchio, C. C., and J. E. Galan. 1995. Functional conservation among members of the *Salmonella typhimurium* InvA family of proteins. *Infect. Immun.* **63**:729–732.

1220. Girardeau, J. P., and Y. Bertin. 1995. Pilins of fimbrial adhesins of different member species of *Enterobacteriaceae* are structurally similar to the C-terminal half of adhesin proteins. *FEBS Lett.* **357**:103–108.

1221. Giridhar, G., A. S. Kreger, Q. N. Myrvik, and A. G. Gristina. 1994. Inhibition of *Staphylococcus* adherence to biomaterials by extracellular slime of *S. epidermidis* RP12. *J. Biomed. Mater. Res.* **28**:1289–1294.

1222. Girod de Bentzmann, S., O. Bajolet-Laudinat, F. Dupuit, D. Pierrot, C. Fuchey, M. C. Plotkowski, and E. Puchelle. 1994. Protection of human respiratory epithelium from *Pseudomonas aeruginosa* adherence by phosphatidylglycerol liposomes. *Infect. Immun.* **62**:704–708.

1223. Girón, J. A., M. S. Donnenberg, W. C. Martin, K. G. Jarvis, and J. B. Kaper. 1993. Distribution of the bundle-forming pilus structural gene (*bfpA*) among enteropathogenic *Escherichia coli*. *J. Infect. Dis.* **168**:1037–1041.

1224. Girón, J. A., A. S. Ho, and G. K. Schoolnik. 1993. Characterization of fimbriae produced by enteropathogenic *Escherichia coli*. *J. Bacteriol.* **175**:7391–7403.

1225. Girón, J. A., M. Lange, and J. B. Baseman. 1996. Adherence, fibronectin binding, and induction of cytoskeleton reorganization in cultured human cells by *Mycoplasma penetrans*. *Infect. Immun.* **64**:197–208.

1226. Girón, J. A., G. I. Viboud, V. Sperandio, O. G. Gomez-Duarte, D. R. Maneval, M. J. Albert, M. M. Levine, and J. B. Kaper. 1995. Prevalence and association of the longus pilus structural gene (*lngA*) with colonization factor antigens, enterotoxin types, and serotypes of enterotoxigenic *Escherichia coli*. *Infect. Immun.* **63**:4195–4198.

1227. Girón, J. A., J. G. Xu, C. R. Gonzalez, D. Hone, J. B. Kaper, and M. M. Levine. 1995. Simultaneous expression of CFA/I and CS3 colonization factor antigens of enterotoxigenic *Escherichia coli* by delta *aroC*, delta *aroD Salmonella typhi* vaccine strain CVD 908. *Vaccine* **13**:939–946.

1228. Girón, J. A., F. Qadri, T. Azim, K. J. Jarvis, J. B. Kaper, and M. J. Albert. 1995. Monoclonal antibodies specific for the bundle-forming pilus of enteropathogenic *Escherichia coli*. *Infect. Immun.* **63**:4949–4952.

1229. Girón, J. A., O. G. Gomez-Duarte, K. G. Jarvis, and J. B. Kaper. 1997. Longus pilus of enterotoxigenic *Escherichia coli* and its relatedness to other type-4 pili—a minireview. *Gene* **192**:39–43.

1230. Gismondo, M. R., L. Drago, C. Fassina, M. L. Garlaschi, M. Rosina, and A. Lombardi. 1994. *Escherichia coli*:effect of fosfomycin trometamol on some urovirulence factors. *J. Chemother.* **6:**167–172.

1231. Gismondo, M. R., L. Drago, A. Lombardi, M. C. Fassina, and B. Mombelli. 1998. Interference on *Helicobacter pylori* growth and adhesion by omeprazole and other drugs. *J. Chemother.* **10:**225–230.

1232. Giugliano, L. G., C. J. Meyer, L. C. Arantes, S. T. Ribeiro, and R. Giugliano. 1993. Mannose-resistant haemagglutination (MRHA) and haemolysin (Hly) production of strains of *Escherichia coli* isolated from children with diarrhoea: effect of breastfeeding. *J. Trop. Pediatr.* **39:**183–187.

1233. Giugliano, L. G., S. T. Ribeiro, M. H. Vainstein, and C. J. Ulhoa. 1995. Free secretory component and lactoferrin of human milk inhibit the adhesion of enterotoxigenic *Escherichia coli. J. Med. Microbiol.* **42:**3–9.

1234. Glancey, G., J. S. Cameron, C. Ogg, and S. Poston. 1993. Adherence of *Staphylococcus aureus* to cultures of human peritoneal mesothelial cells. *Nephrol. Dial. Transpl.* **8:**157–162.

1235. Goebel, W., M. Leimeister-Wachter, M. Kuhn, E. Domann, T. Chakraborty, S. Kohler, A. Bubert, M. Wuenscher, and Z. Sokolovic. 1993. *Listeria monocytogenes*—a model system for studying the pathomechanisms of an intracellular microorganism. *Zentbl. Bakteriol.* **278:**334–347.

1236. Goetz, G. S., A. Mahmood, S. J. Hultgren, M. J. Engle, K. Dodson, and D. H. Alpers. 1999. Binding of pili from uropathogenic *Escherichia coli* to membranes secreted by human colonocytes and enterocytes. *Infect. Immun.* **67:**6161–6163.

1237. Goh, M. S., T. S. Gorton, M. H. Forsyth, K. E. Troy, and S. J. Geary. 1998. Molecular and biochemical analysis of a 105 kDa *Mycoplasma gallisepticum* cytadhesin (GapA). *Microbiology* **144:**2971–2978.

1238. Gold, B. D., M. Dytoc, M. Huesca, D. Philpott, A. Kuksis, S. Czinn, C. A. Lingwood, and P. M. Sherman. 1995. Comparison of *Helicobacter mustelae* and *Helicobacter pylori* adhesion to eukaryotic cells *in vitro. Gastroenterology* **109:**692–700.

1239. Gold, B. D., M. Huesca, P. M. Sherman, and C. A. Lingwood. 1993. *Helicobacter mustelae* and *Helicobacter pylori* bind to common lipid receptors *in vitro. Infect. Immun.* **61:**2632–2638.

1240. Gold, B. D., P. Islur, Z. Policova, S. Czinn, A. W. Neumann, and P. M. Sherman. 1996. Surface properties of *Helicobacter mustelae* and ferret gastrointestinal mucosa. *Clin. Investig. Med.* **19:**92–100.

1241. Goldhar, J. 1994. Bacterial lectinlike adhesins: determination and specificity. *Methods Enzymol.* **236:**211–231.

1242. Goldhar, J. 1995. Erythrocytes as target cells for testing bacterial adhesins. *Methods Enzymol.* **253:**43–50.

1243. Goldmann, D. A., and G. B. Pier. 1993. Pathogenesis of infections related to intravascular catheterization. *Clin. Microbiol. Rev.* **6:**176–192.

1244. Goldstein, R., L. Sun, R. Z. Jiang, U. Sajjan, J. F. Forstner, and C. Campanelli. 1995. Structurally variant classes of pilus appendage fibers coexpressed from *Burkholderia* (*Pseudomonas*) *cepacia. J. Bacteriol.* **177:**1039–1052.

1245. Goluszko, P., S. Nowicki, A. K. Kaul, T. Pham, and B. J. Nowicki. 1995. Dr fimbriae coding region associated hemolytic activity of *Escherichia coli. FEMS Microbiol. Lett.* **130:**13–17.

1246. Goluszko, P., R. Selvarangan, V. Popov, T. Pham, J. W. Wen, and J. Singhal. 1999. Decay-accelerating factor and cytoskeleton redistribution pattern in HeLa cells infected with recombinant *Escherichia coli* strains expressing Dr family of adhesins. *Infect. Immun.* **67:**3989–3997.

1247. Goluszko, P., B. Nowicki, E. Goluszko, S. Nowicki, A. Kaul, and T. Pham. 1995. Association of colony variation in *Serratia marcescens* with the differential expression of protease and type 1 fimbriae. *FEMS Microbiol. Lett.* **133:**41–45.

1248. Goluszko, P., S. L. Moseley, L. D. Truong, A. Kaul, J. R. Williford, R. Selvarangan, S. Nowicki, and B. Nowicki. 1997. Development of experimental model of chronic pyelonephritis with *Escherichia coli* O75:K5:H-bearing Dr fimbriae: mutation in the *dra* region prevented tubulointerstitial nephritis. *J. Clin. Investig.* **99:**1662–1672.

1249. Gomes, T. A., M. A. Vieira, C. M. Abe, D. Rodrigues, P. M. Griffin, and S. R. Ramos. 1998. Adherence patterns and adherence-related DNA sequences in *Escherichia coli* isolates from children with and without diarrhea in Sao Paulo city, Brazil. *J. Clin. Microbiol.* **36:**3609–3613.

1250. Gomez, H. F., J. J. Mathewson, P. C. Johnson, and H. L. DuPont. 1995. Intestinal immune response of volunteers ingesting a strain of enteroadherent (HEp-2 cell-adherent) *Escherichia coli*. *Clin. Diagn. Lab. Immunol.* **2:**10–13.

1251. **Gomez-Duarte, O. G., M. Dehio, C. A. Guzman, G. S. Chhatwal, C. Dehio, and T. F. Meyer.** 1997. Binding of vitronectin to opa-expressing *Neisseria gonorrhoeae* mediates invasion of HeLa cells. *Infect. Immun.* **65:**3857–3866.

1252. **Gomez-Duarte, O. G., and J. B. Kaper.** 1995. A plasmid-encoded regulatory region activates chromosomal eaeA expression in enteropathogenic *Escherichia coli*. *Infect. Immun.* **63:**1767–1776.

1253. **Goncalves, A. G., L. C. Campos, T. A. Gomes, J. Rodrigues, V. Sperandio, T. S. Whittam, and L. R. Trabulsi.** 1997. Virulence properties and clonal structures of strains of *Escherichia coli* O119 serotypes. *Infect. Immun.* **65:**2034–2040.

1254. **Gong, K., D. Y. Wen, T. Ouyang, A. T. Rao, and M. C. Herzberg.** 1995. Platelet receptors for the *Streptococcus sanguis* adhesin and aggregation-associated antigens are distinguished by anti-idiotypical monoclonal antibodies. *Infect. Immun.* **63:**3628–3633.

1255. **Gong, K., and M. C. Herzberg.** 1998. *Streptococcus sanguis* expresses a 150-kilodalton two-domain adhesin: characterization of several independent adhesin epitopes. *Infect. Immun.* **65:**3815–3821.

1256. **Gong, K., T. Ouyang, and M. C. Herzberg.** 1998. A streptococcal adhesion system for salivary pellicle and platelets. *Infect. Immun.* **66:**5388–5392.

1257. **Gong, K., L. Mailloux, and M. C. Herzberg.** 2000. Salivary film expresses a complex, macromolecular binding site for *Streptococcus sanguis*. *J. Biol. Chem.* **275:**8970–8974.

1258. **Gong, M., and L. Makowski.** 1992. Helical structure of P pili from *Escherichia coli*. Evidence from X-ray fiber diffraction and scanning transmission electron microscopy. *J. Mol. Biol.* **228:**735–742.

1259. **Gonzalez, E. A., F. Vazquez, J. Ignacio Garabal, and J. Blanco.** 1995. Isolation of K88 antigen variants (ab, ac, ad) from porcine enterotoxigenic *Escherichia coli* belonging to different serotypes. *Microbiol. Immunol.* **39:**937–942.

1260. **Gonzalez, R., C. Diaz, M. Marino, R. Cloralt, M. Pequeneze, and I. Perez-Schael.** 1997. Age-specific prevalence of *Escherichia coli* with localized and aggregative adherence in Venezuelan infants with acute diarrhea. *J. Clin. Microbiol.* **35:**1103–1107.

1261. **Gophna, U., M. Barlev, R. Seijffers, T. A. Oelschlager, J. Hacker, and E. Z. Ron.** 2001. Curli fibers mediate internalization of *Escherichia coli* by eukaryotic cells. *Infect. Immun.* **69:**2659–2665.

1262. **Gorby, G., D. Simon, and R. F. Rest.** 1994. *Escherichia coli* that express *Neisseria gonorrhoeae* opacity-associated proteins attach to and invade human fallopian tube epithelium. *Ann. N. Y. Acad. Sci.* **730:**286–289.

1263. **Gorby, G. L.** 1994. Digital confocal microscopy allows measurement and three-dimensional multiple spectral reconstruction of *Neisseria gonorrhoeae*/epithelial cell interactions in the human fallopian tube organ culture model. *J. Histochem. Cytochem.* **42:**297–306.

1264. **Gorby, G. L., and G. B. Schaefer.** 1992. Effect of attachment factors (pili plus Opa) on *Neisseria gonorrhoeae* invasion of human fallopian tube tissue *in vitro*: quantitation by computerized image analysis. *Microb. Pathog.* **13:**93–108.

1265. **Gordon, A. E., A. T. Saadi, D. A. MacKenzie, V. S. James, R. A. Elton, D. M. Weir, A. Busuttil, and C. C. Blackwell.** 1999. The protective effect of breast feeding in relation to sudden infant death syndrome (SIDS). II. The effect of human milk and infant formula preparations on binding of *Clostridium perfringens* to epithelial cells. *FEMS Immunol. Med. Microbiol.* **25:**167–173.

1266. **Gori, A. H., K. Ahmed, G. Martinez, H. Masaki, K. Watanabe, and T. Nagatake.** 1999. Mediation of attachment of *Burkholderia pseudomallei* to human pharyngeal epithelial cells by the asialoganglioside GM1-GM2 receptor complex. *Am. J. Trop. Med. Hyg.* **61:**473–475.

1267. **Gorman, S. P., W. M. Mawhinney, C. G. Adair, and M. Issouckis.** 1993. Confocal laser scanning microscopy of peritoneal catheter surfaces. *J. Med. Microbiol.* **38:**411–417.

1268. **Gorter, A. D., P. P. Eijk, S. van Wetering, P. S. Hiemstra, J. Dankert, and L. van Alphen.** 1998. Stimulation of the adherence of *Haemophilus influenzae* to human lung epithelial cells by antimicrobial neutrophil defensins. *J. Infect. Dis.* **178:**1067–1074.

1269. **Gosink, K. K., E. R. Mann, C. Guglielmo, E. I. Tuomanen, and H. R. Masure.** 2000. Role of novel choline binding proteins in virulence of *Streptococcus pneumoniae*. *Infect. Immun.* **68:**5690–5695.

1270. **Gospodarek, E., A. Grzanka, Z.**

Dudziak, and J. Domaniewski. 1998. Electron-microscopic observation of adherence of *Acinetobacter baumannii* to red blood cells. *Acta Microbiol. Pol.* **47:**213–217.

1271. **Goswami, S., S. Sarkar, J. Basu, M. Kundu, and P. Chakrabarti.** 1994. Mycotin: a lectin involved in the adherence of *Mycobacteria* to macrophages. *FEBS Lett.* **355:**183–186.

1272. **Gottenbos, B., D. W. Grijpma, H. C. van der Mei, J. Feijen, and H. J. Busscher.** 2001. Antimicrobial effects of positively charged surfaces on adhering Gram-positive and Gram-negative bacteria. *J. Antimicrob. Chemother.* **48:**7–13.

1273. **Gottenbos, B., H. C. van der Mei, and H. J. Busscher.** 2000. Initial adhesion and surface growth of *Staphylococcus epidermidis* and *Pseudomonas aeruginosa* on biomedical polymers. *J. Biomed. Mater. Res.* **50:**208–214.

1274. **Goulsbra, A. M., C. Edwards, and M. P. Gallagher.** 2001. Surface hygiene monitored using a reporter of *fis* in *Escherichia coli. J. Appl. Microbiol.* **91:**104–109.

1275. **Goyard, S., H. Mireau, and A. Ullmann.** 1995. Mutations which result in constitutive expression of the *Bordetella pertussis* filamentous haemagglutinin gene. *Res. Microbiol.* **146:**363–370.

1276. **Graham, J. C., J. B. Leathart, S. J. Keegan, J. Pearson, A. Bint, and D. L. Gally.** 2001. Analysis of *Escherichia coli* strains causing bacteriuria during pregnancy: selection for strains that do not express type 1 fimbriae. *Infect. Immun.* **69:**794–799.

1277. **Graham, L. L., and K. L. MacDonald.** 1998. The *Campylobacter fetus* S layer is not essential for initial interaction with HEp-2 cells. *Can. J. Microbiol.* **44:**244–250.

1278. **Granato, D., F. Perotti, I. Masserey, M. Rouvet, M. Golliard, A. Servin, and D. Brassart.** 1999. Cell surface-associated lipoteichoic acid acts as an adhesion factor for attachment of *Lactobacillus johnsonii* La1 to human enterocyte-like Caco-2 cells. *Appl. Environ. Microbiol.* **65:**1071–1077.

1279. **Grange, P. A., A. K. Erickson, S. B. Levery, and D. H. Francis.** 1999. Identification of an intestinal neutral glycosphingolipid as a phenotype-specific receptor for the K88ad fimbrial adhesin of *Escherichia coli. Infect. Immun.* **67:**165–172.

1280. **Grange, P. A., and M. Mouricout.** 1996. Susceptibility of infant mice to F5 (K99) *E. coli* infection: differences in glycosyltransferase activities in intestinal mucosa of inbred CBA and DBA/2 strains. *Glycoconj. J.* **13:**45–52.

1281. **Grange, P. A., and M. A. Mouricout.**

1996. Transferrin associated with the porcine intestinal mucosa is a receptor specific for K88ab fimbriae of *Escherichia coli. Infect. Immun.* **64:**606–610.

1282. **Granlund-Edstedt, M., M. Sellin, A. Holm, and S. Hakansson.** 1993. Adherence and surface properties of buoyant density subpopulations of group B streptococci, type III. *APMIS* **101:**141–148.

1283. **Granok, A. B., D. Parsonage, R. P. Ross, and M. G. Caparon.** 2000. The RofA binding site in *Streptococcus pyogenes* is utilized in multiple transcriptional pathways. *J. Bacteriol.* **182:**1529–1540.

1284. **Grant, C. C., M. P. Bos, and R. J. Belland.** 1999. Proteoglycan receptor binding by *Neisseria gonorrhoeae* MS11 is determined by the HV-1 region of OpaA. *Mol. Microbiol.* **32:**233–242.

1285. **Grant, C. C., M. E. Konkel, W. Cieplak, Jr., and L. S. Tompkins.** 1993. Role of flagella in adherence, internalization, and translocation of *Campylobacter jejuni* in nonpolarized and polarized epithelial cell cultures. *Infect. Immun.* **61:**1764–1771.

1286. **Grant, G., S. Bardocz, S. W. Ewen, D. S. Brown, T. J. Duguid, A. Pusztai, D. Avichezer, D. Sudakevitz, A. Belz, N. C. Garber, et al.** 1995. Purified *Pseudomonas aeruginosa* PA-I lectin induces gut growth when orally ingested by rats. *FEMS Immunol. Med. Microbiol.* **11:**191–195.

1287. **Grassmann, B., P. A. Kopp, M. Schmitt, and H. Blobel.** 1997. Adherence of *Borrelia burgdorferi* to granulocytes of different animal species. *Zentbl. Bakteriol.* **285:**501–508.

1288. **Grassmé, H., E. Gulbins, B. Brenner, K. Ferlinz, K. Sandhoff, K. Harzer, F. Lang, and T. F. Meyer.** 1997. Acidic sphingomyelinase mediates entry of *N. gonorrhoeae* into nonphagocytic cells. *Cell* **91:**605–615.

1289. **Grassme, H. U., R. M. Ireland, and J. P. van Putten.** 1996. Gonococcal opacity protein promotes bacterial entry-associated rearrangements of the epithelial cell actin cytoskeleton. *Infect. Immun.* **64:**1621–1630.

1290. **Greco, R., L. De Martino, G. Donnarumma, M. P. Conte, L. Seganti, and P. Valenti.** 1995. Invasion of cultured human cells by *Streptococcus pyogenes. Res. Microbiol.* **146:**551–560.

1291. **Greene, C., D. McDevitt, P. Francois, P. E. Vaudaux, D. P. Lew, and T. J. Foster.** 1995. Adhesion properties of mutants of *Staphylococcus aureus* defective in fibronectin-binding proteins and studies on the expression of *fnb* genes. *Mol. Microbiol.* **17:**1143–1152.

1292. **Greene, C., P. E. Vaudaux, P. Francois, R. A. Proctor, D. McDevitt, and T. J. Foster.** 1996. A low-fibronectin-binding mutant of *Staphylococcus aureus* 879R4S has Tn*918* inserted into its single *fnb* gene. *Microbiology* **142:**2153–2160.

1293. **Greene, J. D., and T. R. Klaenhammer.** 1994. Factors involved in adherence of lactobacilli to human Caco-2 cells. *Appl. Environ. Microbiol.* **60:**4487–4494.

1294. **Greene, R. T., C. Lammler, and M. Schmitt.** 1992. Surface hydrophobicity of *Staphylococcus intermedius* and *Staphylococcus hyicus*. *Res. Vet. Sci.* **52:**90–96.

1295. **Greenfeld, J. I., L. Sampath, S. J. Popilskis, S. R. Brunnert, S. Stylianos, and S. Modak.** 1995. Decreased bacterial adherence and biofilm formation on chlorhexidine and silver sulfadiazine-impregnated central venous catheters implanted in swine. *Crit. Care Med.* **23:**894–900.

1296. **Greiffenberg, L., W. Goebel, K. S. Kim, J. Daniels, and M. Kuhn.** 2000. Interaction of *Listeria monocytogenes* with human brain microvascular endothelial cells: an electron microscopic study. *Infect. Immun.* **68:**3275–3279.

1297. **Grenier, D.** 1992. Demonstration of a bimodal coaggregation reaction between *Porphyromonas gingivalis* and *Treponema denticola*. *Oral Microbiol. Immunol.* **7:**280–284.

1298. **Grenier, D.** 1992. Further evidence for a possible role of trypsin-like activity in the adherence of *Porphyromonas gingivalis*. *Can. J. Microbiol.* **38:**1189–1192.

1299. **Grewal, H. M., W. Gaastra, A.-M. Svennerholm, J. Roli, and H. Sommerfelt.** 1993. Induction of colonization factor antigen I (CFA/I) and coli surface antigen 4 (CS4) of enterotoxigenic *Escherichia coli*:relevance for vaccine production. *Vaccine* **11:**221–226.

1300. **Grewal, H. M. S., H. Valvatne, M. K. Bhan, L. van Dijk, W. Gaastra, and H. Sommerfelt.** 1997. A new putative fimbrial colonization factor, CS19, of human enterotoxigenic *Escherichia coli*. *Infect. Immun.* **65:**507–513.

1301. **Grey, P. A., and S. M. Kirov.** 1993. Adherence to HEp-2 cells and enteropathogenic potential of *Aeromonas* spp. *Epidemiol. Infect.* **110:**279–287.

1302. **Griffiss, J. M., C. J. Lammel, J. Wang, N. P. Dekker, and G. F. Brooks.** 1999. *Neisseria gonorrhoeae* coordinately uses Pili and Opa to activate HEC-1-B cell microvilli, which causes engulfment of the gonococci. *Infect. Immun.* **67:**3469–3480.

1303. **Grimprel, E., P. Begue, I. Anjak, E. Njamkepo, P. Francois, and N. Guiso.** 1996. Long-term human serum antibody responses after immunization with whole-cell pertussis vaccine in France. *Clin. Diagn. Lab. Immunol.* **3:**93–97.

1304. **Gripenberg-Lerche, C., M. Skurnik, and P. Toivanen.** 1995. Role of YadA-mediated collagen binding in arthritogenicity of *Yersinia enterocolitica* serotype O:8: experimental studies with rats. *Infect. Immun.* **63:**3222–3226.

1305. **Gristina, A. G., G. Giridhar, B. L. Gabriel, N. T. Naylor, and Q. N. Myrvik.** 1993. Cell biology and molecular mechanisms in artificial device infections. *Int. J. Artif. Organs* **16:**755–763.

1306. **Grönroos, L., M. Saarela, J. Mättö, U. Tanner-Salo, A. Vuorela, and S. Alaluusua.** 1998. Mutacin production by *Streptococcus mutans* may promote transmission of bacteria from mother to child. *Infect. Immun.* **66:**2595–2600.

1307. **Gross, M., S. E. Cramton, F. Götz, and A. Peschel.** 2001. Key role of teichoic acid net charge in *Staphylococcus aureus* colonization of artificial surfaces. *Infect. Immun.* **69:**3423–3426.

1308. **Grossner-Schreiber, B., M. Griepentrog, I. Haustein, W. D. Muller, K. P. Lange, H. Briedigkeit, and U. B. Gobel.** 2001. Plaque formation on surface modified dental implants. An *in vitro* study. *Clin. Oral Implants Res.* **12:**543–551.

1309. **Grover, V., S. Ghosh, N. Sharma, A. Chakraborti, S. Majumdar, and N. K. Ganguly.** 2001. Characterization of a galactose specific adhesin of enteroaggregative *Escherichia coli*. *Arch. Biochem. Biophys.* **390:**109–118.

1310. **Grund, S., and H. Stolpe.** 1992. Adhesion of *Salmonella typhimurium* var. *copenhagen* in the intestines of pigeons. *Int. J. Food Microbiol.* **15:**299–306.

1311. **Gryllos, I., J. G. Shaw, R. Gavin, S. Merino, and J. M. Tomas.** 2001. Role of *flm* locus in mesophilic *Aeromonas* species adherence. *Infect. Immun.* **69:**65–74.

1312. **Guan, Y. H., T. de Graaf, D. L. Lath, S. M. Humphreys, I. Marlow, and A. H. Brook.** 2001. Selection of oral microbial adhesion antagonists using biotinylated *Streptococcus sanguis* and a human mixed oral microflora. *Arch. Oral Biol.* **46:**129–138.

1313. **Guédin, S., E. Willery, C. Locht, and F. Jacob-Dubuisson.** 1998. Evidence that a globular conformation is not compatible with FhaC-mediated secretion of the *Bordetella per-*

tussis filamentous haemagglutinin. *Mol. Microbiol.* **29**:763–774.

1314. **Guerin, G., Y. Duval-Iflah, M. Bonneau, M. Bertaud, P. Guillaume, and L. Ollivier.** 1993. Evidence for linkage between K88ab, K88ac intestinal receptors to *Escherichia coli* and transferrin loci in pigs. *Anim. Genet.* **24**:393–396.

1315. **Guhathakurta, B., D. Sasmal, and A. Datta.** 1992. Adhesion of *Shigella dysenteriae* type 1 and *Shigella flexneri* to guinea-pig colonic epithelial cells *in vitro*. *J. Med. Microbiol.* **36**:403–405.

1316. **Guhathakurta, B., D. Sasmal, A. N. Ghosh, R. Kumar, P. Saha, D. Biswas, D. Khetawat, and A. Datta.** 1999. Adhesion and invasion of a mutant *Shigella flexneri* to an eukaryotic cell line in absence of the 220-kb virulence plasmid. *FEMS Microbiol. Lett.* **181**:267–275.

1317. **Guhathakurta, B., D. Sasma, A. N. Ghosh, C. R. Pal, and A. Datta.** 1996. Purification of a cell-associated hemagglutinin from *Shigella dysenteriae* type 1. *FEMS Immunol. Med. Microbiol.* **14**:63–66.

1318. **Guhathakurta, B., D. Sasmal, S. Pal, S. Chakraborty, G. B. Nair, and A. Datta.** 1999. Comparative analysis of cytotoxin, hemolysin, hemagglutinin and exocellular enzymes among clinical and environmental isolates of *Vibrio cholerae* O139 and non-O1, non-O139. *FEMS Microbiol. Lett.* **179**:401–407.

1319. **Guhathakurta, B., D. Sasmal, D. Dutta, C. R. Pal, and A. Datta.** 1996. Antibody mediated inhibition of adhesion of *Shigella dysenteriae* 1 to guinea pig colonic epithelial cells *in vitro*. *Microbios* **87**:249–256.

1320. **Guignot, J., M. F. Bernet-Camard, C. Pous, L. Plancon, C. Le Bouguenec, and A. L. Servin.** 2001. Polarized entry of uropathogenic Afa/Dr diffusely adhering *Escherichia coli* strain IH11128 into human epithelial cells: evidence for alpha5beta1 integrin recognition and subsequent internalization through a pathway involving caveolae and dynamic unstable microtubules. *Infect. Immun.* **69**:1856–1868.

1321. **Guignot, J., J. Breard, M. F. Bernet-Camard, I. Peiffer, B. J. Nowicki, A. L. Servin, and A. B. Blanc-Potard.** 2000. Pyelonephritogenic diffusely adhering *Escherichia coli* EC7372 harboring Dr-II adhesin carries classical uropathogenic virulence genes and promotes cell lysis and apoptosis in polarized epithelial Caco-2/TC7 cells. *Infect. Immun.* **68**:7018–7027.

1322. **Guignot, J., I. Peiffer, M. F. Bernet-Camard, D. M. Lublin, C. Carnoy, S. L. Moseley, and A. L. Servin.** 2000. Recruitment of CD55 and CD66e brush border-associated glycosylphosphatidylinositol-anchored proteins by members of the Afa/Dr diffusely adhering family of *Escherichia coli* that infect the human polarized intestinal Caco-2/TC7 cells. *Infect. Immun.* **68**:3554–3563.

1323. **Gulati, B. R., V. K. Sharma, and A. K. Taku.** 1992. Occurrence and enterotoxigenicity of F17 fimbriae bearing *Escherichia coli* from calf diarrhoea. *Vet. Rec.* **131**:348–349.

1324. **Gunther, N. W., IV, V. Lockatell, D. E. Johnson, and H. L. Mobley.** 2001. In vivo dynamics of type 1 fimbria regulation in uropathogenic *Escherichia coli* during experimental urinary tract infection. *Infect. Immun.* **69**:2838–2846.

1325. **Gunzburg, S. T., B. J. Chang, S. J. Elliott, V. Burke, and M. Gracey.** 1993. Diffuse and enteroaggregative patterns of adherence of enteric *Escherichia coli* isolated from aboriginal children from the Kimberley region of Western Australia. *J. Infect. Dis.* **167**:755–758.

1326. **Gunzburg, S. T., N. G. Tornieporth, and L. W. Riley.** 1995. Identification of enteropathogenic *Escherichia coli* by PCR-based detection of the bundle-forming pilus gene. *J. Clin. Microbiol.* **33**:1375–1377.

1327. **Guo, B. P., S. J. Norris, L. C. Rosenberg, and M. Höök.** 1995. Adherence of *Borrelia burgdorferi* to the proteoglycan decorin. *Infect. Immun.* **63**:3467–3472.

1328. **Guo, M., H. Reynolds, M. Stinson, and E. de Nardin.** 2000. Isolation and characterization of a human neutrophil aggregation defective mutant of *Fusobacterium nucleatum*. *FEMS Immunol. Med. Microbiol.* **27**:241–246.

1329. **Guo, W., R. Andersson, R. Odselius, Å. Ljungh, T. Wadström, and S. Bengmark.** 1993. Phospholipid impregnation of abdominal rubber drains: resistance to bacterial adherence but no effect on drain-induced bacterial translocation. *Res. Exp. Med.* **193**:285–296.

1330. **Guo, W., R. Andersson, Å. Ljungh, H. Parsson, K. Johansson, and S. Bengmark.** 1994. Orally administered phospholipids inhibit abdominal rubber-drain-induced bacterial translocation in the rat. *Digestion* **55**:417–424.

1331. **Guo, W., R. Andersson, W. Shen, Å. Ljungh, S. Bengmark, and T. Wadström.** 1993. Binding of *Escherichia coli* to Penrose rubber drains—an *in vitro* study. *Zentbl. Bakteriol.* **278**:73–82.

1332. **Gupta, R., N. K. Ganguly, V. Ahuja, K. Joshi, and S. Sharma.** 1995. An ascending non-obstructive model for chronic pyelonephritis in BALB/c mice. *J. Med. Microbiol.* **43**:33–36.

1333. **Gupta, R., S. Gupta, and N. K. Ganguly.** 1997. Role of type-1 fimbriae in the pathogenesis of chronic pyelonephritis in relation to reactive oxygen species. *J. Med. Microbiol.* **46**:403–406.

1334. **Gupta, S. K., S. A. Masinick, J. A. Hobden, R. S. Berk, and L. D. Hazlett.** 1996. Bacterial proteases and adherence of *Pseudomonas aeruginosa* to mouse cornea. *Exp. Eye Res.* **62**:641–650.

1335. **Gupta, S. K., R. S. Berk, S. Masinick, and L. D. Hazlett.** 1994. Pili and lippolysaccharide of *Pseudomonas aeruginosa* bind to the glycolipid asialo GM1. *Infect. Immun.* **62**:4572–4579.

1336. **Gupta, S. K., S. Masinick, M. Garrett, and L. D. Hazlett.** 1997. *Pseudomonas aeruginosa* lipopolysaccharide binds galectin-3 and other human corneal epithelial proteins. *Infect. Immun.* **65**:2747–2753.

1337. **Gurgan, S., S. Bolay, and R. Alacam.** 1997. *In vitro* adherence of bacteria to bleached or unbleached enamel surfaces. *J. Oral Rehab.* **24**:624–627.

1338. **Guruge, J. L., P. G. Falk, R. G. Lorenz, M. Dans, H. P. Wirth, M. J. Blaser, D. E. Berg, and J. I. Gordon.** 1998. Epithelial attachment alters the outcome of *Helicobacter pylori* infection. *Proc. Natl. Acad. Sci. USA* **95**:3925–3930.

1339. **Gusils, C., J. Palacios, S. Gonzalez, and G. Oliver.** 1999. Lectin-like protein fractions in lactic acid bacteria isolated from chickens. *Biol. Pharm. Bull.* **22**:11–15.

1340. **Guth, B. E., E. G. Aguiar, P. M. Griffin, S. R. Ramos, and T. A. Gomes.** 1994. Prevalence of colonization factor antigens (CFAs) and adherence to HeLa cells in enterotoxigenic *Escherichia coli* isolated from feces of children in Sao Paulo. *Microbiol. Immunol.* **38**:695–701.

1341. **Guy, J. S., L. G. Smith, J. J. Breslin, J. P Vaillancourt, and H. J. Barnes.** 2000. High mortality and growth depression experimentally produced in young turkeys by dual infection with enteropathogenic *Escherichia coli* and turkey coronavirus. *Avian Dis.* **44**:105–113.

1342. **Guzman, C. A., M. J. Walker, M. Rohde, and K. N. Timmis.** 1992. Expression of *Bordetella pertussis* filamentous hemagglutinin in *Escherichia coli* using a two cistron system. *Microb. Pathog.* **12**:383–389.

1343. **Guzman, C. A., M. Rohde, M. Bock, and K. N. Timmis.** 1994. Invasion and intracellular survival of *Bordetella bronchiseptica* in mouse dendritic cells. *Infect. Immun.* **62**:5528–5537.

1344. **Guzman, C. A., M. Rohde, and K. N. Timmis.** 1994. Mechanisms involved in uptake of *Bordetella bronchiseptica* by mouse dendritic cells. *Infect. Immun.* **62**:5538–5544.

1345. **Guzman, C. A., G. Piatti, L. H. Staendner, F. Biavasco, and C. Pruzzo.** 1995. Export of *Bordetella pertussis* serotype 2 and 3 fimbrial subunits by *Escherichia coli.* *FEMS Microbiol. Lett.* **128**:189–194.

1346. **Guzman, C. A., F. Biavasco, and C. Pruzzo.** 1997. News & notes: adhesiveness of *Bacteroides fragilis* strains isolated from feces of healthy donors, abscesses, and blood. *Curr. Microbiol.* **34**:332–334.

1347. **Guzman-Murillo, M., M. L. Merino-Contreras, and F. Ascencio.** 2000. Interaction between *Aeromonas veronii* and epithelial cells of spotted sand bass (*Paralabrax maculatofasciatus*) in culture. *J. Appl. Microbiol.* **88**:897–906.

1348. **Guzman-Murillo, M. A., E. Ruiz-Bustos, B. Ho, and F. Ascencio.** 2001. Involvement of the heparan sulphate-binding proteins of *Helicobacter pylori* in its adherence to HeLa S3 and Kato III cell lines. *J. Med. Microbiol.* **50**:320–329.

1349. **Haapasalo, M., P. Hannam, B. C. McBride, and V. J. Uitto.** 1996. Hyaluronan, a possible ligand mediating *Treponema denticola* binding to periodontal tissue. *Oral Microbiol. Immunol.* **11**:156–160.

1350. **Haapasalo, M., K.-H. Müller, V.-J. Uitto, W. K. Leung, and B. C. McBride.** 1992. Characterization, cloning, and binding properties of the major 53-kilodalton *Treponema denticola* surface antigen. *Infect. Immun.* **60**:2058–2065.

1351. **Haas, R., S. Veit, and T. F. Meyer.** 1992. Silent pilin genes of *Neisseria gonorrhoeae* MS11 and the occurrence of related hypervariant sequences among other gonococcal isolates. *Mol. Microbiol.* **6**:197–208.

1352. **Haataja, S., K. Tikkanen, J. Liukkonen, C. François-Gerard, and J. Finne.** 1993. Characterization of a novel bacterial adhesion specificity of *Streptococcus suis* recognizing blood group P receptor oligosaccharides. *J. Biol. Chem.* **268**:4311–4317.

1353. **Haataja, S., K. Tikkanen, U. Nilsson, G. Magnusson, K.-A. Karlsson, and J. Finne.** 1994. Oligosaccharide-receptor interaction of the Galα1–4Gal binding adhesin of

Streptococcus suis. Combining site architecture and characterization of two variant adhesin specificities. *J. Biol. Chem.* **269:**27466–27472.

1354. **Haataja, S., Z. Zhang, K. Tikkanen, G. Magnusson, and J. Finne.** 1999. Determination of the cell adhesion specificity of *Streptococcus suis* with the complete set of monodeoxy analogues of globotriose. *Glycoconj. J.* **16:**67–71.

1355. **Habash, M. B., H. C. van der Mei, H. J. Busscher, and G. Reid.** 1999. The effect of water, ascorbic acid, and cranberry derived supplementation on human urine and uropathogen adhesion to silicone rubber. *Can. J. Microbiol.* **45:**691–694.

1356. **Habash, M. B., H. C. van der Mei, G. Reid, and H. J. Busscher.** 1997. Adhesion of *Pseudomonas aeruginosa* to silicone rubber in a parallel plate flow chamber in the absence and presence of nutrient broth. *Microbiology* **143:**2569–2574.

1357. **Hachicha, M., P. Rathanaswami, P. H. Naccache, and S. R. McColl.** 1998. Regulation of chemokine gene expression in human peripheral blood neutrophils phagocytosing microbial pathogens. *J. Immunol.* **160:**449–454.

1358. **Hacker, J.** 1992. Role of fimbrial adhesins in the pathogenesis of *Escherichia coli* infections. *Can. J. Microbiol.* **38:**720–727.

1359. **Haesebrouck, F., K. Chiers, I. Van Overbeke, and R. Ducatelle.** 1997. *Actinobacillus pleuropneumoniae* infections in pigs: the role of virulence factors in pathogenesis and protection. *Vet. Microbiol.* **58:**239–249.

1360. **Hagiwara, S. I., M. Takeya, H. Suzuki, T. Kodama, L. J. van der Laan, G. Kraal, N. Kitamura, and K. Takahashi.** 1999. Role of macrophage scavenger receptors in hepatic granuloma formation in mice. *Am. J. Pathol.* **154:**705–720.

1361. **Hagman, M. M., J. B. Dale, and D. L. Stevens.** 1999. Comparison of adherence to and penetration of a human laryngeal epithelial cell line by group A streptococci of various M protein types. *FEMS Immunol. Med. Microbiol.* **23:**195–204.

1362. **Hahn, H. P.** 1997. The type-4 pilus is the major virulence-associated adhesin of *Pseudomonas aeruginosa*—a review. *Gene* **192:**99–108.

1363. **Hahn, T. W., M. J. Willby, and D. C. Krause.** 1998. HMW1 is required for cytadhesin P1 trafficking to the attachment organelle in *Mycoplasma pneumoniae*. *J. Bacteriol.* **180:**1270–1276.

1364. **Haider, K., S. M. Faruque, M. J. Albert, S. Nahar, P. K. Neogi, and A. Hossain.** 1992. Comparison of a modified adherence assay with existing assay methods for identification of enteroaggregative *Escherichia coli*. *J. Clin. Microbiol.* **30:**1614–1616.

1365. **Hajishengallis, G., S. K. Hollingshead, T. Koga, and M. W. Russell.** 1995. Mucosal immunization with a bacterial protein antigen genetically coupled to cholera toxin A2/B subunits. *J. Immunol.* **154:**4322–4332.

1366. **Hajishengallis, G., E. Harokopakis, S. K. Hollingshead, M. W. Russell, and S. M. Michalek.** 1996. Construction and oral immunogenicity of a *Salmonella typhimurium* strain expressing a streptococcal adhesin linked to the A2/B subunits of cholera toxin. *Vaccine* **14:**1545–1548.

1367. **Hajishengallis, G., E. Nikolova, and M. W. Russell.** 1992. Inhibition of *Streptococcus mutans* adherence to saliva-coated hydroxyapatite by human secretory immunoglobulin A (S-IgA) antibodies to cell surface protein antigen I/II: reversal by IgA1 protease cleavage. *Infect. Immun.* **60:**5057–5064.

1368. **Hajishengallis, G., M. W. Russell, and S. M. Michalek.** 1998. Comparison of an adherence domain and a structural region of *Streptococcus mutans* antigen I/II in protective immunity against dental caries in rats after intranasal immunization. *Infect. Immun.* **66:**1740–1743.

1369. **Hajishengallis, G., T. Koga, and M. W. Russell.** 1994. Affinity and specificity of the interactions between *Streptococcus mutans* antigen I/II and salivary components. *J. Dent. Res.* **73:**1493–1502.

1370. **Hakansson, A., I. Carlstedt, J. Davies, A. K. Mossberg, H. Sabharwal, and C. Svanborg.** 1996. Aspects on the interaction of *Streptococcus pneumoniae* and *Haemophilus influenzae* with human respiratory tract mucosa. *Am. J. Respir. Crit. Care Med.* **154:**S187–S191.

1371. **Hakansson, A., A. Kidd, G. Wadell, H. Sabharwal, and C. Svanborg.** 1994. Adenovirus infection enhances in vitro adherence of *Streptococcus pneumoniae*. *Infect. Immun.* **62:**2707–2714.

1372. **Hakansson, S., K. Schesser, C. Persson, E. E. Galyov, R. Rosqvist, F. Homble, and H. Wolf-Watz.** 1996. The YopB protein of *Yersinia pseudotuberculosis* is essential for the translocation of Yop effector proteins across the target cell plasma membrane and displays a contact-dependent membrane disrupting activity. *EMBO J.* **15:**5812–5823.

1373. **Hale, W. B., M. W. van der Woude, B.**

A. Braaten, and D. A. Low. 1998. Regulation of uropathogenic *Escherichia coli* adhesin expression by DNA methylation. *Mol. Genet. Metab.* **65**:191–196.

1374. Hallberg, K., C. Holm, K. J. Hammarstrom, S. Kalfas, and N. Strömberg. 1998. Ribotype diversity of *Actinomyces* with similar intraoral tropism but different types of N-acetyl-β-D-galactosamine binding specificity. *Oral Microbiol. Immunol.* **13**:188–192.

1375. Hallberg, K., C. Holm. U. Öhman, and N. Strömberg. 1998. *Actinomyces naeslundii* displays variant *fimP* and *fimA* fimbrial subunit genes corresponding to different types of acidic proline-rich protein and β-linked galactosamine binding specificity. *Infect. Immun.* **66**:4403–4410.

1376. Hallberg, K., K. J. Hammarstrom, E. Falsen, G. Dahlen, R. J. Gibbons, D. I. Hay, and N. Strömberg. 1998. *Actinomyces naeslundii* genospecies 1 and 2 express different binding specificities to N-acetyl-β-D-galactosamine, whereas *Actinomyces odontolyticus* expresses a different binding specificity in colonizing the human mouth. *Oral Microbiol. Immunol.* **13**:327–336.

1377. Haller, D., C. Bode, W. P. Hammes, A. M. Pfeifer, E. J. Schiffrin, and S. Blum. 2000. Non-pathogenic bacteria elicit a differential cytokine response by intestinal epithelial cell/leucocyte co-cultures. *Gut* **47**:79–87.

1378. Halperin, S. A., L. Barreto, B. Friesen, and W. Meekison. 1994. Immunogenicity of a five-component acellular pertussis vaccine in infants and young children. *Arch. Pediatr. Adolesc. Med.* **148**:495–502.

1379. Halperin, S. A., B. Eastwood, L. Barreto, E. Mills, M. Blatter, K. Reisinger, G. Bader, H. Keyserling, E. A. Roberts, R. Guasparini, et al. 1995. Safety and immunogenicity of two acellular pertussis vaccines with different pertussis toxoid and filamentous hemagglutinin content in infants 2–6 months old. *Scand. J. Infect. Dis.* **27**:279–287.

1380. Halvorsen, T., H. Valvatne, H. M. Grewal, W. Gaastra, and H. Sommerfelt. 1997. Expression of colonization factor antigen I fimbriae by enterotoxigenic *Escherichia coli*: influence of growth conditions and a recombinant positive regulatory gene. *APMIS* **105**:247–254.

1381. Hamada, N., K. Watanabe, C. Sasakawa, M. Yoshikawa, F. Yoshimura, and T. Umemoto. 1994. Construction and characterization of a *fimA* mutant of *Porphyromonas gingivalis*. *Infect. Immun.* **62**:1696–1704.

1382. Hamada, N., H. T. Sojar, M. I. Cho, and R. J. Genco. 1996. Isolation and characterization of a minor fimbria from *Porphyromonas gingivalis*. *Infect. Immun.* **64**:4788–4794.

1383. Hamada, S., T. Ogawa, H. Shimauchi, and Y. Kasumoto. 1992. Induction of mucosal and serum immmune responses to a specific antigen of periodontal bacteria. *Adv. Exp. Med. Biol.* **327**:71–81.

1384. Hamada, S., T. Fujiwara, S. Morishima, I. Takahashi, I. Nakagawa, S. Kimura, and T. Ogawa. 1994. Molecular and immunological characterization of the fimbriae of *Porphyromonas gingivalis*. *Microbiol. Immunol.* **38**:921–930.

1385. Hamada, S., T. Ogawa, and Y. Kusumoto. 1995. Generation and distribution of antibody secreting cells in salivary glands of mice immunized with *Porphyromonas gingivalis* fimbriae. *Adv. Exp. Med. Biol.* **371**:1113–1117.

1386. Hamada S., A. Amano, S. Kimura, I. Nakagawa, S. Kawabata, and I. Morisaki. 1998. The importance of fimbriae in the virulence and ecology of some oral bacteria. *Oral Microbiol. Immunol.* **13**:129–138.

1387. Hamadeh, R. M., M. M. Estabrook, P. Zhou, G. A. Jarvis, and J. M. Griffiss. 1995. Anti-Gal binds to pili of *Neisseria meningitidis*:the immunoglobulin A isotype blocks complement-mediated killing. *Infect. Immun.* **63**:4900–4906.

1388. Hamilton, A. J., L. Jeavons, S. Youngchim, N. Vanittanakom, and R. J. Hay. 1998. Sialic acid-dependent recognition of laminin by *Penicillium marneffei* conidia. *Infect. Immun.* **66**:6024–6026.

1389. Hamilton, R. C., J. Bennet, D. Drane, E. Pietrzykowski, F. Seddon, A. Stefancic, and J. Cox. 1994. Negative staining can cause clumping of *Bordetella pertussis* fimbriae. *Micron* **25**:613–615.

1390. Hammerschmidt, S., R. Hilse, J. P. van Putten, R. Gerardy-Schahn, A. Unkmeir, and M. Frosch. 1996. Modulation of cell surface sialic acid expression in *Neisseria meningitidis* via a transposable genetic element. *EMBO J.* **15**:192–198.

1391. Hamrick, T. S., S. L. Harris, P. A. Spears, E. A. Havell, J. R. Horton, P. W. Russell, and P. E. Orndorff. 2000. Genetic characterization of *Escherichia coli* type 1 pilus adhesin mutants and identification of a novel binding phenotype. *J. Bacteriol.* **182**: 4012–4021.

1392. Hamrick, T. S., E. A. Havell, J. R. Horton, and P. E. Orndorff. 2000. Host and bacterial factors involved in the innate ability of mouse macrophages to eliminate internalized unopsonized *Escherichia coli*. *Infect. Immun.* **68:**125–132.

1393. **Han, N., G. Lepine, J. Whitlock, L. Wojciechowski, and A. Progulske-Fox.** 1998. The *Porphyromonas gingivalis prtP/kgp* homologue exists as two open reading frames in strain 381. *Oral Dis.* **4:**170–179.

1394. **Han, N., J. Whitlock, and A. Progulske-Fox.** 1996. The hemagglutinin gene A (*hagA*) of *Porphyromonas gingivalis* 381 contains four large, contiguous, direct repeats. *Infect. Immun.* **64:**4000–4007.

1395. **Han, Y. W., W. Shi, G. T. Huang, S. Kinder Haake, N. H. Park, H. Kuramitsu, and R. J. Genco.** 2000. Interactions between periodontal bacteria and human oral epithelial cells: *Fusobacterium nucleatum* adheres to and invades epithelial cells. *Infect. Immun.* **68:**3140–3146.

1396. **Han, Y. W., and V. L. Miller.** 1997. Reevaluation of the virulence phenotype of the *inv yadA* double mutants of *Yersinia pseudotuberculosis*. *Infect. Immun.* **65:**327–330.

1397. **Hanazawa, S., Y. Murakami, A. Takeshita, H. Kitami, K. Ohta, S. Amano, and S. Kitano.** 1992. *Porphyromonas gingivalis* fimbriae induce expression of the neutrophil chemotactic factor KC gene of mouse peritoneal macrophages: role of protein kinase C. *Infect. Immun.* **60:**1544–1549.

1398. **Hanazawa, S., Y. Kawata, Y. Murakami, K. Naganuma, S. Amano, Y. Miyata, and S. Kitano.** 1995. *Porphyromonas gingivalis* fimbria-stimulated bone resorption in vitro is inhibited by a tyrosine kinase inhbitor. *Infect. Immun.* **63:**2374–2377.

1399. **Hancox, L. S., K. S. Yeh, and S. Clegg.** 1997. Construction and characterization of type 1 non-fimbriate and non-adhesive mutants of *Salmonella typhimurium*. *FEMS Immunol. Med. Microbiol.* **19:**289–296.

1400. **Hando, S., Y. Nakano, N. Nonomura, H. Niijima, and O. Fujioka.** 1993. Adherence of *Haemophilus influenzae* to middle ear mucosa injured by endotoxin. *In* D. J. Lim, C. D. Bluestone, J. O. Klein, J. D. Nelson, and P. L. Ogra (ed.), *Recent Advances in Otitis Media*. Decker Periodicals, Philadelphia, Pa.

1401. **Hanes, D. E., and D. K. Chandler.** 1993. The role of a 40-megadalton plasmid in the adherence and hemolytic properties of *Aeromonas hydrophila*. *Microb. Pathog.* **15:**313–317.

1402. **Hanisch, F. G., J. Hacker, and H. Schroten.** 1993. Specificity of S fimbriae on recombinant *Escherichia coli*: preferential binding to gangliosides expressing NeuGcα(2–3)Gal and NeuAcα(2–8)NeuAc. *Infect. Immun.* **61:**2108–2115.

1403. Reference deleted.

1404. **Hannah, J. H., F. D. Menozzi, G. Renauld, C. Locht, and M. J. Brennan.** 1994. Sulfated glycoconjugate receptors for the *Bordetella pertussis* adhesin filamentous hemagglutinin (FHA) and mapping of the heparin-binding domain on FHA. *Infect. Immun.* **62:**5010–5019.

1405. **Hannig, M.** 1999. Transmission electron microscopy of early plaque formation on dental materials *in vivo*. *Eur. J. Oral Sci.* **107:**55–64.

1406. **Hansen, H. C., and G. Magnusson.** 1998. Synthesis of some amino and carboxy analogs of galabiose: evaluation as inhibitors of the pilus protein PapGJ96 from *Escherichia coli*. *Carbohydr. Res.* **307:**233–242.

1407. **Hanski, E., and M. Caparon.** 1992. Protein F, a fibronectin-binding protein, is an adhesin of the group A streptococcus *Streptococcus pyogenes*. *Proc. Natl. Acad. Sci. USA* **89:**6172–6176.

1408. **Hanski, E., G. Fogg, A. Tovi, N. Okada, I. Burstein, and M. Caparon.** 1995. Molecular analysis of *Streptococcus pyogenes* adhesion. *Methods Enzymol.* **253:**269–305.

1409. **Hanski, E., P. A. Horwitz, and M. G. Caparon.** 1992. Expression of protein F, the fibronectin-binding protein of *Streptococcus pyogenes* JRS4, in heterologous streptococcal and enterococcal strains promotes their adherence to respiratory epithelial cells. *Infect. Immun.* **60:**5119–5125.

1410. **Hansson, L., P. Wallbrandt, J.-O. Andersson, M. Byström, A. Bäckman, A. Carlstein, K. Enquist, H. Lönn, C. Otter, and M. Strömqvist.** 1995. Carbohydrate specificity of the *Escherichia coli* P-pilus papG protein is mediated by its N-terminal part. *Biochim. Biophys. Acta* **1244:**377–383.

1411. **Haque, M. A., F. Qadri, K. Ohki, and O. Kohashi.** 1995. Surface components of shigellae involved in adhesion and haemagglutination. *J. Appl. Bacteriol.* **79:**186–194.

1412. **Haraldsson, G., and W. P. Holbrook.** 1998. A hemagglutinating variant of *Prevotella melaninogenica* isolated from the oral cavity. *Oral Microbiol. Immunol.* **13:**362–367.

1413. **Harano, K., A. Yamanaka, and K. Okuda.** 1995. An antiserum to a synthetic fimbrial peptide of *Actinobacillus actinomycetemcomitans* blocked adhesion of the microorganism. *FEMS Microbiol. Lett.* **130:**279–286.

1414. **Haraoka, M., T. Matsumoto, Y. Mizunoe, S. Kubo, K. Takahashi, N. Ogata, M. Tanaka, and J. Kumazawa.** 1993. Effect of recombinant human granulo-cyte-colony-stimulating factor on renal scarring following infection with MS-piliated bacteria. *Renal Failure* **15:**141–148.

1415. **Harding, A. S., C. Lakkis, and N. A. Brennan.** 1995. The effects of short-term contact lens wear on adherence of *Pseudomonas aeruginosa* to human corneal cells. *J. Am. Optom. Assoc.* **66:**775–779.

1416. **Hardy, S. J., M. Christodoulides, R. O. Weller, and J. E. Heckels.** 2000. Interactions of *Neisseria meningitidis* with cells of the human meninges. *Mol. Microbiol.* **36:**817–829.

1417. **Harel, J., C. Forget, J. Saint-Amand, F. Daigle, D. Dubreuil, M. Jacques, and J. Fairbrother.** 1992. Molecular cloning of a determinant coding for fimbrial antigen F165(1), a Prs-like fimbrial antigen from porcine septicaemic *Escherichia coli. J. Gen. Microbiol.* **138:**1495–1502.

1418. **Harel, J., C. Forget, M. Ngeleka, M. Jacques, and J. M. Fairbrother.** 1992. Isolation and characterization of adhesin-defective Tn*phoA* mutants of septicaemic porcine *Escherichia coli* of serotype O115:-K⁻:F165. *J. Gen. Microbiol.* **138:**2337–2345.

1419. **Harel, J., M. Jacques, J. M. Fairbrother, M. Bossé, and C. Forget.** 1995. Cloning of determinants encoding F165₂ fimbriae from porcine septicaemic *Escherichia coli* confirms their identity as F1C fimbriae. *Microbiology* **141:**221–228.

1420. **Harkes, G., J. Dankert, and J. Feijen.** 1992. Bacterial migration along solid surfaces. *Appl. Environ. Microbiol.* **58:**1500–1505.

1421. **Harper, J. J., J. G. McCormack, M. H. Tilse.** 1995. Adherence of *Haemophilus influenzae* to dacron fibres: significance of capsule and biotype. *J. Microbiol. Methods* **24:**101–110.

1422. **Hartford, O., L. O'Brien, K. Schofield, J. Wells, and T. J. Foster.** 2001. The Fbe (SdrG) protein of *Staphylococcus epidermidis* HB promotes bacterial adherence to fibrinogen. *Microbiology* **147:**2545–2552.

1423. **Hartford, O., D. McDevitt, and T. J. Foster.** 1999. Matrix-binding proteins of *Staphylococcus aureus:* functional analysis of mutant and hybrid molecules. *Microbiology* **145:**2497–2505.

1424. **Hartland, E. L., M. Batchelor, R. M. Delahay, C. Hale, S. Matthews, G.** Dougan, S. Knutton, I. Connerton, and G. Frankel. 1999. Binding of intimin from enteropathogenic *Escherichia coli* to Tir and to host cells. *Mol. Microbiol.* **32:**151–158.

1425. **Hartland, E. L., S. P. Green, W. A. Phillips, and R. M. Robins-Browne.** 1994. Essential role of YopD in inhibition of the respiratory burst of macrophages by *Yersinia enterocolitica. Infect. Immun.* **62:**4445–4453.

1426. **Hartland, E. L., A. S. Mikosza, R. M. Robins-Browne, and D. J. Hampson.** 1998. Examination of *Serpulina pilosicoli* for attachment and invasion determinants of *Enterobacteria. FEMS Microbiol. Lett.* **165:**59–63.

1427. **Hartleib, J., N. Kohler, R. B. Dickinson, G. S. Chhatwal, J. J. Sixma, O. M. Hartford, T. J. Foster, G. Peters, B. E. Kehrel, and M. Herrmann.** 2000. Protein A is the von Willebrand factor binding protein on *Staphylococcus aureus. Blood* **96:**2149–2156.

1428. **Hartley, M. G., M. J. Hudson, E. T. Swarbrick, A. E. Gent, M. D. Hellier, and R. H. Grace.** 1993. Adhesive and hydrophobic properties of *Escherichia coli* from the rectal mucosa of patients with ulcerative colitis. *Gut* **34:**63–67.

1429. **Hartman, A. B., and M. M. Venkatesan.** 1998. Construction of a stable attenuated *Shigella sonnei* ΔvirG vaccine strain, WRSS1, and protective efficacy and immunogenicity in the guinea pig keratoconjunctivitis model. *Infect. Immun.* **66:**4572–4576.

1430. **Hartmann, E., and C. Lingwood.** 1997. Brief heat shock treatment induces a long-lasting alteration in the glycolipid receptor binding specificity and growth rate of *Haemophilus influenzae. Infect. Immun.* **65:**1729–1733.

1431. **Harty, D. W., H. J. Oakey, M. Patrikakis, E. B. Hume, and K. W. Knox.** 1994. Pathogenic potential of lactobacilli. *Int. J. Food Microbiol.* **24:**179–189.

1432. **Hase, C. C., M. E. Bauer, and R. A. Finkelstein.** 1994. Genetic characterization of mannose-sensitive hemagglutinin (MSHA)-negative mutants of *Vibrio cholerae* derived by Tn5 mutagenesis. *Gene* **150:**17–25.

1433. **Hasman, H., T. Chakraborty, and P. Klemm.** 1999. Antigen-43-mediated autoaggregation of *Escherichia coli* is blocked by fimbriation. *J. Bacteriol.* **181:**4834–4841.

1434. **Hastie, A. T., L. P. Evans, and A. M. Allen.** 1993. Two types of bacteria adherent to bovine respiratory tract ciliated epithelium. *Vet. Pathol.* **30:**12–19.

1435. **Hasty, D. L., I. Ofek, H. S. Courtney,**

and R. J. Doyle. 1992. Multiple adhesins of streptococci. *Infect. Immun.* **60:**2147–2152.

1436. Hata, Y., T. Kita, and M. Murakami. 1999. Bovine milk inhibits both adhesion of *Helicobacter pylori* to sulfatide and *Helicobacter pylori*-induced vacuolation of vero cells. *Dig. Dis. Sci.* **44:**1696–1702.

1437. Hatz, R. A., G. Rieder, M. Stolte, E. Bayerdorffer, G. Meimarakis, F. W. Schildberg, and G. Enders. 1997. Pattern of adhesion molecule expression on vascular endothelium in *Helicobacter pylori*-associated antral gastritis. *Gastroenterology* **112:**1908–1919.

1438. Hauck, C. R., T. F. Meyer, F. Lang, and E. Gulbins. 1998. CD66-mediated phagocytosis of Opa52 *Neisseria gonorrhoeae* requires a Src-like tyrosine kinase- and Rac1-dependent signalling pathway. *EMBO J.* **17:**443–454.

1439. Hauck, C. R., E. Gulbins, F. Lang, and T. F. Meyer. 1999. Tyrosine phosphatase SHP-1 is involved in CD66-mediated phagocytosis of Opa52-expressing *Neisseria gonorrhoeae*. *Infect. Immun.* **67:**5490–5494.

1440. Hauck, C. R., H. Grassme, J. Bock, V. Jendrossek, K. Ferlinz, T. F. Meyer, and E. Gulbins. 2000. Acid sphingomyelinase is involved in CEACAM receptor-mediated phagocytosis of *Neisseria gonorrhoeae*. *FEBS Lett.* **478:**260–266.

1441. Hauck, C. R., D. Lorenzen, J. Saas, and T.F. Meyer. 1997. An in vitro-differentiated human cell line as a model system to study the interaction of *Neisseria gonorrhoeae* with phagocytic cells. *Infect. Immun.* **65:**1863–1869.

1442. Hawkins, B. W., R. D. Cannon, and H. F. Jenkinson. 1993. Interactions of *Actinomyces naeslundii* strains T14V and ATCC 12104 with saliva, collagen and fibrinogen. *Arch. Oral Biol.* **38:**533–535.

1443. Hay, D. I. 1995. Salivary factors in caries models. *Adv. Dent. Res.* **9:**239–243.

1444. Hayashi, H., A. Nagata, D. Hinode, M. Sato, and R. Nakamura. 1992. Survey of a receptor protein in human erythrocytes for hemagglutinin of *Porphyromonas gingivalis*. *Oral Microbiol. Immunol.* **7:**204–211.

1445. Hayashi, H., M. Morioka, S. Ichimiya, K. Yamato, D. Hinode, A. Nagata, and R. Nakamura. 1993. Participation of an arginyl residue of insulin chain B in the inhibition of hemagglutination by *Porphyromonas gingivalis*. *Oral Microbiol. Immunol.* **8:**386–389.

1446. Hayashi, S., T. Sugiyama, A. Yachi, K. Yokota, Y. Hirai, K. Oguma, and N. Fujii. 1997. Effect of ecabet sodium on *Helicobacter pylori* adhesion to gastric epithelial cells. *J. Gastroenterol.* **32:**593–597.

1447. Hayashi, S., T. Sugiyama, A. Yachi, K. Yokota, Y. Hirai, K. Oguma, and N. Fujii. 1997. A rapid and simple method to quantify *Helicobacter pylori* adhesion to human gastric MKN-28 cells. *J. Gastroenterol. Hepatol.* **12:**373–377.

1448. Hayashi, S., T. Sugiyama, K. Amano, H. Isogai, E. Isogai, M. Aihara, M. Kikuchi, M. Asaka, K. Yokota, K. Oguma, N. Fujii, and Y. Hirai. 1998. Effect of rebamipide, a novel antiulcer agent, on *Helicobacter pylori* adhesion to gastric epithelial cells. *Antimicrob. Agents Chemother.* **42:**1895–1899.

1449. Hayashi, S., T. Sugiyama, M. Asaka, K. Yokota, K. Oguma, and Y. Hirai. 1998. Modification of *Helicobacter pylori* adhesion to human gastric epithelial cells by antiadhesion agents. *Dig. Dis. Sci.* **43:**56S–60S.

1450. Hayashi, S., T. Sugiyama, K. Yokota, H. Isogai, E. Isogai, H. Shimomura, K. Oguma, M. Asaka, and Y. Hirai. 2000. Combined effect of rebamipide and ecabet sodium on *Helicobacter pylori* adhesion to gastric epithelial cells. *Microbiol. Immunol.* **44:** 557–562.

1451. Hazenbos, W. L., C. A. Geuijen, B. M. van den Berg, F. R. Mooi, and R. van Furth. 1995. *Bordetella pertussis* fimbriae bind to human monocytes via the minor fimbrial subunit FimD. *J. Infect. Dis.* **171:**924–929.

1452. Hazenbos, W. L., B. M. van den Berg, C. W. Geuijen, F. R. Mooi, and R. van Furth. 1995. Binding of FimD on *Bordetella pertussis* to very late antigen-5 on monocytes activates complement receptor type 3 via protein tyrosine kinases. *J. Immunol.* **155:**3972–3978.

1453. Hazenbos, W. L., B. M. van den Berg, and R. van Furth. 1993. Very late antigen-5 and complement receptor type 3 cooperatively mediate the interaction between *Bordetella pertussis* and human monocytes. *J. Immunol.* **151:**6274–6282.

1454. Hazenbos, W. L., B. M. van den Berg, J. W. van't Wout, F. R. Mooi, and R. van Furth. 1994. Virulence factors determine attachment and ingestion of nonopsonized and opsonized *Bordetella pertussis* by human monocytes. *Infect. Immun.* **62:**4818–4824.

1455. Hazlett, K. R., S. M. Michalek, and J. A. Banas. 1998. Inactivation of the *gbpA* gene of *Streptococcus mutans* increases virulence and promotes in vivo accumulation of recombinations between the glucosyltransferase B and C genes. *Infect. Immun.* **66:**2180–2185.

1456. Hazlett, L., X. Rudner, S. Masinick, M.

Ireland, and S. Gupta. 1995. In the immature mouse, *Pseudomonas aeruginosa* pili bind a 57-kd (α2–6) sialylated corneal epithelial cell surface protein: a first step in infection. *Investig. Ophthalmol. Visual Sci.* 36:634–643.

1457. Hazlett, L. D. 1995. Analysis of ocular microbial adhesion. *Methods Enzymol.* 253:53–66.

1458. Hazlett, L. D., S. Masinick, R. S. Berk, and Z. Zheng. 1992. Proteinase K decreases *Pseudomonas aeruginosa* adhesion to wounded cornea. *Exp. Eye Res.* 55:579–587.

1459. Hazlett, L. D., S. Masinick, R. Barrett, and K. Rosol. 1993. Evidence for asialo GM_1 as a corneal glycolipid receptor for *Pseudomonas aeruginosa* adhesion. *Infect. Immun.* 61:5164–5173.

1460. Hazlett, L. D., and X. L. Rudner. 1994. Investigations on the role of flagella in adhesion of *Pseudomonas aeruginosa* to mouse and human corneal epithelial proteins. *Ophthalmic Res.* 26: 375–379.

1461. Hbabi-Haddioui, L., and C. Roques. 1997. Inhibition of *Streptococcus pneumoniae* adhesion by specific salivary IgA after oral immunisation with a ribosomal immunostimulant. *Drugs* 54:29–32.

1462. He, F., A. C. Ouwehan, H. Hashimoto, E. Isolauri, Y. Benno, and S. Salminen. 2001. Adhesion of *Bifidobacterium* spp. to human intestinal mucus. *Microbiol. Immunol.* 45:259–262.

1463. He, F., A. C. Ouwehand, E. Isolauri, H. Hashimoto, Y. Benno, and S. Salminen. 2001. Comparison of mucosal adhesion and species identification of bifidobacteria isolated from healthy and allergic infants. *FEMS Immunol. Med. Microbiol.* 30:43–47.

1464. He, Q., M. K. Viljanen, R. M. Olander, H. Bogaerts, D. de Grave, O. Ruuskanen, and J. Mertsola. 1994. Antibodies to filamentous hemagglutinin of *Bordetella pertussis* and protection against whooping cough in schoolchildren. *J. Infect. Dis.* 170:705–708.

1465. He, Q., K. Edelman, H. Arvilommi, and J. Mertsola. 1996. Protective role of immunoglobulin G antibodies to filamentous hemagglutinin and pertactin of *Bordetella pertussis* in *Bordetella parapertussis* infection. *Eur. J. Clin. Microbiol. Infect. Dis.* 15:793–798.

1466. He, Q., N. N. Tran Minh, K. Edelman, M. K. Viljanen, H. Arvilommi, and J. Mertsola. 1998. Cytokine mRNA expression and proliferative responses induced by pertussis toxin, filamentous hemagglutinin, and pertactin of *Bordetella pertussis* in the peripheral blood mononuclear cells of infected and immunized schoolchildren and adults. *Infect. Immun.* 66:3796–3801.

1467. Hechemy, K. E., W. A. Samsonoff, H. L. Harris, and M. McKee. 1992. Adherence and entry of *Borrelia burgdorferi* in Vero cells. *J. Med. Microbiol.* 36:229–238.

1468. Hecht, G., and S. D. Savkovic. 1997. Review article: effector role of epithelia in inflammation—interaction with bacteria. *Aliment. Pharmacol. Ther.* 11:64–68.

1469. Heck, D. V., B. L. Trus, and A. C. Steven. 1996. Three-dimensional structure of *Bordetella pertussis* fimbriae. *J. Struct. Biol.* 116:264–269.

1470. Heczko, U., V. C. Smith, R. Mark Meloche, A. M. Buchan, and B. B. Finlay. 2000. Characteristics of *Helicobacter pylori* attachment to human primary antral epithelial cells. *Microbes Infect.* 2:1669–1676.

1471. Hedlund, M., C. Wachtler, E. Johansson, L. Hang, J. E. Somerville, R. P. Darveau, and C. Svanborg. 1999. P fimbriae-dependent, lipopolysaccharide-independent activation of epithelial cytokine responses. *Mol. Microbiol.* 33:693–703.

1472. Hedreyda, C. T., and D. C. Krause. 1995. Identification of a possible cytadherence regulatory locus in *Mycoplasma pneumoniae*. *Infect. Immun.* 63:3479–3483.

1473. Heffernan, E. J., L. Wu, J. Louie, S. Okamoto, J. Fierer, and D. G. Guiney. 1994. Specificity of the complement resistance and cell association phenotypes encoded by the outer membrane protein genes *rck* from *Salmonella typhimurium* and *ail* from *Yersinia enterocolitica*. *Infect. Immun.* 62:5183–5186.

1474. Heilmann, C., C. Gerke, F. Perdreau-Remington, and F. Götz. 1996. Characterization of Tn917 insertion mutants of *Staphylococcus epidermidis* affected in biofilm formation. *Infect. Immun.* 64:277–282.

1475. Heilmann, C., O. Schweitzer, C. Gerke, N. Vanittanakom, D. Mack, and F. Götz. 1996. Molecular basis of intercellular adhesion in the biofilm-forming *Staphylococcus epidermidis*. *Mol. Microbiol.* 20:1083–1091.

1476. Heilmann, C., and F. Götz. 1998. Further characterization of *Staphylococcus epidermidis* transposon mutants deficient in primary attachment or intercellular adhesion. *Zentbl. Bakteriol.* 287:69–83.

1477. Heine, R. P., C. Elkins, P. B. Wyrick, and P. F. Sparling. 1996. Transferrin increases adherence of iron-deprived *Neisseria gonorrhoeae* to human endometrial cells. *Am. J. Obstet. Gynecol.* 174:659–666.

1478. Heinemann, C., J. E. van Hylckama Vlieg, D. B. Janssen, H. J. Busscher, H. C. van der Mei, and G. Reid. 2000. Purification and characterization of a surface-binding protein from *Lactobacillus fermentum* RC-14 that inhibits adhesion of *Enterococcus faecalis* 1131. *FEMS Microbiol. Lett.* **190:**177–180.

1479. Heinzelmann, M., S. A. Gardner, M. Mercer-Jones, A. J. Roll, and H. C. Polk, Jr. 1999. Quantification of phagocytosis in human neutrophils by flow cytometry. *Microbiol. Immunol.* **43:**505–512.

1480. Heithoff, D. M., C. P. Conner, P. C. Hanna, S. M. Julio, U. Hentschel, and M. J. Mahan. 1997. Bacterial infection as assessed by *in vivo* gene expression. *Proc. Natl. Acad. Sci. USA* **94:**934–939.

1481. Hejazi, A., and F. R. Falkiner. 1997. *Serratia marcescens. J. Med. Microbiol.* **46:**903–912.

1482. Hendrickson, B. A., J. Guo, R. Laughlin, Y. Chen, and J. C. Alverdy. 1999. Increased type 1 fimbrial expression among commensal *Escherichia coli* isolates in the murine cecum following catabolic stress. *Infect. Immun.* **67:**745–753.

1483. Hendrixson, D. R., and J. W. St. Geme III. 1998. The *Haemophilus influenzae* Hap serine protease promotes adherence and microcolony formation, potentiated by a soluble host protein. *Mol. Cell* **2:**841–850.

1484. Heneghan, M. A., A. P. Moran, K. M. Feeley, E. L. Egan, J. Goulding, C. E. Connolly, and C. F. McCarthy. 1998. Effect of host Lewis and ABO blood group antigen expression on *Helicobacter pylori* colonisation density and the consequent inflammatory response. *FEMS Immunol. Med. Microbiol.* **20:**257–266.

1485. Hennequin, C., F. Porcheray, A. Waligora-Dupriet, A. Collignon, M. Barc, P. Bourlioux, and T. Karjalainen. 2001. GroEL (Hsp60) of *Clostridium difficile* is involved in cell adherence. *Microbiology* **147:**87–96.

1486. Henrich, B., R.-C. Feldman, and U. Hadding. 1993. Cytoadhesins of *Mycoplasma hominis. Infect. Immun* **61:**2945–2951.

1487. Henrich, B., A. Kitzerow, R. C. Feldmann, H. Schaal, and U. Hadding. 1996. Repetitive elements of the *Mycoplasma hominis* adhesin p50 can be differentiated by monoclonal antibodies. *Infect. Immun.* **64:**4027–4034.

1488. Henrich, B., K. Lang, A. Kitzerow, C. MacKenzie, and U. Hadding. 1998. Truncation as a novel form of variation of the *p50* gene in *Mycoplasma hominis. Microbiology* **144:**2979–2985.

1489. Henrich, B., M. Hopfe, A. Kitzerow, and U. Hadding. 1999. The adherence-associated lipoprotein P100, encoded by an *opp* operon structure, functions as the oligopeptide-binding domain OppA of a putative oligopeptide transport system in *Mycoplasma hominis. J. Bacteriol* **181:**4873–4878.

1490. Henriksson, A., and P. L. Conway. 1992. Adhesion to porcine squamous epithelium of saccharide and protein moieties of *Lactobacillus fermentum* strain 104-S. *J. Gen. Microbiol.* **138:**2657–2661.

1491. Henriksson, A., and P. L. Conway. 1996. Adhesion of *Lactobacillus fermentum* 104-S to porcine stomach mucus. *Curr. Microbiol.* **33:**31–34.

1492. Hensen, S. M., M. J. Pavicic, J. A. Lohuis, and B. Poutrel. 2000. Use of bovine primary mammary epithelial cells for the comparison of adherence and invasion ability of *Staphylococcus aureus* strains. *J. Dairy Sci.* **83:**418–429.

1493. Herias, M. V., T. Midtvedt, L. Å. Hanson, and A. E. Wold. 1995. Role of *Escherichia coli* P fimbriae in intestinal colonization in gnotobiotic rats. *Infect. Immun.* **63:**4781–4789.

1494. Herias, M. V., T. Midtvedt, L. Å. Hanson, and A. E. Wold. 1997. *Escherichia coli* K5 capsule expression enhances colonization of the large intestine in the gnotobiotic rat. *Infect. Immun.* **65:**531–536.

1495. Herndon, B., L. Dall, and P. Suvarnam. 1995. Contribution of the host to test results in assays of *Staphylococcus epidermidis. J. Clin. Lab. Anal.* **9:**81–88.

1496. Herold, B. C., A. Siston, J. Bremer, R. Kirkpatrick, G. Wilbanks, P. Fugedi, C. Peto, and M. Cooper. 1997. Sulfated carbohydrate compounds prevent microbial adherence by sexually transmitted disease pathogens. *Antimicrob. Agents Chemother.* **41:**2776–2780.

1497. Herrera-Insua, I., K. Jacques-Palaz, B. E. Murray, and R. M. Rakita. 1997. The effect of antibiotic exposure on adherence to neutrophils of *Enterococcus faecium* resistant to phagocytosis. *J. Antimicrob. Chemother.* **39:**109–113.

1498. Herrmann, M., Q. J. Lai, R. M. Albrecht, D. F. Mosher, and R. A. Proctor. 1993. Adhesion of *Staphylococcus aureus* to surface-bound platelets: role of fibrinogen/fibrin and platelet integrins. *J. Infect. Dis.* **167:**312–322.

1499. Herrmann, M., J. Hartleib, B. Kehrel, R. R. Montgomery, J. J. Sixma, and G. Peters. 1997. Interaction of von Willebrand factor with *Staphylococcus aureus*. *J. Infect. Dis.* **176**:984–991.

1500. Herzberg, M. C. 1996. Platelet-streptococcal interactions in endocarditis. *Crit. Rev. Oral Biol. Med.* **7**:222–236.

1501. Herzberg, M. C., G. D. MacFarlane, K. Gong, N. N. Armstrong, A. R. Witt, P. R. Erickson, and M. W. Meyer. 1992. The platelet interactivity phenotype of *Streptococcus sanguis* influences the course of experimental endocarditis. *Infect. Immun.* **60**:4809–4818.

1502. Hess, D. J., M. J. Henry-Stanley, E. A. Moore, and C. L. Wells. 2001. Integrin expression, enterocyte maturation, and bacterial internalization. *J. Surg. Res.* **98**:116–122.

1503. Hewlett, E. L. 1997. Pertussis: current concepts of pathogenesis and prevention. *Pediatr. Infect. Dis. J.* **16**:S78–S84.

1504. Heyderman, R. S., N. J. Klein, O. A. Daramola, S. Hammerschmidt, M. Frosch, B. D. Robertson, M. Levin, and C. A. Ison. 1997. Induction of human endothelial tissue factor expression by *Neisseria meningitidis*: the influence of bacterial killing and adherence to the endothelium. *Microb. Pathog.* **22**:265–274.

1505. Hibma, A. M., S. A. Jassim, and M. W. Griffiths. 1996. *In vivo* bioluminescence to detect the attachment of L-forms of *Listeria monocytogenes* to food and clinical contact surfaces. *Int. J. Food Microbiol.* **33**:157–67.

1506. Hickey, T. E., S. Baqar, A. L. Bourgeois, C. P. Ewing, and P. Guerry. 1999. *Campylobacter jejuni*-stimulated secretion of interleukin-8 by INT407 cells. *Infect. Immun.* **67**:88–93.

1507. Hicks, S., D. C. Candy, and A. D. Phillips. 1996. Adhesion of enteroaggregative *Escherichia coli* to formalin-fixed intestinal and ureteric epithelia from children. *J. Med. Microbiol.* **44**:362–371.

1508. Hicks, S., D. C. A. Candy, and A. D. Philips. 1996. Adhesion of enteroaggregative *Escherichia coli* to pediatric intestinal mucosa in vitro. *Infect. Immun.* **64**:4751–4760.

1509. Hicks, S., G. Frankel, J. B. Kaper, G. Dougan, and A. D. Phillips. 1998. Role of intimin and bundle-forming pili in enteropathogenic *Escherichia coli* adhesion to pediatric intestinal tissue in vitro. *Infect. Immun.* **66**:1570–1578.

1510. Hide, E. J., I. D. Connaughton, S. J. Driesen, D. Hasse, R. P. Monckton, and N. G. Sammons. 1995. The prevalence of F107 fimbriae and their association with Shiga-like toxin II in *Escherichia coli* strains from weaned Australian pigs. *Vet. Microbiol.* **47**:235–243.

1511. Higashi, J. M., I. W. Wang, D. M. Shlaes, J. M. Anderson, and R. E. Marchant. 1998. Adhesion of *Staphylococcus epidermidis* and transposon mutant strains to hydrophobic polyethylene. *J. Biomed. Mater. Res.* **39**:341–350.

1512. Higgins, L. M., G. Frankel, I. Connerton, N. S. Goncalves, G. Dougan, and T. T. MacDonald. 1999. Role of bacterial intimin in colonic hyperplasia and inflammation. *Science* **285**:588–591.

1513. High, N., J. Mounier, M. C. Prevost, and P. J. Sansonetti. 1992. IpaB of *Shigella flexneri* causes entry into epithelial cells and escape from the phagocytic vacuole. *EMBO J.* **11**:1991–1999.

1514. Hilbi, H., G. Segal, and H. A. Shuman. 2001. *Icm/dot*-dependent upregulation of phagocytosis by *Legionella pneumophila*. *Mol. Microbiol.* **42**:603–617.

1515. Hill, E. M., A Raji, M. S. Valenzuela, F. Garcia, and R. Hoover. 1992. Adhesion to and invasion of cultured human cells by *Bartonella bacilliformis*. *Infect. Immun.* **60**:4051–4058.

1516. Hill, H. R., N. H. Augustine, P. A. Williams, E. J. Brown, and J. F. Bohnsack. 1993. Mechanism of fibronectin enhancement of group B streptococcal phagocytosis by human neutrophils and culture-derived macrophages. *Infect. Immun.* **61**:2334–2339.

1517. Hirakata, Y., K. Izumikawa, T. Yamaguchi, S. Igimi, N. Furuya, S. Maesaki, K. Tomono, Y. Yamada, S. Kohno, K. Yamaguchi, and S. Kamihira. 1998. Adherence to and penetration of human intestinal Caco-2 epithelial cell monolayers by *Pseudomonas aeruginosa*. *Infect. Immun.* **66**:1748–1751.

1518. Hirano, Y., M. Tamura, Y. Sekine, Y. Nemoto, and K. Hayashi. 1995. Bacterial adsorption to fetuin and mucin pellicle. *J. Nihon Univ. Sch. Dent.* **37**:85–90.

1519. Hirano, Y., K. Kuroda, M. Tamura, H. Oguma, and K. Hayashi. 1998. Citrate promotes attachment of *Prevotella nigrescens* (intermedia) ATCC 25261 to hydroxyapatite. *J. Oral Sci.* **40**:65–69.

1520. Hirano, Y., M. Tamura, and K. Hayashi. 2000. Inhibitory effect of lactoferrin on the adhesion of *Prevotella nigrescens* ATCC 25261 to hydroxyapatite. *J. Oral Sci.* **42**:125–131.

1521. Hiratsuka, K., Y. Abiko, M. Hayakawa, T. Ito, H. Sasahara, and H. Takiguchi. 1992. Role of *Porphyromonas gingivalis* 40-kDa outer membrane protein in the aggregation of *P. gingivalis* vesicles and *Actinomyces viscosus*. *Arch. Oral Biol.* **37**:717–724.

1522. Hirmo, S., S. Kelm, M. Iwersen, K. Hotta, Y. Goso, K. Ishihara, T. Suguri, M. Morita, T. Wadström, and R. Schauer. 1998. Inhibition of *Helicobacter pylori* sialic acid-specific haemagglutination by human gastrointestinal mucins and milk glycoproteins. *FEMS Immunol. Med. Microbiol.* **20**:275–281.

1523. Hirmo, S., S. Kelm, R. Schauer, B. Nilsson, and T. Wadström. 1996. Adhesion of *Helicobacter pylori* strains to alpha-2,3-linked sialic acids. *Glycoconj. J.* **13**:1005–1011.

1524. Hirmo, S., M. Utt, M. Ringner, and T. Wadström. 1995. Inhibition of heparan sulphate and other glycosaminoglycans binding to *Helicobacter pylori* by various polysulphated carbohydrates. *FEMS Immunol. Med. Microbiol.* **10**:301–306.

1525. Hirmo, S., E. Artursson, G. Puu, T. Wadström, and B. Nilsson. 1998. Characterization of *Helicobacter pylori* interactions with sialylglycoconjugates using a resonant mirror biosensor. *Anal. Biochem.* **257**:63–66.

1526. Hirmo, S., S. Kelm, M. Iwersen, K. Hotta, Y. Goso, K. Ishihara, T. Suguri, M. Morita, T. Wadström, and R. Schauer. 1998. Inhibition of *Helicobacter pylori* sialic acid-specific haemagglutination by human gastrointestinal mucins and milk glycoproteins. *FEMS Immunol. Med. Microbiol.* **20**:275–281.

1527. Hiroi, T., K. Fukushima, I. Kantake, Y. Namiki, and T. Ikeda. 1992. De novo glucan synthesis by mutants streptococcal glucosyltransferases present in pellicle promotes firm binding of *Streptococcus gordonii* to tooth surfaces. *FEMS Microbiol. Lett.* **75**:193–198.

1528. Hirose, K., E. Isogai, H. Miura, and I. Ueda. 1997. Levels of *Porphyromonas gingivalis* fimbriae and inflammatory cytokines in gingival crevicular fluid from adult human subjects. *Microbiol. Immunol.* **41**:21–26.

1529. Hirose, K., E. Isogai, H. Mizugai, H. Miura, and I. Ueda. 1996. Inductive effect of *Porphyromonas gingivalis* fimbriae on differentiation of human monocytic tumor cell line U937. *Oral Microbiol. Immunol.* **11**:62–64.

1530. Hirose, K., E. Isogai, H. Mizugai, and I. Ueda. 1996. Adhesion of *Porphyromonas gingivalis* fimbriae to human gingival cell line Ca9–22. *Oral Microbiol. Immunol.* **11**:402–406.

1531. Hirt, H., S. L. Erlandsen, and G. M. Dunny. 2000. Heterologous inducible expression of *Enterococcus faecalis* pCF10 aggregation substance asc10 in *Lactococcus lactis* and *Streptococcus gordonii* contributes to cell hydrophobicity and adhesion to fibrin. *J. Bacteriol.* **182**:2299–2306.

1532. Hnatow, L. L., C. L. Keeler, Jr., L. L. Tessmer, K. Czymmek, and J. E. Dohms. 1998. Characterization of MGC2, a *Mycoplasma gallisepticum* cytadhesin with homology to the *Mycoplasma pneumoniae* 30-kilodalton protein P30 and *Mycoplasma genitalium* P32. *Infect. Immun.* **66**:3436–3442.

1533. Ho, A. S., I. Sohel, and G. K. Schoolnik. 1992. Cloning and characterization of *fxp*, the flexible pilin gene of *Aeromonas hydrophila*. *Mol. Microbiol.* **6**:2725–2732.

1534. Ho, B., and B. Jiang. 1995. The adhesion of *Helicobacter pylori* extract to four mammalian cell lines. *Eur. J. Gastroenterol. Hepatol.* **7**:121–124.

1535. Hobbs, M., E. S. Collie, P. D. Free, S. P. Livingston, and J. S. Mattick. 1993. PilS and PilR, a two-component transcriptional regulatory system controlling expression of type 4 fimbriae in *Pseudomonas aeruginosa*. *Mol. Microbiol.* **7**:669–682.

1536. Hobden, J. A., S. K. Gupta, S. A. Masinick, X. Wu, K. A. Kernacki, R. S. Berk, and L. D. Hazlett. 1996. Anti-receptor antibodies inhibit *Pseudomonas aeruginosa* binding to the cornea and prevent corneal perforation. *Immunol. Cell Biol.* **74**:258–264.

1537. Hoefnagels-Schuermans, A., W. E. Peetermans, M. Jorissen, S. van Lierde, J. van den Oord, R. de Vos, and J. van Eldere. 1999. *Staphylococcus aureus* adherence to nasal epithelial cells in a physiological *in vitro* model. *In Vitro Cell. Dev. Biol.* **35**:472–480.

1538. Hoepelman, A. I., and E. I. Tuomanen. 1992. Consequences of microbial attachment: directing host cell functions with adhesins. *Infect. Immun.* **60**:1729–1733.

1539. Hoffmann, I., E. Eugene, X. Nassif, P. O. Couraud, and S. Bourdoulous. 2001. Activation of ErbB2 receptor tyrosine kinase supports invasion of endothelial cells by *Neisseria meningitidis*. *J. Cell Biol.* **155**:133–143.

1540. Höfner, H., R. Wirth, R. Marré, G. Wanner, and E. Straube. 1995. Subinhibitory concentrations of daptomycin enhance adherence of *Enterococcus faecalis* to in vitro cultivated renal tubuloepithelial cells and induce a sex phermone plasmid-encoded adhesin. *Med. Microbiol. Lett.* **4**:140–149.

1541. **Hofstad, T.** 1992. Virulence factors in anaerobic bacteria. *Eur. J. Clin. Microbiol. Infect. Dis.* **11:**1044–1048.

1542. **Hokama, A., and M. Iwanaga.** 1992. Purification and characterization of *Aeromonas sobria* Ae24 pili: a possible new colonization factor. *Microb. Pathog.* **13:**325–334.

1543. **Holberg-Petersen, M., G. Bukholm, H. Rollag, and M. Degre.** 1994. Infection with human cytomegalovirus enhances bacterial adhesiveness and invasiveness in permissive and semipermissive cells. *APMIS* **102:**703–710.

1544. **Hollyer, T. T., P. A. DeGagne, and M. J. Alfa.** 1994. Characterization of the cytopathic effect of *Haemophilus ducreyi*. *Sex. Transm. Dis.* **21:**247–257.

1545. **Holm, A., S. Kalfas, and S. E. Holm.** 1993. Killing of *Actinobacillus actinomycetemcomitans* and *Haemophilus aphrophilus* by human polymorphonuclear leukocytes in serum and saliva. *Oral Microbiol. Immunol.* **8:**134–140.

1546. **Holmberg, K., K. Bergström, C. Brink, E. Österberg, F. Tiberg, and J. M. Harris.** 1993. Effects of protein adsorption, bacterial adhesion and contact angle of grafting PEG chains to polystyrene. *J. Adhesion Sci. Technol.* **7:**503–517.

1547. **Holmes, A. R., C. Gilbert, J. M. Wells, and H. F. Jenkinson.** 1998. Binding properties of *Streptococcus gordonii* SspA and SspB (antigen I/II family) polypeptides expressed on the cell surface of *Lactococcus lactis* MG1363. *Infect. Immun.* **66:**4633–4639.

1548. **Holmes, K. A., and L. O. Bakaletz.** 1997. Adherence of non-typeable *Haemophilus influenzae* promotes reorganization of the actin cytoskeleton in human or chinchilla epithelial cells *in vitro*. *Microb. Pathog.* **23:**157–166.

1549. **Holmgren, A., M. J. Kuehn, C. I. Branden, and S. J. Hultgren.** 1992. Conserved immunoglobulin-like features in a family of periplasmic pilus chaperones in bacteria. *EMBO J.* **11:**1617–1622.

1550. **Holmstrom, A., R. Rosqvist, H. Wolf-Watz, and A. Forsberg.** 1995. Virulence plasmid-encoded YopK is essential for *Yersinia pseudotuberculosis* to cause systemic infection in mice. *Infect. Immun.* **63:**2269–2276.

1551. **Homma, H., S. Nagaoka, S. Mezawa, T. Matsuyama, E. Masuko, N. Ban, N. Watanabe, and Y. Niitsu.** 1996. Bacterial adhesion on hydrophilic heparinized catheters, with compared with adhesion on silicone catheters, in patients with malignant obstructive jaundice. *J. Gastroenterol.* **31:**836–843.

1552. **Hondalus, M. K., C. R. Sweeney, and D. M. Mosser.** 1992. An assay to quantitate the binding of *Rhodococcus equi* to macrophages. *Vet. Immunol. Immunopathol.* **32:**339–350.

1553. **Hondalus, M. K., M. S. Diamond, L. A. Rosenthal, T. A. Springer, and D. M. Mosser.** 1993. The intracellular bacterium *Rhodococcus equi* requires Mac-1 to bind to mammalian cells. *Infect. Immun.* **61:**2919–2929.

1554. **Honma, K., N. Ishii, T. Kato, K. Ishihara, K. Okuda, and K. Okuda.** 1997. Genetic control of immune response to a synthetic fimbrial antigen of *Actinobacillus actinomycetemcomitans*. *Microbiol. Immunol.* **41:**609–614.

1555. **Hopkins, R. J., J. G. Morris, Jr., P. C. Papadimitriou, C. Drachenberg, D. T. Smoot, S. P. James, and P. Panigrahi.** 1996. Loss of *Helicobacter pylori* hemagglutination with serial laboratory passage and correlation of hemagglutination with gastric epithelial cell adherence. *Pathobiology* **64:**247–254.

1556. **Hoppe, H. C., B. J. M. de Wet, C. Cywes, M. Daffé, and M. R. W. Ehlers.** 1997. Identification of phosphatidylinositol mannoside as a mycobacterial adhesin mediating both direct and opsonic binding to non-phagocytic mammalian cells. *Infect. Immun.* **65:**3896–3905.

1557. **Hopper, S., J. S. Wilbur, B. L. Vasquez, J. Larson, S. Clary, J. J. Mehr, H. S. Seifert, and M. So.** 2000. Isolation of *Neisseria gonorrhoeae* mutants that show enhanced trafficking across polarized T84 epithelial monolayers. *Infect. Immun.* **68:**896–905.

1558. **Horiuchi, S., Y. Inagaki, N. Okamura, R. Nakaya, and N. Yamamoto.** 1992. Type 1 pili enhance the invasion of *Salmonella braenderup* and *Salmonella typhimurium* to HeLa cells. *Microbiol. Immunol.* **36:**593–602.

1559. **Hornick, D. B., B. L. Allen, M. A Horn, and S. Clegg.** 1992. Adherence to respiratory epithelia by recombinant *Escherichia coli* expressing *Klebsiella pneumoniae* type 3 fimbrial gene products. *Infect. Immun.* **60:**1577–1588.

1560. **Hornick, D. B., J. Thommandru, W. Smits, and S. Clegg.** 1995. Adherence properties of an *mrkD*-negative mutant of *Klebsiella pneumoniae*. *Infect. Immun.* **63:**2026–2032.

1561. **Horwitz, M. A.** 1992. Interactions between macrophages and *Legionella pneumophila*. *Curr. Top. Microbiol. Immunol.* **181:**265–282.

1562. **Hosogi, Y., M. Hayakawa, and Y. Abiko.** 2001. Monoclonal antibody against *Porphyromonas gingivalis* hemagglutinin inhibits hemolytic activity. *Eur. J. Oral Sci.* **109:**109–113.

1563. **Hostacka, A.** 1999. Alterations in surface hydrophobicity of *Acinetobacter baumannii* induced by meropenem. *Folia Microbiol.* **44:**267–270.

1564. **Hostacka, A.** 2000. Effect of aminoglycosides on surface hydrophobicity of *Acinetobacter baumannii. Acta Microbiol. Immunol. Hung.* **47:**15–20.

1565. **House-Pompeo, K., Y. Xu, D. Joh, P. Speziale, and M. Höök.** 1996. Conformational changes in the fibronectin binding MSCRAMMs are induced by ligand binding. *J. Biol. Chem.* **271:**1379–1384.

1566. **Howell, A. B., N. Vorsa, A. der Marderosian, and L. Y. Foo.** 1998. Inhibition of the adherence of P-fimbriated *Escherichia coli* to uroepithelial-cell surfaces by proanthrocyanidin. *N. Engl. J. Med.* **339:**1085–1086.

1567. **Hoyne, P. A., R. Haas, T. F. Meyer, J. K. Davies, and T. C. Elleman.** 1992. Production of *Neisseria gonorrhoeae* pili (fimbriae) in *Pseudomonas aeruginosa. J. Bacteriol.* **174:**7321–7327.

1568. **Hsu, S. D., J. O. Cisar, A. L. Sandberg, and M. Kilian.** 1997. Adhesive properties of viridans streptococcal species. *Microb. Ecol. Health Dis.* **7:**125–137.

1569. **Hsu, T., and F. C. Minion.** 1998. Molecular analysis of the 97P cilium adhesin operon of *Mycoplasma hyopneumoniae. Gene* **214:**13–23.

1570. **Hsu, T., and F. C. Minion.** 1998. Identification of the cilium binding epitope of the *Mycoplasma hyopneumoniae* P97 adhesin. *Infect. Immun.* **66:**4762–4766.

1571. **Hsu, T., S. Artiushin, and F. C. Minion.** 1997. Cloning and functional analysis of the P97 swine cilium adhesin gene of *Mycoplasma hyopneumoniae. J. Bacteriol.* **179:**1317–1323.

1572. **Hu, L. T., G. Perides, R. Noring, and M. S. Klempner.** 1995. Binding of human plasminogen to *Borrelia burgdorferi. Infect. Immun.* **63:**3491–3496.

1573. **Hu, Z. L., J. Hasler-Rapacz, S. C. Huang, and J. Rapacz.** 1993. Studies in swine on inheritance and variation in expression of small intestinal receptors mediating adhesion of the K88 enteropathogenic *Escherichia coli* variants. *J. Hered.* **84:**157–165.

1574. **Huang, J., P. W. Keeling, and C. J. Smyth.** 1992. Identification of erythrocyte-binding antigens in *Helicobacter pylori. J. Gen. Microbiol.* **138:**1503–1513.

1575. **Huang, G. T., S. K. Haake, and N. H.**

Park. 1998. Gingival epithelial cells increase interleukin-8 secretion in response to *Actinobacillus actinomycetemcomitans* challenge. *J. Periodontol.* **69:**1105–1110.

1576. **Huang, X.-Z, B. Tall, W. R. Schwan, and D. J. Kopecko.** 1998. Physical limitations of *Salmonella typhi* entry into cultured human intestinal epithelial cells. *Infect. Immun.* **66:**2928–2938.

1577. **Huang, Y., G. Hajishengallis, and S. M. Michalek.** 2000. Construction and characterization of a *Salmonella enterica* serovar Typhimurium clone expressing a salivary adhesin of *Streptococcus mutans* under control of the anaerobically inducible *nirB* promoter. *Infect. Immun.* **68:**1549–1556.

1578. **Huard-Delcourt, A., L. Du, P. Pellen-Mussi, S. Tricot-Doleux, and M. Bonnaure-Mallet.** 1998. Adherence of *Porphyromonas gingivalis* to epithelial cells: analysis by flow cytometry. *Eur. J. Oral Sci.* **106:**938–944.

1579. **Hudson, M. C., W. K. Ramp, N. C. Nicholson, A. S. Williams, and M. T. Nousiainen.** 1995. Internalization of *Staphylococcus aureus* by cultured osteoblasts. *Microb. Pathog.* **19:**409–419.

1580. **Huesca, M., S. Borgia, P. Hoffman, and C. A. Lingwood.** 1996. Acidic pH changes receptor binding specificity of *Helicobacter pylori*: a binary adhesion model in which surface heat shock (stress) proteins mediate sulfatide recognition in gastric colonization. *Infect. Immun.* **64:**2643–2648.

1581. **Huesca, M., B. Gold, P. Sherman, P. Lewin, and C. Lingwood.** 1993. Therapeutics used to alleviate peptic ulcers inhibit *H. pylori* receptor binding *in vitro. Zentbl. Bakteriol* **280:**244–252.

1582. **Huesca, M., A. Goodwin, A. Bhagwansingh, P. Hoffman, and C. A. Lingwood.** 1998. Characterization of an acidic-pH-inducible stress protein (hsp70), a putative sulfatide binding adhesin, from *Helicobacter pylori. Infect. Immun.* **66:**4061–4067.

1583. **Hughes, C. V., R. N. Andersen, and P. E. Kolenbrander.** 1992. Characterization of *Veillonella atypica* PK1910 adhesin-mediated coaggregation with oral *Streptococcus* spp. *Infect. Immun.* **60:**1178–1186.

1584. **Hughes, K. J., K. D. Everiss, C. W. Harkey, and K. M. Peterson.** 1994. Identification of a *Vibrio cholerae* ToxR-activated gene (*tagD*) that is physically linked to the toxin-coregulated pilus (*tcp*) gene cluster. *Gene* **148:**97–100.

1585. **Hugosson, S., J. Angstrom, B. M. Olsson, J. Bergstrom, H. Fredlund, P. Olcen, and S. Teneberg.** 1998. Glycosphingolipid binding specificities of *Neisseria meningitidis* and *Haemophilus influenzae:* detection, isolation, and characterization of a binding-active glycosphingolipid from human oropharyngeal epithelium. *J. Biochem.* **124:**1138–1152.

1586. **Huisman, T. T., E. Pilipcinec, F. Remkes, J. Maaskant, F. K. de Graaf, and B. Oudega.** 1996. Isolation and characterization of chromosomal mTn*10* insertion mutations affecting K88 fimbriae production in *Escherichia coli. Microb. Pathog.* **20:**101–108.

1587. **Hull, R. A., B. Nowicki, A. Kaul, R. Runyan, C. Svanborg, and S. I. Hull.** 1994. Effect of *pap* copy number and receptor specificity on virulence of fimbriated *Escherichia coli* in a murine urinary tract colonization model. *Microb. Pathog.* **17:**79–86.

1588. **Hull, R. A., D. C. Rudy, W. H. Donovan, I. E. Wieser, C. Stewart, and R. O. Darouiche.** 1999. Virulence properties of *Escherichia coli* 83972, a prototype strain associated with asymptomatic bacteriuria. *Infect. Immun.* **67:**429–432.

1589. **Hull, R. A., D. C. Rudy, I. E. Wieser, and W. H. Donovan.** 1998. Virulence factors of *Escherichia coli* isolates from patients with symptomatic and asymptomatic bacteriuria and neuropathic bladders due to spinal cord and brain injuries. *J. Clin. Microbiol.* **36:**115–117.

1590. **Hull, S. I., and R. A. Hull.** 1995. Molecular cloning of adhesion genes. *Methods Enzymol.* **253:**258–269.

1591. **Hulse, M. L., S. Smith, E. Y. Chi, A. Pham, and C. E. Rubens.** 1993. Effect of type III group B streptococcal capsular polysaccharide on invasion of respiratory epithelial cells. *Infect. Immun.* **61:**4835–4841.

1592. **Hultgren, S. J., S. Abraham, M. Caparon, P. Falk, J. W. St. Geme III, and S. Normark.** 1993. Pilus and nonpilus bacterial adhesins: assembly and function in cell recognition. *Cell* **73:**887–901.

1593. **Hultgren, S. J., F. Jacob-Dubuisson, C. H. Jones, and C. I. Branden.** 1993. PapD and superfamily of periplasmic immunoglobulin-like pilus chaperones. *Adv. Protein Chem.* **44:**99–123.

1594. **Hume, E. B., and M. D. Willcox.** 1997. Adhesion and growth of *Serratia marcescens* on artificial closed eye tears soaked hydrogel contact lenses. *Aust. N. Z. J. Ophthalmol.* **25:**S39–S41.

1595. **Humphrey, T. J., S. J. Wilde, and R. J. Rowbury.** 1997. Heat tolerance of *Salmonella typhimurium* DT104 isolates attached to muscle tissue. *Lett. Appl. Microbiol.* **25:**265–268.

1596. **Hung, D. L., and S. J. Hultgren.** 1998. Pilus biogenesis via the chaperone/usher pathway: an integration of structure and function. *J. Struct. Biol.* **124:**201–220.

1597. **Hunt, J. D., D. C. Jackson, P. R. Wood, D. J. Stewart, and L. E. Brown.** 1995. Immunological parameters associated with antigenic competition in a multivalent footrot vaccine. *Vaccine* **13:**1649–1657.

1598. **Hunt, J. D., D. C. Jackson, L. E. Brown, P. R. Wood, and D. J. Stewart.** 1994. Antigenic competition in a multivalent foot rot vaccine. *Vaccine* **12:**457–464.

1599. **Huppertz, H. I., and J. Heesemann.** 1996. The influence of HLA B27 and interferon-gamma on the invasion and persistence of *Yersinia* in primary human fibroblasts. *Med. Microbiol. Immunol.* **185:**163–170.

1600. **Hurtado, A., B. Chahal, R. J. Owen, and A. W. Smith.** 1994. Genetic diversity of the *Helicobacter pylori* haemagglutinin/protease (*hap*) gene. *FEMS Microbiol. Lett.* **123:**173–178.

1601. **Husmann, L. K., and W. Johnson.** 1992. Adherence of *Legionella pneumophila* to guinea pig peritoneal macrophages, J774 mouse macrophages, and undifferentiated U937 human monocytes: role of Fc and complement receptors. *Infect. Immun.* **60:**5212–5218.

1602. **Hussain, M., M. H. Wilcox, P. J. White, M. K. Faulkner, and R. C. Spencer.** 1992. Importance of medium and atmosphere type to both slime production and adherence by coagulase-negative staphylococci. *J. Hosp. Infect.* **20:**173–184.

1603. **Hussain, M., M. H. Wilcox, and P. J. White.** 1993. The slime of coagulase-negative staphylococci: biochemistry and relation to adherence. *FEMS Microbiol Rev.* **10:**191–207.

1604. **Hussain, M., M. Herrmann, C. von Eiff, F. Perdreau-Remington, and G. Peters.** 1997. A 140-kilodalton extracellular protein is essential for the accumulation of *Staphylococcus epidermidis* strains on surfaces. *Infect. Immun.* **65:**519–524.

1605. **Hussain, M., K. Becker, C. von Eiff, J. Schrenzel, G. Peters, and M. Herrmann.** 2001. Identification and characterization of a novel 38.5-kilodalton cell surface protein of *Staphylococcus aureus* with extended-spectrum binding activity for extracellular matrix and plasma proteins. *J. Bacteriol.* **183:**6778–6786.

1606. **Hussain, M., C. Heilmann, G. Peters, and M. Herrmann.** 2001 Teichoic acid enhances adhesion of *Staphylococcus epidermidis* to immobilized fibronectin. *Microb. Pathog.* **31:**261–270.

1607. **Hytonen, J., S. Haataja, P. Isomaki, and J. Finne.** 2000. Identification of a novel glycoprotein-binding activity in *Streptococcus pyogenes* regulated by the *mga* gene. *Microbiology* **146:**31–39.

1608. **Iacono, V. J., K. Demerel, S. M. Zove, S. Holen, and P. N. Baer.** 1992. An augmented regenerative technique for severe osseous defects. *Periodontal Clin. Investig.* **14:**21–27.

1609. **Ibsen, P. H., J. W. Petersen, and I. Heron.** 1993. Quantification of pertussis toxin, filamentous haemagglutinin, 69 kDa outer membrane protein, agglutinogens 2 and 3 and lipopolysaccharide in the Danish whole-cell pertussis vaccine. *Vaccine* **11:**318–322.

1610. **Ichikawa, J. K., A. Norris, M. G. Bangera, G. K. Geiss, A. B. van't Wout, R. E. Bumgarner, and S. Lory.** 2000. Interaction of *Pseudomonas aeruginosa* with epithelial cells: identification of differentially regulated genes by expression microarray analysis of human cDNAs. *Proc. Natl. Acad. Sci. USA* **97:**9659–9664.

1611. **Ichikawa, T., K. Hirota, H. Kanitani, Y. Miyake, and N. Matsumoto.** 1998. *In vitro* adherence of *Streptococcus constellatus* to dense hydroxyapatite and titanium. *J. Oral Rehab.* **25:**125–127.

1612. **Idanpaan-Heikkila, I., P. M. Simon, D. Zopf, T. Vullo, P. Cahill, K. Sokol, and E. Tuomanen.** 1997. Oligosaccharides interfere with the establishment and progression of experimental pneumococcal pneumonia. *J. Infect. Dis.* **176:**704–712.

1613. **Idota, T., and H. Kawakami.** 1995. Inhibitory effects of milk gangliosides on the adhesion of *Escherichia coli* to human intestinal carcinoma cells. *Biosci. Biotechnol. Biochem.* **59:**69–72.

1614. **Ieven, M., D. Ursi, H. van Bever, W. Quint, H. B. Niesters, and H. Goossens.** 1996. Detection of *Mycoplasma pneumoniae* by two polymerase chain reactions and role of *M. pneumoniae* in acute respiratory tract infections in pediatric patients. *J. Infect. Dis.* **173:**1445–1452.

1615. **Ikaheimo, R., A. Siitonen, U. Karkkainen, P. Kuosmanen, and P. H. Makela.** 1993. Characteristics of *Escherichia coli* in acute community-acquired cystitis of adult women. *Scand. J. Infect. Dis.* **25:**705–712.

1616. **Ikaheimo, R., A. Siitonen, U. Karkkainen, and P. H. Makela.** 1993.

1617. **Ikaheimo, R., A. Siitonen, U. Karkkainen, J. Mustonen, T. Heiskanen, and P. H. Makela.** 1994. Community-acquired pyelonephritis in adults: characteristics of *E. coli* isolates in bacteremic and nonbacteremic patients. *Scand. J. Infect. Dis.* **26:**289–296.

1618. **Ikemori, Y., R. C. Peralta, M. Kuroki, H. Yokoyama, and Y. Kodama.** 1993. Research note: avidity of chicken yolk antibodies to enterotoxigenic *Escherichia coli* fimbriae. *Poult. Sci.* **72:**2361–2365.

1619. **Ilver, D., A. Arnqvist, J. Ogren, I. M. Frick, D. Kersulyte, E. T. Incecik, D. E. Berg, A. Covacci, L. Engstrand, and T. Boren.** 1998. *Helicobacter pylori* adhesin binding fucosylated histo-blood group antigens revealed by retagging. *Science* **279:**373–377.

1620. **Ilver, D., H. Källström, S. Normark, and A.-B. Jonsson.** 1998. Transcellular passage of *Neisseria gonorrhoeae* involves pilus phase variation. *Infect. Immun.* **66:**469–473.

1621. **Imayasu, M., W. M. Petroll, J. V. Jester, S. K. Patel, J. Ohashi, and H. D. Cavanagh.** 1994. The relation between contact lens oxygen transmissibility and binding of *Pseudomonas aeruginosa* to the cornea after overnight wear. *Ophthalmology* **101:**371–388.

1622. **Imberechts, H., H. de Greve, C. Schlicker, H. Bouchet, P. Pohl, G. Charlier, H. Bertschinger, P. Wild, J. Vandekerckhove, J. van Damme, M. van Montagu, and P. Lintermans.** 1992. Characterization of F107 fimbriae of *Escherichia coli* 107/86, which causes edema disease in pigs, and nucleotide sequence of the F107 major fimbrial subunit gene, *fedA. Infect. Immun.* **60:**1963–1971.

1623. **Imberechts, H., H. de Greve, J. P. Hernalsteens, C. Schlicker, H. Bouchet, P. Pohl, G. Charlier, H. U. Bertschinger, P. Wild, J. Vandekerckhove, et al.** 1993. The role of adhesive F107 fimbriae and of SLT-IIv toxin in the pathogenesis of edema disease in pigs. *Zentbl. Bakteriol.* **278:**445–450.

1624. **Imberechts, H., H. U. Bertschinger, M. Stamm, T. Sydler, P. Pohl, H. De Greve, J. P. Hernalsteens, M. van Montagu, and P. Lintermans.** 1994. Prevalence of F107 fimbriae on *Escherichia coli* isolated from pigs with oedema disease or postweaning diarrhoea. *Vet. Microbiol.* **40:**219–230.

1625. **Imberechts, H., N. van Pelt, H. de**

Greve, and P. Lintermans. 1994. Sequences related to the major subunit gene *fedA* of F107 fimbriae in porcine *Escherichia coli* strains that express adhesive fimbriae. *FEMS Microbiol. Lett.* **119**:309–314.

1626. Imberechts, H., P. Wild, G. Charlier, H. de Greve, P. Lintermans, and P. Pohl. 1996. Characterization of F18 fimbrial genes *fedE* and *fedF* involved in adhesion and length of enterotoxemic *Escherichia coli* strain 107/86. *Microb. Pathog.* **21**:183–192.

1627. Imberechts, H., H. U. Bertschinger, B. Nagy, P. Deprez, and P. Pohl. 1997. Fimbrial colonisation factors F18ab and F18ac of *Escherichia coli* isolated from pigs with postweaning diarrhea and edema disease. *Adv. Exp. Med. Biol.* **412**:175–183.

1628. Imberechts, H., P. Deprez, E. van Driessche, and P. Pohl. 1997. Chicken egg yolk antibodies against F18ab fimbriae of *Escherichia coli* inhibit shedding of F18 positive *E. coli* by experimentally infected pigs. *Vet. Microbiol.* **54**:329–341.

1629. Imundo, L., J. Barasch, A. Prince, and Q. al-Awqati. 1995. Cystic fibrosis epithelial cells have a receptor for pathogenic bacteria on their apical surface. *Proc. Natl. Acad. Sci. USA* **92**:3019–3023.

1630. Inoue, O. J., J. H. Lee, and R. E. Isaacson. 1993. Transcriptional organization of the *Escherichia coli* pilus adhesin K99. *Mol. Microbiol.* **10**:607–613.

1631. Inoue, T., I. Tanimoto, H. Ohta, K. Kato, Y. Murayama, and K. Fukui. 1998. Molecular characterization of low-molecular-weight component protein, Flp, in *Actinobacillus actinomycetemcomitans* fimbriae. *Microbiol. Immunol.* **42**:253–258.

1632. Iredell, J. R., and P. A. Manning. 1994. Biotype-specific *tcpA* genes in *Vibrio cholerae*. *FEMS Microbiol. Lett.* **121**:47–54.

1633. Iredell, J. R., and P. A. Manning. 1994. The toxin-co-regulated pilus of *Vibrio cholerae* O1: a model for type 4 pilus biogenesis? *Trends Microbiol.* **2**:187–192.

1634. Iredell, J. R., and P. A. Manning. 1997. Translocation failure in a type-4 pilin operon: *rfb* and *tcpT* mutants in *Vibrio cholerae*. *Gene* **192**:71–77.

1635. Iredell, J. R., and P. A. Manning. 1997. Outer membrane translocation arrest of the TcpA pilin subunit in *rfb* mutants of *Vibrio cholerae* O1 strain 569B. *J. Bacteriol.* **179**:2038–2046.

1636. Iredell, J. R., U. H. Stroeher, H. M. Ward, and P. A. Manning. 1998. Lipopolysaccharide O-antigen expression and the effect of its absence on virulence in *rfb* mutants of *Vibrio cholerae* O1. *FEMS Immunol. Med. Microbiol.* **20**:45–54.

1637. Iriarte, M., J. C. Vanooteghem, I. Delor, R. Diaz, S. Knutton, and G. R. Cornelis. 1993. The Myf fibrillae of *Yersinia enterocolitica*. *Mol. Microbiol.* **9**:507–520.

1638. Iriarte, M., and G. R. Cornelis. 1995. MyfF, an element of the network regulating the synthesis of fibrillae in *Yersinia enterocolitica*. *J. Bacteriol.* **177**:738–744.

1639. Iriarte, M., and G. R. Cornelis. 1996. Molecular determinants of *Yersinia* pathogenesis. *Microbiology* **12**:267–280.

1640. Isaacs, R. D. 1994. *Borrelia burgdorferi* bind to epithelial cell proteoglycans. *J. Clin. Investig.* **93**:809–819.

1641. Isaacson, R. E., and S. Patterson. 1994. Analysis of a naturally occurring K99$^+$ enterotoxigenic *Escherichia coli* strain that fails to produce K99. *Infect. Immun.* **62**:4686–4689.

1642. Isaacson, R. E., and E. Trigo. 1995. Pili of *Pasteurella multocida* of porcine origin. *FEMS Microbiol. Lett.* **132**:247–251.

1643. Isacson, J., B. Trollfors, G. Hedvall, J. Taranger, and G. Zackrisson. 1995. Response and decline of serum IgG antibodies to pertussis toxin, filamentous hemagglutinin and pertactin in children with pertussis. *Scand. J. Infect. Dis.* **27**:273–277.

1644. Isacson, J., B. Trollfors, J. Taranger, and T. Lagergard. 1995. Acquisition of IgG serum antibodies against two *Bordetella* antigens (filamentous hemagglutinin and pertactin) in children with no symptoms of pertussis. *Pediatr. Infect. Dis. J.* **14**:517–521.

1645. Isberg, R. R. 1996. Uptake of enteropathogenic *Yersinia* by mammalian cells. *Curr. Top. Microbiol. Immunol.* **209**:1–24.

1646. Isberg, R. R., and G. T. Van Nhieu. 1994. Two mammalian cell internalization strategies used by pathogenic bacteria. *Annu. Rev. Genet.* **28**:395–422.

1647. Ishibashi, Y., D. A. Relman, and A. Nishikawa. 2001. Invasion of human respiratory epithelial cells by *Bordetella pertussis*: possible role for a filamentous hemagglutinin Arg-Gly-Asp sequence and alpha5beta1 integrin. *Microb. Pathog.* **30**:279–288.

1648. Ishibashi, Y., S. Claus, and D. A. Relman. 1994. *Bordetella pertussis* filamentous hemagglutinin interacts with a leukocyte signal transduction complex and stimulates bacterial adherence to monocyte CR3 (CD11b/CD18). *J. Exp. Med.* **180**:1225–1233.

1649. Ishida, A., Y. Yoshikai, S. Murosaki, Y.

Hidaka, and K. Nomoto. 1992. Administration of milk from cows immunized with intestinal bacteria protects mice from radiation-induced lethality. *Biotherapy* **5:**215–225.

1650. **Ishihara, K., K. Honma, T. Miura, T. Kato, and K. Okuda.** 1997. Cloning and sequence analysis of the fimbriae associated protein (*fap*) gene from *Actinobacillus actinomycetemcomitans. Microb. Pathog.* **23:**63–69.

1651. **Ishikawa, H., and W. Sato.** 1997. Role of *Bordetella bronchiseptica* sialic acid-binding hemagglutinin as a putative colonization factor. *J. Vet. Med. Sci.* **59:**43–44.

1652. **Ishiwa, A., and T. Komano.** 2000. The lipopolysaccharide of recipient cells is a specific receptor for PilV proteins, selected by shufflon DNA rearrangement, in liquid matings with donors bearing the R64 plasmid. *Mol. Gen. Genet.* **263:**159–164.

1653. **Ishizuka, S., M. Yamaya, T. Suzuki, K. Nakayama, M. Kamanaka, S. Ida, K. Sekizawa, and H. Sasaki.** 2001. Acid exposure stimulates the adherence of *Streptococcus pneumoniae* to cultured human airway epithelial cells: effects on platelet-activating factor receptor expression. *Am. J. Respir. Cell Mol. Biol.* **24:**459–468.

1654. **Ismaili, A., D. J. Philpott, M. T. Dytoc, and P. M. Sherman.** 1995. Signal transduction responses following adhesion of verocytotoxin-producing *Escherichia coli. Infect. Immun.* **63:**3316–3326.

1655. **Ismaili, A., D. J. Philpott, M. T. Dytoc, R. Soni, S. Ratnam, and P. M. Sherman.** 1995. Alpha-actinin accumulation in epithelial cells infected with attaching and effacing gastrointestinal pathogens. *J. Infect. Dis.* **172:**1393–1396.

1656. **Ismaili, A., B. Bourke, J. C. de Azavedo, S. Ratnam, M. A. Karmali, and P. M. Sherman.** 1996. Heterogeneity in phenotypic and genotypic characteristics among strains of *Hafnia alvei. J. Clin. Microbiol.* **34:**2973–2979.

1657. **Ismaili, A., J. B. Meddings, S. Ratnam, and P. M. Sherman.** 1999. Modulation of host cell membrane fluidity: a novel mechanism for preventing bacterial adhesion. *Am. J. Physiol.* **277:**G201–G208.

1658. **Isogai, E., K. Hirose, N. Fujii, and H. Isogai.** 1992. Three types of binding by *Porphyromonas gingivalis* and oral bacteria to fibronectin, buccal epithelial cells and erythrocytes. *Arch. Oral Biol.* **37:**667–670.

1659. **Isogai, E., H. Isogai, S. Takagi, N. Ishii, N. Fujii, K. Kimura, M. Hayashi, and F. Yoshimura.** 1994. Fimbria-specific immune response in various inbred mice inoculated with *Porphyromonas gingivalis* 381. *Oral Microbiol. Immunol.* **9:**118–122.

1660. **Ista, L. K., H. Fan, O. Baca, and G. P. Lopez.** 1996. Attachment of bacteria to model solid surfaces: oligo(ethylene glycol) surfaces inhibit bacterial attachment. *FEMS Microbiol. Lett.* **142:**59–63.

1661. **Ista, L. K., V. H. Perez-Luna, and G. P. Lopez.** 1999. Surface-grafted, environmentally sensitive polymers for biofilm release. *Appl. Environ. Microbiol.* **65:**1603–1609.

1662. **Itoh, Y., I. Nagano, M. Kunishima, and T. Ezaki.** 1997. Laboratory investigation of enteroaggregative *Escherichia coli* O untypeable:H10 associated with a massive outbreak of gastrointestinal illness. *J. Clin. Microbiol.* **35:**2546–2550.

1663. **Iturralde, M., B. Aguilar, R. Baselga, and B. Amorena.** 1993. Adherence of ruminant mastitis *Staphylococcus aureus* strains to epithelial cells from ovine mammary gland primary cultures and from a rat intestinal cell line. *Vet. Microbiol.* **38:**115–127.

1664. **Iwanaga, M., and A. Hokama.** 1992. Characterization of *Aeromonas sobria* TAP13 pili: a possible new colonization factor. *J. Gen. Microbiol.* **138:**1913–1919.

1665. **Iwanaga, M., N. Nakasone, T. Yamashiro, and N. Higa.** 1993. Pili of *Vibrio cholerae* widely distributed in serogroup O1 strains. *Microbiol. Immunol.* **37:**23–28.

1666. **Izutsu, K. T., C. M. Belton, A. Chan, S. Fatherazi, J. P. Kanter, Y. Park, and R. J. Lamont.** 1996. Involvement of calcium in interactions between gingival epithelial cells and *Porphyromonas gingivalis. FEMS Microbiol. Lett.* **144:**145–150.

1667. **Jabbal-Gill, I., A. N. Fisher, R. Rappuoli, S. S. Davis, and L. Illum.** 1998. Stimulation of mucosal and systemic antibody responses against *Bordetella pertussis* filamentous haemagglutinin and recombinant pertussis toxin after nasal administration with chitosan in mice. *Vaccine* **16:**2039–2046.

1668. **Jackson, A. D., D. Maskell, E. R. Moxon, and R. Wilson.** 1996. The effect of mutations in genes required for lipopolysaccharide synthesis on *Haemophilus influenzae* type b colonization of human nasopharyngeal tissue. *Microb. Pathog.* **21:**463–470.

1669. **Jacob, A., V. B. Sinha, M. K. Sahib, R. Srivastava, J. B. Kaper, and B. S. Srivastava.** 1993. Identification of a 33 kDa antigen associated with an adhesive and colonizing strain of *Vibrio cholerae* El Tor and its role in protection. *Vaccine* **11:**376–382.

1670. **Jacob-Dubuisson, F., J. Heuser, K.**

Dodson, S. Normark, and S. Hultgren. 1993. Initiation of assembly and association of the structural elements of a bacterial pilus depend on two specialized tip proteins. *EMBO J.* **12:**837–847.

1671. Jacob-Dubuisson, F., M. Kuehn, and S. J. Hultgren. 1993. A novel secretion apparatus for the assembly of adhesive bacterial pili. *Trends Microbiol.* **1:**50–55.

1672. Jacob-Dubuisson, F., J. Pinkner, Z. Xu, R. Striker, A. Padmanhaban, and S. J. Hultgren. 1994. PapD chaperone function in pilus biogenesis depends on oxidant and chaperone-like activities of DsbA. *Proc. Natl. Acad. Sci. USA* **91:**11552–11556.

1673. Jacob-Dubuisson, F., R. Striker, and S. J. Hultgren. 1994. Chaperone-assisted self-assembly of pili independent of cellular energy. *J. Biol. Chem.* **269:**12447–12455.

1674. Jacob-Dubuisson, F., C. Buisine, N. Mielcarek, E. Clement, F. D. Menozzi, and C. Locht. 1996. Amino-terminal maturation of the *Bordetella pertussis* filamentous haemagglutinin. *Mol. Microbiol.* **19:**65–78.

1675. Jacob-Dubuisson, F., C. Buisine, E. Willery, G. Renauld-Mongenie, and C. Locht. 1997. Lack of functional complementation between *Bordetella pertussis* filamentous hemagglutinin and *Proteus mirabilis* HpmA hemolysin secretion machineries. *J. Bacteriol.* **179:**775–783.

1676. Jacobs, E., A. Bartl, K. Oberle, and E. Schiltz. 1995. Molecular mimicry by *Mycoplasma pneumoniae* to evade the induction of adherence inhibiting antibodies. *J. Med. Microbiol.* **43:**422–429.

1677. Jacobs, E., M. Vonski, K. Oberle, O. Opitz, and K. Pietsch. 1996. Are outbreaks and sporadic respiratory infections by *Mycoplasma pneumoniae* due to two distinct subtypes? *Eur. J. Clin. Microbiol. Infect. Dis.* **15:**38–44.

1678. Jacobson, S. H., B. Hylander, B. Wretlind, and A. Brauner. 1994. Interleukin-6 and interleukin-8 in serum and urine in patients with acute pyelonephritis in relation to bacterial-virulence-associated traits and renal function. *Nephron* **67:**172–179.

1679. Jacques, M. 1996. Role of lipo-oligosaccharides and lipopolysaccharides in bacterial adherence. *Trends Microbiol.* **4:**408–409.

1680. Jacques, M., B. Blanchard, B. Foiry, C. Girard, and M. Kobisch. 1992. *In vitro* colonization of porcine trachea by *Mycoplasma pneumoniae*. *Ann. Rech. Vet.* **23:**239–247.

1681. Jacques, M., M. Kobisch, M. Belanger, and F. Dugal. 1993. Virulence of capsulated and noncapsulated isolates of *Pasteurella multocida* and their adherence to porcine respiratory tract cells and mucus. *Infect. Immun.* **61:**4785–4792.

1682. Jacques, M., M. Belanger, M. S. Diarra, M. Dargis, and F. Malouin. 1994. Modulation of *Pasteurella multocida* capsular polysaccharide during growth under iron-restricted conditions and *in vivo*. *Microbiology* **140:**263–270.

1683. Jacques, M., and S. E. Paradis. 1998. Adhesin-receptor interactions in *Pasteurellaceae*. *FEMS Microbiol. Rev.* **22:**45–59.

1684. Jadoun, J., and S. Sela. 2000. Mutation in *csrR* global regulator reduces *Streptococcus pyogenes* internalization. *Microb. Pathog.* **29:**311–317.

1685. Jadoun, J., V. Ozeri, E. Burstein, E. Skutelsky, E. Hanski, and S. Sela. 1998. Protein F1 is required for efficient entry of *Streptococcus pyogenes* into epithelial cells. *J. Infect. Dis.* **178:**147–158.

1686. Jafari, A., S. Bouzari, A. A. Farhoudi-Moghaddam, M. Parsi, and F. Shokouhi. 1994. *In vitro* adhesion and invasion of *Salmonella enterica* serovar Havana. *Microb. Pathog.* **16:**65–70.

1687. Jallat, C., A. Darfeuille-Michaud, C. Forestier, B. Joly, and R. Cluzel. 1992. Inhibition of *Haemophilus influenzae* adherence to buccal epithelial cells by cefuroxime. *Chemotherapy* **38:**428–432.

1688. Jallat, C., V. Livrelli, A. Darfeuille-Michaud, C. Rich, and B. Joly. 1993. *Escherichia coli* strains involved in diarrhea in France: high prevalence and heterogeneity of diffusely adhering strains. *J. Clin. Microbiol.* **31:**2031–2037.

1689. Jallat, C., A. Darfeuille-Michaud, C. Rich, and B. Joly. 1994. Survey of clinical isolates of diarrhoeogenic *Escherichia coli*: diffusely adhering *E. coli* strains with multiple adhesive factors. *Res. Microbiol.* **145:**621–632.

1690. James, B. W., and C. W. Keevil. 1999. Influence of oxygen availability on physiology, verocytotoxin expression and adherence of *Escherichia coli* O157. *J. Appl. Microbiol.* **86:**117–124.

1691. Jana, T. K., A. K. Srivastava, K. Csery, and D. K. Arora. 2000. Influence of growth and environmental conditions on cell surface hydrophobicity of *Pseudomonas fluorescens* in non-specific adhesion. *Can. J. Microbiol.* **46:**28–37.

1692. Janakiraman, R. S., and Y. V. Brun. 1999. Cell cycle control of a holdfast attachment gene in *Caulobacter crescentus*. *J. Bacteriol.* **181:**1118–1125.

1693. **Jang, T. J., J. R. Kim, and D. H. Kim.** 1999. Adherence of *Helicobacter pylori* to areas of type II intestinal metaplasia in Korean gastric mucosa. *Yonsei Med. J.* **40:**392–395.

1694. **Jansen, B., J. Beuth, A. Przondo-Mordarska, S. Jansen, H. L. Ko, and G. Pulverer.** 1992. Adhesion of fimbriated and non-fimbriated *Klebsiella* strains to synthetic polymers. *Zentbl. Bakteriol.* **276:**205–212.

1695. **Jansen, B., S. Jansen, G. Peters, and G. Pulverer.** 1992. *In-vitro* efficacy of a central venous catheter ('Hydrocath') loaded with teicoplanin to prevent bacterial colonization. *J. Hosp. Infect.* **22:**93–107.

1696. **Jansen, B., K. G. Kristinsson, S. Jansen, G. Peters, and G. Pulverer.** 1992. *In-vitro* efficacy of a central venous catheter complexed with iodine to prevent bacterial colonization. *J. Antimicrob. Chemother.* **30:**135–139.

1697. **Jansen, B., L. P. Goodman, and D. Ruiten.** 1993. Bacterial adherence to hydrophilic polymer-coated polyurethane stents. *Gastrointest. Endosc.* **39:**670–673.

1698. **Jansen, B., and G. Peters.** 1993. Foreign body associated infection. *J. Antimicrob. Chemother.* **32:**69–75.

1699. **Jansen, B., M. Rinck, P. Wolbring, A. Strohmeier, and T. Jahns.** 1994. *In vitro* evaluation of the antimicrobial efficacy and biocompatibility of a silver-coated central venous catheter. *J. Biomater. Appl.* **9:**55–70.

1700. **Jansen, H. M., A. P. Sachs, and L. van Alphen.** 1995. Predisposing conditions to bacterial infections in chronic obstructive pulmonary disease. *Am. J. Respir. Crit. Care Med.* **151:**2073–2080.

1701. **Jantausch, B. A., and S. I. Hull.** 1996. Restriction fragment length polymorphism of PCR amplified *papE* gene products is correlated with complete serotype among uropathogenic *Escherichia coli* isolates. *Microb. Pathog.* **20:**351–360.

1702. **Jantausch, B. A., B. L. Wiedermann, S. I. Hull, B. Nowicki, P. R. Getson, H. G. Rushton, M. Majd, N. L. Luban, and W. J. Rodriguez.** 1992. *Escherichia coli* virulence factors and 99mTc-dimercaptosuccinic acid renal scan in children with febrile urinary tract infection. *Pediatr. Infect. Dis. J.* **11:**343–349.

1703. **Jaramillo, L., F. Diaz, P. Hernandez, H. Debray, F. Trigo, G. Mendoza, and E. Zenteno.** 2000. Purification and characterization of an adhesin from *Pasteurella haemolytica*. *Glycobiology* **10:**31–37.

1704. **Jarvill-Taylor, K. J., and F. C. Minion.** 1995. The effect of thiol-active compounds and sterols on the membrane-associated hemolysin of *Mycoplasma pulmonis*. *FEMS Microbiol. Lett.* **128:**213–218.

1705. **Jass, J., L. E. Phillips, E. J. Allan, J. W. Costerton, and H. M. Lappin-Scott.** 1994. Growth and adhesion of *Enterococcus faecium* L-forms. *FEMS Microbiol. Lett.* **115:**157–162.

1706. **Jenkinson, H. F.** 1992. Adherence, coaggregation, and hydrophobicity of *Streptococcus gordonii* associated with expression of cell surface lipoproteins. *Infect. Immun.* **60:**1225–1228.

1707. **Jenkinson, H. F.** 1994. Adherence and accumulation of oral streptococci. *Trends Microbiol.* **2:**209–212.

1708. **Jenkinson, H. F.** 1994. Cell surface protein receptors in oral streptococci. *FEMS Microbiol. Lett.* **121:**133–140.

1709. **Jenkinson, H. F.** 1995. Anchorage and release of Gram-positive bacterial cell-surface polypeptides. *Trends Microbiol.* **3:**333–335.

1710. **Jenkinson, H. F.** 1995. Genetic analysis of adherence by oral streptococci. *J. Ind. Microbiol.* **15:**186–192.

1711. **Jenkinson, H. F., S. D. Terry, R. McNab, and G. W. Tannock.** 1993. Inactivation of the gene encoding surface protein SspA in *Streptococcus gordonii* DL1 affects cell interactions with human salivary agglutinin and oral actinomyces. *Infect. Immun.* **61:**3199–3208.

1712. **Jenkinson, H. F., R. McNab, D. M. Loach, and G. W. Tannock.** 1995. Lipoprotein receptors in oral streptococci. *Dev. Biol. Stand.* **85:**333–341.

1713. **Jenkinson, H. F., and R. J. Lamont.** 1997. Streptococcal adhesion and colonization. *Crit. Rev. Oral Biol. Med.* **8:**175–200.

1714. **Jensen, V. B., J. T. Harty, and B. D. Jones.** 1998. Interactions of the invasive pathogens *Salmonella typhimurium*, *Listeria monocytogenes,* and *Shigella flexneri* with M cells and murine Peyer's patches. *Infect. Immun.* **66:**3758–3766.

1715. **Jepson, M. A., M. A. Clark, N. L. Simmons, and B. H. Hirst.** 1993. Actin accumulation at sites of attachment of indigenous apathogenic segmented filamentous bacteria to mouse ileal epithelial cells. *Infect. Immun.* **61:**4001–4004.

1716. **Jerse, A. E., and R. F. Rest.** 1997. Adhesion and invasion by the pathogenic neisseria. *Trends Microbiol.* **5:**217–221.

1717. **Jett, B. D., M. M. Huycke, and M. S. Gilmore.** 1994. Virulence of enterococci. *Clin. Microbiol. Rev.* **7:**462–478.

1718. **Jett, B. D., R. V. Atkuri, and M. S.**

Gilmore. 1998. *Enterococcus faecalis* localization in experimental endophthalmitis: role of plasmid-encoded aggregation substance. *Infect. Immun.* **66**:843–848.

1719. **Jeyasingham, M. D., P. Butty, T. P. King, R. Begbie, and D. Kelly.** 1999. *Escherichia coli* K88 receptor expression in intestine of disease-susceptible weaned pigs. *Vet. Microbiol.* **68**:219–234.

1720. **Ji, Y., N. Schnitzler, E. DeMaster, and P. Cleary.** 1998. Impact of M49, Mrp, Enn, and C5a peptidase proteins on colonization of the mouse oral mucosa by *Streptococcus pyogenes*. *Infect. Immun.* **66**:5399–5405.

1721. **Jiang, Z., N. Nagata, E. Molina, L. O. Bakaletz, H. Hawkins, and J. A. Patel.** 1999. Fimbria-mediated enhanced attachment of nontypeable *Haemophilus influenzae* to respiratory syncytial virus-infected respiratory epithelial cells. *Infect. Immun.* **67**:187–192.

1722. **Jin, L. Z., Y. W. Ho, M. A. Ali, N. Abdullah, and S. Jalaludin.** 1996. Effect of adherent *Lactobacillus* spp. on *in vitro* adherence of salmonellae to the intestinal epithelial cells of chicken. *J. Appl. Bacteriol.* **81**:201–206.

1723. **Jin, L. Z., Y. W. Ho, M. A. Ali, N. Abdullah, K. B. Ong, and S. Jalaludin.** 1996. Adhesion of *Lactobacillus* isolates to intestinal epithelial cells of chicken. *Lett. Appl. Microbiol.* **22**:229–232.

1724. **Jin, L. Z., S. K. Baidoo, R. R. Marquardt, and A. A. Frohlich.** 1998. In vitro inhibition of adhesion of enterotoxigenic *Escherichia coli* K88 to piglet intestinal mucus by egg-yolk antibodies. *FEMS Immunol. Med. Microbiol.* **21**:313–321.

1725. **Jin, L. Z., Y. W. Ho, N. Abdullah, M. A. Ali, and S. Jalaludin.** 1998 Note: Lack of influence of adherent *Lactobacillus* isolates on the attachment of *Escherichia coli* to the intestinal epithelial cells of chicken *in vitro*. *J. Appl. Microbiol.* **84**:1171–1174.

1726. **Jin, L. Z., R. R. Marquardt, S. K. Baidoo, and A. A. Frohlich.** 2000. Characterization and purification of porcine small intestinal mucus receptor for *Escherichia coli* K88ac fimbrial adhesin. *FEMS Immunol. Med. Microbiol.* **27**:17–22.

1727. **Jin, L. Z., R. R. Marquardt, and X. Zhao.** 2000. A strain of *Enterococcus faecium* (18C23) inhibits adhesion of enterotoxigenic *Escherichia coli* K88 to porcine small intestine mucus. *Appl. Environ. Microbiol.* **66**:4200–4204.

1728. **Jin, S., K. S. Ishimoto, and L. Lory.** 1994. PilR, a transcriptional regulator of piliation in *Pseudomonas aeruginosa*: binds to a cis-acting sequence upstream of the pilin gene promoter. *Mol. Microbiol.* **14**:1049–1057.

1729. **Jin, S., A. Joe, J. Lynett, E. K. Hani, P. Sherman, and V. L. Chan.** 2001. JlpA, a novel surface-exposed lipoprotein specific to *Campylobacter jejuni,* mediates adherence to host epithelial cells. *Mol. Microbiol.* **39**:1225–1236.

1730. **Joe, A., R. Verdon, S. Tzipori, G. T. Keusch, and H. D. Ward.** 1998. Attachment of *Cryptosporidium parvum* sporozoites to human intestinal epithelial cells. *Infect. Immun.* **66**:3429–3432.

1731. **Joh, D., P. Speziale, S. Gurusiddappa, J. Manor, and M. Höök.** 1998. Multiple specificities of the staphylococcal and streptococcal fibronectin-binding microbial surface components recognizing adhesive matrix molecules. *Eur. J. Biochem.* **258**:897–905.

1732. **Joh, D., E. R. Wann, B. Kreikemeyer, P. Speziale, and M. Höök.** 1999. Role of fibronectin-binding MSCRAMMs in bacterial adherence and entry into mammalian cells. *Matrix Biol.* **18**:211–223.

1733. **Johanson, I., R. Lindstedt, and C. Svanborg.** 1992. Roles of the *pap*- and *prs*-encoded adhesins in *Escherichia coli* adherence to human uroepithelial cells. *Infect. Immun.* **60**:3416–3422.

1734. **Johansson, A., J. I. Flock, and O. Svensson.** 2001. Collagen and fibronectin binding in experimental staphylococcal osteomyelitis. *Clin. Orthop.* **382**:241–246.

1735. **Johansson, C., T. Nilsson, A. Olsen, and M. J. Wick.** 2001. The influence of curli, a MHC-I-binding bacterial surface structure, on macrophage-T cell interactions. *FEMS Immunol. Med. Microbiol.* **30**:21–29.

1736. **Johansson, L., and K.-A. Karlsson.** 1998. Selective binding by *Helicobacter pylori* of leucocyte gangliosides with 3-linked sialic acid, as identified by a new approach of linkage analysis. *Glycoconj. J.* **15**:713–721.

1737. **John, G., M. Shields, F. Austin, and S. McGinnis.** 1998. Increased *Pseudomonas aeruginosa* adhesion following air drying of etafilcon A soft contact lenses. *CLAO J.* **24**:236–238.

1738. **John, S. F., V. F. Hillier, P. S. Handley, and M. R. Derrick.** 1995. Adhesion of staphylococci to polyurethane and hydrogel-coated polyurethane catheters assayed by an improved radiolabelling technique. *J. Med. Microbiol.* **43**:133–140.

1739. **John, S. F., M. R. Derrick, A. E. Jacob, and P. S. Handley.** 1996. The combined effects of plasma and hydrogel coating on adhesion of *Staphylococcus epidermidis* and *Staphylococcus aureus* to polyurethane catheters. *FEMS Microbiol. Lett.* **144**:241–247.

1740. **John, T., A. B. Kopstein, O. C. John, C. I. Lai, and R. B. Carey.** 2001. *In vitro* adherence of *Staphylococcus epidermidis* to silicone punctual plugs and collagen implants. *J. Cataract Refract. Surg.* **27**:1298–1302.

1741. **Johnson, C. E., J. N. Maslow, D. C. Fattlar, K. S. Adams, and R. D. Arbeit.** 1993. The role of bacterial adhesins in the outcome of childhood urinary tract infections. *Am. J. Dis. Child.* **147**:1090–1093.

1742. **Johnson, C. M.** 1993. *Staphylococcus aureus* binding to cardiac endothelial cells is partly mediated by a 130 kilodalton glycoprotein. *J. Lab. Clin. Med.* **121**:675–682.

1743. **Johnson, D. E., R. G. Russell, C. V. Lockatell, J. C. Zulty, J. W. Warren, and H. L. T. Mobley.** 1993. Contribution of *Proteus mirabilis* urease to persistence, urolithiasis, and acute pyelonephritis in a mouse model of ascending urinary tract infection. *Infect. Immun.* **61**:2748–2754.

1744. **Johnson, D. E., C. V. Lockatell, R. G. Russell, J. R. Hebel, M. D. Island, A. Stapleton, W. E. Stamm, and J. W. Warren.** 1998. Comparison of *Escherichia coli* strains recovered from human cystitis and pyelonephritis infections in transurethrally challenged mice. *Infect. Immun.* **66**:3059–3065.

1745. **Johnson, J. R.** 1995. Epidemiological considerations in studies of microbial adhesion. *Methods Enzymol.* **253**:167–179.

1746. **Johnson, J. R.** 1998. *papG* alleles among *Escherichia coli* strains causing urosepsis: associations with other bacterial characteristics and host compromise. *Infect. Immun.* **66**:4568–4571.

1747. **Johnson, J. R., T. Berrgren, and J. C. Manivel.** 1992. Histopathologic-microbiologic correlates of invasiveness in a mouse model of ascending unobstructed urinary tract infection. *J. Infect. Dis.* **165**:299–305.

1748. **Johnson, J. R., J. L. Swanson, and M. A. Neill.** 1992. Avian P_1 antigens inhibit agglutination mediated by P fimbriae of uropathogenic *Escherichia coli*. *Infect. Immun.* **60**:578–583.

1749. **Johnson, J. R., and A. E. Ross.** 1993. P_1-antigen-containing avian egg whites as inhibitors of P adhesins among wild-type *Escherichia coli* strains from patients with urosepsis. *Infect. Immun.* **61**:4902–4905.

1750. **Johnson, J. R., and T. Berggren.** 1994. Pigeon and dove eggwhite protect mice against renal infection due to P fimbriated *Escherichia coli*. *Am. J. Med. Sci.* **307**:335–339.

1751. **Johnson, J. R., K. M. Skubitz, B. J. Nowicki, K. Jacques-Palaz, and R. M. Rakita.** 1995. Nonlethal adherence to human neutrophils mediated by Dr antigen-specific adhesins of *Escherichia coli*. *Infect. Immun.* **63**:309–316.

1752. **Johnson, J. R., and J. J. Brown.** 1996. A novel multiply primed polymerase chain reaction assay for identification of variant *papG* genes encoding the Gal(alpha 1–4)Gal-binding PapG adhesins of *Escherichia coli*. *J. Infect. Dis.* **173**:920–926.

1753. **Johnson, J. R., A. E. Stapleton, T. A. Russo, F. Scheutz, J. J. Brown, and J. N. Maslow.** 1997. Characteristics and prevalence within serogroup O4 of a J96-like clonal group of uropathogenic *Escherichia coli* O4:H5 containing the class I and class III alleles of *papG*. *Infect. Immun.* **65**:2153–2159.

1754. **Johnson, J. R., J. L. Swanson, T. J. Barela, and J. J. Brown.** 1997. Receptor specificities of variant Gal(alpha1–4)Gal-binding PapG adhesins of uropathogenic *Escherichia coli* as assessed by hemagglutination phenotypes. *J. Infect. Dis.* **175**:373–381.

1755. **Johnson, J. R., T. A. Russo, F. Scheutz, J. J. Brown, L. Zhang, K. Palin, C. Rode, C. Bloch, C. F. Marrs, and B. Foxman.** 1997. Discovery of disseminated J96-like strains of uropathogenic *Escherichia coli* O4:H5 containing genes for both PapG$_{J96}$ (class I) and PrsG$_{J96}$ (class III) Gal(α1–4)Gal-binding adhesins. *J. Infect. Dis.* **175**:983–988.

1756. **Johnson, J. R., J. J. Brown, and J. N. Maslow.** 1998. Clonal distribution of the three alleles of the Gal(alpha1–4)Gal-specific adhesin gene *papG* among *Escherichia coli* strains from patients with bacteremia. *J. Infect. Dis.* **177**:651–661.

1757. **Johnson, J. R., J. J. Brown, and P. Ahmed.** 1998. Diversity of hemagglutination phenotypes among P-fimbriated wild-type strains of *Escherichia coli* in relation to *papG* allele repertoire. *Clin. Diagn. Lab. Immunol.* **5**:160–170.

1758. **Johnson, J. R., T. T. O'Bryan, D. A. Low, G. Ling, P. Delavari, C. Fasching, T. A. Russo, U. Carlino, and A. L. Stell.** 2000. Evidence of commonality between canine and human extraintestinal pathogenic *Escherichia coli* strains that express papG allele III. *Infect. Immun.* **68**:3327–3336.

1759. **Johnson, P. D., F. Oppedisano, V. Bennett-Wood, G. L. Gilbert, and R. M. Robins-Browne.** 1996. Sporadic invasion of cultured epithelial cells by *Haemophilus influenzae* type b. *Infect. Immun.* **64**:1051–1053.

1760. **Johnson-Henry, K., J. L. Wallace, N. S. Basappa, R. Soni, G. K. Wu, and P. M. Sherman.** 2001. Inhibition of attaching and effacing lesion formation following enteropathogenic *Escherichia coli* and Shiga toxin-producing *E. coli* infection. *Infect. Immun.* **69:**7152–7158.

1761. **Johnston, J. L., S. J. Billington, V. Haring, and J. I. Rood.** 1995. Identification of fimbrial assembly genes from *Dichelobacter nodosus:* evidence that *fimP* encodes the type-IV prepilin peptidase. *Gene* **161:**21–26.

1762. **Johnston, J. L., S. J. Billington, V. Haring, and J. I. Rood.** 1998. Complementation analysis of the *Dichelobacter nodosus fimN, fimO,* and *fimP* genes in *Pseudomonas aeruginosa* and transcriptional analysis of the *fimNOP* gene region. *Infect. Immun.* **66:**297–304.

1763. **Jones, A. C., R. P. Logan, S. Foynes, A. Cockayne, B. W. Wren, and C. W. Penn.** 1997. A flagellar sheath protein of *Helicobacter pylori* is identical to HpaA, a putative *N*-acetyl-neuraminyllactose-binding hemagglutinin, but is not an adhesin for AGS cells. *J. Bacteriol.* **179:**5643–5647.

1764. **Jones, A. L., D. DeShazer, and D. E. Woods.** 1997. Identification and characterization of a two-component regulatory system involved in invasion of eukaryotic cells and heavy-metal resistance in *Burkholderia pseudomallei. Infect. Immun.* **65:**4972–4977.

1765. **Jones, C. H., F. Jacob-Dubuisson, K. Dodson, M. Kuehn, L. Slonim, R. Striker, and S. J. Hultgren.** 1992. Adhesin presentation in bacteria requires molecular chaperones and ushers. *Infect. Immun.* **60:**4445–4451.

1766. **Jones, C. H., J. S. Pinkner, A. V. Nicholes, L. N. Slonim, S. N. Abraham, and S. J. Hultgren.** 1993. FimC is a periplasmic PapD-like chaperone that directs assembly of type 1 pili in bacteria. *Proc. Natl. Acad. Sci. USA* **90:**8397–8401.

1767. **Jones, C. H., J. S. Pinkner, R. Roth, J. Heuser, A. V. Nicholes, S. N. Abraham, and S. J. Hultgren.** 1995. FimH adhesin of type 1 pili is assembled into a fibrillar tip structure in the *Enterobacteriaceae. Proc. Natl. Acad. Sci. USA* **92:**2081–2085.

1768. **Jones, C. H., P. N. Danese, J. S. Pinkner, T. J. Silhavy, and S. J. Hultgren.** 1997. The chaperone-assisted membrane release and folding pathway is sensed by two signal transduction systems. *EMBO J.* **16:**6394–6406.

1769. **Jones, D. H., B. W. McBride, H. Jeffery, D. T. O'Hagan, A. Robinson, and G. H. Farrar.** 1995. Protection of mice from *Bordetella pertussis* respiratory infection using microencapsulated pertussis fimbriae. *Vaccine* **13:**675–681.

1770. **Jones, D. H., B. W. McBride, C. Thornton, D. T. O'Hagan, A. Robinson, and G. H. Farrar.** 1996. Orally administered microencapsulated *Bordetella pertussis* fimbriae protect mice from *B. pertussis* respiratory infection. *Infect. Immun.* **64:**489–494.

1771. **Jones, D. S., J. G. McGovern, A. D. Woolfson, and S. P. Gorman.** 1997. Role of physiological conditions in the oropharynx on the adherence of respiratory bacterial isolates to endotracheal tube poly(vinyl chloride). *Biomaterials* **18:**503–510.

1772. **Jones, G. W., D. B. Clewell, L. G. Charles, and M. M. Vickerman.** 1996. Multiple phase variation in haemolytic, adhesive and antigenic properties of *Streptococcus gordonii. Microbiology* **142:**181–189.

1773. **Jones, J. W., R. J. Scott, J. Morgan, and J. V. Pether.** 1992. A study of coagulase-negative staphylococci with reference to slime production, adherence, antibiotic resistance patterns and clinical significance. *J. Hosp. Infect.* **22:**217–227.

1774. **Jonquieres, R., J. Pizarro-Cerda, and P. Cossart.** 2001. Synergy between the N- and C-terminal domains of InlB for efficient invasion of non-phagocytic cells by *Listeria monocytogenes. Mol. Microbiol.* **42:**955–965.

1775. **Jonquières, R., Bierne, J. Mengaud, and P. Cossart.** 1998. The *inlA* gene of *Listeria monocytogenes* LO28 harbors a nonsense mutation resulting in release of internalin. *Infect. Immun.* **66:**3420–3422.

1776. **Jonson, G., J. Holmgren, and A.-M. Svennerholm.** 1992. Analysis of expression of toxin-coregulated pili in classical and El Tor *Vibrio cholerae* O1 in vitro and in vivo. *Infect. Immun.* **60:**4278–4284.

1777. **Jonson, G., M. Lebens, and J. Holmgren.** 1994. Cloning and sequencing of *Vibrio cholerae* mannose-sensitive haemagglutinin pilin gene: localization of *mshA* within a cluster of type 4 pilin genes. *Mol. Microbiol.* **13:**109–118.

1778. **Jonson, G., J. Osek, A.-M. Svennerholm, and J. Holmgren.** 1996. Immune mechanisms and protective antigens of *Vibrio cholerae* serogroup O139 as a basis for vaccine development. *Infect. Immun.* **64:**3778–3785.

1779. **Jonsson, A.-B., J. Pfeifer, and S. Normark.** 1992. *Neisseria gonorrhoeae* PilC expression provides a selective mechanism for structural diversity of pili. *Proc. Natl. Acad. Sci. USA* **89:**3204–3208.

1780. **Jonsson, A.-B., D. Ilver, P. Falk, J. Pepose, and S. Normark.** 1994. Sequence changes in the pilus subunit lead to tropism variation of *Neisseria gonorrhoeae* to human tissue. *Mol. Microbiol.* **13**:403–416.

1781. **Jonsson, A.-B., M. Rahman, and S. Normark.** 1995. Pilus biogenesis gene, *pilC*, of *Neisseria gonorrhoeae: pilC1* and *pilC2* are each part of a larger duplication of the gonococcal genome and share upstream and downstream homologous sequences with *opa* and *pil* loci. *Microbiology* **141**:2367–2377.

1782. **Jonsson, A.-B.** 1998. Identification of a human cDNA clone that mediates adherence of pathogenic *Neisseria* to non-binding cells. *FEMS Microbiol. Lett.* **162**:25–30.

1783. **Jonsson, H., E. Strom, and S. Roos.** 2001. Addition of mucin to the growth medium triggers mucus-binding activity in different strains of *Lactobacillus reuteri in vitro*. *FEMS Microbiol. Lett.* **204**:19–22.

1784. **Jordan, J. L., K. M. Berry, M. F. Balish, and D. C. Krause.** 2001. Stability and subcellular localization of cytadherence-associated protein P65 in *Mycoplasma pneumoniae*. *J. Bacteriol.* **183**:7387–7391.

1785. **Jordi, B. J., B. Dagberg, L. A. de Haan, A. M. Hamers, B. A. van der Zeijst, W. Gaastra, and B. E. Uhlin.** 1992. The positive regulator CfaD overcomes the repression mediated by histone-like protein H-NS (H1) in the CFA/I fimbrial operon of *Escherichia coli*. *EMBO J.* **11**:2627–2632.

1786. **Jordi, B. J., G. A. Willshaw, B. A. van der Zeijst, and W. Gaastra.** 1992. The complete nucleotide sequence of region 1 of the CFA/I fimbrial operon of human enterotoxigenic *Escherichia coli*. *DNA Seq.* **2**:257–263.

1787. **Jordi, B. J., I. I. op den Camp, L. A. de Haan, B. A. van der Zeijst, and W. Gaastra.** 1993. Differential decay of RNA of the CFA/I fimbrial operon and control of relative gene expression. *J. Bacteriol.* **175**:7976–7981.

1788. **Joseph, T. D., and S. K. Bose.** 1992. Surface components of HeLa cells that inhibit cytadherence of *Chlamydia trachomatis*. *FEMS Microbiol Lett.* **70**:177–180.

1789. **Jost, B. H, J. G. Songer, and S. J. Billington.** 2001. Cloning, expression, and characterization of a neuraminidase gene from *Arcanobacterium pyogenes*. *Infect. Immun.* **69**:4430–4437.

1790. **Jouravleva, E. A., G. A. McDonald, J. W. Marsh, R. K. Taylor, M. Boesman-Finkelstein, and R. A. Finkelstein.** 1998. The *Vibrio cholerae* mannose-sensitive hemagglutinin is the receptor for a filamentous bacteriophage from *V. cholerae* O139. *Infect. Immun.* **66**:2535–2539.

1791. **Jouve, M., M. I. Garcia, P. Courcoux, A. Labigne, P. Gounon, and C. le Bouguenec.** 1997. Adhesion to and invasion of HeLa cells by pathogenic *Escherichia coli* carrying the *afa-3* gene cluster are mediated by the AfaE and AfaD proteins, respectively. *Infect. Immun.* **65**:4082–4089.

1792. **Joyanes, P., A. Pascual, L. Martinez-Martinez, A. Hevia, and E. J. Perea.** 2000. *In vitro* adherence of *Enterococcus faecalis* and *Enterococcus faecium* to urinary catheters. *Eur. J. Clin. Microbiol. Infect. Dis.* **19**:124–127.

1793. **Jucker, B. A., H. Harms, and A. J. Zehnder.** 1996. Adhesion of the positively charged bacterium *Stenotrophomonas (Xanthomonas) maltophilia* 70401 to glass and Teflon. *J. Bacteriol.* **178**:5472–5479.

1794. **Jung, K. Y., J. D. Cha, S. H. Lee, W. H. Woo, D. S. Lim, B. K. Choi, and K. J. Kim.** 2001. Involvement of staphylococcal protein A and cytoskeletal actin in *Staphylococcus aureus* invasion of cultured human oral epithelial cells. *J. Med. Microbiol.* **50**:35–41.

1795. **Juntunen, M., P. V. Kirjavainen, A. C. Ouwehand, S. J. Salminen, and E. Isolauri.** 2001. Adherence of probiotic bacteria to human intestinal mucus in healthy infants and during rotavirus infection. *Clin. Diagn. Lab. Immunol.* **8**:293–296.

1796. **Juskova, E., and I. Ciznar.** 1993. Application of biotinylated and ^{32}P probes for detection of P-fimbriae in urinary *E. coli*. *Folia Microbiol.* **38**:259–263.

1797. **Juskova, E., and I. Ciznar.** 1994. Occurrence of genes for P and S fimbriae and hemolysin in urinary *Escherichia coli*. *Folia Microbiol.* **39**:159–161.

1798. **Kaack, M. B., L. N. Martin, S. B. Svenson, G. Baskin, R. H. Steele, and J. A. Roberts.** 1993. Protective anti-idiotype antibodies in the primate model of pyelonephritis. *Infect. Immun.* **61**:2289–2295.

1799. **Kabha, K., L. Nissimov, A. Athamna, Y. Keisari, H. Parolis, L. A. Parolis, R. M. Grue, J. Schlepper-Schafer, A. R. Ezekowitz, D. E. Ohman, and I. Ofek.** 1995. Relationships among capsular structure, phagocytosis, and mouse virulence in *Klebsiella pneumoniae*. *Infect. Immun.* **63**:847–852.

1800. **Kabha, K., J. Schmegner, Y. Keisari, H. Parolis, J. Schlepper-Schaeffer, and I. Ofek.** 1997. SP-A enhances phagocytosis of

Klebsiella by interaction with capsular polysaccharides and alveolar macrophages. *Am. J. Physiol.* **272:**L344–L352.

1801. **Kabir, A. M., Y. Aiba, A. Takagi, S. Kamiya, T. Miwa, and Y. Koga.** 1997. Prevention of *Helicobacter pylori* infection by lactobacilli in a gnotobiotic murine model. *Gut* **41:**49–55.

1802. **Kachlany, S. C., P. J. Planet, R. Desalle, D. H. Fine, D. H. Figurski, and J. B. Kaplan.** 2001. *flp-1,* the first representative of a new pilin gene subfamily, is required for non-specific adherence of *Actinobacillus actinomycetemcomitans*. *Mol. Microbiol.* **40:**542–554.

1803. **Kachlany, S. C., P. J. Planet, M. K. Bhattacharjee, E. Kollia, R. DeSalle, D. H. Fine, and D. H. Figurski.** 2000. Nonspecific adherence by *Actinobacillus actinomycetemcomitans* requires genes widespread in bacteria and archaea. *J. Bacteriol.* **182:**6169–6176.

1804. **Kadioglu, A., J. A. Sharpe, I. Lazou, C. Svanborg, C. Ockleford, T. J. Mitchell, and P. W. Andrew.** 2001. Use of green fluorescent protein in visualisation of pneumococcal invasion of broncho-epithelial cells *in vivo*. *FEMS Microbiol. Lett.* **194:**105–110.

1805. **Kadowaki, T., K. Nakayama, F. Yoshimura, K. Okamoto, N. Abe, and K. Yamamoto.** 1998. Arg-gingipain acts as a major processing enzyme for various cell surface proteins in *Porphyromonas gingivalis*. *J. Biol. Chem.* **273:**29072–29076.

1806. **Kadry, A. A., A. Tawfik, A. A. Abu el-Asrar, and A. M. Shibl.** 1999. Reduction of mucoid *Staphylococcus epidermidis* adherence to intraocular lenses by selected antimicrobial agents. *Chemotherapy* **45:**56–60.

1807. **Kagermeier-Callaway, A. S., B. Willershausen, T. Frank, and E. Stender.** 2000. *In vitro* colonisation of acrylic resin denture base materials by *Streptococcus oralis* and *Actinomyces viscosus*. *Int. Dent. J.* **50:**79–85.

1808. **Kahana, H., J. Grünberg, Y. Bartov, R. Perry, N. Smorodinsky, I. Boldur, E. Weiss, and J. Goldhar.** 1994. Binding of *Escherichia coli* recognizing N-blood group antigen to the erythroleukemic cell line K562 expressing glycophorin A. *Immunol. Infect. Dis.* **4:**161–165.

1809. **Kahane, I., and S. Horowitz.** 1993. Adherence of mycoplasma to cell surfaces. *Sub-Cell. Biochem.* **20:**225–241.

1810. **Kahane, I.** 1995. Adhesion of mycoplasmas. *Methods Enzymol.* **253:**367–373.

1811. **Kahn, A. S., and D. M. Schifferli.** 1994. A minor 987P protein different from the structural fimbrial subunit is the adhesin. *Infect. Immun.* **62:**4233–4243.

1812. **Kalfas, S., Z. Tigyi, M. Wikstrom, and A. S. Naidu.** 1992. Laminin binding to *Prevotella intermedia*. *Oral Microbiol. Immunol.* **7:**235–239.

1813. **Kallman, J., J. Schollin, S. Hakansson, A. Andersson, and E. Kihlstrom.** 1993. Adherence of group B streptococci to human endothelial cells *in vitro*. *APMIS* **101:**403–408.

1814. **Kallstrom, H., M. K. Liszewski, J. P. Atkinson, and A. B. Jonsson.** 1997. Membrane cofactor protein (MCP or CD46) is a cellular pilus receptor for pathogenic *Neisseria*. *Mol. Microbiol.* **25:**639–647.

1815. **Kallstrom, H., M. S. Islam, P. O. Berggren, and A. B. Jonsson.** 1998. Cell signaling by the type IV pili of pathogenic *Neisseria*. *J. Biol. Chem.* **273:**21777–21782.

1816. **Kallstrom, H., P. Hansson-Palo, and A. B. Jonsson.** 2000. Cholera toxin and extracellular Ca^{2+} induce adherence of non-piliated *Neisseria*: evidence for an important role of G-proteins and Rho in the bacteria-cell interaction. *Cell. Microbiol.* **2:**341–351.

1817. **Kallstrom, H., D. Blackmer Gill, B. Albiger, M. K. Liszewski, J. P. Atkinson, and A. B. Jonsson.** 2001. Attachment of *Neisseria gonorrhoeae* to the cellular pilus receptor CD46: identification of domains important for bacterial adherence. *Cell. Microbiol.* **3:**133–143.

1818. **Kalmokoff, M. L., J. W. Austin, X. D. Wan, G. Sanders, S. Banerjee, and J. M. Farber.** 2001. Adsorption, attachment and biofilm formation among isolates of *Listeria monocytogenes* using model conditions. *J. Appl. Microbiol.* **91:**725–734.

1819. **Kaltenbach, L., B. Braaten, J. Tucker, M. Krabbe, and D. Low.** 1998. Use of a two-color genetic screen to identify a domain of the global regulator Lrp that is specifically required for *pap* phase variation. *J. Bacteriol.* **180:**1224–1231.

1820. **Kaltenbach, L. S., B. A. Braaten, and D. A. Low.** 1995. Specific binding of PapI to Lrp-pap DNA complexes. *J. Bacteriol.* **177:**6449–6455.

1821. **Kamaguchi, A., H. Baba, M. Hoshi, and K. Inomata.** 1994. Coaggregation between *Porphyromonas gingivalis* and mutans streptococci. *Microbiol. Immunol.* **38:**457–460.

1822. **Kamaguchi, A., H. Baba, M. Hoshi, and K. Inomata.** 1995. Effect of *Porphyromonas gingivalis* ATCC 33277 vesicle on adherence of *Streptococcus mutans* OMZ 70 to the experimental pellicle. *Microbiol. Immunol.* **39:**521–524.

1823. Kamisago, S., M. Iwamori, T. Tai, K. Mitamura, Y. Yazaki, and K. Sugano. 1996. Role of sulfatides in adhesion of *Helicobacter pylori* to gastric cancer cells. *Infect. Immun.* **64:**624–628.

1824. Kamiya, S., H. Yamaguchi, T. Osaki, H. Taguchi, M. Fukuda, H. Kawakami, and H. Hirano. 1999. Effect of an aluminum hydroxide-magnesium hydroxide combination drug on adhesion, IL-8 inducibility, and expression of HSP60 by *Helicobacter pylori*. *Scand. J. Gastroenterol.* **34:**663–670.

1825. Kamiya, S., T. Osaki, J. Kumada, H. Yamaguchi, and H. Taguchi. 1997. Effect of sofalcone on adherence, production of vacuolating toxin, and induction of interleukin-8 secretion by *Helicobacter pylori*. *J. Clin. Gastroenterol.* **25:**172–178.

1826. Kamperman, L., and S. M. Kirov. 1993. Pili and the interaction of *Aeromonas* species with human peripheral blood polymorphonuclear cells. *FEMS Immunol. Med. Microbiol.* **7:**187–195.

1827. Kaneda, K., T. Masuzawa, K. Yasugami, T. Suzuki, Y. Suzuki, and Y. Yanagihara. 1997. Glycosphingolipid-binding protein of *Borrelia burgdorferi* sensu lato. *Infect. Immun.* **65:**3180–3185.

1828. Kaneda, K., T. Masuzawa, M. M. Simon, E. Isogai, H. Isogai, K. Yasugami, T. Suzuki, Y. Suzuki, and Y. Yanagihara. 1998. Infectivity and arthritis induction of *Borrelia japonica* on SCID mice and immune competent mice: possible role of galactosylceramide binding activity on initiation of infection. *Microbiol. Immunol.* **42:**171–175.

1829. Kang, G., M. M. Mathan, and V. I. Mathan. 1995. Evaluation of a simplified HEp-2 cell adherence assay for *Escherichia coli* isolated from south Indian children with acute diarrhea and controls. *J. Clin. Microbiol.* **33:**2204–2205.

1830. Kang, G., S. Sheela, M. M. Mathan, and V. I. Mathan. 1999. Prevalence of enteroaggregative and other HEp-2 cell adherent *Escherichia coli* in asymptomatic rural south Indians by longitudinal sampling. *Microbios* **100:**57–66.

1831. Kang, G., K. A. Balasubramanian, R. Koshi, M. M. Mathan, and V. I. Mathan. 1998. Salicylate inhibits fimbriae mediated HEp-2 cell adherence of and haemagglutination by enteroaggregative *Escherichia coli*. *FEMS Microbiol. Lett.* **166:**257–265.

1832. Kankaanpaa, P., E. Tuomola, H. el-Nezami, J. Ahokas, and S. J. Salminen. 2000. Binding of aflatoxin B1 alters the adhesion properties of *Lactobacillus rhamnosus* strain GG in a Caco-2 model. *J. Food Prot.* **63:**412–414.

1833. Kankaanpaa, P. E., S. J. Salminen, E. Isolauri, and Y. K. Lee. 2001. The influence of polyunsaturated fatty acids on probiotic growth and adhesion. *FEMS Microbiol. Lett.* **194:**149–153.

1834. Kanoe, M., Y. Koyanagi, C. Kondo, K. Mamba, T. Makita, and K. Kai. 1998. Location of haemagglutinin in bacterial cells of *Fusobacterium necrophorum* subsp. *necrophorum*. *Microbios* **96:**33–38.

1835. Kanzaki, H., and J. Arata. 1992. Role of fibronectin in the adherence of *Staphylococcus aureus* to dermal tissues. *J. Dermatol. Sci.* **4:**87–94.

1836. Kanzaki, H., Y. Morishita, H. Akiyama, and J. Arata. 1996. Adhesion of *Staphylococcus aureus* to horny layer: role of fibrinogen. *J. Dermatol. Sci.* **12:**132–139.

1837. Kapczynski, D. R., R. J. Meinersmann, and M. D. Lee. 2000. Adherence of *Lactobacillus* to intestinal 407 cells in culture correlates with fibronectin binding. *Curr. Microbiol.* **41:**136–141.

1838. Kaper, J. B., T. K. McDaniel, K. G. Jarvis, and O. Gomez-Duarte. 1997. Genetics of virulence of enteropathogenic *E. coli*. *Adv. Exp. Med. Biol.* **412:**279–287.

1839. Karaolis, D. K., T. K. McDaniel, J. B. Kaper, and E. C. Boedeker. 1997. Cloning of the RDEC-1 locus of enterocyte effacement (LEE) and functional analysis of the phenotype on HEp-2 cells. *Adv. Exp. Med. Biol.* **412:**241–245.

1840. Karaolis, D. K., J. A. Johnson, C. C. Bailey, E. C. Boedeker, J. B. Kaper, and P. R. Reeves. 1998 A *Vibrio cholerae* pathogenicity island associated with epidemic and pandemic strains. *Proc. Natl Acad. Sci. USA* **95:**3134–3149.

1841. Karjalainen, T., M. C. Barc, A. Collignon, S. Trolle, H. Boureau, J. Cotte-Laffitte, and P. Bourlioux. 1994. Cloning of a genetic determinant from *Clostridium difficile* involved in adherence to tissue culture cells and mucus. *Infect. Immun.* **62:**4347–4355.

1842. Karlsson, A., M. Markfjall, H. Lundqvist, N. Strømberg, and C. Dahlgren. 1995. Detection of glycoprotein receptors on blotting membranes by binding of live bacteria and amplification by growth. *Anal. Biochem.* **224:**390–394.

1843. Karlsson, A., and C. Dahlgren. 1996. Secretion of type-1-fimbriae binding proteins from human neutrophil granulocytes. *Inflammation* **20**:389–400.

1844. **Karlsson, A., S. R. Carlsson, and C. Dahlgren.** 1994. Identification of the lysosomal membrane glycoprotein Lamp-1 as a receptor for type-1-fimbriated (mannose-specific) *Escherichia coli*. *Biochem. Biophys. Res. Commun.* **219**:168–172.

1845. **Karlsson, K.-A., J. Angstrom, J. Bergstrom, and B. Lanne.** 1992. Microbial interaction with animal cell surface carbohydrates. *APMIS* **27**:71–83.

1846. **Karlsson, K.-A.** 1995. Microbial recognition of target-cell glycoconjugates. *Curr. Opin. Struct. Biol.* **5**:622–635.

1847. **Karmaker, S., A. G. Chaudhuri, and U. Ganguly.** 1996. Comparison of cytosolic levels of calcium and G actin in diffuse and localised adherent *Escherichia coli*-infected HeLa cells. *FEMS Microbiol. Lett.* **135**:245–249.

1848. **Karmaker, S., A. N. Ghosh, D. Dey, G. B. Nair, and U. Ganguly.** 1994. Ultrastructural changes in HeLa cells associated with enteroadherent *Escherichia coli* isolated from infants with diarrhoea in Calcutta. *J. Diarrhoeal Dis. Res.* **12**:274–278.

1849. **Karunasagar, I., B. Senghaas, G. Krohne, and W. Goebel.** 1994. Ultrastructural study of *Listeria monocytogenes* entry into cultured human colonic epithelial cells. *Infect. Immun.* **62**:3554–3558.

1850. **Kataoka, K., A. Amano, S. Kawabata, H. Nagata, S. Hamada, and S. Shizukuishi.** 1999. Secretion of functional salivary peptide by *Streptococcus gordonii* which inhibits fimbria-mediated adhesion of *Porphyromonas gingivalis*. *Infect. Immun.* **67**:3780–3785.

1851. **Kataoka, K., A. Amano, M. Kuboniwa, H. Horie, H. Nagata, and S. Shizukuishi.** 1997. Active sites of salivary proline-rich protein for binding to *Porphyromonas gingivalis* fimbriae. *Infect. Immun.* **65**:3159–3164.

1852. **Katayama, M., X. Xu, R. D. Specian, and E. A. Deitch.** 1997. Role of bacterial adherence and the mucus barrier on bacterial translocation: effects of protein malnutrition and endotoxin in rats. *Ann. Surg.* **225**:317–326.

1853. **Katerov, V., A. Andreev, C. Schalen, and A. A. Totolian.** 1998. Protein F, a fibronectin-binding protein of *Streptococcus pyogenes,* also binds human fibrinogen: isolation of the protein and mapping of the binding region. *Microbiology* **144**:119–126.

1854. **Katouli, M., C. G. Netteblandt, V. Muratov, O. Ljungqvist, T. Bark, T.** Svenberg, and R. Møllby. 1997. Selective translocation of coliform bacteria adhering to caecal epithelium of rats during catabolic stress. *J. Med. Microbiol.* **46:** 571–578.

1855. **Katouli, M., F. Shokouhi, A. A. Farhoudi-Moghaddam, and S. Amini.** 1992. Occurrence of colonization factor antigens I & II in enterotoxigenic *Escherichia coli* associated diarrhoea in Iran & correlation with severity of disease. *Indian J. Med. Res.* **95:**115–120.

1856. **Kaufman, M. R., C. E. Shaw, I. D. Jones, and R. K. Taylor.** 1993. Biogenesis and regulation of the *Vibrio cholerae* toxincoregulated pilus: analogies to other virulence factor secretory systems. *Gene* **126:**43–49.

1857. **Kaul, A., B. J. Nowicki, M. G. Martens, P. Goluszko, A. Hart, M. Nagamani, D. Kumar, T. Q. Pham, and S. Nowicki.** 1994. Decay-accelerating factor is expressed in the human endometrium and may serve as the attachment ligand for Dr pili of *Escherichia coli*. *Am. J. Reprod. Immunol.* **32:**194–199.

1858. **Kaul, A. K., D. Kumar, M. Nagamani, P. Goluszko, S. Nowicki, and BJ. Nowicki.** 1996. Rapid cyclic changes in density and accessibility of endometrial ligands for *Escherichia coli* Dr fimbriae. *Infect. Immun.* **64:**611–615.

1859. **Kawabata, S., M. Torii, T. Minami, T. Fujiwara and S. Hamada.** 1993. Effects of selected surfactants on purified glucosyltransferases from mutans streptococci and cellular adherence to smooth surfaces. *J. Med. Microbiol.* **38:**54–60.

1860. **Kawabata, S., H. Kuwata, I. Nakagawa, S. Morimatsu, K. Sano, and S. Hamada.** 1999. Capsular hyaluronic acid of group A streptococci hampers their invasion into human pharyngeal epithelial cells. *Microb. Pathog.* **27:**71–80.

1861. **Kawai, K., M. Urano, and S. Ebisu.** 2000. Effect of surface roughness of porcelain on adhesion of bacteria and their synthesizing glucans. *J. Prosthet. Dent.* **83:**664–667.

1862. **Kawai, K., and T. Takaoka.** 2001. Inhibition of bacterial and glucan adherence to various light-cured fluoride-releasing restorative materials. *J. Dent.* **29:**119–122.

1863. **Kawasaki, H., G. Sugumaran, and J. E. Silbert.** 1996. Cell surface glycosaminoglycans are not involved in the adherence of *Helicobacter pylori* to cultured Hs 198.St human gastric cells, Hs 746T human gastric adenocarcinoma cells, or HeLa cells. *Glycoconj. J.* **13:**873–877.

1864. **Kawasaki, Y., S. Tazume, K. Shimizu,**

H. Matsuzawa, S. Dosako, H. Isoda, M. Tsukiji, R. Fujimura, Y. Muranaka, and H. Isihida. 2000. Inhibitory effects of bovine lactoferrin on the adherence of enterotoxigenic *Escherichia coli* to host cells. *Biosci. Biotechnol. Biochem.* **64:**348–354.

1865. Kawata, Y., S. Hanazawa, S. Amano, Y. Murakami, T. Matsumoto, K. Nishida, and S. Kitano. 1994. *Porphyromonas gingivalis* fimbriae stimulate bone resorption in vitro. *Infect. Immun.* **62:**3012–3016.

1866. Kawata, Y., H. Iwasaka, S. Kitano, and S. Hanazawa. 1997. *Porphyromonas gingivalis* fimbria-stimulated bone resorption is inhibited through binding of the fimbriae to fibronectin. *Infect. Immun.* **65:**815–817.

1867. Keeler, C. L., Jr., L. L. Hnatow, P. L. Whetzel, and J. E. Dohms. 1996. Cloning and characterization of a putative cytadhesin gene (*mgc1*) from *Mycoplasma gallisepticum*. *Infect. Immun.* **64:**1541–1547.

1868. Keisari, Y., K. Kabha, L. Nissimov, J. Schlepper-Schafer, and I. Ofek. 1997. Phagocyte-bacteria interactions. *Adv. Dent. Res.* **11:**43–49.

1869. Kelle, K., J. M. Pages, and J. M. Bolla. 1998. A putative adhesin gene cloned from *Campylobacter jejuni*. *Res. Microbiol.* **149:** 723–733.

1870. Kellens, J., M. Persoons, M. Vaneechoutte, F. van Tiel, and E. Stobberingh. 1995. Evidence of lectin-mediated adherence of *Moraxella catarrhalis*. *Infection* **23:**37–41.

1871. Keller, R., M. Z. Pedroso, R. Ritchmann, and R. M. Silva. 1998. Occurrence of virulence-associated properties in *Enterobacter cloacae*. *Infect. Immun.* **66:**645–649.

1872. Kelly, C. G., V. Booth, H. Kendal, J. M. Slaney, M. A. Curtis, and T. Lehner. 1997. The relationship between colonization and haemagglutination inhibiting and B cell epitopes of *Porphyromonas gingivalis*. *Clin. Exp. Immunol.* **110:**285–291.

1873. Kelly, C. G., S. Todryk, H. L. Kendal, G. H. Munro, and T. Lehner. 1995. T-cell, adhesion, and B-cell epitopes of the cell surface *Streptococcus mutans* protein antigen I/II. *Infect. Immun.* **63:**3649–3658.

1874. Kelly, C. G., J. S. Younson, B. Y. Hikmat, S. M. Todryk, M. Czisch, P. I. Haris, I. R. Flindall, C. Newby, A. I. Mallet, J. K. Ma, and T. Lehner. 1999. A synthetic peptide adhesion epitope as a novel antimicrobial agent. *Nat. Biotechnol.* **17:**42–47.

1875. Kelly, G., S. Prasannan, S. Daniell, K.

Fleming, G. Frankel, G. Dougan, I. Connerton, and S. Matthews. 1999. Structure of the cell-adhesion fragment of intimin from enteropathogenic *Escherichia coli*. *Nat. Struct. Biol.* **6:**313–318.

1876. Kelly, S. M., B. A. Bosecker, and R. Curtiss, III. 1992. Characterization and protective properties of attenuated mutants of *Salmonella choleraesuis*. *Infect. Immun.* **60:**4881–4890.

1877. Kendall, R. W., C. P. Duncan, J. A. Smith, and J. H. Ngui-Yen. 1996. Persistence of bacteria on antibiotic loaded acrylic depots. *Clin. Orthopaed. Relat. Res.* **329:**273–280.

1878. Kennan, R., O. Soderlind, and P. Conway. 1995. Presence of F107, 2134P and Av24 fimbriae on strains of *Escherichia coli* isolated from Swedish piglets with diarrhoea. *Vet. Microbiol.* **43:**123–129.

1879. Kennan, R. M., R. P. Moncktor, B. M. McDougall, and P. L. Conway. 1995. Confirmation that DNA encoding the major fimbrial subunit of Av24 fimbriae is homologous to DNA encoding the major fimbrial subunit of F107 fimbriae. *Microb. Pathog.* **18:**67–72.

1880. Kenny, B., R. DeVinney, M. Stein, D. J. Reinscheid, E. A. Frey, and B. B. Finlay. 1997. Enteropathogenic *E. coli* (EPEC) transfers its receptor for intimate adherence into mammalian cells. *Cell* **91:**511–520.

1881. Keogh, J. R., and J. W. Eaton. 1994. Albumin binding surfaces for biomaterials. *J. Lab. Clin. Med.* **124:**537–545.

1882. Kernéis, S., G. Chauviere, A. Darfeuille-Michaud, D. Aubel, M. H. Coconnier, B. Joly, and A. L. Servin. 1992. Expression of receptors for enterotoxigenic *Escherichia coli* during enterocytic differentiation of human polarized intestinal epithelial cells in culture. *Infect. Immun.* **60:**2572–2580.

1883. Kernéis, S., M. F. Bernet, M. H. Coconnier, and A. L. Servin. 1994. Adhesion of human enterotoxigenic *Escherichia coli* to human mucus secreting HT-29 cell subpopulations in culture. *Gut* **35:**1449–1454.

1884. Kernéis, S., J. M. Gabastou, M. F. Bernet-Camard, M. H. Coconnier, B. J. Nowicki, and A. L. Servin. 1994. Human cultured intestinal cells express attachment sites for uropathogenic *Escherichia coli* bearing adhesins of the Dr adhesin family. *FEMS Microbiol. Lett.* **119:**27–32.

1885. Kervella, M., J. M. Pages, Z. Pei, G. Grollier, M. J. Blaser, and J. L. Fauchere. 1993. Isolation and characterization of two

Campylobacter glycine-extracted proteins that bind to HeLa cell membranes. *Infect. Immun.* **61:**3440–3448.

1886. **Ketley, J. M.** 1997. Pathogenesis of enteric infection by *Campylobacter. Microbiology* **143:**5–21.

1887. **Ketterer, M. R., J. Q. Shao, D. B. Hornick, B. Buscher, V. K. Bandi, and M. A. Apicella.** 1999. Infection of primary human bronchial epithelial cells by *Haemophilus influenzae:* macropinocytosis as a mechanism of airway epithelial cell entry. *Infect. Immun.* **67:**4161–4170.

1888. **Ketyi, I.** 1994. Effectiveness of antibiotics on the autochthonous *Escherichia coli* of mice in the intestinal biofilm versus its planktonic phase. *Acta Microbiol. Immunol. Hung.* **41:**189–195.

1889. **Keulers, R. A., J. C. Maltha, F. H. Mikx, and J. M. Wolters-Lutgerhorst.** 1993. Attachment of *T. denticola* strains ATCC 33520, ATCC 35405, B11 and Ny541 to a morphologically distinct population of rat palatal epithelial cells. *J. Periodontal Res.* **28:**274–280.

1890. **Keulers, R. A., J. C. Maltha, F. H. Mikx, and J. M. Wolters-Lutgerhorst.** 1993. Involvement of treponemal surface-located protein and carbohydrate moieties in the attachment of *Treponema denticola* ATCC 33520 to cultured rat palatal epithelial cells. *Oral Microbiol. Immunol.* **8:**236–241.

1891. **Khalil, S. B., F. J. Cassels, H. I. Shaheen, L. K. Pannell, N. El-Ghorab, K. Kamal, M. Mansour, S. J. Savarino, and L. F. Peruski, Jr.** 1999. Characterization of an enterotoxigenic *Escherichia coli* strain from Africa expressing a putative colonization factor. *Infect. Immun.* **67:**4019–4026.

1892. **Khan, A. S., B. Kniep, T. A. Oelschlaeger, I. Van Die, T. Korhonen, and J. Hacker.** 2000. Receptor structure for F1C fimbriae of uropathogenic *Escherichia coli. Infect. Immun.* **68:**3541–3547.

1893. **Khan, A. S., I. Muhldorfer, V. Demuth, U. Wallner, T. K. Korhonen, and J. Hacker.** 2000. Functional analysis of the minor subunits of S fimbrial adhesion (SfaI) in pathogenic *Escherichia coli. Mol. Gen. Genet.* **263:**96–105.

1894. **Khan, A. S., and D. M. Schifferli.** 1994. A minor 987P protein different from the structural fimbrial subunit is the adhesin. *Infect. Immun.* **62:**4233–4243.

1895. **Khan, A. S., N. C. Johnston, H. Goldfine, and D. M. Schifferli.** 1996. Porcine 987P glycolipid receptors on intestinal

brush borders and their cognate bacterial ligands. *Infect. Immun.* **64:**3688–3693.

1896. **Khashe, S., D. J. Scales, S. L. Abbott, and J. M. Janda.** 2001. Non-invasive *Providencia alcalifaciens* strains fail to attach to HEp-2 cells. *Curr. Microbiol.* **43:**414–417.

1897. **Khelef, N., C. M. Bachelet, B. B. Vargaftig, and N. Guiso.** 1994. Characterization of murine lung inflammation after infection with parental *Bordetella pertussis* and mutants deficient in adhesins or toxins. *Infect. Immun.* **62:**2893–2900.

1898. **Khoury, A. E., K. Lam, B. Ellis, and J. W. Costerton.** 1992. Prevention and control of bacterial infections associated with medical devices. *ASAIO J.* **38:**M174–M178.

1899. **Kibue, M. A., M. Ehara, and P. G. Waiyaki.** 1992. The fimbriae-like structures on *V. cholerae* isolated in Kenya. *East Afr. Med. J.* **69:**442–444.

1900. **Kienle, Z., L. Emody, C. Svanborg, and P. W. O'Toole.** 1992. Adhesive properties conferred by the plasminogen activator of *Yersinia pestis. J. Gen. Microbiol.* **138:**1679–1687.

1901. **Kihlberg, B. M., J. Cooney, M. G. Caparon, A. Olsen, and L. Bjorck.** 1995. Biological properties of a *Streptococcus pyogenes* mutant generated by Tn916 insertion in *mga. Microb. Pathog.* **19:**299–315.

1902. **Kihlstrom, E., M. Majeed, B. Rozalska, and T. Wadström.** 1992. Binding of *Chlamydia trachomatis* serovar L2 to collagen types I and IV, fibronectin, heparan sulphate, laminin and vitronectin. *Zentbl. Bakteriol.* **277:**329–333.

1903. **Kil, K. S., R. O. Darouiche, R. A. Hull, M. D. Mansouri, and D. M. Musher.** 1997. Identification of a *Klebsiella pneumoniae* strain associated with nosocomial urinary tract infection. *J. Clin. Microbiol.* **35:**2370–2374.

1904. **Kim, J. J., D. Zhou, R. E. Mandrell, and J. M. Griffiss.** 1992. Effect of exogenous sialylation of the lipooligosaccharide of *Neisseria gonorrhoeae* on opsonophagocytosis. *Infect. Immun.* **60:**4439–4442.

1905. **Kim, K. Y., J. F. Frank, and S. E. Craven.** 1996. Three-dimensional visualization of *Salmonella* attachment to poultry skin using confocal scanning laser microscopy. *Lett. Appl. Microbiol.* **22:**280–282.

1906. **Kimoto, H., J. Kurisaki, N. M. Tsuji, S. Ohmomo, and T. Okamoto.** 1999. Lactococci as probiotic strains: adhesion to human enterocyte-like Caco-2 cells and tolerance to low pH and bile. *Lett. Appl. Microbiol.* **29:**313–31.

1907. **Kimsey, H. H., and M. K. Waldor.** 1998. *Vibrio cholerae* hemagglutinin/protease inactivates CTXphi. *Infect. Immun.* **66:**4025–4029.

1908. **Kimura, N., M. Ariga, F. C. Icatlo, Jr., M. Kuroki, M. Ohsugi, Y. Ikemori, K. Umeda, and Y. Kodama.** 1998. A euthymic hairless mouse model of *Helicobacter pylori* colonization and adherence to gastric epithelial cells *in vivo*. *Clin. Diagn. Lab. Immunol.* **5:**578–582.

1909. **Kinder, S. A., and S. C. Holt.** 1994. Coaggregation between bacterial species. *Methods Enzymol.* **236:**254–270.

1910. **Kinder, S. A., and S. C. Holt.** 1993. Localization of the *Fusobacterium nucleatum* T18 adhesin activity mediating coaggregation with *Porphyromonas gingivalis* T22. *J. Bacteriol.* **175:**840–850.

1911. **King, S. S., D. A. Young, L. G. Nequin, and E. M. Carnevale.** 2000. Use of specific sugars to inhibit bacterial adherence to equine endometrium *in vitro*. *Am. J. Vet. Res.* **61:**446–449.

1912. **Kiremitci-Gumusderelioglu, M., and A. Pesmen.** 1996. Microbial adhesion to ionogenic PHEMA, PU and PP implants. *Biomaterials* **17:**443–449.

1913. **Kirjavainen, P. V., A. C. Ouwehand, E. Isolauri, and S. J. Salminen.** 1998. The ability of probiotic bacteria to bind to human intestinal mucus. *FEMS Microbiol. Lett.* **167:**185–189.

1914. **Kirjavainen, P. V., E. M. Tuomola, R. G. Crittenden, A. C. Ouwehand, D. W. Harty, L. F. Morris, H. Rautelin, M. J. Playne, D. C. Donohue, and S. J. Salminen.** 1999. In vitro adhesion and platelet aggregation properties of bacteremia-associated lactobacilli. *Infect. Immun.* **67:**2653–2655.

1915. **Kirkeby, S., and P. E. Hoyer.** 1999. Binding properties of the galactose-detecting lectin *Pseudomonas aeruginosa* agglutinin (PA-IL) to skeletal muscle fibres. Quantitative precipitation and precipitation inhibition assays. *Histochem. J.* **31:**485–493.

1916. **Kirn, T. J., M. J. Lafferty, C. M. Sandoe, and R. K. Taylor.** 2000. Delineation of pilin domains required for bacterial association into microcolonies and intestinal colonization by *Vibrio cholerae*. *Mol. Microbiol.* **35:**896–910.

1917. **Kirov, S. M., T. C. Barnett, C. M. Pepe, M. S. Strom, and M. J. Albert.** 2000. Investigation of the role of type IV *Aeromonas* pilus (Tap) in the pathogenesis of *Aeromonas* gastrointestinal infection. *Infect. Immun.* **68:**4040–4048.

1918. **Kirov, S. M., L. J. Hayward, and M. A. Nerrie.** 1995. Adhesion of *Aeromonas* sp. to cell lines used as models for intestinal adhesion. *Epidemiol. Infect.* **115:**465–473.

1919. **Kirov, S. M., I. Jacobs, L. J. Hayward, and R. H. Hapin.** 1995. Electron microscopic examination of factors influencing the expression of filamentous surface structures on clinical and environmental isolates of *Aeromonas veronii* biotype *sobria*. *Microbiol. Immunol.* **39:**329–338.

1920. **Kirov, S. M., and K. Sanderson.** 1996. Characterization of a type IV bundle-forming pilus (SFP) from a gastroenteritis-associated strain of *Aeromonas veronii* biovar *sobria*. *Microb. Pathog.* **21:**23–34.

1921. **Kirov, S. M., K. Sanderson, and T. C. Dickson.** 1998. Characterisation of a type IV pilus produced by *Aeromonas caviae*. *J. Med. Microbiol.* **47:**527–531.

1922. **Kirov, S. M., L. A. O'Donovan, and K. Sanderson.** 1999. Functional characterization of type IV pili expressed on diarrhea-associated isolates of *Aeromonas* species. *Infect. Immun.* **67:**5447–5454.

1923. **Kirszbaum, L., C. Sotiropoulos, C. Jackson, S. Cleal, N. Slakeski, and E. C. Reynolds.** 1995. Complete nucleotide sequence of a gene *prtR* of *Porphyromonas gingivalis* W50 encoding a 132 kDa protein that contains an arginine-specific thiol endopeptidase domain and a haemagglutinin domain. *Biochem. Biophys. Res. Commun.* **207:**424–431.

1924. **Kist, M., C. Spiegelhalder, T. Moriki, and H. E. Schaefer.** 1993. Interaction of *Helicobacter pylori* (strain 151) and *Campylobacter coli* with human peripheral polymorphonuclear granulocytes. *Zentbl. Bakteriol.* **280:**58–72.

1925. **Kitano, T., Y. Yutani, A. Shimazu, I. Yano, H. Ohashi, and Y. Yamano.** 1996. The role of physicochemical properties of biomaterials and bacterial cell adhesion *in vitro*. *Int. J. Artific. Organs* **19:**353–358.

1926. **Kitzerow, A., U. Hadding, and B. Henrich.** 1999. Cyto-adherence studies of the adhesin P50 of *Mycoplasma hominis*. *J. Med. Microbiol.* **48:**485–493.

1927. **Kjaergaard, K., M. A. Schembri, C. Ramos, S. Molin, and P. Klemm.** 2000. Antigen 43 facilitates formation of multispecies biofilms. *Environ. Microbiol.* **2:**695–702.

1928. **Klann, A. G., R. A. Hull, and S. I. Hull.** 1992. Sequences of the genes encoding the minor tip components of Pap-3 pili of *Escherichia coli*. *Gene* **119:**95–100.

1929. **Klann, A. G., R. A. Hull, T. Palzkill, and S. I. Hull.** 1994. Alanine-scanning mutagenesis reveals residues involved in binding of *pap-3*-encoded pili. *J. Bacteriol.* **176:**2312–2317.

1930. **Klein, N. J., C. A. Ison, M. Peakman, M. Levin, S. Hammerschmidt, M. Frosch, and R. S. Heyderman.** 1996. The influence of capsulation and lipooligosaccharide structure on neutrophil adhesion molecule expression and endothelial injury by *Neisseria meningitidis. J. Infect. Dis.* **173:**172–179.

1931. **Klemm, P., B. J. Jorgensen, B. Kreft, and G. Christiansen.** 1995. The export systems of type 1 and F1C fimbriae are interchangeable but work in parental pairs. *J. Bacteriol.* **177:**621–627.

1932. **Klempner, M. S., R. Noring, and R. A. Rogers.** 1993. Invasion of human skin fibroblasts by the Lyme disease spirochete, *Borrelia burgdorferi. J. Infect. Dis.* **167:**1074–1081.

1933. **Klier, C. M., P. E. Kolenbrander, A. G. Roble, M. L. Marco, S. Cross, and P. S. Handley.** 1997. Identification of a 95 kDa putative adhesin from *Actinomyces* serovar WVA963 strain PK1259 that is distinct from type 2 fimbrial subunits. *Microbiology* **143:**835–846.

1934. **Klier, C. M., A. G. Roble, and P. E. Kolenbrander.** 1998. *Actinomyces* serovar WVA963 coaggregation-defective mutant strain PK2407 secretes lactose-sensitive adhesin that binds to coaggregation partner *Streptococcus oralis* 34. *Oral Microbiol. Immunol.* **13:**337–340.

1935. **Kline, J. B., S. Xu, A. L. Bisno, and C. M. Collins.** 1996. Identification of a fibronectin-binding protein (GfbA) in pathogenic group G streptococci. *Infect. Immun.* **64:**2122–2129.

1936. **Kmet, V., M. L. Callegari, V. Bottazzi, and L. Morelli.** 1995. Aggregation-promoting factor in pig intestinal *Lactobacillus* strains. *Lett. Appl. Microbiol.* **21:**351–353.

1937. **Kmet, V., and F. Lucchini.** 1999. Aggregation of sow lactobacilli with diarrhoeagenic *Escherichia coli. Zentbl. Veterinaermed.* **46:**683–687.

1938. **Knepper, B., I. Heuer, T. F. Meyer, and J. van Putten.** 1997. Differential response of human monocytes to *Neisseria gonorrhoeae* variants expressing pili and opacity proteins. *Infect. Immun.* **65:**4122–4129.

1939. **Knorle, R., and W. Hubner.** 1995. Secondary structure and stability of the bacterial carbohydrate-specific recognition proteins K88ab, AFA-1, NFA-1, and CFA-1. *Biochemistry* **34:**10970–10975.

1940. **Knudsen, T. B., and P. Klemm.** 1998. Probing the receptor recognition site of the FimH adhesin by fimbriae-displayed FimH-FocH hybrids. *Microbiology* **144:**1919–1929.

1941. **Knutton, S.** 1995. Cellular responses to enteropathogenic *Escherichia coli* infection. *Biosci. Rep.* **15:**469–479.

1942. **Knutton, S.** 1995. Electron microscopical methods in adhesion. *Methods Enzymol.* **253:**145–158.

1943. **Knutton, S., R. K. Shaw, M. K. Bhan, H. R. Smith, M. M. McConnell, T. Cheasty, P. H. Williams, and T. J. Baldwin.** 1992. Ability of enteroaggregative *Escherichia coli* strains to adhere in vitro to human intestinal mucosa. *Infect. Immun.* **60:**2083–2091.

1944. **Knutton, S., T. Baldwin, P. Williams, A. Manjarrez-Hernandez, and A. Aitken.** 1993. The attaching and effacing virulence property of enteropathogenic *Escherichia coli. Zentbl. Bakteriol.* **278:**209–217.

1945. **Knutton, S., J. Adu-Bobie, C. Bain, A. D. Phillips, G. Dougan, and G. Frankel.** 1997. Down regulation of intimin expression during attaching and effacing enteropathogenic *Escherichia coli* adhesion. *Infect. Immun.* **65:**1644–1652.

1946. **Knutton, S., I. Rosenshine, M. J. Pallen, I. Nisan, B. C. Neves, C. Bain, C. Wolff, G. Dougan, and G. Frankel.** 1998. A novel EspA-associated surface organelle of enteropathogenic *Escherichia coli* involved in protein translocation into epithelial cells. *EMBO J.* **17:**2166–2176.

1947. **Knutton, S., R. K. Shaw, A. P. Anantha, M. S. Donnenberg, and A. A. Zorgani.** 1999. The type IV bundle-forming pilus of enteropathogenic *Escherichia coli* undergoes dramatic alterations in structure associated with bacterial adherence, aggregation and dispersal. *Mol. Microbiol.* **33:**499–509.

1948. **Kobayashi, Y., K.-I. Okazaki, and K. Murakami.** 1993. Adhesion of *Helicobacter pylori* to gastric epithelial cells in primary cultures obtained from stomachs of various animals. *Infect. Immun.* **61:**4058–4063.

1949. **Kockro, R. A., J. A. Hampl, B. Jansen, G. Peters, M. Scheihing, R. Giacomelli, S. Kunze, and A. Aschoff.** 2000. Use of scanning electron microscopy to investigate the prophylactic efficacy of rifampin-impregnated CSF shunt catheters. *J. Med. Microbiol.* **49:**441–450.

1950. **Koga, T., K. Ishimoto, and S. Lory.** 1993. Genetic and functional characterization of the gene cluster specifying expression of *Pseudomonas aeruginosa* pili. *Infect. Immun.* **61:**1371–1377.

1951. **Koga, T., Y. Yamashita, Y. Nakano, M.**

Kawasaki, T. Oho, H. Yu, M. Nakai, and N. Okahashi. 1995. Surface proteins of *Streptococcus mutans*. *Dev. Biol. Stand.* **85**:363–369.

1952. Kogure, K., E. Ikemoto, and H. Morisaki. 1998. Attachment of *Vibrio alginolyticus* to glass surfaces is dependent on swimming speed. *J. Bacteriol.* **180**:932–937.

1953. Kohnen, W., and B. Jansen. 1995. Polymer materials for the prevention of catheter-related infections. *Zentbl. Bakteriol.* **283**:175–186.

1954. Kohnen, W., J. Schaper, O. Klein, B. Tieke, and B. Jansen. 1998. A silicone ventricular catheter coated with a combination of rifampin and trimethoprim for the prevention of catheter-related infections. *Zentbl. Bakteriol.* **287**:147–156.

1955. Koivusalo, A., H. Makisalo, M. Talja, A. Siitonen, J. Vuopio-Varkila, M. Ruutu, and K. Hockerstedt. 1996. Bacterial adherence and biofilm formation on latex and silicone T-tubes in relation to bacterial contamination of bile. *Scand. J. Gastroenterol.* **31**:398–403.

1956. Kolenbrander, P. E. 1993. Coaggregation of human oral bacteria: potential role in the accretion of dental plaque. *J. Appl. Bacteriol.* **74**:79S–86S.

1957. Kolenbrander, P. E. 1995. Coaggregations among oral bacteria. *Methods Enzymol.* **253**:385–397.

1958. Kolenbrander, P. E., and J. London. 1993. Adhere today, here tomorrow: oral bacterial adherence. *J. Bacteriol.* **175**:3247–3252.

1959. Kolenbrander, P, E., R. N. Andersen, and N. Ganeshkumar. 1994. Nucleotide sequence of the *Streptococcus gordonii* PK488 coaggregation adhesin gene, *scaA,* and ATP-binding cassette. *Infect. Immun.* **62**:4469–448.

1960. Kolenbrander, P. E., K. D. Parrish, R. N. Andersen, and E. P. Greenberg. 1995. Intergeneric coaggregation of oral *Treponema* spp. with *Fusobacterium* spp. and intrageneric coaggregation among *Fusobacterium* spp. *Infect. Immun.* **63**:4584–4588.

1961. Koljalg, S., J. Vuopio-Varkila, O. Lyytikainen, M. Mikelsaar, and T. Wadstöm. 1996. Cell surface properties of *Acinetobacter baumannii*. *APMIS* **104**:659–665.

1962. Konig, B., and W. Konig. 1993. Induction and suppression of cytokine release (tumour necrosis factor-alpha; interleukin-6, interleukin-1 beta) by *Escherichia coli* pathogenicity factors (adhesions, alpha-haemolysin). *Immunology* **78**:526–533.

1963. Konig, D. P., J. M. Schierholz, R. D. Hilgers, C. Bertram, F. Perdreau-Remington, and J. Rutt. 2001. *In vitro* adherence and accumulation of *Staphylococcus epidermidis* RP 62 A and *Staphylococcus epidermidis* M7 on four different bone cements. *Langenbecks Arch. Surg.* **386**:328–332.

1964. Konig, D. P., F. Perdreau-Remington, J. Rutt, R. D. Hilgers, and J. M. Schierholz. 1999. Adherence to and accumulation of *S. epidermidis* on different biomaterials due to extracellular slime production. *In vitro* comparison of a slime-producing strain (Rp 62 A) and its isogenic slime negative mutant (M7). *Zentbl. Bakt.* **289**:355–364.

1965. Konig, D. P., F. Perdreau-Remington, J. Rutt, P. Stossberger, R. D. Hilgers, and G. Plum. 1998. Slime production of *Staphylococcus epidermidis*: increased bacterial adherence and accumulation onto pure titanium. *Acta Orthop. Scand.* **69**:523–526.

1966. Konkel, M. E., and W. Cieplak, Jr. 1992. Altered synthetic response of *Campylobacter jejuni* to cocultivation with human epithelial cells is associated with enhanced internalization. *Infect. Immun.* **60**:4945–4949.

1967. Konkel, M. E., M. D. Corwin, L. A. Joens, and W. Cieplak. 1992. Factors that influence the interaction of *Campylobacter jejuni* with cultured mammalian cells. *J. Med. Microbiol.* **37**:30–37.

1968. Konkel, M. E., B. J. Kim, V. Rivera-Amill, and S. G. Garvis. 1999. Identification of proteins required for the internalization of *Campylobacter jejuni* into cultured mammalian cells. *Adv. Exp. Med. Biol.* **473**:215–224.

1969. Kontani, M., H. Ono, H. Shibata, Y. Okamura, T. Tanaka, T. Fujiwara, S. Kimura, and S. Hamada. 1996. Cysteine protease of *Porphyromonas gingivalis* 381 enhances binding of fimbriae to cultured human fibroblasts and matrix proteins. *Infect. Immun.* **64**:756–762.

1970. Kontani, M., S. Kimura, I. Nakagawa, and S. Hamada. 1997. Adherence of *Porphyromonas gingivalis* to matrix proteins via a fimbrial cryptic receptor exposed by its own arginine-specific protease. *Mol. Microbiol.* **24**:1179–1187.

1971. Kontani, M., A. Amano, T. Nakamura, I. Nakagawa, S. Kawabata, and S. Hamada. 1999. Inhibitory effects of protamines on proteolytic and adhesive activities of *Porphyromonas gingivalis*. *Infect. Immun.* **67**:4917–4920.

1972. Kontiokari, T., M. Uhari, and M. Koskela. 1998. Antiadhesive effects of xylitol on otopathogenic bacteria. *J. Antimicrob. Chemother.* **41**:563–565.

1973. **Kontula, P., M. L. Suihko, A. Von Wright, and T. Mattila-Sandholm.** 1999. The effect of lactose derivatives on intestinal lactic acid bacteria. *J. Dairy Sci.* **82**:249–256.

1974. **Koo, H., B. P. Gomes, P. L. Rosalen, G. M. Ambrosano, Y. K. Park, and J. A. Cury.** 2000. *In vitro* antimicrobial activity of propolis and *Arnica montana* against oral pathogens. *Arch. Oral Biol.* **45**:141–148.

1975. **Koo, H., P. L. Rosalen, J. A. Cury, G. M. Ambrosano, R. M. Murata, R. Yatsuda, M. Ikegaki, S. M. Alencar, and Y. K. Park.** 2000. Effect of a new variety of *Apis mellifera* propolis on mutans streptococci. *Curr. Microbiol.* **41**:192–196.

1976. **Koomey, M.** 1995. Prepilin-like molecules in type 4 pilus biogenesis: minor subunits, chaperones or mediators of organelle translocation? *Trends Microbiol.* **3**:409–410.

1977. **Kopec, L. K., and W. H. Bowen.** 1995. Adherence of microorganisms to rat salivary pellicles. *Caries Res.* **29**:507–512.

1978. **Kopec, L. K., A. M. Vacca Smith, D. Wunder, L. Ng-Evans, and W. H. Bowen.** 2001. Properties of *Streptococcus sanguinis* glucans formed under various conditions. *Caries Res.* **35**:67–74.

1979. **Kopp, P. A., M. Schmitt, H. J. Wellensiek, and H. Blobel.** 1995. Isolation and characterization of fibronectin-binding sites of *Borrelia garinii* N34. *Infect. Immun.* **63**:3804–3808.

1980. **Korhonen, T. K., R. Virkola, K. Lähteenmäki, Y. Björkman, M. Kukkonen, T. Raunio, A.-M. Tarkkanen, and B. Westerlund.** 1992. Penetration of fimbriate enteric bacteria through basement membranes: a hypothesis. *FEMS Microbiol. Lett.* **100**:307–312.

1981. **Korhonen, T. K., K. Lahteenmaki, M. Kukkonen, R. Pouttu, U. Hynonen, K. Savolainen, B. Westerlund-Wikstrom, and R. Virkola.** 1997. Plasminogen receptors. Turning *Salmonella* and *Escherichia coli* into proteolytic organisms. *Adv. Exp. Med. Biol.* **412**:185–192.

1982. **Korneva, H. A., V. A. Grigoriev, E. N. Isaeva, S. M. Kaloshina, and F. S. Barnes.** 1999. Effects of low-level 50 Hz magnetic fields on the level of host defense and on spleen colony formation. *Bioelectromagnetics* **20**:57–63.

1983. **Korth, M. J., J. M. Apostol, Jr., and S. L. Moseley.** 1992. Functional expression of heterologous fimbrial subunits mediated by the F41, K88 and CS31A determinants of *Escherichia coli*. *Infect. Immun.* **60**:2500–2505.

1984. **Korth, M. J., J. C. Lara, and S. L. Moseley.** 1994. Epithelial cell invasion by bovine septicemic *Escherichia coli*. *Infect. Immun.* **62**:41–47.

1985. **Kostrzynska, M., and T. Wadström.** 1992. Binding of laminin, type IV collagen, and vitronectin by *Streptococcus pneumoniae*. *Zentbl. Bakteriol.* **277**:80–83.

1986. **Kotiranta, A., M. Haapasalo, K. Kari, E. Kerosuo, I. Olsen, T. Sorsa, J. H. Jeurman, and K. Lounatmaa.** 1998. Surface structure, hydrophobicity, phagocytosis, and adherence to matrix proteins of *Bacillus cereus* cells with and without the crystalline surface protein layer. *Infect. Immun.* **66**:4895–4902.

1987. **Kottom, T. J., L. K. Nolan, and J. Brown.** 1995. Invasion of Caco-2 cells by *Salmonella typhimurium* (Copenhagen) isolates from healthy and sick chickens. *Avian Dis.* **39**:867–872.

1988. **Kozai, K., J. Suzuki, M. Okada, and N. Nagasaka.** 2000. *In vitro* study of antibacterial and antiadhesive activities of fluoride-containing light-cured fissure sealants and a glass ionomer liner/base against oral bacteria. *ASDC J. Dent. Child.* **67**:117–122.

1989. **Kozarov, E., J. Whitlock, H. Dong, E. Carrasco, and A. Progulske-Fox.** 1998. The number of direct repeats in *hagA* is variable among *Porphyromonas gingivalis* strains. *Infect. Immun.* **66**:4721–4725.

1990. **Krasan, G. P., D. Cutter, S. L. Block, and J. W. St. Geme III.** 1999. Adhesin expression in matched nasopharyngeal and middle ear isolates of nontypeable *Haemophilus influenzae* from children with acute otitis media. *Infect. Immun.* **67**:449–454.

1991. **Krause, D. C., T. Proft, C. T. Hedreyda, H. Hilbert, H. Plagens, and R. Herrmann.** 1997. Transposon mutagenesis reinforces the correlation between *Mycoplasma pneumoniae* cytoskeletal protein HMW2 and cytadherence. *J. Bacteriol.* **179**:2668–2677.

1992. **Krause, D. C.** 1996. *Mycoplasma pneumoniae* cytadherence: unravelling the tie that binds. *Mol. Microbiol.* **20**:247–253.

1993. **Krause, D. C.** 1998. *Mycoplasma pneumoniae* cytadherence: organization and assembly of the attachment organelle. *Trends Microbiol.* **6**:15–18.

1994. **Kreft, B., R. Marre, U. Schramm, and R. Wirth.** 1992. Aggregation substance of *Enterococcus faecalis* mediates adhesion to cultured renal tubular cells. *Infect. Immun.* **60**:25–30.

1995. **Kreft, B., O. Carstensen, E. Straube, S.**

Bohnet, J. Hacker, and R. Marre. 1992. Adherence to and cytotoxicity of *Escherichia coli* for eukaryotic cell lines quantified by MTT (3-[4,5-dimethylthiazol-2-yl]-2,5-diphenyltetrazolium bromide). *Zentbl. Bakteriol.* **276**:231–242.

1996. Kreft, B., S. Bohnet, O. Carstensen, J. Hacker, and R. Marre. 1993. Differential expression of interleukin-6, intracellular adhesion molecule 1, and major histocompatibility complex class II molecules in renal carcinoma cells stimulated with S fimbriae of uropathogenic *Escherichia coli. Infect. Immun.* **61**:3060–3063.

1997. Kreft, B., M. Placzek, C. Doehn, J. Hacker, G. Schmidt, G. Wasenauer, M. R. Daha, F. J. van der Woude, and K. Sack. 1995. S fimbriae of uropathogenic *Escherichia coli* bind to primary human renal proximal tubular epithelial cells but do not induce expression of intercellular adhesion molecule 1. *Infect. Immun.* **63**:3235–3238.

1998. Kreft, B., S. Ilic, W. Ziebuhr, A. Kahl, U. Frei, K. Sack, and M. Trautmann. 1998. Adherence of *Staphylococcus aureus* isolated in peritoneal dialysis-related exit-site infections to HEp-2 cells and silicone peritoneal catheter materials. *Nephrol. Dial. Transpl.* **13**:3160–3164.

1999. Kreikemeyer, B., S. R. Talay, and G. S. Chhatwal. 1995. Characterization of a novel fibronectin-binding surface protein in group A streptococci. *Mol. Microbiol.* **17**:137–145.

2000. Krejany, E. O., T. H. Grant, V. Bennett-Wood, L. M. Adams, and R. M. Robins-Browne. 2000. Contribution of plasmid-encoded fimbriae and intimin to capacity of rabbit-specific enteropathogenic *Escherichia coli* to attach to and colonize rabbit intestine. *Infect. Immun.* **68**:6472–6477.

2001. Kremer, B. H., A. J. Herscheid, W. Papaioannou, M. Quirynen, and T. J. van Steenbergen. 1999. Adherence of *Peptostreptococcus micros* morphotypes to epithelial cells *in vitro. Oral Microbiol. Immunol.* **14**:49–55.

2002. Kremer, B. H., and T. J. van Steenbergen. 2000. *Peptostreptococcus micros* coaggregates with *Fusobacterium nucleatum* and non-encapsulated *Porphyromonas gingivalis. FEMS Microbiol. Lett.* **182**:57–62.

2003. Kresse, A. U., K. Schulze, C. Deibel, F. Ebel, M. Rohde, T. Chakraborty, and C. A. Guzman. 1998. Pas, a novel protein required for protein secretion and attaching and effacing activities of enterohemorrhagic *Escherichia coli. J. Bacteriol.* **180**:4370–4379.

2004. Kreutz, C. 1994. Adherence properties of *Bacteroides vulgatus,* the preponderant colonic organism of adult humans. *Zentbl. Bakteriol.* **281**:225–234.

2005. Krinos, C., A. S. High, and F. G. Rodgers. 1999. Role of the 25 kDa major outer membrane protein of *Legionella pneumophila* in attachment to U-937 cells and its potential as a virulence factor for chick embryos. *J. Appl. Microbiol.* **86**:237–244.

2006. Krishna, M. M., N. B. Powell, and S. P. Borriello. 1996. Cell surface properties of *Clostridium difficile:*haemagglutination, relative hydrophobicity and charge. *J. Med. Microbiol.* **44**:115–123.

2007. Krogfelt, K. A. 1995. Adhesin-dependent isolation and characterization of bacteria from their natural environment. *Methods Enzymol.* **253**:50–53.

2008. Krovacek, K., V. Pasquale, S. B. Baloda, V. Soprano, M. Conte, and S. Dumontet. 1994. Comparison of putative virulence factors in *Aeromonas hydrophila* strains isolated from the marine environment and human diarrheal cases in southern Italy. *Appl. Environ. Microbiol.* **60**:1379–1382.

2009. Kruger, S., E. Brandt, M. Klinger, and B. Kreft. 2000. Interleukin-8 secretion of cortical tubular epithelial cells is directed to the basolateral environment and is not enhanced by apical exposure to *Escherichia coli. Infect. Immun.* **68**:328–334.

2010. Krukonis, E. S., and R. R. Isberg. 1998. SWIM analysis allows rapid identification of residues involved in invasin-mediated bacterial uptake. *Gene* **211**:109–116.

2011. Krukonis, E. S., P. Dersch, J. A. Eble, and R. R. Isberg. 1998. Differential effects of integrin alpha chain mutations on invasin and natural ligand interaction. *J. Biol. Chem.* **273**:31837–31843.

2012. Krzyminska, S., J. Mokracka, M. Laganowska, K. Wlodarczak, E. Guszczynska, J. Liszkowska, E. Popkowska, I. Lima, I. Lemanska, and M. Wendt. 2001. Enhancement of the virulence of *Aeromonas caviae* diarrhoeal strains by serial passages in mice. *J. Med. Microbiol.* **50**:303–312.

2013. Kubiet, M., R. Ramphal, A. Weber, and A. Smith. 2000. Pilus-mediated adherence of *Haemophilus influenzae* to human respiratory mucins. *Infect. Immun.* **68**:3362–3367.

2014. Kubiet, M., and R. Ramphal. 1995. Adhesion of nontypeable *Haemophilus influenzae* from blood and sputum to human tracheobronchial mucins and lactoferrin. *Infect. Immun.* **63**:899–902.

2015. **Kuboniwa, M., A. Amano, S. Shizukuishi, I. Nakagawa, and S. Hamada.** 2001. Specific antibodies to *Porphyromonas gingivalis* Lys-gingipain by DNA vaccination inhibit bacterial binding to hemoglobin and protect mice from infection. *Infect. Immun.* **69:**2972–2979.

2016. **Kuehn, M., K. Lent, J. Haas, J. Hagenzieker, M. Cervin, and A. L. Smith.** 1992. Fimbriation of *Pseudomonas cepacia. Infect. Immun.* **60:**2002–2007.

2017. **Kuehn, M. J.** 1997. Establishing communication via gram-negative bacterial pili. *Trends Microbiol.* **5:**130–132.

2018. **Kuehn, M. J., J. Heuser, S. Normark, and S. J. Hultgren.** 1992. P pili in uropathogenic *E. coli* are composite fibres with distinct fibrillar adhesive tips. *Nature* **356:**252–255.

2019. **Kuehn, M. J., D. J. Ogg, J. Kihlberg, L. N. Slonim, K. Flemmer, T. Bergfors, and S. J. Hultgren.** 1993. Structural basis of pilus subunit recognition by the PapD chaperone. *Science* **262:**1234–1241.

2020. **Kuehn, M. J., F. Jacob-Dubuisson, K. Dodson, L. Slonim, R. Striker, and S. J. Hultgren.** 1994. Genetic, biochemical, and structural studies of biogenesis of adhesive pili in bacteria. *Methods Enzymol.* **236:**282–306.

2021. **Kuhar, I., M. Grabnar, and D. Zgur-Bertok.** 1998. Virulence determinants of uropathogenic *Escherichia coli* in fecal strains from intestinal infections and healthy individuals. *FEMS Microbiol. Lett.* **164:**243–248.

2022. **Kuhn, M.** 1998. The microtubule depolymerizing drugs nocodazole and colchicine inhibit the uptake of *Listeria monocytogenes* by P388D1 macrophages. *FEMS Microbiol. Lett.* **160:**87–90.

2023. **Kuhn, M., and W. Goebel.** 1998. Host cell signalling during *Listeria monocytogenes* infection. *Trends Microbiol.* **6:**11–15.

2024. **Kukkonen, M., T. Raunio, R. Virkola, K. Lahteenmaki, P. H. Makela, P. Klemm, S. Clegg, and T. K. Korhonen.** 1993. Basement membrane carbohydrate as a target for bacterial adhesion: binding of type I fimbriae of *Salmonella enterica* and *Escherichia coli* to laminin. *Mol. Microbiol.* **7:**229–237.

2025. **Kukkonen, M., S. Saarela, K. Lähteenmäki, U. Hynönen, B. Westerlund-Wikström, M. Rhen, and T. K. Korhonen.** 1998. Identification of two laminin-binding fimbriae, the type 1 fimbria of *Salmonella enterica* serovar Typhimurium and the G fimbria of *Escherichia coli,* as plasminogen receptors. *Infect. Immun.* **66:**4965–4970.

2026. **Kumar, K. K., R. Srivastava, V. B.**

Sinha, J. Michalski, J. B. Kaper, and B. S. Srivastava. 1994. *recA* mutations reduce adherence and colonization by classical and El Tor strains of *Vibrio cholerae. Microbiology* **140:**1217–1222.

2027. **Kumar, K. S. N., N. K. Ganguly, I. S. Anand, and P. L. Wahi.** 1992. Adherence of *Streptococcus pyogenes* M type 5 to pharyngeal and buccal cells of patients with rheumatic fever and rheumatic heart disease during a one-year follow-up. *APMIS* **100:**353–359.

2028. **Kumar, S. S., V. Malladi, K. Sankaran, R. Haigh, P. Williams, and A. Balakrishnan.** 2001. Extrusion of actin-positive strands from Hep-2 and Int 407 cells caused by outer membrane preparations of enteropathogenic *Escherichia coli* and specific attachment of wild type bacteria to the strands. *Can. J. Microbiol.* **47:**727–734.

2029. **Kunin, C. M., T. H. Hua, R. L. Guerrant, and L. O. Bakaletz.** 1994. Effect of salicylate, bismuth, osmolytes, and tetracycline resistance on expression of fimbriae by *Escherichia coli. Infect. Immun.* **62:**2178–2186.

2030. **Kuo, C., N. Takahashi, A. F. Swanson, Y. Ozeki, and S. Hakomori.** 1996. An N-linked high-mannose type oligosaccharide, expressed at the major outer membrane protein of *Chlamydia trachomatis,* mediates attachment and infectivity of the microorganism to HeLa cells. *J. Clin. Investig.* **98:**2813–2818.

2031. **Kuramitsu, H., M. Tokuda, M. Yoneda, M. Duncan, and M. I. Cho.** 1997. Multiple colonization defects in a cysteine protease mutant of *Porphyromonas gingivalis. J. Periodontal Res.* **32:**140–142.

2032. **Kuroda, K., E. J. Brown, W. B. Telle, D. G. Russell, and T. L. Ratliff.** 1993. Characterization of the internalization of bacillus Calmette-Guerin by human bladder tumor cells. *J. Clin. Investig.* **91:**69–76.

2033. **Kurono, Y., K. Shimamura, and G. Mogi.** 1993. Role of secretory immunoglobulin A in nasopharyngeal bacterial adherence, p. 222–229. *In* D. J. Lim, C. D. Bluestone, J. O. Klein, J. D. Nelson, and P. L. Ogra (ed.), *Recent Advances in Otitis Media.* Decker Periodicals, Philadelphia, Pa.

2034. **Kushiro, A., T. Takahashi, T. Asahara, H. Tsuji, K. Nomoto, and M. Morotomi.** 2001. *Lactobacillus casei* acquires the binding activity to fibronectin by the expression of the fibronectin binding domain of *Streptococcus pyogenes* on the cell surface. *J. Mol. Microbiol. Biotechnol.* **3:**563–571.

2035. **Kusters, J. G., G. A. W. M. Mulders-Kremers, C. E. M. van Doornik, and B.**

A. M. van der Zeijst. 1993. Effects of multiplicity of infection, bacterial protein synthesis, and growth phase on adhesion to and invasion of human cell lines by *Salmonella typhimurium*. *Infect. Immun.* **61:**5013–5020.

2036. **Kusumoto, Y., T. Ogawa, and S. Hamada.** 1993. Generation of specific antibody-secreting cells in salivary glands of BALB/c mice following parenteral or oral immunization with *Porphyromonas gingivalis* fimbriae. *Arch. Oral Biol.* **38:**361–367.

2037. **Kuusela, P., M. Ullberg, G. Kronvall, T. Tervo, A. Tarkkanen, and O. Saksela.** 1992. Surface-associated activation of plasminogen on gram-positive bacteria. Effect of plasmin on the adherence of *Staphylococcus aureus*. *Acta Ophthalmol.* **202:**42–46.

2038. **Kuykindoll, R. J., and R. G. Holt.** 1996. Characterization of a P1-deficient strain of *Streptococcus mutans* that expresses the SpaA protein of *Streptococcus sobrinus*. *Infect. Immun.* **64:**3652–3658.

2039. **Kwaik, Y. A., C. Venkataraman, O. S. Harb, and L.-Y. Gao.** 1998. Signal transduction in the protozoal host *Hartmannella vermiformis* upon attachment and invasion by *Legionella micdadei*. *Appl. Environ. Microbiol.* **64:**3134–3139.

2040. **Kwiecien, J. M., P. B. Little, and M. A. Hayes.** 1994. Adherence of *Haemophilus somnus* to tumor necrosis factor-alpha-stimulated bovine endothelial cells in culture. *Can. J. Vet. Res.* **58:**211–219.

2041. Reference deleted.

2042. **Labigne, A., and H. de Reuse.** 1996. Determinants of *Helicobacter pylori* pathogenicity. *Infect. Agents Dis.* **5:**191–202.

2043. **Ladage, P. M., K. Yamamoto, D. H. Ren, L. Li, J. V. Jester, W. M. Petroll, and H. D. Cavanagh.** 2001. Effects of rigid and soft contact lens daily wear on corneal epithelium, tear lactate dehydrogenase, and bacterial binding to exfoliated epithelial cells. *Ophthalmology* **108:**1279–1288.

2044. **Lafontaine, E. R., L. D. Cope, C. Aebi, J. L. Latimer, G. H. McCracken Jr., and E. J. Hansen.** 2000. The UspA1 protein and a second type of UspA2 protein mediate adherence of *Moraxella catarrhalis* to human epithelial cells *in vitro*. *J. Bacteriol.* **182:**1364–1373.

2045. **Lagergard, T., M. Purven, and A. Frisk.** 1993. Evidence of *Haemophilus ducreyi* adherence to and cytotoxin destruction of human epithelial cells. *Microb. Pathog.* **14:**417–431.

2046. **Lagrou, K., W. E. Peetermans, M. Jorissen, J. Verhaegen, J. van Damme, and J. van Eldere.** 2000. Subinhibitory concentrations of erythromycin reduce pneumococcal adherence to respiratory epithelial cells *in vitro*. *J. Antimicrob. Chemother.* **46:**717–723.

2047. **Lähteenmäki, K., B. Westerlund, P. Kuusela, and T. K. Korhonen.** 1993. Immobilization of plasminogen on *Escherichia coli* flagella. *FEMS Microbiol. Lett.* **106:**309–314.

2048. **Lähteenmäki, K., R. Virkola, A. Sarén, L. Emödy, and T. K. Korhonen.** 1998. Expression of plasminogen activator Pla of *Yersinia pestis* enhances bacterial attachment to the mammalian extracellular matrix. *Infect. Immun.* **66:**5755–5762.

2049. **Lähteenmäki, K., R. Virkola, R. Pouttu, P. Kuusela, M. Kukkonen, and T. K. Korhonen.** 1998. Bacterial plasminogen receptors: In vitro evidence for a role in degradation of the mammalian extracellular matrix. *Infect. Immun.* **63:**3659–3664.

2050. **Laichakl, L. L., J. M. Danforth, and T. J. Standiford.** 1996. Interleukin-10 inhibits neutrophil phagocytic and bactericidal activity. *FEMS Immunol. Med. Microbiol.* **15:**181–187.

2051. **Laitio, P., M. Virtala, M. Salmi, L. J. Pelliniemi, D. T. Yu, and K. Granfors.** 1997. HLA-B27 modulates intracellular survival of *Salmonella enteritidis* in human monocytic cells. *Eur. J. Immunol.* **27:**1331–1338.

2052. **Lajarin, F., G. Rubio, J. Galvez, and P. Garcia-Penarrubia.** 1996. Adhesion, invasion and intracellular replication of *Salmonella typhimurium* in a murine hepatocyte cell line. Effect of cytokines and LPS on antibacterial activity of hepatocytes. *Microb. Pathog.* **21:**319–329.

2053. **Lalonde, M., M. Segura, S. Lacouture, and M. Gottschalk.** 2000. Interactions between *Streptococcus suis* serotype 2 and different epithelial cell lines. *Microbiol.* **146:**1913–1921.

2054. **Lammel, C. J., N. P. Dekker, J. Palefsky, and G. F. Brooks.** 1993. *In vitro* model of *Haemophilus ducreyi* adherence to and entry into eukaryotic cells of genital origin. *J. Infect. Dis.* **167:**642–650.

2055. **Lammers, A., P. J. Nuijten, E. Kruijt, N. Stockhofe-Zurwieden, U. Vecht, H. E. Smith, and F. G. van Zijderveld.** 1999. Cell tropism of *Staphylococcus aureus* in bovine mammary gland cell cultures. *Vet. Microbiol.* **67:**77–89.

2056. **Lammers, A., P. J. Nuijten, and H. E. Smith.** 1999. The fibronectin binding proteins of *Staphylococcus aureus* are required for adhesion to and invasion of bovine mammary gland cells. *FEMS Microbiol. Lett.* **180:**103–109.

2057. **Lammers, A., C. J. van Vorstenbosch, J. H. Erkens, and H. E. Smith.** 2001. The major bovine mastitis pathogens have different cell tropisms in cultures of bovine mammary gland cells. *Vet. Microbiol.* **80:**255–265.

2058. **Lammler, C., and H. Ding.** 1994. Characterization of fibrinogen-binding properties of *Actinomyces pyogenes*. *Zentbl. Veterinarmed.* **41:**588–596.

2059. **Lamont, J. T.** 1992. Mucus: the front line of intestinal mucosal defense. *Ann. N. Y. Acad. Sci.* **664:**190–201.

2060. **Lamont, R. J., S. G. Hersey, and B. Rosan.** 1992. Characterization of the adherence of *Porphyromonas gingivalis* to oral streptococci. *Oral Microbiol. Immunol.* **7:**193–197.

2061. **Lamont, R. J., D. Oda, R. E. Persson, and G. R. Persson.** 1992. Interaction of *Porphyromonas gingivalis* with gingival epithelial cells maintained in culture. *Oral Microbiol. Immunol.* **7:**364–367.

2062. **Lamont, R. J., C. A. Bevan, S. Gil, R. E. Persson, and B. Rosan.** 1993. Involvement of *Porphyromonas gingivalis* fimbriae in adherence to *Streptococcus gordonii*. *Oral Microbiol. Immunol.* **8:**272–276.

2063. **Lamont, R. J., S. Gil, D. R. Demuth, D. Malamud, and B. Rosan.** 1994. Molecules of *Streptococcus gordonii* that bind to *Porphyromonas gingivalis*. *Microbiology* **140:**867–872.

2064. **Lamont, R. J., G. W. Hsiao, and S. Gil.** 1994. Identification of a molecule of *Porphyromonas gingivalis* that binds to *Streptococcus gordonii*. *Microb. Pathog.* **17:**355–360.

2065. **Lamont, R. J., A. Chan, C. M. Belton, K. T. Izutsu, D. Vasel, and A. Weinberg.** 1995. *Porphyromonas gingivalis* invasion of gingival epithelial cells. *Infect. Immun.* **63:**3878–3885.

2066. **Lan, J., M. D. Willcox, and G. D. Jackson.** 1999. Effect of tear-specific immunoglobulin A on the adhesion of *Pseudomonas aeruginosa* I to contact lenses. *Aust. N. Z. J. Ophthalmol.* **27:**218–220.

2067. **Landa, A. S., H. C. van der Mei, and H. J. Busscher.** 1997. Detachment of linking film bacteria from enamel surfaces by oral rinses and penetration of sodium lauryl sulphate through an artificial oral biofilm. *Adv. Dent. Res.* **11:**528–538.

2068. **Landa, A. S., H. C. van der Mei, G. van Rij, and H. J. Busscher.** 1998. Efficacy of ophthalmic solutions to detach adhering *Pseudomonas aeruginosa* from contact lenses. *Cornea* **17:**293–300.

2069. **Landa, A. S., B. van de Belt-Gritter, H. C. van der Mei, and H. J. Busscher.** 1999. Recalcitrance of *Streptococcus mutans* biofilms towards detergent-stimulated detachment. *Eur. J. Oral Sci.* **107:**236–243.

2070. **Lang, H., G. Jonson, J. Holmgren, and E. T. Palva.** 1994. The maltose regulon of *Vibrio cholerae* affects production and secretion of virulence factors. *Infect. Immun.* **62:**4781–4788.

2071. **Langermann, S., S. Palaszynski, M. Barnhart, G. Auguste, J. S. Pinkner, J. Burlein, P. Barren, S. Koenig, S. Leath, C. H. Jones, and S. J. Hultgren.** 1997. Prevention of mucosal *Escherichia coli* infection by FimH-adhesin-based systemic vaccination. *Science* **276:**607–611.

2072. **Lanne, B., J. Ciopraga, J. Bergstrom, C. Motas, and K. A. Karlsson.** 1994. Binding of the galactose-specific *Pseudomonas aeruginosa* lectin, PA-I, to glycosphingolipids and other glycoconjugates. *Glycoconj. J.* **11:**292–298.

2073. **Lanne, B., B.-M. Olsson, P.-Å Jovall, J. Ångström, H. Linder, B.-I. Marklund, J. Bergström, and K.-A. Karlsson.** 1995. Glycoconjugate receptors for P-fimbriated *Escherichia coli* in the mouse. An animal model of urinary tract infection. *J. Biol. Chem.* **270:**9017–9025.

2074. **La Ragione, R. M., W. A. Cooley, and M. J. Woodward.** 2000. The role of fimbriae and flagella in the adherence of avian strains of *Escherichia coli* O78:K80 to tissue culture cells and tracheal and gut explants. *J. Med. Microbiol.* **49:**327–338.

2075. **La Ragione, R. M., A. R. Sayers, and M. J. Woodward.** 2000. The role of fimbriae and flagella in the colonization, invasion and persistence of *Escherichia coli* O78:K80 in the day-old-chick model. *Epidemiol. Infect.* **124:**351–363.

2076. **Larsson, T., J. Bergstrom, C. Nilsson, and K.-A. Karlsson.** 2000. Use of an affinity proteomics approach for the identification of low-abundant bacterial adhesins as applied on the Lewis(b)-binding adhesin of *Helicobacter pylori*. *FEBS Lett.* **469:**155–158.

2077. **Lassen, B., K. Holmberg, C. Brink, A. Carlén, and J. Olsson.** 1994. Binding of salivary proteins and oral bacteria to hydrophobic and hydrophilic surfaces *in vivo* and *in vitro*. *Colloid Polym. Sci.* **272:**1143–1150.

2078. **Latta, R. K., M. J. Schur, D. L. Tolson, and E. Altman.** 1998. The effect of growth conditions on *in vitro* adherence, invasion, and NAF expression by *Proteus mirabilis* 7570. *Can. J. Microbiol.* **44:**896–904.

2079. Lauer, P., N. H. Albertson, and M. Koomey. 1993. Conservation of genes encoding components of a type IV pilus assembly/two-step protein export pathway in *Neisseria gonorrhoeae. Mol. Microbiol.* **8:**357–368.

2080. Law, D. 1994. Adhesion and its role in the virulence of enteropathogenic *Escherichia coli. Clin. Microbiol. Rev.* **7:**152–173.

2081. Law, D., and H. Chart. 1998. Enteroaggregative *Escherichia coli. J. Appl. Microbiol.* **84:**685–697.

2082. Lawin-Brussel, C. A., M. F. Refojo, and K. R. Kenyon. 1992. *In vitro* adhesion of *Pseudomonas aeruginosa* and *Staphylococcus aureus* to surface passivated poly(methyl methacrylate) intraocular lenses. *J. Cataract Refract. Surg.* **18:**598–601.

2083. Lax, A. J., P. A. Barrow, P. W. Jones, and T. S. Wallis. 1995. Current perspectives in salmonellosis. *Br. Vet. J.* **151:**351–377.

2084. Layh-Schmitt, G., H. Hilbert, and E. Pirkl. 1995. A spontaneous hemadsorption-negative mutant of *Mycoplasma pneumoniae* exhibits a truncated adhesin-related 30 kilodalton protein and lacks the cytadherence-accessory protein HMW1. *J. Bacteriol.* **177:**843–846.

2085. Le Bouguenec, C., M. I. Garcia, V. Ouin, J.-M Desperrier, P. Gounon, and A. Labigne. 1993. Characterization of plasmid-borne *afa-3* gene clusters encoding afimbrial adhesins expressed by *Escherichia coli* strains associated with intestinal or urinary tract infections. *Infect. Immun.* **61:**5106–5114.

2086. Lecuit, M., H. Ohayon, L. Braun, J. Mengaud, and P. Cossart. 1997. Internalin of *Listeria monocytogenes* with an intact leucine-rich repeat region is sufficient to promote internalization. *Infect. Immun.* **65:**5309–5319.

2087. Lee, A. 1995. New microbiological features. *Eur. J. Gastroenterol. Hepatol.* **7:**303–309.

2088. Lee, A., J. O'Rourke, M. C. de Ungria, B. Robertson, G. Daskalopoulos, and M. F. Dixon. 1997. A standardized mouse model of *Helicobacter pylori* infection: introducing the Sydney strain. *Gastroenterology* **112:**1386–1397.

2089. Lee, A., D. Chow, B. Haus, W. Tseng, D. Evans, S. Fleisig, G. Chandy, and T. Machen. 1999. Airway epithelial tight junctions and binding and cytotoxicity of *Pseudomonas aeruginosa. Am. J. Physiol.* **277:**L204–L217.

2090. Lee, B. C., and S. Levesque. 1997. A monoclonal antibody directed against the 97-kilodalton gonococcal hemin-binding protein inhibits hemin utilization by *Neisseria gonorrhoeae. Infect. Immun.* **65:**2970–2974.

2091. Lee, J. H., and R. E. Isaacson. 1995. Expression of the gene cluster associated with the *Escherichia coli* pilus adhesin K99. *Infect. Immun.* **63:**4143–4149.

2092. Lee, J. Y., H. T. Sojar, G. S. Bedi, and R. J. Genco. 1992. Synthetic peptides analogous to the fimbrillin sequence inhibit adherence of *Porphyromonas gingivalis. Infect. Immun.* **60:**1662–1670.

2093. Lee, J. Y., H. T. Sojar, A. Sharma, G. S. Bedi, and R. J. Genco. 1993. Active domains of fimbrillin involved in adherence of *Porphyromonas gingivalis. J. Periodontal Res.* **28:**470–472.

2094. Lee, J. Y., H. T. Sojar, A. Amano, and R. J. Genco. 1995. Purification of major fimbrial proteins of *Porphyromonas gingivalis. Protein Expression Purif.* **6:**496–500.

2095. Lee, J.-Y., and M. Caparon. 1996. An oxygen-induced but protein F-independent fibronectin-binding pathway in *Streptococcus pyogenes. Infect. Immun.* **64:**413–421.

2096. Lee, K. K., W. Y. Wong, H. B. Sheth, R. S. Hodges, W. Paranchych, and R. T. Irvin. 1995. Use of synthetic peptides in characterization of microbial adhesins. *Methods Enzymol.* **253:**115–131.

2097. Lee, K. K., H. B. Sheth, W. Y. Wong, R. Sherburne, W. Paranchych, R. S. Hodges, C. A. Lingwood, H. Krivan, and R. T. Irvin. 1994. The binding of *Pseudomonas aeruginosa* pili to glycosphingolipids is a tip-associated event involving the C-terminal region of the structural pilin subunit. *Mol. Microbiol.* **11:**705–713.

2098. Lee, K. K., L. Yu, D. L. Macdonald, W. Paranchych, R. S. Hodges, and R. T. Irvin. 1996. Anti-adhesin antibodies that recognize a receptor-binding motif (adhesintope) inhibit pilus/fimbrial-mediated adherence of *Pseudomonas aeruginosa* and *Candida albicans* to asialo-GM1 receptors and human buccal epithelial cell surface receptors. *Can. J. Microbiol.* **42:**479–486.

2099. Lee, M. D., R. Curtiss III, and T. Peay. 1996. The effect of bacterial surface structures on the pathogenesis of *Salmonella typhimurium* infection in chickens. *Avian Dis.* **40:**28–36.

2100. Lee, M. D., R. E. Wooley, and J. R. Glisson. 1994. Invasion of epithelial cell monolayers by turkey strains of *Pasteurella multocida. Avian Dis.* **38:**72–77.

2101. Lee, S. F. 1995. Active release of bound antibody by *Streptococcus mutans. Infect. Immun.* **63:**1940–1946.

2102. Lee, S. F., Y. H. Li, and G. H. Bowden. 1996. Detachment of *Streptococcus mutans* biofilm cells by an endogenous enzymatic activity. *Infect. Immun.* **64:**1035–1038.

2103. **Lee, S. G., C. Kim, and Y. C. Ha.** 1997. Successful cultivation of a potentially pathogenic coccoid organism with trophism for gastric mucin. *Infect. Immun.* **65:**49–54.

2104. **Lee, S. W., J. D. Hillman, and A. Progulske-Fox.** 1996. The hemagglutinin genes *hagB* and *hagC* of *Porphyromonas gingivalis* are transcribed in vivo as shown by use of a new expression vector. *Infect. Immun.* **64:**4802–4810.

2105. **Lee, S. Y.** 2001. Effects of chlorhexidine digluconate and hydrogen peroxide on *Porphyromonas gingivalis* hemin binding and coaggregation with oral streptococci. *J. Oral Sci.* **43:**1–7.

2106. **Lee, W., J. L. Farmer, M. Hilty, and Y. B. Kim.** 1998. The protective effects of lactoferrin feeding against endotoxin lethal shock in germfree piglets. *Infect. Immun.* **66:**1421–1426.

2107. **Lee, Y. K., C. Y. Lim, W. L. Teng, A. C. Ouwehand, E. M. Tuomola, and S. Salminen.** 2000. Quantitative approach in the study of adhesion of lactic acid bacteria to intestinal cells and their competition with enterobacteria. *Appl. Environ. Microbiol.* **66:**3692–3697.

2108. **Leffler, H., W. Agace, S. Hedges, R. Lindstedt, M. Svensson, and C. Svanborg.** 1995. Strategies for studying bacterial adhesion *in vivo*. *Methods Enzymol.* **253:**206–220.

2109. **Leher, H., R. Silhavy, H. Alizadeh, J. Huang, and J. Y. Niederkorn.** 1998. Mannose induces the release of cytopathic factors from *Acanthamoeba castellanii*. *Infect. Immun.* **66:**5–10.

2110. **Lehto, E. M., and S. J. Salminen.** 1997. Inhibition of *Salmonella typhimurium* adhesion to Caco-2 cell cultures by *Lactobacillus* strain GG spent culture supernate: only a pH effect? *FEMS Immunol. Med. Microbiol.* **18:**125–132.

2111. **Leininger, E., S. Bowen, G. Renauld-Mongenie, J. H. Rouse, F. D. Menozzi, C. Locht, I. Heron, and M. J. Brennan.** 1997. Immunodominant domains present on the *Bordetella pertussis* vaccine component filamentous hemagglutinin. *J. Infect. Dis.* **175:**1423–1431.

2112. **Leininger, E., C. A. Ewanowich, A. Bhargava, M. S. Peppler, J. G. Kenimer, and M. J. Brennan.** 1992. Comparative roles of the Arg-Gly-Asp sequence present in the *Bordetella pertussis* adhesins pertactin and filamentous hemagglutinin. *Infect. Immun.* **60:**2380–2385.

2113. **Leininger, E., P. G. Probst, M. J. Brennan and J. G. Kenimer.** 1993. Inhibition of *Bordetella pertussis* filamentous hemagglutinin-mediated cell adherence with monoclonal antibodies. *FEMS Microbiol. Lett.* **80:**31–38.

2114. **Leite, D. S., T. Yano, and A. F. de Castro.** 1997. Receptors on chicken erythrocytes for F42 fimbriae of *Escherichia coli* isolated from pigs. *Zentbl. Bakteriol.* **286:**383–388.

2115. **Leitner, G., R. Waiman, and E. D. Heller.** 2001. The effect of apramycin on colonization of pathogenic *Escherichia coli* in the intestinal tract of chicks. *Vet. Q.* **23:**62–66.

2116. **Leknes, K. N.** 1997. The influence of anatomic and iatrogenic root surface characteristics on bacterial colonization and periodontal destruction: a review. *J. Periodontol.* **68:**507–516.

2117. **Lelwala-Guruge, J., Å. Ljungh, and T. Wadström.** 1992. Haemagglutination patterns of *Helicobacter pylori*. Frequency of sialic acid-specific and non-sialic acid-specific haemagglutinins. *APMIS* **100:**908–913.

2118. **Lelwala-Guruge, J., F. Ascencio, Å. Ljungh, and T. Wadström.** 1993. Rapid detection and characterization of sialic acid-specific lectins of *Helicobacter pylori*. *APMIS* **101:**695–702.

2119. **Lelwala-Guruge, J., F. Ascencio, A. S. Kreger, Å. Ljungh, and T. Wadström.** 1993. Isolation of a sialic acid-specific surface haemagglutinin of *Helicobacter pylori* strain NTCC 11637. *Zentbl. Bakteriol.* **280:**93–106.

2120. **Lelwala-Guruge, J., A. S. Kreger, Å. Ljungh, and T. Wadström.** 1995. Immunological properties of the cell surface haemagglutinins (sHAs) of *Helicobacter pylori* strain NCTC 11637. *FEMS Immunol. Med. Microbiol.* **11:**73–77.

2121. **Leong, J. M., L. Moitoso de Vargas, and R. R. Isberg.** 1992. Binding of cultured mammalian cells to immobilized bacteria. *Infect. Immun.* **60:**683–686.

2122. **Leong, J. M., P. E. Morrissey, A. Marra, and R. R. Isberg.** 1995. An aspartate residue of the *Yersinia pseudotuberculosis* invasin protein that is critical for integrin binding. *EMBO J.* **14:**422–431.

2123. **Leong, J. M., P. E. Morrissey, E. Ortega-Barria, M. E. A. Pereira, and J. Coburn.** 1995. Hemagglutination and proteoglycan binding by the Lyme disease spirochete, *Borrelia burgdorferi*. *Infect. Immun.* **63:**874–883.

2124. **Leong, J. M., D. Robbins, L. Rosenfeld, B. Lahiri, and N. Parveen.** 1998. Structural requirements for glycosaminoglycan recognition by the Lyme disease spirochete, *Borrelia burgdorferi*. *Infect. Immun.* **66**:6045–6048.

2125. **Leong, J. M., H. Wang, L. Magoun, J. A. Field, P. E. Morrissey, D. Robbins, J. B. Tatro, J. Coburn, and N. Parveen.** 1998. Different classes of proteoglycans contribute to the attachment of *Borrelia burgdorferi* to cultured endothelial and brain cells. *Infect. Immun.* **66**:994–999.

2126. **Lepine, G., R. P. Ellen, and A. Progulske-Fox.** 1996. Construction and preliminary characterization of three hemagglutinin mutants of *Porphyromonas gingivalis*. *Infect. Immun.* **64**:1467–1472.

2127. **Lepine, G., and A. Progulske-Fox.** 1996. Duplication and differential expression of hemagglutinin genes in *Porphyromonas gingivalis*. *Oral Microbiol. Immunol.* **11**:65–78.

2128. **Lepper, A. W., J. L. Atwell, P. R. Lehrbach, C. L. Schwartzkoff, J. R. Egerton, and J. M. Tennent.** 1995. The protective efficacy of cloned *Moraxella bovis* pili in monovalent and multivalent vaccine formulations against experimentally induced infectious bovine keratoconjunctivitis (IBK). *Vet. Microbiol.* **45**:129–138.

2129. **Lepper, A. W., T. C. Elleman, P. A. Hoyne, P. R. Lehrbach, J. L. Atwell, C. L. Schwartzkoff, J. R. Egerton, and J. M. Tennent.** 1993. A *Moraxella bovis* pili vaccine produced by recombinant DNA technology for the prevention of infectious bovine keratoconjunctivitis. *Vet. Microbiol.* **36**:175–183.

2130. **Lepper, A. W., L. J. Moore, J. L. Atwell, and J. M. Tennent.** 1992. The protective efficacy of pili from different strains of *Moraxella bovis* within the same serogroup against infectious bovine keratoconjunctivitis. *Vet. Microbiol.* **32**:177–187.

2131. **Leppilahti, M., J. Hirvonen, and T. L. Tammela.** 1997. Influence of transient overdistension on bladder wall morphology and enzyme histochemistry. *Scand. J. Urol. Nephrol.* **31**:517–522.

2132. **Leranoz, S., P. Orus, M. Berlanga, F. Dalet, and M. Vinas.** 1997. New fimbrial adhesins of *Serratia marcescens* isolated from urinary tract infections: description and properties. *J. Urol.* **157**:694–698.

2133. **Leriche, V., and B. Carpentier.** 2000. Limitation of adhesion and growth of *Listeria monocytogenes* on stainless steel surfaces by *Staphylococcus sciuri* biofilms. *J. Appl. Microbiol.* **88**:594–605.

2134. **Leung, J. W., G. T. Lau, J. J. Sung, and J. W. Costerton.** 1992. Decreased bacterial adherence to silver-coated stent material: an in vitro study. *Gastrointest. Endosc.* **38:** 338–340.

2135. **Leung, J. W., Y. L. Liu, T. D. Desta, E. D. Libby, J. F. Inciardi, and K. Lam.** 2000. *In vitro* evaluation of antibiotic prophylaxis in the prevention of biliary stent blockage. *Gastrointest. Endosc.* **51**:296–303.

2136. **Leung, J. W., Y. Liu, S. Cheung, R. C. Chan, J. F. Inciardi, and A. F. Cheng.** 2001. Effect of antibiotic-loaded hydrophilic stent in the prevention of bacterial adherence: a study of the charge, discharge, and recharge concept using ciprofloxacin. *Gastrointest. Endosc.* **53**:431–437.

2137. **Leung, K., W. W. Nesbitt, M. Okamoto, and H. Fukushima.** 1999. Identification of a fimbriae-associated haemagglutinin from *Prevotella intermedia*. *Microb. Pathog.* **26**:139–148.

2138. **Leung, K. P., H. Fukushima, W. E. Nesbitt, and W. B. Clark.** 1996. *Prevotella intermedia* fimbriae mediate hemagglutination. *Oral Microbiol. Immunol.* **11**:42–50.

2139. **LeVine, A. M., K. E. Kurak, J. R. Wright, W. T. Watford, M. D. Bruno, G. F. Ross, J. A. Whitsett, and T. R. Korfhagen.** 1999. Surfactant protein-A binds group B streptococcus enhancing phagocytosis and clearance from lungs of surfactant protein-A-deficient mice. *Am. J. Respir. Cell Mol. Biol.* **20**:279–286.

2139a. **L'Hoke, C., S. Berger, S. Bourgerie, Y. Duval-Iflah, R. Julien, and Y. Karamanos.** 1995. Use of porcine fibrinogen as a model glycoprotein to study the binding specificity of the three variants of K88 Lectin. *Infect. Immun.* **63**:1927–1932.

2140. **Li, D. Q., F. Lundberg, and Å. Ljungh.** 2000. Binding of von Willebrand factor by coagulase-negative staphylococci. *J. Med. Microbiol.* **49**:217–225.

2141. **Li, D. Q., F. Lundberg, and Å. Ljungh.** 2001. Characterization of vitronectin-binding proteins of *Staphylococcus epidermidis*. *Curr. Microbiol.* **42**:361–367.

2142. **Li, J., and L. A. McLandsborough.** 1999. The effects of the surface charge and hydrophobicity of *Escherichia coli* on its adhesion to beef muscle. *Int. J. Food Microbiol.* **53**:185–193.

2143. **Li, T., I. Johansson, D. I. Hay, and N. Strömberg.** 1999. Strains of *Actinomyces naeslundii* and *Actinomyces viscosus* exhibit structurally variant fimbrial subunit proteins and bind to different peptide motifs in salivary proteins. *Infect. Immun.* **67**:2053–2059.

2144. **Li, X., H. Zhao, L. Geymonat, F.**

Bahrani, D. E. Johnson, and H. L. T. Mobley. 1997. *Proteus mirabilis* mannose-resistant, *Proteus*-like fimbriae: MrpG is located at the fimbrial tip and is required for fimbrial assembly. *Infect. Immun.* **65:**1327–1334.

2145. Li, X., and H. L. T. Mobley. 1998. MrpB functions as the terminator for assembly of *Proteus mirabilis* mannose-resistant *Proteus*-like fimbriae. *Infect. Immun.* **66:**1759–1763.

2146. Li, X., D. E. Johnson, and H. L. Mobley. 1999. Requirement of MrpH for mannose-resistant *Proteus*-like fimbria-mediated hemagglutination by *Proteus mirabilis*. *Infect. Immun.* **67:**2822–2833.

2147. Li, Y. H., and G. H. Bowden. 1994. Characteristics of accumulation of oral gram-positive bacteria on mucin-conditioned glass surfaces in a model system. *Oral Microbiol. Immunol.* **9:**1–11.

2148. Li, Z., E. Elliott, J. Payne, J. Isaacs, P. Gunning, and E. V. O'Loughlin. 1999. Shiga toxin-producing *Escherichia coli* can impair T84 cell structure and function without inducing attaching/effacing lesions. *Infect. Immun.* **67:**5938–5945.

2149. Li, Z. J., N. Mohamed, and J. M. Ross. 2000. Shear stress affects the kinetics of *Staphylococcus aureus* adhesion to collagen. *Biotechnol. Prog.* **16:**1086–1090.

2150. Liang, O. D., K. T. Preissner, and G. S. Chhatwal. 1997. The hemopexin-type repeats of human vitronectin are recognized by *Streptococcus pyogenes*. *Biochem. Biophys. Res. Commun.* **234:**445–449.

2151. Liang, O. D., M. Maccarana, J.-I. Flock, M. Paulsson, K. T. Preissner, and T. Wadström. 1993. Multiple interactions between human vitronectin and *Staphylococcus aureus*. *Biochim. Biophys. Acta* **1225:**57–63.

2152. Liang, O. D., J.-I. Flock, and T. Wadström. 1994. Evidence that the heparin-binding consensus sequence of vitronectin is recognized by *Staphylococcus aureus*. *J. Biochem.* **116:**457–463.

2153. Liao, J., K. Tomochika, S. Watanabe, and Y. Kanemasa. 1992. Establishment of a mouse model of cystitis and roles of type 1 fimbriated *Escherichia coli* in its pathogenesis. *Microbiol. Immunol.* **36:**243–256.

2154. Ligtenberg, A. J., E. Walgreen-Weterings, E. C. Veerman, J. J. de Soet, J. de Graaff, and A. V. Amerongen. 1992. Influence of saliva on aggregation and adherence of *Streptococcus gordonii* HG 222. *Infect. Immun.* **60:**3878–3884.

2155. Ligtenberg, A. J. M., E. Walgreen-Weterings, E. C. I. Veerman, J. de Graaf, and A. V. Nieuw Amerongen. 1993. Adherence of *Streptococcus gordonii* HG222 in the presence of saliva. *Antonie Leeuwenhoek* **64:**39–45.

2156. Ligtenberg, A. J. M., E. Walgreen-Weterings, E. C. I. Veerman, J. J. De Soet, and A. V. Nieuw Amerongen. 1995. Attachment of *Streptococcus gordonii* HG 222 to *Streptococcus oralis* Ny 586 and the influence of saliva. *Microb. Ecol. Health Dis.* **8:**243–254.

2157. Liles, M. R., V. K. Viswanathan, and N. P. Cianciotto. 1998. Identification and temperature regulation of *Legionella pneumophila* genes involved in type IV pilus biogenesis and type II protein secretion. *Infect. Immun.* **66:**1776–1782.

2158. Lilja, M., J. Silvola, S. Raisanen, and L. E. Stenfors. 1999. Where are the receptors for *Streptococcus pyogenes* located on the tonsillar surface epithelium? *Int. J. Pediatr. Otorhinolaryngol.* **50:**37–43.

2159. Liljemark, W. F., C. G. Bloomquist, and C. H. Lai. 1992. Clustering of an outer membrane adhesin of *Haemophilus parainfluenzae*. *Infect. Immun.* **60:**687–689.

2160. Liljemark, W. F., C. G. Bloomquist, C. L. Bandt, B. L. Pihlstrom, J. E. Hinrichs, and L. F. Wolff. 1993. Comparison of the distribution of *Actinomyces* in dental plaque on inserted enamel and natural tooth surfaces in periodontal health and disease. *Oral Microbiol. Immunol.* **8:**5–15.

2161. Liljemark, W. F., and C. Bloomquist. 1996. Human oral microbial ecology and dental caries and periodontal diseases. *Crit. Rev. Oral Biol. Med.* **7:**180–198.

2162. Lillehoj, E. P., S. W. Hyun, B. T. Kim, X. G. Zhang, D. I. Lee, S. Rowland, and K. C. Kim. 2001. Muc1 mucins on the cell surface are adhesion sites for *Pseudomonas aeruginosa*. *Am. J. Physiol. Lung Cell Mol. Physiol.* **280:**L181–L187.

2163. Lim, D. J., and L. O. Bakaletz. 1992. An investigation of the molecular basis of nontypable *Haemophilus influenzae* adherence to mucosal epithelium. *Rhinology* **14:**37–41.

2164. Lim, J. K., N. W. Gunther IV, H. Zhao, D. E. Johnson, S. K. Keay, and H. L. T. Mobley. 1998. In vivo phase variation of *Escherichia coli* type 1 fimbrial genes in women with urinary tract infection. *Infect. Immun.* **66:**3303–3310.

2165. Lin, T. J., Z. Gao, M. Arock, and S. N. Abraham. 1999. Internalization of FimH[+] *Escherichia coli* by the human mast cell line

(HMC-1 5C6) involves protein kinase C. *J. Leukoc. Biol.* **66:**1031–1038.

2166. **Lin, T. M., L. A. Campbell, M. E. Rosenfeld, and C. C. Kuo.** 2001. Human monocyte-derived insulin-like growth factor-2 enhances the infection of human arterial endothelial cells by *Chlamydia pneumoniae. J. Infect. Dis.* **183:**1368–1372.

2167. **Lindblom, G. B., and B. Kaijser.** 1995. *In vitro* studies of *Campylobacter jejuni/coli* strains from hens and humans regarding adherence, invasiveness, and toxigenicity. *Avian Dis.* **39:**718–722.

2168. **Lindenthal, C., and E. A. Elsinghorst.** 2001. Enterotoxigenic *Escherichia coli* TibA glycoprotein adheres to human intestine epithelial cells. *Infect. Immun.* **69:**52–57.

2169. **Linder, T. E., R. L. Daniels, D. J. Lim, and T. F. DeMaria.** 1994. Effect of intranasal inoculation of *Streptococcus pneumoniae* on the structure of the surface carbohydrates of the chinchilla eustachian tube and middle ear mucosa. *Microb. Pathog.* **16:**435–441.

2170. **Linder, T. E., D. J. Lim, and T. F. DeMaria.** 1992. Changes in the structure of the cell surface carbohydrates of the chinchilla tubotympanum following *Streptococcus pneumoniae*-induced otitis media. *Microb. Pathog.* **13:**293–303.

2171. **Lindgren, P. E., P. Speziale, M. McGavin, H. J. Monstein, M. Höök, L. Visai, T. Kostiainen, S. Bozzini, and M. Lindberg.** 1992. Cloning and expression of two different genes from *Streptococcus dysgalactiae* encoding fibronectin receptors. *J. Biol. Chem.* **267:**1924–1931.

2172. **Lindgren, S. E., H. E. Swaisgood, V. G. Janolini, L. T. Axelsson, C. S. Richter, J. M. Mackenzie, and W. Dobrogosz.** 1992. Binding of *Lactobacillus reuteri* to fibronectin immobilized on glass beads. *Zentbl. Bakteriol. Mikrobiol. Hyg.* **277:**519–528.

2173. **Lindhorst, T. K., C. Kieburg, and U. Krallmann-Wenzel.** 1998. Inhibition of the type 1 fimbriae-mediated adhesion of *Escherichia coli* to erythrocytes by multiantennary alpha-mannosyl clusters: the effect of multivalency. *Glycoconj. J.* **15:**605–613.

2174. **Lindler, L. E., and B. D. Tall.** 1993. *Yersinia pestis* pH 6 antigen forms fimbriae and is induced by intracellular association with macrophages. *Mol. Microbiol.* **8:**311–324.

2175. **Lingwood, C. A., G. Wasfy, H. Han, and M. Huesca.** 1993. Receptor affinity purification of a lipid-binding adhesin from *Helicobacter pylori. Infect. Immun.* **61:**2474–2478.

2176. **Linton, C. J., A. Sherriff, and Millar.** 1999. Use of a modified Robbins device to directly compare the adhesion of *Staphylococcus epidermidis* RP62A to surfaces. *J. Appl. Microbiol.* **86:**194–202.

2177. **Lipman, L. J. A., A. de Nijs, and W. Gaastra.** 1995. Isolation and identification of fimbriae and toxin production by *Escherichia coli* strains from cows with clinical mastitis. *Vet. Microbiol.* **47:**1–7.

2178. **Lis, M., T. Shiroza, and H. K. Kuramitsu.** 1995. Role of C-terminal direct repeating units of the *Streptococcus mutans* glucosyltransferase-S in glucan binding. *Appl. Environ. Microbiol.* **61:**2040–2042.

2179. **Liu, D. F., E. Phillips, T. M. Wizemann, M. M. Siegel, K. Tabei, J. L. Cowell, and E. Tuomanen.** 1997. Characterization of a recombinant fragment that contains a carbohydrate recognition domain of the filamentous hemagglutinin. *Infect. Immun.* **65:**3465–3468.

2180. **Liu, H., L. Magoun, and J. M. Leong.** 1999. β1-chain integrins are not essential for intimin-mediated host cell attachment and enteropathogenic *Escherichia coli*-induced actin condensation. *Infect. Immun.* **67:**2045–2049.

2181. **Liu, W. K., S. E. Tebbs, P. O. Byrne, and T. S. Elliott.** 1993. The effects of electric current on bacteria colonising intravenous catheters. *J. Infect.* **27:**261–269.

2182. **Livrelli, V., C. De Champs, P. Di Martino, A. Darfeuille-Michaud, C. Forestier, and B. Joly.** 1996. Adhesive properties and antibiotic resistance of *Klebsiella, Enterobacter,* and *Serratia* clinical isolates involved in nosocomial infections. *J. Clin. Microbiol.* **34:**1963–1969.

2183. **Ljungh, Å., and T. Wadström.** 1995. Binding of extracellular matrix proteins by microbes. *Methods Enzymol.* **253:**501–514.

2184. **Ljungh, Å., and T. Wadström.** 1995. Growth conditions influence expression of cell surface hydrophobicity of staphylococci and other wound infection pathogens. *Microbiol. Immunol.* **39:**753–757.

2185. **Ljungh, Å., A. P. Moran, and T. Wadström.** 1996. Interactions of bacterial adhesins with extracellular matrix and plasma proteins: pathogenic implications and therapeutic possibilities. *FEMS Immunol. Med. Microbiol.* **16:**117–126.

2186. **Lo, S. C., M. M. Hayes, J. G. Tully, R. Y. Wang, H. Kotani, P. F. Pierce, D. L. Rose, and J. W. Shih.** 1992. *Mycoplasma penetrans* sp. nov., from the urogenital tract of patients with AIDS. *Int. J. Sys. Bacteriol.* **42:**357–364.

2187. Lo, S. C., M. M. Hayes, H. Kotani, P. F. Pierce, D. J. Wear, P. B. Newton III, J. G. Tully, and J. W. Shih. 1993. Adhesion onto and invasion into mammalian cells by *Mycoplasma penetrans:* a newly isolated mycoplasma from patients with AIDS. *Mod. Pathol.* **6:**276–280.

2188. Lobo, A. J., P. M. Sagar, J. Rothwell, P. Quirke, P. Godwin, D. Johnston, and A. T. Axon. 1993. Carriage of adhesive *Escherichia coli* after restorative proctocolectomy and pouch anal anastomosis: relation with functional outcome and inflammation. *Gut* **34:**1379–1383.

2189. Lo Bue, A. M., E. Geremia, C. Castagna, G. Chisari, and G. Nicoletti. 1999. Sub-MIC ciprofloxacin effect on fimbrial production by uropathogenic *Escherichia coli* strains. *J. Chemother.* **11:**357–362.

2190. Locht, C., M. C. Geoffroy, and G. Renauld. 1992. Common accessory genes for the *Bordetella pertussis* filamentous hemagglutinin and fimbriae share sequence similarities with the *papC* and *papD* gene families. *EMBO J.* **11:**3175–3183.

2191. Locht, C., P. Bertin, F. D. Menozzi, and G. Renauld. 1993. The filamentous haemagglutin, a multifaceted adhesin produced by virulent *Bordetella* spp. *Mol. Microbiol.* **9:**653–660.

2192. Lockaby, S. B., F. J. Hoerr, L. H. Lauerman, B. F. Smith, A. M. Samoylov, M. A. Toivio-Kinnucan, and S. H. Kleven. 1999. Factors associated with virulence of *Mycoplasma synoviae. Avian Dis.* **43:**251–261.

2193. Loesche, W. J. 1993. Bacterial mediators in periodontal disease. *Clin. Infect. Dis.* **16:**S203–S210.

2194. Logan, R. P. 1996. Adherence of *Helicobacter pylori. Aliment. Pharmacol. Ther.* **10:**S3–S15.

2195. Logan, R. P., A. Robins, G. A. Turner, A. Cockayne, S. P. Borriello, and C. J. Hawkey. 1998. A novel flow cytometric assay for quantitating adherence of *Helicobacter pylori* to gastric epithelial cells. *J. Immunol. Methods* **213:**19–30.

2196. Loimaranta, V., A. Carlen, J. Olsson, J. Tenovuo, E. L. Syvaoja, and T. H. Korhonen. 1998. Concentrated bovine colostral whey proteins from *Streptococcus mutans/Strep. sobrinus* immunized cows inhibit the adherence of *Strep. mutans* and promote the aggregation of mutans streptococci. *J. Dairy Res.* **65:**599–607.

2197. London, J. 1995. Identifying and isolating fimbrial-associated adhesins of oral gram-negative bacteria. *Methods Enzymol.* **253:**397–403.

2198. Long, C. D., S. F. Hayes, J. P. van Putten, H. A. Harvey, M. A. Apicella, and H. S. Seifert. 2001. Modulation of gonococcal piliation by regulatable transcription of *pilE. J. Bacteriol.* **183:**1600–1609.

2199. Long, C. D., R. N. Madraswala, and H. S. Seifert. 1998. Comparisons between colony phase variation of *Neisseria gonorrhoeae* FA1090 and pilus, pilin, and S-pilin expression. *Infect. Immun.* **66:**1918–1927.

2200. Longhi, C., M. P. Conte, W. Bellamy, L. Seganti, and P. Valenti. 1994. Effect of lactoferricin B, a pepsin-generated peptide of bovine lactoferrin, on *Escherichia coli* HB101(pRI203) entry into HeLa cells. *Med. Microbiol. Immunol.* **183:**77–85.

2201. Longhi, C., M. P. Conte, L. Seganti, M. Polidoro, A. Alfsen, and P. Valenti. 1993. Influence of lactoferrin on the entry process of *Escherichia coli* HB101 (pRI203) in HeLa cells. *Med. Microbiol. Immunol.* **182:**25–35.

2202. Loo, C. Y., M. D. Willcox, and K. W. Knox. 1994. Surface-associated properties of *Actinomyces* strains and their potential relation to pathogenesis. *Oral Microbiol. Immunol.* **9:**12–18.

2203. Loosmore, S. M., R. K. Yacoob, G. R. Zealey, G. E. Jackson, Y. P. Yang, P. S. Chong, J. M. Shortreed, D. C. Coleman, J. D. Cunningham, L. Gisonni, et al. 1995. Hybrid genes over-express pertactin from *Bordetella pertussis. Vaccine* **13:**571–580.

2204. Lopez-Torres, A. J., and V. Stout. 1996. Role of colanic acid polysaccharide in serum resistance *in vivo* and in adherence. *Curr. Microbiol.* **33:**383–389.

2205. Lory, S., and M. S. Strom. 1997. Structure-function relationship of type-IV prepilin peptidase of *Pseudomonas aeruginosa*—a review. *Gene* **192:**117–121.

2206. Lou, Y., W. P. Olson, X. X. Tian, M. E. Klegerman, and M. J. Groves. 1995. Interaction between fibronectin-bearing surfaces and bacillus Calmette-Guérin (BCG) or gelatin microparticles. *J. Pharm. Pharmacol.* **47:**177–181.

2207. Loubeyre, C., J. F. Desnottes, and N. Moreau. 1993. Influence of sub-inhibitory concentrations of antibacterials on the surface properties and adhesion of *Escherichia coli. J. Antimicrob. Chemother.* **31:**37–45.

2208. Louie, M., J. C. de Azavedo, M. Y. Handelsman, C. G. Clark, B. Ally, M. Dytoc, P. Sherman, and J. Brunton. 1993. Expression and characterization of the *eaeA* gene product of *Escherichia coli* serotype O157:H7. *Infect. Immun.* **61:**4085–4092.

2209. **Love, R. M.** 1996. Adherence of *Streptococcus gordonii* to smeared and nonsmeared dentine. *Int. Endodont. J.* **29:**108–112.

2210. **Love, R. M., M. D. McMillan, and H. F. Jenkinson.** 1997. Invasion of dentinal tubules by oral streptococci is associated with collagen recognition mediated by the antigen I/II family of polypeptides. *Infect. Immun.* **65:**5157–5164.

2211. **Love, R. M., M. D. McMillan, Y. Park, and H. F. Jenkinson.** 2000. Coinvasion of dentinal tubules by *Porphyromonas gingivalis* and *Streptococcus gordonii* depends upon binding specificity of streptococcal antigen I/II adhesin. *Infect. Immun.* **68:**1359–1365.

2212. **Lowe, A. B., M. Vamvakaki, M. A. Wassall, L. Wong, N. C. Billingham, S. P. Armes, and A. W. Lloyd.** 2000. Well-defined sulfobetaine-based statistical copolymers as potential antibioadherent coatings. *J. Biomed. Mater. Res.* **52:**88–94.

2213. **Lowe, A. M., P. A. Lambert, and A. W. Smith.** 1995. Cloning of an *Enterococcus faecalis* endocarditis antigen: homology with adhesins from some oral streptococci. *Infect. Immun.* **63:**703–706.

2214. **Lu, D., B. Boyd, and C. A. Lingwood.** 1998. The expression and characterization of a putative adhesin B from *H. influenzae*. *FEMS Microbiol. Lett.* **165:**129–137.

2215. **Lü-Lü, J. S. Singh, M. Y. Galperin, D. Drake, K. G. Taylor, and R. J. Doyle.** 1992. Chelating agents inhibit activity and prevent expression of streptococcal glucan-binding lectins. *Infect. Immun.* **60:**3807–3813.

2216. **Luna, M. G., L. C. Ferreira, D. F. Almeida, and A. Rudin.** 1997. Peptides 14VIDLL18 and 96FEAAAL101 defined as epitopes of antibodies raised against amino acid sequences of enterotoxigenic *Escherichia coli* colonization factor antigen I fused to *Salmonella* flagellin. *Microbiology* **143:**3201–3207.

2217. **Luna, M. G., M. M. Martins, S. M. Newton, S. O. Costa, D. F. Almeida, and L. C. Ferreira.** 1997. Cloning and expression of colonization factor antigen I (CFA/I) epitopes of enterotoxigenic *Escherichia coli* (ETEC) in *Salmonella* flagellin. *Res. Microbiol.* **148:**217–228.

2218. **Lund, S. J., H. A. Rowe, R. Parton, and W. Donachie.** 2001. Adherence of ovine and human *Bordetella parapertussis* to continuous cell lines and ovine tracheal organ culture. *FEMS Microbiol. Lett.* **194:**197–200.

2219. **Lundberg, F., I. Gouda, O. Larm, M. A. Galin, and Å. Ljungh.** 1998. A new model to assess staphylococcal adhesion to intraocular lenses under *in vitro* flow conditions. *Biomaterials* **19:**1727–1733.

2220. **Lundberg, F., S. Schliamser, and Å. Ljungh.** 1997. Vitronectin may mediate staphylococcal adhesion to polymer surfaces in perfusing human cerebrospinal fluid. *J. Med. Microbiol.* **46:**285–296.

2221. **Lundberg, F., T. Lea, and Å. Ljungh.** 1997. Vitronectin-binding staphylococci enhance surface-associated complement activation. *Infect. Immun.* **65:**897–902.

2222. **Lundgren, E., N. Carballeira, R. Vaquez, E. Dubinina, H. Branden, H. Persson, and H. Wolf-Watz.** 1996. Invasin of *Yersinia pseudotuberculosis* activates human peripheral B cells. *Infect. Immun.* **64:**829–835.

2223. **Lutwyche, P., R. Norris-Jones, and D. E. Brooks.** 1995. Aqueous two-phase polymer systems as tools for the study of a recombinant surface-expressed *Escherichia coli* hemagglutinin. *Appl. Environ. Microbiol.* **61:**3251–3255.

2224. **Lutwyche, P., R. Rupps, J. Cavanagh, R. A. Warren, and D. E. Brooks.** 1994. Cloning, sequencing, and viscometric adhesion analysis of heat-resistant agglutinin 1, an integral membrane hemagglutinin from *Escherichia coli* O9:H10:K99. *Infect. Immun.* **62:**5020–5026.

2225. **Lyons, V. O., S. L. Henry, M. Faghiri, and D. Seligson.** 1992. Bacterial adherence to plain and tobramycin-laden polymethylmethacrylate beads. *Clin. Orthopaed. Relat. Res.* **278:**260–264.

2226. **Lyte, M., A. K. Erickson, B. P. Arulanandam, C. D. Frank, M. A. Crawford, and D. H. Francis.** 1997. Norepinephrine-induced expression of the K99 pilus adhesin of enterotoxigenic *Escherichia coli*. *Biochem. Biophys. Res. Commun.* **232:**682–686.

2227. **Ma, Y., M. O. Lassiter, J. A. Banas, M. Y. Galperin, K. G. Taylor, and R. J. Doyle.** 1996. Multiple glucan-binding proteins of *Streptococcus sobrinus*. *J. Bacteriol.* **178:**1572–1577.

2228. **MacBeth, K. J., and C. A. Lee.** 1993. Prolonged inhibition of bacterial protein synthesis abolishes *Salmonella* invasion. *Infect. Immun.* **61:**1544–1546.

2229. **Mack, D., N. Siemssen, and R. Laufs.** 1992. Parallel induction by glucose of adherence and a polysaccharide antigen specific for plastic-adherent *Staphylococcus epidermidis*: evidence for functional relation to intercellular adhesion. *Infect. Immun.* **60:**2048–2057.

2230. **Mack, D., M. Nedelmann, A. Krokotsch,**

A. Schwarzkopf, J. Heesemann, and R. Laufs. 1994. Characterization of transposon mutants of biofilm-producing *Staphylococcus epidermidis* impaired in the accumulative phase of biofilm production: genetic identification of a hexosamine-containing polysaccharide intercellular adhesin. *Infect. Immun.* **62:**3244–3253.

2231. Mack, D., W. Fischer, A. Krokotsch, K. Leopold, R. Hartmann, H. Egge, and R. Laufs. 1996. The intercellular adhesin involved in biofilm accumulation of *Staphylococcus epidermidis* is a linear beta-1,6-linked glucosaminoglycan: purification and structural analysis. *J. Bacteriol.* **178:**175–183.

2232. Mack, D., M. Haeder, N. Siemssen, and R. Laufs. 1996. Association of biofilm production of coagulase-negative staphylococci with expression of a specific polysaccharide intercellular adhesin. *J. Infect. Dis.* **174:**881–884.

2233. Mack, D., J. Riedewald, H. Rohde, T. Magnus, H. H. Feucht, H. A. Elsner, R. Laufs, and M. E. Rupp. 1999. Essential functional role of the polysaccharide intercellular adhesin of *Staphylococcus epidermidis* in hemagglutination. *Infect. Immun.* **67:**1004–1008.

2234. Mack, D., H. Rohde, S. Dobinsky, J. Riedewald, M. Nedelmann, J. K. Knobloch, H. A. Elsner, and H. H. Feucht. 2000. Identification of three essential regulatory gene loci governing expression of *Staphylococcus epidermidis* polysaccharide intercellular adhesin and biofilm formation. *Infect. Immun.* **68:**3799–3807.

2235. Mack, D. R. 1995. The interplay between enteric pathogens and gastrointestinal mucins. *J. Pediatr. Gastroenterol. Nutr.* **21:**116–117.

2236. Mack, D. R., T. S. Gaginella, and P. M. Sherman. 1992. Effect of colonic inflammation on mucin inhibition of *Escherichia coli* RDEC-1 binding *in vitro*. *Gastroenterology* **102:**1199–1211.

2237. Mack, D. R., P. L. Blain-Nelson, and J. W. Mauger. 1993. Lack of inhibition of adhesion of an enteropathogenic *Escherichia coli* by polycarbophil. *J. Biomed. Mater. Res.* **27:**1579–1583.

2238. Mack, D. R., and P. L. Blain-Nelson. 1995. Disparate *in vitro* inhibition of adhesion of enteropathogenic *Escherichia coli* RDEC-1 by mucins isolated from various regions of the intestinal tract. *Pediatr. Res.* **37:**75–80.

2239. Mack, D. R., S. Michail, S. Wei, L. McDougall, and M. A. Hollingsworth. 1999. Probiotics inhibit enteropathogenic *E. coli* adherence *in vitro* by inducing intestinal mucin gene expression. *Am. J. Physiol.* **276:**G941–G950.

2240. Madden, J. C., N. Ruiz, and M. Caparon. 2001. Cytolysin-mediated translocation (CMT): a functional equivalent of type III secretion in gram-positive bacteria. *Cell* **104:**143–152.

2241. Maddox, C. W., K. Kasemsuksukul, W. H. Fales, C. Besch-Williford, C. A. Carson, and K. Wise. 1997. Unique *Salmonella choleraesuis* surface protein affecting invasiveness. Possible *inv* related sequence. *Adv. Exptl. Med. Biol.* **412:**341–348.

2242. Madianos, P. N., P. N. Papapanou, and J. Sandros. 1997. *Porphyromonas gingivalis* infection of oral epithelium inhibits neutrophil transepithelial migration. *Infect. Immun.* **65:**3983–3990.

2243. Madison, B., I. Ofek, S. Clegg, and S. N. Abraham. 1994. Type 1 fimbrial shafts of *Escherichia coli* and *Klebsiella pneumoniae* influence sugar-binding specificities of their FimH adhesins. *Infect. Immun.* **62:**843–848.

2244. Magalhaes, M., R. J. Amorim, Y. Takeda, T. Tsukamoto, M. G. Antas, and S. Tateno. 1992. Localized, diffuse, and aggregative-adhering Escherichia coli from infants with acute diarrhea and matched-controls. *Mem. Inst. Oswaldo Cruz* **87:**93–97.

2245. Maganti, S., M. M. Pierce, A. Hoffmaster, and F. G. Rogers. 1998. The role of sialic acid in opsonin-dependent and opsonin-independent adhesion of *Listeria monocytogenes* to murine peritoneal macrophages. *Infect. Immun.* **66:**620–626.

2246. Magnotta, S., A. Bogucki, R. F. Vieth, and R. W. Coughlin. 1997. Comparative behavior of *E. coli* and *S. aureus* regarding attachment to and removal from a polymeric surface. *J. Biomater. Sci. Polym. Ed.* **8:**683–689.

2247. Magnusson, K. E. 1994. Testing for charge and hydrophobicity correlates in cell-cell adhesion. *Methods Enzymol.* **228:**326–334.

2248. Magnusson, G., S. J. Hultgren, and J. Kihlberg. 1995. Specificity mapping of bacterial lectins by inhibition of hemagglutination using deoxy and deoxyfluoro analogs of receptor-active saccharides. *Methods Enzymol.* **253:**105–114.

2249. Magoun, L., W. R. Zuckert, D. Robbins, N. Parveen, K. R. Alugupalli, T. G. Schwan, A. G. Barbour, and J. M. Leong. 2000. Variable small protein (Vsp)-dependent and Vsp-independent pathways for glycosaminoglycan recognition by relapsing fever spirochaetes. *Mol. Microbiol.* **36:**886–897.

2250. Mahenthiralingham, E., M. E. Campbell,

and D. P. Speert. 1994. Nonmotility and phagocytic resistance of *Pseudomonas aeruginosa* isolates from chronically colonized patients with cystic fibrosis. *Infect. Immun.* **62**:596–605.

2251. **Mahenthiralingam, E., and D. P. Speert.** 1995. Nonopsonic phagocytosis of *Pseudomonas aeruginosa* by macrophages and polymorphonuclear leukocytes requires the presence of the bacterial flagellum. *Infect. Immun.* **63**:4519–4523.

2252. **Maher, D., R. Sherburne, and D. E. Taylor.** 1993. H-pilus assembly kinetics determined by electron microscopy. *J. Bacteriol.* **175**:2175–2183.

2253. **Mahmood, A., M. J. Engle, S. J. Hultgren, G. S. Goetz, K. Dodson, and D. H. Alpers.** 2000. Role of intestinal surfactant-like particles as a potential reservoir of uropathogenic *Escherichia coli*. *Biochim. Biophys. Acta* **1523**:49–55.

2254. **Mai, G. T., J. G. McCormack, W. K. Seow, G. B. Pier, L. A. Jackson, and Y. H. Thong.** 1993. Inhibition of adherence of mucoid *Pseudomonas aeruginosa* by alginase, specific monoclonal antibodies, and antibiotics. *Infect. Immun.* **61**:4338–4343.

2255. **Mainil, J. G., E. Jacquemin, F. Herault, and E. Oswald.** 1997. Presence of *pap-*, *sfa-*, and *afa*-related sequences in necrotoxigenic *Escherichia coli* isolates from cattle: evidence for new variants of the AFA family. *Can. J. Vet. Res.* **61**:193–199.

2256. **Maiti, S. N., J. Harel, and J. M. Fairbrother.** 1993. Structure and copy number analyses of *pap-*, *sfa-*, and *afa*-related gene clusters in F165-positive bovine and porcine *Escherichia coli* isolates. *Infect. Immun.* **61**:2453–2461.

2257. **Maiti, S. N., L. DesGroseillers, J. M. Fairbrother, and J. Harel.** 1994. Analysis of genes coding for the major and minor fimbrial subunits of the Prs-like fimbriae F165(1) of porcine septicemic *Escherichia coli* strain 4787. *Microb. Pathog.* **16**:15–25.

2258. **Majeed, M., M. Gustafsson, E. Kihlstrom, and O. Stendahl.** 1993. Roles of Ca^{2+} and F-actin in intracellular aggregation of *Chlamydia trachomatis* in eucaryotic cells. *Infect. Immun.* **61**:1406–1414.

2259. **Majtan, V., and L. Majtanova.** 1998. Influence of subinhibitory concentrations of antibiotics on surface hydrophobicity of *Salmonella enteritidis*. *Arzneim.-Forsch.* **48**:697–700.

2260. **Majtan, V., and L. Majtanova.** 2000. Effect of new quaternary bisammonium compounds on the growth and cell surface hydrophobicity

of *Enterobacter cloacae*. *Cent. Eur. J. Public Health* **8**:80–82.

2261. **Makakole, S. C., and A. W. Sturm.** 2001. The effect of temperature on the interaction of *Haemophilus ducreyi* with human epithelial cells. *J. Med. Microbiol.* **50**:449–455.

2262. **Makhov, A. M., J. H. Hannah, M. J. Brennan, B. L. Trus, E. Kocsis, J. F. Conway, P. T. Wingfield, M. N. Simon, and A. C. Steven.** 1994. Filamentous hemagglutinin of *Bordetella pertussis*. A bacterial adhesin formed as a 50-nm monomeric rigid rod based on a 19-residue repeat motif rich in beta strands and turns. *J. Mol. Biol.* **241**:110–124.

2263. **Makin, S. A., and T. J. Beveridge.** 1996. The influence of A-band and B-band lipopolysaccharide on the surface characteristics and adhesion of *Pseudomonas aeruginosa* to surfaces. *Microbiology* **142**:299–307.

2264. **Malangoni, M. A., D. H. Livingston, and J. C. Peyton.** 1993. The effect of protein binding on the adherence of staphylococci to prosthetic vascular grafts. *J. Surg. Res.* **54**:168–172.

2265. **Malaviya, R., E. A. Ross, J. I. MacGregor, T. Ikeda, J. R. Little, B. A. Jakschik, and S. N. Abraham.** 1994. Mast cell phagocytosis of FimH-expressing enterobacteria. *J. Immunol.* **152**:1907–1914.

2266. **Malaviya, R., E. Ross, B. A. Jakschik, and S. N. Abraham.** 1994. Mast cell degranulation induced by type 1 fimbriated *Escherichia coli* in mice. *J. Clin. Investig.* **93**:1645–1653.

2267. **Malaviya, R., and S. N. Abraham.** 1995. Interaction of bacteria with mast cells. *Methods Enzymol.* **253**:27–43.

2268. **Malaviya, R., T. Ikeda, E. Ross, and S. N. Abraham.** 1996. Mast cell modulation of neutrophil influx and bacterial clearance at sites of infection through TNF-alpha. *Nature* **381**:77–80.

2269. **Malaviya, R., N. J. Twesten, E. A. Ross, S. N. Abraham, and J. D. Pfeifer.** 1996. Mast cells process bacterial Ags through a phagocytic route for class I MHC presentation to T cells. *J. Immunol.* **156**:1490–1496.

2270. **Malaviya, R., Z. Gao, N. Thankavel, P. A. van der Merwe, and S. N. Abraham.** 1999. The mast cell tumor necrosis factor alpha response to FimH-expressing *Escherichia coli* is mediated by the glycosylphosphatidylinositol-anchored molecule CD48. *Proc. Natl. Acad. Sci. USA* **96**:8110–8115.

2271. **Malek, R., J. G. Fisher, A. Caleca, M. Stinson, C. J. van Oss, J. Y. Lee, M. I. Cho, R. J. Genco, R. T. Evans, and D.**

W. Dyer. 1994. Inactivation of the *Porphyromonas gingivalis fimA* gene blocks periodontal damage in gnotobiotic rats. *J. Bacteriol.* **176**:1052–1059.

2272. **Mammarappallil, J. G., and E. A. Elsinghorst.** 2000. Epithelial cell adherence mediated by the enterotoxigenic *Escherichia coli* Tia protein. *Infect. Immun.* **68**:6595–6601.

2273. **Mamo, W., and G. Froman.** 1994. Adhesion of *Staphylococcus aureus* to bovine mammary epithelial cells induced by growth in milk whey. *Microbiol. Immunol.* **38**:305–308.

2274. **Mamo, W., P. Jonsson, and H. P. Muller.** 1995. Opsonization of *Staphylococcus aureus* with a fibronectin-binding protein antiserum induces protection in mice. *Microb. Pathog.* **19**:49–55.

2275. **Mamo, W., M. Lindahl, and P. Jonsson.** 1992. Binding of fibronectin and type II collagen to *Staphylococcus aureus*: reduction of binding caused by a periodate-sensitive surface structure induced after growth in milk whey. *Microb. Pathog.* **12**:443–449.

2276. **Manch-Citron, J. N., J. Allen, M. Moos Jr., and J. London.** 1992. The gene encoding a *Prevotella loescheii* lectin-like adhesin contains an interrupted sequence which causes a frameshift. *J. Bacteriol.* **174**:7328–7336.

2277. **Manch-Citron, J. N., and J. London.** 1994. Expression of the *Prevotella loescheii* adhesin gene (*plaA*) is mediated by a programmed frameshifting hop. *J. Bacteriol.* **176**:1944–1948.

2278. **Mandell, G. L., and M. O. Frank.** 1992. Microbial defenses against killing by phagocytes. *Trans. Am. Clin. Climatol. Assoc.* **103**:199–209.

2279. **Manganelli, R., and I. van de Rijn.** 1999. Characterization of *emb*, a gene encoding the major adhesin of *Streptococcus defectivus*. *Infect. Immun.* **67**:50–56.

2280. **Manjarrez-Hernandez, A., S. Gavilanes-Parra, M. E. Chavez-Berrocal, J. Molina-Lopez, and A. Cravioto.** 1997. Binding of diarrheagenic *Escherichia coli* to 32- to 33-kilodalton human intestinal brush border proteins. *Infect. Immun.* **65**:4494–4501.

2281. **Manjarrez-Hernandez, H. A., T. J. Baldwin, A. Aitken, S. Knutton, and P. H. Williams.** 1992. Intestinal epithelial cell protein phosphorylation in enteropathogenic *Escherichia coli* diarrhoea. *Lancet* **339**:521–523.

2282. **Mannhardt, W., K. Beutel, P. Habermehl, M. Knuf, and F. Zepp.** 1999. The interaction of buccal mucosal epithelial cells with *E. coli* bacteria enhances the intraepithelial calcium flux and the release of prostaglandin E2 (PgE2). *Int. Urogynecol. J. Pelvic Floor Dysfunct.* **10**:308–315.

2283. **Mannhardt, W., A. Becker, M. Putzer, M. Bork, F. Zepp, J. Hacker, and H. Schulte-Wissermann.** 1996. Host defense within the urinary tract. I. Bacterial adhesion initiates an uroepithelial defense mechanism. *Pediatr. Nephrol.* **10**:568–572.

2284. **Mannhardt, W., M. Putzer, F. Zepp, and H. Schulte-Wissermann.** 1996. Host defense within the urinary tract. II. Signal transducing events activate the uroepithelial defense. *Pediatr. Nephrol.* **10**:573–577.

2285. **Manning, P. A.** 1995. Use of confocal microscopy in studying bacterial adhesion and invasion. *Methods Enzymol.* **253**:159–166.

2286. **Manning, P. A.** 1997. The *tcp* gene cluster of *Vibrio cholerae*. *Gene* **192**:63–70.

2287. **Manning, J. E., E. B. Hume, N. Hunter, and K. W. Knox.** 1994. An appraisal of the virulence factors associated with streptococcal endocarditis. *J. Med. Microbiol.* **40**:110–114.

2288. **Mansouri, E., J. Gabelsberger, B. Knapp, E. Hundt, U. Lenz, K. D. Hungerer, H. E. Gilleland, Jr., J. Staczek, H. Domdey, and B. U. von Specht.** 1999. Safety and immunogenicity of a *Pseudomonas aeruginosa* hybrid outer membrane protein F-I vaccine in human volunteers. *Infect. Immun.* **67**:1461–1470.

2289. **Mantle, M., and S. D. Husar.** 1994. Binding of *Yersinia enterocolitica* to purified, native small intestinal mucins from rabbits and humans involves interactions with the mucin carbohydrate moiety. *Infect. Immun.* **62**:1219–1227.

2290. **Mantle, M., and S. D. Husar.** 1993. Adhesion of *Yersinia enterocolitica* to purified rabbit and human intestinal mucin. *Infect. Immun.* **61**:2340–2346.

2291. **Marc, D., and M. Dho-Moulin.** 1996. Analysis of the *fim* cluster of an avian O2 strain of *Escherichia coli*: serogroup-specific sites within *fimA* and nucleotide sequence of *fimI*. *J. Med. Microbiol.* **44**:444–452.

2292. **Marc, D., P. Arne, A. Bree, and M. Dho-Moulin.** 1998. Colonization ability and pathogenic properties of a *fim⁻* mutant of an avian strain of *Escherichia coli*. *Res. Microbiol.* **149**:473–485.

2293. **Marceau, M., and X. Nassif.** 1999. Role of glycosylation at Ser63 in production of soluble pilin in pathogenic *Neisseria*. *J. Bacteriol.* **181**:656–661.

2294. **Marceau, M., J.-L. Beretti, and X. Nassif.** 1995. High adhesiveness of encapsulated *Neisseria meningitidis* to epithelial cells is associated with the formation of bundles of pili. *Mol. Microbiol.* **17**:855–863.

2295. **Marches, O., J. P. Nougayrede, S. Boullier, J. Mainil, G. Charlier, I. Raymond, P. Pohl, M. Boury, J. De Rycke, A. Milon, and E. Oswald.** 2000. Role of *tir* and intimin in the virulence of rabbit enteropathogenic *Escherichia coli* serotype O103:H2. *Infect. Immun.* **68**:2171–2182.

2296. **Marchese, A., E. A. Debbia, S. Massaro, U. Campora, and G. Schito.** 1996. Brodimoprim: effects of subminimal inhibitory concentrations on virulence traits of respiratory and urinary tract pathogens, and on plasmid transfer and stability. *J. Chemother.* **8**:171–177.

2297. **Mariencheck, W. I., J. Savov, Q. Dong, M. J. Tino, and J. R. Wright.** 1999. Surfactant protein A enhances alveolar macrophage phagocytosis of a live, mucoid strain of *P. aeruginosa. Am. J. Physiol.* **277**:L777–L786.

2298. **Markham, P. F., M. D. Glew, M. R. Brandon, I. D. Walker, and K. G. Whithear.** 1992. Characterization of a major hemagglutinin protein from *Mycoplasma gallisepticum. Infect. Immun.* **60**:3885–3891.

2299. **Markham, P. F., M. D. Glew, K. G. Whithear, and I. D. Walker.** 1993. Molecular cloning of a member of the gene family that encodes pMGA, a hemagglutinin of *Mycoplasma gallisepticum. Infect. Immun.* **61**:903–909.

2300. **Marklund, B. I., J. M. Tennent, E. Garcia, A. Hamers, M. Baga, F. Lindberg, W. Gaastra, and S. Normark.** 1992. Horizontal gene transfer of the *Escherichia coli pap* and *prs* pili operons as a mechanism for the development of tissue-specific adhesive properties. *Mol. Microbiol.* **6**:2225–2242.

2301. **Marone, P., L. Perversi, V. Monzillo, R. Maserati, and E. Antoniazzi.** 1995. Ocular infections: antibiotics and bacterial adhesion on biomaterials used in ocular surgery. *Ophthalmology* **209**:315–318.

2302. **Marques, M. A., V. L. Antonio, E. N. Sarno, P. J. Brennan, and M. C. Pessolani.** 2001. Binding of α2-laminins by pathogenic and non-pathogenic mycobacteria and adherence to Schwann cells. *J. Med. Microbiol.* **50**:23–28.

2303. **Marques, M. A., S. Mahapatra, E. N. Sarno, S. Santos, J. S. Spencer, P. J. Brennan, and M. C. Pessolani.** 2001. Further biochemical characterization of *Mycobacterium leprae* laminin-binding proteins. *Braz. J. Med. Biol. Res.* **34**:463–470.

2304. **Marra, A., and R. R. Isberg.** 1996. Common entry mechanisms. Bacterial pathogenesis. *Curr. Biol.* **6**:1084–1086.

2305. **Marra, A., and R. R. Isberg.** 1997. Invasin-dependent and invasin-independent pathways for translocation of *Yersinia pseudotuberculosis* across the Peyer's patch intestinal epithelium. *Infect. Immun.* **65**:3412–3421.

2306. **Marron, M. B., and C. J. Smyth.** 1995. Molecular analysis of the *cso* operon of enterotoxigenic *Escherichia coli* reveals that CsoA is the adhesin of CS1 fimbriae and that the accessory genes are interchangeable with those of the *cfa* operon. *Microbiology* **141**:2849–2859.

2307. **Marsh, J. W., D. Sun, and R. K. Taylor.** 1996. Physical linkage of the *Vibrio cholerae* mannose-sensitive hemagglutinin secretory and structural subunit gene loci: identification of the *mshG* coding sequence. *Infect. Immun.* **64**:460–465.

2308. **Martin, C., E. Rousset, and H. De Greve.** 1997. Human uropathogenic and bovine septicaemic *Escherichia coli* strains carry an identical F17-related adhesin. *Res. Microbiol.* **148**:55–64.

2309. **Martin, P. R., M. Hobbs, P. D. Free, Y. Jeske, and J. S. Mattick.** 1993. Characterization of *pilQ*, a new gene required for the biogenesis of type 4 fimbriae in *Pseudomonas aeruginosa. Mol. Microbiol.* **9**:857–868.

2310. **Martin, P. R., A. A. Watson, T. F. McCaul, and J. S. Mattick.** 1995. Characterization of a five-gene cluster required for the biogenesis of type 4 fimbriae in *Pseudomonas aeruginosa. Mol. Microbiol.* **16**:497–508.

2311. **Martin, T. R., J. T. Ruzinski, C. E. Rubens, E. Y. Chi, and C. B. Wilson.** 1992. The effect of type-specific polysaccharide capsule on the clearance of group B streptococci from the lungs of infant and adult rats. *J. Infect. Dis.* **165**:306–314.

2312. **Martin, W. J., Jr., J. F. Downing, M. D. Williams, R. Pasula, H. L. Twigg III, and J. R. Wright.** 1995. Role of surfactant protein A in the pathogenesis of tuberculosis in subjects with human immunodeficiency virus infection. *Proc. Assoc. Am. Physicians* **107**:340–345.

2313. **Martinez, J. J., M. A. Mulvey, J. D. Schilling, J. S. Pinkner, and S. J. Hultgren.** 2000. Type 1 pilus-mediated bacterial invasion of bladder epithelial cells. *EMBO J.* **19**:2803–2812.

2314. **Martinez-Moya, M., M. A. de Pedro, H. Schwarz, and F. Garcia-del Portillo.**

1998. Inhibition of *Salmonella* intracellular proliferation by non-phagocytic eucaryotic cells. *Res. Microbiol.* **149**:309–318.

2315. **Martino, M. C., R. A. Stabler, Z. W. Zhang, M. J. Farthing, B. W. Wren, and N. Dorrell.** 2001. *Helicobacter pylori* pore-forming cytolysin orthologue TlyA possesses in vitro hemolytic activity and has a role in colonization of the gastric mucosa. *Infect. Immun.* **69**:1697–1703.

2316. **Marty, N., C. Pasquier, J. L. Dournes, K. Chemin, F. Chavagnat, M. Guinand, G. Chabanon, B. Pipy, and H. Montrozier.** 1998. Effects of characterised *Pseudomonas aeruginosa* exopolysaccharides on adherence to human tracheal cells. *J. Med. Microbiol.* **47**:129–134.

2317. **Maruta, K., H. Miyamoto, T. Hamada, M. Ogawa, H. Taniguchi, and S. Yoshida.** 1998. Entry and intracellular growth of *Legionella dumoffii* in alveolar epithelial cells. *Am. J. Respir. Crit. Care Med.* **157**:1967–1974.

2318. **Maruyama, S., and Y. Katsube.** 1994. Adhesion activity of *Campylobacter jejuni* for intestinal epithelial cells and mucus and erythrocytes. *J. Vet. Med. Sci.* **56**:1123–1127.

2319. **Mascari, L., and J. M. Ross.** 2001. Hydrodynamic shear and collagen receptor density determine the adhesion capacity of *S. aureus* to collagen. *Ann. Biomed. Eng.* **29**:956–962.

2320. **Mase, K., T. Hasegawa, T. Horii, K. Hatakeyama, Y. Kawano, T. Yamashino, and M. Ohta.** 2000. Firm adherence of *Staphylococcus aureus* and *Staphylococcus epidermidis* to human hair and effect of detergent treatment. *Microbiol. Immunol.* **44**:653–656.

2321. **Masinick, S. A., C. P. Montgomery, P. C. Montgomery, and L. D. Hazlett.** 1997. Secretory IgA inhibits *Pseudomonas aeruginosa* binding to cornea and protects against keratitis. *Investig. Ophthalmol. Visual Sci.* **38**:910–918.

2322. **Maslow, J. N., M. E. Mulligan, K. S. Adams, J. C. Justis, and R. D. Arbeit.** 1993. Bacterial adhesins and host factors: Role in the development and outcome of *Escherichia coli* bacteremia. *Clin. Infect. Dis.* **17**:89–97.

2323. **Maslow, J. N., T. S. Whittam, C. F. Gilks, R. A. Wilson, M. E. Mulligan, K. S. Adams, and R. D. Arbeit.** 1995. Clonal relationships among bloodstream isolates of *Escherichia coli*. *Infect. Immun.* **63**:2409–2417.

2324. **Mason, C. M., S. Q. Azizi, and A. R. Dal Nogare.** 1992. Respiratory epithelial carbohydrate levels of rats with Gram-negative bacillary colonization. *J. Lab. Clin. Med.* **120**:740–745.

2325. **Massad, G., F. K. Bahrani, and H. L. Mobley.** 1994. *Proteus mirabilis* fimbriae: identification, isolation, and characterization of a new ambient-temperature fimbria. *Infect. Immun.* **62**:1989–1994.

2326. **Massad, G., and H. L. Mobley.** 1994. Genetic organization and complete sequence of the *Proteus mirabilis pmf* fimbrial operon. *Gene* **150**:101–104.

2327. **Massad, G., C. V. Lockatell, D. E. Johnson, and H. L. T. Mobley.** 1994. *Proteus mirabilis* fimbriae: construction of an isogenic *pmfA* mutant and analysis of virulence in a CBA moust model of ascending urinary tract infection. *Infect. Immun.* **62**:536–542.

2328. **Massad, G., J. F. Fulkerson, Jr., D. C. Watson, and H. L. Mobley.** 1996. *Proteus mirabilis* ambient-temperature fimbriae: cloning and nucleotide sequence of the *aft* gene cluster. *Infect. Immun.* **64**:4390–4395.

2329. **Massey, R. C., M. N. Kantzanou, T. Fowler, N. P. Day, K. Schofield, E. R. Wann, A. R. Berendt, M. Höök, and S. J. Peacock.** 2001. Fibronectin-binding protein A of *Staphylococcus aureus* has multiple, substituting, binding regions that mediate adherence to fibronectin and invasion of endothelial cells. *Cell. Microbiol.* **3**:839–851.

2330. **Masure, H. R., E. A. Campbell, D. R. Cundell, B. J. Pearce, J. Sandros, and B. Spellerberg.** 1995. A new genetic strategy for the analysis of virulence and transformation in *Streptococcus pneumoniae*. *Dev. Biol. Stand.* **85**:251–260.

2331. **Matatov, R., J. Goldhar, E. Skutelsky, I. Sechter, R. Perry, R. Podschun, H. Sahly, K. Thankavel, S. N. Abraham, and I. Ofek.** 1999. Inability of encapsulated *Klebsiella pneumoniae* to assemble functional type 1 fimbriae on their surface. *FEMS Microbiol. Lett.* **179**:123–130.

2332. **Mathan, M. M., V. I. Mathan, and M. J. Albert.** 1993. Electron microscopic study of the attachment and penetration of rabbit intestinal epithelium by *Providencia alcalifaciens*. *J. Pathol.* **171**:67–71.

2333. **Mathewson, J. J., Z. D. Jiang, A. Zumla, C. Chintu, N. Luo, S. R. Calamari, R. M. Genta, A. Steephen, P. Schwartz, and H. L. DuPont.** 1995. HEp-2 cell-adherent *Escherichia coli* in patients with human immunodeficiency virus-associated diarrhea. *J. Infect. Dis.* **171**:1636–1639.

2334. **Mathewson, J. J., B. M. Salameh, H. L. DuPont, Z. D. Jiang, A. C. Nelson, R. Arduino, M. A. Smith, and N. Masozera.** 1998. HEp-2 cell-adherent *Escherichia coli* and intestinal secretory immune response to human immunodeficiency virus (HIV) in outpatients with HIV-associated diarrhea. *Clin. Diagn. Lab. Immunol.* **5**:87–90.

2335. **Matsumoto, M., T. Minami, H. Sasaki, S. Sobue, S. Hamada, and T. Ooshima.** 1999. Inhibitory effects of oolong tea extract on caries-inducing properties of mutans streptococci. *Caries Res.* **33**:441–445.

2336. **Matsumoto, T., Y. Mitzunoe, N. Ogata, M. Tanaka, K. Takahashi, and J. Kumazawa.** 1992. Antioxidant effect on renal scarring following infection of mannose-sensitive-piliated bacteria. *Nephron* **60**:210–215.

2337. **Matsumoto, T., M. Haraoka, Y. Mizunoe, S. Kubo, K. Takahashi, M. Tanaka, and J. Kumazawa.** 1993. Renal scarring is enhanced by phorbol myristate acetate following infection with bacteria with mannose-sensitive pili. *Nephron* **64**:405–409.

2338. **Matsumura, A., T. Saito, M. Arakuni, H. Kitazawa, Y. Kawai, and T. Itoh.** 1999. New binding assay and preparative trial of cell-surface lectin from *Lactobacillus acidophilus* group lactic acid bacteria. *J. Dairy Sci.* **82**:2525–2529.

2339. **Matsuura, T., Y. Miyake, S. Nakashima, H. Komatsuzawa, Y. Akagawa, and H. Suginaka.** 1996. Isolation and characterization of teichoic acid-like substance as an adhesin of *Staphylococcus aureus* to HeLa cells. *Microbiol. Immunol.* **40**:247–254.

2340. **Matthews, K. R., R. A. Almeida, and S. P. Oliver.** 1994. Bovine mammary epithelial cell invasion by *Streptococcus uberis*. *Infect. Immun.* **62**:5641–5646.

2341. **Mattick, J. S., C. B. Whitchurch, and R. A. Alm.** 1996. The molecular genetics of type-4 fimbriae in *Pseudomonas aeruginosa*—a review. *Gene* **179**:147–155.

2342. **Mattoo, S., J. F. Miller, and P. A. Cotter.** 2000. Role of *Bordetella bronchiseptica* fimbriae in tracheal colonization and development of a humoral immune response. *Infect. Immun.* **68**:2024–2033.

2343. **Mattos-Graner, R. O., D. J. Smith, W. F. King, and M. P. Mayer.** 2000. Water-insoluble glucan synthesis by mutans streptococcal strains correlates with caries incidence in 12- to 30-month-old children. *J. Dent. Res.* **79**:1371–1377.

2344. **Mattos-Guaraldi, A. L., L. C. Formiga, and A. F. Andrade.** 1999. Cell surface hydrophobicity of sucrose fermenting and nonfermenting *Corynebacterium diphtheriae* strains evaluated by different methods. *Curr. Microbiol.* **38**:37–42.

2345. **Maurer, J. J., T. P. Brown, W. L. Steffens, and S. G. Thayer.** 1998. The occurrence of ambient temperature-regulated adhesins, curli, and the temperature-sensitive hemagglutinin *tsh* among avian *Escherichia coli*. *Avian Dis.* **42**:106–118.

2346. **May, A. K., C. A. Bloch, R. G. Sawyer, M. D. Spengler, and T. L. Pruett.** 1993. Enhanced virulence of *Escherichia coli* bearing a site-targeted mutation in the major structural subunit of type 1 fimbriae. *Infect. Immun.* **61**:1667–1673.

2347. **May, J. T.** 1994. Antimicrobial factors and microbial contaminants in human milk: recent studies. *J. Paediatr. Child Health* **30**:470–475.

2348. **May, L. L., M. M. Gabriel, R. B. Simmons, L. A. Wilson, and D. G. Ahearn.** 1995. Resistance of adhered bacteria to rigid gas permeable contact lens solutions. *CLAO J.* **21**:242–246.

2349. **McAllister, E. W., L. C. Carey, P. G. Brady, R. Heller, and S. G. Kovacs.** 1993. The role of polymeric surface smoothness of biliary stents in bacterial adherence, biofilm deposition, and stent occlusion. *Gastrointest. Endosc.* **39**:422–425.

2350. **McAllister, T. A., H. D. Bae, G. A. Jones, and K. J. Cheng.** 1994. Microbial attachment and feed digestion in the rumen. *J. Anim. Sci.* **72**:3004–3018.

2351. **McCabe, K., M. D. Mann, and M. D. Bowie.** 1994. pH changes during in vitro adherence of *Escherichia coli* to HeLa cells. *Infect. Immun.* **62**:5164–5167.

2352. **McCabe, K., M. D. Mann, and M. D. Bowie.** 1998. D-Lactate production and [^{14}C]succinic acid uptake by adherent and nonadherent *Escherichia coli*. *Infect. Immun.* **66**:907–911.

2353. **McClain, M. S., I. C. Blomfield, K. J. Eberhardt, and B. I. Eisenstein.** 1993. Inversion-independent phase variation of type 1 fimbriae in *Escherichia coli*. *J. Bacteriol.* **175**:4335–4344.

2354. **McCloskey, J. J., S. Szombathy, A. J. Swift, D. Conrad, and J. A. Winkelstein.** 1993. The binding of pneumococcal lipoteichoic acid to human erythrocytes. *Microb. Pathog.* **14**:23–31.

2355. **McCormick, B. A., P. Klemm, K. A. Krogfelt, R. L. Burghoff, L. Pallesen, D. C. Laux, and P. S. Cohen.** 1993. *Escherichia coli* F-18 phase locked 'on' for expression of type 1 fimbriae is a poor colonizer of the streptomycin-treated mouse large intestine. *Microb. Pathog.* **14**:33–43.

2356. **McCormick, B. A., S. P. Colgan, C.**

Delp-Archer, S. I. Miller, and J. L. Madara. 1993. *Salmonella typhimurium* attachment to human intestinal epithelial monolayers: transcellular signalling to subepithelial neutrophils. *J. Cell Biol.* **123**:895–907.

2357. McCormick, B. A., A. Nusrat, C. A. Parkos, L. D'Andrea, P. M. Hofman, D. Carnes, T. W. Liang, and J. L. Madara. 1997. Unmasking of intestinal epithelial lateral membrane β_1 integrin consequent to transepithelial neutrophil migration in vitro facilitates *inv*-mediated invasion by *Yersinia pseudotuberculosis*. *Infect. Immun.* **65**:1414–1421.

2358. McCormick, B. A., C. A. Parkos, S. P. Colgan, D. K. Carnes, and J. L. Madara. 1998. Apical secretion of a pathogen-elicited epithelial chemoattractant activity in response to surface colonization of intestinal epithelia by *Salmonella typhimurium*. *J. Immunol.* **160**:455–466.

2359. McCrea, K. W., W. J. Watson, J. R. Gilsdorf, and C. F. Marrs. 1997. Identification of two minor subunits in the pilus of *Haemophilus influenzae*. *J. Bacteriol.* **179**:4227–4231.

2360. McCrea, K. W., W. J. Watson, J. R. Gilsdorf, and C. F. Marrs. 1994. Identification of *hifD* and *hifE* in the pilus gene cluster of *Haemophilus influenzae* type b strain Eagan. *Infect. Immun.* **62**:4922–4928.

2361. McCrea, K. W., J. L. St. Sauver, C. F. Marrs, D. Clemans, and J. R. Gilsdorf. 1998. Immunologic and structural relationships of the minor pilus subunits among *Haemophilus influenzae* isolates. *Infect. Immun.* **66**:4788–4796.

2362. McDevitt, D., P. Francois, P. Vaudaux, and T. J. Foster. 1994. Molecular characterization of the clumping factor (fibrinogen receptor) of *Staphylococcus aureus*. *Mol. Microbiol.* **11**:237–248.

2363. McDevitt, D., P. Francois, P. Vaudaux, and T. J. Foster. 1995. Identification of the ligand-binding domain of the surface-located fibrinogen receptor (clumping factor) of *Staphylococcus aureus*. *Mol. Microbiol.* **16**:895–907.

2364. McDevitt, D., T. Nanavaty, K. House-Pompeo, E. Bell, N. Turner, L. McIntire, T. Foster, and M. Höök. 1997. Characterization of the interaction between the *Staphylococcus aureus* clumping factor (ClfA) and fibrinogen. *Eur. J. Biochem.* **247**:416–424.

2365. McDowell, S. G., Y. H. An, R. A. Draughn, and R. J. Friedman. 1995. Application of a fluorescent redox dye for enumeration of metabolically active bacteria on albumin-coated titanium surfaces. *Lett. Appl. Microbiol.* **21**:1–4.

2366. McEwan, N. A. 2000. Adherence by *Staphylococcus intermedius* to canine keratinocytes in atopic dermatitis. *Res. Vet. Sci.* **68**:279–283.

2367. McGavin, M. J., C. Zahradka, K. Rice, and J. E. Scott. 1997. Modification of the *Staphylococcus aureus* fibronectin binding phenotype by V8 protease. *Infect. Immun.* **65**:2621–2628.

2368. McGee, D. J., G.-C. Chen, and R. F. Rest. 1996. Expression of sialyltransferase is not required for interaction of *Neisseria gonorrhoeae* with human epithelial cells and human neutrophils. *Infect. Immun.* **64**:4129–4136.

2369. McGrady, J. A., W. G. Butcher, D. Beighton, and L. M. Switalski. 1995. Specific and charge interactions mediate collagen recognition by oral lactobacilli. *J. Dent. Res.* **74**:649–657.

2370. McGroarty, J. A. 1994. Cell surface appendages of lactobacilli. *FEMS Microbiol. Lett.* **124**:405–409.

2371. McInnes, C., D. Engel, B. J. Moncla, and R. W. Martin. 1992. Reduction in adherence of *Actinomyces viscosus* after exposure to low-frequency acoustic energy. *Oral Microbiol. Immunol.* **7**:171–176.

2372. McInnes, C., D. Engel, and R. W. Martin. 1993. Fimbria damage and removal of adherent bacteria after exposure to acoustic energy. *Oral Microbiol. Immunol.* **8**:277–282.

2373. McInnes, C., C. M. Kay, R. S. Hodges, and B. D. Sykes. 1994. Conformational differences between *cis* and *trans* proline isomers of a peptide antigen representing the receptor binding domain of *Pseudomonas aeruginosa* as studied by ^{1}H-NMR. *Biopolymers* **34**:1221–1230.

2374. McIver, C. J., and J. W. Tapsall. 1995. Virulence and other phenotypic characteristics of urinary isolates of cysteine-requiring *Escherichia coli*. *J. Med. Microbiol.* **42**:39–42.

2375. McKee, M. L., A. R. Melton-Celsa, R. A. Moxley, D. H. Francis, and A. D. O'Brien. Enterohemorrhagic *Escherichia coli* O157:H7 requires intimin to colonize the gnotobiotic pig intestine and to adhere to HEp-2 cells. *Infect. Immun.* **63**:3739–3744.

2376. McKee, M. L., and A. D. O'Brien. 1996. Truncated enterohemorrhagic *Escherichia coli* (EHEC) O157:H7 intimin (EaeA) fusion proteins promote adherence of EHEC strains to HEp-2 cells. *Infect. Immun.* **64**:2225–2233.

2377. McKenney, D., J. Hübner, E. Muller, Y. Wang, D. A. Goldmann, and G. B. Pier.

1998. The *ica* locus of *Staphylococcus epidermidis* encodes production of the capsular polysaccharide/adhesin. *Infect. Immun.* **66:** 4711–4720.

2378. **McLean, N. W., and I. J. Rosenstein.** 2000. Characterisation and selection of a *Lactobacillus* species to re-colonise the vagina of women with recurrent bacterial vaginosis. *J. Med. Microbiol.* **49:**543–552.

2379. **McNab, R., and H. F. Jenkinson.** 1992. Gene disruption identifies a 290 kDa cell-surface polypeptide conferring hydrophobicity and coaggregation properties in *Streptococcus gordonii. Mol. Microbiol.* **6:**2939–2949.

2380. **McNab, R., A. R. Holmes, J. M. Clarke, G. W. Tannock, and H. F. Jenkinson.** 1996. Cell surface polypeptide CshA mediates binding of *Streptococcus gordonii* to other oral bacteria and to immobilized fibronectin. *Infect. Immun.* **64:**4204–4210.

2381. **McNab, R., G. W. Tannock, and H. F. Jenkinson.** 1995. Characterization of CshA, a high molecular mass adhesin of *Streptococcus gordonii. Dev. Biol. Stand.* **85:**371–375.

2382. **McNab, R., and H. F. Jenkinson.** 1998. Altered adherence properties of a *Streptococcus gordonii hppA* (oligopeptide permease) mutant result from transcriptional effects on *cshA* adhesin gene expression. *Microbiology* **144:** 127–136.

2383. **McNab, R., H. Forbes, P. S. Handley, D. M. Loach, G. W. Tannock, and H. F. Jenkinson.** 1999. Cell wall-anchored CshA polypeptide (259 kilodaltons) in *Streptococcus gordonii* forms surface fibrils that confer hydrophobic and adhesive properties. *J. Bacteriol.* **181:**3087–3095.

2384. **McNamara, B. P., and M. S. Donnenberg.** 2000. Evidence for specificity in type 4 pilus biogenesis by enteropathogenic *Escherichia coli. Microbiology* **146:**719–729.

2385. **McNamara, B. P., A. Koutsouris, C. B. O'Connell, J. P. Nougayrede, M. S. Donnenberg, and G. Hecht.** 2001. Translocated EspF protein from enteropathogenic *Escherichia coli* disrupts host intestinal barrier function. *J. Clin. Investig.* **107:**621–629.

2386. **McNeil, G., M. Virji, and E. R. Moxon.** 1994. Interactions of *Neisseria meningitidis* with human monocytes. *Microb. Pathog.* **16:**153–163.

2387. **McNeil, G., and M. Virji.** 1997. Phenotypic variants of meningococci and their potential in phagocytic interactions: the influence of opacity proteins, pili, PilC and surface sialic acids. *Microb. Pathog.* **22:**295–304.

2388. **McQueen, C. E., E. C. Boedeker, R. Reid, D. Jarboe, M. Wolf, M. Le, and W. R. Brown.** 1993. Pili in microspheres protect rabbits from diarrhoea induced by *E. coli* strain RDEC-1. *Vaccine* **11:**201–206.

2389. **Mechin, M. C., E. Rousset, and J. P. Girardeau.** 1996. Identification of surface-exposed linear B-cell epitopes of the nonfimbrial adhesin CS31A of *Escherichia coli* by using overlapping peptides and antipeptide antibodies. *Infect. Immun.* **64:**3555–3564.

2390. **Mecsas, J., R. Welch, J. W. Erickson, and C. A. Gross.** 1995. Identification and characterization of an outer membrane protein, OmpX, in *Escherichia coli* that is homologous to a family of outer membrane proteins including Ail of *Yersinia enterocolitica. J. Bacteriol.* **177:**799–804.

2391. **Medina, E., G. Molinari, M. Rohde, B. Haase, G. S. Chhatwal, and C. A. Guzman.** 1999. Fc-mediated nonspecific binding between fibronectin-binding protein I of *Streptococcus pyogenes* and human immunoglobulins. *J. Immunol.* **163:**3396–3402.

2392. **Medina, M. B.** 2001. Binding of collagen I to *Escherichia coli* O157:H7 and inhibition by carrageenans. *Int. J. Food Microbiol.* **69:** 199–208.

2393. **Megraud, F., V. Neman-Simha, and D. Brugmann.** 1992. Further evidence of the toxic effect of ammonia produced by *Helicobacter pylori* urease on human epithelial cells. *Infect. Immun.* **60:**1858–1863.

2394. **Mehock, J. R., C. E. Greene, F. C. Gherardini, T.-W. Hahn, and D. C. Krause.** 1998. *Bartonella henselae* invasion of feline erythrocytes in vitro. *Infect. Immun.* **66:**3462–3466.

2395. **Meijerink, E., S. Neuenschwander, R. Fries, A. Dinter, H. U. Bertschinger, G. Stranzinger, and P. Vogeli.** 2000. A DNA polymorphism influencing alpha(1,2)fucosyltransferase activity of the pig FUT1 enzyme determines susceptibility of small intestinal epithelium to *Escherichia coli* F18 adhesion. *Immunogenetics* **52:**129–136.

2396. **Meijerink, E., R. Fries, P. Vogeli, J. Masabanda, G. Wigger, C. Stricker, and S. Neuenschwander.** 1997. Two alpha(1,2) fucosyltransferase genes on porcine chromosome 6q11 are closely linked to the blood group inhibitor (S) and *Escherichia coli* F18 receptor (ECF18R) loci. *Mammal. Genome* **8:**736–741.

2397. **Mellies, J., T. Rudel, and T. F. Meyer.** 1997. Transcriptional regulation of *pilC2* in *Neisseria gonorrhoeae:* response to oxygen availability and evidence for growth-phase regulation in *Escherichia coli. Mol. Gen. Genet.* **255**:285–293.

2398. **Mempel, M., T. Schmidt, S. Weidinger, C. Schnopp, T. Foster, J. Ring, and D. Abeck.** 1998. Role of *Staphylococcus aureus* surface-associated proteins in the attachment to cultured HaCaT keratinocytes in a new adhesion assay. *J. Investig. Dermatol.* **111**:452–456.

2399. **Menard, R., P. Sansonetti, and C. Parsot.** 1994. The secretion of the *Shigella flexneri* Ipa invasins is activated by epithelial cells and controlled by IpaB and IpaD. *EMBO J.* **13**:5293–5302.

2400. **Menard, R., P. J. Sansonetti, and C. Parsot.** Nonpolar mutagenesis of the *ipa* genes defines IpaB, IpaC, and IpaD as effectors of *Shigella flexneri* entry into epithelial cells. *J. Bacteriol.* **175**:5899–5906.

2401. **Menard, R., and P. J. Sansonetti.** 1996. *Shigella flexneri:* isolation of noninvasive mutants of gram-negative pathogens. *Methods Enzymol.* **236**:493–509.

2402. **Menard, R., C. Dehio, and P. J. Sansonetti.** 1996. Bacterial entry into epithelial cells: the paradigm of *Shigella. Trends Microbiol.* **4**:220–226.

2403. **Meng, X. Q., K. Yamakawa, H. Ogura, and S. Nakamura.** 1994. Haemagglutination activity of toxigenic and non-toxigenic strains of *Clostridium difficile. FEMS Microbiol. Lett.* **118**:141–144.

2404. **Menozzi, F. D., R. Mutombo, G. Renauld, C. Gantiez, J. H. Hannah, E. Leininger, M. J. Brennan, and C. Locht.** 1994. Heparin-inhibitable lectin activity of the filamentous hemagglutinin adhesin of *Bordetella pertussis. Infect. Immun.* **62**:769–778.

2405. **Menozzi, F. D., P. E. Boucher, G. Riveau, C. Gantiez, and C. Locht.** 1994. Surface-associated filamentous hemagglutinin induces autoagglutination of *Bordetella pertussis. Infect. Immun.* **62**:4261–4269.

2406. **Menozzi, F. D., J. H. Rouse, M. Alavi, M. Laude-Sharp, J. Muller, R. Bischoff, M. J. Brennan, and C. Locht.** 1996. Identification of a heparin-binding hemagglutinin present in *Mycobacteria. J. Exp. Med.* **184**:993–1001.

2407. **Menozzi, F. D., R. Bischoff, E. Fort, M. J. Brennan, and C. Locht.** 1998. Molecular characterization of the mycobacterial heparin-binding hemagglutinin, a mycobacterial adhesin. *Proc. Natl. Acad. Sci. USA* **95**:12625–12630.

2408. **Menzies, B. E., and I. Kourteva.** 1998.

Internalization of *Staphylococcus aureus* by endothelial cells induces apoptosis. *Infect. Immun.* **66**:5994–5998.

2409. **Merien, F., J. Truccolo, G. Baranton, and P. Perolat.** 2000. Identification of a 36-kDa fibronectin-binding protein expressed by a virulent variant of *Leptospira interrogans* serovar *icterohaemorrhagiae. FEMS Microbiol. Lett.* **185**:17–22.

2410. **Merien, F., G. Baranton, and P. Perolat.** 1997. Invasion of Vero cells and induction of apoptosis in macrophages by pathogenic *Leptospira interrogans* are correlated with virulence. *Infect. Immun.* **65**:729–738.

2411. **Merino, S., R. Gavin, M. Altarriba, L. Izquierdo, M. E. Maguire, and J. M. Tomas.** 2001. The MgtE Mg^{2+} transport protein is involved in *Aeromonas hydrophila* adherence. *FEMS Microbiol. Lett.* **198**:189–195.

2412. **Merino, S., X. Rubires, A. Aguilar, and J. M. Tomas.** 1997. The role of O1-antigen in the adhesion to uroepithelial cells of *Klebsiella pneumoniae* grown in urine. *Microb. Pathog.* **23**:49–53.

2413. **Merino, S., X. Rubires, A. Aguilar, and J. M. Tomas.** 1996. The O:34-antigen lipopolysaccharide as an adhesin in *Aeromonas hydrophila. FEMS Microbiol. Lett.* **139**:97–101.

2414. **Merino, S., X. Rubires, S. Knochel, and J. M. Tomas.** 1995. Emerging pathogens: *Aeromonas* spp. *Int. J. Food Microbiol.* **28**:157–168.

2415. **Merkel, T. J., S. Stibitz, J. M. Keith, M. Leef, and R. Shahin.** 1998. Contribution of regulation by the *bvg* locus to respiratory infection of mice by *Bordetella pertussis. Infect. Immun.* **66**:4367–4373.

2416. **Mernaugh, G. R., S. F. Dallo, S. C. Holt, and J. B. Baseman.** 1993. Properties of adhering and nonadhering populations of *Mycoplasma genitalium. Clin. Infect. Dis.* **17**:S69–S78.

2417. **Merritt, K., A. Gaind, and J. M. Anderson.** 1998. Detection of bacterial adherence on biomedical polymers. *J. Biomed. Mater. Res.* **39**:415–422.

2418. **Merz, A. J., and M. So.** 1997. Attachment of piliated, Opa⁻ and Opc⁻ gonococci and meningococci to epithelial cells elicits cortical actin rearrangements and clustering of tyrosine-phosphorylated proteins. *Infect. Immun.* **65**:4341–4349.

2419. **Merz, A. J., D. B. Rifenbery, C. G. Arvidson, and M. So.** 1996. Traversal of a polarized epithelium by pathogenic *Neisseriae:* facilitation by type IV pili and maintenance of epithelial barrier function. *Mol. Med.* **2**:745–754.

2420. **Messick, J. B., and Y. Rikihisa.** 1993. Characterization of *Ehrlichia risticii* binding, internalization, and proliferation in host cells by flow cytometry. *Infect. Immun.* **61**:3803–3810.

2421. **Metzger, Z., L. G. Featherstone, W. W. Ambrose, M. Trope, and R. R. Arnold.** 2001. Kinetics of coaggregation of *Porphyromonas gingivalis* with *Fusobacterium nucleatum* using an automated microtiter plate assay. *Oral Microbiol. Immunol.* **16**:163–169.

2422. **Mey, A., H. Lefflér, Z. Hmama, G. Normier, and J.-P. Revillard.** 1996. The animal lectin galectin-3 interacts with bacterial lipopolysaccharides via two independent sites. *J. Immunol.* **156**:1572–1577.

2423. **Meyer, D. H., and P. M. Fives-Taylor.** 1993. Evidence that extracellular components function in adherence of *Actinobacillus actinomycetemcomitans* to epithelial cells. *Infect. Immun.* **61**:4933–4936.

2424. **Meyer, D. H., and P. M. Fives-Taylor.** 1994. Characteristics of adherence of *Actinobacillus actinomycetemcomitans* to epithelial cells. *Infect. Immun.* **62**:928–935.

2425. **Meyer, D. H., and P. M. Fives-Taylor.** 1995. Adhesion of oral bacteria to soft tissue. *Methods Enzymol.* **253**:373–385.

2426. **Meyer, H. G., and S. Gatermann.** 1994. Surface properties of *Staphylococcus saprophyticus*: hydrophobicity, haemagglutination and *Staphylococcus saprophyticus* surface-associated protein (Ssp) represent distinct entities. *APMIS* **102**:538–544.

2427. **Meyer, H. G., J. Muthing, and S. G. Gatermann.** 1997. The hemagglutinin of *Staphylococcus saprophyticus* binds to a protein receptor on sheep erythrocytes. *Med. Microbiol. Immunol.* **186**:37–43.

2428. **Meyer, H. G., U. Wengler-Becker, and S. G. Gatermann.** 1996. The hemagglutinin of *Staphylococcus saprophyticus* is a major adhesin for uroepithelial cells. *Infect. Immun.* **64**:3893–3896.

2429. **Meyer-Rosberg, K., and T. Berglindh.** 1996. *Helicobacter* colonization of biopsy specimens cultured *in vitro* is dependent on both mucosal type and bacterial strain. *Scand. J. Gastroenterol.* **31**:434–441.

2430. **Meylheuc, T., C. J. van Oss, and M. N. Bellon-Fontaine.** 2001. Adsorption of biosurfactant on solid surfaces and consequences regarding the bioadhesion of *Listeria monocytogenes. J. Appl. Microbiol.* **91**:822–832.

2431. **Mhlanga-Mutangadura, T., G. Morlin,**

A. L. Smith, A. Eisenstark, and M. Golomb. 1998. Evolution of the major pilus gene cluster of *Haemophilus influenzae. J. Bacteriol.* **180**:4693–4703.

2432. **Michaels, R. D., S. C. Whipp, and M. F. Rothschild.** 1994. Resistance of Chinese Meishan, Fengjing, and Minzhu pigs to the K88ac$^+$ strain of *Escherichia coli. Am. J. Vet. Res.* **55**:333–338.

2433. **Michailova, L., N. Markova, T. Radoucheva, S. Stoitsova, V. Kussovski, and M. Jordanova.** 2000. Atypical behaviour and survival of *Streptococcus pyogenes* L forms during intraperitoneal infection in rats. *FEMS Immunol. Med. Microbiol.* **28**:55–65.

2434. **Michetti, P., N. Porta, M. J. Mahan, J. M. Slauch, J. J. Mekalanos, A. L. Blum, J. P. Kraehenbuhl, and M. R. Neutra.** 1994. Monoclonal immunoglobulin A prevents adherence and invasion of polarized epithelial cell monolayers by *Salmonella typhimurium. Gastroenterology* **107**:915–923.

2435. **Middleton, A. M., M. V. Chadwick, A. G. Nicholson, A. Dewar, R. K. Groger, E. J. Brown, and R. Wilson.** 2000. The role of *Mycobacterium avium* complex fibronectin attachment protein in adherence to the human respiratory mucosa. *Mol. Microbiol.* **38**:381–391.

2436. **Midolo, P. D., A. Norton, M. von Itzstein, and J. R. Lambert.** 1997. Novel bismuth compounds have *in vitro* activity against *Helicobacter pylori. FEMS Microbiol. Lett.* **157**:229–232.

2437. **Miettinen, A., B. Westerlund, A. M. Tarkkanen, T. Tornroth, P. Ljungberg, O. V. Renkonen, and T. K. Korhonen.** 1993. Binding of bacterial adhesins to rat glomerular mesangium *in vivo. Kidney Int.* **43**:592–600.

2438. **Mikolajczyk-Pawlinska, J., J. Travis, and J. Potempa.** 1998. Modulation of interleukin-8 activity by gingipains from *Porphyromonas gingivalis*: implications for pathogenicity of periodontal disease. *FEBS Lett.* **440**:282–286.

2439. **Mikx, F. H. M., and R. A. C. Keulers.** 1992. Hemagglutination activity of *Treponema denticola* grown in serum-free medium in continuous culture. *Infect. Immun.* **60**:1761–1766.

2440. **Miliotis, M. D., B. D. Tall, and R. T. Gray.** 1995. Adherence to and invasion of tissue culture cells by *Vibrio hollisae. Infect. Immun.* **63**:4959–4963.

2441. **Miller, V. L.** 1995. Tissue-culture invasion: fact or artefact? *Trends Microbiol.* **3**:69–71.

2442. **Miller, W. G., A. H. Bates, S. T. Horn,**

M. T. Brandl, M. R. Wachtel, and R. E. Mandrell. 2000. Detection on surfaces and in Caco-2 cells of *Campylobacter jejuni* cells transformed with new *gfp*, *yfp*, and *cfp* marker plasmids. *Appl. Environ. Microbiol.* **66:**5426–5436.

2443. Miller-Podraza, H., J. Bergstrom, M. A. Milh, and K.-A. Karlsson. 1997. Recognition of glycoconjugates by *Helicobacter pylori*. Comparison of two sialic acid-dependent specificities based on haemagglutination and binding to human erythrocyte glycoconjugates. *Glycoconj. J.* **14:**467–471.

2444. Mills, S. D., and B. B. Finlay. 1994. Comparison of *Salmonella typhi* and *Salmonella typhimurium* invasion, intracellular growth and localization in cultured human epithelial cells. *Microb. Pathog.* **17:**409–423.

2445. Millsap, K., G. Reid, H. C. van der Mei, and H. J. Busscher. 1994. Displacement of *Enterococcus faecalis* from hydrophobic and hydrophilic substrata by *Lactobacillus* and *Streptococcus* spp. as studied in a parallel plate flow chamber. *Appl. Environ. Microbiol.* **60:**1867–1874.

2446. Millsap, K. W., R. Bos, H. C. van der Mei, and H. J. Busscher. 1999. Influence of aeration of *Candida albicans* during culturing on their surface aggregation in the presence of adhering *Streptococcus gordonii*. *FEMS Immunol. Med. Microbiol.* **26:**69–74.

2447. Millsap, K. W., G. Reid, H. C. van der Mei, and H. J. Busscher. 1997. Adhesion of *Lactobacillus* species in urine and phosphate buffer to silicone rubber and glass under flow. *Biomaterials* **18:**87–91.

2448. Milohanic, E., R. Jonquieres, P. Cossart, P. Berche, and J. L. Gaillard. 2001. The autolysin Ami contributes to the adhesion of *Listeria monocytogenes* to eukaryotic cells via its cell wall anchor. *Mol. Microbiol.* **39:**1212–1224.

2449. Milohanic, E., B. Pron, P. Berche, and J. L. Gaillard. 2000. Identification of new loci involved in adhesion of *Listeria monocytogenes* to eukaryotic cells. *Microbiology* **146:**731–739.

2450. Minion, F. C., C. Adams, and T. Hsu. 2000. R1 region of P97 mediates adherence of *Mycoplasma hyopneumoniae* to swine cilia. *Infect. Immun.* **68:**3056–3060.

2451. Minnick, M. F., S. J. Mitchell, and S. J. McAllister. 1996. Cell entry and the pathogenesis of *Bartonella* infections. *Trends Microbiol.* **4:**343–347.

2452. Mintz, K. P., and P. M. Fives-Taylor. 1999. Binding of the periodontal pathogen *Actinobacillus actinomycetemcomitans* to extracellular matrix proteins. *Oral Microbiol. Immunol.* **14:**109–116.

2453. Mintz, K. P., and P. M. Fives-Taylor. 1994. Adhesion of *Actinobacillus actinomycetemcomitans* to a human oral cell line. *Infect. Immun.* **62:**3672–3678.

2454. Misawa, N., and M. J. Blaser. 2000. Detection and characterization of autoagglutination activity by *Campylobacter jejuni*. *Infect. Immun.* **68:**6168–6175.

2455. Mitsumori, K., A. Terai, S. Yamamoto, and O. Yoshida. 1998. Identification of S, F1C and three PapG fimbrial adhesins in uropathogenic *Escherichia coli* by polymerase chain reaction. *FEMS Immunol. Med. Microbiol.* **21:**261–268.

2456. Miyake, Y., S. Fujiwara, T. Usui, and H. Suginaka. 1992. Simple method for measuring the antibiotic concentration required to kill adherent bacteria. *Chemotherapy* **38:**286–290.

2457. Miyake, M., L. Zhao, T. Ezaki, K. Hirose, A. Q. Khan, Y. Kawamura, R. Shima, and M. Kamijo. 1998. Vi-deficient and nonfimbriated mutants of *Salmonella typhi* agglutinate human blood type antigens and are hyperinvasive. *FEMS Microbiol. Lett.* **161:**75–82.

2458. Miyamoto, N., and L. O. Bakaletz. 1996. Selective adherence of non-typeable *Haemophilus influenzae* (NTHi) to mucus or epithelial cells in the chinchilla eustachian tube and middle ear. *Microb. Pathog.* **21:**343–356.

2459. Miyamoto, N., and L. O. Bakaletz. 1997. Kinetics of the ascension of NTHi from the nasopharynx to the middle ear coincident with adenovirus-induced compromise in the chinchilla. *Microb. Pathog.* **23:**119–126.

2460. Miyata, M., H. Yamamoto, T. Shimizu, A. Uenoyama, C. Citti, and R. Rosengarten. 2000. Gliding mutants of *Mycoplasma mobile*: relationships between motility and cell morphology, cell adhesion and microcolony formation. *Microbiology* **146:**1311–1320.

2461. Miyazaki, S., T. Matsunaga, I. Kobayashi, K. Yamaguchi, and S. Goto. 1992. The other mediator for adherence of *Haemophilus influenzae* organisms without involvement of fimbriae. *Microbiol. Immunol.* **36:**205–212.

2462. Miyoshi, S., K. Kawata, K. Tomochika, and S. Shinoda. 1999. The hemagglutinating action of *Vibrio vulnificus* metalloprotease. *Microbiol. Immunol.* **43:**79–82.

2463. Mizunoe, Y., T. Matsumoto, M. Haraoka, M. Sakumoto, S. Kubo, and J. Kumazawa. 1995. Effect of pili of *Serratia marcescens* on superoxide production and phagocytosis of human polymorphonuclear leukocytes. *J. Urol.* **154**:1227–1230.

2464. **Mizunoe, Y., T. Matsumoto, M. Sakumoto, S. Kubo, O. Mochida, Y. Sakamoto, and J. Kumazawa.** 1997. Renal scarring by mannose-sensitive adhesin of *Escherichia coli* type 1 pili. *Nephron* **77**:412–416.

2465. **Mobley, H. L., G. R. Chippendale, and J. W. Warren.** 1995. *In vitro* adhesion of bacteria to exfoliated uroepithelial cells: criteria for quantitative analysis. *Methods Enzymol.* **253**:360–367.

2466. **Mobley, H. L., M. D. Island, and G. Massad.** 1994. Virulence determinants of uropathogenic *Escherichia coli* and *Proteus mirabilis*. *Kidney Int.* **47**:S129–S136.

2467. **Mobley, H. L., K. G. Jarvis, J. P. Elwood, D. I. Whittle, C. V. Lockatell, R. G. Russell, D. E. Johnson, M. S. Donnenberg, and J. W. Warren.** 1993. Isogenic P-fimbrial deletion mutants of pyelonephritogenic *Escherichia coli*: the role of alpha Gal(1–4) β Gal binding in virulence of a wild-type strain. *Mol. Microbiol.* **10**:143–155.

2468. **Mochida, O., T. Matsumoto, Y. Mizunoe, M. Sakumoto, J. Abe, and J. Kumazawa.** 1998. Preventive effect of dapsone on renal scarring following mannose-sensitive piliated bacterial infection. *Chemotherapy* **44**:36–41.

2469. **Modalsli, K., G. Bukholm, and M. Degre.** 1992. Interferon treatment reduced adherence, invasiveness and intracellular multiplication of *Shigella flexneri* in coxsackie B1 virus-infected cells. *J. Biol. Regul. Homeostatic Agents* **6**:35–45.

2470. **Modalsli, K. R., G. Bukholm, M. Holberg-Petersen, and M. Degre.** 1995. *Shigella flexneri* adherence to and multiplication in coxsackie B1 virus-infected HEp-2 cells. *APMIS* **103**:254–260.

2471. **Mohamed, N., T. R. Rainier, Jr., and J. M. Ross.** 2000. Novel experimental study of receptor-mediated bacterial adhesion under the influence of fluid shear. *Biotechnol. Bioeng.* **68**:628–636.

2472. **Mohamed, N., M. A. Teeters, J. M. Patti, M. Hook, and J. M. Ross.** 1999. Inhibition of *Staphylococcus aureus* adherence to collagen under dynamic conditions. *Infect. Immun.* **67**:589–594.

2473. **Mohamed, N., L. Visai, P. Speziale, and J. M. Ross.** 2000. Quantification of *Staphylococcus aureus* cell surface adhesins using flow cytometry. *Microb. Pathog.* **29**:357–361.

2474. **Mohan, V. P., D. S. Agarwal, K. B. Sharma, P. Ghose, and P. K. Pillai.** 1993.

2475. **Moisset, A., N. Schatz, Y. Lepoivre, S. Amadio, D. Wachsmann, M. Schöller, and J.-P. Klein.** 1994. Conservation of salivary glycoprotein-interacting and human immunoglobulin G-cross-reactive domains of antigen I/II in oral streptococci. *Infect. Immun.* **62**:184–193.

2476. **Mol, O., and B. Oudega.** 1996. Molecular and structural aspects of fimbriae biosynthesis and assembly in *Escherichia coli*. *FEMS Microbiol. Rev.* **19**:25–52.

2477. **Mol, O., W. C. Oudhuis, H. Fokkema, and B. Oudega.** 1996. The N-terminal beta-barrel domain of the *Escherichia coli* K88 periplasmic chaperone FaeE determines fimbrial subunit recognition and dimerization. *Mol. Microbiol.* **22**:379–388.

2478. **Molinari, G., M. Rohde, S. R. Talay, G. S. Chhatwal, S. Beckert, and A. Podbielski.** 2001. The role played by the group A streptococcal negative regulator Nra on bacterial interactions with epithelial cells. *Mol. Microbiol.* **40**:99–114.

2479. **Molinari, G., S. R. Talay, P. Valentin-Weigand, M. Rohde, and G. S. Chhatwal.** 1997. The fibronectin-binding protein of *Streptococcus pyogenes*, SfbI, is involved in the internalization of group A streptococci by epithelial cells. *Infect. Immun.* **65**:1357–1363.

2480. **Molinari, G., and G. S. Chhatwal.** 1998. Invasion and survival of *Streptococcus pyogenes* in eukaryotic cells correlates with the source of the clinical isolates. *J. Infect. Dis.* **177**:1600–1607.

2481. **Moller, P. C., M. J. Evans, R. C. Fader, L. C. Henson, B. Rogers, and J. P. Heggers.** 1994. The effect of anti-exotoxin A on the adherence of *Pseudomonas aeruginosa* to hamster tracheal epithelial cells *in vitro*. *Tissue Cell* **26**:181–188.

2482. **Molnar, C., Z. Hevessy, F. Rozgonyi, and C. G. Gemmell.** 1994. Pathogenicity and virulence of coagulase negative staphylococci in relation to adherence, hydrophobicity, and toxin production *in vitro*. *J. Clin. Pathol.* **47**:743–748.

2483. **Monga, M., and J. A. Roberts.** 1994. Spermagglutination by bacteria: receptor-specific interactions. *J. Androl.* **15**:151–156.

2484. **Monno, R., M. A. Valenza, M. A. Panaro, S. Lisi, L. Marcuccio, D. De Vito, and V. Mitolo.** 1996. Spontaneous binding of *Vibrio cholerae* to human leucocytes. *Microbios* **88**:169–176.

2485. **Montanaro, L., D. Cavedagna, L. Baldassarri, and C. R. Arciola.** 2001. Adhesion of a *Staphylococcus aureus* strain to biomaterials does not select methicillin-resistant mutants. *New Microbiol.* **24**:57–61.

2486. **Montanaro, L., C. R. Arciola, L. Baldassarri, and E. Borsetti.** 1999. Presence and expression of collagen adhesin gene (*cna*) and slime production in *Staphylococcus aureus* strains from orthopaedic prosthesis infections. *Biomaterials* **20**:1945–1949.

2487. **Montanaro, L., C. R. Arciola, E. Borsetti, M. Brigotti, and L. Baldassarri.** 1998. A polymerase chain reaction (PCR) method for the identification of collagen adhesin gene (CNA) in *Staphylococcus*-induced prosthesis infections. *New Microbiol.* **21**:359–363.

2488. **Monteiro-Neto, V., L. C. Campos, A. J. Ferreira, T. A. Gomes, and L. R. Trabulsi.** 1997. Virulence properties of *Escherichia coli* O111:H12 strains. *FEMS Microbiol. Lett.* **146**:123–128.

2489. **Montgomery, R. R., and S. E. Malawista.** 1996. Entry of *Borrelia burgdorferi* into macrophages is end-on and leads to degradation in lysosomes. *Infect. Immun.* **64**:2867–2872.

2490. **Mooi, F. R., H. van Oirschot, K. Heuvelman, H. G. J. van der Heide, W. Gaastra, and R. J. L. Willems.** 1998. Polymorphism in the *Bordetella pertussis* virulence factors P.69/pertactin and pertussis toxin in The Netherlands: temporal trends and evidence for vaccine-driven evolution. *Infect. Immun.* **66**:670–675.

2491. **Mooi, F. R., W. H. Jansen, H. Brunings, H. Gielen, H. G. J. van der Heide, H. C. Walvoort, and P. A. M. Guinee.** 1992. Construction and analysis of *Bordetella pertussis* mutants defective in the production of fimbriae. *Microb. Pathog.* **12**:127–135.

2492. **Moore, K. M., M. W. Jackwood, T. P. Brown, and D. W. Dreesen.** 1994. *Bordetella avium* hemagglutination and motility mutants: isolation, characterization, and pathogenicity. *Avian Dis.* **38**:50–58.

2493. **Moore, K. M., and M. W. Jackwood.** 1994. Production of monoclonal antibodies to the *Bordetella avium* 41-kilodalton surface protein and characterization of the hemagglutinin. *Avian Dis.* **38**:218–224.

2494. **Morabito, S., H. Karch, P. Mariani-Kurkdjian, H. Schmidt, F. Minelli, E. Bingen, and A. Caprioli.** 1998. Enteroaggregative, Shiga toxin-producing *Escherichia coli* O111:H2 associated with an outbreak of hemolytic-uremic syndrome. *J. Clin. Microbiol.* **36**:840–842.

2495. **Moran, A. P.** 1995. Cell surface characteristics of *Helicobacter pylori*. *FEMS Immunol. Med. Microbiol.* **10**:271–280.

2496. **Moran, A. P.** 1996. Pathogenic properties of *Helicobacter pylori*. *Scand. J. Gastroenterol.* **215**:22–31.

2497. **Moran, A. P.** 1996. Bacterial surface structures—an update. *FEMS Immunol. Med. Microbiol.* **16**:61–62.

2498. **Morand, P. C., P. Tattevin, E. Eugene, J. L. Beretti, and X. Nassif.** 2001. The adhesive property of the type IV pilus-associated component PilC1 of pathogenic *Neisseria* is supported by the conformational structure of the N-terminal part of the molecule. *Mol. Microbiol.* **40**:846–856.

2499. **Morata de Ambrosini, V. I., S. N. Gonzalez, and G. Oliver.** 1999. Study of adhesion of *Lactobacillus casei* CRL 431 to ileal intestinal cells of mice. *J. Food Prot.* **62**:1430–1434.

2500. **Moreillon, P., J. M. Entenza, P. Francioli, D. McDevitt, T. J. Foster, P. François, and P. Vaudaux.** 1995. Role of *Staphylococcus aureus* coagulase and clumping factor in pathogenesis of experimental endocarditis. *Infect. Immun.* **63**:4738–4743.

2501. **Morelli, R., L. Baldassarri, V. Falbo, G. Donelli, and A. Caprioli.** 1994. Detection of enteroadherent *Escherichia coli* associated with diarrhoea in Italy. *J. Med. Microbiol.* **41**:399–404.

2502. **Morgan, T. D., and M. Wilson.** Anti-adhesive and antibacterial properties of a proprietary denture cleanser. *J. Appl. Microbiol.* **89**:617–623.

2503. **Mori, N., A. Wada, T. Hirayama, T. P. Parks, C. Stratowa, and N. Yamamoto.** 2000. Activation of intercellular adhesion molecule 1 expression by *Helicobacter pylori* is regulated by NF-kappaB in gastric epithelial cancer cells. *Infect. Immun.* **68**:1806–1814.

2504. **Morioka, H., and M. Tachibana.** 1995. Agglutination of *Staphylococcus saprophyticus*: a structural and cytochemical study. *FEMS Microbiol. Lett.* **132**:101–105.

2505. **Morisaki, H., S. Nagai, H. Ohshima, E. Ikemoto, and K. Kogure.** 1999. The effect of motility and cell-surface polymers on bacterial attachment. *Microbiology* **145**:2797–2802.

2506. **Mork, T., and R. E. Hancock.** 1993. Mechanisms of nonopsonic phagocytosis of

Pseudomonas aeruginosa. Infect. Immun. **61:**3287–3293.

2507. **Morner, A. P., A. Faris, and K. Krovacek.** 1998. Virulence determinants of *Escherichia coli* isolated from milk of sows with coliform mastitis. *Zentbl. Veterinaermed. Reihe B* **45:**287–295.

2508. **Morona, R., J. K. Morona, A. Considine, J. A. Hackett, L. van den Bosch, L. Beyer, and S. R. Attridge.** 1994. Construction of K88- and K99-expressing clones of *Salmonella typhimurium* G30: immunogenicity following oral administration to pigs. *Vaccine* **12:**513–517.

2509. **Morra, M., and C. Cassineli.** 1999. Non-fouling properties of polysaccharide-coated surfaces. *J. Biomater. Sci. Polymer Ed.* **10:**1107–1124.

2510. **Morra, M., and C. Cassinelli.** 1996. *Staphylococcus epidermidis* adhesion to films deposited from hydroxyethylmethacrylate plasma. *J. Biomed. Mater. Res.* **31:**149–155.

2511. **Morra, M., and C. Cassinelli.** 1997. Bacterial adhesion to polymer surfaces: a critical review of surface thermodynamic approaches. *J. Biomater. Sci. Polymer Ed.* **9:**55–74.

2512. **Morrin, M., and D. J. Reen.** 1993. Inhibition of the adherence of *Pseudomonas aeruginosa* to epithelial cells by IgG subclass antibodies. *J. Med. Microbiol.* **39:**459–466.

2513. **Morrin, M., and D. J. Reen.** 1993. Role of IgG subclass response to outer-membrane proteins in inhibiting adhesion of *Pseudomonas aeruginosa* to epithelial cells. *J. Med. Microbiol.* **39:**467–472.

2514. **Morschhauser, J., B. E. Uhlin, and J. Hacker.** 1993. Transcriptional analysis and regulation of the *sfa* determinant coding for S fimbriae of pathogenic *Escherichia coli* strains. *Mol. Gen. Genet.* **238:**97–105.

2515. **Morschhauser, J., V. Vetter, L. Emödy, and J. Hacker.** 1994. Adhesin regulatory genes within large, unstable DNA regions of pathogenic *Escherichia coli:* cross-talk between different adhesin gene clusters. *Mol. Microbiol.* **11:**555–566.

2516. **Morschhauser, J., V. Vetter, T. Korhonen, B. E. Uhlin, and J. Hacker.** 1993. Regulation and binding properties of S fimbriae cloned from *E. coli* strains causing urinary tract infection and meningitis. *Zentbl. Bakteriol.* **278:**165–176.

2517. **Morsy, M. A., V. S. Panangala, V. L. van Santen, and R. C. Bird.** 1993. Cloning and partial sequence analysis of a *Mycoplasma synoviae* DNA fragment encoding epitopes shared with the major adhesin P1 protein of *Mycoplasma pneumoniae. Avian Dis.* **37:**1105–1112.

2518. **Moser, I., and W. Schroder.** 1995. Binding of outer membrane preparations of *Campylobacter jejuni* to INT 457 cell membranes and extracellular matrix proteins. *Med. Microbiol. Immunol.* **184:**147–153.

2519. **Moser, I., and W. Schroder.** 1997. Hydrophobic characterization of thermophilic *Campylobacter* species and adhesion to INT 407 cell membranes and fibronectin. *Microb. Pathog.* **22:**155–164.

2520. **Moser, I., W. Schroeder, and J. Salnikow.** 1997. *Campylobacter jejuni* major outer membrane protein and a 59-kDa protein are involved in binding to fibronectin and INT 407 cell membranes. *FEMS Microbiol. Lett.* **157:**233–238.

2521. **Moses, A. E., M. R. Wessels, K. Zalcman, S. Alberti, S. Natanson-Yaron, T. Menes, and E. Hanski.** 1997. Relative contributions of hyaluronic acid capsule and M protein to virulence in a mucoid strain of the group A *Streptococcus. Infect. Immun.* **65:**64–71.

2522. **Mosleh, I. M., H. J. Boxberger, M. J. Sessler, and T. F. Meyer.** 1997. Experimental infection of native human ureteral tissue with *Neisseria gonorrhoeae:* adhesion, invasion, intracellular fate, exocytosis, and passage through a stratified epithelium. *Infect. Immun.* **65:**3391–3398.

2523. **Mostafavi, M., P. C. Stein, and C. L. Parsons.** 1995. Production of soluble virulence factor by *Escherichia coli. J. Urol.* **153:**1441–1443.

2524. **Mott, M. R., and W. Reilly.** 1994. Double immunogold labelling demonstrating expression of recombinant genes for production of an anti-fertility vaccine. *Micron* **25:**539–545.

2525. **Mounier, J., F. K. Bahrani, and P. J. Sansonetti.** 1997. Secretion of *Shigella flexneri* Ipa invasins on contact with epithelial cells and subsequent entry of the bacterium into cells are growth stage dependent. *Infect. Immun.* **65:**774–782.

2526. **Mouricout, M.** 1997. Interactions between the enteric pathogen and the host. An assortment of bacterial lectins and a set of glycoconjugate receptors. *Adv. Exp. Med. Biol.* **412:**109–123.

2527. **Mouricout, M., M. Milhavet, C. Durie, and P. Grange.** 1995. Characterization of glycoprotein glycan receptors for *Escherichia coli* F17 fimbrial lectin. *Microb. Pathog.* **18:**297–306.

2528. **Moussa, F. W., B. J. Gainor, J. O. Anglen, G. Christensen, and W. A. Simpson.** 1996. Disinfecting agents for removing adherent bacteria from orthopaedic hardware. *Clin. Orthopaed. Relat. Res.* **329:**255–262.

2529. **Muenzner, P., M. Naumann, T. F. Meyer, and S. D. Gray-Owen.** 2001. Pathogenic *Neisseria* trigger expression of their carcinoembryonic antigen-related cellular adhesion molecule 1 (CEACAM1; previously CD66a) receptor on primary endothelial cells by activating the immediate early response transcription factor, nuclear factor-κB. *J. Biol. Chem.* **276:**24331–24340.

2530. **Muenzner, P., C. Dehio, T. Fujiwara, M. Achtman, T. F. Meyer, and S. D. Gray-Owen.** 2000. Carcinoembryonic antigen family receptor specificity of *Neisseria meningitidis* Opa variants influences adherence to and invasion of proinflammatory cytokine-activated endothelial cells. *Infect. Immun.* **68:**3601–3607.

2531. **Muhldorfer, I., and J. Hacker.** 1994. Genetic aspects of *Escherichia coli* virulence. *Microb. Pathog.* **16:**171–181.

2532. **Mukai, T., K. Arihara, and H. Itoh.** 1992. Lectin-like activity of *Lactobacillus acidophilus* strain JCM 1026. *FEMS Microbiol. Lett.* **77:**71–74.

2533. **Mukai, T., S. Kaneko, and H. Ohori.** 1998. Haemagglutination and glycolipid-binding activities of *Lactobacillus reuteri*. *Lett. Appl. Microbiol.* **27:**130–134.

2534. **Mukai, T., T. Toba and H. Ohori.** 1997. Collagen binding of *Bifidobacterium adolescentis*. *Curr. Microbiol.* **34:**326–331.

2535. **Mukherjee, D. P., N. R. Dorairaj, D. K. Mills, D. Graham, and J. T. Krauser.** 2000. Fatigue properties of hydroxyapatite-coated dental implants after exposure to a periodontal pathogen. *J. Biomed. Mater. Res.* **53:**467–474.

2536. **Mukhopadhyay, S., C. Ghosh, and A. C. Ghose.** 1996. Phenotypic expression of a mannose-sensitive hemagglutinin by a *Vibrio cholerae* O1 ElTor strain and evaluation of its role in intestinal adherence and colonization. *FEMS Microbiol. Lett.* **138:**227–232.

2537. **Mukhopadhyay, S., B. Nandi, and A. C. Ghose.** 2000. Antibodies (IgG) to lipopolysaccharide of *Vibrio cholerae* O1 mediate protection through inhibition of intestinal adherence and colonisation in a mouse model. *FEMS Microbiol Lett.* **185:**29–35.

2538. **Muller, E., J. Hubner, N. Gutierrez, S. Takeda, D. A. Goldmann, and G. B. Pier.** 1993. Isolation and characterization of transposon mutants of *Staphylococcus epidermidis* deficient in capsular polysaccharide/adhesin and slime. *Infect. Immun.* **61:**551–558.

2539. **Muller, E., J. Al-Attar, A. G. Wolff, and B. F. Farber.** 1998. Mechanism of salicylate-mediated inhibition of biofilm in *Staphylococcus epidermidis*. *J. Infect. Dis.* **177:**501–503.

2540. **Muller, E., S. Takeda, H. Shiro, D. Goldmann, and G. B. Pier.** 1993. Occurrence of capsular polysaccharide/adhesin among clinical isolates of coagulase-negative staphylococci. *J. Infect. Dis.* **168:**1211–1218.

2541. **Müller, S., T. Hain, P. Pashalidis, A. Lingnau, E. Domann, T. Chakraborty, and J. Wehland.** 1998. Purification of the *inlB* gene product of *Listeria monocytogenes* and demonstration of its biological activity. *Infect. Immun.* **66:**3128–3133.

2542. **Multanen, M., M. Talja, S. Hallanvuo, A. Siitonen, T. Valimaa, T. L. Tammela, J. Seppala, and P. Tormala.** 2000. Bacterial adherence to silver nitrate coated poly-L-lactic acid urological stents *in vitro*. *Urol. Res.* **28:**327–331.

2543. **Multanen, M., M. Talja, S. Hallanvuo, A. Siitonen, T. Valimaa, T. L. Tammela, J. Seppala, and P. Tormala.** 2000. Bacterial adherence to ofloxacin-blended polylactone-coated self-reinforced L-lactic acid polymer urological stents. *BJU Int.* **86:**966–969.

2544. **Mulvey, M. A., Y. S. Lopez-Boado, C. L. Wilson, R. Roth, W. C. Parks, J. Heuser, and S. J. Hultgren.** 1998. Induction and evasion of host defenses by type 1-piliated uropathogenic *Escherichia coli*. *Science* **282:**1494–1497.

2545. **Munch, S., S. Grund and M. Kruger.** 1992. Fimbriae and membranes on *Haemophilus parasuis*. *Zentbl. Veterinaermed. Reihe B* **39:**59–64.

2546. **Munemasa, T., T. Takemoto, G. Dahlen, T. Hino, H. Shiba, T. Ogawa, and H. Kurikara.** 2000. Adherence of *Bacteroides forsythus* to host cells. *Microbios* **101:**115–126.

2547. **Munro, G. H., P. Evans, S. Todryk, P. Buckett, C. G. Kelly, and T. Lehner.** 1993. A protein fragment of streptococcal cell surface antigen I/II which prevents adhesion of *Streptococcus mutans*. *Infect. Immun.* **61:**4590–4598.

2548. **Murai, M., K. Seki, J. Sakurada, A. Usui, and S. Masuda.** 1993. Effects of cytochalasins B and D on *Staphylococcus aureus* adherence to and ingestion by mouse renal cells from primary culture. *Microbiol. Immunol.* **37:**69–73.

2549. **Murai, M., A. Usui, K. Seki, J. Sakurada, and S. Masuda.** 1992. Intracellular localization of *Staphylococcus aureus* within primary cultured mouse kidney cells. *Microbiol. Immunol.* **36:**431–443.

2550. **Murakami, Y., S. Hanazawa, K. Nishida, H. Iwasaka, and S. Kitano.** 1993. *N*-Acetyl-D-galactosamine inhibits TNF-α gene expression induced in mouse peritoneal macrophages by fimbriae of *Porphyromonas* (*Bacteroides*) *gingivalis,* an oral anaerobe. *Biochem. Biophys. Res. Commun.* **192:**826–832.

2551. **Murakami, Y., S. Hanazawa, A. Watanabe, K. Naganuma, H. Iwasaka, K. Kawakami, and S. Kitano.** 1994. *Porphyromonas gingivalis* fimbriae induce a 68-kilodalton phosphorylated protein in macrophages. *Infect. Immun.* **62:**5242–5246.

2552. **Murakami, Y., H. Iwahashi, H. Yasuda, T. Umemoto, I. Namikawa, S. Kitano, and S. Hanazawa.** 1996. *Porphyromonas gingivalis* fimbrillin is one of the fibronectin-binding proteins. *Infect. Immun.* **64:**2571–2576.

2553. **Murakami, Y., H. Tamagawa, S. Shizukuishi, A. Tsunemitsu, and S. Aimoto.** 1992. Biological role of an arginine residue present in a histidine-rich peptide which inhibits hemagglutination of *Porphyromonas gingivalis. FEMS Microbiol. Lett.* **77:**201–204.

2554. **Muratsugu, M., Y. Miyake, N. Ishida, A. Hyodo, and K. Terayama.** 1995. Decrease in surface charge density of *Klebsiella pneumoniae* treated with cefodizime and enhancement of the phagocytic function of human polymorphonuclear leucocytes stimulated by the drug-treated bacteria. *Biol. Pharm. Bull.* **18:**1259–1263.

2555. **Murphree, D., B. Froehlich, and J. R. Scott.** 1997. Transcriptional control of genes encoding CS1 pili: negative regulation by a silencer and positive regulation by Rns. *J. Bacteriol.* **179:**5736–5743.

2556. **Murphy, S., M. Fitzgerald, R. Mulcahy, C. Keane, D. Coakley, and T. Scott.** 1997. Studies on haemagglutination and serum resistance status of strains of *Moraxella catarrhalis* isolated from the elderly. *Gerontology* **43:**277–282.

2557. **Murray, P. A., A. Prakobphol, T. Lee, C. I. Hoover, and S. J. Fisher.** 1992. Adherence of oral streptococci to salivary glycoproteins. *Infect. Immun.* **60:**31–38.

2558. **Muscholl-Silberhorn, A.** 1999. Cloning and functional analysis of Asa373, a novel adhesin unrelated to the other sex pheromone plasmid-encoded aggregation substances of *Enterococcus faecalis. Mol. Microbiol.* **34:**620–630.

2559. **Musmanno, R. A., M. Russi, N. Figura, P. Guglielmetti, A. Zanchi, R. Signori, and A. Rossolini.** 1997. Unusual species of campylobacters isolated in the Siena Tuscany area, Italy. *New Microbiol.* **21:**15–22.

2560. **Myllys, V., T. Honkanen-Buzalski, H. Virtanen, S. Pyorala, and H. P. Muller.** 1994. Effect of abrasion of teat orifice epithelium on development of bovine staphylococcal mastitis. *J. Dairy Sci.* **77:**446–452.

2561. **Mynott, T. L., R. K. Luke, and D. S. Chandler.** 1995. Detection of attachment of enterotoxigenic *Escherichia coli* (ETEC) to human small intestinal cells by enzyme immunoassay. *FEMS Immunol. Med. Microbiol.* **10:**207–218.

2562. **Mynott, T. L., R. K. Luke, and D. L. Chandler.** 1996. Oral administration of protease inhibits enterotoxigenic *Escherichia coli* receptor activity in piglet small intestine. *Gut* **38:**28–32.

2563. **Naaber, P., E. Lehto, S. Salminen, and M. Mikelsaar.** 1996. Inhibition of adhesion of *Clostridium difficile* to Caco-2 cells. *FEMS Immunol. Med. Microbiol.* **14:**205–209.

2564. **Nagamune, K., K. Yamamoto, A. Naka, J. Matsuyama, T. Miwatani, and T. Honda.** 1996. In vitro proteolytic processing and activation of the recombinant precursor of El Tor cytolysin/hemolysin (pro-HlyA) of *Vibrio cholerae* by soluble hemagglutinin/protease of *V. cholerae,* trypsin, and other proteases. *Infect. Immun.* **64:**4655–4658.

2565. **Nagao, P. E., F. Costa e Silva-Filho, L. C. Benchetrit, and L. Barrucand.** 1995. Cell surface hydrophobicity and the net electric surface charge of group B streptococci: the role played in the micro-organism-host cell interaction. *Microbios* **82:**207–216.

2566. **Nagaoka, S., and H. Kawakami.** 1995. Inhibition of bacterial adhesion and biofilm formation by a heparinized hydrophilic polymer. *ASAIO J.* **41:**M365–M368.

2567. **Nagata, H., K. Tazaki, A. Amano, T. Hanioka, H. Tamagawa, and S. Shizukuishi.** 1994. Characterization of coaggregation and fibrinogen-binding by *Porphyromonas gingivalis. J. Osaka Univ. Dent. Sch.* **34:**37–44.

2568. **Nagata, H., A. Amano, M. Ojima, M. Tanaka, K. Kataoka, and S. Shizukuishi.** 1994. Effect of binding of fibrinogen to each bacterium on coaggregation between *Porphyromonas gingivalis* and *Streptococcus oralis. Oral Microbiol. Immunol.* **9:**359–363.

2569. **Nagata, H., A. Sharma, H. T. Sojar, A. Amano, M. J. Levine, and R. J. Genco.**

1997. Role of the carboxyl-terminal region of *Porphyromonas gingivalis* fimbrillin in binding to salivary proteins. *Infect. Immun.* **65**:422–427.

2570. **Nagayama, K., T. Oguchi, M. Arita, and T. Honda.** 1994. Correlation between cell-associated mannose-sensitive hemagglutination by *Vibrio parahaemolyticus* and adherence to a human colonic cell line Caco-2. *FEMS Microbiol. Lett.* **120**:207–210.

2571. **Nagayama, K., T. Oguchi, M. Arita, and T. Honda.** 1995. Purification and characterization of a cell-associated hemagglutinin of *Vibrio parahaemolyticus*. *Infect. Immun.* **63**:1987–1992.

2572. **Nagayama, K., Z. Bi, T. Oguchi, Y. Takarada, S. Shibata, and T. Honda.** 1996. Use of an alkaline phosphatase-conjugated oligonucleotide probe for the gene encoding the bundle-forming pilus of enteropathogenic *Escherichia coli*. *J. Clin. Microbiol.* **34**:2819–2821.

2573. **Nagel, J. A., R. B. Dickinson, and S. L. Cooper.** 1996. Bacterial adhesion to polyurethane surfaces in the presence of pre-adsorbed high molecular weight kininogen. *J. Biomater. Sci. Polym. Ed.* **7**:769–780.

2574. **Nagy, B., L. H. Arp, H. W. Moon, and T. A. Casey.** 1992. Colonization of the small intestine of weaned pigs by enterotoxigenic *Escherichia coli* that lack known colonization factors. *Vet. Pathol.* **29**:239–246.

2575. **Nagy, B., S. C. Whipp, H. Imberechts, H. U. Bertschinger, E. A. Dean-Nystrom, T. A. Casey, and E. Salajka.** 1997. Biological relationship between F18ab and F18ac fimbriae of enterotoxigenic and verotoxigenic *Escherichia coli* from weaned pigs with oedema disease or diarrhoea. *Microb. Pathog.* **22**:1–11.

2576. **Nagy, B., T. A. Casey, S. C. Whipp, and H. W. Moon.** 1992. Susceptibility of porcine intestine to pilus-mediated adhesion by some isolates of piliated enterotoxigenic *Escherichia coli* increases with age. *Infect. Immun.* **60**:1285–1294.

2577. **Nagy, I., G. Fröman, and P.-A. Mårdh.** 1992. Fibronectin binding of *Lactobacillus* species isolated from women with and without bacterial vaginosis. *J. Med. Mcirobiol.* **37**:38–42.

2578. **Nair, S., E. Milohanic, and P. Berche.** 2000. ClpC ATPase is required for cell adhesion and invasion of *Listeria monocytogenes*. *Infect. Immun.* **68**:7061–7068.

2579. **Naito, Y., H. Tohda, K. Okuda, and I. Takazoe.** 1993. Adherence and hydrophobicity of invasive and noninvasive strains of *Porphyromonas gingivalis*. *Oral Microbiol. Immunol.* **8**:195–202.

2580. **Naka, A., K. Yamamoto, T. Miwatani, and T. Honda.** 1992. Characterization of two forms of hemagglutinin/protease produced by *Vibrio cholerae* non-O1. *FEMS Microbiol. Lett.* **77**:197–200.

2581. **Nakae, H., S. Ebisu, and H. Okada.** 1993. Production and characterization of monoclonal antibodies against bacterial lectin of *Eikenella corrodens*. *J. Periodontal Res.* **28**:404–410.

2582. **Nakahara, K., S. Kawabata, H. Ono, K. Ogura, T. Tanaka, T. Ooshima, and S. Hamada.** 1993. Inhibitory effect of oolong tea polyphenols on glycosyltransferases of mutans streptococci. *Appl. Environ. Microbiol.* **59**:968–973.

2583. **Nakai, M., N. Okahashi, H. Ohta, and T. Koga.** 1993. Saliva-binding region of *Streptococcus mutans* surface protein antigen. *Infect. Immun.* **61**:4344–4349.

2584. **Nakamoto, D. A., J. R. Haaga, P. Bove, K. Merritt, and D. Y. Rowland.** 1995. Use of fibrinolytic agents to coat wire implants to decrease infection. An animal model. *Investig. Radiol.* **30**:341–344.

2585. **Nakamura, T., A. Amano, I. Nakagawa, and S. Hamada.** 1999. Specific interactions between *Porphyromonas gingivalis* fimbriae and human extracellular matrix proteins. *FEMS Microbiol. Lett.* **175**:267–272.

2586. **Nakano, Y., M. Fujisawa, T. Matsui, S. Arakawa, and S. Kamidono.** 1999. The significance of the difference in bacterial adherence between bladder and ileum using rat ileal augmented bladder. *J. Urol.* **162**:243–247.

2587. **Nakasone, N., S. Insisengmay, and M. Iwanaga.** 2000. Characterization of the pili isolated from *Vibrio parahaemolyticus* O3:K6. *Southeast Asian J. Trop. Med. Public Health* **31**:360–365.

2588. **Nakasone, N., M. Iwanaga, T. Yamashiro, K. Nakashima, and M. J. Albert.** 1996. *Aeromonas trota* strains, which agglutinate with *Vibrio cholerae* O139 Bengal antiserum, possess a serologically distinct fimbrial colonization factor. *Microbiology* **142**:309–313.

2589. **Nakasone, N., and M. Iwanaga.** 1993. Cell-associated hemagglutinin of classical *Vibrio cholerae* O1 with reference to intestinal adhesion. *FEMS Microbiol. Lett.* **113**:67–70.

2590. **Nakasone, N., and M. Iwanaga.** 1992. The role of pili in colonization of the rabbit intestine by *Vibrio parahaemolyticus* Na2. *Microbiol. Immunol.* **36**:123–130.

2591. **Nakasone, N., T. Yamashiro, M. J.**

Albert, and M. Iwanaga. 1994. Pili of a *Vibrio cholerae* O139. *Microbiol. Immunol.* **38:**225–227.

2592. Nakayama, K., D. B. Ratnayake, T. Tsukuba, T. Kadowaki, K. Yamamoto, and S. Fujimura. 1998. Haemoglobin receptor protein is intragenically encoded by the cysteine proteinase-encoding genes and the haemagglutinin-encoding gene of *Porphyromonas gingivalis*. *Mol. Microbiol.* **27:**51–61.

2593. Nakayama, K., F. Yoshimura, T. Kadowaki, and K. Yamamoto. 1996. Involvement of arginine-specific cysteine proteinase (Arg-gingipain) in fimbriation of *Porphyromonas gingivalis*. *J. Bacteriol.* **178:**2818–2824.

2594. Nallapareddy, S. R., X. Qin, G. M. Weinstock, M. Höök, and B. E. Murray. 2000. *Enterococcus faecalis* adhesin, ace, mediates attachment to extracellular matrix proteins collagen type IV and laminin as well as collagen type I. *Infect. Immun.* **68:**5218–5224.

2595. Namavar, F., M. W. van der Bijl, B. J. Appelmelk, J. De Graaff, and D. M. MacLaren. 1994. The role of neuraminidase in haemagglutination and adherence to colon WiDr cells by *Bacteroides fragilis*. *J. Med. Microbiol.* **40:**393–396.

2596. Namavar, F., M. Sparrius, E. C. I. Veerman, B. J. Appelmelk, and C. M. J. E. Vendenbroucke-Grauls. 1998. Neutrophil-activating protein mediates adhesion of *Helicobacter pylori* to sulfated carbohydrates on high-molecular-weight salivary mucin. *Infect. Immun.* **66:**444–447.

2597. Nanda Kumar, K. S., N. K. Ganguly, I. S. Anand, amd P. L. Wahi. 1992. Adherence of *Streptococcus pyogenes* M type 5 to pharyngeal and buccal cells of patients with rheumatic fever and rheumatic heart disease during a one-year follow-up. *APMIS* **100:**353–359.

2598. Nanda Kumar, K. S., N. K. Ganguly, I. S. Anand, and P. L. Wahi. 1993. Penicillin decreases streptococcal hydrophobicity and streptococcal adherence to human pharyngeal cells. *Indian J. Exp. Biol.* **31:**480–481.

2599. Nandy, R. K., S. K. Barari, and A. C. Ghose. 1994. Expression of antigenically distinct fimbriae with hemagglutination and HeLa cell adherence properties by an enteroaggregative *Escherichia coli* strain belonging to the enteropathogenic serogroup. *FEMS Immunol. Med. Microbiol.* **9:**143–150.

2600. Narat, M., D. Bencina, S. H. Kleven, and F. Habe. 1998. The hemagglutination-positive phenotype of *Mycoplasma synoviae* induces experimental infectious synovitis in chickens more frequently than does the hemagglutination-negative phenotype. *Infect. Immun.* **66:**6004–6009.

2601. Nascimento de Araujo, A., and L. G. Giugliano. 2000. Human milk fractions inhibit the adherence of diffusely adherent *Escherichia coli* (DAEC) and enteroaggregative *E. coli* (EAEC) to HeLa cells. *FEMS Microbiol. Lett.* **184:**91–94.

2602. Nassif, X., J. L. Beretti, J. Lowy, P. Stenberg, P. O'Gaora, J. Pfeifer, S. Normark, and M. So. 1994. Roles of pilin and PilC in adhesion of *Neisseria meningitidis* to human epithelial and endothelial cells. *Proc. Natl. Acad. Sci. USA* **91:**3769–3773.

2603. Nassif, X., J. Lowy, P. Stenberg, P. O'Gaora, A. Ganji, and M. So. 1993. Antigenic variation of pilin regulates adhesion of *Neisseria meningitidis* to human epithelial cells. *Mol. Microbiol.* **8:**719–725.

2604. Nassif, X., and M. So. 1995. Interaction of pathogenic neisseriae with nonphagocytic cells. *Clin. Microbiol. Rev.* **8:**376–388.

2605. Nassif, X., M. Marceau, C. Pujol, B. Pron, J. L. Beretti, and M. K. Taha. 1997. Type-4 pili and meningococcal adhesiveness. *Gene* **192:**149–153.

2606. Natanson, S., S. Sela, A. E. Moses, J. M. Musser, M. G. Caparon, and E. Hanski. 1995. Distribution of fibronectin-binding proteins among group A streptococci of different M types. *J. Infect. Dis.* **171:**871–878.

2607. Nataro, J. P., Y. Deng, S. Cookson, A. Cravioto, S. J. Savarino, L. D. Guers, M. M. Levine, and C. O. Tacket. 1995. Heterogeneity of enteroaggregative *Escherichia coli* virulence demonstrated in volunteers. *J. Infect. Dis.* **171:**465–468.

2608. Nataro, J. P., D. Yikang, J. A. Giron, S. J. Savarino, M. H. Kothary, and R. Hall. 1993. Aggregative adherence fimbria I expression in enteroaggregative *Escherichia coli* requires two unlinked plasmid regions. *Infect. Immun.* **61:**1126–1131.

2609. Nataro, J. P., D. Yikang, D. Yingkang, and K. Walker. 1994. AggR, a transcriptional activator of aggregative adherence fimbria I expression in enteroaggregative *Escherichia coli*. *J. Bacteriol.* **176:**4691–4699.

2610. Nataro, J. P., Y. Deng, K. R. Maneval, A. L. German, W. C. Martin, and M. M. Levine. 1992. Aggregative adherence fimbriae I of enteroaggregative *Escherichia coli* mediate adherence to Hep-2 cells and hemagglutination of human erythrocytes. *Infect. Immun.* **60:**2297–2304.

2611. Nataro, J. P., S. Hicks, A. D. Phillips, P. A. Vial, and C. L. Sears. 1996. T84 cells in culture as a model for enteroaggregative *Escherichia coli* pathogenesis. *Infect. Immun.* **64:**4761–4768.

2612. Naughton, P. J., G. Grant, M. Sojka, S. Bardocz, C. J. Thorns, and A. Pusztai. 2001. Survival and distribution of cell-free SEF 21 of *Salmonella enterica* serovar *Enteritidis* in the stomach and various compartments of the rat gastrointestinal tract *in vivo. J. Med. Microbiol.* **50:**1049–1054.

2613. Navas, E. L., M. F. Venegas, J. L. Duncan, B. E. Anderson, J. S. Chmiel, and A. J. Schaeffer. 1993. Blood group antigen expression on vaginal and buccal epithelial cells and mucus in secretor and nonsecretor women. *J. Urol.* **149:**1492–1498.

2614. Nayak, R., P. B. Kenney, and G. K. Bissonnette. 2001. Inhibition and reversal of *Salmonella typhimurium* attachment to poultry skin using zinc chloride. *J. Food Prot.* **64:**456–461.

2615. Ndour, C. T., K. Ahmed, T. Nakagawa, Y. Nakano, A. Ichinose, G. Tarhan, M. Aikawa, and T. Nagatake. 2001. Modulating effects of mucoregulating drugs on the attachment of *Haemophilus influenzae. Microb. Pathog.* **30:**121–127.

2616. Neeman, R., N. Keller, A. Barzilai, Z. Korenman, and S. Sela. 1998. Prevalence of internalisation-associated gene, *prtF1*, among persisting group-A streptococcus strains isolated from asymptomatic carriers. *Lancet* **352:**1974–1947.

2617. Neeser, J. R., D. Granato, M. Rouvet, A. Servin, S. Teneberg, and K. A. Karlsson. 2000. *Lactobacillus johnsonii* La1 shares carbohydrate-binding specificities with several enteropathogenic bacteria. *Glycobiol.* **10:**1193–1199.

2618. Neeser, J.-R., M. Golliard, A. Woltz, M. Rouvet, M.-L. Dillmann, and B. Guggenheim. 1994. *In vitro* modulation of oral bacterial adhesion to saliva-coated hydroxyapatite beads by milk casein derivatives. *Oral Microbiol. Immunol.* **9:**193–201.

2619. Neeser, J.-R., R. C. Grafström, A. Woltz, D. Brassart, V. Fryder, and B. Guggenheim. 1995. A 23 kDa membrane glycoprotein bearing NeuNAcα2-3Galβ1-3GalNAc O-linked carbohydrate chains acts as a receptor for *Streptococcus sanguis* OMZ 9 on human buccal epithelial cells. *Glycobiology* **5:**97–104.

2620. Negm, R. S., and T. G. Pistole. 1999. The porin OmpC of *Salmonella typhimurium*

mediates adherence to macrophages. *Can. J. Microbiol.* **45:**658–669.

2621. Negm, R. S., and T. G. Pistole. 1998. Macrophages recognize and adhere to an OmpD-like protein of *Salmonella typhimurium. FEMS Immunol. Med. Microbiol.* **20:**191–199.

2622. Nesbitt, W. E., J. E. Beem, K. P. Leung, S. Stroup, R. Swift, W. P. McArthur, and W. B. Clark. 1996. Inhibition of adherence of *Actinomyces naeslundii* (*Actinomyces viscosus*) T14V-J1 to saliva-treated hydroxyapatite by a monoclonal antibody to type 1 fimbriae. *Oral Microbiol. Immunol.* **11:**51–58.

2623. Nesbitt, W. E., H. Fukushima, K. P. Leung, and W. B. Clark. 1993. Coaggregation of *Prevotella intermedia* with oral *Actinomyces* species. *Infect. Immun.* **61:**2011–2014.

2624. Nesbitt, W. W., J. E. Beem, K.-P. Leung, and W. B. Clark. 1992. Isolation and characterization of *Actinomyces viscosus* mutants defective in binding salivary proline-rich proteins. *Infect. Immun.* **60:**1095–1100.

2625. Ness-Greenstein, R. B., M. Rosenberg, R. J. Doyle, and N. Kaplan. 1995. DNA from *Serratia marcescens* confers a hydrophobic character in *Escherichia coli. FEMS Microbiol. Lett.* **125:**71–75.

2626. Neu, T. R. 1996. Significance of bacterial surface-active compounds in interaction of bacteria with interfaces. *Microbiol. Rev.* **60:**151–166.

2627. Neves, M. S., M. P. Nunes, and A. M. Milhomem. 1994. *Aeromonas* species exhibit aggregative adherence to HEp-2 cells. *J. Clin. Microbiol.* **32:**1130–1131.

2628. Newman, F., J. A. Beeley, and T. W. MacFarlane. 1996. Adherence of oral microorganisms to human parotid salivary proteins. *Electrophoresis* **17:**266–270.

2629. Newman, J. V., R. L. Burghoff, L. Pallesen, K. A. Krogfelt, C. S. Kristensen, D. C. Laux, and P. S. Cohen. 1994. Stimulation of *Escherichia coli* F-18Col⁻ type-1 fimbriae synthesis by *leuX. FEMS Microbiol. Lett.* **122:**281–287.

2630. Ng, E. W., G. D. Barrett, and R. Bowman. 1996. *In vitro* bacterial adherence to hydrogel and poly(methyl methacrylate) intraocular lenses. *J. Cataract Refract. Surg.* **22:**1331–1335.

2631. Ngeleka, M., and J. M. Fairbrother. 1999. F165(1) fimbriae of the P fimbrial family inhibit the oxidative response of porcine neutrophils. *FEMS Immunol. Med. Microbiol.* **25:**265–274.

2632. Ngeleka, M., M. Jacques, B. Martineau-

Doize, F. Daigle, J. Harel, and J. M. Fairbrother. 1993. Pathogenicity of an *Escherichia coli* O115:K″V165″ mutant negative for F165(1) fimbriae in septicemia of gnotobiotic pigs. *Infect. Immun.* **61**:836–843.

2633. Ngeleka, M., B. Martineau-Doize, and J. M. Fairbrother. 1994. Septicemia-inducing *Escherichia coli* O115:K″V165″F165(1) resists killing by porcine polymorphonuclear leukocytes in vitro: role of F165(1) fimbriae and K″V165″ O-antigen capsule. *Infect. Immun.* **62**:398–404.

2634. Nguyen, T., B. Ghebrehiwet, and E. I. Peerschke. 2000. *Staphylococcus aureus* protein A recognizes platelet gC1qR/p33: a novel mechanism for staphylococcal interactions with platelets. *Infect. Immun.* **68**:2061–2068.

2635. Niang, M., R. F. Rosenbusch, J. J. Andrews, and M. L. Kaeberle. 1998. Demonstration of a capsule on *Mycoplasma ovipneumoniae*. *Am. J. Vet. Res.* **59**:557–562.

2636. Nicholls, L., T. H. Grant, and R. M. Robins-Browne. 2000. Identification of a novel genetic locus that is required for in vitro adhesion of a clinical isolate of enterohaemorrhagic *Escherichia coli* to epithelial cells. *Mol. Microbiol.* **35**:275–288.

2637. Nicholson, B., and D. A. Low. 2000. DNA methylation-dependent regulation of *pef* expression in *Salmonella typhimurium*. *Mol. Microbiol.* **35**:728–742.

2638. Nickerson, C. A., and R. Curtiss, III. 1997. Role of sigma factor RpoS in initial stages of *Salmonella typhimurium* infection. *Infect. Immun.* **65**:1814–1823.

2639. Niederman, M. S. 1994. The pathogenesis of airway colonization: lessons learned from the study of bacterial adherence. *Eur. Respir. J.* **7**:1737–1740.

2640. Ni Eidhin, D., S. Perkins, P. Francois, P. Vaudaux, M. Höök, and T. J. Foster. 1998. Clumping factor B (ClfB), a new surface-located fibrinogen-binding adhesin of *Staphylococcus aureus*. *Mol. Microbiol.* **30**:245–257.

2641. Nietfeld, J. C., D. E. Tyler, L. R. Harrison, J. R. Cole, K. S. Latimer, and W. A. Crowell. 1992. Invasion of enterocytes in cultured porcine small intestinal mucosal explants by *Salmonella choleraesuis*. *Am. J. Vet. Res.* **53**:1493–1499.

2642. Nieuw Amerongen, A. V., H. Strooker, C. H. Oderkerk, R. A. Bank, Y. M. Henskens, L. C. Schenkels, A. J. Ligtenberg, and E. C. Veerman. 1992. Changes in saliva of epileptic patients. *J. Oral Pathol. Med.* **21**:203–208.

2643. Nikawa, H., T. Hamada, H. Yamashiro, H. Murata, and A. Subiwahjudi. 1998. The effect of saliva or serum on *Streptococcus mutans* and *Candida albicans* colonization of hydroxylapatite beads. *J. Dent.* **26**:31–37.

2644. Nilius, M., G. Bode, M. Buchler, and P. Malfertheiner. 1994. Adhesion of *Helicobacter pylori* and *Escherichia coli* to human and bovine surface mucus cells in vitro. *Eur. J. Clin. Investig.* **24**:454–459.

2645. Nilsson, I. M., J. M. Patti, T. Bremell, M. Höök, and A. Tarkowski. 1998. Vaccination with a recombinant fragment of collagen adhesin provides protection against *Staphylococcus aureus*-mediated septic death. *J. Clin. Investig.* **101**:2640–2649.

2646. Nilsson, K. G., and C. F. Mandenius. 1994. A carbohydrate biosensor surface for the detection of uropathogenic bacteria. *Bio/Technology* **12**:1376–1378.

2647. Nilsson, M., L. Frykberg, J.-I. Flock, L. Pei, M. Lindberg, and B. Guss. 1998. A fibrinogen-binding protein of *Staphylococcus aureus*. *Infect. Immun.* **66**:2666–2673.

2648. Nilsson, P., S. Naureckiene, and B. E. Uhlin. 1996. Mutations affecting mRNA processing and fimbrial biogenesis in the *Escherichia coli pap* operon. *J. Bacteriol.* **178**:683–690.

2649. Nilsson, U., R. T. Striker, S. J. Hultgren, and G. Magnusson. 1996. PapG adhesin from *E. coli* J96 recognizes the same saccharide epitope when present on whole bacteria and as isolated protein. *Bioorg. Med. Chem.* **4**:1809–1817.

2650. Nisapakultorn, K., K. F. Ross, and M. C. Herzberg. 2001. Calprotectin expression inhibits bacterial binding to mucosal epithelial cells. *Infect. Immun.* **69**:3692–3696.

2651. Nishihara, K., Y. Nozawa, S. Nomura, K. Kitazato, and H. Miyake. 1999. Analysis of *Helicobacter pylori* binding site on HEp-2 cells and three cell lines from human gastric carcinoma. *Fundam. Clin. Pharmacol.* **13**:555–561.

2652. Nishikata, M., and F. Yoshimura. 1995. Active site structure of a hemagglutinating protease from *Porphyromonas gingivalis*: similarity to clostripain. *Biochem. Mol. Biol. Int.* **37**:547–553.

2653. Nishikawa, Y., A. Hase, J. Ogawasara, S. M. Scotland, H. R. Smith, and T. Kimura. 1994. Adhesion to and invasion of human colon carcinoma Caco-2 cells by *Aeromonas* strains. *J. Med. Microbiol.* **40**:55–61.

2654. Nishikawa, Y., S. M. Scotland, H. R. Smith, G. A. Willshaw, and B. Rowe.

1995. Catabolite repression of the adhesion of Vero cytotoxin-producing *Escherichia coli* of serogroups 0157 and 0111. *Microb. Pathog.* **18**:223–229.

2655. **Nishimoto, K., S. Maruyama, A. Yasukawa, and M. Ozaki.** 1992. Effects of subminimal inhibitory concentrations of ampicillin on hemagglutination of *Escherichia coli*. *Hiroshima J. Med. Sci.* **41**:13–17.

2656. **Niv, Y., G. Fraser, G. Delpre, A. Neeman, A. Leiser, Z. Samra, E. Scapa, E. Gilon, and S. Bar-Shany.** 1996. *Helicobacter pylori* infection and blood groups. *Am. J. Gastroenterol.* **91**:101–104.

2657. **Njoroge, T., R. J. Genco, H. T. Sojar, N. Hamada, and C. A. Genco.** 1997. A role for fimbriae in *Porphyromonas gingivalis* invasion of oral epithelial cells. *Infect. Immun.* **65**:1980–1984.

2658. **Noach, L. A., T. M. Rolf, and G. N. Tytgat.** 1994. Electron microscopic study of association between *Helicobacter pylori* and gastric and duodenal mucosa. *J. Clin. Pathol.* **47**:699–704.

2659. **Noel, G. J., S. J. Barenkamp, J. W. St Geme III, W. N. Haining, and D. M. Mosser.** 1994. High-molecular-weight surface-exposed proteins of *Haemophilus influenzae* mediate binding to macrophages. *J. Infect. Dis.* **169**:425–429.

2660. **Noel, G. J., D. C. Love, and D. M. Mosser.** 1994. High-molecular-weight proteins of nontypeable *Haemophilus influenzae* mediate bacterial adhesion to cellular proteoglycans. *Infect. Immun.* **62**:4028–4033.

2661. **Noel, J. M., and E. C. Boedeker.** 1997. Enterohemorrhagic *Escherichia coli*: a family of emerging pathogens. *Dig. Dis.* **15**:67–91.

2662. **Nollet, H., P. Deprez, E. van Driessche, and E. Muylle.** 1999. Protection of just weaned pigs against infection with F18+ *Escherichia coli* by non-immune plasma powder. *Vet. Microbiol.* **65**:37–45.

2663. **Nomura, S., A. Kuroiwa, and A. Nagayama.** 1995. Changes of surface hydrophobicity and charge of *Staphylococcus aureus* treated with sub-MIC of antibiotics and their effects on the chemiluminescence response of phagocytic cells. *Chemotherapy* **41**:77–81.

2664. **Nomura, S., F. Lundberg, M. Stollenwerk, K. Nakamura, and Å. Ljungh.** 1997. Adhesion of staphylococci to polymers with and without immobilized heparin in cerebrospinal fluid. *J. Biomed. Mater. Res.* **38**:35–42.

2665. **Nomura, S., K. Murata, and A.**

Nagayama. 1995. Effects of sub-minimal inhibitory concentrations of antimicrobial agents on the cell surface of *Klebsiella pneumoniae* and phagocytic killing activity. *J. Chemother.* **7**:406–413.

2666. **Noormohammadi, A. H., P. F. Markham, K. G. Whithear, I. D. Walker, V. A. Gurevich, D. H. Ley, and G. F. Browning.** 1997. *Mycoplasma synoviae* has two distinct phase-variable major membrane antigens, one of which is a putative hemnagglutinin. *Infect. Immun.* **65**:2542–2547.

2667. **Noormohammadi, A. H., P. F. Markham, M. F. Duffy, K. G. Whithear, and G. F. Browning.** 1998. Multigene families encoding the major hemagglutinins in phylogenetically distinct mycoplasmas. *Infect. Immun.* **66**:3470–3475.

2668. **Nordman, H., T. Boren, J. R. Davies, L. Engstrand, and I. Carlstedt.** 1999. pH-dependent binding of *Helicobacter pylori* to pig gastric mucins. *FEMS Immunol. Med. Microbiol.* **24**:175–181.

2669. **Norris, T. L., R. A. Kingsley, and A. J. Bäumler.** 1998. Expression and transcriptional control of the *Salmonella typhimurium lpf* fimbrial operon by phase variation. *Mol. Microbiol.* **29**:311–320.

2670. **Norwood, D. E., and A. Gilmour.** 1999. Adherence of *Listeria monocytogenes* strains to stainless steel coupons. *J. Appl. Microbiol.* **86**:576–582.

2671. **Nostro, A., M. A. Cannatelli, G. Crisafi, and V. Alonzo.** 2001. The effect of *Nepeta cataria* extract on adherence and enzyme production of *Staphylococcus aureus*. *Int. J. Antimicrob. Agents* **18**:583–585.

2672. **Nou, X., B. Braaten, L. Kaltenbach, and D. A. Low.** 1995. Differential binding of Lrp to two sets of *pap* DNA binding sites mediated by PapI regulates Pap phase variation in *Escherichia coli*. *EMBO J.* **14**:5785–5797.

2673. **Nou, X., B. Skinner, B. Braaten, L. Blyn, D. Hirsch, and D. Low.** 1993. Regulation of pyelonephritis-associated pili phase-variation in *Escherichia coli*: binding of the PapI and the Lrp regulatory proteins is controlled by DNA methylation. *Mol. Microbiol.* **7**:545–553.

2674. **Nougayrede, J. P., O. Marches, M. Boury, J. Mainil, G. Charlier, P. Pohl, J. De Rycke, A. Milon, and E. Oswald.** 1999. The long-term cytoskeletal rearrangement induced by rabbit enteropathogenic *Escherichia coli* is Esp dependent but intimin independent. *Mol. Microbiol.* **31**:19–30.

2675. **Nowicki, B., A. Hart, K. E. Coyne, D.**

M. Lublin, and S. Nowicki. 1993. Short consensus repeat-3 domain of recombinant decay-accelerating factor is recognized by *Escherichia coli* recombinant Dr adhesin in a model of a cell-cell interaction. *J. Exp. Med.* **178:**2115–2121.

2676. Nowicki, B., M. Martens, A. Hart, and S. Nowicki. 1994. Gestational age-dependent distribution of *Escherichia coli* fimbriae in pregnant patients with pyelonephritis. *Ann. N.Y. Acad. Sci.* **730:**290–291.

2677. Nowicki, S., B. Nowicki, M. Martens, A. Kaul, G. Flores, and D. Kumar. 1994. Host factors in the attachment of gonococcal cells to pelvic tissue. *Ann. N. Y. Acad. Sci.* **730:**292–294.

2678. Oberhuber, G., A. Kranz, C. Dejaco, B. Dragosics, I. Mosberger, W. Mayr, and T. Radaszkiewicz. 1997. Blood groups Lewis(b) and ABH expression in gastric mucosa: lack of inter-relation with *Helicobacter pylori* colonisation and occurrence of gastric MALT lymphoma. *Gut* **41:**37–42.

2679. O'Brien-Simpson, N. M., C. L. Black, P. S. Bhogal, S. M. Cleal, N. Slakeski, T. J. Higgins, and E. C. Reynolds. 2000. Serum immunoglobulin G (IgG) and IgG subclass responses to the RgpA-Kgp proteinase-adhesin complex of *Porphyromonas gingivalis* in adult periodontitis. *Infect. Immun.* **68:**2704–2712.

2680. Ochiai, K., K. Kikuchi, K. Fukushima, and T. Kurita-Ochiai. 1999. Coaggregation as a virulent factor of *Streptococcus sanguis* isolated from infective endocarditis. *J. Oral Sci.* **41:**117–122.

2681. Ochiai, K., T. Kurita-Ochiai, Y. Kamino, and T. Ikeda. 1993. Effect of co-aggregation on the pathogenicity of oral bacteria. *J. Med. Microbiol.* **39:**183–190.

2682. O'Connell, W. A., E. K. Hickey, and N. P. Cianciotto. 1996. A *Legionella pneumophila* gene that promotes hemin binding. *Infect. Immun.* **64:**842–848.

2683. Odenbreit, S., M. Till, D. Hofreuter, G. Faller, and R. Haas. 1999. Genetic and functional characterization of the *alpAB* gene locus essential for the adhesion of *Helicobacter pylori* to human gastric tissue. *Mol. Microbiol.* **31:**1537–1548.

2684. Ofek, I. 1995. Enzyme-linked immunosorbent-based adhesion assays. *Methods Enzymol.* **253:**528–536.

2685. Ofek, I., J. Goldhar, Y. Keisari, and N. Sharon. 1995. Nonopsonic phagocytosis of microorganisms. *Annu. Rev. Microbiol.* **49:**239–276.

2686. Oga, M., T. Arizono, and Y. Sugioka. 1993. Bacterial adherence to bioinert and bioactive materials studied *in vitro*. *Acta Orthopaed. Scand.* **64:**273–276.

2687. Oga, M., T. Arizono, and Y. Sugioka. 1992. Inhibition of bacterial adhesion by tobramycin-impregnated PMMA bone cement. *Acta Orthopaed. Scand.* **63:**301–304.

2688. Ogata, S. A., and B. L. Beaman. 1992. Adherence of *Nocardia asteroides* within the murine brain. *Infect. Immun.* **60:**1800–1805.

2689. Ogawa, T., Y. Kusumoto, H. Kiyono, J. R. McGhee, and S. Hamada. 1992. Occurrence of antigen-specific B cells following oral or parenteral immunization with *Porphyromonas gingivalis* fimbriae. *Int. Immunol.* **4:**1003–1010.

2690. Ogawa, T., H. Ogo, and S. Hamada. 1994. Chemotaxis of human monocytes by synthetic peptides that mimic segments of *Porphyromonas gingivalis* fimbrial protein. *Oral Microbiol. Immunol.* **9:**257–261.

2691. Ogawa, T., H. Ogo, H. Uchida, and S. Hamada. 1994. Humoral and cellular immune responses to the fimbriae of *Porphyromonas gingivalis* and their synthetic peptides. *J. Med. Microbiol.* **40:**397–402.

2692. Ogawa, T., H. Uchida, and S. Hamada. 1994. *Porphyromonas gingivalis* fimbriae and their synthetic peptides induce proinflammatory cytokines in human peripheral blood monocyte cultures. *FEMS Microbiol. Lett.* **116:**237–242.

2693. Ogawa, T., H. Uchida, and K. Yasuda. 1995. Mapping of murine Th1 and Th2 helper T-cell epitopes on fimbriae from *Porphyromonas gingivalis*. *J. Med. Microbiol.* **42:**165–170.

2694. Ogawa, T., and H. Uchida. 1995. A peptide, ALTTE, within the fimbrial subunit protein from *Porphyromonas gingivalis,* induces production of interleukin 6, gene expression and protein phosphorylation in human peripheral blood mononuclear cells. *FEMS Immunol. Med. Microbiol.* **11:**197–205.

2695. Ogawa, T., K. Yasuda, K. Yamada, H. Mori, K. Ochiai, and M. Hasegawa. 1995. Immunochemical characterisation and epitope mapping of a novel fimbrial protein (Pg-II fimbria) of *Porphyromonas gingivalis*. *FEMS Immunol. Med. Microbiol.* **11:**247–255.

2696. Ogawa, T. 1994. The potential protective immune responses to synthetic peptides containing conserved epitopes of *Porphyromonas gingivalis* fimbrial protein. *J. Med. Microbiol.* **41:**349–358.

2697. Ogawa, T., and S. Hamada. 1994.

Hemagglutinating and chemotactic properties of synthetic peptide segments of fimbrial protein from *Porphyromonas gingivalis*. *Infect. Immun.* **62:**3305–3310.

2698. **Ogierman, M. A., and P. A. Manning.** 1992. Homology of TcpN, a putative regulatory protein of *Vibrio cholerae,* to the AraC family of transcriptional activators. *Gene* **116:**93–97.

2699. **Ogierman, M. A., and P. A. Manning.** 1992. TCP pilus biosynthesis in *Vibrio cholerae* O1: gene sequence of *tcpC* encoding an outer membrane lipoprotein. *FEMS Microbiol. Lett.* **76:**179–184.

2700. **Ogierman, M. A., A. W. Paton, and J. C. Paton.** 2000. Up-regulation of both intimin- and *eae*-independent adherence of Shiga toxigenic *Escherichia coli* O157 by *ler* and phenotypic impact of a naturally occurring *ler* mutation. *Infect. Immun.* **68:**5344–5353.

2701. **Ogierman M. A., E. Voss, C. Meaney, R. Faast, S. R. Attridge, and P. A. Manning.** 1996. Comparison of the promoter proximal regions of the toxin-coregulated *tcp* gene cluster in classical and El Tor strains of *Vibrio cholerae* O1. *Gene* **170:**9–16.

2702. **Ogierman, M. A., S. Zabihi, L. Mourtzios, and P. A. Manning.** 1993. Genetic organization and sequence of the promoter-distal region of the tcp gene cluster of *Vibrio cholerae. Gene* **126:**51–60.

2703. **Ogle, K. F., K. K. Lee, and D. C. Krause.** 1992. Nucleotide sequence analysis reveals novel features of the phase-variable cytadherence accessory protein HMW3 of *Mycoplasma pneumoniae. Infect. Immun.* **60:**1633–1641.

2704. **O'Gorman, L. E., E. O. Krejany, V. R. Bennett-Wood, and R. M. Robins-Browne.** 1996. Transfer of attaching and effacing from a strain of enteropathogenic *Escherichia coli* to *E. coli* K-12. *Microbiol. Res.* **151:**379–385.

2705. **Ogunniyi, A. D., I. Kotlarski, R. Morona, and P. A. Manning.** 1997. Role of SefA subunit protein of SEF14 fimbriae in the pathogenesis of *Salmonella enterica* serovar Enteritidis. *Infect. Immun.* **65:**708–717.

2706. **Ogunniyi, A. D., P. A. Manning, and I. Kotlarski.** 1994. A *Salmonella enteritidis* 11RX pilin induces strong T-lymphocyte responses. *Infect. Immun.* **62:**5376–5383.

2707. **Ohman, L., K. Tullus, M. Katouli, L. G. Burman, and O. Stendahl.** 1995. Correlation between susceptibility of infants to infections and interaction with neutrophils of *Escherichia coli* strains causing neonatal and

infantile septicemia. *J. Infect. Dis.* **171:** 128–133.

2708. **Oho, T., H. Yu, Y. Yamashita, and T. Koga.** 1998. Binding of salivary glycoprotein-secretory immunoglobulin A complex to the surface protein antigen of *Streptococcus mutans. Infect. Immun.* **66:**115–121.

2709. **Ojeniyi, B., P. Ahrens, and A. Meyling.** 1994. Detection of fimbrial and toxin genes in *Escherichia coli* and their prevalence in piglets with diarrhoea. The application of colony hybridization assay, polymerase chain reaction and phenotypic assays. *Zentbl. Veterinaermed. Reihe B* **41:**49–59.

2710. **Okada, N., A. P. Pentland, P. Falk, and M. G. Caparon.** 1994. M protein and protein F act as important determinants of cell-specific tropism of *Streptococcus pyogenes* in skin tissue. *J. Clin. Investig.* **94:**965–977.

2711. **Okada, N., I. Tatsuno, E. Hanski, M. Caparon, and C. Sasakawa.** 1998. *Streptococcus pyogenes* protein F promotes invasion of HeLa cells. *Microbiology* **144:** 3079–3086.

2712. **Okada, N., M. K. Liszewski, J. P. Atkinson, and M. Caparon.** 1995. Membrane cofactor protein (CD46) is a keratinocyte receptor for the M protein of the group A streptococcus. *Proc. Natl. Acad. Sci. USA* **92:**2489–2493.

2713. **Okada, N., M. Watarai, V. Ozeri, E. Hanski, M. Caparon, and C. Sasakawa.** 1997. A matrix form of fibronectin mediates enhanced binding of *Streptococcus pyogenes* to host tissue. *J. Biol. Chem.* **272:**26978–82694.

2714. **Okada, Y., M. Kanoe, Y. Yaguchi, T. Watanabe, H. Ohmi, and K. Okamoto.** 1999. Adherence of *Fusobacterium necrophorum* subspecies *necrophorum* to different animal cells. *Microbios* **99:**95–104.

2715. **Okada, Y., K. Kitada, M. Takagaki, H. O. Ito, and M. Inoue.** 2000. Endocardiac infectivity and binding to extracellular matrix proteins of oral *Abiotrophia* species. *FEMS Immunol. Med. Microbiol.* **27:**257–261.

2716. **Okamoto, M., N. Maeda, K. Kondo, and K. P. Leung.** 1999. Hemolytic and hemagglutinating activities of *Prevotella intermedia* and *Prevotella nigrescens. FEMS Microbiol. Lett.* **178:**299–304.

2717. **Okte, E., N. Sultan, B. Dogan, and S. Asikainen.** 1999. Bacterial adhesion of *Actinobacillus actinomycetemcomitans* serotypes to titanium implants: SEM evaluation. A preliminary report. *J. Periodontol.* **70:**1376–1382.

2718. **Oligino, L., and P. Fives-Taylor.** 1993. Overexpression and purification of a fimbria-associated adhesin of *Streptococcus parasanguis*. *Infect. Immun.* **61:**1016–1022.

2719. **Olmsted, S. B., G. M. Dunny, S. L. Erlandsen, and C. L. Wells.** 1994. A plasmid-encoded surface protein on *Enterococcus faecalis* augments its internalization by cultured intestinal epithelial cells. *J. Infect. Dis.* **170:**1549–1556.

2720. **Olmsted, S. B., and N. I. Norcross.** 1992. Effect of specific antibody on adherence of *Staphylococcus aureus* to bovine mammary epithelial cells. *Infect. Immun.* **60:**249–256.

2721. **Olsen, A., A. Arnqvist, M. Hammar, and S. Normark.** 1993. Environmental regulation of curli production in *Escherichia coli*. *Infect. Agents Dis.* **2:**272–274.

2722. **Olsen, A., A. Arnqvist, M. Hammar, S. Sukupolvi, and S. Normark.** 1993. The RpoS sigma factor relieves H-NS-mediated transcriptional repression of *csgA*, the subunit gene of fibronectin-binding curli in *Escherichia coli*. *Mol. Microbiol.* **7:**523–536.

2723. **Olsén, A., M. J. Wick, M. Mörgelin, and L. Björck.** 1998. Curli, fibrous surface proteins of *Escherichia coli*, interact with major histocompatibility complex class I molecules. *Infect. Immun.* **66:**944–949.

2724. **Olsen, P. B., and P. Klemm.** 1994. Localization of promoters in the *fim* gene cluster and the effect of H-NS on the transcription of *fimB* and *fimE*. *FEMS Microbiol. Lett.* **116:**95–100.

2725. **Olson, J. C., E. M. McGuffie, and D. W. Frank.** 1997. Effects of differential expression of the 49-kilodalton exoenzyme S by *Pseudomonas aeruginosa* on cultured eukaryotic cells. *Infect. Immun.* **65:**248–256.

2726. **Olson, L. D., and A. A. Gilbert.** 1993. Characteristics of *Mycoplasma hominis* adhesion. *J. Bacteriol.* **175:**3224–3227.

2727. **Olsson, J., Y. van der Heijde, and K. Holmberg.** Plaque formation *in vivo* and bacterial attachment *in vitro* on permanently hydrophobic and hydrophilic surfaces. *Caries Res.* **26:**428–433.

2728. **O'Meara, T. J., J. R. Egerton, and H. W. Raadsma.** 1993. Recombinant vaccines against ovine footrot. *Immunol. Cell Biol.* **71:**473–488.

2729. **Onagawa, M., K. Ishihara, and K. Okuda.** 1994. Coaggregation between *Porphyromonas gingivalis* and *Treponema denticola*. *Bull. Tokyo Dent. Coll.* **35:**171–181.

2730. **Onaolapo, J. A., and J. O. Salami.** 1995. Effect of subminimum inhibitory concentration of ceftriaxone on adherence of *Pseudomonas aeruginosa* to inert surfaces in an experimental model. *Afr. J. Med. Med. Sci.* **24:**275–281.

2731. **Ondarza, M. A., and F. Sotelo.** 1996. Neutral glycolipids in adult rabbit blood and analysis of their function as specific receptors for micro-organisms. *Biomed. Chromatogr.* **10:**6–10.

2732. **Onoe, T., C. I. Hoover, K. Nakayama, T. Ideka, H. Nakamura, and F. Yoshimura.** 1995. Identification of *Porphyromonas gingivalis* prefimbrilin possessing a long leader peptide: possible involvement of trypsin-like protease in fimbrilin maturation. *Microb. Pathog.* **19:**351–364.

2733. **Ooshima, T., M. Matsumura, T. Hoshino, S. Kawabata, S. Sobue, and T. Fujiwara.** 2001. Contributions of three glycosyltransferases to sucrose-dependent adherence of *Streptococcus mutans*. *J. Dent. Res.* **80:**1672–1677.

2734. **Opitz, O., and E. Jacobs.** 1992. Adherence epitopes of *Mycoplasma genitalium* adhesin. *J. Gen. Microbiol.* **138:**1785–1790.

2735. **Orden, J. A., J. A. Ruiz-Santa-Quiteria, D. Cid, S. Garcia, and R. de la Fuente.** 1999. Prevalence and characteristics of necrotoxigenic *Escherichia coli* (NTEC) strains isolated from diarrhoeic dairy calves. *Vet. Microbiol.* **66:**265–273.

2736. **Orenstein, J. M., and D. P. Kotler.** 1995. Diarrheogenic bacterial enteritis in acquired immune deficiency syndrome: a light and electron microscopy study of 52 cases. *Hum. Pathol.* **26:**481–492.

2737. **Orihuela, C. J., R. Janssen, C. W. Robb, D. A. Watson, and D. W. Niesel.** 2000. Peritoneal culture alters *Streptococcus pneumoniae* protein profiles and virulence properties. *Infect. Immun.* **68:**6082–6086.

2738. **Ormonde, P., P. Horstedt, R. O'Toole, and D. L. Milton.** 2000. Role of motility in adherence to and invasion of a fish cell line by *Vibrio anguillarum*. *J. Bacteriol.* **182:**2326–2328.

2739. **O'Rourke, J., A. Lee, and J. G. Fox.** 1992. An ultrastructural study of *Helicobacter mustelae* and evidence of a specific association with gastric mucosa. *J. Med. Microbiol.* **36:**420–427.

2740. **Ortega-Barria, E., and M. E. Pereira.** 1992. Identification of a lectin activity in *Pneumocystis carinii*. *Trop. Med. Parasitol.* **43:**186–190.

2741. **Osaki, T., H. Yamaguchi, H. Taguchi, J. Kumada, S. Ogata, and S. Kamiya.** 1997. Studies on the relationship between adhesive activity and haemagglutination by *Helicobacter pylori*. *J. Med. Microbiol.* **46:**117–121.

2742. Osaki, T., H. Yamaguchi, H. Taguchi, M. Fukuda, H. Kawakami, H. Hirano, S. Watanabe, A. Takagi, and S. Kamiya. 1998. Establishment and characterisation of a monoclonal antibody to inhibit adhesion of *Helicobacter pylori* to gastric epithelial cells. *J. Med. Microbiol.* **47:**505–512.

2743. Osek, J., G. Jonson, A.-M. Svennerholm, and J. Holmgren. 1994. Role of antibodies against biotype-specific *Vibrio cholerae* pili in protection against experimental classical and El Tor cholera. *Infect. Immun.* **62:**2901–2907.

2744. Osek, J. 1999. Prevalence of virulence factors of *Escherichia coli* strains isolated from diarrheic and healthy piglets after weaning. *Vet. Microbiol.* **68:**209–217.

2745. Osek, J., M. Truszczynski, K. Tarasiuk, and Z. Pejsak. 1995. Evaluation of different vaccines to control of pig colibacillosis under large-scale farm conditions. *Comp. Immunol. Microbiol. Infect. Dis.* **18:**1–8.

2746. Osek, J., A.-M. Svennerholm, and J. Holmgren. 1992. Protection against *Vibrio cholerae* El Tor infection by specific antibodies against mannose-binding hemagglutinin pili. *Infect. Immun.* **60:**4961–4964.

2747. Osek, J., and M. Truszczynski. 1992. Occurrence of fimbriae and enterotoxins in *Escherichia coli* strains isolated from piglets in Poland. *Comp. Immunol. Microbiol. Infect. Dis.* **15:**285–292.

2748. Osset, J., R. M. Bartolome, E. Garcia, and A. Andreu. 2001. Assessment of the capacity of *Lactobacillus* to inhibit the growth of uropathogens and block their adhesion to vaginal epithelial cells. *J. Infect. Dis.* **183:**485–491.

2749. Ota, H., T. Katsuyama, S. Nakajima, H. El-Zimaity, J. G. Kim, D. Y. Graham, and R. M. Genta. 1998. Intestinal metaplasia with adherent *Helicobacter pylori:* a hybrid epithelium with both gastric and intestinal features. *Hum. Pathol.* **29:**846–850.

2750. Otoi, T., S. Tachikawa, S. Kondo, and T. Suzuki. 1992. Effect of antibiotics treatment of *in vitro* fertilized bovine embryos to remove adhering bacteria. *J. Vet. Med. Sci.* **54:**763–765.

2751. Otoi, T., S. Tachikawa, S. Kondo, and T. Suzuki. 1993. Effect of washing, antibiotics and trypsin treatment of bovine embryos on the removal of adhering K99+ *Escherichia coli. J. Vet. Med. Sci.* **55:**1053–1055.

2752. O'Toole, P. W., L. Janzon, P. Doig, J. Huang, M. Kostrzynska, and T. J. Trust. 1995. The putative neuraminyllactose-binding hemagglutinin HpaA of *Helicobacter pylori*

2753. O'Toole, G. A., and R. Kolter. 1998. Flagellar and twitching motility are necessary for *Pseudomonas aeruginosa* biofilm development. *Mol. Microbiol.* **30:**295–304.

2754. Otto, K., J. Norbeck, T. Larsson, K.-A. Karlsson, and M. Hermansson. 2001. Adhesion of type 1-fimbriated *Escherichia coli* to abiotic surfaces leads to altered composition of outer membrane proteins. *J. Bacteriol.* **183:**2445–2453.

2755. Ouadia, A., Y. Karamanos, and R. Julien. 1992. Detection of the ganglioside *N*-glycolyl-neuraminyl-lactosyl-ceramide by biotinylated *Escherichia coli* K99 lectin. *Glycoconj. J.* **9:**21–26.

2756. Ouwehand, A. C., E. M. Tuomola, S. Tolkko, and S. Salminen. 2001. Assessment of adhesion properties of novel probiotic strains to human intestinal mucus. *Int. J. Food Microbiol.* **64:**119–126.

2757. Ouwehand, A. C., and P. L. Conway. 1996. Purification and characterization of a component produced by *Lactobacillus fermentum* that inhibits the adhesion of K88 expressing *Escherichia coli* to porcine ileal mucus. *J. Appl. Bacteriol.* **80:**311–318.

2758. Ouwehand, A. C., E. Isolauri, P. V. Kirjavainen, and S. J. Salminen. 1999. Adhesion of four *Bifidobacterium* strains to human intestinal mucus from subjects in different age groups. *FEMS Microbiol. Lett.* **172:**61–64.

2759. Ouwehand, A. C., E. Isolauri, P. V. Kirjavainen, S. Tolkko, and S. J. Salminen. 2000. The mucus binding of *Bifidobacterium lactis* Bb12 is enhanced in the presence of *Lactobacillus* GG and *Lact. delbrueckii* subsp. *bulgaricus. Lett. Appl. Microbiol.* **30:**10–13.

2760. Ouwehand, A. C., S. Tolkko, J. Kulmala, S. Salminen, and E. Salminen. 2000. Adhesion of inactivated probiotic strains to intestinal mucus. *Lett. Appl. Microbiol.* **31:**82–86.

2761. Ouwehand, A. C., P. L. Conway, and S. J. Salminen. 1995. Inhibition of S-fimbriae-mediated adhesion to human ileostomy glycoproteins by a protein isolated from bovine colostrum. *Infect. Immun.* **63:**4917–4920.

2762. Ouwehand, A. C., and S. J. Salminen. 1998. Adhesion inhibitory activity of beta-lactoglobulin isolated from infant formulae. *Acta Paediatr.* **87:**491–493.

2763. Owen, R. J., M. Desai, N. Figura, B. F. Bayeli, L. Di Gregorio, M. Russi, and R.

A. Musmanno. 1993. Comparisons between degree of histological gastritis and DNA fingerprints, cytotoxicity and adhesivity of *Helicobacter pylori* from different gastric sites. *Eur. J. Epidemiol.* **9:**315–321.

2764. Oyofo, B. A., S. H. el-Etr, M. O. Wasfy, L. Peruski, B. Kay, M. Mansour, and J. R. Campbell. 1995. Colonization factors of enterotoxigenic *E. coli* (ETEC) from residents of northern Egypt. *Microbiol. Res.* **150:**429–436.

2765. Ozaki, C. K., M. D. Phaneuf, M. J. Bide, W. C. Quist, J. M. Alessi, and F. W. LoGerfo. 1993. *In vivo* testing of an infection-resistant vascular graft material. *J. Surg. Res.* **55:**543–547.

2766. Ozeri, V., I. Rosenshine, A. Ben-Ze'Ev, G. M. Bokoch, T. S. Jou, and E. Hanski. 2001. De novo formation of focal complex-like structures in host cells by invading streptococci. *Mol. Microbiol.* **41:**561–573.

2767. Pace, J. L., and J. E. Galán. 1994. Measurement of free intracellular calcium levels in epithelial cells as consequence of bacterial invasion. *Methods Enzymol.* **236:**482–490.

2768. Pacheco-Soares, C., L. C. Gaziri, W. Loyola, and I. Felipe. 1992. Phagocytosis of enteropathogenic *Escherichia coli* and *Candida albicans* by lectin-like receptors. *Brazil. J. Med. Biol. Res.* **25:**1015–1024.

2769. Padilla-Vaca, F., S. Ankri, R. Bracha, L. A. Koole, and D. Mirelman. 1999. Down regulation of *Entamoeba histolytica* virulence by monoxenic cultivation with *Escherichia coli* O55 is related to a decrease in expression of the light (35-kilodalton) subunit of the Gal/GalNAc lectin. *Infect. Immun.* **67:**2096–2102.

2770. Paerregaard, A. 1992. Interactions between *Yersinia enterocolitica* and the host with special reference to virulence plasmid encoded adhesion and humoral immunity. *Dan. Med. Bull.* **39:**155–172.

2771. Pagé, D., and R. Roy. 1996. Synthesis of divalent α-D-mannopyranosylated clusters having enhanced binding affinities towards concanavalin A and pea lectins. *J. Bioorg. Med. Chem. Lett.* **6:**1765–1770.

2772. Pal, S., D. Sasmal, B. Guhathakurta, R. Mallick, and A. Datta. 1995. Haemagglutinating property & cell surface hydrophobicity of *Vibrio cholerae* O139. *Indian J. Med. Res.* **101:**175–178.

2773. Pal, U., A. M. de Silva, R. R. Montgomery, D. Fish, J. Anguita, J. F. Anderson, Y. Lobet, and E. Fikrig. 2000. Attachment of *Borrelia burgdorferi* within *Ixodes*

scapularis mediated by outer surface protein A. *J. Clin. Investig.* **106:**561–569.

2774. Palenik, C. J., M. J. Behnen, J. C. Setcos, and C. H. Miller. 1992. Inhibition of microbial adherence and growth by various glass ionomers *in vitro*. *Dent. Mater.* **8:**16–20.

2775. Pallesen, L., L. K. Poulsen, G. Christiansen, and P. Klemm. 1995. Chimeric FimH adhesin of type 1 fimbriae: a bacterial surface display system for heterologous sequences. *Microbiology* **141:**2839–2848.

2776. Palma, M., A. Haggar, and J. I. Flock. 1999. Adherence of *Staphylococcus aureus* is enhanced by an endogenous secreted protein with broad binding activity. *J. Bacteriol.* **181:**2840–2845.

2777. Palma, M., S. Nozohoor, T. Shennings, A. Heimdahl, and J.-I. Flock. 1996. Lack of the extracellular 19-kilodalton fibrinogen-binding protein from *Staphylococcus aureus* decreases virulence in experimental wound infection. *Infect. Immun.* **64:**5284–5289.

2778. Palmeira, P., S. B. Carbonare, M. L. Silva, L. R. Trabulsi, and M. M. Carneiro-Sampaio. 2001. Inhibition of enteropathogenic *Escherichia coli* (EPEC) adherence to HEp-2 cells by bovine colostrum and milk. *Allergol. Immunopathol.* **29:**229–237.

2779. Palmer, K. L., and R. S. Munson, Jr. 1992. Construction of chimaeric genes for mapping a surface-exposed epitope on the pilus of non-typable *Haemophilus influenzae* strain M37. *Mol. Microbiol.* **6:**2583–2588.

2780. Palmer, L. M., T. J. Reilly, S. J. Utsalo, and M. S. Donenberg. 1997. Internalization of *Escherichia coli* by human renal epithelial cells is associated with tyrosine phosphorylation of specific host cell proteins. *Infect. Immun.* **65:**2570–2575.

2781. Palmer, R. J., Jr., and D. C. White. 1997. Developmental biology of biofilms: implications for treatment and control. *Trends Microbiol.* **5:**435–440.

2782. Palomar, J., A. M. Leranoz, and M. Vinas. 1995. *Serratia marcescens* adherence: the effect of O-antigen presence. *Microbios* **81:**107–113.

2783. Palovuori, R., A. Perttu, Y. Yan, R. Karttunen, S. Eskelinen, and T. J. Karttunen. 2000. *Helicobacter pylori* induces formation of stress fibers and membrane ruffles in AGS cells by *rac* activation. *Biochem. Biophys. Res. Commun.* **269:**247–253.

2784. Pan, Y. T., B. Xu, K. Rice, S. Smith, R. Jackson, and A. D. Elbein. 1997. Specificity of the high-mannose recognition site between *Enterobacter cloacae* pili adhesin and

HT-29 cell membranes. *Infect. Immun.* **65**:4199–4206.

2785. **Pandiripally, V. K., D. G. Westbrook, G. R. Sunki, and A. K. Bhunia.** 1999. Surface protein p104 is involved in adhesion of *Listeria monocytogenes* to human intestinal cell line, Caco-2. *J. Med. Microbiol.* **48**:117–124.

2786. **Panigrahi, P., S. Gupta, I. H. Gewolb, and J. G. Morris, Jr.** 1994. Occurrence of necrotizing enterocolitis may be dependent on patterns of bacterial adherence and intestinal colonization: studies in Caco-2 tissue culture and weanling rabbit models. *Pediatr. Res.* **36**:115–121.

2787. **Panjwani, N., Z. Zhao, M. B. Raizman, and F. Jungalwala.** 1996. Pathogenesis of corneal infection: binding of *Pseudomonas aeruginosa* to specific phospholipids. *Infect. Immun.* **64**:1819–1825.

2788. **Papadogiannakis, N., R. Willen, B. Carlen, S. Sjostedt, T. Wadström, and A. Gad.** 2000. Modes of adherence of *Helicobacter pylori* to gastric surface epithelium in gastroduodenal disease: a possible sequence of events leading to internalisation. *APMIS* **108**:439–447.

2789. **Papazisi, L., K. E. Troy, T. S. Gorton, X. Liao, and S. J. Geary.** 2000. Analysis of cytadherence-deficient, GapA-negative *Mycoplasma gallisepticum* strain R. *Infect. Immun.* **68**:6643–6649.

2790. **Paradis, S.-É., J. D. Dubreuil, M. Gottschalk, M. Archambault, and M. Jacques.** 1999. Inhibition of adherence of *Actinobacillus pleuropneumoniae* to porcine respiratory tract cells by monoclonal antibodies directed against LPS and partial characterization of the LPS receptors. *Curr. Microbiol.* **39**:313–320.

2791. **Paradis, S.-É., D. Dubreuil, S. Rioux, M. Gottschalk, and M. Jacques.** 1994. High-molecular-mass lipopolysaccharides are involved in *Actinobacillus pleuropneumoniae* adherence to porcine respiratory tract cells. *Infect. Immun.* **62**:3311–3319.

2792. **Paranjpye, R. N., J. C. Lara, J. C. Pepe, C. M. Pepe, and M. S. Strom.** 1998. The type IV leader peptidase/N-methyltransferase of *Vibrio vulnificus* controls factors required for adherence to HEp-2 cells and virulence in iron-overloaded mice. *Infect. Immun.* **66**:5659–5668.

2793. **Parge, H. E., K. T. Forest, M. J. Hickey, D. A. Christensen, E. D. Getzoff, and J. A. Tainer.** 1995. Structure of the fibre-forming protein pilin at 2.6 Å resolution. *Nature* **378**:32–38.

2794. **Park, J. H., K. B. Lee, I. C. Kwon, and Y. H. Bae.** 2001. PDMS-based poly-urethanes with MPEG grafts: mechanical properties, bacterial repellency, and release behavior of rifampicin. *J. Biomater. Sci. Polym. Ed.* **12**:629–645.

2795. **Park, K. D., Y. S. Kim, D. K. Han, Y. H. Kim, E. H. Lee, H. Suh, and K. S. Choi.** 1998. Bacterial adhesion on PEG modified polyurethane surfaces. *Biomaterials* **19**:851–859.

2796. **Park, Y., and R. J. Lamont.** 1998. Contact-dependent protein secretion in *Porphyromonas gingivalis*. *Infect. Immun.* **66**:4777–4782.

2797. **Parker, C. T., A. W. Kloser, C. A. Schnaitman, M. A. Stein, S. Gottesman, and B. W. Gibson.** 1992. Role of the *rfaG* and *rfaP* genes in determining the lipopolysaccharide core structure and cell surface properties of *Escherichia coli* K-12. *J. Bacteriol.* **174**:2525–2538.

2798. **Parment, P. A., M. Gabriel, G. W. Bruse, S. Stegall, and D. G. Ahearn.** 1993. Adherence of *Serratia marcescens*, *Serratia liquefaciens*, *Pseudomonas aeruginosa* and *Staphylococcus epidermidis* to blood transfusion bags (CPD-SAGMAN sets). *Scand. J. Infect. Dis.* **25**:721–724.

2799. **Parment, P. A.** 1997. The role of *Serratia marcescens* in soft contact lens associated ocular infections. A review. *Acta Ophthalmol. Scand.* **75**:67–71.

2800. **Parment, P. A., C. Svanborg-Edén, M. J. Chaknis, A. D. Sawant, L. Hagberg, L. A. Wilson, and D. G. Ahearn.** 1992. Hemagglutination (fimbriae) and hydrophobicity in acherence of *Serratia marcescens* to urinary tract epithelium and contact lenses. *Curr. Microbiol.* **25**:113–118.

2801. **Parsons, L. M., R. J. Limberger, and M. Shayegani.** 1997. Alterations in levels of DnaK and GroEL result in diminished survival and adherence of stressed *Haemophilus ducreyi*. *Infect. Immun.* **65**:2413–2419.

2802. **Partridge, S. R., M. S. Baker, M. J. Walker, and M. R. Wilson.** 1996. Clusterin, a putative complement regulator, binds to the cell surface of *Staphylococcus aureus* clinical isolates. *Infect. Immun.* **64**:4324–4329.

2803. **Parveen, N., and J. M. Leong.** 2000. Identification of a candidate glycosaminoglycan-binding adhesin of the Lyme disease spirochete *Borrelia burgdorferi*. *Mol. Microbiol.* **35**:1220–1234.

2804. **Parveen, N., D. Robbins, and J. M. Leong.** 1999. Strain variation in glycosaminoglycan recognition influences cell-type-specific binding by lyme disease spirochetes. *Infect. Immun.* **67:**1743–1749.

2805. **Pascual, A., E. Ramirez de Arellano, L. Martinez Martinez, and E. J. Perea.** 1993. Effect of polyurethane catheters and bacterial biofilms on the *in-vitro* activity of antimicrobials against *Staphylococcus epidermidis. J. Hosp. Infect.* **24:**211–218.

2806. **Patel, J., H. Faden, S. Sharma, and P. L. Ogra.** 1992. Effect of respiratory syncytial virus on adherence, colonization and immunity of non-typable *Haemophilus influenzae:* implications for otitis media. *Int. J. Pediatr. Otorhinolaryngol.* **23:**15–23.

2807. **Paton, A. W., and J. C. Paton.** 1999. Molecular characterization of the locus encoding biosynthesis of the lipopolysaccharide O antigen of *Escherichia coli* serotype O113. *Infect. Immun.* **67:**5930–5937.

2808. **Paton, A. W., E. Voss, P. A. Manning, and J. C. Paton.** 1997. Shiga toxin-producing *Escherichia coli* isolates from cases of human disease show enhanced adherence to intestinal epithelial (Henle 407) cells. *Infect. Immun.* **65:**3799–3805.

2809. **Paton, A. W., E. Voss, P. A. Manning, and J. C. Paton.** 1998. Antibodies to lipopolysaccharide block adherence of Shiga toxin-producing *Escherichia coli* to human intestinal epithelial (Henle 407) cells. *Microb. Pathog.* **24:**57–63.

2810. **Patrick, S., J. P. McKenna, S. O'Hagan, and E. Dermott.** 1996. A comparison of the haemagglutinating and enzymic activities of *Bacteroides fragilis* whole cells and outer membrane vesicles. *Microb. Pathog.* **20:**191–202.

2811. **Patti, J. M., B. L. Allen, M. J. McGavin, and M. Höök.** 1994. MSCRAMM-mediated adherence of microorganisms to host tissues. *Annu. Rev. Microbiol.* **48:**585–617.

2812. **Patti, J. M., and M. Höök.** 1994. Microbial adhesins recognizing extracellular matrix macromolecules. *Curr. Opin. Cell Biol.* **6:**752–758.

2813. **Patti, J. M., T. Bremell, D. Krajewska-Pietrasik, A. Abdelnour, A. Tarkowski, C. Rydén, and M. Höök.** 1994. The *Staphylococcus aureus* collagen adhesin is a virulence determinant in experimental septic arthritis. *Infect. Immun.* **62:**152–161.

2814. **Paul, M., T. Tsukamoto, A. R. Ghosh, S. K. Bhattacharya, B. Manna, S. Chakrabarti, G. B. Nair, D. A. Sack, D. Sen, and Y. Takeda.** 1994. The significance of enteroaggregative *Escherichia coli* in the etiology of hospitalized diarrhoea in Calcutta, India

and the demonstration of a new honeycombed pattern of aggregative adherence. *FEMS Microbiol. Lett.* **117:**319–325.

2815. **Paulsson, M., I. Gouda, O. Larm, and Å. Ljungh.** 1994. Adherence of coagulase-negative staphylococci to heparin and other glycosaminoglycans immobilized on polymer surfaces. *J. Biomed. Mater. Res.* **28:**311–317.

2816. **Paulsson, M., M. Kober, C. Freij-Larsson, M. Stollenwerk, B. Wesslen, and Å. Ljungh.** 1993. Adhesion of staphylococci to chemically modified and native polymers, and the influence of preadsorbed fibronectin, vitronectin and fibrinogen. *Biomaterials* **14:**845–853.

2817. **Paulsson, M., Å Ljungh, and T. Wadström.** 1994. Inhibition of lactoferrin and vitronectin binding to *Staphylococcus aureus* by heparin. *Curr. Microbiol.* **29:**113–117.

2818. **Paulsson, M., A. C. Petersson, and Å. Ljungh.** 1993. Serum and tissue protein binding and cell surface properties of *Staphylococcus lugdunensis. J. Med. Microbiol.* **38:**96–102.

2819. **Pavloff, N., P. A. Pemberton, J. Potempa, W. C. Chen, R. N. Pike, V. Prochazka, M. C. Kiefer, J. Travis, and P. J. Barr.** 1997. Molecular cloning and characterization of *Porphyromonas gingivalis* lysine-specific gingipain. A new member of an emerging family of pathogenic bacterial cysteine proteinases. *J. Biol. Chem.* **272:**1595–1600.

2820. **Payne, D., M. O'Reilly, and D. Williamson.** 1993. The K88 fimbrial adhesin of enterotoxigenic *Escherichia coli* binds to β1-linked galactosyl residues in glycosphingolipids. *Infect. Immun.* **61:**3673–3677.

2821. **Payne, D.** 1994. A study of K88-mediated haemagglutination by enterotoxigenic *Escherichia coli* (ETEC). *New Microbiol.* **17:**99–110.

2822. **Payne, D., D. Tatham, E. D. Williamson, and R. W. Titball.** 1998. The pH antigen of *Yersinia pestis* binds to β1-linked galactosyl residues in glycosphingolipids. *Infect. Immun.* **66:**4545–4548.

2823. **Payne, D., M. O'Reilly, and D. Williamson.** 1993. The K88 fimbrial adhesin of enterotoxigenic *Escherichia coli* binds to β1-linked galactosyl residues in glycosphingolipids. *Infect. Immun.* **61:**3673–3677.

2824. **Peacock, S. J., N. P. Day, M. G. Thomas, A. R. Berendt, and T. J. Foster.** 2000. Clinical isolates of *Staphylococcus aureus* exhibit diversity in *fnb* genes and adhesion to human fibronectin. *J. Infect.* **41:**23–31.

2825. **Peacock, S. J., T. J. Foster, B. J.**

Cameron, and A. R. Berendt. 1999. Bacterial fibronectin-binding proteins and endothelial cell surface fibronectin mediate adherence of *Staphylococcus aureus* to resting human endothelial cells. *Microbiology* 145:3477–3486.

2826. Peake, P., A. Gooley, and W. J. Britton. 1993. Mechanism of interaction of the 85B secreted protein of *Mycobacterium bovis* with fibronectin. *Infect. Immun.* 61:4828–4834.

2827. Pearce, A. M., R. N. Seabrook, L. I. Irons, L. A. Ashworth, T. Atkinson, and A. Robinson. 1994. Localization of antigenic domains on the major subunits of *Bordetella pertussis* serotype 2 and 3 fimbriae. *Microbiology* 140:205–211.

2828. Pece, S., A. B. Maffione, R. Petruzzelli, B. Greco, G. Giuliani, M. R. Partipilo, S. Amarri, F. Schettini, E. Jirillo, and D. Fumarola. 1994. *Rochalimaea henselae* organisms possess an elevated capacity of binding to peripheral blood lymphocytes from patients with cat scratch disease. *Microbios* 77:95–100.

2829. Pederson, K. J., and D. E. Pierson. 1995. Ail expression in *Yersinia enterocolitica* is affected by oxygen tension. *Infect. Immun.* 63:4199–4201.

2830. Pei, L., and J. I. Flock. 2001. Functional study of antibodies against a fibrogenin-binding protein in *Staphylococcus epidermidis* adherence to polyethylene catheters. *J. Infect. Dis.* 184:52–55.

2831. Pei, L., and J. I. Flock. 2001. Lack of *fbe*, the gene for a fibrinogen-binding protein from *Staphylococcus epidermidis*, reduces its adherence to fibrinogen coated surfaces. *Microb. Pathog.* 31:185–193.

2832. Pei, L., M. Palma, M. Nilsson, B. Guss, and J. I. Flock. 1999. Functional studies of a fibrinogen binding protein from *Staphylococcus epidermidis*. *Infect. Immun.* 67:4525–4530.

2833. Pei, Z., C. Burucoa, B. Grignon, S. Baqar, X.-Z. Huang, D. J. Kopecko, A. L. Bourgeois, J.-L. Fauchere, and M. J. Blaser. 1998. Mutation in the *peb1A* locus of *Campylobacter jejuni* reduces interactions with epithelial cells and intestinal colonization of mice. *Infect. Immun.* 66:938–943.

2834. Peiffer, I., M. F. Bernet-Camard, M. Rousset, and A. L. Servin. 2001. Impairments in enzyme activity and biosynthesis of brush border-associated hydrolases in human intestinal Caco-2/TC7 cells infected by members of the Afa/Dr family of diffusely adhering *Escherichia coli*. *Cell. Microbiol.* 3:341–357.

2835. Peiffer, I., A. B. Blanc-Potard, M.-F.

2835. Bernet-Camard, J. Guignot, A. Barbat, and A. L. Servin. 2000. Afa/Dr diffusely adhering *Escherichia coli* C1845 infection promotes selective injuries in the junctional domain of polarized human intestinal Caco-2/TC7 cells. *Infect. Immun.* 68:3431–3442.

2836. Peiffer, I., J. Guignot, A. Barbat, C. Carnoy, S. L. Moseley, B. J. Nowicki, A. L. Servin, and M.-F. Bernet-Camard. 2000. Structural and functional lesions in brush border of human polarized intestinal Caco-2/TC7 cells infected by members of the Afa/Dr diffusely adhering family of *Escherichia coli*. *Infect. Immun.* 68:5979–5990.

2837. Peiffer, I., A. L. Servin, and M.-F. Bernet-Camard. 1998. Piracy of decay-accelerating factor (CD55) signal transduction by the diffusely adhering strain *Escherichia coli* C1845 promotes cytoskeletal F-actin rearrangements in cultured human intestinal INT407 cells. *Infect. Immun.* 66:4036–4042.

2838. Peiser, L., P. J. Gough, T. Kodama, and S. Gordon. 2000. Macrophage class A scavenger receptor-mediated phagocytosis of *Escherichia coli*: role of cell heterogeneity, microbial strain, and culture conditions in vitro. *Infect. Immun.* 68:1953–1963.

2839. Pelayo, J. S., I. C. Scaletsky, M. Z. Pedroso, V. Sperandio, J. A. Giron, G. Frankel, and L. R. Trabulsi. 1999. Virulence properties of atypical EPEC strains. *J. Med. Microbiol.* 48:41–49.

2840. Pelicic, V., J. M. Reyrat, L. Sartori, C. Pagliaccia, R. Rappuoli, J. L. Telford, C. Montecucco, and E. Papini. 1999. *Helicobacter pylori* VacA cytotoxin associated with the bacteria increases epithelial permeability independently of its vacuolating activity. *Microbiology* 145:2043–2050.

2841. Peng, J. S., W. C. Tsai, and C. C. Chou. 2001. Surface characteristics of *Bacillus cereus* and its adhesion to stainless steel. *Int. J. Food Microbiol.* 65:105–111.

2842. Penteado, A. S., L. Aidar, A. F. Pestana de Castro, A. Yamada, J. R. Andrade, J. Blanco, M. Blanco, and J. E. Blanco. 2001. *eae*-Negative attaching and effacing *Escherichia coli* from piglets with diarrhea. *Res. Microbiol.* 152:75–81.

2843. Peotta, V. A., F. L. Gomes, A. C. Arnholdt, and P. E. Nagao. 2001. Human monocytes and monocyte-derived macrophage phagocytosis of serotype III group B streptococci strains. *Curr. Microbiol.* 43:64–68.

2844. Pepe, C. M., M. W. Eklund, and M. S. Strom. 1996. Cloning of an *Aeromonas hydrophila* type IV pilus biogenesis gene cluster: complementation of pilus assembly functions and characterization of a type IV leader peptidase/N-methyltransferase required for extracellular protein secretion. *Mol. Microbiol.* **19:**857–869.

2845. **Pepe, J. C., and S. Lory.** 1998. Amino acid substitutions in PilD, a bifunctional enzyme of *Pseudomonas aeruginosa.* Effect on leader peptidase and N-methyltransferase activities *in vitro* and *in vivo. J. Biol. Chem.* **273:**19120–19129.

2846. **Peralta, R. C., H. Yokoyama, Y. Ikemori, M. Kuroki, and Y. Kodama.** 1995. Passive immunisation against experimental salmonellosis in mice by orally administered hen egg-yolk antibodies specific for 14-kDa fimbriae of *Salmonella enteritidis. J. Med. Microbiol.* **41:**29–35.

2847. **Percival, R. S., P. D. Marsh, D. A. Devine, M. Rangarajan, J. Aduse-Opoku, P. Shepherd, and M. A. Curtis.** 1999. Effect of temperature on growth, hemagglutination, and protease activity of *Porphyromonas gingivalis. Infect. Immun.* **67:**1917–1921.

2848. **Perdreau-Remington, F., M. A. Sande, G. Peters, and H. F. Chambers.** 1998. The abilities of a *Staphylococcus epidermidis* wild-type strain and its slime-negative mutant to induce endocarditis in rabbits are comparable. *Infect. Immun.* **66:**2778–2781.

2849. **Pereira, S. H., M. P. Cervante, S. Bentzmann, and M. C. Plotkowski.** 1997. *Pseudomonas aeruginosa* entry into Caco-2 cells is enhanced in repairing wounded monolayers. *Microb. Pathog.* **23:**249–255.

2850. **Perez, P. F., Y. Minnaard, E. A. Disalvo, and G. L. De Antoni.** 1998. Surface properties of bifidobacterial strains of human origin. *Appl. Environ. Microbiol.* **64:**21–26.

2851. **Perez-Casal, J., N. Okada, M. G. Caparon, and J. R. Scott.** 1995. Role of the conserved C-repeat region of the M protein of *Streptococcus pyogenes. Mol. Microbiol.* **15:**907–916.

2852. **Perkins, S., E. J. Walsh, C. C. Deivanayagam, S. V. Narayana, T. J. Foster, and M. Höök.** 2001. Structural organization of the fibrinogen-binding region of the clumping factor B MSCRAMM of *Staphylococcus aureus. J. Biol. Chem.* **276:**44721–44728.

2853. **Perrone, M., L. E. Gfell, M. Fontana, and R. L. Gregory.** 1997. Antigenic characterization of fimbria preparations from *Streptococcus mutans* isolates from caries-free and caries-susceptible subjects. *Clin. Diagn. Lab. Immunol.* **4:**291–296.

2854. **Persson, C., R. Nordfelth, A. Holmstrom, S. Hakansson, R. Rosqvist, and H. Wolf-Watz.** 1995. Cell-surface-bound *Yersinia* translocate the protein tyrosine phosphatase YopH by a polarized mechanism into the target cell. *Mol. Microbiol.* **18:**135–150.

2855. **Perugini, M. R., and M. C. Vidotto.** 1996. Frequency of *pap* and *pil* operons in *Escherichia coli* strains associated with urinary infections. *Braz. J. Med. Biol. Res.* **29:**351–357.

2856. **Pessina, A., A. Raimondi, and M. G. Neri.** 1995. Different biological conditions influencing bacterial adherence assay. *J. Chemother.* **7:**8–11.

2857. **Petas, A., J. Vuopio-Varkila, A. Siitonen, T. Valimaa, M. Talja, and K. Taari.** 1998. Bacterial adherence to self-reinforced polyglycolic acid and self-reinforced polylactic acid 96 urological spiral stents *in vitro. Biomaterials* **19:**677–681.

2858. **Peters, D. J., and J. L. Benach.** 1997. *Borrelia burgdorferi* adherence and injury to undifferentiated and differentiated neural cells *in vitro. J. Infect. Dis.* **176:**470–477.

2859. **Peters, R. R., R. A. Le Dane, L. W. Douglass, and M. J. Paape.** 1992. Intramammary response to modified intramammary devices. *J. Dairy Sci.* **75:**85–95.

2860. **Petersen, A. M., J. Blom, L. P. Andersen, and K. A. Krogfelt.** 2000. Role of strain type, AGS cells and fetal calf serum in *Helicobacter pylori* adhesion and invasion assays. *FEMS Immunol. Med. Microbiol.* **29:**59–67.

2861. **Petersen, A. M., V. Fussing, H. Colding, J. Blom, A. Norgaard, L. P. Andersen, and K. A. Krogfelt.** 2000. Phenotypic and genotypic characterization of *Helicobacter pylori* from patients with and without peptic ulcer disease. *Scand. J. Gastroenterol.* **35:**359–367.

2862. **Peterson, P. K., G. Gekker, S. Hu, W. S. Sheng, W. R. Anderson, R. J. Ulevitch, P. S. Tobias, K. V. Gustafson, T. W. Molitor, and C. C. Chao.** 1995. CD14 receptor-mediated uptake of nonopsonized *Mycobacterium tuberculosis* by human microglia. *Infect. Immun.* **63:**1598–1602.

2863. **Pethe, K., S. Alonso, F. Biet, G. Delogu, M. J. Brennan, C. Locht, and F. D. Menozzi.** 2001. The heparin-binding haemagglutinin of *M. tuberculosis* is required for extrapulmonary dissemination. *Nature* **412:**190–194.

2864. **Pethe, K., V. Puech, M. Daffe, C. Josenhans, H. Drobecq, C. Locht, and F. D. Menozzi.** 2001. *Mycobacterium smegmatis* laminin-binding glycoprotein shares epitopes with *Mycobacterium tuberculosis* heparin-binding haemagglutinin. *Mol. Microbiol.* **39**:89–99.

2865. **Pettersson, J., A. Holmstrom, J. Hill, S. Leary, E. Frithz-Lindsten, A. von Euler-Matell, E. Carlsson, R. Titball, A. Forsberg, and H. Wolf-Watz.** 1999. The V-antigen of *Yersinia* is surface exposed before target cell contact and involved in virulence protein translocation. *Mol. Microbiol.* **32**:961–976.

2866. **Pettersson, J., R. Nordfelth, E. Dubinina, T. Bergman, M. Gustafsson, K. E. Magnusson, and H. Wolf-Watz.** 1996. Modulation of virulence factor expression by pathogen target cell contact. *Science* **273**:1231–1233.

2867. **Pham, T., A. Kaul, A. Hart, P. Goluszko, J. Moulds, S. Nowicki, D. M. Lublin, and B. J. Nowicki.** 1995. *dra*-related X adhesins of gestational pyelonephritis-associated *Escherichia coli* recognize SCR-3 and SCR-4 domains of recombinant decay-accelerating factor. *Infect. Immun.* **63**:1663–1668.

2868. **Pham, T. Q., P. Goluszko, V. Popov, S. Nowicki, and B. J. Nowicki.** 1997. Molecular cloning and characterization of Dr-II, a nonfimbrial adhesin-I-like adhesin isolated from gestational pyelonepritis-associated *Escherichia coli* that binds to decay-accelerating factor. *Infect. Immun.* **65**:4309–4318.

2869. **Philippon, S., H. J. Streckert, and K. Morgenroth.** 1993. *In vitro* study of the bronchial mucosa during *Pseudomonas aeruginosa* infection. *Virchows Arch. A* **423**:39–43.

2870. **Phillips, A. D., J. Giron, S. Hicks, G. Dougan, and G. Frankel.** 2000. Intimin from enteropathogenic *Escherichia coli* mediates remodelling of the eukaryotic cell surface. *Microbiology* **146**:1333–1344.

2871. **Phillips, A. D., S. Navabpour, S. Hicks, G. Dougan, T. Wallis, and G. Frankel.** 2000. Enterohaemorrhagic *Escherichia coli* O157:H7 target Peyer's patches in humans and cause attaching/effacing lesions in both human and bovine intestine. *Gut* **47**:377–381.

2872. **Phillips, A. D.** 1998. The medium is the messenger. *Gut* **43**:456–457.

2873. **Philpott, D. J., A. Ismaili, M. T. Dytoc, J. R. Cantey, and P. M. Sherman.** 1995. Increased adherence of *Escherichia coli* RDEC-1 to human tissue culture cells results in the activation of host signaling pathways. *J. Infect. Dis.* **172**:136–143.

2874. **Piatti, G.** 1994. Bacterial adhesion to respiratory mucosa and its modulation by antibiotics at sub-inhibitory concentrations. *Pharmacol. Res.* **30**:289–299.

2875. **Piatti, G., T. Gazzola, and L. Allegra.** 1997. Bacterial adherence in smokers and nonsmokers. *Pharmacol. Res.* **36**:481–484.

2876. **Pichel, M., N. Binsztein, and G. Viboud.** 2000. CS22, a novel human enterotoxigenic *Escherichia coli* adhesin, is related to CS15. *Infect. Immun.* **68**:3280–3285.

2877. **Picot, L., S. M. Abdelmoula, A. Merieau, P. Leroux, I. Cazin, N. Orange, and M. G. Feuilloley.** 2001. *Pseudomonas fluorescens* as a potential pathogen: adherence to nerve cells. *Microbes Infect.* **3**:985–995.

2878. **Pier, G. B., M. Grout, T. S. Zaidi, and J. B. Goldberg.** 1996. How mutant CFTR may contribute to *Pseudomonas aeruginosa* infection in cystic fibrosis. *Am. J. Respir. Crit. Care Med.* **154**:S175–S182.

2879. **Pier, G. B., G. Meluleni, and E. Neuger.** 1992. A murine model of chronic mucosal colonization by *Pseudomonas aeruginosa*. *Infect. Immun.* **60**:4768–4776.

2880. **Pierce, M. M., R. E. Gibson, and F. G. Rodgers.** 1996. Opsonin-independent adherence and phagocytosis of *Listeria monocytogenes* by murine peritoneal macrophages. *J. Med. Microbiol.* **45**:258–262.

2881. **Pierson, D. E.** 1994. Mutations affecting lipopolysaccharide enhance ail-mediated entry of *Yersinia enterocolitica* into mammalian cells. *J. Bacteriol.* **176**:4043–4051.

2882. **Pietsch, K., and E. Jacobs.** 1993. Characterization of the cellular response of spleen cells in BALB/c mice inoculated with *Mycoplasma pneumoniae* or the P1 protein. *Med. Microbiol. Immunol.* **182**:77–85.

2883. **Pike, R., W. McGraw, J. Potempa, and J. Travis.** 1994. Lysine- and arginine-specific proteinases from *Porphyromonas gingivalis*. Isolation, characterization, and evidence for the existence of complexes with hemagglutinins. *J. Biol. Chem.* **269**:406–411.

2884. **Pike, R. N., J. Potempa, W. McGraw, T. H. Coetzer, and J. Travis.** 1996. Characterization of the binding activities of proteinase-adhesin complexes from *Porphyromonas gingivalis*. *J. Bacteriol.* **178**:2876–2882.

2885. **Pillar, C. M., L. D. Hazlett, and J. A. Hobden.** 2000. Alkaline protease-deficient mutants of *Pseudomonas aeruginosa* are virulent in the eye. *Curr. Eye Res.* **21**:730–739.

2886. **Pillien, F., C. Chalareng, M. Boury, C. Tasca, J. de Rycke, and A. Milon.** 1996. Role of adhesive factor/rabbit 2 in experimental enteropathogenic *Escherichia coli* O103 diarrhea of weaned rabbit. *Vet. Microbiol.* **50**:105–115.

2887. **Pin, C., M. L. Marin, M. D. Selgas, M.**

L. Garcia, J. Tormo, and C. Casas. 1994. Virulence factors in clinical and food isolates of *Aeromonas* species. *Folia Microbiol.* **39:**331–336.

2888. Pinna, A., L. A. Sechi, S. Zanetti, D. Delogu, and F. Carta. 2000. Adherence of ocular isolates of *Staphylococcus epidermidis* to ACRYSOF intraocular lenses. A scanning electron microscopy and molecular biology study. *Ophthalmology* **107:**2162–2166.

2889. Pinna, A., S. Zanetti, L. A. Sechi, D. Usai, M. P. Falchi, and F. Carta. 2000. *In vitro* adherence of *Staphylococcus epidermidis* to polymethyl methacrylate and acrysof intraocular lenses. *Ophthalmology* **107:**1042–1046.

2890. Piotrowski, J., V. L. Murty, A. Czajkowski, A. Slomiany, F. Yotsumoto, J. Majka, and B. L. Slomiany. 1994. Association of salivary bacterial aggregating activity with sulfomucin. *Biochem. Mol. Biol. Int.* **32:**713–721.

2891. Pitt, W. G., M. O. McBride, A. J. Barton, and R. D. Sagers. 1993. Air-water interface displaces adsorbed bacteria. *Biomaterials* **14:**605–608.

2892. Planet, P. J., S. C. Kachlany, R. DeSalle, and D. H. Figurski. 2001. Phylogeny of genes for secretion NTPases: identification of the widespread *tadA* subfamily and development of a diagnostic key for gene classification. *Proc. Natl. Acad. Sci. USA* **98:**2503–2508.

2893. Plant, L., and P. Conway. 2001. Association of *Lactobacillus* spp. with Peyer's patches in mice. *Clin. Diagn. Lab. Immunol.* **8:**320–324.

2894. Plos, K., H. Connell, U. Jodal, B. I. Marklund, S. Marild, B. Wettergren, and C. Svanborg. 1995. Intestinal carriage of P fimbriated *Escherichia coli* and the susceptibility to urinary tract infection in young children. *J. Infect. Dis.* **171:**625–631.

2895. Plotkin, B. J., and D. A. Bemis. 1998. Carbon source utilisation by *Bordetella bronchiseptica*. *J. Med. Microbiol.* **47:**761–765.

2896. Plotkowski, M.-C., A. O. Costa, V. Morandi, H. S. Barbosa, H. B. Nader, S. de Bentzmann, and E. Puchelle. 2001. Role of heparan sulphate proteoglycans as potential receptors for non-piliated *Pseudomonas aeruginosa* adherence to non-polarised airway epithelial cells. *J. Med. Microbiol.* **50:**183–190.

2897. Plotkowski, M.-C., O. Bajolet-Laudinat, and E. Puchelle. 1993. Cellular and molecular mechanisms of bacterial adhesion to respiratory mucosa. *Eur. Respir. J.* **6:**903–916.

2898. Plotkowski, M.-C., M. Chevillard, D. Pierrot, D. Altemayer, and E. Puchelle. 1992. Epithelial respiratory cells from cystic fibrosis patients do not possess specific *Pseudomonas aeruginosa*-adhesive properties. *J. Med. Microbiol.* **36:**104–111.

2899. Plotkowski, M.-C., A. M. Saliba, S. H. Pereira, M. P. Cervante, and O. Bajolet-Laudinat. 1994. *Pseudomonas aeruginosa* selective adherence to and entry into human endothelial cells. *Infect. Immun.* **62:**5456–5463.

2900. Plotkowski, M.-C., J. M. Zahm, J.-M. Tournier, and E. Puchelle. 1992. *Pseudomonas aeruginosa* adhesion to normal and injured respiratory mucosa. *Mem. Inst. Oswaldo Cruz* **87:**61–68.

2901. Plotkowski, M.-C., J.-M. Tournier, and E. Puchelle. 1996. *Pseudomonas aeruginosa* strains possess specific adhesins for laminin. *Infect. Immun.* **64:**600–605.

2902. Plummer, F. A., H. Chubb, J. N. Simonsen, M. Bosire, L. Slaney, N. J. Nagelkerke, I. Maclean, J. O. Ndinya-Achola, P. Waiyaki, and R. C. Brunham. 1994. Antibodies to opacity proteins (Opa) correlate with a reduced risk of gonococcal salpingitis. *J. Clin. Investig.* **93:**1748–1755.

2903. Podschun, R., D. Sievers, A. Fischer, and U. Ullmann. 1993. Serotypes, hemagglutinins, siderophore synthesis, and serum resistance of *Klebsiella* isolates causing human urinary tract infections. *J. Infect. Dis.* **168:**1415–1421.

2904. Podschun, R., and U. Ullmann. 1992. *Klebsiella* capsular type K7 in relation to toxicity, susceptibility to phagocytosis and resistance to serum. *J. Med. Microbiol.* **36:**250–254.

2905. Poelstra, K. A., H. C. van der Mei, B. Gottenbos, D. W. Grainger, J. R. van Horn, and H. J. Busscher. 2000. Pooled human immunoglobulins reduce adhesion of *Pseudomonas aeruginosa* in a parallel plate flow chamber. *J. Biomed. Mater. Res.* **51:**224–232.

2906. Pohlmann-Dietze, P., M. Ulrich, K. B. Kiser, G. Doring, J. C. Lee, J. M. Fournier, K. Botzenhart, and C. Wolz. 2000. Adherence of *Staphylococcus aureus* to endothelial cells: influence of capsular polysaccharide, global regulator *agr,* and bacterial growth phase. *Infect. Immun.* **68:**4865–4871.

2907. Poitrineau, P., C. Forestier, M. Meyer, C. Jallat, C. Rich, G. Malpuech, and C. De Champs. 1995. Retrospective case-control study of diffusely adhering *Escherichia coli* and clinical features in children with diarrhea. *J. Clin. Microbiol.* **33:**1961–1962.

2908. Polotsky, Y., J. P. Nataro, D. Kotler, T. J. Barrett, and J. M. Orenstein. 1997. HEp-2 cell adherence patterns, serotyping, and DNA analysis of *Escherichia coli* isolates from eight patients with AIDS and chronic diarrhea. *J. Clin. Microbiol.* **35:**1952–1958.

2909. **Polotsky, Y., E. Dragunsky, and T. Khavkin.** 1994. Morphologic evaluation of the pathogenesis of bacterial enteric infections. *Crit. Rev. Microbiol.* **20:**161–208.

2910. **Ponniah, S., S. N. Abraham, and R. O. Endres.** 1992. T-cell-independent stimulation of immunoglobulin secretion in resting human B lymphocytes by the mannose-specific adhesin of *Escherichia coli* type 1 fimbriae. *Infect. Immun.* **60:**5197–5203.

2911. **Poortinga, A. T., R. Bos, and H. J. Busscher.** 2001. Charge transfer during staphylococcal adhesion to TiNOX coatings with different specific resistivity. *Biophys. Chem.* **91:**273–279.

2912. **Poortinga, A. T., R. Bos, and H. J. Busscher.** 2001. Lack of effect of an externally applied electric field on bacterial adhesion to glass. *Colloids Surf. Ser. B* **20:**189–194.

2913. **Poortinga, A. T., J. Smit, H. C. van der Mei, and H. J. Busscher.** 2001. Electric field induced desorption of bacteria from a conditioning film covered substratum. *Biotechnol. Bioeng.* **76:**395–399.

2914. **Pope, L. M., K. E. Reed, and S. M. Payne.** 1995. Increased protein secretion and adherence to HeLa cells by *Shigella* spp. following growth in the presence of bile salts. *Infect. Immun.* **63:**3642–3648.

2915. **Popham, P. L., T.-W. Hahn, K. A. Krebes, and D. C. Krause.** 1997. Loss of HMW1 and HMW3 in noncytadhering mutants of *Mycoplasma pneumoniae* occurs posttranslationally. *Proc. Natl. Acad. Sci. USA* **94:**13979–13984.

2916. **Popov, V. L., X. Yu, and D. H. Walker.** 2000. The 120 kDa outer membrane protein of *Ehrlichia chaffeensis:* preferential expression on dense-core cells and gene expression in *Escherichia coli* associated with attachment and entry. *Microb. Pathog.* **28:**71–80.

2917. **Porat, N., M. A. Apicella, and M. S. Blake.** 1995. *Neisseria gonorrhoeae* utilizes and enhances the biosynthesis of the asialoglycoprotein receptor expressed on the surface of the hepatic HepG2 cell line. *Infect. Immun.* **63:**1498–1506.

2918. **Porat, N., M. A. Apicella, and M. S. Blake.** 1995. A lipooligosaccharide-binding site on HepG2 cells similar to the gonococcal opacity-associated surface protein Opa. *Infect. Immun.* **63:**2164–2172.

2919. **Porter, J. F., K. Connor, A. van der Zee, F. Reubsaet, P. Ibsen, I. Heron, R. Chaby, K. Le Blay, and W. Donachie.** 1995. Characterisation of ovine *Bordetella parapertussis* isolates by analysis of specific endotoxin (lipopolysaccharide) epitopes, filamentous haemagglutinin production, cellular fatty acid composition and antibiotic sensitivity. *FEMS Microbiol. Lett.* **132:**195–201.

2920. **Porter, M. E., S. G. Smith, and C. J. Dorman.** 1998. Two highly related regulatory proteins, *Shigella flexneri* VirF and enterotoxigenic *Escherichia coli* Rns, have common and distinct regulatory properties. *FEMS Microbiol. Lett.* **162:**303–309.

2921. **Portoles, M., F. Austin, S. Nos-Barbera, C. Paterson, and M. F. Refojo.** 1995. Effect of Poloxamer 407 on the adherence of *Pseudomonas aeruginosa* to corneal epithelial cells. *Cornea* **14:**56–61.

2922. **Portolés, M., M. F. Refojo, and F.-L. Leong.** 1994. Poloxamer 407 as a bacterial adhesive for hydrogel contact lenses. *J. Biomed. Mater. Res.* **28:**303–309.

2923. **Portolés, M., M. F. Refojo, and F.-L. Leong.** 1993. Reduced bacterial adhesion to heparin-surface-modified intraocular lenses. *J. Cataract. Refract. Surg.* **19:**755–759.

2924. **Postnova, T., O. G. Gomez-Duarte, and K. Richardson.** 1996. Motility mutants of *Vibrio cholerae* O1 have reduced adherence *in vitro* to human small intestinal epithelial cells as demonstrated by ELISA. *Microbiology* **142:**2767–2776.

2925. **Poston, S. M., G. R. Glancey, J. E. Wyatt, T. Hogan, and T. J. Foster.** Coelimination of *mec* and *spa* genes in *Staphylococcus aureus* and the effect of *agr* and protein A production on bacterial adherence to cell monolayers. *J. Med. Microbiol.* **39:**422–428.

2926. **Poulain-Godefroy, O., N. Mielcarek, N. Ivanoff, F. Remoue, A. M. Schacht, N. Phillips, C. Locht, A. Capron, and G. Riveau.** 1998. *Bordetella pertussis* filamentous hemagglutinin enhances the immunogenicity of liposome-delivered antigen administered intranasally. *Infect. Immun.* **66:**1764–1767.

2927. **Pourbakhsh, S. A., M. Dho-Moulin, A. Bree, C. Desautels, B. Martineau-Doize, and J. M. Fairbrother.** 1997. Localization of the *in vivo* expression of P and F1 fimbriae in chickens experimentally inoculated with pathogenic *Escherichia coli*. *Microb. Pathog.* **22:**331–341.

2928. **Pourbakhsh, S. A., and J. M. Fairbrother.** 1994. Purification and characterization of P fimbriae from an *Escherichia coli* strain isolated from a septicemic turkey. *FEMS Microbiol. Lett.* **122:**313–318.

2929. **Pourbakhsh, S. A., M. Boulianne, B. Martineau-Doize, and J. M. Fairbrother.** 1997. Virulence mechanisms of avian fimbriated *Escherichia coli* in experimentally inoculated chickens. *Vet. Microbiol.* **58:**195–213.

2930. **Pouttu, R., T. Puustinen, R. Virkola, J. Hacker, P. Klemm, and T. K. Korhonen.** 1999. Amino acid residue Ala-62 in the FimH fimbrial adhesin is critical for the adhesiveness of meningitis-associated *Escherichia coli* to collagens. *Mol. Microbiol.* **31:**1747–1757.

2931. **Pouvelle, B., T. Fusaï, C. Lépolard, and J. Gysin.** 1998. Biological and biochemical characteristics of cytoadhesion of *Plasmodium falciparum*-infected erythrocytes to chondroitin-4-sulfate. *Infect. Immun.* **66:**4950–4956.

2932. **Prakobphol, A., C. A. Burdsal, and S. J. Fisher.** 1995. Quantifying the strength of bacterial adhesive interactions with salivary glycoproteins. *J. Dent. Res.* **74:**1212–1218.

2933. **Prakobphol, A., F. Xu, V. M. Hoang, T. Larsson, J. Bergstrom, I. Johansson, L. Frangsmyr, U. Holmskov, H. Leffler, C. Nilsson, T. Boren, J. R. Wright, N. Strömberg, and S. J. Fisher.** 2000. Salivary agglutinin, which binds *Streptococcus mutans* and *Helicobacter pylori*, is the lung scavenger receptor cysteine-rich protein gp-340. *J. Biol. Chem.* **275:**39860–39866.

2934. **Prakobphol, A., H. Leffler, and S. J. Fisher.** 1995. Identifying bacterial receptor proteins and quantifying strength of interactions they mediate. *Methods Enzymol.* **253:**132–142.

2935. **Prasad, K. N., T. N. Dhole, and A. Ayyagari.** 1996. Adherence, invasion and cytotoxin assay of *Campylobacter jejuni* in HeLa and HEp-2 cells. *J. Diarrhoeal Dis. Res.* **14:**255–259.

2936. **Prasad, K. N., S. K. Mathur, A. Ayyagari, S. Singhal, and T. N. Dhole.** 1992. Mannose-resistant haemagglutination by *Campylobacter jejuni*—a preliminary communication. *Indian J. Med. Res.* **95:**184–186.

2937. **Prasad, S. M., Y. Yin, E. Rodzinski, E. I. Tuomanen, and H. R. Masure.** 1993. Identification of a carbohydrate recognition domain in filamentous hemagglutinin from *Bordetella pertussis*. *Infect. Immun.* **61:**2780–2785.

2938. **Prasadarao, N. V., E. Lysenko, C. A. Wass, K. S. Kim, and J. N. Weiser.** 1999. Opacity-associated protein A contributes to the binding of *Haemophilus influenzae* to Chang epithelial cells. *Infect. Immun.* **67:**4153–4160.

2939. **Prasadarao, N. V., C. A. Wass, J. Hacker, K. Jann, and K. S. Kim.** 1993. Adhesion of S-fimbriated *Escherichia coli* to brain glycolipids mediated by *sfaA* gene-encoded protein of S-fimbriae. *J. Biol. Chem.* **268:**10356–10363.

2940. **Prasadarao, N. V., C. A. Wass, and K. S. Kim.** 1997. Identification and characterization of S fimbriae-binding sialoglycoproteins on brain microvascular endothelial cells. *Infect. Immun.* **65:**2852–2860.

2941. **Pratten, J., S. J. Foster, P. F. Chan, M. Wilson, and S. P. Nair.** 2001. *Staphylococcus aureus* accessory regulators: expression within biofilms and effect on adhesion. *Microbes Infect.* **3:**633–637.

2942. **Preston, A., R. E. Mandrell, B. W. Gibson, and M. A. Apicella.** 1996. The lipooligosaccharides of pathogenic gram-negative bacteria. *Crit. Rev. Microbiol.* **22:**139–180.

2943. **Prigent-Combaret, C., E. Brombacher, O. Vidal, A. Ambert, P. Lejeune, P. Landini, and C. Dorel.** 2001. Complex regulatory network controls initial adhesion and biofilm formation in *Escherichia coli* via regulation of the *csgD* gene. *J. Bacteriol.* **183:**7213–7223.

2944. **Prince, A.** 1992. Adhesins and receptors of *Pseudomonas aeruginosa* associated with infection of the respiratory tract. *Microb. Pathog.* **13:**251–260.

2945. **Prins, J. M., C. Schultsz, P. Speelman, and S. J. van Deventer.** 1996. Known bacterial virulence factors do not explain the variation in urinary cytokine levels in patients with urosepsis. *FEMS Immunol. Med. Microbiol.* **16:**283–289.

2946. **Proft, T., H. Hilbert, H. Plagens, and R. Herrmann.** 1996. The P200 protein of *Mycoplasma pneumoniae* shows common features with the cytadherence-associated proteins HMW1 and HMW3. *Gene* **171:**79–82.

2947. **Progulske-Fox, A., E. Kozarov, B. Dorn, W. Dunn, Jr., J. Burks, and Y. Wu.** 1999. *Porphyromonas gingivalis* virulence factors and invasion of cells of the cardiovascular system. *J. Periodontal Res.* **34:**393–399.

2948. **Progulske-Fox, A., V. Rao, N. Han, G. Lepine, J. Witlock, and M. Lantz.** 1993. Molecular characterization of hemagglutinin genes of periodontopathic bacteria. *J. Periodontal Res.* **28:**473–474.

2949. **Progulske-Fox, A., S. Tumwasorn, G. Lepine, J. Whitlock, D. Savett, J. J. Ferretti, and J. A. Banas.** 1995. The cloning, expression and sequence analysis of a second *Porphyromonas gingivalis* gene that codes for a protein involved in hemagglutination. *Oral Microbiol. Immunol.* **10**:311–318.

2950. **Pron, B., M. K. Taha, C. Rambaud, J. C. Fournet, N. Pattey, J. P. Monnet, M. Musilek, J. L. Beretti, and X. Nassif.** 1997. Interaction of *Neisseria maningitidis* with the components of the blood-brain barrier correlates with an increased expression of PilC. *J. Infect. Dis.* **176**:1285–1292.

2951. **Prosdocimo, G., S. Grandesso, and G. Amici.** 1997. Influence of optic and haptic materials on the adherence of *Staphylococcus epidermidis* to intraocular lenses: a pilot study. *Eur. J. Ophthalmol.* **7**:241–244.

2952. **Provence, D. L., and R. Curtiss III.** 1994. Isolation and characterization of a gene involved in hemagglutination by an avian pathogenic *Escherichia coli* strain. *Infect. Immun.* **62**:1369–1380.

2953. **Provenzano, D., D. A. Schuhmacher, J. L. Barker, and K. E. Klose.** 2000. The virulence regulatory protein ToxR mediates enhanced bile resistance in *Vibrio cholerae* and other pathogenic *Vibrio* species. *Infect. Immun.* **68**:1491–1497.

2954. **Pruimboom, I. M., R. B. Rimler, M. R. Ackermann, and K. A. Brogden.** 1996. Capsular hyaluronic acid-mediated adhesion of *Pasteurella multocida* to turkey air sac macrophages. *Avian Dis.* **40**:887–893.

2955. **Pruimboom, I. M., R. B. Rimler, and M. R. Ackermann.** 1999. Enhanced adhesion of *Pasteurella multocida* to cultured turkey peripheral blood monocytes. *Infect. Immun.* **67**:1292–1296.

2956. **Przondo-Mordarska, A., D. Smutnicka, H. L. Ko, J. Beuth, and G. Pulverer.** 1996. Adhesive properties of P-like fimbriae in *Klebsiella* species. *Zentbl. Bakteriol.* **284**:372–377.

2957. **Puente, J. L., D. Bieber, S. W. Ramer, W. Murray, and G. K. Schoolnik.** 1996. The bundle-forming pili of enteropathogenic *Escherichia coli*: transcriptional regulation by environmental signals. *Mol. Microbiol.* **20**:87–100.

2958. **Pujol, C., E. Eugene, M. Marceau, and X. Nassif.** 1999. The meningococcal PilT protein is required for induction of intimate attachment to epithelial cells following pilus-mediated adhesion. *Proc. Natl. Acad. Sci. USA* **96**:4017–4022.

2959. **Pujol, C., E. Eugène, L. de Saint Martin, and X. Nassif.** 1997. Interaction of *Neisseria meningitidis* with a polarized monolayer of epithelial cells. *Infect. Immun.* **65**:4836–4842.

2960. **Pusztai, A., G. Grant, R. J. Spencer, T.**

J. Duguid, D. S. Brown, S. W. B. Ewen, W. J. Peumans, E. J. M. van Damme, and S. Bardocz.** 1993. Kidney bean lectin-induced *Escherichia coli* overgrowth in the small intestine is blocked by GNA, a mannose specific lectin. *J. Appl. Bacteriol.* **75**:360–368.

2961. **Puzova, H., L. Siegfried, M. Kmetova, J. Filka, V. Takacova, and J. Durovicova.** 1994. Fimbriation, surface hydrophobicity and serum resistance in uropathogenic strains of *Escherichia coli*. *FEMS Immunol. Med. Microbiol.* **9**:223–229.

2962. **Qadri, F., G. Jonson, Y. A. Begum, C. Wenneras, M. J. Albert, M. A. Salam, and A.-M. Svennerholm.** 1997. Immune response to the mannose-sensitive hemagglutinin in patients with cholera due to *Vibrio cholerae* O1 and O0139. *Clin. Diagn. Lab. Immunol.* **4**:429–434.

2963. **Qadri, F., A. Haque, S. M. Faruque, K. A. Bettelheim, R. Robins-Browne, and M. J. Albert.** 1994. Hemagglutinating properties of enteroaggregative *Escherichia coli*. *J. Clin. Microbiol.* **32**:510–514.

2964. **Qian, H., and M. L. Dao.** 1993. Inactivation of the *Streptococcus mutans* wall-associated protein A gene (*wapA*) results in a decrease in sucrose-dependent adherence and aggregation. *Infect. Immun.* **61**:5021–5028.

2965. **Qiu, J., D. R. Hendrixson, E. N. Baker, T. F. Murphy, J. W. St. Geme III, and A. G. Plaut.** 1998. Human milk lactoferrin inactivates two putative colonization factors expressed by *Haemophilus influenzae*. *Proc. Natl. Acad. Sci. USA* **95**:12641–12646.

2966. **Que, Y. A., J. A. Haefliger, P. Francioli, and P. Moreillon.** 2000. Expression of *Staphylococcus aureus* clumping factor A in *Lactococcus lactis* subsp. *cremoris* using a new shuttle vector. *Infect. Immun.* **68**:3516–3522.

2967. **Quinn, F. D., R. S. Weyant, M. J. Worley, E. H. White, E. A. Utt, and E. A. Ades.** 1995. Human microvascular endothelial tissue culture cell model for studying pathogenesis of Brazilian purpuric fever. *Infect. Immun.* **63**:2317–2322.

2968. **Quinn, D. M., C. Y. F. Wong, H. M. Atkinson, and R. L. P. Flower.** 1993. Isolation of carbohydrate-reactive outer membrane proteins of *Aeromonas hydrophila*. *Infect. Immun.* **61**:371–377.

2969. **Quinn, M. O., V. E. Miller, and A. R. Dal Nogare.** 1994. Increased salivary exoglycosidase activity during critical illness. *Am. J. Respir. Crit. Care Med.* **150**:179–183.

2970. **Quirynen, M., W. Papaioannou, T. J. van Steenbergen, K. Dierickx, J. J.**

Cassiman, and D. van Steenberghe. 2001. Adhesion of *Porphyromonas gingivalis* strains to cultured epithelial cells from patients with a history of chronic adult periodontitis or from patients less susceptible to periodontitis. *J. Periodontol.* **72:**626–633.

2971. **Quirynen, M.** 1994. The clinical meaning of the surface roughness and the surface free energy of intra-oral hard substrata on the microbiology of the supra- and subgingival plaque: results of *in vitro* and *in vivo* experiments. *J. Dent.* **22:**S13–S16.

2972. **Raad, I., R. Darouiche, R. Hachem, M. Mansouri, and G. P. Bodey.** 1996. The broad-spectrum activity and efficacy of catheters coated with minocycline and rifampin. *J. Infect. Dis.* **173:**418–424.

2973. **Raadsma, H. W., J. R. Egerton, F. W. Nicholas, and S. C. Brown.** 1992. Disease resistance in Merino sheep. I. Traits indicating resistance to footrot following experimental challenge and subsequent vaccination with an homologous rDNA pilus vaccine. *J. Anim. Breed. Genet.* **110:**281–300.

2974. **Raadsma, H. W., T. J. O'Meara, J. R. Egerton, P. R. Lehrbach, and C. L. Schwartzkoff.** 1994. Protective antibody titres and antigenic competition in multivalent *Dichelobacter nodosus* fimbrial vaccines using characterised rDNA antigens. *Vet. Immunol. Immunopathol.* **40:**253–274.

2975. **Rabaan, A. A., I. Gryllos, J. M. Tomas, and J. G. Shaw.** 2001. Motility and the polar flagellum are required for *Aeromonas caviae* adherence to HEp-2 cells. *Infect. Immun.* **69:**4257–4267.

2976. **Rabinowitz, R. P., L. C. Lai, K. Jarvis, T. K. McDaniel, J. B. Kaper, K. D. Stone, and M. S. Donnenberg.** 1996. Attaching and effacing of host cells by enteropathogenic *Escherichia coli* in the absence of detectable tyrosine kinase mediated signal transduction. *Microb. Pathog.* **21:**157–171.

2977. **Rachid, S., K. Ohlsen, W. Witte, J. Hacker, and W. Ziebuhr.** 2000. Effect of subinhibitory antibiotic concentrations on polysaccharide intercellular adhesin expression in biofilm-forming *Staphylococcus epidermidis*. *Antimicrob. Agents Chemother.* **44:**3357–3363.

2978. **Rahman, I., M. Shahamat, M. A. Chowdhury, and R. R. Colwell.** 1996. Potential virulence of viable but nonculturable *Shigella dysenteriae* type 1. *Appl. Environ. Microbiol.* **62:**115–120.

2979. **Rahman, M., A. B. Jonsson, and T. Holme.** 1998. Monoclonal antibodies to the epitope α-Gal-(1–4)-β-Gal-(1- of *Moraxella catarrhalis* LPS react with a similar epitope in type IV pili of *Neisseria meningitidis*. *Microb. Pathog.* **24:**299–308.

2980. **Raimondi, A., A. Pessina, and M. G. Neri.** 1995. Loss of porins following carbapenem-resistance selection and adherence modification in enterobacteria. *J. Chemother.* **7:**171–174.

2981. **Rainard, P.** 1993. Binding of bovine fibronectin to mastitis-causing *Streptococcus agalactiae* induces adherence to solid substrate but not phagocytosis by polymorphonuclear cells. *Microb. Pathog.* **14:**239–248.

2982. **Rainho, C. S., E. A. de Sa, L. C. Jabur Gaziri, H. Ostrensky Saridakis, and I. Felipe.** Modulation of lectinophagocytosis of *Escherichia coli* by variation of pH and temperature. *FEMS Immunol. Med. Microbiol.* **24:**91–95.

2983. **Raj, P. A., M. Johnsson, M. J. Levine, and G. H. Nancollas.** 1992. Salivary statherin. Dependence on sequence, charge, hydrogen bonding potency, and helical conformation for adsorption to hydroxyapatite and inhibition of mineralization. *J. Biol. Chem.* **267:**5968–5976.

2984. **Rajan, N., Q. Cao, B. E. Anderson, D. L. Pruden, J. Sensibar, J. L. Duncan, and A. J. Schaeffer.** 1999. Roles of glycoproteins and oligosaccharides found in human vaginal fluid in bacterial adherence. *Infect. Immun.* **67:**5027–5032.

2985. **Rajashekara, G., S. Munir, C. M. Lamichhane, A. Back, V. Kapur, D. A. Halvorson, and K. V. Nagaraja.** 1998. Application of recombinant fimbrial protein for the specific detection of *Salmonella enteritidis* infection in poultry. *Diagn. Microbiol. Infect. Dis.* **32:**147–157.

2986. **Rajashekara, G., S. Munir, M. F. Alexeyev, D. A. Halvorson, C. L. Wells, and K. V. Nagaraja.** 2000. Pathogenic role of SEF14, SEF17, and SEF21 fimbriae in *Salmonella enterica* serovar Enteritidis infection of chickens. *Appl. Environ. Microbiol.* **66:**1759–1763.

2987. **Rakita, R. M., N. N. Vanek, K. Jacques-Palaz, M. Mee, M. M. Mariscalco, G. M. Dunny, M. Snuggs, W. B. Van Winkle, and S. I. Simon.** 1999. *Enterococcus faecalis* bearing aggregation substance is resistant to killing by human neutrophils despite phagocytosis and neutrophil activation. *Infect. Immun.* **67:**6067–6075.

2988. **Rambukkana, A., H. Yamada, G. Zanazzi, T. Mathus, J. L. Salzer, P. D. Yurchenco, K. P. Campbell, and V. A.**

Fischetti. 1998. Role of α-dystroglycan as a Schwann cell receptor for *Mycobacterium leprae*. *Science* **282**:2076–2079.

2989. **Rambukkana, A., J. L. Salzer, P. D. Yurchenko, and E. I. Tuomanen.** 1997. Neural targeting of *Mycobacterium leprae* mediated by the G domain of the laminin-α2 chain. *Cell* **88**:811–821.

2990. **Ramer, S. W., D. Bieber, and G. K. Schoolnik.** 1996. BfpB, an outer membrane lipoprotein required for the biogenesis of bundle-forming pili in enteropathogenic *Escherichia coli*. *J. Bacteriol.* **178**: 6555–6563.

2991. **Ramphal, R., S. K. Arora, and B. W. Ritchings.** 1996. Recognition of mucin by the adhesin-flagellar system of *Pseudomonas aeruginosa*. *Am. J. Respir. Crit. Care Med.* **154**:S170–S174.

2992. **Rani, D. B., M. E. Bayer, and D. M. Schifferli.** 1999. Polymeric display of immunogenic epitopes from herpes simplex virus and transmissible gastroenteritis virus surface proteins on an enteroadherent fimbria. *Clin. Diagn. Lab. Immunol.* **6**:30–40.

2993. **Rankin, S., G. Tran Van Nhieu, and R. R. Isberg.** 1994. Use of *Staphylococcus aureus* coated with invasin derivatives to assay invasin function. *Methods Enzymol.* **236**:566–577.

2994. **Rao, S. P., K. R. Gehlsen, and A. Catanzaro.** 1992. Identification of a β1-integrin on *Mycobacterium avium-Mycobacterium intracellulare*. *Infect. Immun.* **60**:3652–3657.

2995. **Rao, S. P., K. Ogata, and A. Catanzaro.** 1993. *Mycobacterium avium-M. intracellulare* binds to the integrin receptor αvβ3 on human monocytes and monocyte-derived macrophages. *Infect. Immun.* **61**:663–670.

2996. **Rao, S. P., K. Ogata, S. L. Morris, and A. Catanzaro.** 1993. Identification of a 68 kd surface antigen of *Mycobacterium avium* that binds to human macrophages. *J. Lab. Clin. Med.* **123**:526–535.

2997. **Rao, V. K., J. A. Whitlock, and A. Progulske-Fox.** 1993. Coning, characterization and sequencing of two haemagglutinin genes from *Eikenella corrodens*. *J. Gen. Microbiol.* **139**:639–650.

2998. **Raoof, S., M. M. Grant, M. S. Niederman. M. A. Poehlman, A. F. Matin, F. A. Khan, and A. M. Fein.** 1995. Cytokines affect pseudomonas binding to tracheal cells via a neutrophil-mediated process. *Am. J. Respir. Crit. Care Med.* **152**:921–926.

2999. **Rappuoli, R.** 1994. Pathogenicity mechanisms of *Bordetella*. *Curr. Top. Microbiol. Immunol.* **192**:319–336.

3000. **Rappuoli, R., A. Covacci, P. Ghiara, and J. Telford.** 1994. Pathogenesis of *Helicobacter pylori* and perspectives of vaccine development against an emerging pathogen. *Behring Inst. Mitt.* **95**:42–48.

3001. **Raskin, E. M., M. G. Speaker, S. A. McCormick, D. Wong, J. A. Menikoff, and K. Pelton-Henrion.** 1993. Influence of haptic materials on the adherence of staphylococci to intraocular lenses. *Arch. Ophthalmol.* **111**:250–253.

3002. **Rastawicki, W., R. Raty, and M. Kleemola.** 1996. Detection of antibodies to *Mycoplasma pneumoniae* adhesion P1 in serum specimens from infected and non-infected subjects by immunoblotting. *Diagn. Microbiol. Infect. Dis.* **26**:141–143.

3003. **Ratliff, T. L., R. McCarthy, W. B. Telle, and E. J. Brown.** 1993. Purification of a mycobacterial adhesin for fibronectin. *Infect. Immun.* **61**:1889–1894.

3004. **Ratnakar, P., S. P. Rao, and A. Catanzaro.** 1996. Isolation and characterization of a 70 kDa protein from *Mycobacterium avium*. *Microb. Pathog.* **21**:471–486.

3005. **Rautelin, H., B. Blomberg, H. Fredlund, G. Jarnerot, and D. Danielsson.** 1993. Incidence of *Helicobacter pylori* strains activating neutrophils in patients with peptic ulcer disease. *Gut* **34**:599–603.

3006. **Rautelin, H., E. Kihlstrom, M. Jurstrand, and D. Danielsson.** 1995 Adhesion to and invasion of HeLa cells by *Helicobacter pylori*. *Zentbl. Bakteriol.* **282**:50–53.

3007. **Ravins, M., J. Jaffe, E. Hanski, I. Shetzigovski, S. Natanson-Yaron, and A. E. Moses.** 2000. Characterization of a mouse-passaged, highly encapsulated variant of group A streptococcus in *in vitro* and *in vivo* studies. *J. Infect. Dis.* **182**:1702–1711.

3008. **Ravizzola, G., M. Longo, P. Pollara, M. De Francesco, and R. Pizzi.** 1994. Effect of subinhibitory concentrations of lomefloxacin on bacterial adherence. *New Microbiol.* **17**:211–216.

3009. **Ray, C. A., L. E. Gfell, T. L. Buller, and R. L. Gregory.** 1999. Interactions of *Streptococcus mutans* fimbria-associated surface proteins with salivary components. *Clin. Diagn. Lab. Immunol.* **6**:400–404.

3010. **Rayner, C. F., A. Dewar, E. R. Moxon, M. Virji, and R. Wilson.** 1995. The effect of variations in the expression of pili on the interaction of *Neisseria meningitidis* with human nasopharyngeal epithelium. *J. Infect. Dis.* **171**:113–121.

3011. **Rayner, C. F., A. D. Jackson, A.**

Rutman, A. Dewar, T. J. Mitchell, P. W. Andrew, P. J. Cole, and R. Wilson. 1995. Interaction of pneumolysin-sufficient and -deficient isogenic variants of *Streptococcus pneumoniae* with human respiratory mucosa. *Infect. Immun.* **63**:442–447.

3012. **Rayner, C. F. J., A. Dewar, E. R. Moxon, M. Virji, and R. Wilson.** 1994. The effect of variations in the expression of pili on the interaction of *Neisseria meningitidis* with human nasopharyngeal epithelium. *J. Infect. Dis.* **171**:113–121.

3013. **Raza, M. W., M. M. Ogilvie, C. C. Blackwell, J. Stewart, R. A. Elton, and D. M. Weir.** 1993. Effect of respiratory syncytial virus infection on binding of *Neisseria meningitidis* and *Haemophilus influenzae* type b to a human epithelial cell line (HEp-2). *Epidemiol. Infect.* **110**:339–347.

3014. **Raza, M. W., C. C. Blackwell, M. M. Ogilvie, A. T. Saadi, J. Stewart, R. A. Elton, and D. M. Weir.** 1994. Evidence for the role of glycoprotein G of respiratory syncytial virus in binding of *Neisseria meningitidis* to HEp-2 cells. *FEMS Immunol. Med. Microbiol.* **10**:25–30.

3015. **Raza, M. W., O. R. el Ahmer, M. M. Ogilvie, C. C. Blackwell, A. T. Saadi, R. A. Elton, and D. M. Weir.** 1999. Infection with respiratory syncytial virus enhances expression of native receptors for non-pilate *Neisseria meningitidis* on HEp-2 cells. *FEMS Immunol. Med. Microbiol.* **23**:115–124.

3016. **Razatos, A., Y. L. Ong, M. M. Sharma, and G. Georgiou.** 1998. Evaluating the interaction of bacteria with biomaterials using atomic force microscopy. *J. Biomater. Sci.* **9**:1361–1373.

3017. **Razatos, A., Y. L. Ong, M. M. Sharma, and G. Georgiou.** 1998. Molecular determinants of bacterial adhesion monitored by atomic force microscopy. *Proc. Natl. Acad. Sci. USA* **95**:11059–11064.

3018. **Read, R. C., A. A. Rutman, P. K. Jeffery, V. J. Lund, A. P. Brain, E. R. Moxon, P. J. Cole, and R. Wilson.** 1992. Interaction of capsulate *Haemophilus influenzae* with human airway mucosa in vitro. *Infect. Immun.* **60**:3244–3252.

3019. **Read, R. C., S. Zimmerli, C. Broaddus, D. A. Sanan, D. S. Stephens, and J. D. Ernst.** 1996. The (α2→8)-linked polysialic acid capsule of group B *Neisseria meningitidis* modifies multiple steps during interaction with human macrophages. *Infect. Immun.* **64**:3210–3217.

3020. **Read, T. D., M. Dowdell, S. W. Satola,** and M. M. Farley. 1996. Duplication of pilus gene complexes of *Haemophilus influenzae* biogroup *aegyptius*. *J. Bacteriol.* **178**:6564–6570.

3021. **Read, T. D., S. W. Satola, J. A. Opdyke, and M. M. Farley.** 1998. Copy number of pilus gene clusters in *Haemophilus influenzae* and variation in the *hifE* pilin gene. *Infect. Immun.* **66**:1622–1631.

3022. **Reddy, K., and J. M. Ross.** 2001. Shear stress prevents fibronectin binding protein-mediated *Staphylococcus aureus* adhesion to resting endothelial cells. *Infect. Immun.* **69**:3472–3475.

3023. **Reddy, M. S.** 1996. Binding between *Pseudomonas aeruginosa* adhesins and human salivary, tracheobronchial and nasopharyngeal mucins. *Biochem. Mol. Biol. Int.* **40**:403–408.

3024. **Reddy, M. S.** 1992. Human tracheobronchial mucin: purification and binding to *Pseudomonas aeruginosa*. *Infect. Immun.* **60**:1530–1535.

3025. **Reddy, M. S., M. J. Levine, and W. Paranchych.** 1993. Low-molecular-mass human salivary mucin, MG2: structure and binding of *Pseudomonas aeruginosa*. *Crit. Rev. Oral Biol. Med.* **4**:315–323.

3026. **Reddy, M. S., T. F. Murphy, H. S. Faden, and J. M. Bernstein.** 1997. Middle ear mucin glycoprotein: purification and interaction with nontypable *Haemophilus influenzae* and *Moraxella catarrhalis*. *Otolaryngol. Head Neck Surg.* **116**:175–180.

3027. **Reddy, M. S.** 1998. Binding of the pili of *Pseudomonas aeruginosa* to a low-molecular-weight mucin and neutral cystatin of human submandibular-sublingual saliva. *Curr. Microbiol.* **37**:395–402.

3028. **Reddy, S. P., W. G. Rasmussen, and J. B. Baseman.** 1995. Molecular cloning and characterization of an adherence-related operon of *Mycoplasma genitalium*. *J. Bacteriol.* **177**:5943–5951.

3029. **Reddy, S. P., W. G. Rasmussen, and J. B. Baseman.** 1996. Correlations between *Mycoplasma pneumoniae* sensitivity to cyclosporin A and cyclophilin-mediated regulation of mycoplasma cytadherence. *Microb. Pathog.* **20**:155–169.

3030. **Reddy, S. P., W. G. Rasmussen, and J. B. Baseman.** 1996. Isolation and characterization of transposon Tn*4001*-generated, cytadherence-deficient transformants of *Mycoplasma pneumoniae* and *Mycoplasma genitalium*. *FEMS Immunol. Med. Microbiol.* **15**:199–211.

3031. **Reece, S., C. P. Simmons, R. J. Fitzhenry, S. Matthews, A. D. Phillips,**

G. Dougan, and G. Frankel. 2001. Site-directed mutagenesis of intimin alpha modulates intimin-mediated tissue tropism and host specificity. *Mol. Microbiol.* **40:**86–98.

3032. **Reed, K. A., T. A. Booth, B. H. Hirst, and M. A. Jepson.** 1996. Promotion of *Salmonella typhimurium* adherence and membrane ruffling in MDCK epithelia by staurosporine. *FEMS Microbiol. Lett.* **145:**233–238.

3033. **Reed, K. A., M. A. Clark, T. A. Booth, C. J. Hueck, S. I. Miller, B. H. Hirst, and M. A. Jepson.** 1998. Cell-contact-stimulated formation of filamentous appendages by *Salmonella typhimurium* does not depend on the type III secretion system encoded by *Salmonella* pathogenicity island 1. *Infect. Immun.* **66:**2007–2017.

3034. **Reed, W. P., C. Metzler, and E. Albright.** 1994. Streptococcal adherence to Langerhans cells: a possible step in the pathogenesis of streptococcal pharyngitis. *Clin. Immunol. Immunopathol.* **70:**28–31.

3035. **Register, K. B., M. R. Ackermann, and M. E. Kehrli, Jr.** 1994. Non-opsonic attachment of *Bordetella bronchiseptica* mediated by CD11/CD18 and cell surface carbohydrates. *Microb. Pathog.* **17:**375–385.

3036. **Register, K. B., and M. R. Ackermann.** 1997. A highly adherent phenotype associated with virulent Bvg$^+$-phase swine isolates of *Bordetella bronchiseptica* grown under modulating conditions. *Infect. Immun.* **65:**5295–5300.

3037. **Reid, G., R. Charbonneau-Smith, D. Lam, Y. S. Kang, M. Lacerte, and K. C. Hayes.** 1992. Bacterial biofilm formation in the urinary bladder of spinal cord injured patients. *Paraplegia* **30:**711–717.

3038. **Reid, G., J. D. Denstedt, Y. S. Kang, D. Lam, and C. Nause.** 1992. Microbial adhesion and biofilm formation on ureteral stents *in vitro* and *in vivo*. *J. Urol.* **148:**1592–1594.

3039. **Reid, G., and H. J. Busscher.** 1992. Importance of surface properties in bacterial adhesion to biomaterials, with particular reference to the urinary tract. *Int. Biodeterior. Biodegrad.* **30:**105–122.

3040. **Reid, G., P. L. Cuperus, A. W. Bruce, H. C. van der Mei, L. Tomeczek, A. H. Khoury, and H. J. Busscher.** 1992. Comparison of contact angles and adhesion to hexadecane of urogenital, dairy, and poultry lactobacilli: effect of serial culture passages. *Appl. Environ. Microbiol.* **58:**1549–1553.

3041. **Reid, G., A. L. Servin, A. W. Bruce, and H. J. Busscher.** 1993. Adhesion of three *Lactobacillus* strains to human urinary and intestinal epithelial cells. *Microbios* **75:**57–65.

3042. **Reid, G., and C. Tieszer.** 1993. Preferential adhesion of urethral bacteria from a mixed population to a urinary catheter. *Cells Mater.* **3:**171–176.

3043. **Reid, G., D. Lam, A. W. Bruce, H. C. van der Mei, and H. J. Busscher.** 1994. Adhesion of lactobacilli to urinary catheters and diapers: effect of surface properties. *J. Biomed. Mater. Res.* **28:**731–734.

3044. **Reid, G., L. Dafoe, G. Delaney, M. Lacerte, M. Valvano, and K. C. Hayes.** 1994. Use of adhesion counts to help predict symptomatic infection and the ability of fluoroquinolones to penetrate bacterial biofilms on the bladder cells of spinal cord injured patients. *Paraplegia* **32:**468–472.

3045. **Reid, G., S. Sharma, K. Advikolanu, C. Tieszer, R. A. Martin, and A. W. Bruce.** 1994. Effects of ciprofloxacin, norfloxacin, and ofloxacin on in vitro adhesion and survival of *Pseudomonas aeruginosa* AK1 on urinary catheters. *Antimicrob. Agents Chemother.* **38:**1490–1495.

3046. **Reid, G.** 1994. Applications from bacterial adhesion and biofilm studies in relation to urogenital tissues and biomaterials: a review. *J. Ind. Microbiol.* **13:**90–96.

3047. **Reid, G., A. E. Khoury, C. A. K. Preston, and J. W. Costerton.** 1994. Influence of dextrose dialysis solutions on adhesion of *Staphylococcus aureus* and *Pseudomonas aeruginosa* to three catheter surfaces. *Am. J. Nephrol.* **14:**37–40.

3048. **Reid, G., C. Tieszer, and D. Lam.** 1995. Influence of lactobacilli on the adhesion of *Staphylococcus aureus* and *Candida albicans* to fibers and epithelial cells. *J. Ind. Microbiol.* **15:**248–253.

3049. **Reid, G., H. C. van der Mei, C. Tieszer, and H. J. Busscher.** 1996. Uropathogenic *Escherichia coli* adhere to urinary catheters without using fimbriae. *FEMS Immunol. Med. Microbiol.* **16:**159–162.

3050. **Reinhard, J., C. Basset, J. Holton, M. Binks, P. Youinou, and D. Vaira.** 2000. Image analysis method to assess adhesion of *Helicobacter pylori* to gastric epithelium using confocal laser scanning microscopy. *J. Microbiol. Methods* **39:**179–187.

3051. **Ren, D. H., W. M. Petroll, J. V. Jester, J. Ho-Fan, and H. H. Cavanagh.** 1999. The relationship between contact lens oxygen permeability and binding of *Pseudomonas aeruginosa* to human corneal epithelial cells after overnight and extended wear. *CLAO J.* **25:**80–100.

3052. **Ren, D. H., W. M. Petroll, J. V. Jester, J.**

Ho-Fan, and H. D. Cavanagh. 1999. Short-term hypoxia downregulates epithelial cell desquamation in vivo, but does not increase *Pseudomonas aeruginosa* adherence to exfoliated human corneal epithelial cells. *CLAO J.* **25:**73–79.

3053. **Ren, H., W. M. Petroll, J. V. Jester, H. D. Cavanagh, W. D. Mathers, J. A. Bonanno, and R. H. Kennedy.** 1997. Adherence of *Pseudomonas aeruginosa* to shed rabbit corneal epithelial cells after overnight wear of contact lenses. *CLAO J.* **23:**63–68.

3054. **Renauld-Mongenie, G., J. Cornette, N. Mielcarek, F. D. Menozzi, and C. Locht.** 1996. Distinct roles of the N-terminal and C-terminal precursor domains in the biogenesis of the *Bordetella pertussis* filamentous hemagglutinin. *J. Bacteriol.* **178:**1053–1060.

3055. **Renauld-Mongenie, G., N. Mielcarek, J. Cornette, A. M. Schacht, A. Capron, G. Riveau, and C. Locht.** 1996. Induction of mucosal immune responses against a heterologous antigen fused to filamentous hemagglutinin after intranasal immunization with recombinant *Bordetella pertussis. Proc. Natl. Acad. Sci. USA* **93:**7944–7949.

3056. **Renesto, P., J. Mounier, and P. J. Sansonetti.** 1996. Induction of adherence and degranulation of polymorphonuclear leukocytes: a new expression of the invasive phenotype of *Shigella flexneri. Infect. Immun.* **64:**719–723.

3057. **Rest, R. F., and J. V. Frangipane.** 1992. Growth of *Neisseria gonorrhoeae* in CMP-*N*-acetylneuraminic acid inhibits nonopsonic (opacity-associated outer membrane protein-mediated) interactions with human neutrophils. *Infect. Immun.* **60:**989–997.

3058. **Rest, R. F.** 1995. Association of bacteria with human phagocytes. *Methods Enzymol.* **253:**12–26.

3059. **Rest, R. F., and D. P Speert.** 1994. Measurement of nonopsonic phagocytic killing by human and mouse phagocytes. *Methods Enzymol.* **236:**91–108.

3060. **Restrepo, C. I., Q. Dong, J. Savov, W. I. Mariencheck, and J. R. Wright.** 1999. Surfactant protein D stimulates phagocytosis of *Pseudomonas aeruginosa* by alveolar macrophages. *Am. J. Respir. Cell Mol. Biol.* **21:**576–585.

3061. **Reyes, L., M. K. Davidson, L. C. Thomas, and J. K. Davis.** 1999. Effects of *Mycoplasma fermentans incognitus* on differentiation of THP-1 cells. *Infect. Immun.* **67:**3188–3192.

3062. **Rhem, M. N., E. M. Lech, J. M. Patti, D.**

3063. **Rhen, M., P. Riikonen, and S. Taira.** 1993. Transcriptional regulation of *Salmonella enterica* virulence plasmid genes in cultured macrophages. *Mol. Microbiol.* **10:**45–56.

3064. **Rhine, J. A., and R. K. Taylor.** 1994. TcpA pilin sequences and colonization requirements for O1 and O139 *Vibrio cholerae. Mol. Microbiol.* **13:**1013–1020.

3065. **Riber, U., F. Espersen, and A. Kharazmi.** 1995. Comparison of adherent and non-adherent staphylococci in the induction of polymorphonuclear leukocyte activation in vitro. *APMIS* **103:**439–446.

3066. **Riber, U., and P. Lind.** 1999. Interaction between *Salmonella typhimurium* and phagocytic cells in pigs. Phagocytosis, oxidative burst and killing in polymorphonuclear leukocytes and monocytes. *Vet. Immunol. Immunopathol.* **67:**259–270.

3067. **Rich, R. L., B. Demeler, K. Ashby, C. C. Deivanayagam, J. W. Petrich, J. M. Patti, S. V. Narayana, and M. Höök.** 1998. Domain structure of the *Staphylococcus aureus* collagen adhesin. *Biochemistry* **37:**15423–15433.

3068. **Rich, R. L., C. C. Deivanayagam, R. T. Owens, M. Carson, A. Höök, D. Moore, J. Symersky, V. W. Yang, S. V. Narayana, and M. Höök.** 1999. Trench-shaped binding sites promote multiple classes of interactions between collagen and the adherence receptors, $\alpha_1\beta_1$ integrin and *Staphylococcus aureus* cna MSCRAMM. *J. Biol. Chem.* **274:**24906–24913.

3069. **Rich, R. L., B. Kreikemeyer, R. T. Owens. S. LaBrenz. S. V. Narayana, G. M. Weinstock, B. E. Murray, and M. Höök.** 1999. Ace is a collagen-binding MSCRAMM from *Enterococcus faecalis. J. Biol. Chem.* **274:**26939–26945.

3070. **Rieder, G., R. A. Hatz, A. P. Moran, A. Walz, M. Stolte, and G. Enders.** 1997. Role of adherence in interleukin-8 induction in *Helicobacter pylori*-associated gastritis. *Infect. Immun.* **65:**3622–3630.

3071. **Riise, G. C., S. Larsson, and B. A. Andersson.** 1994. Bacterial adhesion to oropharyngeal and bronchial epithelial cells in smokers with chronic bronchitis and in healthy nonsmokers. *Eur. Respir. J.* **7:**1759–1764.

3072. **Rijnaarts, H. H. M., W. Norde, E. J. Bouwer, J. Lyklema, and A. J. B.**

Zehnder. 1993. Bacterial adhesion under static and dynamic conditions. *Appl. Environ. Microbiol.* **59:**3255–3265.

3073. **Rikitomi, N., K. Ahmed, and T. Nagatake.** 1997. *Moraxella (Branhamella) catarrhalis* adherence to human bronchial and oropharyngeal cells: the role of adherence in lower respiratory tract infections. *Microbiol. Immunol.* **41:**487–494.

3074. **Rikitomi, N., T. Nagatake, T. Sakamoto, and K. Matsumoto.** 1994. The role of MRSA (methicillin-resistant *Staphylococcus aureus*) adherence and colonization in the upper respiratory tract of geriatric patients in nosocomial pulmonary infections. *Microbiol. Immunol.* **38:**607–614.

3075. **Rimler, R. B., K. B. Register, T. Magyar, and M. R. Ackermann.** Influence of chondroitinase on indirect hemagglutination titers and phagocytosis of *Pasteurella multocida* serogroups A, D and F. *Vet. Microbiol.* **47:**287–294.

3076. **Ringnér, M., P. Aleljung, and T. Wadström.** 1993. Adherence of haemagglutinating *Helicobacter pylori* to five cell lines. *Zentbl. Bakteriol.* **280:**107–112.

3077. **Ringnér, M., K. H. Valkonen, and T. Wadström.** 1994. Binding of vitronectin and plasminogen to *Helicobacteri pylori*. *FEMS Immunol. Med. Microbiol.* **9:**29–34.

3078. **Rioux, S., C. Galarneau, J. Harel, J. Frey, J. Nicolet, M. Kobisch, J. D. Dubreuil, and M. Jacques.** 1999. Isolation and characterization of mini-Tn*10* lipopolysaccharide mutants of *Actinobacillus pleuropneumoniae* serotype 1. *Can. J. Microbiol.* **45:**1017–1026.

3079. **Rioux, S., C. Galarneau, J. Harel, M. Kobisch, J. Frey, M. Gottschalk, and M. Jacques.** 2000. Isolation and characterization of a capsule-deficient mutant of *Actinobacillus pleuropneumoniae* serotype 1. *Microb. Pathog.* **28:**279–289.

3080. **Rippinger, P., H. U. Bertschinger, H. Imberechts, B. Nagy, I. Sorg, M. Stamm, P. Wild, and W. Wittig.** 1995. Designations F18ab and F18ac for the related fimbrial types F107, 2134P and 8813 of *Escherichia coli* isolated from porcine postweaning diarrhoea and from oedema disease. *Vet. Microbiol.* **45:**281–295.

3081. **Ritchings, B. W., E. C. Almira, S. Lory, and R. Ramphal.** 1995. Cloning and phenotypic charactrerization of *fleS* and *fleR,* new response regulators of *Pseudomonas aeruginosa* which regulate motility and adhesion to mucin. *Infect. Immun.* **63:**4868–4876.

3082. **Ritter, A., D. L. Gally, P. B. Olsen, U. Dobrindt, A. Friedrich, P. Klemm, and J. Hacker.** 1997. The Pai-associated *leuX* specific tRNA$_5$^{Leu} affects type 1 fimbriation in pathogenic *Escherichia coli* by conrtol of FimB recombinase expression. *Mol. Microbiol.* **25:**871–882.

3083. **Ritter, A., G. Blum, L. Emody, M. Kerenyi, A. Bock, B. Neuhierl, W. Rabsch, F. Scheutz, and J. Hacker.** 1995. tRNA genes and pathogenicity islands: influence on virulence and metabolic properties of uropathogenic *Escherichia coli*. *Mol. Microbiol.* **17:**109–121.

3084. **Rivera-Amill, V., B. J. Kim, J. Seshu, and M. E. Konkel.** 2001. Secretion of the virulence-associated *Campylobacter* invasion antigens from *Campylobacter jejuni* requires a stimulatory signal. *J. Infect. Dis.* **183:**1607–1616.

3085. **Robert, A., A. Silva, J. A. Benitez, B. L. Rodriguez, R. Fando, J. Campos, D. K. Sengupta, M. Boesman-Finkelstein, and R. A. Finkelstein.** 1996. Tagging a *Vibrio cholerae* El Tor candidate vaccine strain by disruption of its hemagglutinin/protease gene using a novel reporter enzyme: *Clostridium thermocellum* endoglucanase A. *Vaccine* **14:**1517–1522.

3086. **Roberts, J. A.** 1996. Factors predisposing to urinary tract infections in children. *Pediatr. Nephrol.* **10:**517–522.

3087. **Roberts, J. A.** 1996. Neonatal circumcision: an end to the controversy? *South. Med. J.* **89:**167–171.

3088. **Roberts, J. A.** 1996. Tropism in bacterial infections: urinary tract infections. *J. Urol.* **156:**1552–1559.

3089. **Roberts, J. A., M. B. Kaack, and E. N. Fussell.** 1993. Adherence to urethral catheters by bacteria causing nosocomial infections. *Urology* **41:**338–342.

3090. **Roberts, J. A., B. I. Marklund, D. Ilver, D. Haslam, M. B. Kaack, G. Baskin, M. Louis, R. Möllby, J. Winberg, and S. Normark.** 1994. The Gal(α1–4)Gal-specific tip adhesin of *Escherichia coli* P-fimbriae is needed for pyelonephritis to occur in the normal urinary tract. *Proc. Natl. Acad. Sci. USA* **91:**11889–11893.

3091. **Roberts, J. A., M. B. Kaack, G. Baskin, and S. B. Svenson.** 1995. Vaccination with a formalin-killed P-fimbriated *E. coli* whole-cell vaccine prevents renal scarring from pyelonephritis in the non-human primate. *Vaccine* **13:**11–16.

3092. **Roberts, J. A., M. B. Kaack, G. Baskin,**

B. I. Marklund, and S. Normark. 1997. Epitopes of the P-fimbrial adhesin of *E. coli* cause different urinary tract infections. *J. Urol.* **158:**1610–1613.

3093. Robertson, J. M., G. Grant, E. Allen-Vercoe, M. J. Woodward, A. Pusztai, and H. J. Flint. 2000. Adhesion of *Salmonella enterica* var *Enteritidis* strains lacking fimbriae and flagella to rat ileal explants cultured at the air interface or submerged in tissue culture medium. *J. Med. Microbiol.* **49:**691–696.

3094. Robins-Browne, R. M., and V. Bennett-Wood. 1992. Quantitative assessment of the ability of *Escherichia coli* to invade cultured animal cells. *Microb. Pathog.* **12:**159–164.

3095. Robins-Browne, R. M., A. M. Tokhi, L. M. Adams, V. Bennett-Wood, A. V. Moisidis, E. O. Krejany, and L. E. O'Gorman. 1994. Adherence characteristics of attaching and effacing strains of *Escherichia coli* from rabbits. *Infect. Immun.* **62:**1584–1592.

3096. Robins-Browne, R. M., A. M. Tokhi, L. M. Adams, and V. Bennett-Wood. 1994. Host specificity of enteropathogenic *Escherichia coli* from rabbits: lack of correlation between adherence in vitro and pathogenicity for laboratory animals. *Infect. Immun.* **62:**3329–3336.

3097. Robinson, A., and S. G. Funnell. 1992. Potency testing of acellular pertussis vaccines. *Vaccine* **10:**139–141.

3098. Rocha, C. L., and V. A. Fischetti. 1999. Identification and characterization of a novel fibronectin-binding protein on the surface of group A streptococci. *Infect. Immun.* **67:**2720–2728.

3099. Rocha-De-Souza, C. M., A. V. Colombo, R. Hirata, A. L. Mattos-Guaraldi, L. H. Monteiro-Leal, J. O. Previato, A. C. Freitas, and A. F. Andrade. 2001. Identification of a 43-kDa outer-membrane protein as an adhesin in *Aeromonas caviae. J. Med. Microbiol.* **50:**313–319.

3100. Roche, N., T. Larsson, J. Angstrom, and S. Teneberg. 2001. *Helicobacter pylori*-binding gangliosides of human gastric adenocarcinoma. *Glycobiology* **11:**935–944.

3101. Roche, R. J., and E. R. Moxon. 1992. The molecular study of bacterial virulence: a review of current approaches, illustrated by the study of adhesion in uropathogenic *Escherichia coli. Pediatr. Nephrol.* **6:**587–596.

3102. Rodgers, F. G., and F. C. Gibson III. 1993. Opsonin-independent adherence and intracellular development of *Legionella pneumophila* within U-937 cells. *Can. J. Microbiol.* **39:**718–722.

3103. Rodgers, J., F. Phillips, and C. Olliff.

1994. The effects of extracellular slime from *Staphylococcus epidermidis* on phagocytic ingestion and killing. *FEMS Immunol. Med. Microbiol.* **9:**109–115.

3104. Rodrigues, V. S., M. C. Vidotto, I. Felipe, D. S. Santos, and L. C. Gaziri. 1999. Apoptosis of murine peritoneal macrophages induced by an avian pathogenic strain of *Escherichia coli. FEMS Microbiol. Lett.* **179:**73–78.

3105. Rodriguez, M. E., A. L. Samo, D. F. Hozbor, and O. M. Yantorno. 1993. Effect of hydromechanical forces on the production of filamentous haemagglutinin and pertussis toxin of *Bordetella pertussis. J. Ind. Microbiol.* **12:**103–108.

3106. Rodriguez, M. E., D. F. Hozbor, and O. M. Yantorno. 1996. Effect of hydromechanical stress on cellular antigens of *Bordetella pertussis. J. Ind. Microbiol.* **17:**53–55.

3107. Roe, A. J., C. Currie, D. G. Smith, and D. L. Gally. 2001. Analysis of type 1 fimbriae expression in verotoxigenic *Escherichia coli*:a comparison between serotypes O157 and O26. *Microbiology* **147:**145–152.

3108. Roger, P., E. Puchelle, O. Bajolet-Laudinat, J. M. Tournier, C. Debordeaux, M. C. Plotkowski, J. H. Cohen, D. Sheppard, and S. de Bentzmann. 1999. Fibronectin and $\alpha_5\beta_1$ integrin mediate binding of *Pseudomonas aeruginosa* to repairing airway epithelium. *Eur. Respir. J.* **13:**1301–1309.

3109. Roger, V., J. Tenovuo, M. Lenander-Lumikari, E. Söderling, and P. Vilja. 1994. Lysozyme and lactoperoxidase inhibit the adherence of *Streptococcus mutans* NCTC 10449 (serotype c) to saliva-treated hydroxyapatite in vitro. *Caries Res.* **28:**421–428.

3110. Rogers, J. D., R. J. Palmer, Jr., P. E. Kolenbrander, and F. A. Scannapieco. 2001. Role of *Streptococcus gordonii* amylase-binding protein A in adhesion to hydroxyapatite, starch metabolism, and biofilm formation. *Infect. Immun.* **69:**7046–7056.

3111. Roggenkamp, A., H. R. Neuberger, A. Flugel, T. Schmoll, and J. Heesemann. 1995. Substitution of two histidine residues in YadA protein of *Yersinia enterocolitica* abrogates collagen binding, cell adherence and mouse virulence. *Mol. Microbiol.* **16:**1207–1219.

3112. Roggenkamp, A., K. Ruckdeschel, L. Leitritz, R. Schmitt, and J. Heesemann. 1996. Deletion of amino acids 29 to 81 in adhesion protein YadA of *Yersinia enterocolitica* serotype O:8 results in selective abrogation of adherence to neutrophils. *Infect. Immun.* **64:**2506–2514.

3113. **Rojas, I. A., J. B. Slunt, and D. W. Grainger.** 2000. Polyurethane coatings release bioactive antibodies to reduce bacterial adhesion. *J. Controlled Release* **63:**175–189.

3114. **Rojas, M., and P. L. Conway.** 1996. Colonization by lactobacilli of piglet small intestinal mucus. *J. Appl. Bacteriol.* **81:**474–480.

3115. **Romalde, J. L., and A. E. Toranzo.** 1993. Pathological activities of *Yersinia ruckeri,* the enteric redmouth (ERM) bacterium. *FEMS Microbiol Lett.* **112:**291–299.

3116. **Romero-Arroyo, C. E., J. Jordan, S. J. Peacock, M. J. Willby, M. A. Farmer, and D. C. Krause.** 1999. *Mycoplasma pneumoniae* protein P30 is required for cytadherence and associated with proper cell development. *J. Bacteriol.* **181:**1079–1087.

3117. **Römling, U., W. D. Sierralta, K. Eriksson, and S. Normark.** 1998. Multicellular and aggregative behavior of *Salmonella typhimurium* strains is controlled by mutations in the *agfD* promoter. *Mol. Microbiol.* **28:**249–264.

3118. **Roof, M. B., T. T. Kramer, and J. A. Roth.** 1992. A comparison of virulent and avirulent strains of *Salmonella choleraesuis* and their ability to invade Vero cell monolayers. *Vet. Microbiol.* **30:**355–368.

3119. **Roos, S., S. Lindgren, and H. Jonsson.** 1999. Autoaggregation of *Lactobacillus reuteri* is mediated by a putative DEAD-box helicase. *Mol. Microbiol.* **32:**427–436.

3120. **Roos, S., P. Aleljung, N. Robert, B. Lee, T. Wadström, M. Lindberg, and H. Jonsson.** 1997. A collagen binding protein from *Lactobacillus reuteri* is part of an ABC transporter system? *FEMS Microbiol. Lett.* **144:**33–38.

3121. **Roques, C. G., S. el Kaddouri, P. Barthet, J. F. Duffort, and M. Arellano.** 2000. *Fusobacterium nucleatum* involvement in adult periodontitis and possible modification of strain classification. *J. Periodontol.* **71:**1144–1150.

3122. **Rosa, A. C., A. T. Mariano, A. M. Pereira, A. Tibana, T. A. Gomes, and J. R. Andrade.** 1998 Enteropathogenicity markers in *Escherichia coli* isolated from infants with acute diarrhoea and healthy controls in Rio de Janeiro, Brazil. *J. Med. Microbiol.* **47:**781–790.

3123. **Rosa, A. C., M. A. Vieira, A. Tibana, T. A. Gomes, and J. R. Andrade.** 2001. Interactions of *Escherichia coli* strains of non-EPEC serogroups that carry *eae* and lack the EAF and *stx* gene sequences with undifferentiated and differentiated intestinal human Caco-2 cells. *FEMS Microbiol. Lett.* **200:**117–122.

3124. **Rose, R. K.** 2000. The role of calcium in oral streptococcal aggregation and the implications for biofilm formation and retention. *Biochim. Biophys. Acta* **1475:**76–82.

3125. **Rose, R. K., G. H. Dibdin, and R. P. Shellis.** 1993. A quantitative study of calcium binding and aggregation in selected oral bacteria. *J. Dent. Res.* **72:**78–84.

3126. **Rosenau, A., P. Y. Sizaret, J. M. Musser, A. Goudeau, and R. Quentin.** 1993. Adherence to human cells of a cryptic *Haemophilus* genospecies responsible for genital and neonatal infections. *Infect. Immun.* **61:**4112–4118.

3127. **Rosenow, C., P. Ryan, J. N. Weiser, S. Johnson, P. Fontan, A. Ortqvist, and H. R. Masure.** 1997. Contribution of novel choline-binding proteins to adherence, colonization and immunogenicity of *Streptococcus pneumoniae*. *Mol. Microbiol.* **25:**819–829.

3128. **Rosenshine, I., S. Ruschkowski, M. Stein, D. J. Reinscheid, S. D. Mills, and B. B. Finlay.** 1996. A pathogenic bacterium triggers epithelial signals to form a functional bacterial receptor that mediates actin pseudopod formation. *EMBO J.* **15:**2613–2624.

3129. **Rosenstein, I. J., C. T. Yuen, M. S. Stoll, and T. Feizi.** 1992. Differences in the binding specificities of *Pseudomonas aeruginosa* M35 and *Escherichia coli* C600 for lipid-linked oligosaccharides with lactose-related core regions. *Infect. Immun.* **60:**5078–5084.

3130. **Rosenstein, I. J.** 1994. The use of lipid-linked oligosaccharides (neoglycolipids) in the identification of carbohydrate receptors for microbial pathogens. *Biomed. Pharmacother.* **48:**319–326.

3131. **Rosocha, J., I. Mikula, V. Kalinacova, and Z. Kollarova.** 1995. Purification and partial immunochemical characterization of proteins of fimbriae F107 from *Escherichia coli* isolated from edema disease of pigs. *Folia Microbiol.* **40:**541–546.

3132. **Rosqvist, R., D.-E. Magnusson, and H. Wolf-Watz.** 1994. Target cell contact triggers expression and polarized transfer of *Yersinia* YopE cytotoxin into mammalian cells. *EMBO J.* **13:**964–972.

3133. **Rossoni, E. M., and C. C. Gaylarde.** 2000. Comparison of sodium hypochlorite and peracetic acid as sanitising agents for stainless steel food processing surfaces using epifluorescence microscopy. *Int. J. Food Microbiol.* **61:**81–85.

3134. **Rostand, K. S., and J. D. Esko.** 1993.

Cholesterol and cholesterol esters: host receptors for *Pseudomonas aeruginosa* adherence. *J. Biol. Chem.* **268**:24053–24059.

3135. **Rousset, É., J. Harel, and J. D. Dubreuil.** 1998. Sulfatide from the pig jejunum brush border epithelial cell surface is involved in binding of *Escherichia coli* enterotoxin b. *Infect. Immun.* **66**:5650–5658.

3136. **Rozalska, B., and T. Wadström.** 1993. Protective opsonic activity of antibodies against fibronectin-binding proteins (FnBPs) of *Staphylococcus aureus*. *Scand. J. Immunol.* **37**:575–580.

3137. **Rozalska, B., and Å. Ljungh.** 1995. Biomaterial-associated staphylococcal peritoneal infections in a neutropaenic mouse model. *FEMS Immunol. Med. Microbiol.* **11**:307–319.

3138. **Rozalski, A., Z. Sidorczyk, and K. Kotelko.** 1997. Potential virulence factors of *Proteus* bacilli. *Microbiol. Mol. Biol. Rev.* **61**:65–89.

3139. **Rozdzinski, E., and E. Tuomanen.** 1994. Interactions of bacteria with leukocyte integrins. *Methods Enzymol.* **236**:333–345.

3140. **Rozdzinski, E., J. Sandros, M. van der Flier, A. Young, B. Spellerberg, C. Bhattacharyya, J. Straub, G. Musso, S. Putney, R. Starzyk, et al.** 1995. Inhibition of leukocyte-endothelial cell interactions and inflammation by peptides from a bacterial adhesin which mimic coagulation factor X. *J. Clin. Investig.* **95**:1078–1085.

3141. **Rozdzinski, E., B. Spellerberg, M. van der Flier, C. Bhattacharyya, A. I. Hoepelman, M. A. Moranm, A. Jarpe, S. D. Putney, R. M. Starzyk, and E. Tuomanen.** 1995. Peptide from a prokaryotic adhesin blocks leukocyte migration in vitro and in vivo. *J. Infect. Dis.* **172**:785–793.

3142. **Rozdzinski, E., and E. Tuomanen.** 1995. Adhesion of microbial pathogens to leukocyte integrins: methods to study ligand mimicry. *Methods Enzymol.* **253**:3–12.

3143. **Rozen, R., G. Bachrach, M. Bronshteyn, I. Gedalia, and D. Steinberg.** 2001. The role of fructans on dental biofilm formation by *Streptococcus sobrinus, Streptococcus mutans, Streptococcus gordonii* and *Actinomyces viscosus*. *FEMS Microbiol. Lett.* **195**:205–210.

3144. **Rubino, S., G. Leori, P. Rizzu, G. Erre, M. M. Colombo, S. Uzzau, G. Masala, and P. Cappuccinelli.** 1993. TnphoA *Salmonella abortusovis* mutants unable to adhere to epithelial cells and with reduced virulence in mice. *Infect. Immun.* **61**:1786–1792.

3145. **Rubins, J. B., A. H. Paddock, D.**

Charboneau, A. M. Berry, J. C. Paton, and E. N. Janoff.** 1998. Pneumolysin in pneumococcal adherence and colonization. *Microb. Pathog.* **25**:337–342.

3146. **Rubinstein, A., M. Ezra, and J. S. Rokem.** 1992. Adhesion of bacteria on pectin casted films. *Microbios* **70**:163–170.

3147. **Rubinstein, A., R. Radai, M. Friedman, P. Fischer, and J. S. Rokem.** 1997. The effect of intestinal bacteria adherence on drug diffusion through solid films under stationary conditions. *Pharm. Res.* **14**:503–507.

3148. **Rudel, T., J. P. van Putten, C. P. Gibbs, R. Haas, and T. F. Meyer.** 1992. Interaction of two variable proteins (PilE and PilC) required for pilus-mediated adherence of *Neisseria gonorrhoeae* to human epithelial cells. *Mol. Microbiol.* **6**:3439–3450.

3149. **Rudel, T., H. J. Boxberger, and T. F. Meyer.** 1995. Pilus biogenesis and epithelial cell adherence of *Neisseria gonorrhoeae pilC* double knock-out mutants. *Mol. Microbiol.* **17**:1057–1071.

3150. **Rudel, T., D. Facius, R. Barten, I. Scheuerpflug, E. Nonnenmacher, and T. F. Meyer.** 1995. Role of pili and the phase-variable PilC protein in natural competence for transformation of *Neisseria gonorrhoeae*. *Proc. Natl. Acad. Sci. USA* **92**:7986–7990.

3151. **Rudel T., I. Scheuerpflug, and T. F. Meyer.** 1995. *Neisseria* PilC protein identified as type-4 pilus tip-located adhesin. *Nature* **373**:357–359.

3152. **Ruder, H., C. Thurn, and J. P. Guggenbichler.** 1992. Adherence of *E. coli* to uroepithelia in children and adolescents after renal transplantation. *Transplant Proc.* **24**:2576–2577.

3153. **Rudin, A., M. M. McConnell, and A.-M. Svennerholm.** 1994. Monoclonal antibodies against enterotoxigenic *Escherichia coli* colonization factor antigen I (CFA/I) that cross-react immunologically with heterologous CFAs. *Infect. Immun.* **62**:4339–4346.

3154. **Rudin, A., L. Olbe, and A.-M. Svennerholm.** 1996. Monoclonal antibodies against fimbrial subunits of colonization factor antigen I (CFA/I) inhibit binding to human enterocytes and protect against enterotoxigenic *Escherichia coli* expressing heterologous colonization factors. *Microb. Pathog.* **21**:35–45.

3155. **Rudin, A., and A.-M. Svennerholm.** 1996. Identification of a cross-reactive continuous B-cell epitope in enterotoxigenic *Escherichia coli* colonization factor antigen I. *Infect. Immun.* **64**:4508–4513.

3156. **Rudmann, D. G., K. A. Eaton, and S. Krakowka.** 1992. Ultrastructural study of

Helicobacter pylori adherence properties in gnotobiotic piglets. *Infect. Immun.* **60:**2121–2124.

3157. **Rudner, X. L., L. D. Hazlett, and R. S. Berk.** 1992. Systemic and topical protection studies using *Pseudomonas aeruginosa* flagella in an ocular model of infection. *Curr. Eye Res.* **11:**727–738.

3158. **Rudner, X. L., R. S. Berk, and L. D. Hazlett.** 1993. Immunization with homologous *Pseudomonas aeruginosa* pili protects against ocular disease. *Reg. Immunol.* **5:**245–252.

3159. **Rudney, J. D., Z. Ji, C. J. Larson, W. F. Liljemark, and K. L. Hickey.** 1995. Saliva protein binding to layers of oral streptococci *in vitro* and *in vivo. J. Dent. Res.* **74:**1280–1288.

3160. **Rudney, J. D., K. L. Hickey, and Z. Ji.** 1999. Cumulative correlations of lysozyme, lactoferrin, peroxidase, S-IgA, amylase, and total protein concentrations with adherence of oral viridans streptococci to microplates coated with human saliva. *J. Dent. Res.* **78:**759–768.

3161. **Ruehl, W. W., C. F. Marrs, L. George, S. J. Banks, and G. K. Schoolnik.** 1993. Infection rates, disease frequency, pilin gene rearrangement, and pilin expression in calves inoculated with *Moraxella bovis* pilin-specific isogenic variants. *Am. J. Vet. Res.* **54:**248–253.

3162. **Ruehl, W. W., C. Marrs, M. K. Beard, V. Shokooki, J. R. Hinojoza, S. Banks, D. Bieber, and J. S. Mattick.** 1993. Q pili enhance the attachment of *Moraxella bovis* to bovine corneas *in vitro. Mol. Microbiol.* **7:**285–288.

3163. **Ruffolo, C. G., J. M. Tennent, W. P. Michalski, and B. Adler.** 1997. Identification, purification, and characterization of the type 4 fimbriae of *Pasteurella multocida. Infect. Immun.* **65:**339–343.

3164. **Ruggieri, M. R., R. K. Balagani, J. J. Rajter, and P. M. Hanno.** 1992. Characterization of bovine bladder mucin fractions that inhibit *Escherichia coli* adherence to the mucin deficient rabbit bladder. *J. Urol.* **148:**173–178.

3165. **Ruhl, S., J. O. Cisar, and A. L. Sandberg.** 2000. Identification of polymorphonuclear leukocyte and HL-60 cell receptors for adhesins of *Streptococcus gordonii* and *Actinomyces naeslundii. Infect. Immun.* **68:** 6346–6354.

3166. **Ruiz, C., J. M. Sabio, J. L. Santos, M. J. Montes, and E. G. Olivares.** 1993. Application of flow cytometry to the study of antiphagocytic properties of *Klebsiella pneumoniae* capsular polysaccharide. *FEMS Immunol. Med. Microbiol.* **7:**63–66.

3167. **Ruiz, N., B. Wang, A. Pentland, and M.**

Caparon. 1998. Streptolysin O and adherence synergistically modulate proinflammatory responses of keratinocytes to group A streptococci. *Mol. Microbiol.* **27:**337–346.

3168. **Rumyantsev, S. N., N. P. Shabalov, M. F. Pyasetskaya, N. M. Rogacheva, and L. I. Bardakova.** 2000. Species, population and age diversity in cell resistance to adhesion of *Neisseria meningitidis* serogroups A, B and C. *Microbes Infect.* **2:**447–453.

3169. **Rundegren, J., T. Simonsson, L. Petersson, and E. Hansson.** 1992. Effect of delmopinol on the cohesion of glucan-containing plaque formed by *Streptococcus mutans* in a flow cell system. *J. Dent. Res.* **71:**1792–1796.

3170. **Rupp, M. E., and G. L. Archer.** 1992. Hemagglutination and adherence to plastic by *Staphylococcus epidermidis. Infect. Immun.* **60:**4322–4327.

3171. **Rupp, M. E., and G. L. Archer.** 1994. Coagulase-negative staphylococci: pathogens associated with medical progress. *Clin. Infect. Dis.* **19:**231–243.

3172. **Rupp, M. E., J. Han, and S. Gatermann.** 1995. Hemagglutination by *Staphylococcus aureus* strains responsible for human bacteremia or bovine mastitis. *Med. Microbiol. Immunol.* **184:**33–36.

3173. **Rupp, M. E., N. Sloot, H. G. Meyer, J. Han, and S. Gatermann.** 1995. Characterization of the hemagglutinin of *Staphylococcus epidermidis. J. Infect. Dis.* **172:**1509–1518.

3174. **Rupp, M. E., and K. E. Hamer.** 1998. Effect of subinhibitory concentrations of vancomycin, cefazolin, ofloxacin, L-ofloxacin and D-ofloxacin on adherence to intravascular catheters and biofilm formation by *Staphylococcus epidermidis. J. Antimicrob. Chemother.* **41:**155–161.

3175. **Rupp, M. E., J. S. Ulphani, P. D. Fey, K. Bartscht, and D. Mack.** 1999. Characterization of the importance of polysaccharide intercellular adhesin/hemagglutinin of *Staphylococcus epidermidis* in the pathogenesis of biomaterial-based infection in a mouse foreign body infection model. *Infect. Immun.* **67:**2627–2632.

3176. **Rupp, M. E., J. S. Ulphani, P. D. Fey, and D. Mack.** 1999. Characterization of *Staphylococcus epidermidis* polysaccharide intercellular adhesin/hemagglutinin in the pathogenesis of intravascular catheter-associated infection in a rat model. *Infect. Immun.* **67:**2656–2659.

3177. **Rupp, M. E., P. D. Fey, C. Heilmann, and F. Götz.** 2001. Characterization of the importance of *Staphylococcus epidermidis* autolysin and polysaccharide intercellular adhesin in the pathogenesis of intravascular catheter-associated infection in a rat model. *J. Infect. Dis.* **183**:1038–1042.

3178. **Russell, M. A., and A. Darzins.** 1994. The *pilE* gene product of *Pseudomonas aeruginosa*, required for pilus biogenesis, shares amino acid sequence identity with the N-termini of type 4 prepilin proteins. *Mol. Microbiol.* **13**:973–985.

3179. **Russell, R. G., B. D. Tall, and J. G. Morris, Jr.** 1992. Non-O1 *Vibrio cholerae* intestinal pathology and invasion in the removable intestinal tie adult rabbit diarrhea model. *Infect. Immun.* **60**:435–442.

3180. **Russell, R. R.** 1994. Control of specific plaque bacteria. *Adv. Dent. Res.* **8**:285–290.

3181. **Ryan, P. A., V. Pancholi, and V. A. Fischetti.** 2001. Group A streptococci bind to mucin and human pharyngeal cells through sialic acid-containing receptors. *Infect. Immun.* **69**:7402–7412.

3182. **Ryd, M., T. Schennings, M. Flock, A. Heimdahl, and J. I. Flock.** 1996. *Streptococcus mutans* major adhesion surface protein, P1 (I/II), does not contribute to attachment to valvular vegetations or to the development of endocarditis in a rat model. *Arch. Oral Biol.* **41**:999–1002.

3183. **Ryll, R. R., T. Rudel, I. Scheuerpflug, R. Barten, and T. F. Meyer.** 1997. PilC of *Neisseria meningitidis* is involved in class II pilus formation and restores pilus assembly, natural transformation competence and adherence to epithelial cells in PilC-deficient gonococci. *Mol. Microbiol.* **23**:879–892.

3184. **Ryu, H., Y. S. Kim, P. A. Grange, and F. J. Cassels.** 2001. *Escherichia coli* strain RDEC-1 AF/R1 endogenous fimbrial glycoconjugate receptor molecules in rabbit small intestine. *Infect. Immun.* **69**:640–649.

3185. **Saadi, A. T., C. C. Blackwell, M. W. Raza, V. S. James, J. Stewart, R. A. Elton, and D. M. Weir.** 1993. Factors enhancing adherence of toxigenic *Staphylococcus aureus* to epithelial cells and their possible role in sudden infant death syndrome. *Epidemiol. Infect.* **110**:507–517.

3186. **Saadi, A. T., D. M Weir, I. R. Poxton, J. Stewart, S. D. Essery, C. C. Blackwell, M. W. Raza, and A. Busuttil.** 1994. Isolation of an adhesin from *Staphylococcus aureus* that binds Lewis[a] blood group antigen and its relevance to sudden infant death syndrome. *FEMS Immunol. Med. Microbiol.* **8**:315–320.

3187. **Saadi, A. T., C. C. Blackwell, S. D. Essery, M. W. Raza, O. R. el Ahmer, D. A. MacKenzie, V. S. James, D. M. Weir, M. M. Ogilvie, R. A. Elton, A. Busuttil, and J. W. Keeling.** 1996. Development and environmental factors that enhance binding of *Bordetella pertussis* to human epithelial cells in relation to sudden infant death syndrome (SIDS). *FEMS Immunol. Med. Microbiol.* **16**:51–59.

3188. **Saadi, A. T., A. E. Gordon, D. A. MacKenzie, V. S. James, R. A. Elton, D. M. Weir, A. Busuttil, and C. C. Blackwell.** 1999. The protective effect of breast feeding in relation to sudden infant death syndrome (SIDS). I. The effect of human milk and infant formula preparations on binding of toxigenic *Staphylococcus aureus* to epithelial cells. *FEMS Immunol. Med. Microbiol.* **25**:155–165.

3189. **Saarela, S., S. Taira, E. L. Nurmiaho-Lassila, A. Makkonen, and M. Rhen.** 1995. The *Escherichia coli* G-fimbrial lectin protein participates both in fimbrial biogenesis and in recognition of the receptor N-acetyl-D-glucosamine. *J. Bacteriol.* **177**:1477–1484.

3190. **Saarela, S., B. Westerlund-Wikstrom, M. Rhen, and T. K. Korhonen.** 1996. The GafD protein of the G (F17) fimbrial complex confers adhesiveness of *Escherichia coli* to laminin. *Infect. Immun.* **64**:2857–2860.

3191. **Sachse, K., H. Pfutzner, M. Heller, and I. Hanel.** 1993. Inhibition of *Mycoplasma bovis* cytadherence by a monoclonal antibody and various carbohydrate substances. *Vet. Microbiol.* **36**:307–316.

3192. **Sachse, K., C. Grajetzki, R. Rosengarten, I. Hanel, M. Heller, and H. Pfutzner.** 1996. Mechanisms and factors involved in *Mycoplasma bovis* adhesion to host cells. *Zentbl. Bakteriol.* **284**:80–92.

3193. **Sadosky, A. B., L. A. Wiater, and H. A. Shuman.** 1993. Identification of *Legionella pneumophila* genes required for growth within and killing of human macrophages. *Infect. Immun.* **61**:5361–5373.

3194. **Saeki, Y., T. Kato, Y. Naito, I. Takazoe, and K. Okuda.** 1996. Inhibitory effects of funoran on the adherence and colonization of mutans streptococci. *Caries Res.* **30**:119–125.

3195. **Saeki, Y.** 1994. Effect of seaweed extracts on *Streptococcus sobrinus* adsorption to saliva-coated hydroxyapatite. *Bull. Tokyo Dent. Coll.* **35**:9–15.

3196. **Sahly, H., R. Podschun, T. A. Oelschlaeger, M. Greiwe, H. Parolis, D. Hasty, J. Kekow, U. Ullmann, I. Ofek,** and S. Sela. 2000. Capsule impedes adhesion to and invasion of epithelial cells by *Klebsiella pneumoniae*. *Infect. Immun.* **68:**6744–6749.

3197. Saiman, L., G. Cacalano, D. Gruenert, and A. Prince. 1992. Comparison of adherence of *Pseudomonas aeruginosa* to respiratory epithelial cells from cystic fibrosis patients and healthy subjects. *Infect. Immun.* **60:**2808–2814.

3198. Saiman, L., and A. Prince. 1993. *Pseudomonas aeruginosa* pili bind to asialoGM1 which is increased on the surface of cystic fibrosis epithelial cells. *J. Clin. Investig.* **92:**1875–1880.

3199. Sainz, T., C. Wacher, J. Espinoza, D. Centurion, A. Navarro, J. Molina, A. Inzunza, A. Cravioto, and C. Eslava. 2001. Survival and characterization of *Escherichia coli* strains in a typical Mexican acid-fermented food. *Int. J. Food Microbiol.* **71:**169–176.

3200. Saito, S., K. Hiratsuka, M. Hayakawa, H. Takiguchi, and Y. Abiko. 1997. Inhibition of a *Porphyromonas gingivalis* colonizing factor between *Actinomyces viscosus* ATCC 19246 by monoclonal antibodies against recombinant 40-kDa outer-membrane protein. *Gen. Pharmacol.* **28:**675–680.

3201. Saito, T., T. Takatsuka, T. Kato, K. Ishihara, and K. Okuda. 1997. Adherence of oral streptococci to an immobilized antimicrobial agent. *Arch. Oral Biol.* **42:**539–545.

3202. Sajjan, U. S., and J. F. Forstner. 1992. Identification of the mucin-binding adhesin of *Pseudomonas cepacia* isolated from patients with cystic fibrosis. *Infect. Immun.* **60:**1434–1440.

3203. Sajjan, U., J. Reisman, P. Doig, R. T. Irvin, G. Forstner, and J. Forstner. 1992. Binding of nonmucoid *Pseudomonas aeruginosa* to normal human intestinal mucin and respiratory mucin from patients with cystic fibrosis. *J. Clin. Investig.* **89:**657–665.

3204. Sajjan, U., Y. Wu, G. Kent, and J. Forstner. 2000. Preferential adherence of cable-piliated *Burkholderia cepacia* to respiratory epithelia of CF knockout mice and human cystic fibrosis lung explants. *J. Med. Microbiol.* **49:**875–885.

3205. Sajjan, U. S., M. Corey, M. A. Karmali, and J. F. Forstner. 1992. Binding of *Pseudomonas cepacia* to normal human intestinal mucin and respiratory mucin from patients with cystic fibrosis. *J. Clin. Investig.* **89:**648–656.

3206. Sajjan, U. S., and J. F. Forstner. 1993. Role of a 22-kilodalton pilin protein in binding of *Pseudomonas cepacia* to buccal epithelial cells. *Infect. Immun.* **61:**3157–3163.

3207. Sajjan, U. S., L. Sun, R. Goldstein, and J. F. Forstner. 1995. Cable (*cbl*) type II pili of cystic fibrosis-associated *Burkholderia* (*Pseudomonas*) *cepacia*: nucleotide sequence of the *cblA* major subunit pilin gene and novel morphology of the assembled appendage fibers. *J. Bacteriol.* **177:**1030–1038.

3208. Sajjan, U. S., F. A. Sylvester, and J. F. Forstner. 2000. Cable-piliated *Burkholderia cepacia* binds to cytokeratin 13 of epithelial cells. *Infect. Immun.* **68:**1787–1795.

3209. Sakai, Y., M. Fujisawa, Y. Nakano, S. Miyazaki, S. Arakawa, and S. Kamidono. 2000. Bacterial adherence in a rat bladder augmentation model: ileocystoplasty versus colocystoplasty. *J. Urol.* **164:**2104–2107.

3210. Sakanaka, S., M. Aizawa, M. Kim, and T. Yamamoto. 1996. Inhibitory effects of green tea polyphenols on growth and cellular adherence of an oral bacterium, *Porphyromonas gingivalis*. *Biosci. Biotechnol. Biochem.* **60:**745–749.

3211. Sakellaris, H., D. P. Balding, and J. R. Scott. 1996. Assembly proteins of CS1 pili of enterotoxigenic *Escherichia coli*. *Mol. Microbiol.* **21:**529–541.

3212. Sakellaris, H., and J. R. Scott. 1998. New tools in an old trade: CS1 pilus morphogenesis. *Mol. Microbiol.* **30:**681–687.

3213. Sakellaris, H., G. P. Munson, and J. R. Scott. 1999. A conserved residue in the tip proteins of CS1 and CFA/I pili of enterotoxigenic *Escherichia coli* that is essential for adherence. *Proc. Natl. Acad. Sci. USA* **96:**12828–12832.

3214. Sakurai, S., N. Shinagawa, T. Fukui, and J. Yura. 1992. Bacterial adherence to human gallbladder epithelium. *Surg. Today* **22:**504–507.

3215. Sakurai, Y., H. Suzuki, and E. Terada. 1993. Purification and characterisation of haemagglutinin from *Bordetella bronchiseptica*. *J. Med. Microbiol.* **39:**388–392.

3216. Salajka, E., Z. Salajkova, P. Alexa, and M. Hornich. 1992. Colonization factor different from K88, K99, F41 and 987P in enterotoxigenic *Escherichia coli* strains isolated from postweaning diarrhoea in pigs. *Vet. Microbiol.* **32:**163–175.

3217. Salasia, S. I., C. Lammler, and L. A. Devriese. 1994. Serotypes and putative virulence markers of *Streptococcus suis* isolates from cats and dogs. *Res. Vet. Sci.* **57:**259–261.

3218. Salasia, S. I., C. Lammler, and G. Herrmann. 1995. Properties of a *Streptococcus suis* isolate of serotype 2 and two capsular mutants. *Vet. Microbiol.* **45:**151–156.

3219. **Salih, B. A., and R. F. Rosenbusch.** 1999. Interactions of *Mycoplasma bovoculi* with erythrocytes: role of p94 surface protein. *Zentbl. Veterinaermed. Reihe B* **46:**323–329.

3220. **Salminen, S., E. Isolauri, and E. Salminen.** 1996. Clinical uses of probiotics for stabilizing the gut mucosal barrier: successful strains and future challenges. *Antonie Leeuwenhoek* **70:**347–358.

3221. **Salmond, G. P.** 1996. Pili, peptidases and protein secretion: curious connections. *Trends Microbiol.* **4:**474–476.

3222. **Saltman, L. H., Y. Lu, E. M. Zaharias, and R. R. Isberg.** 1996. A region of the *Yersinia pseudotuberculosis* invasin protein that contributes to high affinity binding to integrin receptors. *J. Biol. Chem.* **271:**23438–23444.

3223. **Samaranayake, L. P., D. Hamilton, and T. W. MacFarlane.** 1994. The effect of indigenous bacterial populations on buccal epithelial cells on subsequent microbial adhesion *in vitro*. *Oral Microbiol. Immunol.* **9:**236–240.

3224. **Sambri, V., F. Basso, F. Massaria, M. Ardizzoni, and R. Cevenini.** 1993. Adherence of *Borrelia burgdorferi* and *Borrelia hermsii* to mammalian cells *in vitro*. *New Microbiol.* **16:**43–49.

3225. **Sampson, J. S., S. P. O'Connor, A. R. Stinson, J. A. Tharpe, and H. Russell.** 1994. Cloning and nucleotide sequence analysis of *psaA*, the *Streptococcus pneumoniae* gene encoding a 37-kilodalton protein homologous to previously reported *Streptococcus* sp. adhesins. *Infect. Immun.* **62:**319–324.

3226. **Sanchez, R., L. Kanarek, J. Koninkx, H. Hendriks, P. Lintermans, A. Bertels, G. Charlier, and E. Van Driessche.** 1993. Inhibition of adhesion of enterotoxigenic *Escherichia coli* cells expressing F17 fimbriae to small intestinal mucus and brush-border membranes of young calves. *Microb. Pathog.* **15:**207–219.

3227. **Sandberg, A. L., S. Ruhl, R. A. Joralmon, M. J. Brennan, M. J. Sutphin, and J. O. Cisar.** 1995. Putative glycoprotein and glycolipid polymorphonuclear leukocyte receptors for the *Actinomyces naeslundii* WVU45 fimbrial lectin. *Infect. Immun.* **63:**2625–2631.

3228. **Sandros, J., P. Papapanou, and G. Dahlen.** 1993. *Porphyromonas gingivalis* invades oral epithelial cells *in vitro*. *J. Periodontal. Res.* **28:**219–226.

3229. **Sandros, J., and E. Tuomanen.** 1993. Attachment factors of *Bordetella pertussis*:mimicry of eukaryotic cell recognition molecules. *Trends Microbiol.* **1:**192–196.

3230. **Sandros, J., P. N. Papapanou, U. Nannmark, and G. Dahlen.** 1994. *Porphyromonas gingivalis* invades human pocket epithelium *in vitro*. *J. Periodontal Res.* **29:**62–69.

3231. **Sandros, J., C. Karlsson, D. F. Lappin, P. N. Madianos, D. F. Kinane, and P. N. Papapanou.** 2000. Cytokine responses of oral epithelial cells to *Porphyromonas gingivalis* infection. *J. Dent. Res.* **79:**1808–1814.

3232. **Sandt, C. H., Y.-D. Wang, R. A. Wilson, and C. W. Hill.** 1997. *Escherichia coli* strains with nonimmune immunoglobulin-binding activity. *Infect. Immun.* **65:**4572–4579.

3233. **Sandt, C. H., and C. W. Hill.** 2000. Four different genes responsible for nonimmune immunoglobulin-binding activities within a single strain of *Escherichia coli*. *Infect. Immun.* **68:**2205–2214.

3234. **Sansonetti, P. J.** 1992. Molecular and cellular biology of *Shigella flexneri* invasiveness: from cell assay systems to shigellosis. *Curr. Top. Microbiol. Immunol.* **180:**1–19.

3235. **Sansonetti, P. J.** 1993. Bacterial pathogens, from adherence to invasion: comparative strategies. *Med. Microbiol. Immunol.* **182:**223–232.

3236. **Sansonetti, P. J., J. Arondel, J. R. Cantey, M. C. Prevost, and M. Huerre.** 1996. Infection of rabbit Peyer's patches by *Shigella flexneri*: effect of adhesive or invasive bacterial phenotypes on follicle-associated epithelium. *Infect. Immun.* **64:**2752–2764.

3237. **Santiago, N. I., A. Zipf, and A. K. Bhunia.** 1999. Influence of temperature and growth phase on expression of a 104-kilodalton *Listeria* adhesion protein in *Listeria monocytogenes*. *Appl. Environ. Microbiol.* **65:**2765–2769.

3238. **Sapatnekar, S., K. M. Kieswetter, K. Merritt, J. M. Anderson, L. Cahalan, M. Verhoeven, M. Hendriks, B. Fouache, and P. Cahalan.** 1995. Blood-biomaterial interactions in a flow system in the presence of bacteria: effect of protein adsorption. *J. Biomed. Mater. Res.* **29:**247–256.

3239. **Sarem, F., L. O. Sarem-Damerdji, and J. P. Nicolas.** 1996. Comparison of the adherence of three *Lactobacillus* strains to Caco-2 and Int-407 human intestinal cell lines. *Lett. Appl. Microbiol.* **22:**439–442.

3240. **Saren, A., R. Virkola, J. Hacker, and T. K. Korhonen.** 1999. The cellular form of human fibronectin as an adhesion target for the S fimbriae of meningitis-associated *Escherichia coli*. *Infect. Immun.* **67:**2671–2676.

3241. **Saridakis, H. O., S. A. el Gared, M. C.**

Vidotto, and B. E. Guth. 1997. Virulence properties of *Escherichia coli* strains belonging to enteropathogenic (EPEC) serogroups isolated from calves with diarrhea. *Vet. Microbiol.* **54**:145–153.

3242. **Sarrazin, E., and H. U. Bertschinger.** 1997. Role of fimbriae F18 for actively acquired immunity against porcine enterotoxigenic *Escherichia coli*. *Vet. Microbiol.* **54**:133–144.

3243. **Sartingen, S., E. Rozdzinski, A. Muscholl-Silberhorn, and R. Marre.** 2000. Aggregation substance increases adherence and internalization, but not translocation, of *Enterococcus faecalis* through different intestinal epithelial cells in vitro. *Infect. Immun.* **68**:6044–6047.

3244. **Sasaki, T., T. Kenri, N. Okazaki, M. Iseki, R. Yamashita, M. Shintani, Y. Sasaki, and M. Yayoshi.** 1996. Epidemiological study of *Mycoplasma pneumoniae* infections in Japan based on PCR-restriction fragment length polymorphism of the P1 cytadhesin gene. *J. Clin. Microbiol.* **34**:447–449.

3245. **Sasmal, D., B. Guhathakurta, A. N. Ghosh, C. R. Pal, and A. Datta.** 1992. *N*-acetyl-D-glucosamine-specific lectin purified from *Vibrio cholerae* 01. *FEMS Microbiol. Lett.* **77**:217–224.

3246. **Sasmal, D., B. Guhathakurta, A. N. Ghosh, C. R. Pal, and A. Datta.** 1995. Studies on adhesion, haemagglutination and other biological properties of *Vibrio cholerae* O139. *FEMS Immunol. Med. Microbiol.* **10**:199–205.

3247. **Sasmal, D., B. Guhathakurta, S. K. Bhattacharya, C. R. Pal, and A. Datta.** 1996. Role of cell-associated *N*-acetyl-D-glucosamine specific haemagglutinin in the adhesion of *Vibrio cholerae* O1 to rabbit intestinal epithelial cells in vitro. *FEMS Immunol. Med. Microbiol.* **13**:101–105.

3248. **Sasmal, D., B. Guhathakurta, A. N. Ghosh, C. R. Pal, and A. Datta.** 1999. Purification of a mannose/glucose-specific hemagglutinin/lectin from a *Vibrio cholerae* O1 strain. *FEMS Immunol. Med. Microbiol.* **23**:221–227.

3249. **Sato, Y., Y. Yamamoto, and H. Kizaki.** 1997. Cloning and sequence analysis of the *gbpC* gene encoding a novel glucan-binding protein of *Streptococcus mutans*. *Infect. Immun.* **65**:668–675.

3250. **Sato, Y., Y. Yamamoto, and H. Kizaki.** 2000. Xylitol-induced elevated expression of the *gbpC* gene in a population of *Streptococcus mutans* cells. *Eur. J. Oral Sci.* **108**:538–545.

3251. **Saulino, E. T., D. G. Thanassi, J. S.**

Pinkner, and S. J. Hultgren. 1998. Ramifications of kinetic partitioning on usher-mediated pilus biogenesis. *EMBO J.* **17**:2177–2185.

3252. **Saunders, J. R., H. O'Sullivan, J. Wakeman, G. Sims, C. A. Hart, M. Virji, J. E. Heckels, C. Winstanley, J. A. Morgan, and R. W. Pickup.** 1993. Flagella and pili as antigenically variable structures on the bacterial surface. *J. Appl. Bacteriol.* **74**:33S–42S.

3253. **Sauter, S. L., S. M. Rutherfurd, C. Wagener, J. E. Shively, and S. A. Hefta.** 1993. Identification of the specific oligosaccharide sites recognized by type 1 fimbriae from *Escherichia coli* on nonspecific cross-reacting antigen, a CD66 cluster granulocyte glycoprotein. *J. Biol. Chem.* **268**:15510–15516.

3254. **Savage, D. C.** 1992. Growth phase, cellular hydrophobicity, and adhesion in vitro of lactobacilli colonizing the keratinizing gastric epithelium in the mouse. *Appl. Environ. Microbiol.* **58**:1992–1995.

3255. **Savarino, S. J.** 1993. Diarrheal disease: current concepts and future challenges. Enteroadherent *Escherichia coli*: a heterogeneous group of *E. coli* implicated as diarrheal pathogens. *Trans. R. Soc. Trop. Med. Hyg.* **87**:49S–53S.

3256. **Savarino, S. J., P. Fox, Y. Deng, and J. P. Nataro.** 1994. Identification and characterization of a gene cluster mediating enteroaggregative *Escherichia coli* aggregative adherence fimbria I biogenesis. *J. Bacteriol.* **176**:4949–4957.

3257. **Savelkoul, P. H., B. Kremer, J. G. Kusters, B. A. van der Zeijst, and W. Gaastra.** 1993. Invasion of HeLa cells by *Bordetella bronchiseptica*. *Microb. Pathog.* **14**:161–168.

3258. **Savelkoul, P. H. M., D. P. G. de Kerf, R. J. Willems, F. R. Mooi, B. A. M. van der Zeijst, and W. Gaastra.** 1996. Characterization of the *fim2* and *fim3* fimbrial subunit genes of *Bordetella bronchiseptica*: roles of Fim2 and Fim3 fimbriae and flagella in adhesion. *Infect. Immun.* **64**:5098–5105.

3259. **Savett, D. A., and A. Progulske-Fox.** 1995. Restriction fragment length polymorphism analysis of two hemagglutinin loci, serotyping and agglutinating activity of *Porphyromonas gingivalis* isolates. *Oral Microbiol. Immunol.* **10**:1–7.

3260. **Savkovic, S. D., A. Koutsouris, and G. Hecht.** 1996. Attachment of a noninvasive enteric pathogen, enteropathogenic *Escherichia coli*, to cultured human intestinal epithelial

monolayers induces transmigration of neutrophils. *Infect. Immun.* **64:**4480–4487.

3261. **Savoia, D., M. Millesimo, and G.
Fontana.** 1994. Evaluation of virulence factors and the adhesive capability of *Escherichia
coli* strains. *Microbios* **80:**73–81.

3262. **Savolainen, K., L. Paulin, B. Westerlund-
Wikstrom, T. J. Foster, T. K. Korhonen,
and P. Kuusela.** 2001. Expression of *pls,* a
gene closely associated with the *mecA* gene of
methicillin-resistant *Staphylococcus aureus,* prevents bacterial adhesion in vitro. *Infect. Immun.*
69:3013–3020.

3263. **Sawyer, R. T., D. A. Drevets, P. A.
Campbell, and T. A. Potter.** 1996.
Internalin A can mediate phagocytosis of
Listeria monocytogenes by mouse macrophage
cell lines. *J. Leukoc. Biol.* **60:**603–610.

3264. **Scaglione, F., G. Demartini, S. Dugnani,
F. Ferrara, G. Maccarinelli, C. Cocuzza,
and F. Fraschini.** 1994. Effect of antibiotics
on *Bordetella pertussis* adhering activity:
hypothesis regarding mechanism of action.
Chemotherapy **40:**215–220.

3265. **Scaletsky, I. C., M. S. Gatti, J. F. da
Silveira, I. M. DeLuca, E. Freymuller,
and L. R. Travassos.** 1995. Plasmid coding
for drug resistance and invasion of epithelial
cells in enteropathogenic *Escherichia coli*
0111:H. *Microb. Pathog.* **18:**387–399.

3266. **Scaletsky, I. C., M. Z. Pedroso, and U.
Fagundes-Neto.** 1996. Attaching and effacing enteropathogenic *Escherichia coli* O18ab
invades epithelial cells and causes persistent
diarrhea. *Infect. Immun.* **64:**4876–4881.

3267. **Scaletsky, I. C., M. Z. Pedroso, C. A.
Oliva, R. L. Carvalho, M. B. Morais, and
U. Fagundes-Neto.** 1999. A localized adher-
ence-like pattern as a second pattern of adherence of classic enteropathogenic *Escherichia coli*
to HEp-2 cells that is associated with infantile
diarrhea. *Infect. Immun.* **67:**3410–3415.

3268. **Scaletsky, I. C., M. Z. Pedroso, and R.
M. Silva.** 1999. Phenotypic and genetic features of *Escherichia coli* strains showing simultaneous expression of localized and diffuse
adherence. *FEMS Immunol. Med. Microbiol.*
23:181–188.

3269. **Scannapieco, F. A., G. G. Haraszthy,
M.-I. Cho, and M. J. Levine.** 1992.
Characterization of an amylase-binding com-
ponent of *Streptococcus gordonii* G9B. *Infect.
Immun.* **60:**4726–4733.

3270. **Scannapieco, F. A., G. Torres, and M. J.
Levine.** 1993. Salivary alpha-amylase: role in
dental plaque and caries formation. *Crit. Rev.
Oral Biol. Med.* **4:**301–307.

3271. **Scannapieco, F. A.** 1994. Saliva-bacterium
interactions in oral microbial ecology. *Crit.
Rev. Oral Biol. Med.* **5:**203–248.

3272. **Scannapieco, F. A., L. Solomon, and R.
O. Wadenya.** 1994. Emergence in human
dental plaque and host distribution of amylase-binding streptococci. *J. Dent. Res.* **73:**
1627–1635.

3273. **Scannapieco, F. A., G. I. Torres, and M.
J. Levine.** 1995. Salivary amylase promotes
adhesion of oral streptococci to hydroxyapatite. *J. Dent. Res.* **74:**1360–1366.

3274. **Schamhart, D. H., E. C. de Boer, R. F.
Bevers, K. H. Kurth, and P. A.
Steerenberg.** 1992. Mycobacterial adherence
and BCG treatment of superficial bladder cancer. *Prog. Clin. Biol. Res.* **378:**75–80.

3275. **Schamhart, D. H., E. C. de Boer, and K.
H. Kurth.** 1994. Interaction between bacteria
and the lumenal bladder surface: modulation
by pentosan polysulfate, an experimental and
theoretical approach with clinical implication.
World J. Urol. **12:**27–37.

3276. **Scharfman, A., G. Lamblin, and P.
Roussel.** 1995. Interactions between human
respiratory mucins and pathogens. *Biochem.
Soc. Trans.* **23:**836–839.

3277. **Scharfman, A., E. Van Brussel, N.
Houdret, G. Lamblin, and P. Roussel.**
1996. Interactions between glycoconjugates
from human respiratory airways and *Pseudomonas aeruginosa. Am. J. Respir. Crit. Care Med.*
154:S163–S169.

3278. **Scharfman, A., H. Kroczynski, C.
Carnoy, E. van Brussel, G. Lamblin, R.
Ramphal, and P. Roussel.** 1996. Adhesion
of *Pseudomonas aeruginosa* to respiratory mucins
and expression of mucin-binding proteins are
increased by limiting iron during growth.
Infect. Immun. **64:**5417–5420.

3279. **Scharfman, A., S. Degroote, J. Beau,
G. Lamblin, P. Roussel, and J.
Mazurier.** 1999. *Pseudomonas aeruginosa*
binds to neoglycoconjugates bearing mucin
carbohydrate determinants and predominantly to sialyl-Lewis x conjugates. *Glycobiology* **9:**757–764.

3280. **Schauer, D. B., and S. Falkow.** 1993. The
eae gene of *Citrobacter freundii* biotype 4280 is
necessary for colonization in transmissible
murine colonic hyperplasia. *Infect. Immun.*
61:4654–4661.

3281. **Schauer, D. B., S. N. McCathey, B. M.
Daft, S. S. Jha, L. E. Tatterson, N. S.
Taylor, and J. G. Fox.** 1998. Proliferative
enterocolitis associated with dual infection
with enteropathogenic *Escherichia coli* and

Lawsonia intracellularis in rabbits. *J. Clin. Microbiol.* **36:**1700–1703.

3282. **Scheie, A. A.** 1994. Mechanisms of dental plaque formation. *Adv. Dent. Res.* **8:**246–253.

3283. **Schembri, M. A., K. Kjaergaard, and P. Klemm.** 1999. Bioaccumulation of heavy metals by fimbrial designer adhesins. *FEMS Microbiol. Lett.* **170:**363–371.

3284. **Schembri, M. A., E. V. Sokurenko, and P. Klemm.** 2000. Functional flexibility of the FimH adhesin: insights from a random mutant library. *Infect. Immun.* **68:**2638–2646.

3285. **Schembri, M. A., and P. Klemm.** 2001. Biofilm formation in a hydrodynamic environment by novel *fimH* variants and ramifications for virulence. *Infect. Immun.* **69:**1322–1328.

3286. **Schembri, M. A., G. Christiansen, and P. Klemm.** 2001. FimH-mediated autoaggregation of *Escherichia coli. Mol. Microbiol.* **41:**1419–1430.

3287. **Schenkels, L. C. P. M., A. J. M. Ligtenberg, E. C. I. Veerman, and A. van Niew Amerongen.** 1993. Interaction of the salivary glycoprotein EP-GP with the bacterium *Streptococcus salivarius* HB. *J. Dent. Res.* **72:**1559–1565.

3288. **Schennings, T., A. Heimdahl, K. Coste, and J. I. Flock.** 1993. Immunization with fibronectin binding protein from *Staphylococcus aureus* protects against experimental endocarditis in rats. *Microb. Pathog.* **15:**227–236.

3289. **Scheuerpflug, I., T. Rudel, R. Ryll, J. Pandit, and T. F. Meyer.** 1999. Roles of PilC and PilE proteins in pilus-mediated adherence of *Neisseria gonorrhoeae* and *Neisseria meningitidis* to human erythrocytes and endothelial and epithelial cells. *Infect. Immun.* **67:**834–843.

3290. **Schiavano, G. F., F. Bruscolini, A. Albano, and, G. Brandi.** 1998. Virulence factors in *Aeromonas* spp and their association with gastrointestinal disease. *New Microbiol.* **21:**23–30.

3291. **Schiemann, D. A.** 1995. Association with MDCK epithelial cells by *Salmonella typhimurium* is reduced during utilization of carbohydrates. *Infect. Immun.* **63:**1462–1467.

3292. **Schierholz, J., B. Jansen, L. Jaenicke, and G. Pulverer.** 1994. *In-vitro* efficacy of an antibiotic releasing silicone ventricle catheter to prevent shunt infection. *Biomaterials* **15:**996–1000.

3293. **Schifferli, D. M., and M. A. Alrutz.** 1994. Permissive linker insertion sites in the outer membrane protein of 987P fimbriae of *Escherichia coli. J. Bacteriol.* **176:**1099–1110.

3294. **Schifferli, D. M.** 1995. Use of TnphoA and

T7 RNA polymerase to study fimbrial proteins. *Methods Enzymol.* **253:**242–258.

3295. **Schilling, K. M., and W. H. Bowen.** 1992. Glucans synthesized in situ in experimental salivary pellicle function as specific binding sites for *Streptococcus mutans. Infect. Immun.* **60:**284–295.

3296. **Schilling, K. M., R. G. Carson, C. A. Bosko, G. D. Golikeri, A. Bruinooge, K. Hoyberg, A. M. Wller, and N. P. Hughes.** 1994. A microassay for bacterial adherence to hydroxyapatite. *Colloid Surf. Ser. B* **3:**31–38.

3297. **Schilling, K. M., and R. J. Doyle.** 1995. Bacterial adhesion to hydroxylapatite. *Methods Enzymol.* **253:**536–542.

3298. **Schlager, T. A., J. A. Lohr, and J. O. Hendley.** 1993. Antibacterial activity of the bladder mucosa. *Urol. Res.* **21:**313–317.

3299. **Schlager, T. A., T. S. Whittam, J. O. Hendley, R. J. Hollis, M. A. Pfaller, R. A. Wilson, and A. Stapleton.** 1995. Comparison of expression of virulence factors by *Escherichia coli* causing cystitis and *E. coli* colonizing the periurethra of healthy girls. *J. Infect. Dis.* **172:**772–777.

3300. **Schlech, W.F., III, Q. Luo, G. Faulkner, and S. Galsworthy.** 1994. Interaction of *Listeria* species with human cell monolayers. *Clin. Investig. Med.* **17:**9–17.

3301. **Schlesinger, L. S., S. R. Hull, and T. M. Kaufman.** 1994. Binding of the terminal mannosyl units of lipoarabinomannan from a virulent strain of *Mycobacterium tuberculosis* to human macrophages. *J. Immunol.* **152:**4070–4079.

3302. **Schmidt, H., C. Knop, S. Franke, S. Aleksic, J. Heesemann, and H. Karch.** 1995. Development of PCR for screening of enteroaggregative *Escherichia coli. J. Clin. Microbiol.* **33:**701–705.

3303. **Schmidt, H., E. Schloricke, R. Fislage, H. A. Schulze, and R. Guthoff.** 1998. Effect of surface modifications of intraocular lenses on the adherence of *Staphylococcus epidermidis. Zentbl. Bakteriol.* **287:**135–145.

3304. **Schmidt, M. A.** 2001. Glycosylation with heptose residues mediated by the *aah* gene product is essential for adherence of the AIDA-I adhesin. *Mol. Microbiol.* **40:**1403–1413.

3305. **Schneider, B., A. Thanhauser, D. Jocham, H. Loppnow, E. Vollmer, J. Galle, H. D. Flad, A. J. Ulmer, and A. Bohle.** 1994. Specific binding of bacillus Calmette-Guerin to urothelial tumor cells *in vitro. World J. Urol.* **12:**337–344.

3306. **Schneitz, C., L. Nuotio, and K. Lounatma.** 1993. Adhesion of *Lactobacillus acidophilus* to avian intestinal epithelial cells mediated by the crystalline bacterial cell surface layer (S-layer). *J. Appl. Bacteriol.* **74:**290–294.

3307. **Schnitzler, N., G. Haase, A. Podbielski, R. Lutticken, and K. G. Schweizer.** 1999. A co-stimulatory signal through ICAM-beta2 integrin-binding potentiates neutrophil phagocytosis. *Nat. Med.* **5:**231–235.

3308. **Schoeb, T. R., M. M. Juliana, P. W. Nichols, J. K. Davis, and J. R. Lindsey.** 1993. Effects of viral and mycoplasmal infections, ammonia exposure, vitamin A deficiency, host age, and organism strain on adherence of *Mycoplasma pulmonis* in cultured rat tracheas. *Lab. Anim. Sci.* **43:**417–424.

3309. **Schonian, G., W. Sokolowska-Kohler, R. Bollmann, A. Schubert, Y. Graser, and W. Presber.** 1992. Determination of S fimbriae among *Escherichia coli* strains from extraintestinal infections by colony hybridization and dot enzyme immunoassay. *Zentbl. Bakteriol.* **276:**273–279.

3310. **Schoolnik, G. K.** 1994. Purification of somatic pili. *Methods Enzymol.* **236:**271–282.

3311. **Schoor, R. A., B. Anderson, D. J. Klumpp, and A. J. Schaeffer.** 2001. Secretory IGA differentially promotes adherence of type 1-piliated *Escherichia coli* to immortalized vaginal epithelial cell lines. *Urology* **57:**556–561.

3312. **Schorey, J. S., Q. Li, D. W. McCourt, M. Bong-Mastek, J. E. Clark-Curtiss, T. L. Ratliff, and E. J. Brown.** 1995. A *Mycobacterium leprae* gene encoding a fibronectin binding protein is used for efficient invasion of epithelial cells and Schwann cells. *Infect. Immun.* **63:**2652–2657.

3313. **Schrager, H. M., S. Alberti, C. Cywes, G. J. Dougherty, and M. R. Wessels.** 1998. Hyaluronic acid capsule modulates M protein-mediated adherence and acts as a ligand for attachment of group A streptococcus to CD44 on human keratinocytes. *J. Clin. Investig.* **101:**1708–1716.

3314. **Schroder, W., and I. Moser.** 1997. Primary structure analysis and adhesion studies on the major outer membrane protein of *Campylobacter jejuni. FEMS Microbiol. Lett.* **150:**141–147.

3315. **Schroeder, T. H., N. Reiniger, G. Meluleni, M. Grout, F. T. Coleman, and G. B. Pier.** 2001. Transgenic cystic fibrosis mice exhibit reduced early clearance of *Pseudomonas aeruginosa* from the respiratory tract. *J. Immunol.* **166:**7410–7418.

3316. **Schroeder, T. H., T. Zaidi, and G. B. Pier.** 2001. Lack of adherence of clinical isolates of *Pseudomonas aeruginosa* to asialo-GM$_1$ on epithelial cells. *Infect. Immun.* **69:**719–729.

3317. **Schroten, H., F. G. Hanisch, R. Plogmann, J. Hacker, G. Uhlenbruck, R. Nobis-Bosch, and V. Wahn.** 1992. Inhibition of adhesion of S-fimbriated *Escherichia coli* to buccal epithelial cells by human milk fat globule membrane components: a novel aspect of the protective function of mucins in the nonimmunoglobulin fraction. *Infect. Immun.* **60:**2893–2899.

3318. **Schroten, H., A. Lethen, F. G. Hanisch, R. Plogmann, J. Hacker, R. Nobis-Bosch, and V. Wahn.** 1992. Inhibition of adhesion of S-fimbriated *Escherichia coli* to epithelial cells by meconium and feces of breast-fed and formula-fed newborns: mucins are the major inhibitory component. *J. Pediatr. Gastroenterol. Nutr.* **15:**150–158.

3319. **Schroten, H., M. Steinig, R. Plogmann, F. G. Hanisch, J. Hacker, P. Herzig, and V. Wahn.** 1992. S-fimbriae mediated adhesion of *Escherichia coli* to human buccal epithelial cells is age independent. *Infection* **20:**273–275.

3320. **Schroten, H., R. Plogmann, F. G. Hanisch, J. Hacker, R. Nobis-Bosch, and V. Wahn.** 1993. Inhibition of adhesion of S-fimbriated *E. coli* to buccal epithelial cells by human skim milk is predominantly mediated by mucins and depends on the period of lactation. *Acta Paediatr.* **82:**6–11.

3321. **Schubert, R. H., and A. Holz-Bremer.** 1999. Cell adhesion of *Plesiomonas shigelloides. Zentbl. Hyg. Umweltmed.* **202:**383–388.

3322. **Schulte, R., and I. B. Autenrieth.** 1998. *Yersinia enterocolitica*-induced interleukin-8 secretion by human intestinal epithelial cells depends on cell differentiation. *Infect. Immun.* **66:**1216–1224.

3323. **Schulte, R., R. Zumbihl, D. Kampik, A. Fauconnier, and I. B. Autenrieth.** 1998. Wortmannin blocks *Yersinia* invasin-triggered internalization, but not interleukin-8 production by epithelial cells. *Med. Microbiol. Immunol.* **187:**53–60.

3324. **Schultsz, C., M. Moussa, R. van Ketel, G. N. Tytgat, and J. Dankert.** 1997. Frequency of pathogenic and enteroadherent *Escherichia coli* in patients with inflammatory bowel disease and controls. *J. Clin. Pathol.* **50:**573–579.

3325. **Schultz, C. L., M. R. Pezzutti, D. Silor, and R. White.** 1995. Bacterial adhesion measurements on soft contact lenses using a

Modified Vortex Device and a Modified Robbins Device. *J. Ind. Microbiol.* **15**:243–247.

3326. **Schulze-Koops, H., H. Burkhardt, J. Heesemann, K. von der Mark, and F. Emmrich.** 1992. Plasmid-encoded outer membrane protein YadA mediates specific binding of enteropathogenic yersiniae to various types of collagen. *Infect. Immun.* **60**:2153–2159.

3327. **Schulze-Koops, H., H. Burkhardt, J. Heesemann, T. Kirsch, B. Swoboda, C. Bull, S. Goodman, and F. Emmrich.** 1993. Outer membrane protein YadA of enteropathogenic yersiniae mediates specific binding to cellular but not plasma fibronectin. *Infect. Immun.* **61**:2513–2519.

3328. **Schulze-Koops, H., H. Burkhardt, J. Heesemann, K. von der Mark, and F. Emmrich.** 1995. Characterization of the binding region for the *Yersinia enterocolitica* adhesin YadA on types I and II collagen. *Arthritis Rheum.* **38**:1283–1289.

3329. **Schumacher-Perdreau, F., C. Heilmann, G. Peters, F. Götz, and G. Pulverer.** 1994. Comparative analysis of a biofilm-forming *Staphylococcus epidermidis* strain and its adhesion-positive, accumulation-negative mutant M7. *FEMS Microbiol. Lett.* **117**:71–78.

3330. **Schupbach, P., J. R. Neeser, M. Golliard, M. Rouvet, and B. Guggenheim.** 1996. Incorporation of caseinoglycomacropeptide and caseinophosphopeptide into the salivary pellicle inhibits adherence of mutans streptococci. *J. Dent. Res.* **75**:1779–1788.

3331. **Schurtz, T. A., D. B. Hornick, T. K. Korhonen, and S. Clegg.** 1994. The type 3 fimbrial adhesin gene (*mrkD*) of *Klebsiella* species is not conserved among all fimbriate strains. *Infect. Immun.* **62**:4186–4191.

3332. **Schwab, U. E., A. E. Wold, J. L. Carson, M. W. Leigh, P. W. Cheng, P. H. Gilligan, and T. F. Boat.** 1993. Increased adherence of *Staphylococcus aureus* from cystic fibrosis lungs to airway epithelial cells. *Am. Rev. Respir. Dis.* **148**:365–369.

3333. **Schwab, U., H. J. Thiel, K. P. Steuhl, and G. Doering.** 1996. Binding of *Staphylococcus aureus* to fibronectin and glycolipids on corneal surfaces. *German J. Ophthalmol.* **5**:417–421.

3334. **Schwan, W. R., H. S. Seifert, and J. L. Duncan.** 1992. Growth conditions mediate differential transcription of *fim* genes involved in phase variation of type 1 pili. *J. Bacteriol.* **174**:2367–2375.

3335. **Schwank, S., Z. Rajacic, W. Zimmerli, and J. Blaser.** 1998. Impact of bacterial biofilm formation on *in vitro* and *in vivo* activities of antibiotics. *Antimicrob. Agents Chemother.* **42**:895–898.

3336. **Schwartz, D., R. Perry, D. M. Dombroski, J. M. Merrick, and J. Goldhar.** 1996. Invasive ability of *C. jejuni/coli* isolates from children with diarrhea and the effect of iron-regulated proteins. *Zentbl. Bakteriol.* **283**:485–491.

3337. **Schwartzkoff, C. L., J. R. Egerton, D. J. Stewart, P. R. Lehrbach, T. C. Elleman, and P. A. Hoyne.** 1993. The effects of antigenic competition on the efficacy of multivalent footrot vaccines. *Aust. Vet. J.* **70**:123–126.

3338. **Schwartzkoff, C. L., P. R. Lehrbach, M. L. Ng, and A. Poi.** 1993. The effect of time between doses on serological response to a recombinant multivalent pilus vaccine against footrot in sheep. *Aust. Vet. J.* **70**:127–129.

3339. **Schwarz-Linek, U., M. J. Plevin, A. R. Pickford, M. Höök, I. D. Campbell, and J. R. Potts.** 2001. Binding of a peptide from a *Streptococcus dysgalactiae* MSCRAMM to the N-terminal F1 module pair of human fibronectin involves both modules. *FEBS Lett.* **497**:137–140.

3340. **Schweizer, F., H. Jiao. O. Hindsgaul, W. Y. Wong, and R. T. Irvin.** 1998. Interaction between the pili of *Pseudomonas aeruginosa* PAK and its carbohydrate receptor beta-D-GalNAc(1–4)beta-D-Gal analogs. *Can. J. Microbiol.* **44**:307–311.

3341. **Schwertmann, A., H. Schroten, J. Hacker, and C. Kunz.** 1999. S-fimbriae from *Escherichia coli* bind to soluble glycoproteins from human milk. *J. Pediatr. Gastroenterol. Nutr.* **28**:257–263.

3342. **Sciotti, M. A., I. Yamodo, J. P. Klein, and J. A. Ogier.** 1997. The N-terminal half part of the oral streptococcal antigen I/IIf contains two distinct binding domains. *FEMS Microbiol. Lett.* **153**:439–445.

3343. **Scotland, S. M., G. A. Willshaw, H. R. Smith, B. Said, N. Stokes, and B. Rowe.** 1993. Virulence properties of *Escherichia coli* strains belonging to serogroups O26, O55, O111 and O128 isolated in the United Kingdom in 1991 from patients with diarrhoea. *Epidemiol. Infect.* **111**:429–438.

3344. **Scotland, S. M., H. R. Smith, T. Cheasty, B. Said, G. A. Willshaw, N. Stokes, and B. Rowe.** 1996. Use of gene probes and adhesion tests to characterise *Escherichia coli* belonging to enteropathogenic serogroups isolated in the United Kingdom. *J. Med. Microbiol.* **44**:438–443.

3345. **Scott, J. R., J. C. Wakefield, P. W.**

Russell, P. E. Orndorff, and B. J. Froehlich. 1992. CooB is required for assembly but not transport of CS1 pilin. *Mol. Microbiol.* **6:**293–300.

3346. Sebghati, T. A., and S. Clegg. 1999. Construction and characterization of mutations within the *Klebsiella mrkD1P* gene that affect binding to collagen type V. *Infect. Immun.* **67:**1672–1676.

3347. Sebghati, T. A. S., T. K. Korhonen, D. B. Hornick, and S. Clegg. 1998. Characterization of the type 3 fimbrial adhesins of *Klebsiella* strains. *Infect. Immun.* **66:**2887–2894.

3348. Secott, T. E., T. L. Lin, and C. C. Wu. 2001. Fibronectin attachment protein homologue mediates fibronectin binding by *Mycobacterium avium* subsp. *paratuberculosis*. *Infect. Immun.* **69:**2075–2082.

3349. Sedgley, C. M., L. P. Samaranayake, and B. W. Darvell. 1996. The influence of incubation conditions on the adherence of oral *Enterobacteriaceae* to HeLa cells. *APMIS* **104:**583–590.

3350. See, W. A., and J. L. Smith. 1992. Urinary trypsin levels observed in pancreas transplant patients with duodenocystostomies promote *in vitro* fibrinolysis and *in vivo* bacterial adherence to urothelial surfaces. *Urol. Res.* **20:**409–413.

3351. See, W. A., and R. D. Williams. 1992. Urothelial injury and clotting cascade activation: common denominators in particulate adherence to urothelial surfaces. *J. Urol.* **147:**541–548.

3352. Segal, E. D., S. Falkow, and L. S. Tompkins. 1996. *Helicobacter pylori* attachment to gastric cells induces cytoskeletal rearrangements and tyrosine phosphorylation of host cell proteins. *Proc. Natl. Acad. Sci. USA* **93:**1259–1264.

3353. Seganti, L., M. P. Conte. C. Longhi, M. Marchetti, M. Nicoletti, and N. Orsi. 1994. Invasiveness of *Shigella flexneri* in poliovirus infected HT-29 cells. *New Microbiol.* **17:**29–36.

3354. Segura, M., J. Stankova, and M. Gottschalk. 1999. Heat-killed *Streptococcus suis* capsular type 2 strains stimulate tumor necrosis factor alpha and interleukin-6 production by murine macrophages. *Infect. Immun.* **67:**4646–4654.

3355. Segura, M. A., P. Cleroux, and M. Gottschalk. 1998. *Streptococcus suis* and group B *Streptococcus* differ in their interactions with murine macrophages. *FEMS Immunol. Med. Microbiol.* **21:**189–195.

3356. Seifert, H. S. 1996. Questions about gonococcal pilus phase- and antigenic variation. *Mol. Microbiol.* **21:**433–440.

3357. Seifert, H. S., C. J. Wright, A. E. Jerse, M. S. Cohen, and J. G. Cannon. 1994. Multiple gonococcal pilin antigenic variants are produced during experimental human infections. *J. Clin. Investig.* **93:**2744–2749.

3358. Seignole, D., P. Grange, Y. Duval-Iflah, and M. Mouricout. 1994. Characterization of *O*-glycan moieties of the 210 and 240 kDa pig intestinal receptors for *Escherichia coli* K88ac fimbriae. *Microbiology* **140:**2467–2473.

3359. Sekizaki, T., Y. Nakasato, and I. Nonomura. 1992. Acid-induced autoagglutination found in chicken pathogenic *Escherichia coli* strain. *J. Vet. Med. Sci.* **54:**493–499.

3360. Sela, M. N., D. Steinberg, A. Klinger, A. A. Krausz, and D. Kohavi. 1999. Adherence of periodontopathic bacteria to bioabsorbable and non-absorbable barrier membranes in vitro. *Clin. Oral Implant Res.* **10:**445–452.

3361. Sela, S., A. Aviv, A. Tovi, I. Burstein, M. G. Caparon, and E. Hanski. 1993. Protein F: an adhesin of *Streptococcus pyogenes* binds fibronectin via two distinct domains. *Mol. Microbiol.* **10:**1049–1055.

3362. Sela, S., M. J. Marouni, R. Perry, and A. Barzilai. 2000. Effect of lipoteichoic acid on the uptake of *Streptococcus pyogenes* by HEp-2 cells. *FEMS Microbiol. Lett.* **193:**187–193.

3363. Sela, S., R. Neeman, N. Keller, and A. Barzilai. 2000. Relationship between asymptomatic carriage of *Streptococcus pyogenes* and the ability of the strains to adhere to and be internalised by cultured epithelial cells. *J. Med. Microbiol.* **49:**499–502.

3364. Selvarangan, R., P. Goluszko, V. Popov, J. Singhal, T. Pham, D. M. Lublin, S. Nowicki, and B. Nowicki. 2000. Role of decay-accelerating factor domains and anchorage in internalization of Dr-fimbriated *Escherichia coli*. *Infect. Immun.* **68:**1391–1399.

3365. Sengupta, T. K., D. K. Sengupta, and A. C. Ghose. 1993. A 20-kDa pilus protein with haemagglutination and intestinal adherence properties expressed by a clinical isolate of non-01 *Vibrio cholerae*. *FEMS Microbiol. Lett.* **112:**237–242.

3366. Sengupta, T. K., D. K. Sengupta, G. B. Nair, and A. C. Ghose. 1994. Epidemic isolates of *Vibrio cholerae* 0139 express antigenically distinct types of colonization pili. *FEMS Microbiol. Lett.* **118:**265–271.

3367. Sengupta, T. K., R. K. Nandy, S. Mukhopadhyay, R. H. Hall, V. Sathyamoorthy, and A. C. Ghose. 1998.

Characterization of a 20-kDa pilus protein expressed by a diarrheogenic strain of non-O1/non-O139 *Vibrio cholerae*. *FEMS Microbiol. Lett.* **160**:183–189.

3368. **Senior, D. F., P. deMan, and C. Svanborg.** 1992. Serotype, hemolysin production, and adherence characteristics of strains of *Escherichia coli* causing urinary tract infection in dogs. *Am. J. Vet. Res.* **53**:494–498.

3369. **Senpuku, H., K. Matin, S. M. Abdus, I. Kurauchi, S. Sakurai, M. Kawashima, T. Murata, H. Miyazaki, and N. Hanada.** 2001. Inhibitory effects of MoAbs against a surface protein antigen in real-time adherence *in vitro* and recolonization *in vivo* of *Streptococcus mutans*. *Scand. J. Immunol.* **54**:109–116.

3370. **Senpuku, H., H. Kato, M. Todoroki, N. Hanada, and T. Nisizawa.** 1996. Interaction of lysozyme with a surface protein antigen of *Streptococcus mutans*. *FEMS Microbiol. Lett.* **139**:195–201.

3371. **Seo, K. H., and J. F. Frank.** 1999. Attachment of *Escherichia coli* O157:H7 to lettuce leaf surface and bacterial viability in response to chlorine treatment as demonstrated by using confocal scanning laser microscopy. *J. Food Prot.* **62**:3–9.

3372. **Sepulveda, M., M. Ruiz, H. Bello, M. Dominguez, M. A. Martinez, M. E. Pinto, G. Gonzalez, S. Mella, and R. Zemelman.** 1998. Adherence of *Acinetobacter baumannii* to rat bladder tissue. *Microbios* **95**:45–53.

3373. **Seto, S., G. Layh-Schmitt, T. Kenri, and M. Miyata.** 2001. Visualization of the attachment organelle and cytadherence proteins of *Mycoplasma pneumoniae* by immunofluorescence microscopy. *J. Bacteriol.* **183**:1621–1630.

3374. **Sexton, M., and D. J. Reen.** 1992. Characterization of antibody-mediated inhibition of *Pseudomonas aeruginosa* adhesion to epithelial cells. *Infect. Immun.* **60**:3332–3338.

3375. **Shahal, Y., D. Steinberg, Z. Hirschfeld, M. Bronshteyn, and K. Kopolovic.** 1998. *In vitro* bacterial adherence onto pellicle-coated aesthetic restorative materials. *J. Oral Rehab.* **25**:52–58.

3376. **Shaia, C. I., J. Voyich, S. J. Gillis, B. N. Singh, and D. E. Burgess.** 1998. Purification and expression of the Tf190 adhesin in *Tritrichomonas foetus*. *Infect. Immun.* **66**:1100–1105.

3377. **Shani, S., M. Friedman, and D. Steinberg.** 2000. The anticariogenic effect of amine fluorides on *Streptococcus sobrinus* and glucosyltransferase in biofilms. *Caries Res.* **34**:260–267.

3378. **Shaniztki, B., D. Hurwitz, N. Smorodinsky, N. Ganeshkumar, and E. I. Weiss.** 1997. Identification of a *Fusobacterium nucleatum* PK1594 galactose-binding adhesin which mediates coaggregation with periopathogenic bacteria and hemagglutination. *Infect. Immun.* **65**:5231–5237.

3379. **Shaniztki, B., N. Ganeshkumar, and E. I. Weiss.** 1998. Characterization of a novel *N*-acetylneuraminic acid-specific *Fusobacterium nucleatum* PK1594 adhesin. *Oral Microbiol. Immunol.* **13**:47–50.

3380. **Sharma, A., H. T. Sojar, J. Y. Lee, and R. J. Genco.** 1993. Expression of a functional *Porphyromonas gingivalis* fimbrillin polypeptide in *Escherichia coli*: purification, physicochemical and immunochemical characterization, and binding characteristics. *Infect. Immun.* **61**:3570–3573.

3381. **Sharma, A., H. Nagata, N. Hamada, H. T. Sojar, D. E. Hruby, H. K. Kuramitsu, and R. J. Genco.** 1996. Expression of functional *Porphyromonas gingivalis* fimbrillin polypeptide domains on the surface of *Streptococcus gordonii*. *Appl. Environ. Microbiol.* **62**:3933–3938.

3382. **Sharma, A., H. T. Sojar, I. Glurich, K. Honma, H. K. Kuramitsu, and R. J. Genco.** 1998. Cloning, expression, and sequencing of a cell surface antigen containing a leucine-rich repeat motif from *Bacteroides forsythus* ATCC 43037. *Infect. Immun.* **66**:5703–5710.

3383. **Sharma, A., E. K. Novak, H. T. Sojar, R. T. Swank, H. K. Kuramitsu, and R. J. Genco.** 2000. *Porphyromonas gingivalis* platelet aggregation activity: outer membrane vesicles are potent activators of murine platelets. *Oral Microbiol. Immunol.* **15**:393–396.

3384. **Sharon, N., and I. Ofek.** 1995. Identification of receptors for bacterial lectins by blotting techniques. *Methods Enzymol.* **253**:91–98.

3385. **Shaw, R. K., S. Daniell, F. Ebel, G. Frankel, and S. Knutton.** 2001. EspA filament-mediated protein translocation into red blood cells. *Cell Microbiol.* **3**:213–222.

3386. **Sheikh, J., S. Hicks, M. Dall'Agnol, A. D. Phillips, and J. P. Nataro.** 2001. Roles for Fis and YafK in biofilm formation by enteroaggregative *Escherichia coli*. *Mol. Microbiol.* **41**:983–997.

3387. **Shen, W., H. Steinruck, and Å Ljungh.** 1995. Expression of binding of plasminogen, thrombospondin, vitronectin, and fibrinogen, and adhesive properties by *Escherichia coli* strains isolated from patients with colonic diseases. *Gut* **36**:401–406.

3388. **Shenkman, B., E. Rubinstein, I. Tamarin, R. Dardik, N. Savion, and D. Varon.** 2000. *Staphylococcus aureus* adherence to thrombin-treated endothelial cells is mediated by fibrinogen but not by platelets. *J. Lab. Clin. Med.* **135**:43–51.

3389. **Shenkman, B., E. Rubinstein, A. L. Cheung, G. E. Brill, R. Dardik, I. Tamarin, N. Savion, and D. Varon.** 2001. Adherence properties of *Staphylococcus aureus* under static and flow conditions: roles of *agr* and *sar* loci, platelets, and plasma ligands. *Infect. Immun.* **69**:4473–4478.

3390. **Shepel, M., J. Boyd, J. Luider, and A. P. Gibb.** 2001. Interaction of *Yersinia enterocolitica* and *Y. pseudotuberculosis* with platelets. *J. Med. Microbiol.* **50**:1030–1038.

3391. **Sherman, M. P., L. A. Campbell, T. A. Merritt, W. A. Long, J. H. Gunkel, T. Curstedt, and B. Robertson.** 1994. Effect of different surfactants on pulmonary group B streptococcal infection in premature rabbits. *J. Pediatr.* **125**:939–947.

3392. **Sheth, H. B., L. M. Glasier, N. W. Ellert, P. Cachia, W. Kohn, K. K. Lee, W. Paranchych, R. S. Hodges, and R. T. Irvin.** 1995. Development of an anti-adhesive vaccine for *Pseudomonas aeruginosa* targeting the C-terminal region of the pilin structural protein. *Biomed. Pept. Proteins Nucleic Acids* **1**:141–148.

3393. **Sheth, H. B., K. K. Lee, W. Y. Wong, G. Srivastava, O. Hindsgaul, R. S. Hodges, W. Paranchych, and R. T. Irvin.** 1994. The pili of *Pseudomonas aeruginosa* strains PAK and PAO bind specifically to the carbohydrate sequence beta GalNAc(1–4)beta Gal found in glycosphingolipids asialo-GM_1 and asialo-GM_2. *Mol. Microbiol.* **11**:715–723.

3394. **Shi, Y., D. B. Ratnayake, K. Okamoto, Y. Abe, K. Yamamoto, and K. Nakayama.** 1999. Genetic analyses of proteolysis, hemoglobin binding, and hemagglutination of *Porphyromonas gingivalis*. Construction of mutants with a combination of *rgpA*, *rgpB*, *kgp*, and *hagA*. *J. Biol. Chem.* **274**:17955–17960.

3395. **Shiau, A. L., and C. L. Wu.** 1998. The inhibitory effect of *Staphylococcus epidermidis* slime on the phagocytosis of murine peritoneal macrophages is interferon-independent. *Microbiol. Immunol.* **42**:33–40.

3396. **Shibata, H., I. Kimura-Takagi, M. Nagaoka, S. Hashimoto, H. Sawada, S.** Ueyama, and T. Yokokura. 1999. Inhibitory effect of *Cladosiphon* fucoidan on the adhesion of *Helicobacter pylori* to human gastric cells. *J. Nutr. Sci. Vitaminol.* **45**:325–336.

3397. **Shibata, Y., K. Kurihara, H. Takiguchi, and Y. Abiko.** 1998. Construction of a functional single-chain variable fragment antibody against hemagglutinin from *Porphyromonas gingivalis*. *Infect. Immun.* **66**:2207–2212.

3398. **Shibata, Y., M. Hayakawa, H. Takiguchi, T. Shiroza, and Y. Abiko.** 1999. Determination and characterization of the hemagglutinin-associated short motifs found in *Porphyromonas gingivalis* multiple gene products. *J. Biol. Chem.* **274**:5012–5020.

3399. **Shibl, A. M., M. A. Ramadan, and A. F. Tawfik.** 1994. Differential inhibition by clindamycin on slime formation, adherence to teflon catheters and hemolysin production by *Staphylococcus epidermidis*. *J. Chemother.* **6**:107–110.

3400. **Shimaoka, T., N. Kume, M. Minami, K. Hayashida, T. Sawamura, T. Kita, and S. Yonehara.** 2001. LOX-1 supports adhesion of Gram-positive and Gram-negative bacteria. *J. Immunol.* **166**:5108–5114.

3401. **Shimizu, T., W. Ba-Thein, M. Tamaki, and H. Hayashi.** 1994. The *virR* gene, a member of a class of two-component response regulators, regulates the production of perfringolysin O, collagenase, and hemagglutinin in *Clostridium perfringens*. *J. Bacteriol.* **176**:1616–1623.

3402. **Shimodori, S., K. Iida, F. Kojima, A. Takade, M. Ehara, and K. Amako.** 1997. Morphological features of a filamentous phage of *Vibrio cholerae* O139 Bengal. *Microbiol. Immunol.* **41**:757–763.

3403. **Shimoji, Y., V. Ng, K. Matsumura, V. A. Fischetti, and A. Rambukkana.** 1999. A 21-kDa surface protein of *Mycobacterium leprae* binds peripheral nerve laminin-2 and mediates Schwann cell invasion. *Proc. Natl. Acad. Sci. USA* **96**:9857–9862.

3404. **Shin, J. S., Z. Gao, and S. N. Abraham.** 2000. Involvement of cellular caveolae in bacterial entry into mast cells. *Science* **289**:785–788.

3405. **Shinagawa, H., T. Taniguchi, O. Yamaguchi, K. Yamamoto, and T. Honda.** 1993. Cloning of the genes that control formation of the fimbrial colonization factor antigen III (CFA/III) from an enterotoxigenic *Escherichia coli*. *Microbiol. Immunol.* **37**:689–694.

3406. **Shinji, H., J. Sakurada, K. Seki, M.**

Murai, and S. Masuda. 1998. Different effects of fibronectin on the phagocytosis of *Staphylococcus aureus* and coagulase-negative staphylococci by murine peritoneal macrophages. *Microbiol. Immunol.* **42**:851–861.

3407. **Shiono, A., and Y. Ike.** 1999. Isolation of *Enterococcus faecalis* clinical isolates that efficiently adhere to human bladder carcinoma T24 cells and inhibition of adhesion by fibronectin and trypsin treatment. *Infect. Immun.* **67**:1585–1592.

3408. **Shiro, H., E. Muller, N. Gutierrez, S. Boisot, M. Grout, T. D. Tosteson, D. Goldmann, and G. B. Pier.** 1995. Transposon mutants of *Staphylococcus epidermidis* deficient in elaboration of capsular polysaccharide/adhesin and slime are avirulent in a rabbit model of endocarditis. *J. Infect. Dis.* **169**:1042–1049.

3409. **Shoberg, R. J., and D. D. Thomas.** 1993. Specific adherence of *Borrelia burgdorferi* extracellular vesicles to human endothelial cells in culture. *Infect. Immun.* **61**:3892–3900.

3410. **Shuman, H. A., and M. A. Horwitz.** 1996. *Legionella pneumophila* invasion of mononuclear phagocytes. *Curr. Topics Microbiol. Immunol.* **209**:99–112.

3411. **Shuter, J., V. B. Hatcher, and F. D. Lowy.** 1996. *Staphylococcus aureus* binding to human nasal mucin. *Infect. Immun.* **64**:310–318.

3412. **Siboo, I. R., A. L. Cheung, A. S. Bayer, and P. M. Sullam.** 2001. Clumping factor A mediates binding of *Staphylococcus aureus* to human platelets. *Infect. Immun.* **69**:3120–3127.

3413. **Siegfried, L., M. Kmetova, H. Puzova, M. Molokacova, and J. Filka.** 1994. Virulence-associated factors in *Escherichia coli* strains isolated from children with urinary tract infections. *J. Med. Microbiol.* **41**:127–132.

3414. **Siitonen, A., R. Martikainen, R. Ikaheimo, J. Palmgren, and P. H. Makela.** 1993. Virulence-associated characteristics of *Escherichia coli* in urinary tract infection: a statistical analysis with special attention to type 1C fimbriation. *Microb. Pathog.* **15**:65–75.

3415. **Siitonen, A., A. Takala, Y. A. Ratiner, A. Pere, and P. H. Makela.** 1993. Invasive *Escherichia coli* infections in children: bacterial characteristics in different age groups and clinical entities. *Pediatr. Infect. Dis. J.* **12**:606–612.

3416. **Silva, M. L., and C. M. Giampaglia.** 1992. Colostrum and human milk inhibit localized adherence of enteropathogenic *Escherichia coli* to HeLa cells. *Acta Paediatr.* **81**:266–267.

3417. **Silva, R. M., R. Giraldi, R. Keller, L. C. Campos, and B. E. Guth.** 1996. Diffuse

adherence, ST-I enterotoxin and CFA/IV colonization factor are encoded by the same plasmid in the *Escherichia coli* O29:H21 strain. *Braz. J. Med. Biol. Res.* **29**:969–976.

3418. **Simala-Grant, J. L., D. Zopf, and D. E. Taylor.** 2001. Antibiotic susceptibility of attached and free-floating *Helicobacter pylori*. *J. Antimicrob. Chemother.* **47**:555–563.

3419. **Simhi, E., H. C. van der Mei, E. Z. Ron, E. Rosenberg, and H. J. Busscher.** 2000. Effect of the adhesive antibiotic TA on adhesion and initial growth of *E. coli* on silicone rubber. *FEMS Microbiol. Lett.* **192**:97–100.

3420. **Simionato, M. R. L., M. P. A. Mayer, S. Cai, J. L. de Lorenzo, and F. Zelante.** 1994. Influence of lectins on adhesion of *Streptococcus salivarius* to buccal epithelial cells. *Rev. Microbiol. Sao Paulo* **25**:83–85.

3421. **Simon, D., and R. F. Rest.** 1992. *Escherichia coli* expressing a *Neisseria gonorrhoeae* opacity-associated outer membrane protein invade human cervical and endometrial epithelial cell lines. *Proc. Natl. Acad. Sci. USA* **89**:5512–5516.

3422. **Simon, D., J. T. Liu, M. S. Blake, C. R. Blake, and R. F. Rest.** 1996. Structure-function studies with *Neisseria gonorrhoeae* Opa outer membrane proteins expressed in *Escherichia coli*. *Ann. N. Y. Acad. Sci.* **797**:253–254.

3423. **Simon, P. M., P. L. Goode, A. Mobasseri, and D. Zopf.** 1997. Inhibition of *Helicobacter pylori* binding to gastrointestinal epithelial cells by sialic acid-containing oligosacchrides. *Infect. Immun.* **65**:750–757.

3424. **Simonet, M., P. Triadou, C. Frehel, M. C. Morel-Kopp, C. Kaplan, and P. Berche.** 1992. Human platelet aggregation by *Yersinia pseudotuberculosis* is mediated by invasin. *Infect. Immun.* **60**:366–373.

3425. **Simonet, M., B. Riot, N. Fortineau, and P. Berche.** 1996. Invasin production by *Yersinia pestis* is abolished by insertion of an IS200-like element within the *inv* gene. *Infect. Immun.* **64**:375–379.

3426. **Simons, B. L., O. Mol, J. F. L. van Breemen, and B. Oudega.** 1994. Morphological appearances of K88ab fimbriae and optical diffraction analysis of K88 paracrystalline structures. *FEMS Microbiol. Lett.* **118**:83–88.

3427. **Simpson, D. A., R. Ramphal, and S. Lory.** 1992. Genetic analysis of *Pseudomonas aeruginosa* adherence: distinct genetic loci control attachment to epithelial cells and mucins. *Infect. Immun.* **60**:3771–3779.

3428. **Simpson, D. A., R. Ramphal, and S.**

Lory. 1995. Characterization of *Pseudomonas aeruginosa fliO*, a gene involved in flagellar biosynthesis and adherence. *Infect. Immun.* 63:2950–2957.

3429. **Singh, D. V., R. S. Dubey, and S. C. Sanyal.** 1993. Adherence of haemagglutinating and non-haemagglutinating clinical and environmental isolates of *Aeromonas. J. Diarrhoeal Dis. Res.* 11:157–160.

3430. **Singh, R. D., M. Khullar, and N. K. Ganguly.** Role of anaerobiosis in virulence of *Salmonella typhimurium. Mol. Cell. Biochem.* 215:39–46.

3431. **Singh, S. N., R. Srivastava, V. B. Sinha, and B. S. Srivastava.** 1994. A 53 kDa protein of *Vibrio cholerae* classical strain O395 involved in intestinal colonization. *Microb. Pathog.* 17:69–78.

3432. **Sirakova, T., P. E. Kolattukudy, D. Murwin, J. Billy, E. Leake, D. Lim, T. DeMaria, and L. Bakaletz.** 1994. Role of fimbriae expressed by nontypeable *Haemophilus influenzae* in pathogenesis of and protection against otitis media and relatedness of the fimbrin subunit to outer membrane protein A. *Infect. Immun.* 62:2002–2020.

3433. (See reference 3432.)

3434. **Sjöbring, U., G. Pohl, and A. Olsén.** 1994. Plasminogen, absorbed by *Escherichia coli* expressing curli or by *Salmonella enteritidis* expressing thin aggregative fimbriae, can be activated by simultaneously captured tissue-type plasminogen activator (t-PA). *Mol. Microbiol.* 14:443–452.

3435. **Skopek, R. J., W. F. Liljemark, C. G. Bloomquist, and J. D. Rudney.** 1993. Dental plaque development on defined streptococcal surfaces. *Oral Microbiol. Immunol.* 8:16–23.

3436. **Skopek, R. J., and W. F. Liljemark.** 1994. The influence of saliva on interbacterial adherence. *Oral Microbiol. Immunol.* 9:19–24.

3437. **Skorupski, K., and R. K. Taylor.** 1997. Cyclic AMP and its receptor protein negatively regulate the coordinate expression of cholera toxin and toxin-coregulated pilus in *Vibrio cholerae. Proc. Natl. Acad. Sci. USA* 94:265–270.

3438. **Skurnik, M., Y. el Tahir, M. Saarinen, S. Jalkanen, and P. Toivanen.** 1994. YadA mediates specific binding of enteropathogenic *Yersinia enterocolitica* to human intestinal submucosa. *Infect. Immun.* 62:1252–1261.

3439. **Slakeski, N., S. M. Cleal, and E. C. Reynolds.** 1996. Characterization of a *Porphyromonas gingivalis* gene *prtR* that encodes an arginine-specific thiol proteinase and multiple adhesins. *Biochem. Biophys. Res. Commun.* 224:605–610.

3440. **Slakeski, N., P. S. Bhogal, N. M. O'Brien-Simpson, and E. C. Reynolds.** 1998. Characterization of a second cell-associated Arg-specific cysteine proteinase of *Porphyromonas gingivalis* and identification of an adhesin-binding motif involved in association of the *prtR* and *prtK* proteinases and adhesins into large complexes. *Microbiology* 144:1583–1592.

3441. **Slakeski, N., S. M. Cleal, P. S. Bhogal, and E. C. Reynolds.** 1999. Characterization of a *Porphyromonas gingivalis* gene *prtK* that encodes a lysine-specific cysteine proteinase and three sequence-related adhesins. *Oral Microbiol. Immunol.* 14:92–97.

3442. **Slavikova, M., R. Lodinova-Zadnikova, I. Adlerberth, L. A. Hanson, C. Svanborg, and A. E. Wold.** 1995. Increased mannose-specific adherence and colonizing ability of *Escherichia coli* O83 in breast-fed infants. *Adv. Exp. Med. Biol.* 371A:421–423.

3443. **Sloan, A. R., and T. G. Pistole.** 1992. A quantitative method for measuring the adherence of group B streptococci to murine peritoneal exudate macrophages. *J. Immunol. Methods* 154:217–223.

3444. **Sloan, A. R., and T. G. Pistole.** 1993. Characterization of the murine macrophage receptor for group B streptococci. *Zentbl. Bakteriol.* 278:541–552.

3445. **Slomiany, B. L., and A. Slomiany.** 1992. Mechanism of *Helicobacter pylori* pathogenesis: focus on mucus. *J. Clin. Gastroenterol.* 14:S114–S121.

3446. **Slomiany, B. L., J. Piotrowski, A. Czajkowski, F. E. Shovlin, and A. Slomiany.** 1993. Differential expression of salivary mucin bacterial aggregating activity with caries status. *Int. J. Biochem.* 25:935–940.

3447. **Slomiany, B. L., V. L. Murty, J. Piotrowski, and A. Slomiany.** 1994. Gastroprotective agents in mucosal defense against *Helicobacter pylori. Gen. Pharmacol.* 25:833–841.

3448. **Slonim, L. N., J. S. Pinkner, C. I. Branden, and S. J. Hultgren.** 1992. Interactive surface in the PapD chaperone cleft is conserved in pilus chaperone superfamily and essential in subunit recognition and assembly. *EMBO J.* 11:4747–4756.

3449. **Smart, W., P. A. Sastry, W. Paranchych, and B. Singh.** 1993. Immune recognition of polar pili from *Pseudomonas aeruginosa O. Infect. Immun.* 61:3527–3529.

3450. **Smeds, A., K. Hemmann, M. Jakava-Viljanen, S. Pelkonen, H. Imberechts, and A. Palva.** 2001. Characterization of the adhesin of *Escherichia coli* F18 fimbriae. *Infect. Immun.* **69:**7941–7945.

3451. **Smeltzer, M. S., A. F. Gillaspy, F. L. Pratt, Jr., M. D. Thames, and J. J. Iandolo.** 1997. Prevalence and chromosomal map location of *Staphylococcus aureus* adhesin genes. *Gene* **196:**249–259.

3452. **Smith, B. L., and M. K. Hostetter.** 2000. C3 as substrate for adhesion of *Streptococcus pneumoniae. J. Infect Dis.* **182:**497–508.

3453. **Smith, C. J., J. B. Kaper, and D. R. Mack.** 1995. Intestinal mucin inhibits adhesion of human enteropathogenic *Escherichia coli* to HEp-2 cells. *J. Pediatr. Gastroenterol. Nutr.* **21:**269–276.

3454. **Smith, D. G., W. C. Russell, and D. Thirkell.** 1994. Adherence of *Ureaplasma urealyticum* to human epithelial cells. *Microbiology* **140:**2893–2498.

3455. **Smith, H.** 1995. The revival of interest in mechanisms of bacterial pathogenicity. *Biol. Rev. Camb. Philos. Soc.* **70:**277–316.

3456. **Smith, H. R., S. M. Scotland, G. A. Willshaw, B. Rowe, A. Cravioto, and C. Eslava.** 1994. Isolates of *Escherichia coli* O44:H18 of diverse origin are enteroaggregative. *J. Infect. Dis.* **170:**1610–1613.

3457. **Smith, L. M., V. Laganas, and T. G. Pistole.** 1998. Attachment of group B streptococci to macrophages is mediated by a 21-kDa protein. *FEMS Immunol. Med. Microbiol.* **20:**89–97.

3458. **Smith, S. H., R. G. Murray, and M. Hall.** 1994. The surface structure of *Leptotrichia buccalis. Can. J. Microbiol.* **40:**90–98.

3459. **Smoot, D. T., J. H. Resau, T. Naab, B. C. Desbordes, T. Gilliam, K. Bull-Henry, S. B. Curry, J. Nidirym, J. Sewchand, K. Mills-Robertson, K. Frontin, E. Abebe, M. Dillon, G. R. Chippendale, P. C. Phelps, V. F. Scott, and H. L. T. Mobley.** 1993. Adherence of *Helicobacter pylori* to cultured human gastric epithelial cells. *Infect. Immun.* **61:**350–355.

3460. **Smoot, D. T.** 1997. How does *Helicobacter pylori* cause mucosal damage? Direct mechanisms. *Gastroenterology* **113:**S31–S34.

3461. **Smyth, C. J., M. B. Marron, J. M. Twohig, and S. G. Smith.** 1996. Fimbrial adhesins: similarities and variations in structure and biogenesis. *FEMS Immunol. Med. Microbiol.* **16:**127–139.

3462. **Snodgrass, J. L., N. Mohamed, J. M. Ross, S. Sau, C. Y. Lee, and M. S. Smeltzer.** 1999. Functional analysis of the *Staphylococcus aureus* collagen adhesin B domain. *Infect. Immun.* **67:**3952–3959.

3463. **Soane, M. C., A. Jackson, D. Maskell, A. Allen, P. Keig, A. Dewar, G. Dougan, and R. Wilson.** 2000. Interaction of *Bordetella pertussis* with human respiratory mucosa *in vitro. Respir. Med.* **94:**791–799.

3464. **Soderhall, M., U. S. Bergerheim, S. H. Jacobson, J. Lundahl, R. Möllby, S. Normark, and J. Winberg.** 1997. Molecular evidence for pap-G specific adhesion of *Escherichia coli* to human renal cells. *J. Urol.* **157:**346–350.

3465. **Soderhall, M., S. Normark, K. Ishikawa, K. Karlsson, S. Teneberg, J. Winberg, and R. Möllby.** 1997. Induction of protective immunity after *Escherichia coli* bladder infection in primates. Dependence of the globoside-specific P-fimbrial tip adhesin and its cognate receptor. *J. Clin. Investig.* **100:**364–372.

3466. **Soell, M., M. Diab, G. Haan-Archipoff, A. Beretz, C. Herbelin, B. Poutrel, and J. P. Klein.** 1995. Capsular polysaccharide types 5 and 8 of *Staphylococcus aureus* bind specifically to human epithelial (KB) cells, endothelial cells, and monocytes and induce release of cytokines. *Infect. Immun.* **63:**1380–1386.

3467. **Sohel, I., J. L. Puente, W. J. Murray, J. Vuopio-Varkila, and G. K. Schoolnik.** 1993. Cloning and characterization of the bundle-forming pilin gene of enteropathogenic *Escherichia coli* and its distribution in *Salmonella* serotypes. *Mol. Microbiol.* **7:**563–575.

3468. **Sohel, I., J. L. Puente, S. W. Ramer, D. Bieber, C. Y. Wu, and G. K. Schoolnik.** 1996. Enteropathogenic *Escherichia coli:* identification of a gene cluster coding for bundle-forming pilus morphogenesis. *J. Bacteriol.* **178:**2613–2628.

3469. **Sojar, H. T., J.-Y. Lee, and R. J. Genco.** 1995 Fibronectin binding domain of *P. gingivalis* fimbriae. *Biochem. Biophys. Res. Commun.* **216:**785–792.

3470. **Sojar, H. T., N. Hamada, and R. J. Genco.** 1997. High-performance liquid chromatographic separation of *Porphyromonas gingivalis* fimbriae. *Protein Expression Purif.* **9:**49–52.

3471. **Sojar, H. T., N. Hamada, and R. J. Genco.** 1997. Isolation and characterization of fimbriae from a sparsely fimbriated strain of *Porphyromonas gingivalis. Appl. Environ. Microbiol.* **63:**2318–2323.

3472. **Sojar, H. T., N. Hamada, and R. J. Genco.** 1998. Structures involved in the interaction of *Porphyromonas gingivalis* fimbriae and human lactoferrin. *FEBS Lett.* **422**:205–208.

3473. **Sojar, H. T., Y. Han, N. Hamada, A. Sharma, and R. J. Genco.** 1999. Role of the amino-terminal region of *Porphyromonas gingivalis* fimbriae in adherence to epithelial cells. *Infect. Immun.* **67**:6173–6176.

3474. **Sojka, M. G., M. Dibb-Fuller, and C. J. Thorns.** 1996. Characterisation of monoclonal antibodies specific to SEF 21 fimbriae of *Salmonella enteritidis* and their reactivity with other *Salmonellae* and *Enterobacteria*. *Vet. Microbiol.* **48**:207–221.

3475. **Sojka, M. G., M. A. Carter, and C. J. Thorns.** 1998. Characterisation of epitopes of type 1 fimbriae of *Salmonella* using monoclonal antibodies specific for SEF21 fimbriae of *Salmonella enteritidis*. *Vet. Microbiol.* **59**:157–174.

3476. **Sokolowska-Kohler, W., G. Schonian, R. Bollmann, A. Schubert, J. Parschau, A. Seeberg, and W. Presber.** 1997. Occurrence of S and F1C/S-related fimbrial determinants and their expression in *Escherichia coli* strains isolated from extraintestinal infections. *FEMS Immunol. Med. Microbiol.* **18**:1–6.

3477. **Sokurenko, E. V., H. S. Courtney, S. N. Abraham, P. Klemm, and D. L. Hasty.** 1992. Functional heterogeneity of type 1 fimbriae of *Escherichia coli*. *Infect. Immun.* **60**:4709–4719.

3478. **Sokurenko, E. V., H. S. Courtney, D. E. Ohman, P. Klemm, and D. L. Hasty.** 1994. FimH family of type 1 fimbrial adhesins: Functional heterogeneity due to minor sequence variations among *fimH* genes. *J. Bacteriol.* **176**:748–755.

3479. **Sokurenko, E. V., H. S. Courtney, J. Maslow, A. Siitonen, and D. L. Hasty.** 1995. Quantitative differences in adhesiveness of type 1 fimbriated *Escherichia coli* due to structural differences in *fimH* genes. *J. Bacteriol.* **177**:3680–3686.

3480. **Sokurenko, E. V., V. A. McMackin, and D. L. Hasty.** 1995. Bacterial adhesion measured by growth of adherent organisms. *Methods Enzymol.* **253**:519–528.

3481. **Sokurenko, E. V., and D. L. Hasty.** 1995. Assay for adhesion of host cells to immobilized bacteria. *Methods Enzymol.* **253**:220–226.

3482. **Sokurenko, E. V., V. Chesnokova, R. J. Doyle, and D. L. Hasty.** 1997. Diversity of the *Escherichia coli* type 1 fimbrial lectin. Differential binding to mannosides and uroepithelial cells. *J. Biol. Chem.* **272**:17880–17886.

3483. **Sokurenko, E. V., V. Chesnokova, D. E. Dykhuizen, I. Ofek, X.-R. Wu, K. A.**

Krogfelt, C. Struve, M. A. Schembri, and D. L. Hasty. 1998. Pathogenic adaptation of *Escherichia coli* by natural variation of the FimH adhesin. *Proc. Natl. Acad. Sci. USA* **95**:8922–8926.

3484. **Sokurenko, E. V., M. S. Schembri, E. Trintchina, K. Kjaergaard, D. L. Hasty, and P. Klemm.** 2001. Valency conversion in the type 1 fimbrial adhesin of *Escherichia coli*. *Mol. Microbiol.* **41**:675–686.

3485. **Solano, C., B. Sesma, M. Alvarez, T. J. Humphrey, C. J. Thorns, and C. Gamazo.** 1998. Discrimination of strains of *Salmonella enteritidis* with differing levels of virulence by an in vitro glass adherence test. *J. Clin. Microbiol.* **36**:674–678.

3486. **Somers, E. B., J. L. Schoeni, and A. C. Wong.** 1994. Effect of trisodium phosphate on biofilm and planktonic cells of *Campylobacter jejuni*, *Escherichia coli* O157:H7, *Listeria monocytogenes* and *Salmonella typhimurium*. *Int. J. Food Microbiol.* **22**:269–276.

3487. **Sommerfelt, H., H. M. Grewal, A.-M. Svennerholm, W. Gaastra, P. R. Flood, G. Viboud, and M. K. Bhan.** 1992. Genetic relationship of putative colonization factor O166 to colonization factor antigen I and coli surface antigen 4 of enterotoxigenic *Escherichia coli*. *Infect. Immun.* **60**: 3799–3806.

3488. **Song, W., L. Ma, R. Chen, and D. C. Stein.** 2000. Role of lipooligosaccharide in Opa-independent invasion of *Neisseria gonorrhoeae* into human epithelial cells. *J. Exp. Med.* **191**:949–960.

3489. **Sonstein, S. A., and J. C. Burnham.** 1993. Effect of low concentrations of quinolone antibiotics on bacterial virulence mechanisms. *Diagn. Microbiol. Infect. Dis.* **16**:277–289.

3490. **Soriano, F., C. Ponte, and M. J. Galiano.** 1993. Adherence of *Corynebacterium urealyticum* (CDC group D2) and *Corynebacterium jeikeium* to intravascular and urinary catheters. *Eur. J. Clin. Microbiol. Infect. Dis.* **12**:453–456.

3491. **Sory, M. P., and G. R. Cornelis.** 1994. Translocation of a hybrid YopE-adenylate cyclase from *Yersinia enterocolitica* into HeLa cells. *Mol. Microbiol.* **14**:583–594.

3492. **Soto, G. E., K. W. Dodson, D. Ogg, C. Liu, J. Heuser, S. Knight, J. Kihlberg, C. H. Jones, and S. J. Hultgren.** 1998. Periplasmic chaperone recognition motif of subunits mediates quaternary interactions in the pilus. *EMBO J.* **17**:6155–6167.

3493. **Soukka, T., J. Tenovuo, and J. Rundegren.** 1993. Agglutination of *Streptococcus mutans* serotype C cells but inhibition of *Porphyromonas gingivalis* autoaggregation by human lactoferrin. *Arch. Oral Biol.* **38:**227–232.

3494. **Southwick, F. S., and D. L. Purich.** 1994. Dynamic remodeling of the actin cytoskeleton: lessons learned from *Listeria* locomotion. *Bioessays* **16:**885–891.

3495. **Speert, D. P., and S. Gordon.** 1992. Phagocytosis of unopsonized *Pseudomonas aeruginosa* by murine macrophages is a two-step process requiring glucose. *J. Clin. Investig.* **90:**1085–1092.

3496. **Spellerberg, B., E. Rozdzinski, S. Martin, J. Weber-Heynemann, N. Schnitzler, R. Lutticken, and A. Podbielski.** 1999. Lmb, a protein with similarities to the LraI adhesin family, mediates attachment of *Streptococcus agalactiae* to human laminin. *Infect. Immun.* **67:**871–878.

3497. **Spence, J. M., J. C. Chen, and V. L. Clark.** 1997. A proposed role for the lutropin receptor in contact-inducible gonococcal invasion of Hec1B cells. *Infect. Immun.* **65:**3736–3742.

3498. **Spence, J. M., and V. L. Clark.** 2000. Role of ribosomal protein L12 in gonococcal invasion of Hec1B cells. *Infect. Immun.* **68:**5002–5010.

3499. **Spencer, J., H. Chart, H. R. Smith, and B. Rowe.** 1997. Improved detection of enteroaggregative *Escherichia coli* using formalin-fixed HEp-2 cells. *Lett. Appl. Microbiol.* **25:**325–326.

3500. **Spencer, J., H. Chart, H. R. Smith, and B. Rowe.** 1998. Expression of membrane-associated proteins by strains of enteroaggregative *Escherichia coli.* *FEMS Microbiol. Lett.* **161:**325–330.

3501. **Spencer, R. J., and A. Chesson.** 1994. The effect of *Lactobacillus* spp. on the attachment of enterotoxigenic *Escherichia coli* to isolated porcine enterocytes. *J. Appl. Bacteriol.* **77:**215–220.

3502. **Sperandio, V., and W. D. da Silveira.** 1993. Comparison between enterotoxigenic *Escherichia coli* strains expressing "F42," F41 and K99 colonization factors. *Microbiol. Immunol.* **37:**869–875.

3503. **Sperandio, V., J. A. Giron, W. D. Silveira, and J. B. Kaper.** 1995. The OmpU outer membrane protein, a potential adherence factor of *Vibrio cholerae.* *Infect. Immun.* **63:**4433–4438.

3504. **Sperandio, V., J. B. Kaper, M. R. Bortolini, B. C. Neves, R. Keller, and L. R. Trabulsi.** 1998. Characterization of the locus of enterocyte effacement (LEE) in different enteropathogenic *Escherichia coli* (EPEC) and Shiga-toxin producing *Escherichia coli* (STEC) serotypes. *FEMS Microbiol. Lett.* **164:**133–139.

3505. **Speziale, P., D. Joh, L. Visai, S. Bozzini, K. House-Pompeo, M. Lindberg, and M. Höök.** 1996. A monoclonal antibody enhances ligand binding of fibronectin MSCRAMM (adhesin) from *Streptococcus dysgalactiae.* *J. Biol. Chem.* **271:**1371–1378.

3506. **Spitz, J., G. Hecht, M. Taveras, E. Aoys, and J. Alverdy.** 1994. The effect of dexamethasone administration on rat intestinal permeability: the role of bacterial adherence. *Gastroenterology* **106:**35–41.

3507. **Spitz, J., R. Yuhan, A. Koutsouris, C. Blatt, J. Alverdy, and G. Hecht.** 1995. Enteropathogenic *Escherichia coli* adherence to intestinal epithelial monolayers diminishes barrier function. *Am. J. Physiol.* **268:**G374–G379.

3508. **Spitz, J. C., S. Ghandi, M. Taveras, E. Aoys, and J. C. Alverdy.** 1996. Characteristics of the intestinal epithelial barrier during dietary manipulation and glucocorticoid stress. *Crit. Care Med.* **24:**635–641.

3509. **Sreenivasan, P. K., D. H. Meyer, and P. M. Fives-Taylor.** 1993. Requirements for invasion of epithelial cells by *Actinobacillus actinomycetemcomitans.* *Infect. Immun.* **61:**1239–1245.

3510. **Stalhammar-Carlemalm, M., T. Areschoug, C. Larsson, and G. Lindahl.** 1999. The R28 protein of *Streptococcus pyogenes* is related to several group B streptococcal surface proteins, confers protective immunity and promotes binding to human epithelial cells. *Mol. Microbiol.* **33:**208–219.

3511. **Stalheim, T., and P. E. Granum.** 2001. Characterization of spore appendages from *Bacillus cereus* strains. *J. Appl. Microbiol.* **91:**839–845.

3512. **Stapleton, A., E. Nudelman, H. Clausen, S. Hakomori, and W. E. Stamm.** 1992. Binding of uropathogenic *Escherichia coli* R45 to glycolipids extracted from vaginal epithelial cells is dependent on histo-blood group secretor status. *J. Clin. Investig.* **90:**965–972.

3513. **Stapleton, A., T. M. Hooton, C. Fennell, P. L. Roberts, and W. E. Stamm.** 1995. Effect of secretor status on vaginal and rectal colonization with fimbriated *Escherichia coli* in women with and without recurrent urinary tract infection. *J. Infect. Dis.* **171:**717–720.

3514. **Stapleton, A. E., M. R. Stroud, S. I. Hakomori, and W. E. Stamm.** 1998. The globoseries glycosphingolipid sialosyl galactosyl globoside is found in urinary tract tissues and is a preferred binding receptor in vitro for uropathogenic *Escherichia coli* expressing *pap*-encoded adhesins. *Infect. Immun.* **66**:3856–3861.

3515. **Starzyk, R. M., C. Rosenow, J. Frye, M. Leismann, E. Rodzinski, S. Putney, and E. I. Tuomanen.** 2000. Cerebral cell adhesion molecule: a novel leukocyte adhesion determinant on blood-brain barrier capillary endothelium. *J. Infect. Dis.* **181**:181–187.

3516. **Stathopoulos, C., D. L. Provence, and R. Curtiss III.** 1999. Characterization of the avian pathogenic *Escherichia coli* hemagglutinin Tsh, a member of the immunoglobulin A protease-type family of autotransporters. *Infect. Immun.* **67**:772–781.

3517. **Steer, J. A., G. B. Hill, S. Srinivasan, J. Southern, and A. P. Wilson.** 1997. Slime production, adherence and hydrophobicity in coagulase-negative staphylococci causing peritonitis in peritoneal dialysis. *J. Hosp. Infect.* **37**:305–316.

3518. **Steidler, L., E. Remaut, and W. Fiers.** 1993. Pap pili as a vector system for surface exposition of an immunoglobulin G-binding domain of protein A of *Staphylococcus aureus* in *Escherichia coli. J. Bacteriol.* **175**:7639–7643.

3519. **Stein, M. A., S. D. Mills, and B. B. Finlay.** 1994. *Salmonella*: now you see it, now you don't. *Bioessays* **16**:537–538.

3520. **Steinberg, D., L. K. Kopec, and W. H. Bowen.** 1993. Adhesion of *Actinomyces* isolates to experimental pellicles. *J. Dent. Res.* **72**:1015–1020.

3521. **Steinberg, D., C. Mor, H. Dogan, B. Zacks, and I. Rotstein.** 1999. Effect of salivary biofilm on the adherence of oral bacteria to bleached and non-bleached restorative material. *Dent. Mater.* **15**:14–20.

3522. **Steinberg, D., and M. Rothman.** 1996. Antibacterial effect of chlorhexidine on bacteria adsorbed onto experimental dental plaque. *Diagn. Microbiol. Infect. Dis.* **26**:109–115.

3523. **Steinberg, D., M. N. Sela, A. Klinger, and D. Kohavi.** 1998. Adhesion of periodontal bacteria to titanium, and titanium alloy powders. *Clin. Oral Implant Res.* **9**:67–72.

3524. **Stenfors, L. E., and S. Raisanen.** 1992. Abundant attachment of bacteria to nasopharyngeal epithelium in otitis-prone children. *J. Infect. Dis.* **165**:1148–1150.

3525. **Stenfors, L. E., and S. Raisanen.** 1992. Bacterial attachment in vivo to epithelial cells of the nasopharynx during otitis media with effusion. *J. Laryngol. Otol.* **106**:111–115.

3526. **Stenfors, L. E., and S. Raisanen.** 1992. *In vivo* attachment of *Streptococcus pneumoniae* and *Haemophilus influenzae* to nasopharyngeal

epithelium in children. *J. Oto-Rhino-Laryngol. Relat. Spec.* **54**:25–28.

3527. **Stentebjerg-Olesen, B., L. Pallesen, L. B. Jensen, G. Christiansen, and P. Klemm.** 1997. Authentic display of a cholera toxin epitope by chimeric type 1 fimbriae: effects of insert position and host background. *Microbiology* **143**:2027–2038.

3528. **Stepanovic, S., D. Vukovic, I. Dakic, B. Savic, and M. Svabic-Vlahovic.** 2000. A modified microtiter-plate test for quantification of staphylococcal biofilm formation. *J. Microbiol. Methods* **40**:175–179.

3529. **Stephens, C., and L. Shapiro.** 1996. Delivering the payload. Bacterial pathogenesis. *Curr. Biol.* **6**:927–930.

3530. **Stephens, D. S., P. A. Spellman, and J. S. Swartley.** 1993. Effect of the (α2→8)-linked polysialic acid capsule on adherence of *Neisseria meningitidis* to human mucosal cells. *J. Infect. Dis.* **167**:475–479.

3531. **Stephens, R. S., F. S. Fawaz, K. A. Kennedy, K. Koshiyama, B. Nichols, C. van Ooij, and J. N. Engel.** 2000. Eukaryotic cell uptake of heparin-coated microspheres: a model of host cell invasion by *Chlamydia trachomatis. Infect. Immun.* **68**:1080–1085.

3532. **Stepinska, M., and E. A. Trafny.** 1995. Modulation of *Pseudomonas aeruginosa* adherence to collagen type I and type II by carbohydrates. *FEMS Immunol. Med. Microbiol.* **12**:187–194.

3533. **Steuer, M. K., J. Beuth, F. Hofstadter, L. Probster, H. L. Ko, G. Pulverer, and J. Strutz.** 1995. Blood group phenotype determines lectin-mediated adhesion of *Pseudomonas aeruginosa* to human outer ear canal epithelium. *Zentbl. Bakteriol.* **282**:287–295.

3534. **Stevens, M. K., and D. C. Krause.** 1992. *Mycoplasma pneumoniae* cytadherence phase-variable protein HMW3 is a component of the attachment organelle. *J. Bacteriol.* **174**:4265–4274.

3535. **St. Geme, J. W., III, and S. Falkow.** 1992. Capsule loss by *Haemophilus influenzae* type b results in enhanced adherence to and entry into human cells. *J. Infect. Dis.* **165**:S117–S118.

3536. **St. Geme, J. W., III, S. Falkow, and S. J. Barenkamp.** 1993. High-molecular-weight proteins of nontypable *Haemophilus influenzae* mediate attachment to human epithelial cells. *Proc. Natl. Acad. Sci. USA* **90**:2875–2879.

3537. **St. Geme, J. W., III, M. L. de la Morena, and S. Falkow.** 1994. A *Haemophilus influen-*

zae IgA protease-like protein promotes intimate interaction with human epithelial cells. *Mol. Microbiol.* **14:**217–233.

3538. **St. Geme, J. W., III.** 1994. The HMW1 adhesin of nontypeable *Haemophilus influenzae* recognizes sialylated glycoprotein receptors on cultured human epithelial cells. *Infect. Immun.* **62:**3881–3889.

3539. **St. Geme J. W., III.** 1996. Molecular determinants of the interaction between *Haemophilus influenzae* and human cells. *Am. J. Respir. Crit. Care Med.* **154:**S192–S196.

3540. **St. Geme, J. W., III, D. Cutter, and S. J. Barenkamp.** 1996. Characterization of the genetic locus encoding *Haemophilus influenzae* type b surface fibrils. *J. Bacteriol.* **178:**6281–6287.

3541. **St. Geme, J. W., III.** 1997. Bacterial adhesins: determinants of microbial colonization and pathogenicity. *Adv. Pediatr.* **44:**43–72.

3542. **St. Geme, J. W., III.** 1997. Insights into the mechanism of respiratory tract colonization by nontypable *Haemophilus influenzae*. *Pediatr. Infect. Dis. J.* **16:**931–935.

3543. **St. Geme, J. W., III, V. V. Kumar, D. Cutter, and S. J. Barenkamp.** 1998. Prevalence and distribution of the hmw and hia genes and the HMW and Hia adhesins among genetically diverse strains of nontypeable *Haemophilus influenzae*. *Infect. Immun.* **66:**364–368.

3544. **Stimson, E., M. Virji, K. Makepeace, A. Dell, H. R. Morris, G. Payne, J. R. Saunders, M. P. Jennings, S. Barker, M. Panico, et al.** 1995. Meningococcal pilin: a glycoprotein substituted with digalactosyl 2,4-diacetamido-2,4,6-trideoxyhexose. *Mol. Microbiol.* **17:**1201–1214.

3545. **Stins, M. F., N. V. Prasadarao, L. Ibric, C. A. Wass, P. Luckett, and K. S. Kim.** 1994. Binding characteristics of S fimbriated *Escherichia coli* to isolated brain microvascular endothelial cells. *Am. J. Pathol.* **145:**1228–1236.

3546. **Stins, M. F., P. V. Nemani, C. Wass, and K. S. Kim.** 1999. *Escherichia coli* binding to and invasion of brain microvascular endothelial cells derived from humans and rats of different ages. *Infect. Immun.* **67:**5522–5525.

3547. **Stinson, M. W., and M. J. Levine.** 193. Modulation of intergeneric adhesion of oral bacteria by human saliva. *Crit. Rev. Oral Biol. Med.* **4:**309–314.

3548. **Stinson, M. W., G. G. Haraszthy, X. L. Zhang, and M. J. Levine.** 1992. Inhibition of *Porphyromonas gingivalis* adhesion to *Streptococcus gordonii* by human submandibular-sublingual saliva. *Infect. Immun.* **60:**2598–2604.

3549. **Stockbauer, K. E., B. Fuchslocher, J. F. Miller, and P. A. Cotter.** 2001. Identification and characterization of BipA, a *Bordetella* Bvg-intermediate phase protein. *Mol. Microbiol.* **39:**65–78.

3550. **Stokes, R. W., and D. P. Speert.** 1995. Lipoarabinomannan inhibits nonopsonic binding of *Mycobacterium tuberculosis* to murine macrophages. *J. Immunol.* **155:**1361–1369.

3551. **Stokes, R. W., and D. Doxsee.** 1999. The receptor-mediated uptake, survival, replication, and drug sensitivity of *Mycobacterium tuberculosis* within the macrophage-like cell line THP-1: a comparison with human monocyte-derived macrophages. *Cell. Immunol.* **197:**1–9.

3552. **Stollenwerk, M., C. Fallgren, F. Lundberg, J. O. Tegenfeldt, L. Montelius, and Å. Ljungh.** 1998. Quantitation of bacterial adhesion to polymer surfaces by bioluminescence. *Zentbl. Bakteriol.* **287:**7–18.

3553. **Stollerman, G. H., A. L. Bisno, and J. M. Sullivan.** 1993. *Helicobacter pylori* infection and gastric cancer. *Hosp. Pract.* **17:**21–22.

3554. **Stolpe, H., S. Grund, and W. Schröder.** 1994. Purification and partial characterization of type 3 fimbriae from *Salmonella typhimurium* var. copenhagen. *Zentbl. Bakteriol.* **281:**8–15.

3555. **Stone, B. J., C. M. Garcia, J. L. Badger, T. Hassett, R. I. Smith, and V. L. Miller.** 1992. Identification of novel loci affecting entry of *Salmonella enteritidis* into eukaryotic cells. *J. Bacteriol.* **174:**3945–3952.

3556. **Stone, B. J., and Y. Abu Kwaik.** 1998. Expression of multiple pili by *Legionella pneumophila*: identification and characterization of a type IV pilin gene and its role in adherence to mammalian and protozoan cells. *Infect. Immun.* **66:**1768–1775.

3557. **Stone, K. D., H. Z. Zhang, L. K. Carlson, and M. S. Donnenberg.** 1996. A cluster of fourteen genes from enteropathogenic *Escherichia coli* is sufficient for the biogenesis of a type IV pilus. *Mol. Microbiol.* **20:**325–337.

3558. **Stoor, P., E. Soderling, and R. Grenman.** 1999. Interactions between the bioactive glass S53P4 and the atrophic rhinitis-associated microorganism *Klebsiella ozaenae*. *J. Biomed. Mater. Res.* **48:**869–874.

3559. **Stoor, P., E. Soderling, and R. Grenman.** 2001. Bioactive glass S53P4 in repair of septal perforations and its interactions with the respiratory infection-associated microorganisms *Haemophilus influenzae* and *Streptococcus pneumoniae*. *J. Biomed. Mater. Res.* **58:**113–120.

3560. **Straley, S. C.** 1993. Adhesins in *Yersinia pestis*. *Trends Microbiol.* **1:**285–286.

3561. Straley, S. C., E. Skrzypek, G. V. Plano, and J. B. Bliska. 1993. Yops of Yersinia spp. pathogenic for humans. *Infect. Immun.* **61:**3105–3110.

3562. Straube, E., M. Kretschmar, G. Schmidt, R. Marre, B. Kreft, and G. Zingler. 1992. Is *Escherichia coli* invading tubuloepithelial cells? *Zentbl. Bakteriol.* **277:**193–203.

3563. Straube, E., G. Schmidt, R. Marre, and J. Hacker. 1993. Adhesion and internalization of *E. coli* strains expressing various pathogenicity determinants. *Zentbl. Bakteriol.* **278:**218–228.

3564. Strauss, E. J., N. Ghori, and S. Falkow. 1997. An *Edwardsiella tarda* strain containing a mutation in a gene with homology to *shlB* and *hpmB* is defective for entry into epithelial cells in culture. *Infect. Immun.* **65:**3924–3932.

3565. Striker, R., F. Jacob-Dubuisson, C. Freiden, and S. J. Hultgren. 1994. Stable fiber-forming and nonfiber-forming chaperone-subunit complexes in pilus biogenesis. *J. Biol. Chem.* **269:**12233–12239.

3566. Striker, R., U. Nilsson, A. Stonecipher, G. Magnusson, and S. J. Hultgren. 1995. Structural requirements for the glycolipid receptor of human uropathogenic *Escherichia coli. Mol. Microbiol.* **16:**1021–1029.

3567. Strom, M. S., and S. Lory. 1993. Structure-function and biogenesis of the type IV pili. *Annu. Rev. Microbiol.* **47:**565–596.

3568. Strom, M. S., D. N. Nunn, and S. Lory. 1993. A single bifunctional enzyme, PilD, catalyzes cleavage and N-methylation of proteins belonging to the type IV pilin family. *Proc. Natl. Acad. Sci. USA* **90:**2404–2408.

3569. Strömberg, N., T. Boren, A. Carlen, and J. Olsson. 1992. Salivary receptors for GalNAc beta-sensitive adherence of *Actinomyces* spp.: evidence for heterogeneous GalNAc beta and proline-rich protein receptor properties. *Infect. Immun.* **60:**3278–3286.

3570. Strömberg, N., and T. Boren. 1992. *Actinomyces* tissue specificity may depend on differences in receptor specificity for GalNAc beta-containing glycoconjugates. *Infect. Immun.* **60:**3268–3277.

3571. Strömberg, N., S. Ahlfors, T. Boren, P. Bratt, K. Hallberg, K. J. Hammarstrom, C. Holm, I. Johansson, M. Jarvholm, J. Kihlberg, T. Li, M. Ryberg, and G. Zand. 1996. Anti-adhesion and diagnostic strategies for oro-intestinal bacterial pathogens. *Adv. Exp. Med. Biol.* **408:**9–24.

3572. Stromqvist, M., P. Falk, S. Bergstrom, L. Hansson, B. Lonnerdal, S. Normark, and O. Hernell. 1995. Human milk kappa-casein and inhibition of *Helicobacter pylori* adhesion to human gastric mucosa. *J. Pediatr. Gastroenterol. Nutr.* **21:**288–296.

3573. Struve, C., and K. A. Krogfelt. 1999. In vivo detection of *Escherichia coli* type 1 fimbrial expression and phase variation during experimental urinary tract infection. *Microbiology* **145:**2683–2690.

3574. Sturegård, E., H. Sjunnesson, B. Ho, R. Willén, P. Aleljung, H. C. Ng, and T. Wadström. 1998. Severe gastritis in guinea-pigs infected with *Helicobacter pylori. J. Med. Microbiol.* **47:**1–7.

3575. Stutzmann Meier, P., J. M. Entenza, P. Vaudaux, P. Francioli, M. P. Glauser, and P. Moreillon. 2001. Study of *Staphylococcus aureus* pathogenic genes by transfer and expression in the less virulent organism *Streptococcus gordonii. Infect. Immun.* **69:**657–664.

3576. Styriak, I., P. Galfi, and V. Kmet. 1992. Adherence of ruminal *Streptococcus bovis* and *Lactobacillus* strains to primary and secondary cultures of rumen epithelium. *Acta Microbiol. Hung.* **39:**323–325.

3577. Styriak, I., P. Galfi, and V. Kmet. 1994. The adherence of three *Streptococcus bovis* strains to cells of rumen epithelium primoculture under various conditions. *Arch. Tierernahr.* **46:**357–365.

3578. Styriak, I., A. Laukova, C. Fallgren, and T. Wadström. 1999. Binding of selected extracellular matrix proteins to enterococci and *Streptococcus bovis* of animal origin. *Curr. Microbiol.* **39:**327–335.

3579. Styriak, I., A. Laukova, C. Fallgren, and T. Wadström. 1999. Binding of extracellular matrix proteins by animal strains of staphylococcal species. *Vet. Microbiol.* **67:**99–112.

3580. Styriak, I., B. Zatkovic, V. Kmet. 2001. Binding of *E. coli* isolates from pigs with postweaning diarrhea or edema disease to crude intestinal mucin of a weaned pig. *Dtsch. Tieraerztl. Wochenschr.* **108:**454–458.

3581. Su, B., P. M. Hellstrom, C. Rubio, J. Celik, M. Granstrom, and S. Normark. 1998. Type I *Helicobacter pylori* shows Lewis(b)-independent adherence to gastric cells requiring de novo protein synthesis in both host and bacteria. *J. Infect. Dis.* **178:**1379–1390.

3582. Su, B., S. Johansson, M. Fallman, M. Patarroyo, M. Granstrom, and S. Normark. 1999. Signal transduction-mediated adherence and entry of *Helicobacter pylori* into cultured cells. *Gastroenterology* **117:**595–604.

3583. **Su, C. J., S. F. Dallo, A. Chavoya, and J. B. Baseman.** 1993. Possible origin of sequence divergence in the P1 cytadhesin gene of *Mycoplasma pneumoniae*. *Infect. Immun.* **61**:816–822.

3584. **Su, H., L. Raymond, D. D. Rockey, E. Fischer, T. Hackstadt, and H. D. Caldwell.** 1996. A recombinant *Chlamydia trachomatis* major outer membrane protein binds to heparan sulfate receptors on epithelial cells. *Proc. Natl. Acad. Sci. USA* **93**:11143–11148.

3585. **Su, H., and H. D. Caldwell.** 1998. Sulfated polysaccharides and a synthetic sulfated polymer are potent inhibitors of *Chlamydia trachomatis* infectivity in vitro but lack protective efficacy in an in vivo murine model of chlamydial genital tract infection. *Infect. Immun.* **66**:1258–1260.

3586. **Su, S. H., J. W. Eaton, R. A. Venezia, and L. Tang.** 1998. Interactions of vancomycin resistant enterococci with biomaterial surfaces. *ASAIO J.* **44**:770–775.

3587. **Sudha, P. S., H. J. Devaraj, and N. Devaraj.** 2001 Adherence of *Shigella dysenteriae* 1 to human colonic mucin. *Curr. Microbiol.* **42**:381–387.

3588. **Sueyoshi, M., H. Fukui, S. Tanaka, M. Nakazawa, and K. Ito.** 1996. A new adherent form of an attaching and effacing *Escherichia coli* (eaeA$^+$, bfp$^-$) to the intestinal epithelial cells of chicks. *J. Vet. Med. Sci.* **58**:1145–1147.

3589. **Sugano, N., H. Tanaka, K. Ito, and S. Murai.** 1997. Arg-Gly-Asp (RGD) peptides inhibit *Streptococcus mitis* to adhere to fibronectin. *J. Nihon Univ. Sch. Dent.* **39**:154–155.

3590. **Sugita-Konishi, Y., K. Shibata, S. S. Yun, Y. Hara-Kudo, K. Yamaguchi, and S. Kumagai.** 1996. Immune functions of immunoglobulin Y isolated from egg yolk of hens immunized with various infectious bacteria. *Biosci. Biotechnol. Biochem.* **60**:886–888.

3591. **Sugiyama, A., T. Ogawa, Y. Daikuhara, and H. Takada.** 2000. Enhancement of hepatocyte growth factor (scatter factor) production by human gingival fibroblasts in culture stimulated with *Porphyromonas gingivalis* fimbriae. *J. Med. Microbiol.* **49**:319–325.

3592. **Suhonen, J., K. Hartiala, and M. K. Viljanen.** 1998. Tube phagocytosis, a novel way for neutrophils to phagocytize *Borrelia burgdorferi*. *Infect. Immun.* **66**:3433–3435.

3593. **Suhr, M., I. Benz, and M. A. Schmidt.** 1996. Processing of the AIDA-I precursor: removal of AIDAc and evidence for the outer membrane anchoring as a beta-barrel structure. *Mol. Microbiol.* **22**:31–42.

3594. **Sukupolvi, S., R. G. Lorenz, J. I. Gordon, Z. Bian, J. D. Pfeifer, S. J. Normark, and M. Rhen.** 1997. Expression of thin aggregative fimbriae promotes interaction of *Salmonella typhimurium* SR-11 with mouse small intestinal epithelial cells. *Infect. Immun.* **65**:5320–5325.

3595. **Suljak, J. P., G. Reid, S. M. Wood, R. J. McConnell, H. C. van der Mei, and H. J. Busscher.** 1995. Bacterial adhesion to dental amalgam and three resin composites. *J. Dent.* **23**:171–176.

3596. **Sullam, P. M., A. S. Bayer, W. M. Foss, and A. L. Cheung.** 1996. Diminished platelet binding *in vitro* by *Staphylococcus aureus* is associated with reduced virulence in a rabbit model of infective endocarditis. *Infect. Immun.* **64**:4915–4921.

3597. **Suman, E., K. Gopalkrishna Bhat, and B. M. Hegde.** 2001. Bacterial adherence and immune response in recurrent urinary tract infection. *Int. J. Gynaecol. Obstet.* **75**:263–268.

3598. **Sun, D., M. J. Lafferty, J. A. Peek, and R. K. Taylor.** 1997. Domains within the *Vibrio cholerae* toxin coregulated pilin subunit that mediate bacterial colonization. *Gene* **192**:79–85.

3599. **Sun, R., T. J. Anderson, A. K. Erickson, E. A. Nelson, and D. H. Francis.** 2000. Inhibition of adhesion of *Escherichia coli* k88ac fimbria to its receptor, intestinal mucin-type glycoproteins, by a monoclonal antibody directed against a variable domain of the fimbria. *Infect. Immun.* **68**:3509–3515.

3600. **Sun, Q., G. M. Smith, C. Zahradka, and M. J. McGavin.** 1997. Identification of D motif epitopes in *Staphylococcus aureus* fibronectin-binding protein for the production of antibody inhibitors of fibronectin binding. *Infect. Immun.* **65**:537–543.

3601. **Sundberg-Kovamees, M., T. Holme, and A. Sjogren.** 1994. Specific binding of *Streptococcus pneumoniae* to two receptor saccharide structures. *Microb. Pathog.* **17**:63–68.

3602. **Sundberg-Kovamees, M., T. Holme, and A. Sjogren.** 1996. Interaction of the C-polysaccharide of *Streptococcus pneumoniae* with the receptor asialo-GM$_1$. *Microb. Pathog.* **21**:223–234.

3603. **Sunderland, J., N. Brenwald, E. Smith, J. M. Andrews, R. Wise, and D. Honeybourne.** 1995. Enumeration of phagocytosed *Staphylococcus aureus* inside cells of the mouse macrophage cell line J774 by bioluminescence, fluorescence and viable counting techniques. *J. Biolum. Chemilum.* **10**:291–299.

3604. **Sundqvist, G.** 1993. Pathogenicity and virulence of black-pigmented gram-negative anaerobes. *FEMS Immunol. Med. Microbiol.* **6**:125–137.

3605. **Sundstrom, J., C. Agrup, G. Kronvall, and B. Wretlind.** 1997. *Pseudomonas aeruginosa* adherence to external auditory canal epithelium. *Arch. Otolaryngol. Head Neck Surg.* **123**:1287–1292.

3606. **Sung, J. Y., E. A. Shaffer, K. Lam, I. Rususka, and J. W. Costerton.** 1994. Hydrophobic bile salt inhibits bacterial adhesion on biliary stent material. *Dig. Dis. Sci.* **39**:999–1006.

3607. **Sung, J. Y., E. A. Shaffer, K. Lam, I. Rususka, and J. W. Costerton.** 1994. Hydrophobic bile salt inhibits bacterial adhesion on biliary stent material. *Dig. Dis. Sci.* **39**:999–1006.

3608. **Suresh, P., and L. H. Arp.** 1995. Effect of passively administered immunoglobulin G on the colonization and clearance of *Bordetella avium* in turkeys. *Vet. Immunol. Immunopathol.* **49**:229–239.

3609. **Susa, M., B. Kreft, G. Wasenauer, A. Ritter, J. Hacker, and R. Marre.** 1996. Influence of cloned tRNA genes from a uropathogenic *Escherichia coli* strain on adherence to primary human renal tubular epithelial cells and nephropathogenicity in rats. *Infect. Immun.* **64**:5390–5394.

3610. **Sussmuth, S. D., A. Muscholl-Silberhorn, R. Wirth, M. Susa, R. Marre, and E. Rozdzinski.** 2000. Aggregation substance promotes adherence, phagocytosis, and intracellular survival of *Enterococcus faecalis* within human macrophages and suppresses respiratory burst. *Infect. Immun.* **68**:4900–4906.

3611. **Svanborg, C., W. Agace, S. Hedges, H. Linder, and M. Svensson.** 1993. Bacterial adherence and epithelial cell cytokine production. *Zentbl. Bakteriol.* **278**:359–364.

3612. **Svanborg, C., W. Agace, S. Hedges, R. Lindstedt, and M. L. Svensson.** 1994. Bacterial adherence and mucosal cytokine production. *Ann. N. Y. Acad. Sci.* **730**:162–181.

3613. **Svanborg, C., B. Andersson, I. Andersson von Rosen, G. Aniansson, O. Nylén, and H. K. Sabharwal.** 1993. Bacterial adherence and acute otitis media, p. 367–373. *In* D. J. Lim, C. D. Bluestone, J. O. Klein, J. D. Nelson, and P. L. Ogra (ed.), *Recent Advances in Otitis Media.* Decker Periodicals, Philadelphia, Pa.

3614. **Svanborg, C., M. Hedlund, H. Connell,**

W. Agace, R. D. Duan, A. Nilsson, and B. Wullt. 1996. Bacterial adherence and mucosal cytokine responses. Receptors and transmembrane signaling. *Ann. N. Y. Acad. Sci.* **797**:177–190.

3615. **Svensson, M., R. Lindstedt, N. S. Radin, and C. Svanborg.** 1994. Epithelial glucosphingolipid expression as a determinant of bacterial adherence and cytokine production. *Infect. Immun.* **62**:4404–4410.

3616. **Swanson, A. F., and C.-C. Kuo.** 1994. The 32-kDa glycoprotein of *Chlamydia trachomatis* is an acidic protein that may be involved in the attachment process. *FEMS Microbiol. Lett.* **123**:113–117.

3617. **Swanson, A. F., and C.-C. Kuo.** 1994. Binding of the glycan of the major outer membrane protein of *Chlamydia trachomatis* to HeLa cells. *Infect. Immun.* **62**:24–28.

3618. **Swanson, A. F., R. A. Ezekowitz, A. Lee, and C. C. Kuo.** 1998. Human mannose-binding protein inhibits infection of HeLa cells by *Chlamydia trachomatis.* *Infect. Immun.* **66**:1607–1612.

3619. **Swanson, J., R. J. Belland, and S. A. Hill.** 1992. *Neisserial* surface variation: how and why? *Curr. Opin. Genet. Dev.* **2**:805–811.

3620. **Swanson, K. V., G. A. Jarvis, G. F. Brooks, B. J. Barham, M. D. Cooper, and J. M. Griffiss.** 2001. CEACAM is not necessary for *Neisseria gonorrhoeae* to adhere to and invade female genital epithelial cells. *Cell. Microbiol.* **3**:681–691.

3621. **Swanson, M. S., and R. R. Isberg.** 1995. Association of *Legionella pneumophila* with the macrophage endoplasmic reticulum. *Infect. Immun.* **63**:3609–3620.

3622. **Swartley, J. S., J. T. Balthazar, J. Coleman, W. M. Shafer, and D. S. Stephens.** 1995. Membrane glycerophospholipid biosynthesis in *Neisseria meningitidis* and *Neisseria gonorrhoeae*: identification, characterization, and mutagenesis of a lysophosphatidic acid acyltransferase. *Mol. Microbiol.* **18**:401–412.

3623. **Swenson, D. L., and S. Clegg.** 1992. Identification of ancillary *fim* genes affecting *fimA* expression in *Salmonella typhimurium.* *J. Bacteriol.* **174**:7697–7704.

3624. **Switalski, L. M., W. G. Butcher, P. C. Caufield, and M. S. Lantz.** 1993. Collagen mediates adhesion of *Streptococcus mutans* to human dentin. *Infect. Immun.* **61**:4119–4125.

3625. **Switalski, L. M., J. M. Patti, W. Butcher, A. G. Gristina, P. Speziale, and M. Höök.** 1993. A collagen receptor on *Staphylococcus aureus* strains isolated from patients with septic arthritis mediates adhesion to cartilage. *Mol. Microbiol.* **7:**99–107.

3626. **Switalski, L. M., and W. G. Butcher.** 1994. An *in vitro* model for adhesion of bacteria to human tooth root surfaces. *Arch. Oral Biol.* **39:**155–161.

3627. **Swords, W. E., B. A. Buscher, I. K. Ver Steeg, A. Prestonk, W. A. Nichols, J. N. Weiser, B. W. Gibson, and M. A. Apicella.** 2000. Non-typeable *Haemophilus influenzae* adhere to and invade human bronchial epithelial cells via an interaction of lipooligosaccharide with the PAF receptor. *Mol. Microbiol.* **37:**13–27.

3628. **Syder, A. J., J. L. Guruge, Q. Li, Y. Hu, C. M. Oleksiewicz, R. G. Lorenz, S. M. Karam, P. G. Falk, and J. I. Gordon.** 1999. *Helicobacter pylori* attaches to NeuAc α2,3Gal β1,4 glycoconjugates produced in the stomach of transgenic mice lacking parietal cells. *Mol. Cell* **3:**263–274.

3629. **Sylvester, F. A., D. Philpott, B. Gold, A. Lastovica, and J. F. Forstner.** 1996. Adherence to lipids and intestinal mucin by a recently recognized human pathogen, *Campylobacter upsaliensis*. *Infect. Immun.* **64:**4060–4066.

3630. **Symersky, J., J. M. Patti, M. Carson, K. House-Pompeo, M. Teale, D. Moore, L. Jin, A. Schneider, L. J. DeLucas, M. Höök, and S. V. Narayana.** 1997. Structure of the collagen-binding domain from a *Staphylococcus aureus* adhesin. *Nat. Struct. Biol.* **4:**833–838.

3631. **Szöke, I., C. Pascu, E. Nagy, Å. Ljung, and T. Wadström.** 1996. Binding of extracellular matrix proteins to the surface of anaerobic bacteria. *J. Med. Microbiol.* **45:**338–343.

3632. **Szöke, I., C. Pascu, E. Nagy, Å. Ljungh, and T. Wadström.** 1996. Binding of extracellular matrix proteins to the surface of anaerobic bacteria. *J. Med. Microbiol.* **45:**338–343.

3633. **Szymanski, C. M., M. King, M. Haardt, and G. D. Armstrong.** 1995. *Campylobacter jejuni* motility and invasion of Caco-2 cells. *Infect. Immun.* **63:**4295–4300.

3634. **Szymanski, C. M., and G. D. Armstrong.** 1996. Interactions between *Campylobacter jejuni* and lipids. *Infect. Immun.* **64:**3467–3474.

3635. **Tacket, C. O., R. K. Taylor, G. Losonsky, Y. Lim, J. P. Nataro, J. B. Kaper, and M. M. Levine.** 1998. Investigation of the roles of toxin-coregulated pili and mannose-sensitive hemagglutinin pili in the pathogenesis of *Vibrio cholerae* O139 infection. *Infect. Immun.* **66:**692–695.

3636. **Tadikonda, K. R., and R. H. Davis.** 1994. Cell separations using targeted monoclonal antibodies against overproduced surface proteins. *Appl. Biochem. Biotechnol.* **45–46:**233–244.

3637. **Tafazoli, F., A. Holmstrom, A. Forsberg, and K. E. Magnusson.** 2000. Apically exposed, tight junction-associated beta1-integrins allow binding and YopE-mediated perturbation of epithelial barriers by wild-type *Yersinia* bacteria. *Infect. Immun.* **68:**5335–5343.

3638. **Tagashira, M., K. Uchiyama, T. Yoshimura, M. Shirota, and N. Uemitsu.** 1997. Inhibition by hop bract polyphenols of cellular adherence and water-insoluble glucan synthesis of mutans streptococci. *Biosci. Biotechnol. Biochem.* **61:**332–335.

3639. **Taguchi, H., T. Osaki, H. Yamaguchi, and S. Kamiya.** 1995. Flow cytometric analysis using lipophilic dye PKH-2 for adhesion of *Vibrio cholerae* to Intestine 407 cells. *Microbiol. Immunol.* **39:**891–894.

3640. **Taguchi, H., H. Yamaguchi, T. Y. Osaki, T. Yamamoto, S. Ogata, and S. Kamiya.** 1997. Flow cytometric analysis for adhesion of *Vibrio cholerae* to human intestinal epithelial cell. *Eur. J. Epidemiol.* **13:**719–724.

3641. **Taha, M. K., D. Giorgini, and X. Nassif.** 1996. The *pilA* regulatory gene modulates the pilus-mediated adhesion of *Neisseria meningitidis* by controlling the transcription of *pilC1*. *Mol. Microbiol.* **19:**1073–1084.

3642. **Taha, M. K., P. C. Morand, Y. Pereira, E. Eugene, D. Giorgini, M. Larribe, and X. Nassif.** 1998. Pilus-mediated adhesion of *Neisseria meningitidis*: the essential role of cell contact-dependent transcriptional upregulation of the PilC1 protein. *Mol. Microbiol.* **28:**1153–1163.

3643. **Tahir, Y. E., P. Kuusela, and M. Skurnik.** 2000. Functional mapping of the *Yersinia enterocolitica* adhesin YadA. Identification of eight NSVAIG-S motifs in the amino-terminal half of the protein involved in collagen binding. *Mol. Microbiol.* **37:**192–206.

3644. **Tai, S. S., T. R. Wang, and C.-J. Lee.** 1997. Characterization of hemin binding activity of *Streptococcus pneumoniae*. *Infect. Immun.* **65:**1083–1087.

3645. **Takagi, A., Y. Koga, Y. Aiba, A. M. Kabir, S. Watanabe, U. Ohta-Tada, T. Osaki, S. Kamiya, and T. Miwa.** 2000. Plaunotol suppresses interleukin-8 secretion induced by *Helicobacter pylori*: therapeutic effect of plaunotol on *H. pylori* infection. *J. Gastroenterol. Hepatol.* **15:**374–380.

3646. Takagi, M., N. Hirayama, T. Simazaki, K. Taguchi, R. Yamaoka, and S. Ohta. 1993. Purification of hemagglutinin from *Haemophilus paragallinarum* using monoclonal antibody. *Vet. Microbiol.* **34**:191–197.

3647. Takahashi, M., H. Taguchi, H. Yamaguchi, T. Osaki, and S. Kamiya. 2000. Studies of the effect of *Clostridium butyricum* on *Helicobacter pylori* in several test models including gnotobiotic mice. *J. Med. Microbiol.* **49**:635–642.

3648. Takahashi, N., T. Saito, S. Ohwada, H. Ota, H. Hashiba, and T. Itoh. 1996. A new screening method for the selection of *Lactobacillus acidophilus* group lactic acid bacteria with high adhesion to human colonic mucosa. *Biosci. Biotechnol. Biochem.* **60**:1434–1438.

3649. Takahashi, Y., F. Yoshimura, M. Kawanami, and H. Kato. 1992. Detection of fimbrilin gene (*fimA*) in *Porphyromonas* (*Bacteroides*) *gingivalis* by Southern blot analysis. *J. Periodontal Res.* **27**:599–603.

3650. Takahashi, Y., A. L. Sandberg, S. Ruhl, J. Muller, and J. O. Cisar. 1997. A specific cell surface antigen of *Streptococcus gordonii* is associated with bacterial hemagglutination and adhesion to α2-3-linked sialic acid-containing receptors. *Infect. Immun.* **65**:5042–5051.

3651. Takahashi, Y., H. Yoshimoto, D. Kato, N. Hamada, M. Arai, and T. Umemoto. 2001. Reduced fimbria-associated activities of *Porphyromonas gingivalis* induced by recombinant fimbrial expression. *FEMS Microbiol. Lett.* **195**:217–222.

3652. Takayama, Y., M. Kanoe, K. Maeda, Y. Okada, and K. Kai. 2000. Adherence of *Fusobacterium necrophorum* subsp. *necrophorum* to ruminal cells derived from bovine rumenitis. *Lett. Appl. Microbiol.* **30**:308–311.

3653. Takemoto, T., M. Ozaki, M. Shirakawa, T. Hino, and H. Okamoto. 1992. Purification of arginine-sensitive hemagglutinin from *Fusobactrerium nucleatum* and its role in coaggregation. *J. Periodontal Res.* **28**:21–26.

3654. Takemoto, T., T. Hino, M. Yoshida, K. Nakanishi, M. Shirakawa, and H. Okamoto. 1995. Characteristics of multimodal co-aggregation between *Fusobacterium nucleatum* and streptococci. *J. Periodontal Res.* **30**:252–257.

3655. Takeshita, A., Y. Murakami, Y. Yamashita, M. Ishida, S. Fujisawa, S. Kitano, and S. Hanazawa. 1998. *Porphyromonas gingivalis* fimbriae use β²-integrin (CD11/CD18) on mouse peritoneal macrophages as a cellular receptor, and the CD18 beta chain plays a functional role in fimbrial signaling. *Infect. Immun.* **66**:4056–4060.

3656. Takumi, K., T. Koga, T. Oka, H. Tsuji, M. Tsubokura, and Y. J. Sheng. 1992. Extracellular localization of a nonfimbrial hemagglutinin of *Yersinia pseudotuberculosis*. *Microbiol. Immunol.* **36**:767–771.

3657. Talay, S. R., P. Valentin-Weigand, K. N. Timmis, and G. S. Chhatwal. 1994. Domain structure and conserved epitopes of Sfb protein, the fibronectin-binding adhesin of *Streptococcus pyogenes*. *Mol. Microbiol.* **13**:531–539.

3658. Talbot, U. M., A. W. Paton, and J. C. Paton. 1996. Uptake of *Streptococcus pneumoniae* by respiratory epithelial cells. *Infect. Immun.* **64**:3772–3777.

3659. Tamamoto, T., K. Nakashima, N. Nakasone, Y. Honma, N. Higa, and T. Yamashiro. 1998. Adhesive property of toxin-coregulated pilus of *Vibrio cholerae* O1. *Microbiol. Immunol.* **42**:41–45.

3660. Tambic, T., V. Oberiter, J. Delmis, and A. Tambic. 1992. Diagnostic value of a P-fimbriation test in determining duration of therapy in children with urinary tract infections. *Clin. Ther.* **14**:667–671.

3661. Tamura, G. S., J. M. Kuypers, S. Smith, H. Raff, and C. E. Rubens. 1994. Adherence of group B streptococci to cultured epithelial cells: roles of environmental factors and bacterial surface components. *Infect. Immun.* **62**:2450–2458.

3662. Tamura, G. S., and C. E. Rubens. 1995. Group B streptococci adhere to a variant of fibronectin attached to a solid phase. *Mol. Microbiol.* **15**:581–589.

3663. Tamura, G. S., and A. Nittayajarn. 2000. Group B streptococci and other gram-positive cocci bind to cytokeratin 8. *Infect. Immun.* **68**:2129–2134.

3664. Tamura, M., K. Kuroda, Y. Ueda, N. Saito, Y. Hirano, and K. Hayashi. 1995. Adsorption of saliva-coated and just-harvested *Streptococcus sanguis* to saliva-coated hydroxyapatite beads. *J. Nihon Univ. Sch. Dent.* **37**:170–177.

3665. Tamura, M., Y. Hirano, and K. Hayashi. 1999. The phenomenon of salivary protein adsorption onto *Streptococcus mitis* ATCC 903 cells. *J. Oral Sci.* **41**:169–172.

3666. Tanaka, H., S. Ebara, K. Otsuka, and K. Hayashi. 1996. Adsorption of saliva-coated and plain streptococcal cells to the surfaces of hydroxyapatite beads. *Arch. Oral Biol.* **41**:505–508.

3667. Tanemoto, K., and H. Fujinami. 1994.

Experimental study on bacterial colonization of fibrin glue and its prevention. *Clin. Ther.* **16**:1016–1027.

3668. **Tang, H., M. Kays, and A. Prince.** 1995. Role of *Pseudomonas aeruginosa* pili in acute pulmonary infection. *Infect. Immun.* **63**:1278–1285.

3669. **Tang, H. B., E. DiMango, R. Bryan, M. Gambello, B. H. Iglewski, J. B. Goldberg, and A. Prince.** 1996. Contribution of specific *Pseudomonas aeruginosa* virulence factors to pathogenesis of pneumonia in a neonatal mouse model of infection. *Infect. Immun.* **64**:37–43.

3670. **Tang, P., I. Rosenshine, and B. B. Finlay.** 1994. *Listeria monocytogenes,* an invasive bacterium, stimulates MAP kinase upon attachment to epithelial cells. *Mol. Biol. Cell* **5**:455–464.

3671. **Tang, P., C. L. Sutherland, M. R. Gold, and B. B. Finlay.** 1998. *Listeria monocytogenes* invasion of epithelial cells requires the MEK-1/ERK-2 mitogen-activated protein kinase pathway. *Infect. Immun.* **66**:1106–1112.

3672. **Taniguchi, T., Y. Fujino, K. Yamamoto, T. Miwatani, and T. Honda.** 1995. Sequencing of the gene encoding the major pilin of pilus colonization factor antigen III (CFA/III) of human enterotoxigenic *Escherichia coli* and evidence that CFA/III is related to type IV pili. *Infect. Immun.* **63**:724–728.

3673. **Tanner, J., P. K. Vallittu, and E. Soderling.** 2001. Effect of water storage of E-glass fiber-reinforced composite on adhesion of *Streptococcus mutans. Biomaterials* **22**:1613–1618.

3674. **Tanner, J., P. K. Vallittu, and E. Soderling.** 2000. Adherence of *Streptococcus mutans* to an E-glass fiber-reinforced composite and conventional restorative materials used in prosthetic dentistry. *J. Biomed. Mater. Res.* **49**:250–256.

3675. **Tarkkanen, A.-M., B. L. Allen, P. H. Williams, M. Kauppi, K. Haahtela, A. Siitonen, I. Ørskov, F. Ørskov, S. Clegg, and T. K. Korhonen.** 1992. Fimbriation, capsulation, and iron-scavenging systems of *Klebsiella* strains associated with human urinary tract infection. *Infect. Immun.* **60**:1187–1192.

3676. **Tarkkanen, A.-M., R. Virkola, S. Clegg, and T. K. Korhonen.** 1997. Binding of the type 3 fimbriae of *Klebsiella pneumoniae* to human endothelial and urinary bladder cells. *Infect. Immun.* **65**:1546–1549.

3677. **Tarkkanen, A.-M., B. Westerlund-Wikstrom, L. Erkkila, and T. K. Korhonen.** 1998. Immunohistological localization of the MrkD adhesin in the type 3 fim-

briae of *Klebsiella pneumoniae. Infect. Immun.* **66**:2356–2361.

3678. **Tarsi, R., R. A. Muzzarelli, C. A. Guzman, and C. Pruzzo.** 1997. Inhibition of *Streptococcus mutans* adsorption to hydroxyapatite by low-molecular-weight chitosans. *J. Dent. Res.* **76**:665–672.

3679. **Tarsi, R., B. Corbin, C. Pruzzo, and R. A. Muzzarelli.** 1998. Effect of low-molecular-weight chitosans on the adhesive properties of oral streptococci. *Oral Microbiol. Immunol.* **13**:217–224.

3680. **Tarsi, R., and C. Pruzzo.** 1999. Role of surface proteins in *Vibrio cholerae* attachment to chitin. *Appl. Environ. Microbiol.* **65**:1348–1351.

3681. **Tart, R. C., and I. van de Rijn.** 1993. Identification of the surface component of *Streptococcus defectivus* that mediates extracellular matrix adherence. *Infect. Immun.* **61**:4994–5000.

3682. **Tartera, C., and E. S. Metcalf.** 1993. Osmolarity and growth phase overlap in regulation of *Salmonella typhi* adherence to and invasion of human intestinal cells. *Infect. Immun.* **61**:3084–3089.

3683. **Tasteyre, A., M. C. Barc, A. Collignon, H. Boureau, and T. Karjalainen.** 2001. Role of FliC and FliD flagellar proteins of *Clostridium difficile* in adherence and gut colonization. *Infect. Immun.* **69**:7937–7940.

3684. **Tatsuno, I., M. Horie, H. Abe, T. Miki, K. Makino, H. Shinagawa, H. Taguchi, S. Kamiya, T. Hayashi, and C. Sasakawa.** 2001. *toxB* gene on pO157 of enterohemorrhagic *Escherichia coli* O157:H7 is required for full epithelial cell adherence phenotype. *Infect. Immun.* **69**:6660–6669.

3685. **Tatsuno, I., H. Kimura, A. Okutani, K. Kanamaru, H. Abe, S. Nagai, K. Makino, H. Shinagawa, M. Yoshida, K. Sato, J. Nakamoto, T. Tobe, and C. Sasakawa.** 2000. Isolation and characterization of mini-Tn5Km2 insertion mutants of enterohemorrhagic *Escherichia coli* O157:H7 deficient in adherence to Caco-2 cells. *Infect. Immun.* **68**:5943–5952.

3686. **Tawfik, A. F., F. A. Al-Zamil, M. A. Ramadan, and A. M. Shibl.** 1996. Effect of beta-lactamase inhibitors on normal immune capabilities and their interactions with staphylococcal pathogenicity. *J. Chemother.* **8**:102–106.

3687. **Tawfik, A. F., M. A. Ramadan, and A. M. Shibl.** 1997. Inhibition of motility and adherence of *Proteus mirabilis* to uroepithelial cells by subinhibitory concentrations of amikacin. *Chemotherapy* **43**:424–429.

3688. **Tay, S. T., S. Devi, S. Puthucheary, and I. Kautner.** 1996. In vitro demonstration of the invasive ability of Campylobacters. Zentbl. Bakterol. **283:**306–313.

3689. **Taylor, D. C., R. L. Clancy, A. W. Cripps, H. Butt, L. Bartlett, and K. Murree-Allen.** 1994. An alteration in the host-parasite relationship in subjects with chronic bronchitis prone to recurrent episodes of acute bronchitis. Immunol. Cell Biol. **72:**143–151.

3690. **Taylor, K. A., P. W. Luther, and M. S. Donnenberg.** 1999. Expression of the EspB protein of enteropathogenic Escherichia coli within HeLa cells affects stress fibers and cellular morphology. Infect. Immun. **67:**120–135.

3691. **Taylor, N. S., A. T. Hasubski, J. G. Fox, and A. Lee.** 1992. Haemagglutination profiles of Helicobacter species that cause gastritis in man and animals. J. Med. Microbiol. **37:**299–303.

3692. **Taylor, R. L., M. D. Willcox, T. J. Williams, and J. Verran.** 1998. Modulation of bacterial adhesion to hydrogel contact lenses by albumin. Optom. Vision Sci. **75:**23–29.

3693. **Tebbs, S. E., and T. S. Elliott.** 1993. A novel antimicrobial central venous catheter impregnated with benzalkonium chloride. J. Antimicrob. Chemother. **31:**261–271.

3694. **Tebbs, S. E., and T. S. Elliott.** 1994. Modification of central venous catheter polymers to prevent in vitro microbial colonisation. Eur. J. Clin. Microbiol. Infect. Dis. **13:**111–117.

3695. **Tebbs, S. E., A. Sawyer, and T. S. Elliott.** 1994. Influence of surface morphology on in vitro bacterial adherence to central venous catheters. Br. J. Anaesth. **72:**587–591.

3696. **Temple, L. M., A. A. Weiss, K. E. Walker, H. J. Barnes, V. L. Christensen, D. M. Miyamoto, C. B. Shelton, and P. E. Orndorff.** 1998. Bordetella avium virulence measured in vivo and in vitro. Infect. Immun. **66:**5244–5251.

3697. **ten Cate, J. M., and P. D. Marsh.** 1994. Procedures for establishing efficacy of antimicrobial agents for chemotherapeutic caries prevention. J. Dent. Res. **73:**695–703.

3698. **Teneberg, S., P. T. Willemsen, F. K. de Graaf, and K. A. Karlsson.** 1993. Calf small intestine receptors for K99 fimbriated enterotoxigenic Escherichia coli. FEMS Microbiol. Lett. **109:**107–112.

3699. **Teppema, J. S., E. C. de Boer, P. A. Steerenberg, and A. P. van der Meijden.** 1992. Morphological aspects of the interaction of Bacillus Calmette-Guerin with urothelial bladder cells in vivo and in vitro: relevance for antitumor activity? Urol. Res. **20:**219–228.

3700. **Teraguchi, S., K. Shin, Y. Fukuwatari, and S. Shimamura.** 1996. Glycans of bovine lactoferrin function as receptors for the type 1 fimbrial lectin of Escherichia coli. Infect. Immun. **64:**1075–1077.

3701. **Terai, A., S. Yamamoto, K. Mitsumori, Y. Okada, H. Kurazono, Y. Takeda, and O. Yoshida.** 1997. Escherichia coli virulence factors and serotypes in acute bacterial prostatitis. Int. J. Urol. **4:**289–294.

3702. **Tertii, R., M. Skurnik, T. Vartio, and P. Kuusela.** 1992. Adhesion protein YadA of Yersinia species mediates binding of bacteria to fibronectin. Infect. Immun. **60:**3021–3024.

3703. (See reference 3702.)

3704. **Tesh, V. L., and A. D. O'Brien.** 1992. Adherence and colonization mechanisms of enteropathogenic and enterohemorrhagic Escherichia coli. Microb. Pathog. **12:**245–254.

3705. **Tetz, V. V., and L. L. Norman.** 1994. Effect of subinhibitory concentrations of antimicrobial agents on virulence factors of Shigella flexneri 2a and Escherichia coli O124. J. Med. Microbiol. **41:**279–281.

3706. **Tewari, R., J. I. MacGregor, T. Ikeda, J. R. Little, S. J. Hultgren, and S. N. Abraham.** 1993. Neutrophil activation by nascent FimH subunits of type 1 fimbriae purified from the periplasm of Escherichia coli. J. Biol. Chem. **268:**3009–3015.

3707. **Tewari, R., T. Ikeda, R. Malaviya, J. I. MacGregor, J. R. Little, S. J. Hultgren, and S. N. Abraham.** 1994. The PapG tip adhesin of P fimbriae protects Escherichia coli from neutrophil bactericidal activity. Infect. Immun. **62:**5296–5304.

3708. **Thakker, M., J. S. Park, V. Carey, and J. C. Lee.** 1998. Staphylococcus aureus serotype 5 capsular polysaccharide is antiphagocytic and enhances bacterial virulence in a murine bacteremia model. Infect. Immun. **66:**5183–5189.

3709. **Thakur, A., A. Chauhan, and M. D. Willcox.** 1999. Effect of lysozyme on adhesion and toxin release by Staphylococcus aureus. Aust. N. Z. J. Ophthalmol. **27:**224–227.

3710. **Tham, T. N., S. Ferris, E. Bahraoui, S. Canarelli, L. Montagnier, and A. Blanchard.** 1994. Molecular characterization of the P1-like adhesin gene from Mycoplasma pirum. J. Bacteriol. **176:**781–788.

3711. **Tham, T. N., S. Ferris, E. Bahraoui, S. Canarelli, L. Montagnier, and A. Blanchard.** 1994. Identification of an adhesin-like gene of Mycoplasma pirum isolated from AIDS patients. *Ann. N. Y. Acad. Sci.* **730**:279–282.

3712. **Thankavel, K., B. Madison, T. Ideda, R. Malaviya, A. H. Shah, P. M. Arumugam, and S. N. Abraham.** 1997. Localization of a domain in the FimH adhesin of *Escherichia coli* type 1 fimbriae capable of receptor recognition and use of a domain-specific antibody to confer protection against experimental urinary tract infection. *J. Clin. Investig.* **100**:1123–1136.

3713. **Thankavel, K., A. H. Shah, M. S. Cohen, T. Ikeda, R. G. Lorenz, R. Curtiss III, and S. N. Abraham.** 1999. Molecular basis for the enterocyte tropism exhibited by *Salmonella typhimurium* type 1 fimbriae. *J. Biol. Chem.* **274**:5797–5809.

3714. **Tharpe, J. A., and H. Russell.** 1996. Purification and seroreactivity of pneumococcal surface adhesin A (PsaA). *Clin. Diagn. Lab. Immunol.* **3**:227–229.

3715. **Thelin, K. H., and R. K. Taylor.** 1996. Toxin-coregulated pilus, but not mannose-sensitive hemagglutinin, is required for colonization by *Vibrio cholerae* O1 El Tor biotype and O139 strains. *Infect. Immun.* **64**:2853–2856.

3716. **Theriot, J. A.** 1995. The cell biology of infection by intracellular bacterial pathogens. *Annu. Rev. Cell Dev. Biol.* **11**:213–239.

3717. **Thiagarajan, D., A. M. Saeed, and E. K. Asem.** 1994. Mechanism of transovarian transmission of *Salmonella enteritidis* in laying hens. *Poult. Sci.* **73**:89–98.

3718. **Thiagarajan, D., M. Saeed, J. Turek, and E. Asem.** 1996. *In vitro* attachment and invasion of chicken ovarian granulosa cells by *Salmonella enteritidis* phage type 8. *Infect. Immun.* **64**:5015–5021.

3719. **Thiagarajan, D., H. L. Thacker, and A. M. Saeed.** 1996. Experimental infection of laying hens with *Salmonella enteritidis* strains that express different types of fimbriae. *Poult. Sci.* **75**:1365–1372.

3720. **Thomas, L. H., J. A. Leigh, A. P. Bland, and R. S. Cook.** 1992. Adherence and colonization by bacterial pathogens in explant cultures of bovine mammary tissue. *Vet. Res. Commun.* **16**:87–96.

3721. **Thomas, V. L., B. A. Sanford, and M. A. Ramsay.** 1993. Calcium- and mucin-binding proteins of staphylococci. *J. Gen. Microbiol.* **139**:623–629.

3722. **Thornley, J. P., J. G. Shaw, I. A. Gryllos, and A. Eley.** 1996. Adherence of *Aeromonas caviae* to human cell lines Hep-2 and Caco-2. *J. Med. Microbiol.* **45**:445–451.

3723. **Thorns, C. J.** 1995. *Salmonella* fimbriae:

novel antigens in the detection and control of salmonella infections. *Br. Vet. J.* **151**:643–658.

3724. **Thorns, C. J., M. G. Sojka, I. M. McLaren, and M. Dibb-Fuller.** 1992. Characterisation of monoclonal antibodies against a fimbrial structure of *Salmonella enteritidis* and certain other serogroup D salmonellae and their application as serotyping reagents. *Res. Vet. Sci.* **53**:300–308.

3725. **Thorns, C. J., C. Turcotte, C. G. Gemmell, and M. J. Woodward.** 1996. Studies into the role of the SEF14 fimbrial antigen in the pathogenesis of *Salmonella enteritidis*. *Microb Pathog.* **20**:235–246.

3726. **Thorson, L. M., D. Doxsee, M. G. Scott, P. Wheeler, and R. W. Stokes.** 2001. Effect of mycobacterial phospholipids on interaction of *Mycobacterium tuberculosis* with macrophages. *Infect. Immun.* **69**:2172–2179.

3727. **Thuruthyil, S. J., H. Zhu, and M. D. Willcox.** 2001. Serotype and adhesion of *Pseudomonas aeruginosa* isolated from contact lens wearers. *Clin. Exp. Ophthalmol.* **29**:147–149.

3728. **Tickoo, S. K., M. K. Bhan, R. Srivastava, B. K. Dass, M. Shariff, S. Saini, and R. Kumar.** 1992. Intestinal colonization & production of diarrhoea by enteroadherent-aggregative *Escherichia coli*. *Indian J. Med. Res.* **95**:278–283.

3729. **Tikkanen, K., S. Haataja, C. François-Gerard, and J. Finne.** 1995. Purification of a galactosyl-α1-4-galactose-binding adhesin from the gram-positive meningitis-associated bacterium *Streptococcus suis*. *J. Biol. Chem.* **270**:28874–28878.

3730. **Tikkanen, K., S. Haataja, and J. Finne.** 1996. The galactosyl-(α1–4)-galactose-binding adhesion of *Streptococcus suis*: occurrence in strains of different hemagglutination activities and induction of opsonic antibodies. *Infect. Immun.* **64**:3659–3665.

3731. **Toba, T., R. Virkola, B. Westerlund, Y. Björkman, J. Sillanpää, T. Vartio, N. Kalkkinen, and T. K. Korhonen.** 1995. A collagen-binding S-layer protein in *Lactobacillus crispatus*. *Appl. Environ. Microbiol.* **61**:2467–2471.

3732. **Tobe, T., and C. Sasakawa.** 2001. Role of bundle-forming pilus of enteropathogenic *Escherichia coli* in host cell adherence and in microcolony development. *Cell. Microbiol.* **3**:579–585.

3733. **Tobe, T., G. K. Schoolnik, I. Sohel, V. H. Bustamante, and J. L. Puente.** 1996. Cloning and characterization of *bfpTVW*, genes required for the transcriptional activation of *bfpA* in enteropathogenic *Escherichia coli*. *Mol. Microbiol.* **21**:963–975.

3734. **Todoriki, K., T. Mukai, S. Sato, and T. Toba.** 2001. Inhibition of adhesion of food-borne pathogens to Caco-2 cells by *Lactobacillus* strains. *J. Appl. Microbiol.* **91**:154–159.

3735. **Tokuda, M., M. Duncan, M. I. Cho, and H. K. Kuramitsu.** 1996. Role of *Porphyromonas gingivalis* protease activity in colonization of oral surfaces. *Infect. Immun.* **64**:4067–4073.

3736. **Tokuda, M., T. Karunakaran, M. Duncan, N. Hamada, and H. Kuramitsu.** 1998. Role of Arg-gingipain A in virulence of *Porphyromonas gingivalis*. *Infect. Immun.* **66**:1159–1166.

3737. **Toledo-Arana, A., J. Valle, C. Solano, M. J. Arrizubieta, C. Cucarella, M. Lamata, B. Amorena, J. Leiva, J. R. Penades, and I. Lasa.** 2001. The enterococcal surface protein, Esp, is involved in *Enterococcus faecalis* biofilm formation. *Appl. Environ. Microbiol.* **67**:4538–4545.

3738. **Toleman, M., E. Aho, and M. Virji.** 2001. Expression of pathogen-like Opa adhesins in commensal *Neisseria*: genetic and functional analysis. *Cell. Microbiol.* **3**:33–44.

3739. **Tolson, D. L., B. A. Harrison, R. K. Latta, K. K. Lee, and E. Altman.** 1997. The expression of nonagglutinating fimbriae and its role in *Proteus mirabilis* adherence to epithelial cells. *Can. J. Microbiol.* **43**:709–717.

3740. **Tompkins, D. C., L. J. Blackwell, V. B. Hatcher, D. A. Elliott, C. O'Hagan-Sotsky, and F. D. Lowy.** 1992. *Staphylococcus aureus* proteins that bind to human endothelial cells. *Infect. Immun.* **60**:965–969.

3741. **Tong, H. H., L. M. Fisher, G. M. Kosunick, and T. F. Demaria.** 1999. Effect of tumor necrosis factor alpha and interleukin 1-alpha on the adherence of *Streptococcus pneumoniae* to chinchilla tracheal epithelium. *Acta Oto-Laryngol.* **119**:78–82.

3742. **Tong, H. H., M. A. McIver, L. M. Fisher, and T. F. DeMaria.** 1999. Effect of lacto-N-neotetraose, asialoganglioside-GM_1 and neuraminidase on adherence of otitis media-associated serotypes of *Streptococcus pneumoniae* to chinchilla tracheal epithelium. *Microb. Pathog.* **26**:111–119.

3743. **Tønjum, T., W. Weir, K. Bovre, A. Progulske-Fox, and C. F. Marrs.** 1993. Sequence divergence in two tandemly located pilin genes of *Eikenella corrodens*. *Infect. Immun.* **61**:1909–1916.

3744. **Tønjum, T., N. E. Freitag, E. Namork,**

and M. Koomey. 1995. Identification and characterization of *pilG*, a highly conserved pilus-assembly gene in pathogenic *Neisseria*. *Mol. Microbiol.* **16**:451–464.

3745. **Tønjum, T., and M. Koomey.** 1997. The pilus colonization factor of pathogenic neisserial species: organelle biogenesis and structure/function relationships—a review. *Gene* **192**:155–163.

3746. **Tønjum, T., D. A. Caugant, S. A. Dunham, and M. Koomey.** 1998. Structure and function of repetitive sequence elements associated with a highly polymorphic domain of the *Neisseria meningitidis* PilQ protein. *Mol. Microbiol.* **29**: 111–124.

3747. **Tosh, F. D., and L. J. Douglas.** 1992. Charactrerization of a fucoside-binding adhesin of *Candida albicans*. *Infect. Immun.* **60**:4734–4739.

3748. **Totten, P. A., L. C. Lara, D. V. Norn, and W. E. Stamm.** 1994. *Haemophilus ducreyi* attaches to and invades human epithelial cells in vitro. *Infect. Immun.* **62**:5632–5640.

3749. **Trafny, E. A., M. Antos-Bielska, and J. Grzybowski.** 1999. Antibacterial activity of liposome-encapsulated antibiotics against *Pseudomonas aeruginosa* attached to the matrix of human dermis. *J. Microencapsulation* **16**:419–429.

3750. **Trafny, E. A., M. Stepinska, M. Antos, and J. Grzybowski.** 1995. Effects of free and liposome-encapsulated antibiotics on adherence of *Pseudomonas aeruginosa* to collagen type I. *Antimicrob. Agents Chemother.* **39**:2645–2649.

3751. **Trafny, E. A.** 1998. Susceptibility of adherent organisms from *Pseudomonas aeruginosa* and *Staphylococcus aureus* strains isolated from burn wounds to antimicrobial agents. *Int. J. Antimicrob. Agents* **10**:223–228.

3752. **Trafny, E. A., K. Kowalska, and J. Grzybowski.** 1998. Adhesion of *Pseudomonas aeruginosa* to collagen biomaterials: effect of amikacin and ciprofloxacin on the colonization and survival of the adherent organisms. *J. Biomed. Mater. Res.* **41**:593–599.

3753. **Tran Van Nhieu, G., and R. R. Isberg.** 1993. Bacterial internalization mediated by beta 1 chain integrins is determined by ligand affinity and receptor density. *EMBO J.* **12**:1887–1895.

3754. **Tran Van Nhieu, G., A. Ben-Ze'ev, and P. J. Sansonetti.** 1997. Modulation of bacterial entry into epithelial cells by association between vinculin and the *Shigella* IpaA invasin. *EMBO J.* **16**:2717–2729.

3755. **Trifillis, A. L., M. S. Donnenberg, X.**

Cui, R. G. Russell, S. J. Utsalo, H. L. Mobley, and J. W. Warren. 1994. Binding to and killing of human renal epithelial cells by hemolytic P-fimbriated *E. coli. Kidney Int.* **46:**1083–1091.

3756. Trust, T. J., M. Kostrzynska, L. Emody, and T. Wadström. 1993. High-affinity binding of the basement membrane protein collagen type IV to the crystalline virulence surface protein array of *Aeromonas salmonicida. Mol. Microbiol.* **7:**593–600.

3757. Tsai, C. L., T. K. Liu, and M. H. Hung. 1992. Glycocalyx production and adherence of *Staphylococcus* to biomaterials. *Acta Med. Okayama* **46:**11–16.

3758. Tsai, P. J., C. F. Kuo, K. Y. Lin, Y. S. Lin, H. Y. Lei, F. F. Chen, J. R. Wang, and J. J. Wu. 1998. Effect of group A streptococcal cysteine protease on invasion of epithelial cells. *Infect. Immun.* **66:**1460–1466.

3759. Tsang, K. W., A. Rutman, E. Tanaka, V. Lund, A. Dewar, P. J. Cole, and R. Wilson. 1994. Interaction of *Pseudomonas aeruginosa* with human respiratory mucosa *in vitro. Eur. Respir. J.* **7:**1746–1753.

3760. Tsang, R. S., J. M. Luk, D. L. Woodward, and W. M. Johnson. 1996. Immunochemical characterization of a haemagglutinating antigen of *Arcobacter* spp. *FEMS Microbiol. Lett.* **136:**209–213.

3761. Tsumori, H., and H. Kuramitsu. 1997. The role of the *Streptococcus mutans* glucosyltransferases in the sucrose-dependent attachment to smooth surfaces: essential role of the GtfC enzyme. *Oral Microbiol. Immunol.* **12:**274–280.

3762. Tue, C. J., R. A. Bojar, and K. T. Holland. 1998. The adhesion of cutaneous micro-organisms to human skin lipids. *Dermatology* **196:**71–72.

3763. Tufano, M. A., F. Rossano, P. Catalanotti, G. Liguori, A. Marinelli, A. Baroni, and P. Marinelli. 1994. Properties of *Yersinia enterocolitica* porins: interference with biological functions of phagocytes, nitric oxide production and selective cytokine release. *Res. Microbiol.* **145:**297–307.

3764. Tullio, V., N. A. Carlone, A. M. Cuffini, and S. Fazari. 1992. Adhesion of *Proteus mirabilis* and *Proteus vulgaris* to uroepithelial cells following exposure to various antimicrobial agents. *Microbios* **69:**67–75.

3765. Tullio, V., A. M. Cuffini, A. Bonino, A. I. Palarchio, V. Rossi, and N. A. Carlone. 1999. Cellular uptake and intraphagocytic activity of the new fluoro-quinolone AF 3013 against *Klebsiella pneumoniae. Drugs Exp. Clin. Res.* **25:**1–11.

3766. Tullus, K., A. Brauner, B. Fryklund, T. Munkhammar, W. Rabsch, R. Reissbrodt, and L. G. Burman. 1992. Host factors versus virulence-associated bacterial characteristics in neonatal and infantile bacteraemia and meningitis caused by *Escherichia coli. J. Med. Microbiol.* **36:**203–208.

3767. Tullus, K., I. Kühn, I. Ørskov, F. Ørskov, and R. Möllby. 1992. The importance of P and type 1 fimbriae for the persistence of *Escherichia coli* in the human gut. *Epidemiol. Infect.* **108:**415–421.

3768. Tumber-Saini, S. K., B. F. Habbick, A. M. Oles, J. P. Schaefer, and K. Komiyama. 1992. The role of saliva in aggregation and adherence of *Pseudomonas aeruginosa* in patients with cystic fibrosis. *J. Oral Pathol. Med.* **21:**299–304.

3769. Tuomanen, E. I. 1996. Molecular and cellular mechanisms of pneumococcal meningitis. *Ann. N. Y. Acad. Sci.* **797:**42–52.

3770. Tuomola, E. M., and S. J. Salminen. 1998. Adhesion of some probiotic and dairy *Lactobacillus* strains to Caco-2 cell cultures. *Int. J. Food Microbiol.* **41:**45–51.

3771. Tuomola, E. M., A. C. Ouwehand, and S. J. Salminen. 1999. Human ileostomy glycoproteins as a model for small intestinal mucus to investigate adhesion of probiotics. *Lett. Appl. Microbiol.* **28:**159–163.

3772. Tuomola, E. M., A. C. Ouwehand, and S. J. Salminen. 1999. The effect of probiotic bacteria on the adhesion of pathogens to human intestinal mucus. *FEMS Immunol. Med. Microbiol.* **26:**137–142.

3773. Tuomola, E. M., A. C. Ouwehand, and S. J. Salminen. 2000. Chemical, physical and enzymatic pre-treatments of probiotic lactobacilli alter their adhesion to human intestinal mucus glycoproteins. *Int. J. Food Microbiol.* **60:**75–81.

3774. Turi, M., E. Turi, S. Koljalg, and M. Mikelsaar. 1997. Influence of aqueous extracts of medicinal plants on surface hydrophobicity of *Escherichia coli* strains of different origin. *APMIS* **105:**956–962.

3775. Turner, L. R., J. C. Lara, D. N. Nunn, and S. Lory. 1993. Mutations in the consensus ATP-binding sites of XcpR and PilB eliminate extracellular protein secretion and pilus biogenesis in *Pseudomonas aeruginosa. J. Bacteriol.* **175:**4962–4969.

3776. Tuttle, R. S., N. A. Strubel, J. Mourad, and D. F. Mangan. 1992. A non-lectin-like mechanism by which *Fusobacterium nucleatum*

10953 adheres to and activates human lymphocytes. *Oral Microbiol. Immunol.* **7**:78–83.

3777. **Tzipori, S., F. Gunzer, M. S. Donnenberg, L. de Montigny, J. B. Kaper, and A. Donohue-Rolfe.** 1995. The role of the *eaeA* gene in diarrhea and neurological complications in a gnotobiotic piglet model of enterohemorrhagic *Escherichia coli* infection. *Infect. Immun.* **63**: 3621–3627.

3778. **Tzouvelekis, L. S., A. F. Mentis, E. Tzelpi, A. Tsakris, K. Kyriakis, and N. J. Legakis.** 1992. Exposure effects of netilmicin on *Enterobacter cloacae* clinical isolates. *Chemotherapy* **38**:405–409.

3779. **Uberos, J., C. Augustin, J. Liebana, A. Molina, and A. Munoz-Hoyos.** 2001. Comparative study of the influence of melatonin and vitamin E on the surface characteristics of *Escherichia coli*. *Lett. Appl. Microbiol.* **32**:303–306.

3780. **Uchimura, M., and T. Yamamoto.** 1992. Production of hemagglutinins and pili by *Vibrio mimicus* and its adherence to human and rabbit small intestines *in vitro*. *FEMS Microbiol. Lett.* **91**:73–78.

3781. **Uehara, Y., H. Nakama, K. Agematsu, M. Uchida, Y. Kawakami, A. S. Abdul Fattah, and N. Maruchi.** 2000. Bacterial interference among nasal inhabitants: eradication of *Staphylococcus aureus* from nasal cavities by artificial implantation of *Corynebacterium* sp. *J. Hosp. Infect.* **44**:127–133.

3782. **Ulleryd, P., K. Lincoln, F. Scheutz, and T. Sandberg.** 1994. Virulence characteristics of *Escherichia coli* in relation to host response in men with symptomatic urinary tract infection. *Clin. Infect. Dis.* **18**:579–584.

3783. **Ulrich, M., S. Herbert, J. Berger, G. Bellon, D. Louis, G. Munker, and G. Doring.** 1998. Localization of *Staphylococcus aureus* in infected airways of patients with cystic fibrosis and in a cell culture model of *S. aureus* adherence. *Am. J. Respir. Cell Mol. Biol.* **19**:83–91.

3784. **Umemoto, T., Y. Nakatani, Y. Nakamura, and I. Namikawa.** 1993. Fibronectin-binding proteins of a human oral spirochete *Treponema denticola*. *Microbiol. Immunol.* **37**:75–78.

3785. **Umemoto, T., and I. Namikawa.** 1994. Binding of host-associated treponeme proteins to collagens and laminin: a possible mechanism of spirochetal adherence to host tissues. *Microbiol. Immunol.* **38**:655–663.

3786. **Umemoto, T., M. Li, and I. Namikawa.** 1997. Adherence of human oral spirochetes by collagen-binding proteins. *Microbiol. Immunol.* **41**:917–923.

3787. **Umemoto, T., F. Yoshimura, H. Kureshiro, J. Hayashi, T. Noguchi, and T. Ogawa.** 1999. Fimbria-mediated coaggregation between human oral anaerobes *Treponema medium* and *Porphyromonas gingivalis*. *Microbiol. Immunol.* **43**:837–845.

3788. **Unhanand, M., I. Maciver, O. Ramilo, O. Arencibia-Mireles, J. C. Argyle, G. H. McCracken, Jr., and E. J. Hansen.** 1992. Pulmonary clearance of *Moraxella catarrhalis* in an animal model. *J. Infect. Dis.* **165**:644–650.

3789. **Usui, A., M. Murai, K. Seki, J. Sakurada, and S. Masuda.** 1992. Conspicuous ingestion of *Staphylococcus aureus* organisms by murine fibroblasts *in vitro*. *Microbiol. Immunol.* **36**:545–550.

3790. **Usui, Y., Y. Ohshima, Y. Ichiman, T. Ohtomo, and J. Shimada.** 1997. Some biochemical properties of the components of *Staphylococcus aureus* binding to human platelets. *Zentbl. Bakteriol.* **286**:56–62.

3791. **Utsunomiya, A., T. Naito, M. Ehara, Y. Ichinose, and A. Hamamoto.** 1992. Studies on novel pili from *Shigella flexneri*. I. Detection of pili and hemagglutination activity. *Microbiol. Immunol.* **36**:803–813.

3792. **Utsunomiya, A., M. Nakamura, and A. Hamamoto.** 2000. Expression of fimbriae and hemagglutination activity in *Shigella boydii*. *Microbiol. Immunol.* **44**:529–531.

3793. **Vaahtontemi, L. H., S. Raisanen, and L. E. Stenfors.** 1992. The age-dependence of bacterial presence on oral epithelial surfaces *in vivo*. *Oral Microbiol. Immunol.* **7**:263–266.

3794. **Vacca-Smith, A. M., C. A. Jones, M. J. Levine, and M. W. Stinson.** 1994. Glucosyltransferase mediates adhesion of *Streptococcus gordonii* to human endothelial cells in vitro. *Infect. Immun.* **62**:2187–2194.

3795. **Vacca-Smith, A. M., B. C. Van Wuyckhuyse, L. A. Tabak, and W. H. Bowen.** 1994. The effect of milk and casein proteins on the adherence of *Streptococcus mutans* to saliva-coated hydroxyapatite. *Arch. Oral Biol.* **39**:1063–1069.

3796. **Vacca-Smith, A. M., A. R. Venkitaraman, R. G. Quivey, Jr., and W. H. Bowen.** 1996. Interactions of streptococcal glucosyltransferases with alpha-amylase and starch on the surface of saliva-coated hydroxyapatite. *Arch. Oral Biol.* **41**:291–198.

3797. **Vacca-Smith, A. M., L. Ng-Evans, D. Wunder, and W. H. Bowen.** 2000. Studies concerning the glucosyltransferase of *Streptococcus sanguis*. *Caries Res.* **34**:295–302.

3798. **Vacheethasanee, K., J. S. Temenoff, J. M. Higashi, A. Gary, J. M. Anderson, R. Bayston, and R. E. Marchant.** 1998. Bacterial surface properties of clinically isolated *Staphylococcus epidermidis* strains determine adhesion on polyethylene. *J. Biomed. Mater. Res.* **42:**425–432.

3799. **Vacheethasanee, K., and R. E. Marchant.** 2000. Surfactant polymers designed to suppress bacterial (*Staphylococcus epidermidis*) adhesion on biomaterials. *J. Biomed. Mater. Res.* **50:**302–312.

3800. **Valentin-Weigand, P., K. N. Timmis, and G. S. Chhatwal.** 1993. Role of fibronectin in staphylococcal colonisation of fibrin thrombi and plastic surfaces. *J. Med. Microbiol.* **38:**90–95.

3801. **Valentin-Weigand, P., S. R. Talay, K. N. Timmis, and G. S. Chhatwal.** 1993. Identification of a fibronectin-binding protein as adhesin of *Streptococcus pyogenes*. *Zentbl. Bakteriol.* **278:**238–245.

3802. **Valentin-Weigand, P., S. R. Talay, A. Kaufhold, K. N. Timmis, and G. S. Chhatwal.** 1994. The fibronectin binding domain of the Sfb protein adhesin of *Streptococcus pyogenes* occurs in many group A streptococci and does not cross-react with heart myosin. *Microb. Pathog.* **17:**111–120.

3803. **Valentin-Weigand, P., P. Benkel, M. Rohde, and G. S. Chhatwal.** 1996. Entry and intracellular survival of group B streptococci in J774 macrophages. *Infect. Immun.* **64:**2467–2473.

3804. **Valkonen, K. H., M. Ringner, Å. Ljungh, and T. Wadström.** 1993. High-affinity binding of laminin by *Helicobacter pylori*: evidence for a lectin-like interaction. *FEMS Immunol. Med. Microbiol.* **7:**29–37.

3805. **Valkonen, K. H., T. Wadström, and A. P. Moran.** 1994. Interaction of lipopolysaccharides of *Helicobacter pylori* with basement membrane protein laminin. *Infect. Immun.* **62:**3640–3648.

3806. **Valkonen, K. H., T. Wadström, and A. P. Moran.** 1997. Identification of the N-acetylneuraminyllactose-specific laminin-binding protein of *Helicobacter pylori*. *Infect. Immun.* **65:**916–923.

3807. **Valpotic, I., M. Frankovic, and I. Vrbanac.** 1992. Identification of infant and adult swine susceptible to enterotoxigenic *Escherichia coli* by detection of receptors for F4(K88)ac fimbriae in brush borders or feces. *Comp. Immunol. Microbiol. Infect. Dis.* **15:**271–279.

3808. **Valvatne, H., H. Sommerfelt, W. Gaastra, M. K. Bhan, and H. M. Grewal.** 1996. Identification and characterization of CS20, a new putative colonization factor of enterotoxigenic *Escherichia coli*. *Infect. Immun.* **64:**2635–2642.

3809. **van Alphen, L., P. Eijk, H. Kayhty, J. van Marle, and J. Dankert.** 1996. Antibodies to *Haemophilus influenzae* type b polysaccharide affect bacterial adherence and multiplication. *Infect. Immun.* **64:**995–1001.

3810. **van Asten, F. J., H. G. Hendriks, J. F. Koninkx, B. A. van der Zeijst, and W. Gaastra.** Inactivation of the flagellin gene of *Salmonella enterica* serotype enteritidis strongly reduces invasion into differentiated Caco-2 cells. *FEMS Microbiol. Lett.* **185:**175–179.

3811. **van den Akker, W. M.** 1997. *Bordetella bronchiseptica* has a BvgAS-controlled cytotoxic effect upon interaction with epithelial cells. *FEMS Microbiol. Lett.* **156:**239–244.

3812. **van den Akker, W. M.** 1998. The filamentous hemagglutinin of *Bordetella parapertussis* is the major adhesin in the phase-dependent interaction with NCI-H292 human lung epithelial cells. *Biochem. Biophys. Res. Commun.* **252:**128–133.

3813. **van den Berg, B. M., H. Beekhuizen, F. R. Mooi, and R. van Furth.** 1999. Role of antibodies against *Bordetella pertussis* virulence factors in adherence of *Bordetella pertussis* and *Bordetella parapertussis* to human bronchial epithelial cells. *Infect. Immun.* **67:**1050–1055.

3814. **van den Berg, B. M., H. Beekhuizen, R. J. Willems, F. R. Mooi, and R. van Furth.** 1999. Role of *Bordetella pertussis* virulence factors in adherence to epithelial cell lines derived from the human respiratory tract. *Infect. Immun.* **67:**1056–1062.

3815. **van den Bosch, J. F., J. H. Hendriks, I. Gladigau, H. M. Willems, P. K. Storm, and F. K. de Graaf.** 1993. Identification of F11 fimbriae on chicken *Escherichia coli* strains. *Infect. Immun.* **61:**800–806.

3816. **van den Brink, G. R., K. M. Tytgat, R. W. van der Hulst, C. M. van der Loos, A. W. Einerhand, H. A. Buller, and J. Dekker.** 2000. *H. pylori* colocalises with MUC5AC in the human stomach. *Gut* **46:**601–607.

3817. **van den Broeck, W., E. Cox, and B. M. Goddeeris.** 1999. Receptor-specific binding of purified F4 to isolated villi. *Vet. Microbiol.* **68:**255–263.

3818. **van der Flier, M., N. Chhun, T. M. Wizemann, J. Min, J. B. McCarthy, and E. I. Tuomanen.** 1995. Adherence of *Streptococcus pneumoniae* to immobilized fibronectin. *Infect. Immun.* **63:**4317–4322.

3819. **van der Mei, H. C., S. D. Cox, G. I.**

Geertsema-Doornbusch, R. J. Doyle, and H. J. Busscher. 1993. A critical appraisal of positive cooperativity in oral streptococcal adhesion: Scatchard analyses of adhesion data versus analyses of the spatial arrangement of adhering bacteria. *J. Gen. Microbiol.* **139:**937–948.

3820. van der Mei, H. C., J. de Vries, and H. J. Busscher. 1993. Hydrophobic and electrostatic cell surface properties of thermophilic dairy streptococci. *Appl. Environ. Microbiol.* **59:**4305–4312.

3821. van der Mei, H. C., and H. J. Busscher. 1996. Detection by physico-chemical techniques of an amphiphilic surface component on *Streptococcus mitis* strains involved in non-electrostatic binding to surfaces. *Eur. J. Oral Sci.* **104:**48–55.

3822. van der Mei, H. C., and H. J. Busscher. 1997. The use of X-ray photoelectron spectroscopy for the study of oral streptococcal cell surfaces. *Adv. Dent. Res.* **11:**388–394.

3823. van der Mei, H. C., B. van de Belt-Gritter, G. Reid, H. Bialkowska-Hobrzanska, and H. J. Busscher. 1997. Adhesion of coagulase-negative staphylococci grouped according to physico-chemical surface properties. *Microbiology* **143:**3861–3870.

3824. van der Velden, A. W. M., A. J. Bäumler, R. M. Tsolis, and F. Heffron. 1998. Multiple fimbrial adhesins are required for full virulence of *Salmonella typhimurium* in mice. *Infect. Immun.* **66:**2803–2808.

3825. van der Woude, M. W., B. A. Braaten, and D. A. Low. 1992. Evidence for global regulatory control of pilus expression in *Escherichia coli* by Lrp and DNA methylation: model building based on analysis of pap. *Mol. Microbiol.* **6:**2429–2435.

3826. van der Woude, M. W., and D. A. Low. 1994. Leucine-responsive regulatory protein and deoxyadenosine methylase control the phase variation and expression of the *sfa* and *daa* pili operons in *Escherichia coli*. *Mol. Microbiol.* **11:**605–618.

3827. van der Woude, M. W., L. S. Kaltenbach, and D. A. Low. 1995. Leucine-responsive regulatory protein plays dual roles as both an activator and a repressor of the *Escherichia coli* pap fimbrial operon. *Mol. Microbiol.* **17:**303–312.

3828. van der Woude, M., B. Braaten, and D. Low. 1996. Epigenetic phase variation of the *pap* operon in *Escherichia coli*. *Trends Microbiol.* **4:**5–9.

3829. van der Zee, A., C. V. Noordegraaf, H. van den Bosch, J. Gielen, H. Bergmans, W. Hoekstra, and I. van Die. 1995. P-fim-

briae of *Escherichia coli* as carriers for gonadotropin releasing hormone: development of a recombinant contraceptive vaccine. *Vaccine* **13:**753–758.

3830. Vandevoorde, L., H. Christiaens, and W. Verstraete. 1992. Prevalence of coaggregation reactions among chicken lactobacilli. *J. Appl. Bacteriol.* **72:**214–219.

3831. van Doorn, J., B Oudega, and D. M. MacLaren. 1992. Characterization and detection of the 40 kDa fimbrial subunit of *Bacteroides fragilis* BE1. *Microb. Pathog.* **13:**75–79.

3832. Vanek, N. N., S. I. Simon, K. Jacques-Palaz, M. M. Mariscalco, G. M. Dunny, and R. M. Rakita. 1999. *Enterococcus faecalis* aggregation substance promotes opsonin-independent binding to human neutrophils via a complement receptor type 3-mediated mechanism. *FEMS Immunol. Med. Microbiol.* **26:**49–60.

3833. van Golde, L. M. 1995. Potential role of surfactant proteins A and D in innate lung defense against pathogens. *Biol. Neonate* **67:**2–17.

3834. van Ham, S. M., L. van Alphen, and F. R. Mooi. 1992. Fimbria-mediated adherence and hemagglutination of *Haemophilus influenzae*. *J. Infect. Dis.* **165:**S97–S99.

3835. van Ham, S. M., L. van Alphen, F. R. Mooi, and J. P. van Putten. 1994. The fimbrial gene cluster of *Haemophilus influenzae* type b. *Mol. Microbiol.* **13:**673–684.

3836. van Ham, S. M., L. van Alphen, F. R. Mooi, and J. P. van Putten. 1995. Contribution of the major and minor subunits to fimbria-mediated adherence of *Haemophilus influenzae* to human epithelial cells and erythrocytes. *Infect. Immun.* **63:**4883–4889.

3837. (See reference 3836.)

3838. van Heyningen, T., G. Fogg, D. Yates, E. Hanski, and M. Caparon. 1993. Adherence and fibronectin binding are environmentally regulated in the group A streptococci. *Mol. Microbiol.* **9:**1213–1222.

3839. van Hoogmoed, C. G., M. van der Kuijl-Booij, H. C. van der Mei, and H. J. Busscher. 2000. Inhibition of *Streptococcus mutans* NS adhesion to glass with and without a salivary conditioning film by biosurfactant-releasing *Streptococcus mitis* strains. *Appl. Environ. Microbiol.* **66:**659–663.

3840. Vanmaele, R. P., L. D. Heerze, and G. D. Armstrong. 1999. Role of lactosyl glycan sequences in inhibiting enteropathogenic *Escherichia coli* attachment. *Infect. Immun.* **67:**3302–3307.

3841. Vanmaele, R. P., and G. D. Armstrong.

1997. Effect of carbon source on localized adherence of enteropathogenic *Escherichia coli*. *Infect. Immun.* **65:**1408–1413.

3842. **van Putten, J. P. M., and S. M. Paul.** 1995. Binding of syndecan-like cell surface proteoglycan receptors is required for *Neisseria gonorrhoeae* entry into human mucosal cells. *EMBO J.* **14:**2144–2154.

3843. **van Putten, J. P. M., S. F. Hayes, and T. D. Duensing.** 1997. Natural proteoglycan receptor analogs determine the dynamics of Opa adhesin-mediated gonococcal infection of Chang epithelial cells. *Infect. Immun.* **65:**5028–5034.

3844. **van Raamsdonk, M., H. C. van der Mei, J. J. de Soet, H. J. Busscher, and J. de Graaff.** 1995. Effect of polyclonal and monoclonal antibodies on surface properties of *Streptococcus sobrinus*. *Infect. Immun.* **63:**1698–1702.

3845. **van Raamsdonk, M., H. C. van der Mei, G. I. Geertsema-Doornbusch, J. J. de Soet, H. J. Busscher, and J. de Graaff.** 1995. Physicochemical aspects of microbial adhesion—influence of antibody adsorption on the deposition of *Streptococcus sobrinus* in a parallel-plate flow chamber. *Colloids Surf. Ser. B* **4:**401–410.

3846. **van Rosmalen, M., and M. H. Saier, Jr.** 1993. Structural and evolutionary relationships between two families of bacterial extracytoplasmic chaperone proteins which function cooperatively in fimbrial assembly. *Res. Microbiol.* **144:**507–527.

3847. **van Schie, P. M., and M. Fletcher.** 1999. Adhesion of biodegradative anaerobic bacteria to solid surfaces. *Appl. Environ. Microbiol.* **65:**5082–5088.

3848. **van Schilfgaarde, M., P. van Ulsen, P. Eijk, M. Brand, M. Stam, J. Kouame, L. van Alphen, and J. Dankert.** 2000. Characterization of adherence of nontypeable *Haemophilus influenzae* to human epithelial cells. *Infect. Immun.* **68:**4658–4665.

3849. **van't Wout, J., W. N. Burnette, V. L. Mar, E. Rozdzinski, S. D. Wright, and E. I. Tuomanen.** 1992. Role of carbohydrate recognition domains of pertussis toxin in adherence of *Bordetella pertussis* to human macrophages. *Infect. Immun.* **60:**3303–3308.

3850. **van Wamel, W. J., C. M. Vandenbroucke-Grauls, J. Verhoef, and A. C. Fluit.** 1998. The effect of culture conditions on the *in-vitro* adherence of methicillin-resistant *Staphylococcus aureus*. *J. Med. Microbiol.* **47:**705–709.

3851. **van Zijderveld, F. G., A. M. van Zijderveld-van Bemmel, and D. Bakker.** 1998. The F41 adhesin of enterotoxigenic

Escherichia coli: inhibition of adhesion by monoclonal antibodies. *Vet. Q.* **20:**S73–S78.

3852. **Vats, N., and S. F. Lee.** 2000. Active detachment of *Streptococcus mutans* cells adhered to epon-hydroxylapatite surfaces coated with salivary proteins *in vitro*. *Arch. Oral Biol.* **45:**305–314.

3853. **Vaudaux, P., T. Avramoglou, D. Letourneur, D. P. Lew, and J. Jozefonvicz.** 1992. Inhibition by heparin and derivatized dextrans of *Staphylococcus aureus* adhesion to fibronectin-coated biomaterials. *J. Biomater. Sci. Polym. Ed.* **4:**89–97.

3854. **Vaudaux, P., D. Pittet, A. Haeberli, P. G. Lerch, J.-J. Morgenthaler, R. A. Proctor, F. A Waldvogel, and D. P. Lew.** 1993. Fibronectin is more active than fibrin or fibrinogen in promoting *Staphylococcus aureus* adherence to inserted intravascular catheters. *J. Infect. Dis.* **167:**633–641.

3855. **Vaudaux, P. E., P. François, R. A. Proctor, D. McDevitt, T. J. Foster, R. M. Albrecht, D. P. Lew, H. Wabers, and S. L. Cooper.** 1995. Use of adhesion-defective mutants of *Staphylococcus aureus* to define the role of specific plasma proteins in promoting bacterial adhesion to canine arteriovenus shunts. *Infect. Immun.* **63:**585–590.

3856. **Vaudaux, P. E., V. Monzillo, P. Francois, D. P. Lew, T. J. Foster, and B. Berger-Bachi.** 1998. Introduction of the *mec* element (methicillin resistance) into *Staphylococcus aureus* alters in vitro functional activities of fibrinogen and fibronectin adhesins. *Antimicrob. Agents Chemother.* **42:**564–570.

3857. **Vazquez, F., E. A. Gonzalez, J. I. Garabal, S. Valderrama, J. Blanco, and S. B. Baloda.** 1996. Development and evaluation of an ELISA to detect *Escherichia coli* K88 (F4) fimbrial antibody levels. *J. Med. Microbiol.* **44:**453–463.

3858. **Vazquez, F., E. A. Gonzalez, J. I. Garabal, and J. Blanco.** 1996. Fimbriae extracts from enterotoxigenic *Escherichia coli* strains of bovine and porcine origin with K99 and/or F41 antigens. *Vet. Microbiol.* **48:**231–241.

3859. **Vazquez-Torres, A., J. Jones-Carson, A. J. Baumler, S. Falkow, R. Valdivia, W. Brown, M. Le, R. Berggren, W. T. Parks, and F. C. Fang.** 1999. Extraintestinal dissemination of *Salmonella* by CD18-expressing phagocytes. *Nature* **401:**804–808.

3860. **Veenstra, G. J., F. F. Cremers, H. van Dijk, and A. Fleer.** 1996. Ultrastructural organization and regulation of a biomaterial adhesin of *Staphylococcus epidermidis*. *J. Bacteriol.* **178**:537–541.

3861. **Veerman, E. C., A. J. Ligtenberg, L. C. Schenkels, E. Walgreen-Weterings, and A. V. Nieuw Amerongen.** 1995. Binding of human high-molecular-weight salivary mucins (MG1) to *Hemophilus parainfluenzae*. *J. Dent. Res.* **74**:351–357.

3862. **Veerman, E. C., C. M. Bank, F. Namavar, B. J. Appelmelk, J. G. Bolscher, and A. V. Nieuw Amerongen.** 1997. Sulfated glycans on oral mucin as receptors for *Helicobacter pylori*. *Glycobiology* **7**:737–473.

3863. **Velge, P., E. Bottreau, B. Kaeffer, N. Yurdusev, P. Pardon, and N. Van Langendonck.** 1994. Protein tyrosine kinase inhibitors block the entries of *Listeria monocytogenes* and *Listeria ivanovii* into epithelial cells. *Microb. Pathog.* **17**:37–50.

3864. **Velge, P., B. Kaeffer, E. Bottreau, and N. Van Langendonck.** 1995. The loss of contact inhibition and anchorage-dependent growth are key steps in the acquisition of *Listeria monocytogenes* susceptibility phenotype by non-phagocytic cells. *Biol. Cell* **85**:55–66.

3865. **Velge, P., E. Bottreau, N. Van Langendonck, and B. Kaeffer.** 1997. Cell proliferation enhances entry of *Listeria monocytogenes* into intestinal epithelial cells by two proliferation-dependent entry pathways. *J. Med. Microbiol.* **46**:681–692.

3866. **Velraeds, M. M., H. C. van der Mei, G. Reid, and H. J. Busscher.** 1996. Inhibition of initial adhesion of uropathogenic *Enterococcus faecalis* by biosurfactants from *Lactobacillus* isolates. *Appl. Environ. Microbiol.* **62**:1958–1963.

3867. **Velraeds, M. M., H. C. van der Mei, G. Reid, and H. J. Busscher.** 1997. Inhibition of initial adhesion of uropathogenic *Enterococcus faecalis* to solid substrata by an adsorbed biosurfactant layer from *Lactobacillus acidophilus*. *Urology* **49**:790–794.

3868. **Velraeds, M. M., B. van de Belt-Gritter, H. C. van der Mei, G. Reid, and H. J. Busscher.** 1998. Interference in initial adhesion of uropathogenic bacteria and yeasts to silicone rubber by a *Lactobacillus acidophilus* biosurfactant. *J. Med. Microbiol.* **47**:1081–1085.

3869. **Venegas, M. F., E. L. Navas, R. A. Gaffney, J. L. Duncan, B. E. Anderson, and A. J. Schaeffer.** 1995. Binding of type 1-piliated *Escherichia coli* to vaginal mucus. *Infect. Immun.* **63**:416–422.

3870. **Venisse, A., J. J. Fournie, and G. Puzo.** 1995. Mannosylated lipoarabinomannan interacts with phagocytes. *Eur. J. Biochem.* **231**:440–447.

3871. **Venkataraman, C., B. J. Haack, S. Bondada, and Y. A. Kwaik.** 1997. Identification of a Gal/GalNAc lectin in the protozoan *Hartmannella vermiformis* as a potential receptor for attachment and invasion by the Legionnaires' disease bacterium. *J. Exp. Med.* **186**:537–547.

3872. **Venkataraman, C., L.-Y. Gao, S. Bondada, and Y. A. Kwaik.** 1998. Identification of putative cytoskeletal protein homologues in the protozoan host *Hartmanella vermiformis* as substrates for induced tyrosine phosphatase activity upon attachment to the Legionnaires' disease bacterium, *Legionella pneumophila*. *J. Exp. Med.* **188**:505–514.

3873. **Verdon, R., D. Mirelman, and P. J. Sansonetti.** 1992. A model of interaction between *Entamoeba histolytica* and *Shigella flexneri*. *Res. Microbiol.* **143**:67–74.

3874. **Vergeres, P., and J. Blaser.** 1992. Amikacin, ceftazidime, and flucloxacillin against suspended and adherent *Pseudomonas aeruginosa* and *Staphylococcus epidermidis* in an *in vitro* model of infection. *J. Infect. Dis.* **165**:281–289.

3875. **Verheyen, C. C., W. J. Dhert, J. M. de Blieck-Hogervorst, T. J. van der Reijden, P. L. Petit, and K. de Groot.** 1993. Adherence to a metal, polymer and composite by *Staphylococcus aureus* and *Staphylococcus epidermidis*. *Biomaterials* **14**:383–391.

3876. **Verjans, G. M., J. H. Ringrose, L. van Alphen, T. E. Feltkamp, and J. G. Kusters.** 1994. Entrance and survival of *Salmonella typhimurium* and *Yersinia enterocolitica* within human B- and T-cell lines. *Infect. Immun.* **62**:2229–2235.

3877. **Vetter, V., and J. Hacker.** 1995. Strategies for employing molecular genetics to study tip adhesions. *Methods Enzymol.* **253**:229–241.

3878. **Veyries, M. L., F. Faurisson, M. L. Joly-Guillou, and B. Rouveix.** 2000. Control of staphylococcal adhesion to polymethylmethacrylate and enhancement of susceptibility to antibiotics by poloxamer 407. *Antimicrob. Agents Chemother.* **44**:1093–1096.

3879. **Viboud, G. I., N. Binsztein, and A.-M. Svennerholm.** 1993. Characterization of monoclonal antibodies against putative colonization factors of enterotoxigenic *Escherichia coli* and their use in an epidemiological study. *J. Clin. Microbiol.* **31**:558–564.

3880. **Viboud, G. I., N. Binsztein, and A.-M. Svennerholm.** 1993. A new fimbrial putative colonization factor, PCFO20, in human enterotoxigenic *Escherichia coli. Infect. Immun.* **61**:5190–5197.

3881. **Viboud, G. I., G. Jonson, E. Dean-Nystrom, and A. M. Svennerholm.** 1996. The structural gene encoding human enterotoxigenic *Escherichia coli* PCFO20 is homologous to that for porcine 987P. *Infect. Immun.* **64**:1233–1239.

3882. **Viboud, G. I., M. M. McConnell, A. Helander, and A.-M. Svennerholm.** 1996. Binding of enterotoxigenic *Escherichia coli* expressing different colonization factors to tissue-cultured Caco-2 cells and to isolated human enterocytes. *Microb. Pathog.* **21**:139–147.

3883. **Vica Pacheco, S., O. Garcia Gonzalez, and G. L. Paniagua Contreras.** 1997. The *lom* gene of bacteriophage lambda is involved in *Escherichia coli* K12 adhesion to human buccal epithelial cells. *FEMS Microbiol. Lett.* **156**:129–132.

3884. **Vickerman, M. M., D. B. Clewell, and G. W. Jones.** 1992. Glucosyltransferase phase variation in *Streptococcus gordonii* modifies adhesion to saliva-coated hydroxyapatite surfaces in a sucrose-independent manner. *Oral Microbiol. Immunol.* **7**:118–120.

3885. **Vickerman, M. M., and G. W. Jones.** 1992. Adhesion of glucosyltransferase phase variants to *Streptococcus gordonii* bacterium-glucan substrata may involve lipoteichoic acid. *Infect. Immun.* **60**:4301–4308.

3886. **Vickerman, M. M., and G. W. Jones.** 1995. Sucrose-dependent accumulation of oral streptococci and their adhesion-defective mutants on saliva-coated hydroxyapatite. *Oral Microbiol. Immunol.* **10**:175–182.

3887. **Vidal, O., R. Longin, C. Prigent-Combaret, C. Dorel, M. Hooreman, and P. Lejeune.** 1998. Isolation of an *Escherichia coli* K-12 mutant strain able to form biofilms on inert surfaces: involvement of a new *ompR* allele that increases curli expression. *J. Bacteriol.* **180**:2442–2449.

3888. **Vidotto, M. C., H. R. Navarro, and L. C. Gaziri.** 1997. Adherence pili of pathogenic strains of avian *Escherichia coli. Vet. Microbiol.* **59**:79–87.

3889. **Vieira, M. A., J. R. Andrade, L. R. Trabulsi, A. C. Rosa, A. M. Dias, S. R. Ramos, G. Frankel, and T. A. Gomes.** 2001. Phenotypic and genotypic characteristics of *Escherichia coli* strains of non-enteropathogenic *E. coli* (EPEC) serogroups that carry EAE and lack the EPEC adherence factor and Shiga toxin DNA probe sequences. *J. Infect. Dis.* **183**:762–772.

3890. **Vijayakumari, S., M. M. Khin. B. Jiang,**
and **B. Ho.** 1995. The pathogenic role of the coccoid form of *Helicobacter pylori. Cytobios* **82**:251–260.

3891. **Virji, M., K. Makepeace, D. J. Ferguson, M. Achtman, J. Sarkari, and E. R. Moxon.** 1992. Expression of the Opc protein correlates with invasion of epithelial and endothelial cells by *Neisseria meningitidis. Mol. Microbiol.* **6**:2785–2795.

3892. **Virji, M., C. Alexandrescu, D. J. P. Ferguson, J. R. Saunders, and E. R. Moxon.** 1992. Variations in the expression of pili: the effect of adherence of *Neisseria meningitidis* to human epithelial and endothelial cells. *Mol. Microbiol.* **6**:1271–1279.

3893. **Virji, M., J. R. Saunders, G. Sims, K. Makepeace, D. Maskell, and D. J. P. Ferguson.** 1993. Pilus-facilitated adherence of *Neisseria meningitidis* to human epithelial and endothelial cells: modulation of adherence phenotype occurs concurrently with changes in primary amino acid sequence and the glycosylation status of pilin. *Mol. Microbiol.* **10**:1013–1028.

3894. **Virji, M., K. Makepeace, D. J. P. Ferguson, M. Achtman, and E. R. Moxon.** 1993. Meningococcal Opa and Opc proteins: their role in colonization and invasion of human epithelial and endothelial cells. *Mol. Microbiol.* **10**:499–510.

3895. **Virji, M., K. Makepeace, and E. R. Moxon.** 1994. Distinct mechanisms of interactions of Opc-expressing meningococci at apical and basolateral surfaces of human endothelial cells; the role of integrins in apical interactions. *Mol. Microbiol.* **14**:173–184.

3896. **Virji, M., K. Makepeace, I. Peak, G. Payne, J. R. Saunders, D. J. Ferguson, and E. R. Moxon.** 1995. Functional implications of the expression of PilC proteins in meningococci. *Mol. Microbiol.* **16**:1087–1097.

3897. **Virji, M., K. Makepeace, I. R. Peak, D. J. Ferguson, M. P. Jennings, and E. R. Moxon.** 1995. Opc- and pilus-dependent interactions of meningococci with human endothelial cells: molecular mechanisms and modulation by surface polysaccharides. *Mol. Microbiol.* **18**:741–754.

3898. **Virji, M., K. Makepeace, D. J. Ferguson, and S. M. Watt.** 1996. Carcinoembryonic antigens (CD66) on epithelial cells and neutrophils are receptors for Opa proteins of pathogenic neisseriae. *Mol. Microbiol.* **22**:941–950.

3899. **Virji, M., K. Makepeace, I. R. Peak, D. J. Ferguson, and E. R. Moxon.** 1996. Pathogenic mechanisms of *Neisseria meningitidis. Ann. N. Y. Acad. Sci.* **797**:273–276.

3900. **Virji, M., E. Stimson, K. Makepeace, A. Dell, H. R. Morris, G. Payne, J. R. Saunders, and E. R. Moxon.** 1996. Posttranslational modifications of meningococcal pili. Identification of a common trisaccharide substitution on variant pilins of strain C311. *Ann. N. Y. Acad. Sci.* **797:**53–64.

3901. **Virji, M.** 1997. Post-translational modifications of meningococcal pili. Identification of common substituents: glycans and alphaglycerophosphate—a review. *Gene* **192:**141–147.

3902. **Virji, M., D. Evans, A. Hadfield, F. Grunert, A. M. Teixeira, and S. M. Watt.** 1999. Critical determinants of host receptor targeting by *Neisseria meningitidis* and *Neisseria gonorrhoeae*: identification of Opa adhesiotopes on the N-domain of CD66 molecules. *Mol. Microbiol.* **34:**538–551.

3903. **Virkola, R., J. Parkkinen, J. Hacker, and T. K. Korhonen.** 1993. Sialyloligosaccharide chains of laminin as an extracellular matrix target for S fimbriae of *Escherichia coli*. *Infect. Immun.* **61:**4480–4484.

3904. **Virkola, R., K. Lahteenmaki, T. Eberhard, P. Kuusela, L. van Alphen, M. Ullberg, and T. K. Korhonen.** 1996. Interaction of *Haemophilus influenzae* with the mammalian extracellular matrix. *J. Infect. Dis.* **173:**1137–1147.

3905. **Virkola, R., M. Brummer, H. Rauvala, L. van Alphen, and T. K. Korhonen.** 2000. Interaction of fimbriae of *Haemophilus influenzae* type B with heparin-binding extracellular matrix proteins. *Infect. Immun.* **68:**5696–5701.

3906. **Visai, L., Y. Xu, F. Casolini, S. Rindi, M. Höök, and P. Speziale.** 2000. Monoclonal antibodies to CNA, a collagenbinding microbial surface component recognizing adhesive matrix molecules, detach *Staphylococcus aureus* from a collagen substrate. *J. Biol. Chem.* **275:**39837–39845.

3907. **Viscount, H. B., C. L. Munro, D. Burnette-Curley, D. L. Peterson, and F. L. Macrina.** 1997. Immunization with FimA protects against *Streptococcus parasanguis* endocarditis in rats. *Infect. Immun.* **65:**994–1002.

3908. **Visser, L. G., E. Seijmonsbergen, P. H. Nibbering, P. J. van den Broek, and R. van Furth.** 1999. Yops of *Yersinia enterocolitica* inhibit receptor-dependent superoxide anion production by human granulocytes. *Infect. Immun.* **67:**1245–1250.

3909. **Visser, M. R., H. Beumer, A. I. Hoepelman, M. Rozenberg-Arska, and J. Verhoef.** 1993. Changes in adherence of respiratory pathogens to HEp-2 cells induced by subinhibitory concentrations of sparfloxacin, ciprofloxacin, and trimethoprim. *Antimicrob. Agents Chemother.* **37:**885–888.

3910. **Voegele, K., H. Sakellaris, and J. R. Scott.** 1997. CooB plays a chaperone-like role for the proteins involved in formation of CS1 pili of enterotoxigenic *Escherichia coli*. *Proc. Natl. Acad. Sci. USA* **94:**13257–13261.

3911. **Vogeli, P., H. U. Bertschinger, M. Stamm, C. Stricker, C. Hagger. R. Fries, J. Rapacz, and G. Stranzinger.** 1996. Genes specifying receptors for F18 fimbriated *Escherichia coli,* causing oedema disease and postweaning diarrhoea in pigs, map to chromosome 6. *Anim. Genet.* **27:**321–328.

3912. **Vogeli, P., B. Kuhn, R. Kuhne, R. Obrist, G. Stranzinger, S. C. Huang, Z. L. Hu, J. Hasler-Rapacz, and J. Rapacz.** 1992. Evidence for linkage between the swine L blood group and the loci specifying the receptors mediating adhesion of K88 *Escherichia coli* pilus antigens. *Anim. Genet.* **23:**19–29.

3913. **Vogely, H. C., C. J. Oosterbos. E. W. Puts, M. W. Nijhof, P. G. Nikkels, A. Fleer. A. J. Tonino, W. J. Dhert, and A. J. Verbout.** 2000. Effects of hydrosyapatite coating on Ti-6A1-4V implant-site infection in a rabbit tibial model. *J. Orthop. Res.* **18:**485–493.

3914. **von Hunolstein, C., M. L. Ricci, and G. Orefici.** 1993. Adherence of glucan-positive and glucan-negative strains of *Streptococcus bovis* to human epithelial cells. *J. Med. Microbiol.* **39:**53–57.

3915. **von Moll, L. K., and J. R. Cantey.** 1997. Peyer's patch adherence of enteropathogenic *Escherichia coli* strains in rabbits. *Infect. Immun.* **65:**3788–3793.

3916. **Vorobjova, T., K. Kisand, A. Haukanomm, H.-I. Maaroos, T. Wadström, and R. Uibo.** 1994. The prevalence of *Helicobacter pylori* antibodies in a population from southern Estonia. *Eur. J. Gastroenterol. Hepatol.* **6:**529–533.

3917. **Vranes, J.** 1996. Effect of subinhibitory concentrations of ceftazidime, ciprofloxacin, and azithromycin on the hemagglutination and adherence of uropathogenic *Escherichia coli* strains. *Chemotherapy* **42:**177–185.

3918. **Vranes, J.** 2000. Effect of subminimal inhibitory concentrations of azithromycin on adherence of *Pseudomonas aeruginosa* to polystyrene. *J. Chemother.* **12:**280–285.

3919. **Vranes, J.** 1997. Hemagglutination ability and adherence to the Buffalo green monkey kidney cell line of uropathogenic *Escherichia coli. APMIS* **105**:831–837.

3920. **Vranes, J., Z. Zagar, and S. Kurbel.** 1996. Influence of subinhibitory concentrations of ceftazidime, ciprofloxacin and azithromycin on the morphology and adherence of P-fimbriated *Escherichia coli. J. Chemother.* **8**:254–260.

3921. **Vuong, C., H. L. Saenz, F. Götz, and M. Otto.** 2000. Impact of the *agr* quorum-sensing system on adherence to polystyrene in *Staphylococcus aureus. J. Infect. Dis.* **182**:1688–1693.

3922. **Wada, Y., Y. Nakaoka, H. Kondo, M. Nakazawa, and M. Kubo.** 1996. Dual infection with attaching and effacing *Escherichia coli* and enterotoxigenic *Escherichia coli* in post-weaning pigs. *J. Comp. Pathol.* **114**:93–99.

3923. **Wada, Y., H. Kondo, M. Sueyoshi, M. Kubo, and Y. Adachi.** 1997. A novel developmental process of intestinal epithelial lesions in a calf infected with attaching and effacing *Escherichia coli. J. Vet. Med. Sci.* **59**:401–403.

3924. **Wadström, T., F. Ascencio, Å. Ljungh, J. Lelwala-Guruge, M. Ringnér, M. Utt, and K. Valkonen.** 1993. *Helicobacter pylori* adhesins. *Eur. J. Gastroenterol. Hepatol.* **5**:SD12–S15.

3925. **Wadström, T.** 1996. Biochemical aspects of *Helicobacter pylori* colonization of the human gastric mucosa. *Aliment. Pharmacol. Ther.* **10**:1–10.

3926. **Wadström, T., M. Ringnér, and K. H. Valkonen.** 1994. Interactions of microbial lectins and extracellular matrix, p. 73–77. *In* J. Beuth, and G. Pulverer (ed.), *Lectin Blocking: New Strategies for the Prevention and Therapy of Tumor Metastasis and Infectious Diseases.* Gustav Fischer Verlag, New York, N.Y.

3927. **Wadström, T., S. Hirmo, and T. Boren.** 1996. Biochemical aspects of *Helicobacter pylori* colonization of the human gastric mucosa. *Aliment. Pharmacol. Ther.* **10**:S17–S27.

3928. **Wadström, T., S. Hirmo, H. Novak, A. Guzman, M. Ringner-Pantzar, M. Utt, and P. Aleljung.** 1997. Sulfatides inhibit binding of *Helicobacter pylori* to the gastric cancer Kato III cell line. *Curr. Microbiol.* **34**:267–272.

3929. **Wadström, T., S. Hirmo, and B. Nilsson.** 1997. Biochemical aspects of *H. pylori* adhesion. *J. Physiol. Pharmacol.* **48**:325–331.

3930. **Wagner, S., W. Beil, U. E. Mai, C. Bokemeyer, H. J. Meyer, and M. P. Manns.** 1994. Interaction between *Helicobacter pylori* and human gastric epithelial cells in culture: effect of antiulcer drugs. *Pharmacology* **49**:226–237.

3931. **Wai, S. N., A. Takade, and K. Amako.** 1996. The hydrophobic surface protein layer of enteroaggregative *Escherichia coli* strains. *FEMS Microbiol. Lett.* **135**:17–22.

3932. **Wainwright, L. A., K. H. Pritchard, and H. S. Seifert.** 1994. A conserved DNA sequence is required for efficient gonococcal pilin antigenic variation. *Mol. Microbiol.* **13**:75–87.

3933. **Waldbeser, L. S., R. S. Ajioka, A. J. Merz, D. Puaoi, L. Lin, M. Thomas, and M. So.** 1994. The *opaH* locus of *Neisseria gonorrhoeae* MS11A is involved in epithelial cell invasion. *Mol. Microbiol.* **13**:919–928.

3934. **Walduck, A. K., and J. P. Opdebeeck.** 1996. Effect of adjuvants on antibody responses of sheep immunised with recombinant pili from *Dichelobacter nodosus. Aust. Vet. J.* **74**:451–455.

3935. **Wales, A. D., G. R. Pearson, A. M. Skuse, J. M. Roe, C. M. Hayes, A. L. Cookson, and M. J. Woodward.** 2001. Attaching and effacing lesions caused by *Escherichia coli* O157:H7 in experimentally inoculated neonatal lambs. *J. Med. Microbiol.* **50**:752–758.

3936. **Waligora, A. J., M. C. Barc, P. Bourlioux, A. Collignon, and T. Karjalainen.** 1999. *Clostridium difficile* cell attachment is modified by environmental factors. *Appl. Environ. Microbiol.* **65**:4234–4238.

3937. **Waligora, A. J., C. Hennequin, P. Mullany, P. Bourlioux, A. Collignon, and T. Karjalainen.** 2001. Characterization of a cell surface protein of *Clostridium difficile* with adhesive properties. *Infect. Immun.* **69**:2144–2153.

3938. **Walker, S. G., J. L. Ebersole, and S. C. Holt.** 1999. Studies on the binding of *Treponema pectinovorum* to HEp-2 epithelial cells. *Oral Microbiol. Immunol.* **14**:165–171.

3939. **Wallace, A., and M. C. M. Pérombelon.** 1992. Haemagglutinins and fimbriae of soft rot *Erwinias. J. Appl. Bacteriol.* **73**:114–119.

3940. **Walse, B., J. Kihlberg, K. F. Karlsson, M. Nilsson, K. G. Wahlund, J. S. Pinkner, S. J. Hultgren, and T. Drakenberg.** 1997. Transferred nuclear Overhauser effect spectroscopy study of a peptide from the PapG pilus subunit bound by the *Escherichia coli* PapD chaperone. *FEBS Lett.* **412**:115–120.

3941. **Walser, B. L., R. D. Newman, A. A. M. Lima, and R. L. Guerrant.** 1992. Pathogen and host differences in bacterial adherence to human buccal epithelial cells in a northeast Brazilian community. *Infect. Immun.* **60:** 4793–4800.

3942. **Wang, B., E. Kraig, and K. Kolodrubetz.** 2000. Use of defined mutants to assess the role of the *Campylobacter rectus* S-layer in bacterium-epithelial cell interactions. *Infect. Immun.* **68:**1465–1473.

3943. **Wang, B., N. Ruiz, A. Pentland, and M. Caparon.** 1997. Keratinocyte proinflammatory responses to adherent and nonadherent group A streptococci. *Infect. Immun.* **65:**2119–2126.

3944. **Wang, H. L., K. Yuan, F. Burgett, Y. Shyr, and S. Syed.** 1994. Adherence of oral microorganisms to guided tissue membranes: an *in vitro* study. *J. Periodontol.* **65:**211–218.

3945. **Wang, I. W., J. M. Anderson, and R. E. Marchant.** 1993. Platelet-mediated adhesion of *Staphylococcus epidermidis* to hydrophobic NHLBI reference polyethylene. *J. Biomed. Mater. Res.* **27:**1119–1128.

3946. **Wang, I. W., J. M. Anderson, M. R. Jacobs, and R. E. Marchant.** 1995. Adhesion of *Staphylococcus epidermidis* to biomedical polymers: contributions of surface thermodynamics and hemodynamic shear conditions. *J. Biomed. Mater. Res.* **29:**485–493.

3947. **Wang, I. W., J. M. Anderson, and R. E. Marchant.** 1993. *Staphylococcus epidermidis* adhesion to hydrophobic biomedical polymer is mediated by platelets. *J. Infect. Dis.* **167:**329–336.

3948. **Wang, J., S. Singh, K. G. Taylor, and R. J. Doyle.** 1995. Streptococcal glucan-binding lectins do not recognize methylated alpha-1,6 glucans. *Glycoconj. J.* **12:**109–112.

3949. **Wang, J., S. D. Gray-Owen, A. Knorre, T. F. Meyer, and C. Dehio.** 1998. Opa binding to cellular CD66 receptors mediates the transcellular traversal of *Neisseria gonorrhoeae* across polarized T84 epithelial cell monolayers. *Mol. Microbiol.* **30:**657–671.

3950. **Wang, J. R., and M. W. Stinson.** 1994. M protein mediates streptococcal adhesion to HEp-2 cells. *Infect. Immun.* **62:**442–448.

3951. **Wang, X., V. Soltesz, W. Guo, and R. Andersson.** 1994. Water-soluble ethylhydroxyethyl cellulose: a new agent against bacterial translocation from the gut after major liver resection. *Scand. J. Gastroenterol.* **29:**833–840.

3952. **Wang, X. D., H. Parsson, R. Andersson, V. Soltesz, K. Johansson, and S. Bengmark.** 1994. Bacterial translocation, intestinal ultrastructure and cell membrane permeability early after major liver resection in the rat. *Bri. J. Surg.* **81:**579–584.

3953. **Wanke, C. A., H. Mayer, R. Weber, R. Zbinden, D. A. Watson, and D. Acheson.** 1998. Enteroaggregative *Escherichia coli* as a potential cause of diarrheal disease in adults infected with human immunodeficiency virus. *J. Infect. Dis.* **178:**185–190.

3954. **Wann, E. R., S. Gurusiddappa, and M. Höök.** 2000. The fibronectin-binding MSCRAMM FnbpA of *Staphylococcus aureus* is a bifunctional protein that also binds to fibrinogen. *J. Biol. Chem.* **275:**13863–13871.

3955. **Ward, C. K., S. R. Lumbley, J. L. Latimer, L. D. Cope, and E. J. Hansen.** 1998. *Haemophilus ducreyi* secretes a filamentous hemagglutinin-like protein. *J. Bacteriol.* **180:**6013–6022.

3956. **Ward, C. K., M. L. Lawrence, H. P. Veit, and T. J. Inzana.** 1998. Cloning and mutagenesis of a serotype-specific DNA region involved in encapsulation and virulence of *Actinobacillus pleuropneumoniae* serotype 5a: concomitant expression of serotype 5a and 1 capsular polysaccharides in recombinant *A. pleuropneumoniae* serotype 1. *Infect. Immun.* **66:**3326–3336.

3957. **Ward, T. T.** 1992. Comparison of *in vitro* adherence of methicillin-sensitive and methicillin-resistant *Staphylococcus aureus* to human nasal epithelial cells. *J. Infect. Dis.* **166:**400–404.

3958. **Washburn, L. R., S. Hirsch, and L. L. Voelker.** 1993. Mechanisms of attachment of *Mycoplasma arthritidis* to host cells in vitro *Infect. Immun.* **61:**2670–2680.

3959. **Washburn, L. R., and E. J. Weaver.** 1997. Protection of rats against *Mycoplasma arthritidis*-induced arthritis by active and passive immunizations with two surface antigens. *Clin. Diagn. Lab. Immunol.* **4:**321–327.

3960. **Washington, O. R., M. Deslauriers, D. P. Stevens, L. K. Lyford, S. Haque, Y. Yan, and P. M. Flood.** 1993. Generation and purification of recombinant fimbrillin from *Porphyromonas (Bacteroides) gingivalis* 381. *Infect. Immun.* **61:**1040–1047.

3961. **Wassall, M. A., M. Santin, C. Isalberti, M. Cannas, and S. P. Denyer.** 1997. Adhesion of bacteria to stainless steel and silver-coated orthopedic external fixation pins. *J. Biomed. Mater. Res.* **36:**325–330.

3962. **Wassall, M. A., M. Santin, G. Peluso, and S. P. Denyer.** 1998. Possible role of alpha-1-microglobulin in mediating bacterial attachment to model surfaces. *J. Biomed. Mater. Res.* **40:**365–370.

3963. **Watanabe, K., Y. Yamaji, and T. Umemoto.** 1992. Correlation between cell-adherent activity and surface structure in *Porphyromonas gingivalis*. *Oral Microbiol. Immunol.* **7:**357–363.

3964. **Watanabe, K., H. Kumada, F. Yoshimura, and T. Umemoto.** 1996. The induction of polyclonal B-cell activation and interleukin-1 production by the 75-kDa cell surface protein from *Porphyromonas gingivalis* in mice. *Arch. Oral Biol.* **41:**725–731.

3965. **Watanabe, K., T. Onoe, M. Ozeki, Y. Shimizu, T. Sakayori, H. Nakamura, and F. Yoshimura.** 1996. Sequence and product analyses of the four genes downstream from the fimbrilin gene (*fimA*) of the oral anaerobe *Porphyromonas gingivalis*. *Microbiol. Immunol.* **40:**725–734.

3966. **Watanabe-Kato, T., J. I. Hayashi, Y. Terazawa, C. I. Hoover, K. Nakayama, E. Hibi, N. Kawakami, T. Ikeda, H. Nakamura, T. Noguchi, and F. Yoshimura.** 1998. Isolation and characterization of transposon-induced mutants of *Porphyromonas gingivalis* deficient in fimbriation. *Microb. Pathog.* **24:**25–35.

3967. **Watarai, M., T. Tobe, M. Yoshikawa, and C. Sasakawa.** 1995. Contact of *Shigella* with host cells triggers release of Ipa invasins and is an essential function of invasiveness. *EMBO J.* **14:**2461–2470.

3968. **Watnick, P. I., K. J. Fullner, and R. Kolter.** 1999. A role for the mannose-sensitive hemagglutinin in biofilm formation by *Vibrio cholerae* El Tor. *J. Bacteriol.* **181:**3606–3609.

3969. **Watnick, P. I., and R. Kolter.** 1999. Steps in the development of a *Vibrio cholerae* El Tor biofilm. *Mol. Microbiol.* **34:**586–595.

3970. **Watson, A. A., R. A. Alm, and J. S. Mattick.** 1996. Identification of a gene, *pilF*, required for type 4 fimbrial biogenesis and twitching motility in *Pseudomonas aeruginosa*. *Gene* **180:**49–56.

3971. **Watson, A. A., J. S. Mattick, and R. A. Alm.** 1996. Functional expression of heterologous type 4 fimbriae in *Pseudomonas aeruginosa*. *Gene* **175:**143–150.

3972. **Watson, R. W., H. P. Redmond, J. H. Wang, C. Condron, and D. Bouchier-Hayes.** 1996. Neutrophils undergo apoptosis following ingestion of *Escherichia coli*. *J. Immunol.* **156:**3986–3992.

3973. **Watson, W. J., J. R. Gilsdorf, M. A. Tucci, K. W. McCrea, L. J. Forney, and C. F. Marrs.** 1994. Identification of a gene essential for piliation in *Haemophilus influenzae*

type b with homology to the pilus assembly platform genes of gram-negative bacteria. *Infect. Immun.* **62:**468–475.

3974. **Webb, L. X., J. Holman, B. de Araujo, D. J. Zaccaro, and E. S. Gordon.** 1994. Antibiotic resistance in staphylococci adherent to cortical bone. *J. Orthopaed. Trauma* **8:**28–33.

3975. **Weiger, R., E. M. Decker, G. Krastl, and M. Brecx.** 1999. Deposition and retention of vital and dead *Streptococcus sanguinis* cells on glass surfaces in a flow-chamber system. *Arch. Oral Biol.* **44:**621–628.

3976. **Weiger, R., L. Netuschil, T. Wester-Ebbinghaus, and M. Brecx.** 1998. An approach to differentiate between antibacterial and antiadhesive effects of mouthrinses *in vivo*. *Arch. Oral Biol.* **43:**559–565.

3977. **Weikert, L. F., K. Edwards, Z. C. Chroneos, C. Hager, L. Hoffman, and V. L. Shepherd.** 1997. SP-A enhances uptake of bacillus Calmette-Guerin by macrophages through a specific SP-A receptor. *Am. J. Physiol.* **272:**L989–995.

3978. **Weinberg, A., C. M. Belton, Y. Park, and R. J. Lamont.** 1997. Role of fimbriae in *Porphyromonas gingivalis* invasion of gingival epithelial cells. *Infect. Immun.* **65:**313–316.

3979. **Weingart, C. L., G. Broitman-Maduro, G. Dean, S. Newman, M. Peppler, and A. A. Weiss.** 1999. Fluorescent labels influence phagocytosis of *Bordetella pertussis* by human neutrophils. *Infect. Immun.* **67:**4264–4267.

3980. **Weingart, C. L., and A. A. Weiss.** 2000. *Bordetella pertussis* virulence factors affect phagocytosis by human neutrophils. *Infect. Immun.* **68:**1735–1739.

3981. **Weinstein, D. L., B. L. O'Neill, and E. S. Metcalf.** 1997. *Salmonella typhi* stimulation of human intestinal epithelial cells induces secretion of epithelial cell-derived interleukin-6. *Infect. Immun.* **65:**395–404.

3982. **Weinstein, D. L., B. L. O'Neill, D. M Hone, and E. S. Metcalf.** 1998. Differential early interactions between *Salmonella enterica* serovar Typhi and two other pathogenic *Salmonella* serovars with intestinal epithelial cells. *Infect. Immun.* **66:**2310–2318.

3983. **Weir, S., and C. F. Marrs.** 1992. Identification of type 4 pili in *Kingella denitrificans*. *Infect. Immun.* **60:**3437–3441.

3984. **Weir, S., L. W. Lee, and C. F. Marrs.** 1997. Type-4 pili of *Kingella denitrificans*. *Gene* **192:**171–617.

3985. **Weiss, E. I., R. Lev-Dor, Y. Kashamn, J. Goldhar, N. Sharon, and I. Ofek.** 1998.

Inhibiting interspecies coaggregation of plaque bacteria with a cranberry juice constituent. *J. Am. Dent. Assoc.* **129**:1719–1723.

3986. **Weiss, E. I., B. Shaniztki, M. Dotan, N. Ganeshkumar, P. E. Kolenbrander, and Z. Metzger.** 2000. Attachment of *Fusobacterium nucleatum* PK1594 to mammalian cells and its coaggregation with periodontopathogenic bacteria are mediated by the same galactose-binding adhesin. *Oral Microbiol. Immunol.* **15**:371–377.

3987. **Wells, C. L., R. P. Jechorek, S. B. Olmsted, and S. L. Erlandsen.** 1994. Bacterial translocation in cultured enterocytes: magnitude, specificity, and electron microscopic observations of endocytosis. *Shock* **1**:443–451.

3988. **Wells, C. L., R. P. Jechorek, K. M. Kinneberg, S. M. Debol, and S. L. Erlandsen.** 1999. The isoflavone genistein inhibits internalization of enteric bacteria by cultured Caco-2 and HT-29 enterocytes. *J. Nutr.* **129**:634–640.

3989. **Wells, C. L., E. M. van de Westerlo, R. P. Jechorek, and S. L. Erlandsen.** 1995. Exposure of the lateral enterocyte membrane by dissociation of calcium-dependent junctional complex augments endocytosis of enteric bacteria. *Shock* **4**:204–210.

3990. **Wells, C. L., E. M. van de Westerlo, R. P. Jechorek, B. A. Feltis, T. D. Wilkins, and S. L. Erlandsen.** 1996. *Bacteroides fragilis* enterotoxin modulates epithelial permeability and bacterial internalization by HT-29 enterocytes. *Gastroenterology* **110**:1429–1437.

3991. **Wells, C. L., E. M. A. van de Westerlo, R. P. Jechorek, H. M. Haines, and S. L. Erlandsen.** 1998. Cytochalasin-induced actin disruption of polarized enterocytes can augment internalization of bacteria. *Infect. Immun.* **66**:2410–2419.

3992. **Wennerås, C., J. R. Neeser, and A.-M. Svennerholm.** 1995. Binding of the fibrillar CS3 adhesin of enterotoxigenic *Escherichia coli* to rabbit intestinal glycoproteins is competitively prevented by GalNAcβ1–4Gal-containing glycoconjugates. *Infect. Immun.* **63**:640–646.

3993. Reference deleted.

3994. **Westerlund, B., and T. K. Korhonen.** 1993. Bacterial proteins binding to the mammalian extracellular matrix. *Mol. Microbiol.* **9**:687–694.

3995. **Westerlund, B., I. Van Die, W. Hoekstra, R. Virkola, and T. K. Korhonen.** 1993. P fimbriae of uropathogenic *Escherichia coli* as multifunctional adherence organelles. *Zentbl. Bakteriol.* **278**:229–237.

3996. **Westerlund-Wikstrom, B., J. Tanskanen, R. Virkola, J. Hacker, M. Lindberg, M. Skurnik, and T. K. Korhonen.** 1997. Functional expression of adhesive peptides as fusions to *Escherichia coli* flagellin. *Protein Eng.* **10**:1319–1326.

3997. **Westerman, R. B., G. W. Fortner, K. W. Mills, R. M. Phillips, and J. M. Greenwood.** 1993. Use of monoclonal antibodies specific for the a determinant of K88 pili for detection of enterotoxigenic *Escherichia coli* in pigs. *J. Clin. Microbiol.* **31**:311–314.

3998. **Whitchurch, C. B., and J. S. Mattick.** 1994. *Escherichia coli* contains a set of genes homologous to those involved in protein secretion, DNA uptake and the assembly of type-4 fimbriae in other bacteria. *Gene* **150**:9–15.

3999. **White-Ziegler, C. A., and D. A. Low.** 1992. Thermoregulation of the *pap* operon: evidence for the involvement of RimJ, the N-terminal acetylase of ribosomal protein S5. *J. Bacteriol.* **174**:7003–7012.

4000. **Whittaker, C. J., D. L. Clemans, and P. E. Kolenbrander.** 1996. Insertional inactivation of an intrageneric coaggregation-relevant adhesin locus from *Streptococcus gordonii* DL1 (Challis). *Infect. Immun.* **64**:4137–4142.

4001. **Whittaker, C. J., C. M. Klier, and P. E. Kolenbrander.** 1996. Mechanisms of adhesion by oral bacteria. *Annu. Rev. Microbiol.* **50**:513–552.

4002. **Wibawan, I. W. T., C. Lämmler, R. S. Seleim, and F. H. Pasaribu.** 1993. A haemagglutinating adhesin of group B streptococci isolated from cases of bovine mastitis mediates adherence to HeLa cells. *J. Gen. Microbiol.* **139**:2173–2178.

4003. **Wibawan, I. W., and C. Lammler.** 1994. Relation between encapsulation and various properties of *Streptococcus suis*. *Zentbl. Veterinaermed. Reihe B* **41**:453–459.

4004. **Wibawan, I. W., F. H. Pasaribu, I. H. Utama, A. Abdulmawjood, and C. Lammler.** 1999. The role of hyaluronic acid capsular material of *Streptococcus equi* subsp. *zooepidemicus* in mediating adherence to HeLa cells and in resisting phagocytosis. *Res. Vet. Sci.* **67**:131–135.

4005. **Widdicombe, J.** 1995. Relationships among the composition of mucus, epithelial lining liquid, and adhesion of microorganisms. *Am. J. Respir. Crit. Care Med.* **151**: 2088–2092.

4006. **Wieler, L. H., E. Vieler, C. Erpenstein, T. Schlapp, H. Steinruck, R. Bauerfeind, A. Byomi, and G. Baljer.** 1996. Shiga toxin-producing *Escherichia coli* strains from bovines: association of adhesion with carriage of *eae* and other genes. *J. Clin. Microbiol.* **34:**2980–2984.

4007. **Wieler, L. H., T. K. McDaniel, T. S. Whittam, and J. B. Kaper.** 1997. Insertion site of the locus of enterocyte effacement in enteropathogenic and enterohemorrhagic *Escherichia coli* differs in relation to the clonal phylogeny of the strains. *FEMS Microbiol. Lett.* **156:**49–53.

4008. **Wieler, L. H., A. Schwanitz, E. Vieler, B. Busse, H. Steinruck, J. B. Kaper, and G. Baljer.** 1998. Virulence properties of Shiga toxin-producing *Escherichia coli* (STEC) strains of serogroup O118, a major group of STEC pathogens in calves. *J. Clin. Microbiol.* **36:**1604–1607.

4009. **Wilcox, M. H., A. Cook, I. Geary, and A. Eley.** 1994. Toxin production, adherence and protein expression by clinical *Aeromonas* spp. isolates in broth and human pooled ileostomy fluid. *Epidemiol. Infect.* **113:**235–245.

4010. **Wilcox, M. H., and F. Schumacher-Perdreau.** 1994. Lack of evidence for increased adherent growth in broth or human serum of clinically significant coagulase-negative staphylococci. *J. Hosp. Infect.* **26:**239–250.

4011. **Wilcox, M. H., T. G. Winstanley, and R. C. Spencer.** 1994. Binding of teicoplanin and vancomycin to polymer surfaces. *J. Antimicrob. Chemother.* **33:**431–441.

4012. **Wilkinson, S. M., J. R. Uhl, B. C. Kline, and F. R. Cockerill III.** 1998. Assessment of invasion frequencies of cultured HEp-2 cells by clinical isolates of *Helicobacter pylori* using an acridine orange assay. *J. Clin. Pathol.* **51:**127–133.

4013. **Willcox, M. D.** 1995. Potential pathogenic properties of members of the "*Streptococcus milleri*" group in relation to the production of endocarditis and abscesses. *J. Med. Microbiol.* **43:**405–410.

4014. **Willcox, M. D., R. J. Fitzgerald, B. O. Adams, M. Patrikakis, and K. W. Knox.** 1993. Biochemical properties of *Streptococcus sobrinus* reisolates from the gastrointestinal tract of a gnotobiotic rat. *J. Gen. Microbiol.* **139:**929–935.

4015. **Willcox, M. D., M. Patrikakis, D. W. Harty, C. Y. Loo, and K. W. Knox.** 1993. Coaggregation of oral lactobacilli with streptococci from the oral cavity. *Oral Microbiol. Immunol.* **8:**319–321.

4016. **Willcox, M. D., C. Y. Loo, D. W. Harty, and K. W. Knox.** 1995. Fibronectin binding by *Streptococcus milleri* group strains and partial

characterisation of the fibronectin receptor of *Streptococcus anginosus* F4. *Microb. Pathog.* **19:**129–137.

4017. **Willems, R. J., H. G. van der Heide, and F. R. Mooi.** 1992. Characterization of a *Bordetella pertussis* fimbrial gene cluster which is located directly downstream of the filamentous haemagglutinin gene. *Mol. Microbiol.* **6:**2661–2671.

4018. **Willems, R. J., C. Geuijen, H. G. van der Heide, G. Renauld, P. Bertin, W. M. van den Akker, C. Locht, and F. R. Mooi.** 1994. Mutational analysis of the *Bordetella pertussis fim/fha* gene cluster: identification of a gene with sequence similarities to haemolysin accessory genes involved in export of FHA. *Mol. Microbiol.* **11:**337–347.

4019. **Willemsen, P. T. J., and F. K. de Graaf.** 1993. Multivalent binding of K99 fimbriae to the N-glycolyl-GM$_3$ ganglioside receptor. *Infect. Immun.* **61:**4518–4522.

4020. **Williams, I., W. A. Venables, D. Lloyd, F. Paul, and I. Critchley.** 1997. The effects of adherence to silicone surfaces on antibiotic susceptibility in *Staphylococcus aureus.* *Microbiology* **143:**2407–2413.

4021. **Williams, I., F. Paul, D. Lloyd, R. Jepras, I. Critchley, M. Newman, J. Warrack, T. Giokarini, A. J. Hayes, P. F. Randerson, and W. A. Venables.** 1999. Flow cytometry and other techniques show that *Staphylococcus aureus* undergoes significant physiological changes in the early stages of surface-attached culture. *Microbiology* **145:**1325–1333.

4022. **Williams, T. J., M. D. Willcox, and R. P. Schneider.** 1998. Interactions of bacteria with contact lenses: the effect of soluble protein and carbohydrate on bacterial adhesion to contact lenses. *Optom. Vision Sci.* **75:**266–271.

4023. **Willshaw, G. A., S. M. Scotland, H. R. Smith, and B. Rowe.** 1992. Properties of Vero cytotoxin-producing *Escherichia coli* of human origin of O serogroups other than O157. *J. Infect. Dis.* **166:**797–802.

4024. **Wilson, R., R. B. Dowling, and A. D. Jackson.** 1996. The biology of bacterial colonization and invasion of the respiratory mucosa. *Eur. Respir. J.* **9:**1523–1530.

4025. **Wilson, R. L., J. Elthon, S. Clegg, and B. D. Jones.** 2000. *Salmonella enterica* serovars Gallinarum and Pullorum expressing *Salmonella enterica* serovar Typhimurium type 1 fimbriae exhibit increased invasiveness for mammalian cells. *Infect. Immun.* **68:**4782–4785.

4026. **Wilson, D. R., A. Siebers, and B. B. Finlay.** 1998. Antigenic analysis of *Bordetella* *pertussis* filamentous hemagglutinin with phage display libraries and rabbit anti-filamentous hemagglutinin polyclonal antibodies. *Infect. Immun.* **66**:4884–4894.

4027. **Wilson, S. L., and D. A. Drevets.** 1998. *Listeria monocytogenes* infection and activation of human brain microvascular endothelial cells. *J. Infect. Dis.* **178**:1658–1666.

4028. **Wilton, J. L., A. L. Scarman, M. J. Walker, and S. P. Djordjevic.** 1998. Reiterated repeat region variability in the ciliary adhesin gene of *Mycoplasma hyopneumoniae. Microbiology* **144**:1931–1943.

4029. **Winberg, J., M. Herthelius-Elman, R. Möllby, and C. E. Nord.** 1993. Pathogenesis of urinary tract infection—experimental studies of vaginal resistance to colonization. *Pediatr. Nephrol.* **7**:509–514.

4030. **Winberg, J., R. Möllby, J. Bergström, K.-A. Karlsson, I. Leonardsson, M. A. Milh, S. Teneberg, D. Haslam, B.-I. Marklund, and S. Normark.** 1995. The PapG-adhesin at the tip of P-fimbriae provides *Escherichia coli* with a competitive edge in experimental bladder infections of *Cynomolgus* monkeys. *J. Exp. Med.* **182**:1695–1702.

4031. **Winram, S. B., M. Jonas, E. Chi, and C. E. Rubens.** 1998. Characterization of group B streptococcal invasion of human chorion and amnion epithelial cells in vitro. *Infect. Immun.* **66**:4932–4941.

4032. **Winsor, D. K., Jr., S. Ashkenazi, R. Chiovetti, and T. G. Cleary.** 1992. Adherence of enterohemorrhagic *Escherichia coli* strains to a human colonic epithelial cell line (T$_{84}$). *Infect. Immun.* **60**:1613–1617.

4033. **Winther-Larsen, H. C., F. T. Hegge, M. Wolfgang, S. F. Hayes, J. P. van Putten, and M. Koomey.** 2001. *Neisseria gonorrhoeae* PilV, a type IV pilus-associated protein essential to human epithelial cell adherence. *Proc. Natl. Acad. Sci. USA* **98**:15276–15281.

4034. **Wittig, W., R. Prager, M. Stamm, W. Streckel, and H. Tschape.** 1994. Expression and plasmid transfer of genes coding for the fimbrial antigen F107 in porcine *Escherichia coli* strains. *Zentbl. Bakteriol.* **281**:130–139.

4035. **Wittig, W., H. Klie, P. Gallien, S. Lehmann, M. Timm, and H. Tschape.** 1995. Prevalence of the fimbrial antigens F18 and K88 and of enterotoxins and verotoxins among *Escherichia coli* isolated from weaned pigs. *Zentbl. Bakteriol.* **283**:95–104.

4036. **Wizemann, T. M., J. Moskovitz, B. J. Pearce, D. Cundell, C. G. Arvidson, M. So, H. Weissbach, N. Brot, and H. R. Masure.** 1996. Peptide methionine sulfoxide reductase contributes to the maintenance of adhesins in three major pathogens. *Proc. Natl. Acad. Sci. USA* **93**:7985–7990.

4037. **Wold, A. E., D. A. Caugant, G. Lidin-Janson, P. de Man, and C. Svanborg.** 1992. Resident colonic *Escherichia coli* strains frequently display uropathogenic characteristics. *J. Infect. Dis.* **165**:46–52.

4038. **Wolff, H., A. Panhans, W. Stolz, and M. Meurer.** 1993. Adherence of *Escherichia coli* to sperm: a mannose mediated phenomenon leading to agglutination of sperm and *E. coli. Fertil. Steril.* **60**:154–158.

4039. **Wolfgang, M., H. S. Park, S. F. Hayes, J. P. van Putten, and M. Koomey.** 1998. Suppression of an absolute defect in type IV pilus biogenesis by loss-of-function mutations in *pilT,* a twitching motility gene in *Neisseria gonorrhoeae. Proc. Natl. Acad. Sci. USA* **95**:14973–14978.

4040. **Wolinsky, L. E., S. Mania, S. Nachnani, and S. Ling.** 1996. The inhibiting effect of aqueous *Azadirachta indica* (Neem) extract upon bacterial properties influencing in vitro plaque formation. *J. Dent. Res.* **75**:816–822.

4041. **Wolter, J. M., and J. G. McCormack.** 1998. The effect of subinhibitory concentrations of antibiotics on adherence of *Pseudomonas aeruginosa* to cystic fibrosis (CF) and non-CF-affected tracheal epithelial cells. *J. Infect.* **37**:217–223.

4042. **Wolz, C., D. McDevitt, T. J. Foster, and A. L. Chung.** 1996. Influence of *agr* on fibrinogen binding in *Staphylococcus aureus* Newman. *Infect. Immun.* **64**:3142–3147.

4043. **Wong, S. Y., L. M. Guerdoud, A. Cantin, and D. P. Speert.** 1999. Glucose stimulates phagocytosis of unopsonized *Pseudomonas aeruginosa* by cultivated human alveolar macrophages. *Infect. Immun.* **67**:16–21.

4044. **Wong, W. Y., A. P. Campbell, C. McInnes, B. D. Sykes, W. Paranchych, R. T. Irvin, and R. S. Hodges.** 1995. Structure-function analysis of the adherence-binding domain on the pilin of *Pseudomonas aeruginosa* strains PAK and KB7. *Biochemistry* **34**:12963–12972.

4045. **Wong, W. Y., R. T. Irvin, W. Paranchych, and R. S. Hodges.** 1992. Antigen-antibody interactions: elucidation of the epitope and strain-specificity of a monoclonal antibody directed against the pilin protein adherence binding domain of *Pseudomonas aeruginosa* strain K. *Protein Sci.* **1**:1308–1318.

4046. **Wood, G. E., S. M. Dutro, and P. A.**

Totten. 2001. *Haemophilus ducreyi* inhibits phagocytosis by U-937 cells, a human macrophage-like cell line. *Infect. Immun.* **69:**4726–4733.

4047. **Wood, M. W., R. Rosqvist, P. B. Mullan, M. H. Edwards, and E. E. Galyov.** 1996. SopE, a secreted protein of *Salmonella dublin,* is translocated into the target eukaryotic cell via a sip-dependent mechanism and promotes bacterial entry. *Mol. Microbiol.* **22:**327–338.

4048. **Woodall, L. D., P. W. Russell, S. L. Harris, and P. E. Orndorff.** 1993. Rapid, synchronous, and stable induction of type 1 piliation in *Escherichia coli* by using a chromosomal *lacUV5* promoter. *J. Bacteriol.* **175:**2770–2778.

4049. **Woods, C. R., Jr., E. O. Mason, Jr., and S. L. Kaplan.** 1992. Interaction of *Citrobacter diversus* strains with HEp-2 epithelial and human umbilical vein endothelial cells. *J. Infect. Dis.* **166:**1035–1044.

4050. **Woods, D. E., A. L. Jones, and P. J. Hill.** 1993. Interaction of insulin with *Pseudomonas pseudomallei.* *Infect. Immun.* **61:**4045–4050.

4051. **Woodward, J. M., I. D. Connaughton, V. A. Fahy, A. J. Lymbery, and D. J. Hampson.** 1993. Clonal analysis of *Escherichia coli* of serogroups O9, O20, and O101 isolated from Australian pigs with neonatal diarrhea. *J. Clin. Microbiol.* **31:**1185–1188.

4052. **Woodward, M. J., E. Allen-Vercoe, and J. S. Redstone.** 1996. Distribution, gene sequence and expression in vivo of the plasmid encoded fimbrial antigen of *Salmonella* serotype Enteritidis. *Epidemiol. Infect.* **117:**17–28.

4053. **Woodward, M. J., M. Sojka, K. A. Sprigings, and T. J. Humphrey.** 2000. The role of SEF14 and SEF17 fimbriae in the adherence of *Salmonella enterica* serotype Enteritidis to inanimate surfaces. *J. Med. Microbiol.* **49:**481–487.

4054. **Wooley, R. E., P. S. Gibbs, T. P. Brown, J. R. Glisson, W. L. Steffens, and J. J. Maurer.** 1998. Colonization of the chicken trachea by an avirulent avian *Escherichia coli* transformed with plasmid pHK11. *Avian Dis.* **42:**194–198.

4055. **Wu, Q., Q. Wang, K. G. Taylor, and R. J. Doyle.** 1995. Subinhibitory concentrations of antibiotics affect cell surface properties of *Streptococcus sobrinus.* *J. Bacteriol.* **177:**1399–1401.

4056. **Wu, X., S. K. Gupta, and L. D. Hazlett.** 1995. Characterization of *P. aeruginosa* pili binding human corneal epithelial proteins. *Curr. Eye Res.* **14:**969–977.

4057. **Wu, X., M. Kurpakus, and L. D. Hazlett.** 1996. Some *P. aeruginosa* pilus-binding proteins of corneal epithelium are cytokeratins. *Curr. Eye Res.* **15:**782–791.

4058. **Wu, X.-R., T.-T. Sun, and J. J. Medina.** 1996. *In vitro* binding of type 1-fimbriated *Escherichia coli* to uroplakins Ia and Ib: relation to urinary tract infections. *Proc. Natl. Acad. Sci. USA* **93:**9630–9635.

4059. **Wu, Z., D. Milton, P. Nybom, A. Sjo, and K. E. Magnusson.** 1996. *Vibrio cholerae* hemagglutinin/protease (HA/protease) causes morphological changes in cultured epithelial cells and perturbs their paracellular barrier function. *Microb. Pathog.* **21:**111–123.

4060. **Wullenweber, M., L. Beutin, S. Zimmermann, and C. Jonas.** 1993. Influence of some bacterial and host factors on colonization and invasiveness of *Escherichia coli* K1 in neonatal rats. *Infect. Immun.* **61:**2138–2144.

4061. **Wullt, B., G. Bergsten, H. Connell, P. Rollano, N. Gebretsadik, R. Hull, and C. Svanborg.** 2000. P fimbriae enhance the early establishment of *Escherichia coli* in the human urinary tract. *Mol. Microbiol.* **38:**456–464.

4062. **Wu-Yuan, C. D., K. J. Eganhouse, J. C. Keller, and K. S. Walters.** 1995. Oral bacterial attachment to titanium surfaces: a scanning electron microscopy study. *J. Oral Implantol.* **21:**207–213.

4063. **Wu-Yuan, C. D., and M. Scheinost.** 1994. Glucan-binding proteins of *Streptococcus sobrinus* B13 grown in high glucose media. *Comp. Biochem. Physiol. Ser B* **108:**237–240.

4064. **Wu-Yuan, C. D., and R. E. Gill.** 1992. An 87-kilodalton glucan-binding protein of *Streptococcus sobrinus* B13. *Infect. Immun.* **60:**5291–5293.

4065. **Wyrick, P. B., C. H. Davis, and E. A. Wayner.** 1994. *Chlamydia trachomatis* does not bind to $\alpha\beta_1$ integrins to colonize a human endometrial epithelial cell line cultured *in vitro*. *Microb. Pathog.* **17:**159–166.

4066. **Xiong, H., Y. Li, M. F. Slavik, and J. T. Walker.** 1998. Spraying chicken skin with selected chemicals to reduce attached *Salmonella typhimurium*. *J. Food Prot.* **61:**272–275.

4067. **Xu, H., T. Storch, M. Yu, S. P. Elliott, and D. B. Haslam.** 1999. Characterization of the human Forssman synthetase gene. An evolving association between glycolipid synthesis and host-microbial interactions. *J. Biol. Chem.* **274:**29390–29398.

4068. **Xu, J. G., B. Q. Cheng, Y. P. Wu, L. B. Huang, X. H. Lai, B. Y. Liu, X. Z. Lo,** and H. F. Li. 1996. Adherence patterns and DNA probe types of *Escherichia coli* isolated from diarrheal patients in China. *Microbiol. Immunol.* **40:**89–97.

4069. Xu, S., R. D. Arbeit, and J. C. Lee. 1992. Phagocytic killing of encapsulated and microencapsulated *Staphylococcus aureus* by human polymorphonuclear leukocytes. *Infect. Immun.* **60:**1358–1362.

4070. Xu, Z., C. H. Jones, D. Haslam, J. S. Pinkner, K. Dodson, J. Kihlberg, and S. J. Hultgren. 1995. Molecular dissection of PapD interaction with PapG reveals two chaperone-binding sites. *Mol. Microbiol.* **16:**1011–1020.

4071. Yamada, M., T. Saito, T. Toba, H. Kitazawa, J. Uemura, and T. Itoh. 1994. Hemagglutination activity of *Lactobacillus acidophilus* group lactic acid bacteria. *Biosci. Biotechnol. Biochem.* **58:**910–915.

4072. Yamaguchi, H., T. Osaki, H. Taguchi, T. Hanawa, T. Yamamoto, and S. Kamiya. 1998. Relationship between expression of HSP60, urease activity, production of vacuolating toxin, and adherence activity of *Helicobacter pylori*. *J. Gastroenterol.* **33:**6–9.

4073. Yamaguchi, H., T. Osaki, N. Kurihara, H. Taguchi, T. Hanawa, T. Yamamoto, and S. Kamiya. 1997. Heat-shock protein 60 homologue of *Helicobacter pylori* is associated with adhesion of *H. pylori* to human gastric epithelial cells. *J. Med. Microbiol.* **46:**825–831.

4074. Yamaguchi, M., M. Kanoe, K. Kai, and Y. Okada. 1999. Actin degradation concomitant with *Fusobacterium necrophorum* subsp. *necrophorum* adhesion to bovine portal cells. *Microbios* **98:**87–94.

4075. Yamaguchi, T., K. Kasamo, M. Chuman, M. Machigashira, M. Inoue, and T. Sueda. 1998. Preparation and characterization of an *Actinomyces naeslundii* aggregation factor that mediates coaggregation with *Porphyromonas gingivalis*. *J Periodontal Res.* **33:**460–468.

4076. Yamamoto, K., T. Miwa, H. Taniguchi, T. Nagano, K. Shimamura, T. Tanaka, and H. Kumagai. 1996. Binding specificity of *Lactobacillus* to glycolipids. *Biochem. Biophys. Res. Commun.* **228:**148–152.

4077. Yamamoto, K., S. Ohashi, E. Taki, and K. Hirata. 1996. Adherence of oral streptococci to composite resin of varying surface roughness. *Dent. Mater. J.* **15:**201–204.

4078. Yamamoto, S., K. Nakata, K. Yuri, H. Katae, A. Terai, H. Kurazono, Y. Takeda, and O Yoshida. 1996. Assessment of the significance of virulence factors of uropathogenic *Escherichia coli* in experimental urinary tract infection in mice. *Microbiol. Immunol.* **40:**607–610.

4079. Yamamoto, T., P. Echeverria, and T. Yokota. 1992. Drug resistance and adherence to human intestines of enteroaggregative *Escherichia coli*. *J. Infect. Dis.* **165:**744–749.

4080. Yamamoto, T., Y. Koyama, M. Matsumoto, E. Sonoda, S. Nakayama, M. Uchimura, W. Paveenkittiporn, K. Tamura, T. Yokota, and P. Echeverria. 1992. Localized, aggregative, and diffuse adherence to HeLa cells, plastic, and human small intestines by *Escherichia coli* isolated from patients with diarrhea. *J. Infect. Dis.* **166:**1295–1310.

4081. Yamamoto, T., M. J. Albert, and R. B. Sack. 1994. Adherence to human small intestines of capsulated *Vibrio cholerae* O139. *FEMS Microbiol. Lett.* **119:**229–235.

4082. Yamamoto, T., M. Kaneko, S. Changchawalit, O. Serichantalergs, S. Ijuin, and P. Echeverria. 1994. Actin accumulation associated with clustered and localized adherence in *Escherichia coli* isolated from patients with diarrhea. *Infect. Immun.* **62:**2917–2929.

4083. Yamamoto, T., N. Wakisaka, T. Nakae, T. Kamano, O. Serichantalergs, and P. Echeverria. 1996. Characterization of a novel hemagglutinin of diarrhea-associated *Escherichia coli* that has characteristics of diffusely adhering *E. coli* and enteroaggregative *E. coli*. *Infect. Immun.* **64:**3694–3702.

4084. (See reference 4083.)

4085. Yamamoto, T., N. Wakisaka, and T. Nakae. 1997. A novel cryohemagglutinin associated with adherence of enteroaggregative *Escherichia coli*. *Infect. Immun.* **65:**3478–3484.

4086. Yamamoto, Y., S. Okubo, T. W. Klein, K. Onozaki, T. Saito, and H. Friedman. 1994. Binding of *Legionella pneumophila* to macrophages increases cellular cytokine mRNA. *Infect. Immun.* **62:**3947–3956.

4087. Yamamoto, Y., T. W. Klein, and H. Friedman. 1996. Induction of cytokine granulocyte-macrophage colony-stimulating factor and chemokine macrophage inflammatory protein 2 mRNAs in macrophages by *Legionella pneumophila* or *Salmonella typhimurium* attachment requires different ligand-receptor systems. *Infect. Immun.* **64:**3062–3068.

4088. Yamamoto-Osaki, T., H. Yamaguchi, H. Taguchi, S. Ogata, and S. Kamiya. 1995. Adherence of *Helicobacter pylori* to cultured human gastric carcinoma cells. *Eur. J. Gastroenterol. Hepatol.* **7:**S89–S92.

4089. **Yamasaki, T., T. Ichimiya, K. Hirai, K. Hiramatsu, and M. Nasu.** 1997. Effect of antimicrobial agents on the piliation of *Pseudomonas aeruginosa* and adherence to mouse tracheal epithelium. *J. Chemother.* **9:**32–37.

4090. **Yamashiro, T., N. Nakasone, and M. Iwanaga.** 1993. Purification and characterization of pili of a *Vibrio cholerae* non-O1 strain. *Infect. Immun.* **61:**5398–5400.

4091. **Yamashiro, T., N. Nakasone, Y. Honma, M. J. Albert, and M. Iwanaga.** 1994. Purification and characterization of *Vibrio cholerae* O139 fimbriae. *FEMS Microbiol. Lett.* **115:**247–252.

4092. **Yamashiro, T., and M. Iwanaga.** 1996. Purification and characterization of a pilus of a *Vibrio cholerae* strain: a possible colonization factor. *Infect. Immun.* **64:**5233–5238.

4093. **Yang, G. H., S. D. Rhee, H. H. Jung, and K. H. Yang.** 1996. Organization and nucleotide sequence of genes for hemagglutinin components of *Clostridium botulinum* type B progenitor toxin. *Biochem. Mol. Biol. Int.* **39:**1141–1146.

4094. **Yang, J., R. Bos, G. F. Belder, and H. J. Busscher.** 2001. Co-adhesion and removal of adhering bacteria from salivary pellicles by three different modes of brushing. *Eur. J. Oral Sci.* **109:**325–329.

4095. **Yang, Y., and R. R. Isberg.** 1993. Cellular internalization in the absence of invasin expression is promoted by the *Yersinia pseudotuberculosis yadA* product. *Infect. Immun.* **61:**3907–3913.

4096. **Yang, Y., and R. R. Isberg.** 1997. Transcriptional regulation of the *Yersinia pseudotuberculosis* pH6 antigen adhesin by two envelope-associated components. *Mol. Microbiol.* **24:**499–510.

4097. **Yang, Y., J. J. Merriam, J. P. Mueller, and R. R. Isberg.** 1996. The *psa* locus is responsible for thermoinducible binding of *Yersinia pseudotuberculosis* to cultured cells. *Infect. Immun.* **64:**2483–2489.

4098. **Yano, T., M. Garcia, D. S. Leite, A. F. Pestana-de-Castro, and M. A. Shenk.** 1995. Determination of the efficiency of K99-F41 fimbrial antigen vaccine in newborn calves. *Braz. J. Med. Biol. Res.* **28:**651–654.

4099. **Yao, E. S., R. J. Lamont, S. P. Leu, and A. Weinberg.** 1996. Interbacterial binding among strains of pathogenic and commensal oral bacterial species. *Oral Microbiol. Immunol.* **11:**35–41.

4100. **Yao, L., V. Bengualid, F. D. Lowy, J. J. Gibbons, V. B. Hatcher, and J. W. Berman.** 1995. Internalization of *Staphylococcus aureus* by endothelial cells induces cytokine gene expression. *Infect. Immun.* **63:**1835–1839.

4101. **Yao, L., V. Bengualid, J. W. Berman, and F. D. Lowy.** 2000. Prevention of endothelial cell cytokine induction by a *Staphylococcus aureus* lipoprotein. *FEMS Immunol. Med. Microbiol.* **28:**301–305.

4102. **Yao, R., D. H. Burr, P. Doig, T. J. Trust, H. Niu, and P. Guerry.** 1994. Isolation of motile and non-motile insertional mutants of *Campylobacter jejuni*:the role of motility in adherence and invasion of eukaryotic cells. *Mol. Microbiol.* **14:**883–893.

4103. **Yassien, M., N. Khardori, A. Ahmedy, and M. Toama.** 1995. Modulation of biofilms of *Pseudomonas aeruginosa* by quinolones. *Antimicrob. Agents Chemother.* **39:**2262–2268.

4104. **Yasuda, H., Y. Ajiki, J. Aoyama, and T. Yokota.** 1994. Interaction between human polymorphonuclear leucocytes and bacteria released from in-vitro bacterial biofilm models. *J. Med. Microbiol.* **41:**359–367.

4105. **Yeaman, M. R., P. M. Sullam, P. F. Dazin, and A. S. Bayer.** 1994. Platelet microbicidal protein alone and in combination with antibiotics reduces *Staphylococcus aureus* adherence to platelets in vitro. *Infect. Immun.* **62:**3416–3423.

4106. **Yeh, K. S., L. S. Hancox, and S. Clegg.** 1995. Construction and characterization of a *fimZ* mutant of *Salmonella typhimurium*. *J. Bacteriol.* **177:**6861–6865.

4107. **Yeung, M. K.** 1992. Conservation of an *Actinomyces viscosus* T14V type 1 fimbrial subunit homolog among divergent groups of *Actinomyces* spp. *Infect. Immun.* **60:**1047–1054.

4108. **Yeung, M. K.** 1995. Construction and use of integration plasmids to generate site-specific mutations in the *Actinomyces viscosus* T14V chromosome. *Infect. Immun.* **63:** 2924–2930.

4109. **Yeung, M. K., and P. A. Ragsdale.** 1997. Synthesis and function of *Actinomyces naeslundii* T14V type 1 fimbriae require the expression of additional fimbria-associated genes. *Infect. Immun.* **65:**2629–2639.

4110. **Yeung, M. K., J. A. Donkersloot, J. O. Cisar, and P. A. Ragsdale.** 1998. Identification of a gene involved in assembly of *Actinomyces naeslundii* T14V type 2 fimbriae. *Infect. Immun.* **66:**1482–1491.

4111. **Yokoyama, H., T. Hashi, K. Umeda, F. C. Icatlo, Jr., M. Kuroki, Y. Ikemori, and Y. Kodama.** 1997. Effect of oral egg antibody in experimental F18[+] *Escherichia coli* infection in weaned pigs. *J. Vet. Med. Sci.* **59:**917–921.

4112. **Yokoyama, H., R. C. Peralta, R. Diaz, S. Sendo, Y. Ikemori, and Y. Kodama.** 1992. Passive protective effect of chicken egg yolk immunoglobulins against experimental enterotoxigenic *Escherichia coli* infection in neonatal piglets. *Infect. Immun.* **60:** 998–1007.

4113. **Yoneda, M., and H. K. Kuramitsu.** 1996. Genetic evidence for the relationship of *Porphyromonas gingivalis* cysteine protease and hemagglutinin activities. *Oral Microbiol. Immunol.* **11:**129–134.

4114. **Yonezawa, H., K. Ishihara, and K. Okuda.** 2001. Arg-gingipain a DNA vaccine induces protective immunity against infection by *Porphyromonas gingivalis* in a murine model. *Infect. Immun.* **69:**2858–2864.

4115. **Yoshida, T., N. Furuya, M. Ishikura, T. Isobe, K. Haino-Fukushima, T. Ogawa, and T. Komano.** 1998. Purification and characterization of thin pili of IncI1 plasmids Collb-P9 and R64: formation of PilV-specific cell aggregates by type IV pili. *J. Bacteriol.* **180:**2842–2848.

4116. **Yoshimura, F., Y. Takahashi, E. Hibi, T. Takasawa, H. Kato, and D. P. Dickinson.** 1993. Proteins with molecular masses of 50 and 80 kildaltons encoded by genes downstream from the fimbrilin gene (*fimA*) are components associated with fimbriae in the oral anaerobe *Porphyromonas gingivalis. Infect. Immun.* **61:**5181–5189.

4117. **Yoshinari, M., Y. Oda, T. Kato, K. Okuda, and A. Hirayama.** 2000. Influence of surface modifications to titanium on oral bacterial adhesion *in vitro. J. Biomed. Mater. Res.* **52:**388–394.

4118. **Yoshinari, M., Y. Oda, H. Ueki, and S. Yokose.** 2001. Immobilization of bisphosphonates on surface modified titanium. *Biomaterials* **22:**709–715.

4119. **Young, K. A., R. P. Allaker, J. M. Hardie, and R. A. Whiley.** 1996. Interactions between *Eikenella corrodens* and 'Streptococcus milleri-group' organisms: possible mechanisms of pathogenicity in mixed infections. *Antonie Leeuwenhoek* **69:**371–373.

4120. **Young, T. F., E. L. Thacker, B. Z. Erickson, and R. F. Ross.** 2000. A tissue culture system to study respiratory ciliary epithelial adherence of selected swine mycoplasmas. *Vet. Microbiol.* **71:**269–279.

4121. **Young, V.B., S. Falkow, and G. K. Schoolnik.** 1992. The invasin protein of *Yersinia enterocolitica:*internalization of invasin-

bearing bacteria by eukaryotic cells is associated with reorganization of the cytoskeleton. *J. Cell Biol.* **116:**197–207.

4122. **Yousefi Rad, A., H. Ayhan, and E. Piskin.** 1998. Adhesion of different bacterial strains to low-temperature plasma-treated sutures. *J. Biomed. Mater. Res.* **41:**349–358.

4123. **Yousefi Rad, A., H. Ayhan, U. Kisa, and E. Piskin.** 1998. Adhesion of different bacterial strains to low-temperature plasma treated biomedical PVC catheter surfaces. *J. Biomater. Sci.* **9:**915–929.

4124. **Yu, B., and H. Y. Tsen.** 1993. *Lactobacillus* cells in the rabbit digestive tract and the factors affecting their distribution. *J. Appl. Bacteriol.* **75:**269–275.

4125. **Yu, H., Y. Nakano, Y. Yamashita, T. Oho, and T. Koga.** 1997. Effects of antibodies against cell surface protein antigen PAc-glucosyltransferase fusion proteins on glucan synthesis and cell adhesion of *Streptococcus mutans. Infect. Immun.* **65:**2292–2298.

4126. **Yu, J., M. N. Montelius, M. Paulsson, I. Gouda, O. Larm, L. Montelius, and Å. Ljungh.** 1994. Adhesion of coagulase-negative staphylococci and adsorption of plasma proteins to heparinized polymer surfaces. *Biomaterials* **15:**805–814.

4127. **Yu, J. L., R. Andersson, Å. Ljungh, L. Q. Wang, E. Jakab, B. G. Persson, T. Wadström, and S. Bengmark.** 1993. Reduction of *E. coli* adherence to rubber slices treated with phospholipids. *APMIS* **101:**582–586.

4128. **Yu, J. L., Å. Ljungh, R. Andersson, E. Jakab, S. Bengmark, and T. Wadström.** 1994. Promotion of *Escherichia coli* adherence to rubber slices by adsorbed fibronectin. *J. Med. Microbiol.* **41:**133–138.

4129. **Yu, J. L., R. Andersson, L. Q. Wang, S. Bengmark, and Å. Ljungh.** 1995. Fibronectin on the surface of biliary drain materials—a role in bacterial adherence. *J. Surg. Res.* **59:**596–600.

4130. **Yu, J. L., R. Andersson, and Å. Ljungh.** 1996. Protein adsorption and bacterial adhesion to biliary stent materials. *J. Surg. Res.* **62:**69–73.

4131. **Yu, J. L., R. Andersson, H. Parsson, E. Hallberg, Å. Ljungh, and S. Bengmark.** 1996. A bacteriologic and scanning electron microscope study after implantation of foreign bodies in the biliary tract in rats. *Scand. J. Gastroenterol.* **31:**175–181.

4132. **Yu, J. L., R. Andersson, and Å. Ljungh.** 1998. Binding of immobilized fibronectin by biliary drain isolates. *Zentbl. Bakteriol.* **287:**461–473.

4133. **Yu, L., K. K. Lee, R. S. Hodges, W. Paranchych, and R. T. Irvin.** 1994. Adherence of *Pseudomonas aeruginosa* and *Candida albicans* to glycosphingolipid (asialo-GM$_1$) receptors is achieved by a conserved receptor-binding domain present on their adhesins. *Infect. Immun.* **62:**5213–5219.

4134. **Yu, L., K. K. Lee, W. Paranchych, R. S. Hodges, and R. T. Irvin.** 1996. Use of synthetic peptides to confirm that the *Pseudomonas aeruginosa* PAK pilus adhesin and the *Candida albicans* fimbrial adhesin possess a homologous receptor-binding domain. *Mol. Microbiol.* **19:**1107–1116.

4135. **Yuki, N., T. Shimazaki, A. Kushiro, K. Watanabe, K. Uchida, T. Yuyama, and M. Morotomi.** 2000. Colonization of the stratified squamous epithelium of the nonsecreting area of horse stomach by lactobacilli. *Appl. Environ. Microbiol.* **66:**5030–5034.

4136. **Yumoto, H., H. Azakami, H. Nakae, T. Matsuo, and S. Ebisu.** 1996. Cloning, sequencing and expression of an *Eikenella corrodens* gene encoding a component protein of the lectin-like adhesin complex. *Gene* **183:**115–121.

4137. **Yumoto, H., H. Nakae, K. Fujinaka, S. Ebisu, and T. Matsuo.** 1999. Interleukin-6 (IL-6) and IL-8 are induced in human oral epithelial cells in response to exposure to periodontopathic *Eikenella corrodens. Infect. Immun.* **67:**384–394.

4138. **Yuri, K., K. Nakata, H. Katae, S. Yamamoto, and A. Hasegawa.** 1998. Distribution of uropathogenic virulence factors among *Escherichia coli* strains isolated from dogs and cats. *J. Vet. Med. Sci.* **60:**287–290.

4139. **Zabaleta, J., M. Arias, J. R. Maya, and L. F. Garcia.** 1998. Diminished adherence and/or ingestion of virulent *Mycobacterium tuberculosis* by monocyte-derived macrophages from patients with tuberculosis. *Clin. Diagn. Lab. Immunol.* **5:**690–694.

4140. **Zabel, L. T., A. Neuer, and B. Manncke.** 1996. Fibronectin binding and cell surface hydrophobicity contribute to adherence properties of group B streptococci. *Zentbl. Bakteriol.* **285:**35–43.

4141. **Zaffran, Y., L. Zhang, and J. J. Ellner.** 1998. Role of CR4 in *Mycobacterium tuberculosis*-human macrophages binding and signal transduction in the absence of serum. *Infect. Immun.* **66:**4541–4544.

4142. **Zaidi, T. S., M. J. Preston, and G. B. Pier.** 1997. Inhibition of bacterial adherence to host tissue does not markedly affect disease in the murine model of *Pseudomonas aeruginosa* corneal infection. *Infect. Immun.* **65:**1370–1376.

4143. **Zaidi, T. S., S. M. J. Fleiszig, M. J. Preston, J. B. Goldberg, and G. B Pier.** 1996. Lipopolysaccharide outer core is a ligand for corneal cell binding and ingestion of *Pseudomonas aeruginosa. Investig. Opthalmol. Visual Sci.* **37:**76–986.

4144. **Zanetti, S., A. Angioi, P.L. Fiori, and G. Fadda.** 1994. Adhesion of mucoid uropathogenic strains of *Pseudomonas aeruginosa* to HEp-2 cells. *New Microbiol.* **17:**297–305.

4145. **Zanetti, S., A. Deriu, L. Volterra, M. P. Falchi, P. Molicotti, G. Fadda, and L. Sechi.** 2000. Virulence factors in *Vibrio alginolyticus* strains isolated from aquatic environments. *Ann. Igiene* **12:**487–491.

4146. **Zar, H., L. Saiman, L. Quittell, and A. Prince.** 1995. Binding of *Pseudomonas aeruginosa* to respiratory epithelial cells from patients with various mutations in the cystic fibrosis transmembrane regulator. *J. Pediatr.* **126:**230–233.

4147. **Zareba, T. W., C. Pascu, W. Hryniewicz, and T. Wadström.** 1997. Binding of extracellular matrix proteins by enterococci. *Curr. Microbiol.* **34:**6–11.

4148. **Zaretzky, F. R., R. Pearce-Pratt, and D. M. Phillips.** 1995. Sulfated polyanions block *Chlamydia trachomatis* infection of cervix-derived human epithelia. *Infect. Immun.* **63:**3520–3526.

4149. **Zaretzky, F. R., and T. H. Kawula.** 1999. Examination of early interactions between *Haemophilus ducreyi* and host cells by using cocultured HaCaT keratinocytes and foreskin fibroblasts. *Infect. Immun.* **67:**5352–5360.

4150. **Zdanowski, Z., E. Ribbe, and C. Schalen.** 1993. Bacterial adherence to synthetic vascular prostheses and influence of human plasma. An *in vitro* study. *Eur. J. Vasc. Surg.* **7:**277–282.

4151. **Zdanowski, Z., E. Ribbe, and C. Schalen.** 1993. Influence of some plasma proteins on *in vitro* bacterial adherence to PTFE and Dacron vascular prostheses. *APMIS* **101:**926–932.

4152. **Zdanowski, Z., B. Koul, E. Hallberg, and C. Schalen.** 1997. Influence of heparin coating on *in vitro* bacterial adherence to poly(vinyl chloride) segments. *J. Biomater. Sci. Polym. Ed.* **8:**825–832.

4153. **Zepeda-Lopez, H. M., and G. M. Gonzalez-Lugo.** 1995. *Escherichia coli* adherence to HEp-2 cells with prefixed cells. *J. Clin. Microbiol.* **33:**1414–1417.

4154. **Zhan, Y., and C. Cheers.** 1995. Differential

induction of macrophage-derived cytokines by live and dead intracellular bacteria in vitro. *Infect. Immun.* **63:**720–723.

4155. **Zhanel, G. G., S. O. Kim, R. J. Davidson, D. J. Hoban, and L. E. Nicolle.** 1993. Effect of subinhibitory concentrations of Ciprofloxacin and gentamicin on the adherence of *Pseudomonas aeruginosa* to Vero cells and voided uroepithelial cells. *Chemotherapy* **39:**105–111.

4156. **Zhang, H. Z., and M. S. Donnenberg.** 1996. DsbA is required for stability of the type IV pilin of enteropathogenic *Escherichia coli. Mol. Microbiol.* **21:**787–797.

4157. **Zhang, J. P., and R. S. Stephens.** 1992. Mechanism of *C. trachomatis* attachment to eukaryotic host cells. *Cell* **69:**861–869.

4158. **Zhang, J. P., and S. Normark.** 1996. Induction of gene expression in *Escherichia coli* after pilus-mediated adherence. *Science* **273:**1234–1236.

4159. **Zhang, J. R., K. E. Mostov, M. E. Lamm, M. Nanno, S. Shimida, M. Ohwaki, and E. Tuomanen.** 2000. The polymeric immunoglobulin receptor translocates pneumococci across human nasopharyngeal epithelial cells. *Cell* **102:**827–837.

4160. **Zhang, L., B. Foxman, P. Tallman, E. Cladera, C. Le Bouguenec, and C. F. Marrs.** 1997. Distribution of *drb* genes coding for Dr binding adhesins among uropathogenic and fecal *Escherichia coli* isolates and identification of new subtypes. *Infect. Immun.* **65:**2011–2018.

4161. **Zhang, Q., T. F. Young, and R. F. Ross.** 1994. Microtiter plate adherence assay and receptor analogs for *Mycoplasma hyopneumoniae. Infect. Immun.* **62:**1616–1622.

4162. **Zhang, Q., and K. S. Wise.** 1996. Molecular basis of size and antigenic variation of a *Mycoplasma hominis* adhesin encoded by divergent *vaa* genes. *Infect. Immun.* **64:**2737–2744.

4163. **Zhang, Q, T. F. Young, and R. F. Ross.** 1995. Identification and characterization of a *Mycoplasma hyopneumoniae* adhesin. *Infect. Immun.* **63:**1013–1019.

4164. **Zhang, Q., T. F. Young, and R. F. Ross.** 1994. Glycolipid receptors for attachment of *Mycoplasma hyopneumoniae* to porcine respiratory ciliated cells. *Infect. Immun.* **62:**4367–4373.

4165. **Zhang, Q. Y., D. DeRyckere, P. Lauer, and M. Koomey.** 1992. Gene conversion in *Neisseria gonorrhoeae:* evidence for its role in pilus antigenic variation. *Proc. Natl. Acad. Sci. USA* **89:**5366–5370.

4166. **Zhang, X. L., I. S. Tsui, C. M. Yip, A.** W. Fung, D. K. Wong, X. Dai, Y. Yang, J. Hackett, and C. Morris. 2000. *Salmonella enterica* serovar Typhi uses type IVB pili to enter human intestinal epithelial cells. *Infect. Immun.* **68:**3067–3073.

4167. **Zhang, Y. J.** 1993. Heterogeneity of *Porphyromonas gingivalis* strains on fimbrillin gene locus by restriction fragment length polymorphism analysis. *Bull. Tokyo Med. Dent. Univ.* **40:**113–123.

4168. **Zhao, S., J. Meng, M. P. Doyle, R. Meinersman, G. Wang, and P. Zhao.** 1996. A low molecular weight outer-membrane protein of *Escherichia coli* O157:H7 associated with adherence to INT407 cells and chicken caeca. *J. Med. Microbiol.* **45:**90–96.

4169. **Zhao, W., J. S. Schorey, M. Bong-Mastek, J. Ritchey, E. J. Brown, and T. L. Ratliff.** 2000. Role of a bacillus Calmette-Guerin fibronectin attachment protein in BCG-induced antitumor activity. *Int. J. Cancer* **86:**83–88.

4170. **Zhao, W., J. S. Schorey, R. Groger, P. M. Allen, E. J. Brown, and T. L. Ratliff.** 1999. Characterization of the fibronectin binding motif for a unique mycobacterial fibronectin attachment protein, FAP. *J. Biol. Chem.* **274:**4521–4526.

4171. **Zhao, Z., and N. Panjwani.** 1995. *Pseudomonas aeruginosa* infection of the cornea and asialo-GM₁. *Infect. Immun.* **63:**353–355.

4172. **Zheng, C. H., K. Ahmed, N. Rikitomi, G. Martinez, and T. Nagatake.** 1999. The effects of *S*-carboxymethylcysteine and *N*-acetylcysteine on the adherence of *Moraxella catarrhalis* to human pharyngeal epithelial cells. *Microbiol. Immunol.* **43:**107–113.

4173. **Zhou, Z. X., Z. P. Deng, and J. Y. Ding.** 1995. Role of glycoconjugates in adherence of *Salmonella pullorum* to the intestinal epithelium of chicks. *Br. Poult. Sci.* **26:**79–86.

4174. **Zhu, C., J. Harel, M. Jacques, and J. M. Fairbrother.** 1995. Interaction with pig ileal explants of *Escherichia coli* O45 isolates from swine with postweaning diarrhea. *Can. J. Vet. Res.* **59:**118–123.

4175. **Zhu, C., J. Harel, M. Jacques, C. Desautels, M. S. Donnenberg, M. Beaudry, and J. M. Fairbrother.** 1995. Virulence properties and attaching-effacing activity of *Escherichia coli* O45 from swine postweaning diarrhea. *Infect. Immun.* **62:**4153–4159.

4176. **Ziebuhr, W., C. Heilmann, F. Götz, P. Meyer, K. Wilms, E. Straube, and J. Hacker.** 1997. Detection of the intercellular adhesion gene cluster (*ica*) and phase variation in *Staphylococcus epidermidis* blood culture strains and mucosal isolates. *Infect. Immun.* **65**:890–896.

4177. **Ziebuhr, W., V. Krimmer, S. Rachid, I. Lossner, F. Götz, and J. Hacker.** 1999. A novel mechanism of phase variation of virulence in *Staphylococcus epidermidis*: evidence for control of the polysaccharide intercellular adhesin synthesis by alternating insertion and excision of the insertion sequence element IS*256*. *Mol. Microbiol.* **32**:345–356.

4178. **Zielinski, G. C., and R. F. Ross.** 1993. Adherence of *Mycoplasma hyopneumoniae* to porcine ciliated respiratory tract cells. *Am. J. Vet. Res.* **54**:1262–1269.

4179. **Zierler, M. K., and J. E. Galan.** 1995. Contact with cultured epithelial cells stimulates secretion of *Salmonella typhimurium* invasion protein InvJ. *Infect. Immun.* **63**:4024–4028.

4180. **Zingler, G., G. Blum, U. Falkenhagen, I. Orskov, F. Orskov, J. Hacker, and M. Ott.** 1993. Clonal differentiation of uropathogenic *Escherichia coli* isolates of serotype O6:K5 by fimbrial antigen typing and DNA long-range mapping techniques. *Med. Microbiol. Immunol.* **182**:13–24.

4181. **Zingler, G., M. Ott, G. Blum, U. Falkenhagen, G. Naumann, W. Sokolowska-Kohler, and J. Hacker.** 1992. Clonal analysis of *Escherichia coli* serotype O6 strains from urinary tract infections. *Microb. Pathog.* **12**:299–310.

4182. **Ziprin, R. L., C. R. Young, L. H. Stanker, M. E. Hume, and M. E. Konkel.** 1999. The absence of cecal colonization of chicks by a mutant of *Campylobacter jejuni* not expressing bacterial fibronectin-binding protein. *Avian Dis.* **43**:586–589.

4183. **Zopf, D., and S. Roth.** 1996. Oligosaccharide anti-infective agents. *Lancet* **347**:1017–1021.

4184. **Zorgani, A. A., J. Stewart, C. C. Blackwell, R. A. Elton, and D. M. Weir.** 1994. Inhibitory effect of saliva from secretors and non-secretors on binding of meningococci to epithelial cells. *FEMS Immunol. Med. Microbiol.* **9**:135–142.

4185. **Zucchelli, G., F. Pollini, C. Clauser, and M. De Sanctis.** 2000. The effect of chlorhexidine mouthrinses on early bacterial colonization of guided tissue regeneration membranes. An *in vivo* study. *J. Periodontol.* **71**:263–271.

4186. **Zuniga, A., H. Yokoyama, P. Albicker-Rippinger, E. Eggenberger, and H. U. Bertschinger.** 1997. Reduced intestinal colonisation with F18-positive enterotoxigenic *Escherichia coli* in weaned pigs fed chicken egg antibody against the fimbriae. *FEMS Immunol. Med. Microbiol.* **18**:153–161.

4187. **Zunino, P., C. Piccini, and C. Legnani-Fajardo.** 1999. Growth, cellular differentiation and virulence factor expression by *Proteus mirabilis in vitro* and *in vivo. J. Med. Microbiol.* **48**:527–534.

INDEX